1982

University of St. Francis
GEN 599.0133 H559
Principles of metabolic contro
3 0301 00076778 6

D1566494

Principles of Metabolic Control in Mammalian Systems

Principles of
Metabolic Control in
Mammalian Systems

Edited by
Robert H. Herman
Letterman Army Institute of Research
San Francisco, California

Robert M. Cohn
and
Pamela D. McNamara
University of Pennsylvania
Philadelphia, Pennsylvania

PLENUM PRESS · NEW YORK AND LONDON

Library of Congress Cataloging in Publication Data

Main entry under title:

Principles of metabolic control in mammalian systems.

Includes index.
1. Metabolic regulation. 2. Mammals—Physiology. I. Herman, Robert H. II. Cohn, Robert M. III. McNamara, Pamela D.
QP171.M3773   574.1'33   79-9177
ISBN 0-306-40261-0

© 1980 Plenum Press, New York
A Division of Plenum Publishing Corporation
227 West 17th Street, New York, N.Y. 10011

All rights reserved

No part of this book may be reproduced, stored in a retrieval system, or transmitted, in any form or by any means, electronic, mechanical, photocopying, microfilming, recording, or otherwise without written permission from the Publisher

Printed in the United States of America

Contributors

Ifeanyi J. Arinze • Department of Biochemistry and Nutrition, Meharry Medical College, Nashville, Tennessee

F. John Ballard • CSIRO Division of Human Nutrition, Adelaide, Australia

Robert M. Cohn • Department of Metabolic Research, Children's Hospital of Philadelphia, University of Pennsylvania, Philadelphia, Pennsylvania

Harold I. Friedman • Department of Surgery, Letterman Army Medical Center, Presidio of San Francisco, California

Richard W. Hanson • Department of Biochemistry, Case Western Reserve University School of Medicine, Cleveland, Ohio

Robert H. Herman • Endocrine-Metabolic Service, Letterman Army Medical Center, Presidio of San Francisco, California

Pamela D. McNamara • Department of Metabolic Research, Children's Hospital of Philadelphia, University of Pennsylvania, Philadelphia, Pennsylvania

Božena Ožegović • Laboratory for Experimental Medicine, University of Zagreb, Yugoslavia

Michael J. Palmieri • Department of Metabolic Research, Children's Hospital of Philadelphia, University of Pennsylvania, Philadelphia, Pennsylvania

O. David Taunton • Department of Medicine, University of Alabama School of Medicine, Birmingham, Alabama

Donald A. Vessey • Departments of Medicine and Pharmacology, University of California, San Francisco, California; and Liver Study Unit, Department of Medicine, Division of Molecular Biology, Veterans Administration Hospital, San Francisco, California

John R. Yandrasitz • Department of Metabolic Research, Children's Hospital of Philadelphia, University of Pennsylvania, Philadelphia, Pennsylvania

Marc Yudkoff • Department of Metabolic Research, Children's Hospital of Philadelphia, University of Pennsylvania, Philadelphia, Pennsylvania

David Zakim • Departments of Medicine and Pharmacology, University of California, San Francisco, California; and Liver Study Unit, Department of Medicine, Division of Molecular Biology, Veterans Administration Hospital, San Francisco, California

Preface

In this work we present the basic principles of metabolic control which we hope will serve as a foundation for the vast array of factual matter which the biochemist and the physician engaged in metabolic research must accumulate. Accordingly, we attempt to set forth these principles, along with sufficient explanation, so that the reader may apply them to the ever-expanding literature of biochemistry. If we are successful, this will provide a theoretical approach which can be applied to any given set of metabolic reactions.

It is impossible to enumerate each and every biochemical reaction and pathway since such a work would be too unwieldy for efficient use. Rather, we hope our presentation of the principles of metabolic control will be sufficiently basic to be of lasting usefulness no matter how detailed biochemistry may become. We would like to be able to condense biochemistry into a theoretical biology that will not only allow for the general treatment of any given reaction but will enable predictions to be made as to the existence of necessary pathways and the consequences of altered control. Such is not possible today, but this may be accomplished in the future. We believe it is now possible to institute the beginnings of such a theoretical biology.

We hope this book will fulfill the varied needs of the graduate student, physician, and biochemist, as well as selected undergraduates and medical students who wish to augment their understanding of the principles that underlie metabolic regulation. There is a tendency to become so enmeshed in the accumulation of factual knowledge that a firm grasp of the concept and nature of metabolic interrelationships and control mechanisms suffers; we think that our approach may act as a corrective. Perhaps, after the beginning student is introduced to the structural details of carbohydrates, lipids, proteins, enzymes, and nucleic acids, the next logical presentation should be the principles of metabolic control and the means by which the supramolecular entities of life are constructed and maintained.

Obviously, a book such as this cannot hope to give examples of all types of control phenomena, but it can serve as a starting point for the newcomer to biochemistry as well as for individuals engaged in closely related areas who may have hitherto felt somewhat chary about opening up the Pandora's box of metabolic control. Any discussion of regulation must, of necessity, depend on arbitrary and somewhat artificial categories, but an attempt at discrete redundancy has been made in order to present an integrated picture. In this way, one can follow in logical fashion the various principles of metabolic control or, if desired, may read only that selection which is appropriate to his needs.

We have limited our viewpoint to the control principles of mammalian systems although many of the mechanisms have been derived from nonmammalian systems. It appears that many metabolic control systems may be shared by both mammalian and nonmammalian organisms, although certain specific details may vary. Thus, it may be most useful to consider control of metabolic reactions in microorganisms, recognizing that these controls may apply only partially to mammalian organisms. It is assumed that a majority of metabolic control principles are biologically universal and therefore can be derived at any level of biological complexity and can be applied to all organisms. Certain metabolic controls that are present in mammals are lacking in unicellular organisms so that knowledge of these principles must be derived directly from the study of mammalian organisms.

In general, the principles of metabolic control that will be covered will be those principles which govern intracellular reactions and not ordinarily those that deal with intercellular regulation or physiological mechanisms. It is assumed that physiological mechanisms result from intracellular functioning and thus are derivative phenomena. The borderline between intra- and extracellular events is not sharp. There exist areas where one merges into the other. Yet, we would like to make the distinction between intra- and extracellular events and thus have made certain arbitrary choices for our conveniences.

We take the responsibility for our selection and organization of topics. If we have omitted important principles or included those of lesser importance, the fault is ours.

With an endeavor which embraces so many interlocking areas of biochemistry, it is essential for the editors to turn to someone who can organize references and proofread text and reference lists, as well as expedite the thousand little aspects of the compilation process. We have been fortunate to have the services of Louise Pepe who gave so unselfishly of her free time to serve these and many other functions. Our thanks to her and to Steven Weiss, who, in the last frantic stages of manuscript preparation, have done so much in so little time. Additionally, we have been fortunate to have available a cadre of dedicated secretaries, including Clara Polaski and Marie L. Carlson, who have been with us since work on the book began, as well as Ronnie Payton, Linda Braxton, Edith Mayes, Dee Daly, Marie Marcacci, and Sheri Silberg, who worked enthusiastically toward the completion of this book. A large portion of the artwork has been contributed by Nancy Cohn, who graciously met every deadline, no matter how imminent. Stephen Updegrove, Andrew Eisen, Bonnie Yost, and Steven Weiss read several chapters and made suggestions which we feel have improved the comprehensibility of the book.

Two of us (RMC and PDM) wish to thank Dr. Stanton Segal for generously providing the intellectual environment conducive to accomplishing this undertaking.

The editors at Plenum have been understanding and encouraging, but it is our spouses who have shown us a measure of forbearance and support which even we sometimes doubted we deserved. We hope this book merits their confidence in us.

Robert H. Herman
Robert M. Cohn
Pamela D. McNamara

Contents

Chapter 1
The Principles of Metabolic Control
Robert H. Herman

I. The First Fundamental Theorem of Theoretical Biology 1
   A. The Nature of the Life Process 2
   B. The Products of Protein Biosynthesis 32
   C. Cellular Replacement and Replication 39
II. The Second Fundamental Theorem of Theoretical Biology 42
   A. The Order of Acquisition of Function 44
   B. The Functional Specialization of DNA, RNA, and Protein 44
   C. Biochemical Implications of Evolution 46
III. Other Fundamental Theorems of Theoretical Biology 46
   A. Theorem 3 46
   B. Theorem 4 46
   C. Theorem 5 46
IV. Summary 47
References 48

Chapter 2
Nonequilibrium Thermodynamics, Noncovalent Forces, and Water
Robert M. Cohn, Michael J. Palmieri, and Pamela D. McNamara

I. Introduction 63
II. Stability, Thermodynamics, and Biological Organization 65
   A. The Development of a General Theory of Thermodynamics 67
   B. Order through Fluctuations 70
   C. Stability Criteria, Instability, and Thermodynamic Theory 72
   D. Model Dissipative Structures 76
   E. Evolutionary Feedback 80

III. Noncovalent Forces   81
   A. Electrostatic Interactions   82
   B. Van der Waals Forces   82
   C. Hydrogen Bonding   83
   D. Hydrophobic Interactions   85
IV. Water   85
V. Conclusions and Implications   86
Appendix I   88
Appendix II: Glossary   89
References   91

Chapter 3
Enzymes and Coenzymes: A Mechanistic View
Robert M. Cohn

I. Introduction   93
II. Chemical Bonding   94
   A. Bond Energy   95
   B. Noncovalent Interactions   96
III. Chemical Reactions   97
   A. Reaction Intermediates   97
IV. The Protein Nature of Enzymes   99
   A. The Amino Acid Side Chains   100
   B. The Active Site   101
V. Enzyme Mechanisms   102
   A. Approximation and Orientation   102
   B. The Transition State   104
   C. Other Factors in Catalysis   105
VI. Coenzymes   113
   A. Adenosine Triphosphate   113
   B. Nicotinamide Nucleotides   115
   C. Coenzyme A   117
   D. Pyridoxal Phosphate   118
   E. Lipoic Acid   120
   F. Thiamine Pyrophosphate   120
   G. Flavins   122
   H. Biotin   123
   I. Folate Coenzymes   124
   J. Metal Ions in Catalysis   126
   K. Coenzyme B_{12} 126
VII. Evolution of Enzyme Function   128
References   129

Chapter 4
Modulation of Enzyme Activity
Robert M. Cohn and John R. Yandrasitz

I. Introduction 135
II. Noncovalent Regulatory Mechanisms 136
  A. Modulation by Substrate and Product Concentration in Enzymes Following Classical Michaelis–Menten Kinetics 136
  B. Cooperativity 139
  C. Modulation of Enzyme Activity by Binding of Small Molecules to Regulatory Sites 140
  D. Kinetics of Interacting Enzyme Sites 142
  E. Modulation by Metabolite Ratios 144
  F. Modulation by Protein–Protein Interaction 147
  G. Other Regulatory Phenomena 150
  H. Summary of Noncovalent Regulation of Enzyme Activity 152
III. Covalent Regulatory Mechanisms 153
  A. Covalent Modification by Irreversible Interconversions 154
  B. Modification by Reversible Covalent Action 154
  C. Substrate Cycles 159
IV. Enzyme Synthesis and Degradation 160
V. Evaluation of the Physiologic Importance of Regulatory Mechanisms 161
  A. Identification of Potentially Rate-Controlling Steps 161
VI. Conclusion: An Overview of Regulation 163
References 164

Chapter 5
Regulation of Protein Biosynthesis
Robert M. Cohn, Pamela D. McNamara, and Robert H. Herman

I. Introduction 171
  A. The Reason for Protein Biosynthesis 171
  B. The Complexity of Protein Biosynthesis 172
  C. The Third Fundamental Theorem of Theoretical Biology 173
II. The Mechanism of Protein Biosynthesis 173
  A. A General Description of Protein Biosynthesis 173
  B. Gene Structure and Protein–Nucleic Acid Interactions 175
  C. Genetic Code 180
  D. DNA-Dependent RNA Polymerase 184
  E. Posttranscriptional Modification of Ribonucleic Acids 187
  F. Protein Biosynthesis 197

III. The Regulation of Protein Biosynthesis   205
    A. Regulation at the Gene Level   205
    B. Translational Control of Protein Synthesis   214
References   216

Chapter 6
Degradation of Enzymes
F. John Ballard

I. Introduction   221
    A. Why Enzyme Degradation?   221
    B. Partial or Complete Degradation?   224
II. Kinetics of Enzyme Degradation   225
    A. Kinetic Order   225
    B. First-Order Kinetics   228
III. Techniques for the Measurement of Enzyme Degradation   229
    A. Steady-State Methods for the Determination of Degradation Rate Constants   229
    B. Non-Steady-State Methods   232
IV. Variability of Enzyme Half-Lives   234
    A. Enzymes with Short Half-Lives   234
    B. Stable Enzymes   236
    C. Abnormal Enzymes   238
    D. Lysosomal Enzymes   239
    E. Mitochondrial Enzymes   240
    F. Protein Properties Correlating with Half-Lives   240
    G. Half-Lives of the Same Enzymes in Different Tissues   241
V. Changes to Degradation Rate Constants   242
    A. Effects of Ligands   242
    B. Effects of Hormones and Nutrients   244
    C. Effects of Growth and Development   245
    D. Relative Contribution of Changes in k_d to Alterations in Enzyme Content   247
VI. Intracellular Localization of Degradative Pathways   249
    A. Lysosomes and Autophagy   249
    B. Degradation of Proteins within Organelles   252
    C. Possible Experimental Approaches for Defining the Intracellular Localization of Protein Breakdown   253
VII. Initial Reactions in Enzyme Degradation   254
    A. Inactivation of Enzymes in Vitro   254
    B. Sulfhydryl Reactions and Protein Catabolism   256
    C. Coenzyme Dissociation   257
    D. Specific Proteolytic Enzymes   257

VIII. Conclusions 258
References 258

Chapter 7
DNA Replication and the Cell Cycle
Robert H. Herman

   I. Introduction 265
  II. Chromatin Structure 265
 III. The Cell Cycle 269
  IV. DNA Synthesis 272
   V. Mitosis 277
  VI. Gene Activation and Inactivation 281
 VII. Summary 284
References 285

Chapter 8
Servomechanisms and Oscillatory Phenomena
Robert M. Cohn, Mark Yudkoff, and Pamela D. McNamara

   I. Introduction 295
  II. Feedback and Feedforward Phenomena 296
      A. Glycolysis: The Pasteur Effect 298
      B. Fatty Acid Synthesis 300
      C. Cholesterol Synthesis 301
      D. Other Examples of Feedback Control 302
 III. Oscillatory Phenomena 304
      A. Oscillations in Open Systems 304
      B. Biological Examples 306
      C. Involvement of Oscillatory Behavior in Collective Phenomena 307
  IV. Proposed Physiological Significance of Oscillatory Phenomena 309
References 310

Chapter 9
Membrane-Bound Enzymes
Robert M. Cohn and Pamela D. McNamara

   I. Introduction 313
  II. Membrane Composition and Structure 314
      A. Isolation and Solubilization of Membrane-Bound Enzymes 315

III. Endoplasmic Reticulum   316
   A. Microsomal Acyl-CoA Desaturation System   317
   B. Microsomal Hydroxylation System   319
   C. Sarcoplasmic Reticulum   319
IV. Golgi Apparatus   320
V. Mitochondria   321
   A. Respiratory Chain and Electron Transport   322
   B. H^+–ATPase   324
VI. Plasma Membrane   328
VII. Temperature Effects   329
   A. Lipid Liquid Crystals   329
   B. Lipid–Protein Interactions   329
VIII. Conclusion: Effects of Lipids on Enzymatic Activity   330
References   331

Chapter 10
The Importance of Phospholipid–Protein Interactions for Regulation of the Activities of Membrane-Bound Enzymes
David Zakim and Donald A. Vessey

I. Introduction   337
II. Effect of Lipid Composition on the Properties of Membranes and Membrane-Bound Proteins   339
   A. Influence of Chain Length and Unsaturation of Phospholipid Fatty Acids on Membrane Structure and Function   339
   B. Influence of Phospholipid Headgroups on Membrane Structure and Function   342
   C. Influence of Cholesterol on the Properties of Phospholipid Membranes   344
   D. Inhomogeneous Nature of the Lipid Phase of Biological Membranes   345
   E. Sensitivity of Membrane-Bound Proteins to Temperature-Induced Changes in Membrane Lipids   346
III. The Effect of Proteins on the Properties of Membrane Lipids   349
IV. The Effect of Phospholipids on the Activities of Soluble Enzymes and Proteins   350
V. Reconstituted Systems   352
VI. Alteration of the Properties of Tightly-Bound Membrane Enzymes by Perturbation of Their Membrane Lipid Environment   353
   A. Glucose-6-Phosphatase   354
   B. UDP-Glucuronyltransferase   357

VII. Model for Lipid–Protein Interactions   361
VIII. Consideration of Factors Regulating the Activities of Membrane-Bound Enzymes in Vivo   362
IX. Conclusions   364
References   365

Chapter 11
Membrane Structure and Transport Systems
Pamela D. McNamara and Božena Ožegović

I. Introduction   373
II. Contact Inhibition and Intercellular Communication   375
III. Antigenic and Receptor Sites   377
IV. Membrane Structure   381
V. Membrane Composition   387
   A. Carbohydrates   391
   B. Proteins   393
   C. Lipids   395
VI. Transport Systems   400
   A. Free Diffusion   401
   B. Diffusion through Pores   402
   C. Pinocytosis   402
   D. Carrier-Mediated Transport and Ion Pumps   403
   E. Transport of Amino Acids and Sugars   412
   F. Water Transport   418
VII. Summary   419
References   419

Chapter 12
Cellular Mechanisms of Secretion
Harold I. Friedman

I. Introduction   439
II. Representative Secretory Cells   440
   A. The Pancreatic Exocrine Cell   440
   B. The Intestinal Goblet Cell   446
   C. The Intestinal Absorptive Cell   451
III. Membrane Flow and Differentiation   468
   A. The Hypothesis   468
   B. Supportive Experimental Data   469
IV. Membrane Reutilization   472
   A. Supportive Experimental Data   472

B. Reutilization versus Flow and Differentiation of
      Membranes   474
   C. Unidirectional Secretory Product Transport   476
V. The Clinical Importance of Intracellular Membranes for
   Secretion   476
VI. Microtubules   478
   A. Structure   478
   B. Function   479
   C. Microtubules and Secretion   479
VII. Mechanisms of Secretory Activation   483
   A. Calcium and cAMP   483
   B. A Hypothesis of Secretory Activation   486
VIII. Types of Secretory Discharge   486
  IX. Summary   489
References   489

Chapter 13
Compartmentation and Its Role in Metabolic Regulation
Ifeanyi J. Arinze and Richard W. Hanson

   I. Introduction   495
  II. Nature of Intracellular Compartments   496
 III. Zymogen Activation and Compartmentation   497
      A. Enzymes as Zymogens in Their Own Degradation   498
      B. Compartmentation of Secretory Proteins—The Signal
         Peptide Theory   498
  IV. Membrane Permeability and the Movement of Molecules in the
      Cell   499
      A. Compartmentation and Oxidative Phosphorylation   500
      B. ATP Translocation across the Inner Mitochondrial
         Membrane   503
      C. Compartmentation of Citric-Acid-Cycle Intermediates and
         Other Anions   506
      D. Species Differences in Compartmentation   509
      E. Measurement of the Compartmentation of Intermediates—
         Limitations of Available Methods   512
   V. Examples of the Role of Compartmentation in the Regulation
      of Energy Metabolism   517
      A. Gluconeogenesis   518
      B. Lipid Biosynthesis   525
      C. Fatty Acid Oxidation   527
  VI. Conclusions   529
References   530

Chapter 14
The Mechanism of Action of Hormones
Robert H. Herman and O. David Taunton

I. Introduction 535
   A. Definitions 536
   B. Classification of Hormones 537
   C. The Criteria for Establishing a Substance as a Hormone 538
   D. The Need for Hormones and Other Signal Molecules 539
II. Hormones 540
   A. The Basic Principles of Hormone Function 540
   B. The Membrane Receptor–Adenylate Cyclase–cAMP–Cyclic Nucleotide Systems 564
   C. The Membrane Receptor–Non-Adenylate-Cyclase System 587
   D. Intracellular Hormone-Binding Proteins 591
References 597

Chapter 15
The Biochemical Basis of Disease
Robert H. Herman and Robert M. Cohn

I. Introduction 621
   A. The Biochemical Nature of Disease 621
   B. The Definition of Intermediary Metabolism 627
II. The Biochemical Basis of Disease 627
   A. Implications of the Basic Thesis 627
   B. The Adaptive Response 629
   C. The Biochemical Abnormality 630
   D. Enzyme Deficiency States 640
III. Therapeutic Approaches to Disease 643
   A. Treatment of Inborn Errors of Metabolism 643
   B. The Classical Modes of Therapy in Medicine 649
References 650

Index 659

Principles of Metabolic Control in Mammalian Systems

1

The Principles of Metabolic Control

Robert H. Herman

I. The First Fundamental Theorem of Theoretical Biology

Theoretical biology attempts to construct basic principles of biology and, by derivation therefrom, to describe the behavior of biological systems. As such, the principles of metabolic control form a subset of the basic principles of theoretical biology.

Both theoretical biology and metabolic control are dependent on certain fundamental theorems. The first fundamental theorem of theoretical biology is that the life process is explicable in terms of chemistry and physics, i.e., in terms of matter and energy. There is a large body of knowledge that demonstrates that the life process involves chemical transformations of substrates mediated by organic catalysts (enzymes), within a complex organized entity (the cell), with the elaboration of products and energy. Biochemistry treats the transformations of matter in living systems and the assimilation of nonliving matter into the life process. This assimilation of nonliving matter into the life process requires a cybernetic or information component (see Chapter 2, Section 2, The Reductionist View).

The basic principles of chemistry and physics govern the chemical reactions and related processes that occur within the cell. Water occupies an axial role in the life processes, serving both as solvent for the myriad substrates within the cell and as the major factor in directing the ultimate, three-dimensional structure of the informational and functional polymers of life (cf. Chapter 2). The energy generated by the

---

Robert H. Herman • Endocrine-Metabolic Service, Letterman Army Medical Center, Presidio of San Francisco, California 94129.

cell is used to maintain the system in a far-from-equilibrium state (see Chapter 2) and hence, in the modern view, constitutes the driving force of the life process.

While it is beyond the scope of this chapter to enter into the details concerning the physical and chemical principles involved or to consider all of the biochemical reactions which occur according to such principles, the reader may wish to consult some standard references to enhance his appreciation of this theorem (e.g., Metzler, 1977; Florkin and Stotz, 1962–1977; Mahler and Cordes, 1971). Our focus in this chapter will be to explore what constitutes a principle of intracellular metabolic control and to enumerate the various principles so far as we perceive them.

A. The Nature of the Life Process

1. The Maintenance of the Constancy of the Intracellular Environment: Homeostasis

The intracellular environment of mammalian cells must be kept relatively constant. When alterations in the environment occur (e.g., pH, temperature, ionic concentration, substrate and product concentrations, etc.), the intracellular function of the cell must be altered in order to adjust to the changed or changing conditions and to help return the environment to its optimum state. Within mammalian systems, the extracellular milieu may be altered by exposure of the whole organism to environmental factors (nutrients, oxygen, temperature, irradiation, foreign protein, etc.) or as a consequence of the internal state of being (consciousness, sleep, exercise, rest, emotion, disease, etc.). There is a range of extracellular changes within which the cell can survive and a large range within which cellular function is impaired; beyond the latter limits, cell death occurs. The intracellular machinery must be able to adapt to various physiological and pathological perturbations if cell death is to be avoided. Clearly, metabolic control mechanisms must exist to regulate changes if the cell is to be able to withstand an altered extracellular environment. The necessity of maintaining the constancy of the intracellular environment was recognized as early as 1878 by Claude Bernard and more recently by Walter Cannon (1929). The term "homeostasis" was coined by Cannon to denote the steady state of the intracellular and extracellular environments maintained by coordinated, complex physiological reactions.

Assuming that all life is explicable in terms of matter and energy, it follows that metabolic control mechanisms also are explicable in terms of the chemistry and physics of the living process. The metabolic control mechanisms constitute an information system which translates environmental change into biochemical response.

2. Physiological Chemical and Physical Reactions

The life process comprises the totality of enzymatic and nonenzymatic chemical and physical reactions occurring within the cell and which are integrated in a delicately balanced, organized state. In this context, chemical reactions involve the transformation of substrates into products through covalent bond formation, while physical reactions involve alterations in molecular state that do not involve the making and breaking of covalent bonds. This latter category includes translocations and binding reactions. Table 1 provides a classification of physiological chemical and physical reactions.

The great majority of chemical reactions that occur within the cell are enzymatically catalyzed reactions. These may be called complete enzyme-dependent reactions since the entire reaction depends primarily on the enzyme. We will not consider the general enzymatic reaction any further here since it is treated in Chapters 3 and 4. Less well appreciated are the partial enzyme-dependent reactions, an example of which is the enzymatically mediated photochemical reaction. Here, in addition to the enzyme, energy from a radiation source is necessary for the chemical reaction to occur. Table 2 lists examples of several such reactions.

A minority of physiologically occurring chemical reactions are nonenzymatic. One category of nonenzymatic reactions comprises spontaneous chemical reactions. These reactions often would not occur physiologically were it not for the enzymatic generation of the chemical compounds which then react spontaneously. This is not always the case, however. In some instances, the spontaneous reaction is governed by the concentrations of the substances involved. Some examples of spontaneous chemical reactions are listed in Table 3. Another category of nonenzymatic chemical reactions includes those that are strictly photochemical in nature, i.e., that involve only the input of radiant energy. Two examples of photochemical and irradiation reactions are the transformation of 7-dehydrocholesterol into vitamin D_3 (cholecalciferol) upon exposure to ultraviolet radiation (Holick et al., 1977) and the transformation of rhodopsin into a series of products upon exposure to visible light (Ostroy, 1977), with the generation of a nerve impulse. Some examples of photochemical reactions are given in Table 4.

Turning our attention now to physiological nonenzymatic physical

Table 1. Classification of Physiological Chemical and Physical Reactions

A. Chemical reactions
  1. Enzymatic reactions
     a. Complete enzyme-dependent reactions: The vast majority of enzymatic reactions
     b. Partial enzyme-dependent reactions: Photoenzymatic reactions (Table 2)
  2. Nonenzymatic reactions
     a. Spontaneous biochemical reactions (Table 3)
     b. Photobiochemical reactions (Table 4)
        1) In vivo
        2) In vitro
B. Physical reactions
  1. Translocation of molecules
     a. Nonmembranous physical reactions: Diffusion
     b. Membranous physical reactions: Osmosis
  2. Binding physical reactions (Table 5)
     a. Extracellular binding
     b. Intracellular binding
        1) Membrane binding by insoluble proteins
        2) Cytoplasmic binding by soluble substances
           a) Carrier protein
           b) Mucopolysaccharides (Table 6)
           c) Protein aggregation by noncovalent forces (Section I.B.2)
        3) Nuclear binding by nucleic acids (Table 6)
C. Physicochemical reactions
  1. Electron transport (oxido–reductive reactions)

reactions that occur within the cell, we are particularly interested in diffusion and binding reactions (Table 1). Diffusion of molecules occurs ubiquitously within the cytoplasm and other body fluids. Transmembrane movement of water is governed by osmotic force, a phenomenon treated in Chapter 11.

Binding reactions occur both intra- and extracellularly. Some binding reactions occur on insoluble receptors located within cell membranes. Some of these receptors are involved in membrane transport and may act as carriers, while others mediate hormone activity. Certain membrane proteins, such as the high-molecular-weight cell surface glycoprotein termed LETS (large, external, transformation-sensitive) glycoprotein, are involved in mediating cell–substratum adhesion and cell–cell interaction (Hynes, 1976), while other membrane proteins are able to bind immunoglobulins (Newman et al., 1977). These topics are considered in Chapters 11 and 14. Among the functionally diverse components of the cell membrane are those which bind enzymes as well as those which bind a variety of substances important for the structural integrity of the membrane.

Binding reactions between soluble proteins and their ligands also occur both intra- and extracellularly. With certain proteins, the binding process is reversible so that the proteins serve as carrier substances, binding a ligand at one site and releasing it at a distant site. Other proteins utilize the binding process as part of a storage function. The nonenzymatic binding of substances constitutes a large category of physiologically important physical reactions. Table 5 lists many extracellular and intracellular soluble proteins which bind various substances.

Certain examples illustrate the importance of nonenzymatic binding reactions. These include hemoglobin, protein–phospholipid interactions, and the nonenzymatic binding abilities of proteins, nucleic acids, and mucopolysaccharides.

Table 2. Partial Enzyme-Dependent Reactions: Photoenzymatic Reactions

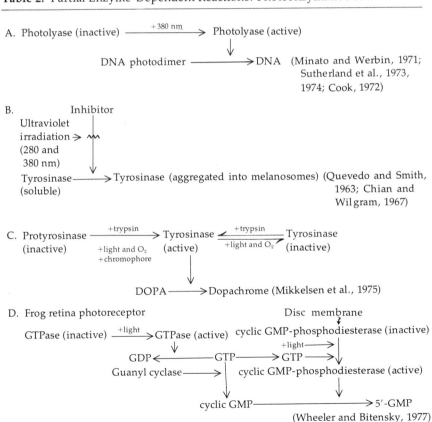

Table 3. Spontaneous Biochemical Reactions

A. o-Aminobenzoyl pyruvic acid $\xrightarrow{-HOH}$ Kynurenic acid (Musajo et al., 1950; Jakoby and Bonner, 1956; Long et al., 1954)

B. Hydroxy o-aminobenzoyl pyruvic acid $\xrightarrow{-HOH}$ Xanthurenic acid (Jakoby and Bonner, 1956)

C. α-Amino-β-carboxymuconic-δ-semialdehyde → Quinolinic acid (Bokman and Schweigert, 1951; Mehler, 1956)

D. Serine $\xrightarrow[-H_2O]{+serine\ dehydrase}$ α-Amino acrylic acid

α-Amino acrylic acid → α-Iminopropionic acid (spont.)

α-Iminopropionic acid $\xrightarrow{+HOH,\ -NH_3}$ Pyruvic acid (spont.) (Longnecker and Snell, 1957)

E. Threonine $\xrightarrow[-H_2O]{+threonine\ dehydrase}$ α-Amino crotonic acid

α-Amino crotonic acid → α-Iminobutyric acid (spont.)

α-Iminobutyric acid $\xrightarrow{-HOH,\ -NH_3}$ α-Ketobutyric acid (spont.) (Longnecker and Snell, 1957)

F. Formation of crosslinks in elastin forming desmosine and isodesmosine (Davis and Anwar, 1970; Rucker and Murray, 1978)

G. Hemoglobin (reduced) (Fe^{2+}) → Methemoglobin (Fe^{3+}) (Goldberg and Stern, 1976)

H. Methemoglobin + Ascorbic acid → Hemoglobin (reduced) + Dehydroascorbic acid (Vestling, 1942)

I. Hemoglobin A + Glucose → Hemoglobin A_{1c} (McDonald et al., 1978)

J. Hemoglobin A + Glucose-6-phosphate → Hemoglobin A_{1a2} (McDonald et al., 1978)

Hemoglobin is able to bind a variety of substances such as O_2, CO_2, and D-2,3-diphosphoglycerate (2,3-DPG). Oxygen binding is influenced by O_2 pressure, pH changes, 2,3-DPG concentration (Arnone, 1974; Brewer and Eaton, 1971; Chanutin and Curnish, 1967), temperature, and certain organic phosphates (Brewer, 1974) such as inositol hexaphosphate (Nelson et al., 1974), ATP, ADP, and AMP (Benesch and Benesch, 1967; Lo and Schimmel, 1969). The affinity of hemoglobin for O_2 is altered if the amino acid structure of the hemoglobin is changed. For example, hemoglobin Kansas, in which threonine replaces aspartic acid in residue 102 of the β-chain, has decreased oxygen affinity, while hemoglobin Abruzzo, in which arginine replaces histidine in residue 143 of the β-chain, has high oxygen affinity (Riggs and Gibson, 1973; Bonaventura et al., 1975). Oxygen is displaced from hemoglobin by 2,3-DPG, which has a greater affinity for deoxyhemoglobin than for oxyhemoglobin (Benesch et al., 1968, 1969). Hemoglobin Hiroshima (143 His → Asp) has a high O_2 affinity, possibly owing to a lack of affinity for 2,3-DPG. 2,3-DPG can bind to hexokinase and inhibit its action (Dische, 1941; Brewer, 1969), and this is reversed by ATP or Mg^{2+}, in increasing concentration. Increased pH decreases 2,3-DPG binding to hemoglobin.

There are species differences with regard to the action of 2,3-DPG on hemoglobin. The hemoglobins of man, horse, dog, rabbit, guinea pig, and rat have high oxygen affinity when free of organic phosphates and react strongly with 2,3-DPG. The red cells of these species have high concentrations of 2,3-DPG. The hemoglobins of sheep, goat, cow, and cat have low oxygen affinity and react weakly with 2,3-DPG. The red cells of these latter species have low concentrations of 2,3-DPG (Bunn, 1971).

Both NADH-cytochrome b_5 reductase (a flavoprotein) and cytochrome b_5 bind to preformed phospholipid vesicles and microsomes.

Table 4. Photobiochemical Reactions

A. In vivo reactions

1.
$$7\text{-Dehydrocholesterol} \xrightarrow{+UV} \text{(Intermediate)} \xrightarrow{+UV} \text{Previt. D}_3 \longrightarrow \text{Vitamin D}_3$$
with Tachysterol$_3$ and Lumisterol$_3$ as $+UV$ branches from (Intermediate). (Holick et al. 1977)

2. Rhodopsin $\xrightarrow{+\text{light}}$ Bathorhodopsin $\rightarrow$ Lumirhodopsin $\rightarrow$ Metarhodopsin I

$\uparrow$ +Opsin $\downarrow$
11-cis-Retinal   Metarhodopsin II
$\uparrow$ +E^a $\downarrow$
11-cis-Retinol $\xleftarrow{+E}$ Retinol $\xleftarrow{+E}$ trans-Retinal $\xleftarrow{-\text{Opsin}}$ Metarhodopsin III

(Ostroy, 1977)

3. DNA $\xrightarrow{+UV}$ Photodimers, etc. (Varghese, 1972; Setlow, 1968; Cleaver, 1972)

4. Cholesterol $\xrightarrow{+UV}$ Cholesterol-5α,6α-Epoxide (Black and Lo, 1971; Lo and Black, 1972; Black and Douglas, 1972)

5. Protoporphyrin $\xrightarrow{+\text{Light}}$ Cutaneous photosensitivity (Magnus et al., 1961; Kosenow and Treibs, 1953; Peterka et al., 1965)

6. Bilirubin $\xrightarrow{+400-450 \text{ nm}}$ Photoproducts (McDonagh, 1976)

B. In vitro reactions

1. Protoporphyrin $\xrightarrow{+410 \text{ nm, } +O_2}$ Photohemolysis (Schothorst et al., 1970; Haining et al., 1969)

2. Oxyhemoglobin $\xrightarrow{+\text{white light}}$ Methemoglobin + O_2^- (Demma and Salhany, 1977)

^{a}E = Enzyme.

Table 5. Binding Physical Reactions

Substance (ligand)	Extracellular	Reference	Intracellular	Reference
O_2, CO_2, CO, 2,3-DPG, organic phosphates			Hemoglobin (red cells)	Kilmartin and Rossi-Bernardi (1973); Arnone (1974); Brewer and Eaton (1971)
O_2			Myoglobin (red muscle)	Wittenberg (1970)
CO_2			Hemoglobin β-chain	Bauer and Kurtz (1977)
2,3-DPG Hemoglobin			Deoxyhemoglobin	Benesch et al. (1968, 1969)
Heme	Haptoglobin	Polonovski and Jayle (1938); Sutton (1970)		
Calcium	Hemopexin	Muller-Eberhard (1970)		
	Albumin	Walser (1961)	Calcium binding protein	Huang et al. (1975)
Copper	Albumin (portal blood)	Evans (1973); Dixon and Sarkar (1974)	Metallothionein	Evans (1973)
	Ceruloplasmin (nonportal blood)	Evans (1973)		
	IgG$_1$ myeloma protein	Baker and Hultquist (1978)		
	Diglycyl-histidine (synthetic peptide)	Lau et al. (1974)		
Iron	Ferritin	Jacobs and Worwood (1975)	Ferritin	Jacobs and Worwood (1975)
	Transferrin (Siderophilin)	Schade and Caroline (1946); Wang and Sutton (1956)	Conalbumin	Schade and Caroline (1944)
	Albumin	Giroux et al. (1976)		

Iron, copper	Lactoferrin	Masson and Heremans (1968)	Lactoferrin	Leffell and Spitznagel (1972); Masson et al. (1969)
Lactoferrin	Plasma proteins	Hekman (1971)		
Lead			Hemoglobin	Raghavan and Gonick (1977)
			Low-molecular-weight protein	Raghavan and Gonick (1977)
Zinc	α_2-Macroglobulin	Parisi and Vallee (1970)		
	Albumin	Giroux et al. (1976)		
Zinc, cadmium			Zinc-binding protein (metallothionein)	Evans et al. (1970); Richards and Cousins (1977)
Fatty acids	Albumin	Spector et al. (1969)		
Cholesterol	Lipoproteins[a]	Jackson et al. (1976)	Cholesterol-binding protein	Scallen et al. (1975)
Triglycerides				
Phospholipids				
Squalene			Sterol carrier protein	Srikantaiah et al. (1976)
Lanosterol				
Estrogen	α_1-Fetoprotein	Swartz et al. (1974)		
Cysteine	Albumin	Koch-Weser and Sellers (1976a,b)		
Tryptophan				
Multiple drugs				
Vitamin A (retinol)	Prealbumin	Goodman (1974)	Intracellular binding protein	Glover et al. (1974); Saari and Futterman (1976)
	Retinol binding protein (RBP)	Goodman (1974)		
Vitamin A-RBP complex	Prealbumin	Goodman (1974)		
Retinoic acid			Intracellular binding protein	Saari and Futterman (1976)
11-cis-Retinal			11-cis-Retinal-binding protein	Futterman et al. (1977)

[a] Solubility depends on particle size.

(Continued)

Table 5 (*continued*)

Substance (ligand)	Soluble binding proteins				
	Extracellular	Reference	Intracellular	Reference	
Vitamin B$_{12}$	Intrinsic factor (in gastrointestinal tract)	Castle and Townsend (1929); Ellenbogen et al. (1958); Lien et al. (1973); Glass (1963)	PMN-Vitamin-B$_{12}$-binding protein	Burger et al. (1975)
	Transcobalamin I (R binder)	Hall (1975)		
	Transcobalamin II	Hall (1975)		
	Transcobalamin III (R binder)	Burger et al. (1975)		
	Milk vitamin-B$_{12}$-binding protein	Kirk et al. (1972)		
Folic acid			Folate binding protein	Zamierowski and Wagner (1977)
1,25-Dihydroxy-cholecalciferol			1,25-Dihydroxy-cholecalciferol-binding protein	Tsai and Norman (1973); Brumbaugh and Haussler (1974)
25-Hydroxycholecalciferol, 1,25-Dihydroxy-cholecalciferol, vitamin D$_2$, vitamin D$_3$	Group-specific component (Gc); Vitamin-D-binding globulin (VDBG)	Brissenden and Cox (1978); Svasti and Bowman (1978)		

Ligand	Binding protein	Reference
α-Tocopherol	α-Tocopherol binding protein	Catignani (1975)
Thyroxine	Prealbumin	Goodman (1974)
Triiodothyronine	Thyroxine-binding protein	Ingbar (1963); Oppenheimer (1968)
Bilirubin	Albumin	Klatskin and Bungards (1956)
	α-Fetoprotein	Verseé and Barel (1978)
Organic anions }	Y and Z proteins	Levi et al. (1969)
BSP }	Y and Z proteins	Levi et al. (1969)
β-Glucuronidase	Egasyn	Tomino and Paigen (1975)
Adenosine deaminase	Adenosine deaminase-binding protein	Daddona and Kelley (1978)
Actin, smooth muscle	Filamin	Wallach et al. (1978)

Each protein contains a hydrophilic, catalytic segment and a hydrophobic segment which is required for binding (Enoch et al., 1977). Only when the reductase and cytochrome b_5 were bound to the same type of vesicle did the rate of electron transfer from NADH to the cytochrome b_5 approach the maximum turnover rate of the reductase. When reductase–phospholipid vesicles were mixed with cytochrome b_5 vesicles, a third type of vesicle appeared (not due to vesicle fusion) which contained both proteins.

Proteins, nucleic acids, and mucopolysaccharides are able to bind substances in nonenzymatic reactions. It is the ability of proteins to bind one another, nucleic acids, and other substances that makes possible the complex protein biosynthetic system (see Chapter 5). Mucopolysaccharides (glycosaminoglycans) can bind water, ions (e.g., Na^+, Ca^{2+}, La^{3+}) (Rees, 1975; Comper and Laurent, 1978), collagen, α-elastin, tropoelastin, hyaluronate, antithrombin III, platelet factor 4, low-density and very-low-density lipoproteins, high-density lipoproteins, lipoprotein lipase and hepatic lipase (Lindahl and Høøk, 1978; Comper and Laurent, 1978). Glycosaminoglycans do not occur as free polysaccharide chains in vivo but as proteoglycans, where several polysaccharide chains are covalently linked to a polypeptide core. Proteoglycan molecules are flexible and polyanionic. They give rise to an internal osmotic pressure. The chains are "soluble" but cannot go into true solution because of the molecular structure. This results in a swollen gel, with the forces of chain separation being balanced by the elastic resistance to further swelling (Rees, 1975; Comper and Laurent, 1978).

Noncovalent (nonchemical) forces (hydrogen bonds, Van der Waals forces, electrostatic bonds, etc.; see Chapter 2) may cause aggregation of protein and enzyme molecules. The formation of multienzyme complexes is one example of enzyme aggregation, while the formation of enzyme polymers is another example (see Section I.B.2, which discusses the special functional categories of proteins).

Not all enzymatic reactions mediate covalent changes. For example, electron transport is enzyme mediated, although covalent bonds, per se, are not changed. However, electron transport leads to oxido–reductive changes that are the equivalents of covalent bonds. Electron transport represents a physicochemical category of physiological reaction. Nor are all oxido–reductive reactions enzyme mediated. Certain oxidations take place at a continual rate, and the reduced state is maintained by specific reductases. Hemoglobin and myoglobin, for example, bind oxygen in the reduced state. When both are oxidized to the metheme form (ferric form) they are no longer able to bind oxygen. The metheme forms are then reduced by their specific reductases. The change in the oxido–reductive state generates a functionally different compound, even though the actual structure is not altered. In the absence of methemoglobin reductase, methemoglobin accumulates (Hsieh and Jaffe, 1971). Since, in methemoglobin reductase deficiency, methemoglobin can be reduced by ascorbic acid, the amount of methemoglobin in the deficient patient may depend on the amount of ascorbic acid in the diet (Tomoda et al., 1976). In methemoglobin reductase deficiency, the regulation of the level of methemoglobin in the red blood cell, as a function of dietary ascorbic acid, exemplifies the interaction between the environment and the regulatory mechanisms within the cell. It is of interest that oxyhemoglobin autooxidizes to methemoglobin and superoxide (O_2^-) (McCord and Fridovich, 1978). This represents a physiological, nonenzymatic, spontaneous reaction (see Table 3). Table 6 gives some examples of redox reactions.

3. The Steady State

Physiological chemical and physical reactions operate in a steady-state system. While none of the reactions reach equilibrium (von Bertalanffy, 1950), the tendency to equilibrium is the driving force of spontaneous reactions (Krebs, 1975). In any reaction where a substrate, A, is transformed into a product, B, equilibrium will occur, given sufficient time, as denoted by the double arrows in equation (1).

$$A \rightleftharpoons B \tag{1}$$

At equilibrium, the ratio of the concentration of A ($a - x_1$) and B (x_1) are constant [equation (2)], as indicated by an equilibrium constant, $K_{eq} = K_1$,

$$\frac{(x_1)}{(a - x_1)} = K_1 \tag{2}$$

Table 6. Oxido–Reductive (Redox) Reactions

A. Hemoglobin (reduced) (Fe^{2+}) → Methemoglobin (Fe^{3+}) (Goldberg and Stern, 1976)
B. Myoglobin (reduced) (Fe^{2+}) → Metmyoglobin (Fe^{3+}) (Hagler et al., 1976)
C. Ferricytochrome b (Fe^{3+}) → Ferrocytochrome b (Fe^{2+})
D. Ferricytochrome c_1 (Fe^{3+}) → Ferrocytochrome c_1 (Fe^{2+})
E. Ferricytochrome c (Fe^{3+}) → Ferrocytochrome c (Fe^{2+}) (Harrison, 1974; Stellwagen and Cass, 1975)
F. Ferricytochrome b_5 (Fe^{3+}) → Ferrocytochrome b_5 (Fe^{2+}) (Strittmatter and Velick, 1956)
G. Ferrocytochrome b_5 (Fe^{2+}) + Ferricytochrome c (Fe^{3+}) → Ferricytochrome b_5 (Fe^{3+}) + Ferrocytochrome c (Fe^{2+}) (Strittmatter and Velick, 1956)

where (x_1) is the concentration of B at time t_1, (a) is the concentration of A at $t = 0$, and $(a - x_1)$ is the concentration of A at t_1. If another reaction is added to the system, a new equilibrium occurs,

$$A \underset{}{\overset{K_1}{\rightleftharpoons}} B \underset{}{\overset{K_2}{\rightleftharpoons}} C \tag{3}$$

where (x_2) is the concentration of C at t_2, $(x_1 - x_2)$ is the concentration of B at t_2, and $(a - x_1 - x_2)$ is the concentration of A at t_2. Thus,

$$\frac{(x_2)}{(x_1 - x_2)} = K_2 \tag{4}$$

and

$$\frac{(x_1 - x_2)}{(a - x_1 - x_2)} = K_1 \tag{5}$$

so that

$$\frac{(x_2)}{(a - x_1 - x_2)} = K_1 K_2 \tag{6}$$

For any extended sequence, equilibrium becomes complex:

$$A \rightleftharpoons B \rightleftharpoons C \rightleftharpoons \ldots \rightleftharpoons N \tag{7}$$

and

$$\frac{(x_n)}{(a - x_1 - x_2 - \ldots - x_n)} = K_1 K_2 \ldots K_n = K \tag{8}$$

If the system can be isolated so that A can be added and N can be removed,

$$A \rightarrow / \rightarrow A \rightleftharpoons B \rightleftharpoons C \rightleftharpoons \ldots \rightleftharpoons N \rightarrow / \rightarrow N \tag{9}$$

then equilibrium can never be attained. In equation (9), the symbol "/" denotes a barrier which permits only the unidirectional movement of A and N and is impermeable to all other substances. If the rate of input of A into the system equals the rate of output of N from the system, a steady state ensues. This holds true no matter how many substances enter the system, how many products leave, and no matter what the number and complexity of the intermediate steps. The steady state may be viewed as an example of the local equilibrium condition in a far-from-equilibrium system (see Chapter 2). The steady state prevents intermediates from accumulating to toxic levels and provides a carefully poised or balanced state which is susceptible to metabolic control (see, for example, Fichera et al., 1977a,b). Oscillatory behavior around the steady state, and limit-cycle formation, are less commonly recognized expressions of control in biochemical pathways and, again, represent examples of dissipative structures in a far-from-equilibrium system (see Chapter 7 on servomechanisms and oscillatory phenomena, and also Chapter 2).

The complexities of the steady-state system are treated in depth by Engasser et al. (1977). Stadtman and Chock (1977) and Chock and Stadtman (1977) provide an analysis of the steady state that occurs in monocyclic and multicyclic cascade systems. Several mathematical treatments of steady-state pathways have been developed by various investigators (Park, 1974, 1975; Rapoport et al., 1976; Sel'kov, 1975; Tschudy and Bonkowsky, 1973). As has been pointed out (Tschudy and Bonkowsky, 1973), superimposed control mechanisms on a metabolic pathway help maintain the steady state and permit the pathway to evolve to a new steady state when the need arises. Such evolutionary potential in metabolic pathways exemplifies the behavior of an open system which exists far-from-equilibrium and evidences nonlinear kinetics.

4. Cellular Structure

The physiological chemical reactions, both enzymatic and nonenzymatic, and binding reactions alone do not constitute the life process. All of the physiological chemical reactions must occur within a given, bounded volume which is termed the cell. The cell is an ordered structure which contains subcellular organelles and substances composed of simpler subunits. The cell is bounded by a plasma membrane and contains cytoplasm (cytosol), a nucleus, mitochondria, endoplasmic reticulum (microsomes in broken-cell preparations), nuclear membrane, nucleolus, chromatin, Golgi apparatus or membrane, lysosomes, peroxisomes, microtubules, secretory granules, lipid vacuoles, ribosomes, phi bodies, and other special structures depending on the type of cell. The formation of the cell and its subcellular organelles is possible only with the development of the various cell membranes. This depends on the physicochemical properties of hydrophobic phospholipids, with the insertion of hydrophobic proteins to permit communication through the membrane (Tanford, 1973, 1978).

The nucleus is surrounded by a nuclear membrane and contains chromatin, which is composed of deoxyribonucleic acid (DNA), wherein reside the genes, histones (basic protein), nuclear acidic protein, and one or more nucleoli which contain ribonucleic acid (RNA). The histones and DNA are organized into nucleosomes (see Chapter 7). It has been estimated that each nucleus contains 6.2 pg of DNA (Vendrely and Vendrely, 1956). The mitochondria have an inner and outer membrane and contain the pyruvate dehydrogenase multienzyme complex, the Krebs tricarboxylic acid cycle (citric acid cycle) enzymes, the cytochrome system which mediates electron transport and oxidative phosphorylation, and portions of the urea cycle and porphyrin synthetic pathways. The endoplasmic reticulum serves as a series of internal membranes which carry out secretory and absorptive functions in conjunction with the Golgi membranes, secretory granules, and microtubules. The endoplasmic reticulum can serve as a point of attachment for ribosomes, which are the subcellular structures involved in protein synthesis. Lysosomes contain a large set of hydrolytic enzymes which are involved in intracellular degradative processes (Dean and Barrett, 1976). Peroxisomes are more recently recognized intracellular structures that contain catalase, D-amino acid oxidase, and urate oxidase (Masters and Holmes, 1977). The cytoplasm contains some or all of the enzymes of the glycolytic, pentose phosphate, galactose, uronic acid, glycogen, fructose, mucopolysaccharide, pyrimidine, purine, amino acid, fatty acid biosynthetic, and other metabolic pathways.

The plasma membrane helps maintain the appropriate intracellular environment by confining the reactants within the small volume that constitutes the cell [about 4000 μm^3 (20 × 20 × 10) for a liver cell]. The plasma membrane is semipermeable so that escape of the reactants from the cell is minimized.

Other special structures are present in certain cells but not all. These include cilia, flagella, brush border structures, centrioles, filaments, glycogen granules, melanin granules and melanosomes, contractile proteins (actomyosin and related proteins), intercellular junctions, and water vacuoles. Peroxisomes and azurophilic granules of leukocytes (which are thought to be lysosomes) can be transformed into new crystalloid structures called phi bodies (Hanker and Romanovicz, 1977), which are rich in catalase.

The different cell structures serve to compartmentalize the cellular interior. Various types of inclusions are found within cells and represent storage components, secretory granules, pigments, crystals, or substances formed in reaction to some environmental stimulus. There are a number of reference books which treat subcellular structure in great detail and which may be consulted by the interested reader (Fawcett, 1966; Lentz, 1971).

5. The Hierarchical Organization of Intracellular Metabolic Processes

When we consider that nonenzymatic and enzymatic proteins bind substances, that enzymatic proteins not only bind substrates but alter covalent bonds, and that enzymes catalyze chemical reactions that can be linked sequentially, we realize that it is possible to conceptualize intracellular metabolic processes as organized in an ascending order of acquisition of function. Such an ascending order is depicted in Table 7. A sequential order of enzymatically catalyzed chemical reactions results in a metabolic pathway. The product of each biochemical reaction serves as the substrate for the next, thus generating a series of reactions, which constitutes the metabolic pathway. Multiple interacting metabolic pathways form a network. The output of each pathway can become the input for the next pathway, serving to link otherwise disparate pathways. In many cases, the linking reactions must traverse membrane barriers, so that the metabolic network has several compartmented areas. The metabolic network, together with ligand binding and other nonenzymatic reactions, represents the totality of intracellular chemical and physical reactions which constitutes the life process.

Although it is tempting to speculate that the acquisition of function by hydrophilic and hydrophobic proteins is the evolutionary sequence of events in the development of living systems, there is no direct proof that such is the case. In what follows, we utilize this concept to develop some of the basic principles of metabolic control as we understand them.

The emergence of catalytic function can transform a carrier-type protein into an enzyme. And, so long as its enzymatic action does not preclude its carrier function, the enzyme can continue to bind certain substances and act as a carrier. The paradigm of such a protein is hemoglobin which, having acquired a primordial enzymatic capability, exercises the crucial carrier function associated with oxygen transport. Other receptor proteins may develop enzymatic abilities which preclude their basic carrier function.

If the enzymes and resulting intermediates in a metabolic pathway coexist in a mixture, the number of times each enzyme and substrate

Table 7. The Hierarchical Organization of Intracellular Metabolic Processes

Level of function	Acquisition of function
1. Protein	1. Hydrophilic or hydrophobic solubility
2. Carrier or receptor protein	2. Binding site for ligand
3. Enzyme	3. Binding site for substrate and/or cofactor and/or allosteric modulator and catalytic site for substrate
4. Enzymatic reaction	4. Binding and catalytic sites and release of product
5. Metabolic pathway	5. Sequential order of enzymatic reactions
6. Metabolic network	6. Multiple interacting metabolic pathways

can come into contact depends statistically upon their concentrations. For optimal flux through a pathway, each substrate must find its enzyme quickly and effectively. This can be accomplished by positioning the enzyme and substrate in an organized fashion within the cell.

Although the glycolytic pathway was once considered to consist of freely diffusible soluble enzymes, it is now believed likely that these enzymes exist in ordered sequence in the cytoplasm (Ottaway and Mowbray, 1977; Foemmel et al., 1975). The rate of formation of lactic acid from glucose, as short a time as 5 s in Ehrlich ascites tumor cells (Lee et al., 1967), appears to be too rapid to be explained by random distribution of substrates and enzymes. In the reverse pathway of gluconeogenesis, glucose is formed from lactate in 75 s (Exton and Park, 1967).

6. The Principles of Metabolic Control

The ordered sequence of chemical reactions that forms a metabolic pathway presupposes metabolic control. Clearly, the metabolic pathways cannot operate at maximum rates, since substrates would become exhausted, there might be accumulation of toxic levels of one or more intermediates and various pathways might be rendered ineffectual through mutual competition. Similarly, metabolic pathways cannot operate at minimal rates, since this would be incompatible with life because of the failure to generate enough of the products and energy necessary for critical reactions, and there would be no way to adjust to changing conditions. Above certain concentrations, specific intermediates can inhibit particular reactions. This becomes of critical concern in the case of enzyme deficiencies, as discussed more fully in Chapter 15, The Biochemical Basis of Disease.

We can define a principle of metabolic control as a generalized function which is derived from the hierarchical organization of metabolic processes and which is involved in altering the rate of a chemical (enzymatic) or binding reaction. Such a function may alter the rate of a reaction by changing the amount of substrate (e.g., by regulating a transport system), cofactor, enzyme (by synthesis, degradation, activation–inactivation, aggregation, membrane-binding, etc.), or product (via a transport system, secretion, a servomechanism, etc.). Any factor which modulates function in a secondary manner (i.e., hormones, regulation of energy production, alteration of solubility, etc.) also becomes a principle of metabolic control.

Although the number of principles must be finite, we cannot be certain that we have considered all of the major principles of metabolic control. Undoubtedly, additional variations on the theme of a principle remain to be observed. Our intent in using the hierarchical organization of metabolic processes has been to develop a logical system from which the important principles can be derived.

The *basic principle of metabolic control* is the regulation of at least one rate-limiting step—the metabolic control point—in a given metabolic pathway. The specific principle is determined by the particular component of the reaction that is rate limiting, i.e., the substrate, cofactor, enzyme, etc., and how it is made rate limiting. One of the goals of theoretical biology is to determine the number of control principles, the minimum number of necessary principles, and the consequences of failure of each of the principles. A definition of the basic principle of metabolic control is that which regulates a rate-limiting step in a dose-responsive manner under otherwise constant conditions. The amount and activity of the regulatory enzymes in the system determine the direction and rates of flow of the substrates. The activities of the enzymes are dependent on their inherent catalytic abilities and activation–inactivation mechanisms. The inherent catalytic ability of an enzyme depends on its primary amino acid sequence (which dictates its three-dimensional configuration), the specific substrates, and the conditions of the molecular environment.

In principle, inherent catalytic function should be susceptible to quantum biochemical analysis. However, such an endeavor is beyond the ability of present methods (Scheiner and Kern, 1978). The activation–inactivation mechanisms are, at least in part, hormone mediated. The amount of enzyme present is dependent on the rate of protein biosynthesis versus the rate of degradation. These processes, in turn, are dependent on the effect of substrate on the protein biosynthetic mechanism necessary for the manufacture of the glycolytic, gluconeogenetic, and degradative enzymes.

The principles of metabolic control can be derived from a consideration of the hierarchical organization of intracellular processes. And by analyzing the organization of intracellular metabolic processes, one can decide what category of metabolic control constitutes a fundamental principle.

Table 8 relates the specific principles of metabolic control to the hierarchical organization of metabolic control processes. The principles of metabolic control ascend through echelons of organizational complexity within hydrophobic and hydrophilic systems. As cumulative acquisitions of functions occur in the hydrophobic and hydrophilic systems, various metabolic functions emerge, with their attendent control mechanisms. In the hydrophobic system, there appear membrane systems, membrane transport, hormonal membrane receptors coupled to signal-molecule generating systems, membrane-bound enzymes, electron transport systems, cellular secretory mechanisms, and compartmentation of functions. The physicochemical basis for the development of hydrophobic phospholipid membranes has been developed by Tanford (1973, 1978). As has been proposed, such membranes totally isolate the interior of the membrane from the exterior. However, by insertion of hydrophobic proteins into the phospholipid membrane, controlled access through the otherwise impermeable membrane becomes possible. The hydrophobic membrane and its proteins form the essential basis for membrane transport, compartmentation, and other membrane functions. In the hydrophilic system, there is the appearance of carrier proteins, modulation of enzyme action, servomechanisms, enzyme activation and inactivation, multienzyme complexes, and hormonal regulation. With the development of the metabolic network, the functions of the hydrophobic and hydrophilic systems merge, so that both are involved in protein biosynthesis, cellular secretory mechanisms, DNA replication, and mitosis.

Certain basic postulates of metabolic control can be enumerated.

1. Metabolic control is necessary to permit intracellular adjustments to extracellular changes.
2. Control mechanisms vary as to time of onset, speed of action, duration of action, magnitude of action, and responsiveness to various stimuli.
3. For every control mechanism there is a reciprocal or inverse mechanism, which subserves the same reaction but is opposite in direction. It is likely, but not certain, that there would be the same time of onset of action, speed, duration, magnitude, and responsiveness to stimuli for this reciprocal mechanism.
4. Failure of a metabolic control mechanism in a given tissue leads to impairment of the metabolic pathway, and may result in illness owing to accumulation of precursors or absence of an essential metabolite(s).

7. The Ordered State of Living Systems

The hierarchical organization of metabolic processes, and hence the ordered state of the life process, is maintained through an elaborate system of information transfer. The components of the metabolic network are continually generated and degraded at varying rates, so that a relatively constant net concentration of components is present at any given time. A certain amount of information is required, dependent on the complexity of the system, in order to maintain the concentration of components at the necessary level for each state of the system. The continued production and degradation of enzymes and the continued operation of metabolic pathways at the appropriate rates constitute the ordered system. In essence, this is a thermodynamic state of negative entropy. There are many factors which operate to disrupt the ordered state and thus increase the entropy. In order to counter these disruptive forces and to protect the metabolic network and its information system

Table 8. Relation of Specific Principles of Metabolic Control to the Hierarchial Organization of Metabolic Processes

Hydrophobic state			Hydrophilic state	
Metabolic control principle	Acquisition of function	Level of function	Acquisition of function	Metabolic control principle
Membrane structure	Hydrophobic solubility	Protein	Hydrophilic solubility	Tertiary structure
Membrane transport; hormone receptors; cell adhesion	Binding site for ligand	Carrier or receptor protein	Binding site for ligand	Soluble carrier proteins
Oxidoreduction reactions; electron transport	Binding sites for substrate and/or cofactor and/or allosteric modulator	Primordial enzyme	Binding sites for substrate and/or cofactor and/or allosteric modulator	Inhibited biochemical reaction
Oxidative phosphorylation; generation of hormone-dependent signal molecules; membrane-bound enzymes	Binding and catalytic sites and release of products	Enzymatic biochemical reaction	Binding and catalytic sites and release of products	Modulation of enzyme activity
Hormonal regulation; compartmentation	Ordered sequence of biochemical reactions	Metabolic pathway	Ordered sequence of biochemical reactions	Servomechanisms; enzyme activation–inactivation; multienzyme complexes; hormonal regulation
Protein biosynthesis and degradation; cellular secretory mechanisms; mitosis and DNA replication	Multiple interacting pathways	Metabolic network	Multiple interacting pathways	Protein biosynthesis and degradation; cellular secretory mechanisms; DNA replication and mitosis

(which is also part of the metabolic network), there exists a stabilizing system for the various components of the network.

a. Inherent Instability of Metabolic Systems. Since all living systems ultimately age and die, this implies that the metabolic network ultimately deteriorates, according to the constraints of the second law of thermodynamics. Hence, the entropy of the system increases as it goes from an ordered to a more disordered state. On reaching equilibrium, the system dies. There are multiple factors which cause the loss of function of metabolic systems, i.e., destabilize metabolic systems. Various of the components of the system may be affected, resulting in chemical change with diminution or loss of function. Proteins or lipids in cell membranes may be oxidized, reduced, or in some way affected so that the membrane becomes more permeable. Proteins in solution may be denatured by O_2, O_2^-, H_2O_2, H^+, ionizing radiation, hydroxyl radicals, singlet oxygen (see Fig. 1), intracellular proteolytic enzymes, or as a consequence of catalytic activity itself. Proteins kept in solution may slowly denature with time. Enzymes, while in solution, may slowly lose activity, possibly through interaction with water, with the oxygen dissolved in the water, or with superoxide and other free radicals generated by irradiation and endogenous reactions, or through intrinsic instability, thermal agitation, or by exposure to trace metals. Repeated enzymatic reactions may lead to gradual denaturation. Denaturation of enzymes may occur because of other factors, such as interaction with poor substrates, adherence to cell membranes, and chemical alterations (phosphorylation, acetylation, methylation, sulfation, glycosylation, reduction, oxidation, etc.). Denaturation of proteins is a type of spontaneous reaction. Denaturation of protein, especially enzymes, is the main reason for instability of biological systems. The very nature of enzyme structure and the mechanism of enzyme action (the conformability of the enzyme) renders an enzyme susceptible to denaturation. Denaturation involves the disruption of the higher-order structure of protein (unfolding), with concomitant loss of function. Up to a point, the peptide is still soluble, since its hydrophilic groups predominate over the hydrophobic ones. But past a given point, the hydrophobic groups predominate and the denatured protein precipitates. Destabilizing factors promote unfolding, while stabilizing factors help to maintain the protein in its folded condition.

For example, red blood cells gradually lose enzyme activity with time. Glucose-6-phosphate dehydrogenase (G-6-PD) activity gradually declines as the cells age (Marks, 1958); and the enzymes glutamic–oxaloacetic transaminase, 6-phosphogluconic acid dehydrogenase, lactic acid dehydrogenase, purine nucleoside phosphorylase, hexokinase,

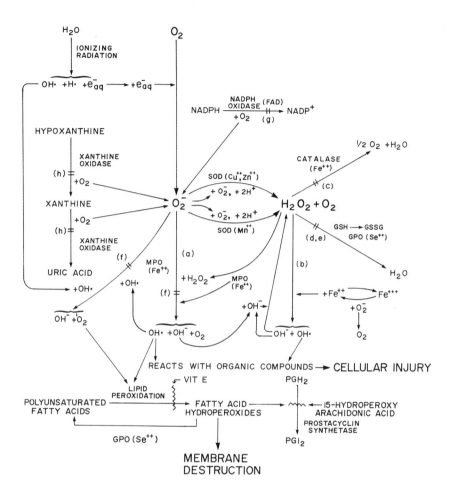

Figure 1. Reactions for the generation and disposal of superoxide (O_2^-), peroxides, singlet oxygen ($^1O_2$), and hydroxyl radicals (OH^-, $OH\cdot$). (a) Haber-Weiss reaction, (b) Fenton reaction, (c) catalase deficiency in red blood cells (acatalasemia), (d) glutathione peroxidase deficiency in red blood cells (causes a hemolytic anemia), (e) glutathione peroxidase deficiency in platelets (Glanzmann's thrombocytopenia), (f) myeloperoxidase deficiency in white blood cells, (g) chronic granulomatous disease, (h) xanthine oxidase deficiency. SOD, superoxide dismutase; MPO, myeloperoxidase; GPO, glutathione peroxidase; GSH, reduced glutathione; GSSG, oxidized glutathione; PGH_2, prostaglandin endoperoxide; PGI_2, prostacyclin; e^-_{aq}, hydrated electron; ⟿, inhibition of a reaction, ⊣⊢, enzyme deficiency (see Fridovich, 1977; Koppenol and Butler, 1977; Babior, 1978).

aldolase, and pyruvate kinase all decrease in activity as the cells age (Sass et al., 1964; Brok et al., 1966; Walter and Selby, 1966; Chapman and Schaumburg, 1967). However, no significant change occurs in glyceraldehyde-3-phosphate dehydrogenase activity (Chapman and Schaumburg, 1967). Since red blood cells do not have protein synthetic mechanisms or intracellular proteases, the loss of enzyme activity as a function of time may be due to the intrinsic instability of the enzymes. There is a gradual incremental loss of normal enzyme activity. The rate of loss of activity may increase if the enzyme structure is altered because of a mutation or by exposure to abnormal intracellular conditions.

In cell cultures of human fibroblasts, an increasing amount of altered protein occurs as the cells age (Holliday and Tarrant, 1972; Lewis and Tarrant, 1972; Goldstein and Moerman, 1975). As much as 25% of G-6-PD, 6-phosphogluconic acid dehydrogenase, and hypoxanthine-guanine-phosphoribosyltransferase are altered. It has been found that three lysosomal enzymes (N-acetyl-β-D-glucosaminidase, α-D-glucosidase, N-acetyl-α-D-galactosaminidase) and one mitochondrial enzyme (sulfite cytochrome c reductase) are not altered in aged fibroblasts (Houben and Remacle, 1978). The results were interpreted to mean that decreases in cytoplasmic enzyme resistance to heat inactivation are the result of posttranslational modification (see concept of secondary isozyme in Section I.B.2.a).

In order to minimize the instability of protein systems, there is a continual renewal of proteins and enzymes through biosynthesis and removal of proteins and enzymes by degradative systems. Thus, as denaturation and degradation occur, fresh enzymes are supplied while older molecules are eliminated. Chapter 6, on enzyme degradation, gives data on the turnover rates of various enzymes.

Despite the orderly process of enzyme biosynthesis and the denaturation and degradation of older enzymes, there are problems of stability of proteins in solution. There is continual endogenous generation of peroxides, ammonia, hydrogen ions, carbon monoxide, superoxide, singlet oxygen, and hydroxyl radicals. Oxygen, as well as cations, anions, and trace metals are present in all cellular fluids. These various substances may inhibit or destabilize various enzymes.

A variety of systems are available to minimize the different, potentially destabilizing factors. For example, histones appear to stabilize DNA in the nucleus. Buffer systems minimize the impact of organic acids. Proteins are capable of binding trace metals, anions, and cations. Enzymatic systems are present which metabolize cyanide, sulfate, ammonia, peroxides, and superoxide. Mechanisms are available to dispose of carbon dioxide, and the heat generated by metabolic processes is dissipated in aqueous solution by virtue of the specific heat of water and a complex thermoregulatory system.

Figure 1 depicts some of the probable reactions for generating and removing superoxide, peroxide, hydroxyl radicals, and singlet oxygen. These free radicals are used by granulocytes to destroy bacteria (Babior, 1978). In the process, however, the granulocyte itself is destroyed. Granulocytes are expendable and are replaced by cell renewal mechanisms. Certain enzyme deficiencies in the system impair the bactericidal action of granulocytes or the function of other cells (Fridovich, 1977; McCord and Fridovich, 1978; Babior, 1978). Superoxide dismutase can activate guanylate cyclase, presumably through the production of hydrogen peroxide, which then reacts with superoxide to produce hydroxyl radicals which are the primary stimulants of guanylate cyclase (Mittal and Murad, 1977). Superoxide can cause lysis of the erythrocyte membrane (Lynch and Fridovich, 1978). Hydrogen peroxide generated in tissue culture under the influence of fluorescent light has been seen to cause chromosome damage in adult mouse lung cells (Parshad et al., 1978). Certain aspects of the inflammatory response seem to be related to superoxide (McCord and Fridovich, 1978).

Many proteins become unstable in purified form, particularly in dilute solution, but are stable in cruder preparations. Mixtures of proteins may act to stabilize enzymes from the destabilizing effects that occur when enzymes are below a critical threshold concentration or when other proteins are not present to provide weak binding forces which stabilize the enzymes. A certain ionic strength may be necessary to maintain enzyme stability. Sulfhydryl groups or linkages may have to be protected in order to maintain enzyme function. Substrates and coenzymes can often stabilize enzymes. For example, pyruvate carboxylase is a tetramer composed of four identical protomers of molecular weight 125,000, each containing biotin, which is rapidly inactivated at low temperatures (0°C), with dissociation of the tetramer to protomers. Acetyl-CoA, which is a potent activator of pyruvate carboxylase, protects this enzyme against cold inactivation, urea, or pH changes (Frey and Utter, 1977). The isozymes of L-α-hydroxyacid oxidase have different stabilities in their purified forms. The A isozyme, when stored in ammonium sulfate at 4°C for several months, loses 70% of its activity and a considerable portion of its flavin content, while the B isozyme is more stable under the same conditions (Duley and Holmes, 1976). In general, in cold-sensitive enzymes, low temperatures destabilize hydrophobic bonds, which provide molecular stability at higher temperatures (Peat and Soderwall, 1972). It has been found that the adenine phosphoribosyltransferase of red blood cells declines markedly as the cells age (Rubin et al., 1969) but increases in activity in cells which accumulate 5-phosphoribosyl-1-pyrophosphate (PRPP) (Green et al., 1970) as a consequence of a deficiency of hypoxanthine-guanine phosphoribosyltransferase (Lesch–Nyhan syndrome). The increased concentration of free PRPP protects the adenine phosphoribosyltransferase from heat inactivation and prolongs the half-life of the enzyme.

In rat liver there are two types of glutaminase, a phosphate-dependent and a phosphate-independent form. The phosphate-dependent form polymerizes at phosphate concentrations of 10^{-1} M and is thereby activated. In contrast, the phosphate-independent form of the glutaminase is polymerized to its active state by its substrate, glutamate, and this action of glutamate is facilitated by maleate, N-acetyl-glutamate, or tricarboxylic acids. Although phosphate does not activate or enhance activation by the glutamate substrate, 10^{-3} M phosphate does stabilize the enzyme. In the absence of phosphate, the enzyme is very labile. Concentrations greater than 10^{-3} M phosphate, however, inhibit the polymerization of the enzyme (Katsunuma et al., 1968).

Argininosuccinase loses its activity in the cold except in the presence of substrate or phosphate. Activity is regained by warming for a brief period (Havir et al., 1965). Cold inactivation is caused by dissociation of the enzyme into catalytically inactive subunits.

Purified rat liver carbamyl phosphate synthetase is unstable, but this can be overcome by storage at $-20°C$ in a buffered ammonium sulfate solution containing 20% glycerol (Guthöhrlein and Knappe, 1968).

Kynurenine transaminase of rat kidney is inactivated at a pH less than 7.0 in the presence of phosphate, but this is prevented by the coenzyme, pyridoxal phosphate.

As noted earlier, alterations in protein structure may increase the instability of a protein. Although certain changes in amino acid sequence may lead to no discernible difference in stability or function of an enzyme, as compared to the "normal" structure, other changes may drastically alter function, stability, or both. A number of mutant species of different enzymes are known which lead to different enzyme abnormalities with different clinical manifestations; however, the exact molecular mechanisms that are involved are unknown in most cases, since complete amino acid structures of mutant enzymes generally are unknown. For example, there are more than 80 species of G-6-PD which can be distinguished from one another by biochemical criteria (Yoshida, 1973), but the structures of these variant enzymes are not known. Thus, the correlation between amino acid structure and biochemical function or biochemical properties or both cannot be made. However, the structure of the various mutant species of hemoglobin are known. It is instructive, therefore, to look at the structures of the various unstable hemoglobins and determine how the changes in their amino acid sequences lead to alterations in their functions, stabilities, or both. Changes in primary structure which lead to the instability of certain hemoglobins exemplify the forces which destabilize biological substances.

Some 56 hemoglobin variants are known which denature and precipitate within the red blood cell, forming a so-called Heinz body. There is an associated hemolytic anemia of varying severity, which is termed congenital Heinz body hemolytic anemia (Bunn et al., 1977). Examination of the structures of the abnormal hemoglobins suggests that instability can be attributed to five mechanisms (Perutz and Lehmann, 1968; Morimoto et al., 1971). There may be (1) amino acid substitution in the vicinity of the heme pocket; (2) disruption of secondary structure; (3) substitution in the interior of the subunit; (4) amino acid deletions; and (5) elongation of the subunit.

Heme is inserted into a hydrophobic cleft on the surface of the hemoglobin subunit, interacting with invariant, nonpolar amino acids. Substitution of a polar for a nonpolar residue may allow water into the hydrophobic heme pocket, weakening the heme–globin linkage. Unstable hemoglobins which have amino acid substitutions and deletions in the heme pocket are hemoglobins Hammersmith, Zurich, Koln (Jacob et al., 1968a,b), Sabine (Schneider et al., 1969), Bristol, Boras, Olmsted, and Shepherds Bush.

About 75% of hemoglobin is in the form of an α-helix and 25% is a random coil form. Certain amino acid substitutions could shift the equilibrium between the α-helix and the random coil, thereby altering tertiary structure (Chou and Fasman, 1974). One such substitution is proline, which cannot participate in an α-helix (except as one of the first three residues). Hemoglobins Duarte and Madrid have a proline substitution for alanine, while eight other hemoglobins substitute proline for leucine.

Substitution of polar for nonpolar amino acids in the interior of the subunit could destabilize the hydrophobic interactions within the interior (Bunn et al., 1977). The polar amino acid substitutions could allow water into the hydrophobic interior, leading to alteration of the tertiary structure. Neutral amino acid substitutions for one of the interior amino acids may alter the stereochemical congruity of the interior amino acids. Several hemoglobins have amino acid substitutions in the subunit interior: Sogn, Riverdale-Bronx, Russ, Ann Arbor, Wien, Bristol, Shepherds Bush, Boras, and Olmsted. Of the unstable hemoglobin variants, 17 involve substitution of one noncharged amino acid for another (Bunn et al., 1977).

Deletion of one or more amino acids may affect conformation of the hemoglobin structure (Bunn et al., 1977). There are 10 variants known with from one to five amino acid deletions: Leiden, Lyon, Freiburg,

Leslie, Coventry, St. Antoine, Tours, Niteroi, Tochigi, and Gun Hill. The deletions occur at or near interhelical corners. In the case of Hemoglobin Gun Hill, heme binding to the β-chain is not possible because of deletion of five residues in the F–FG region.

Hemoglobin Cranston is unstable, probably because of a hydrophobic segment that is attached to the C-terminal end of the β-chain.

It is proposed that the changes in the amino acid structure (substitutions, deletions, additions, or elongations) alter the tertiary structure of the hemoglobin. This alters the heme–globin linkage, which is dependent on the precise stereochemical fit by the amino acids in the heme pocket. The oxidation state of the heme iron also affects the heme–globin linkage, since ferriheme is less tightly bound to the globin than ferroheme. Thus, the iron of the ferriheme group binds to certain groups of the globin chain, forming ferrichromes of the reversible and then finally of the irreversible type (Rachmilewitz, 1974). This results in precipitation of the hemoglobin and formation of the Heinz body, which becomes hydrophobically bonded to the cell membrane. The deoxygenation of the hemoglobin to form oxidized hemoglobin results in the formation of superoxide, which enhances the hemoglobin precipitation. The hydrogen peroxide formed through the action of superoxide dismutase (=erythrocuprein) interacts with the superoxide, generating hydroxyl radicals, which destroy the cell membrane, causing hemolysis (Winterbourn and Carrell, 1974; McCord and Fridovich, 1978) (Fig. 2).

Many proteins and enzymes within the cell appear to have a half-life that depends upon their amino acid structures. Alterations in the amino acid sequence may increase the liability to denaturation. Various mechanisms exist to stabilize the enzymes and protect them from the destabilizing factors in their environment. Destabilization becomes more evident when protein biosynthesis, which maintains the enzymes at a steady-state concentration, decreases for whatever reason. Thus, when there is decreased replacement of protein and enzymes, destabilization is aggravated by the loss of enzymes necessary to maintain the stabilizing systems.

Degradation of proteins and enzymes is accomplished by a set of proteolytic enzymes (Schimke, 1975). Degradation is a specific category of the more general class of proteolytic regulation of metabolic systems. Proteolytic enzymes are involved in the process of zymogen activation, i.e., conversion of inactive precursors (zymogens) into active forms by selective enzymatic cleavage (limited proteolysis). Zymogen activation is involved, for example, in blood clotting, fibrinolysis, complement activation, proteolytic digestion, hormone biosynthesis, enzyme activation, and fertilization (Neurath and Walsh, 1976). Degradation is

The Principles of Metabolic Control

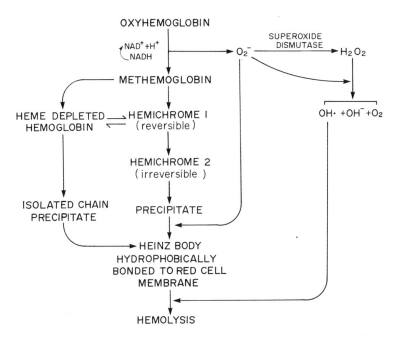

Figure 2. Mechanism of hemolysis by unstable hemoglobin. The figure depicts the postulated sequence of events in the precipitation of hemoglobin and its attachment to the red cell membrane rendering the membrane susceptible to attack by hydroxyl radicals (OH·, OH$^-$) generated from superoxide (O$_2^-$). Modified from Winterbourn and Carrell (1974), *J. Clin. Invest.* **54**:678, with kind permission from the authors and the Rockefeller University Press.

accomplished intracellularly by various lysosomal cathepsins (Dean and Barrett, 1976). Control of proteolysis is necessary to prevent destruction of important proteins and enzymes. One way in which such control is effected is that intracellular lysosomal enzymes are compartmentalized. In the vascular system, the serine proteases of the blood clotting mechanism are inactivated by various proteolytic inhibitors such C'1 esterase inhibitor, α_2-macroglobulin, inter α-trypsin inhibitor, α_1-antitrypsin, α_1-chymotrypsin, and anti-thrombin III. Deficiencies of proteolytic inhibitors can lead to destruction of various proteins and enzymes, resulting in a number of pathological processes (Ulevitch et al., 1975).

Nucleic acids also can be affected by the intracellular environment. Biochemical changes can occur in DNA. The instability of DNA can lead to profound genetic abnormalities including base substitutions, additions and deletions, crossovers, duplications, inversions, translocations, and ultimately to a variety of gross chromosomal aberrations

(Drake and Baltz, 1976). The instability of DNA and the occurrence of mutations is considered briefly in Section II. The ability of DNA to denature (unwind) is essential, on the one hand, for replication and in order to serve as a template for RNA synthesis, and, on the other hand, may lead to irreversible changes. This ability is similar to the conformability of enzymes, which is essential for enzymatic action but which may lead to irreversible denaturation.

Why must a living system renew itself? Evidently this is the price biological systems must pay for their inherent instability, since a totally stable system is a dead system, existing in chemical and thermodynamic equilibrium. In stark contrast, a living system is not completely covalent and hence is unstable, existing in a far-from-equilibrium state which is responsive to fluctuations in the environment.

b. The Hierarchical Organization of Metabolic Control. Analysis of the various principles of metabolic control suggests that they can be arranged in an hierarchical fashion. Figure 3 illustrates the sites of control and the hierarchical organization. Control may take place at the plasma membrane level (transport and hormonal receptor systems), the cytoplasmic level (generation of hormonal signal molecules, intracellular carrier proteins), the enzyme level (local environmental factors, activation–inactivation of enzymes, enzyme degradation, servomechanisms), the ribosomal level (protein biosynthesis), and the nuclear level (hormonal regulation, substrate induction, product repression, and gene control of mRNA and other RNA production).

c. The Time-Order Rates of Regulatory Mechanisms. An advantage that accrues from the hierarchical organization of metabolic control is the possibility of a range of responses which differ in their time of onset and duration of expression. Regulatory reactions occurring at the cell membrane level (substrate or product transport, hormone action) result in a rapid response, occurring in minutes to hours. Regulatory reactions occurring at the enzyme level (enzyme activation–inactivation, servomechanisms, degradation) may occur rapidly in minutes to hours, or, in the case of degradation, more slowly, in hours to days. For example, the effect of glucagon on gluconeogenesis in the perfused rat liver occurs in about 70 s (Exton and Park, 1968). Regulatory reactions occurring at the nuclear level (hormone action, substrate induction–product repression) also occur more slowly, in hours to days. Cell replication and cell renewal (see Section I.C) is the slowest process of all, taking days or even months. The exact time values for the time-order rates of the regulatory mechanisms are discussed in the appropriate chapters on protein biosynthesis, protein degradation, the mechanisms of hormone action, servomechanisms, etc. Thus, each type of regulatory mechanism covers a period of time and sets in operation a series of

The Principles of Metabolic Control

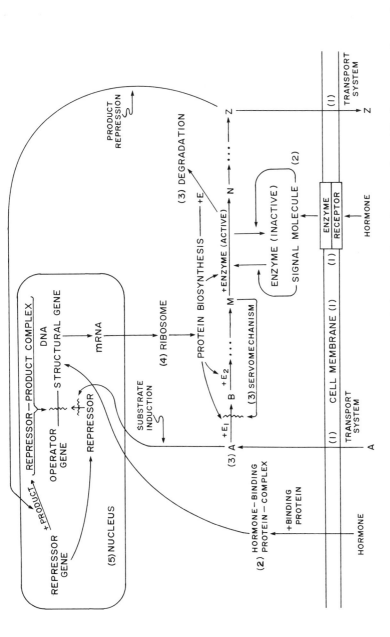

Figure 3. Hierarchical levels of metabolic control. Sites of metabolic control are designated as: (1) plasma membrane level: active transport systems, hormone receptors; (2) cytoplasmic level: hormone binding protein complex, signal molecule generation; (3) enzymatic level: steady-state enzymatic pathway, servomechanisms, enzyme degradation; (4) ribosomal level: protein biosynthesis; (5) nuclear level: hormonal control of gene action, operon control of gene action (substrate induction, product repression). The symbol, ∿, indicates inhibition of a reaction.

biochemical events which allows the cell to adjust to an environmental stimulus for that period of time. The period of time during which biochemical changes are necessary depends on the nature of the environmental stimulus, its persistence and strength, and the type(s) of regulatory control(s) involved.

B. The Products of Protein Biosynthesis

1. General Functional Categories of Proteins

The synthesis of protein is the ultimate result of the transcription and translation of the information contained in DNA. Various RNA species are formed via transcription (e.g., mRNA, tRNA, rRNA, etc.) and are then utilized in protein biosynthesis. Proteins can be classified in terms of function or in terms of their subcellular distribution in the cytoplasm, mitochondria, lysosomes, multienzyme complexes, brush border, nuclei, peroxisomes, etc. The nature of the specific tissues of the body is determined by the type and distribution of the various proteins of that tissue. Failure to produce the normal protein structure, sufficient amount of protein, or to regulate the production of protein in response to the appropriate stimulus results in disease in one or more tissues or organs, dependent on the distribution of that protein. The manifestations of the disease will depend on the function of the involved protein. These problems are considered in Chapter 15, The Biochemical Basis of Disease.

2. Special Functional Categories of Proteins

a. Categories of Function. While most enzymes have a single catalytic function and exist in a single milieu (soluble system, membrane system, etc.), certain enzymes have multiple functions, including nonenzymatic functions, and exist in more than one milieu or in a combined state. Table 9 lists examples of enzymes that have special functions. There are sets of enzymes that function sequentially in a multienzyme complex, which constitutes a special metabolic pathway that is not quite particulate and not quite soluble. The multienzyme complex allows for repetitive enzymatic reactions (e.g., fatty acid synthesis) or for metabolism of compounds that require special conditions (α-ketoacids).

Certain enzymes exist in both a reversibly membrane-bound state and a soluble state and, depending on intracellular conditions, will exist predominantly in one or the other state. While many enzymes show group specificity, using a number of related substrates, there is generally a preferred major physiological substrate that is metabolized by a specific enzyme to the general exclusion of other substrates. But certain single-enzyme molecules have specificity for two or more major physiological substrates. The multifunctional enzymes utilize different substrates, all of which are physiologically and metabolically important.

It is important to distinguish between enzymes of broad specificity, multienzyme complexes, and multifunctional enzymes. Multifunctional enzymes "are [those] characterized . . . by multiple binding or by mixed catalytic and binding functions and for which the existence of domains has been established . . ." (Kirschner and Bisswanger, 1976). Enzymes with broad specificity catalyze a number of reactions at a single active center, and the reactions are usually chemically similar, e.g., glyceraldehyde-3-phosphate dehydrogenase (Francis et al., 1973) and glucose-6-phosphatase (Nordlie, 1974), while multienzyme complexes are composed of distinct enzyme subunits that associate with each other by noncovalent interactions. In some cases, enzyme activities which had been attributed to multienzyme complexes have been found to reside in multifunctional enzymes. Such is the case for fatty acid synthetase (Lornitzo et al., 1975; Muensing et al., 1975; Qureshi et al., 1974).

Multiple forms of the same enzymes, each with their distinct structures, often exist in the same or different tissues. Many of these isoenzymes or isozymes, as they are called, consist of separate subunits arranged in all possible combinations (Purich and Fromm, 1972). Isozymes may be classified in accordance with their genesis. Primary isozymes arise from multiple gene loci or from multiple alleles at a single locus (Harris, 1969). Secondary isozymes are those thought to arise through posttranslational modifications of protein structures (Turner et al., 1975) (see Sections I.C.1 and I.A.7.a). Five of 16 human red cell enzymes were shown to undergo electrophoretic changes with increasing red cell age. Identical changes to those occurring in vivo to purine–nucleoside phosphorylase could be demonstrated in vitro using the partially purified enzyme from cultured human lymphoid cells (Turner et al., 1975). An extensively studied isozyme is lactate dehydrogenase, a tetramer composed of two types of subunits. Various arrangements of the subunits give rise to five types of lactate dehydrogenase which vary in electrophoretic mobility. These types of isozymes also belong to the class of aggregated enzymes. Examples of isozymes occurring in different tissues are listed in Table 9, D. Certain isozymes are differentially localized in the same cell (Table 9, E).

Some proteins have primary functions that are nonenzymatic, yet under certain conditions they exhibit enzymatic activity (Table 9, F). Whether such enzymatic function is physiologically significant is not always clear.

Table 9. Special Functional Categories of Enzymes and Certain Proteins

A. Aggregated enzymes
  1. Multienzyme complexes
     a. Fatty acid synthetase complex (Volpe and Vagelos, 1973)
     b. Pyruvate dehydrogenase complex (Reed and Cox, 1966)
     c. α-Ketoglutarate dehydrogenase complex (Reed and Cox, 1966)
     d. Branched chain amino acid ketoacid analogue dehydrogenase complex (Goedde and Keller, 1967)
     e. Orotate phosphoribosyltransferase: Orotidylate decarboxylase complex (Kavipurapu and Jones, 1976)
  2. Polymer forms: Monomer ⇌ Dimer ⇌ . . . ⇌ Polymer
     a. Phosphorylase a, inactive tetramer ⇌ Phosphorylase b, active dimer (Metzger et al., 1967)
     b. Ceruloplasmin monomer ⇌ Dimer (Poillon and Bearn, 1967)
     c. Ferritin ⇌ Dimer ⇌ Trimer ⇌ Polymer (Jacobs and Worwood, 1975)
     d. Haptoglobin ⇌ Homopolymers (Hp^2/Hp^2) ⇌ Heteropolymers (Hp^2/Hp^1) (Sutton, 1970)
     e. Hemoglobin S (sol) ⇌ Polymer (gel) (Waterman and Cottam, 1976)
     f. Tyrosinase (soluble) → Tyrosinase (aggregate) (Chian and Wilgram, 1967)
     g. L-Threonine dehydrase (monomer) $\underset{}{\overset{+AMP}{\rightleftharpoons}}$ Tetramer (Phillips and Wood, 1964; Whanger et al., 1968)
     h. Hydroxyacid oxidase monomer ⇌ Dimer ⇌ Tetramer (Phillips et al., 1976)
     i. Acetyl-CoA carboxylase inactive protomer $\underset{}{\overset{+citrate}{\rightleftharpoons}}$ Active polymer (Halestrap and Denton, 1974)
     j. Mitochondrial carbamyl phosphate synthetase monomer ⇌ Dimer (Clarke, 1976)
     k. Glutaminase, phosphate-dependent $\underset{}{\overset{+phosphate}{\rightleftharpoons}}$ Active polymer (Katsunuma et al., 1968)
     l. Glutaminase, phosphate-independent $\underset{}{\overset{+substrate}{\rightleftharpoons}}$ Active polymer (Katsunuma et al., 1968)
     m. Argininosuccinase active tetramer $\underset{+warm}{\overset{+cold}{\rightleftharpoons}}$ Inactive dimer (Ratner, 1973)
     n. Rat liver pyruvate carboxylase tetramer ⇌ Dimer ⇌ Monomer (Taylor et al., 1978)
B. Reversibly bound enzymes
  1. Hexokinase: Cytoplasm ⇌ Rat jejunal mitochondrial membrane (Anderson and King, 1975)
  2. Various enzymes: Cytoplasm ⇌ Skeletal muscle F-actin (Arnold and Pette, 1968, 1970)
     a. Aldolase
     b. Triosephosphate dehydrogenase
     c. Glyceraldehyde-3-phosphate dehydrogenase
     d. Fructose-6-phosphate kinase
     e. Phosphoglycerate kinase
     f. Pyruvate kinase
     g. Lactate dehydrogenase
  3. Various enzymes: Cytoplasm ⇌ Beef red blood cell membrane (Green et al., 1965)
     a. Triosephosphate dehydrogenase
     b. Aldolase
     c. 3-Phosphoglyceraldehyde dehydrogenase

(continued)

Table 9 (*continued*)

   d. 3-Phosphoglyceric acid kinase
   e. Pyruvate kinase
   f. Lactate dehydrogenase
   g. Glucose-6-phosphate dehydrogenase
   h. 6-Phosphogluconic acid dehydrogenase
  4. Glyceraldehyde-3-phosphate dehydrogenase: Cytoplasm ⇌ Human red blood cell membrane (Shin and Carraway, 1973; McDaniel et al., 1974)
  5. Hexokinase: Cytoplasm ⇌ Liver mitochondria (Ruchti and McLaren, 1965)
  6. Ribonuclease: Cytoplasm ⇌ Liver mitochondria (Beard and Razzell, 1964)
  7. Diaphorase: Cytoplasm ⇌ Liver mitochondria (Conover and Ernster, 1962)
  8. Hexokinase: Cytoplasm ⇌ Ascites tumor mitochondria (Rose and Warms, 1967)
  9. Hexokinase: Cytoplasm ⇌ Heart particulate state (Hernandez and Crane, 1966)
  10. Fructose 1,6-bisphosphate aldolase: Cytoplasm ⇌ Rat liver endoplasmic reticulum (Foemmel et al., 1975)
C. Multifunctional enzymes
  1. Sucrase-isomaltase (Conklin et al., 1975)
  2. Fructose-1-phosphate aldolase—fructose-1,6-diphosphate aldolase (Peanasky and Lardy, 1958)
  3. Formyl-methenyl-methylenetetrahydrofolate synthetase (combined) (Paukert et al., 1976)
  4. Formiminoglutamate: Tetrahydrofolate formiminotransferase—formiminotetrahydrofolate deaminase (Drury et al., 1975)
  5. Delta^5-3β-hydroxysteroid dehydrogenase-isomerase (Ford and Engel, 1974)
  6. Aromatic amino-acid-synthesizing enzyme with 5 separate sequential enzymatic activities (Gaertner and Cole, 1977)
  7. Acetyl-CoA carboxylase (Tanabe et al., 1975)
  8. Fatty acid synthase (Lornitzo et al., 1975; Muensing et al., 1975; Stoops et al., 1975; Qureshi et al., 1974)
  9. Glycogen debranching enzyme (Taylor et al., 1975; White and Nelson, 1975)
  10. First three enzymes of the pyrimidine nucleotide pathway: Carbamylphosphate synthetase; aspartate transcarbamylase; dihydro-orotase (Coleman et al., 1977)
D. Isozymes

Enzyme	Types	Subunit structure	Tissue distribution
1. Lactate dehydrogenase (tetramer) (Kitamura and Nishina, 1975; Eventoff et al., 1977)	1	H_4	Heart, fetal muscle
	2	H_3M	
	3	H_2M_2	
	4	HM_3	
	5	M_4	Skeletal muscle
2. Creatine phosphokinase (dimer) (Hooton, 1968)	I	MM	Skeletal muscle
	II	MB	
	III	BB	Fetal muscle, brain, heart
3. Fructose 1,6-diphosphate aldolase (tetramer) (Horecker, 1975)		A_4	Muscle, brain, all tissues
		B_4	Liver, kidney
		C_4	Brain, testes, heart, spleen
		A_3B	Kidney
		A_2B_2	Kidney
		AB_3	Kidney

(*continued*)

Table 9 (*continued*)

Enzyme	Types	Subunit structure	Tissue distribution
		AC_3	Brain
		A_2C_2	Brain
		A_3C	Brain
4. Isocitric dehydrogenase (Bell and Baron, 1961)	1		Heart, muscle
	2		Liver, kidney
	3		Liver, kidney, heart, muscle
	4		Muscle, heart
5. Glutamic–pyruvic transaminase (dimer) (Chen and Giblett, 1971)	1	AA	Red blood cell
	2-1	AB	Red blood cell
	2	BB	Red blood cell
6. AMP deaminase (tetramer) (Ogasawara et al., 1975, 1977)	A	A_4	Rat muscle
	B (V)	B_4	Rat liver, kidney, brain
	C (I)	C_4	Rat heart, brain
	II	C_3B	Rat brain
	III	C_2B_2	Rat brain
	IV	CB_3	Rat brain
7. Phosphofructokinase (tetramer) (Gonzalez and Kemp, 1978)		A_4	Rabbit muscle, heart, adipose tissue
		B_4	Rabbit liver, red blood cell, adipose tissue
		A_2B_2	Rabbit adipose tissue
		A_3B	Rabbit adipose tissue
		AB_3	Rabbit adipose tissue
8. Pyruvate kinase (Berglund et al., 1977)	M		Muscle, brain
	L		Liver, kidney, small intestine
	K (M_2)		Kidney, other tissues

E. Isozymes differentially localized within the same cell

Enzyme	Intracellular localization	Function/reference
1. Carbamyl phosphate synthetase (*N*-acetyl glutamate dependent)	Inner mitochondrial membrane	Urea cycle (Clarke, 1976)
Carbamyl Phosphate Synthetase (Glutamine Dependent)	Cytosol	Pyrimidine biosynthesis (Ratner, 1973)
2. Aconitate [citrate(isocitrate)] hydro-lyase	Mitochondria Cytosol	Interconversion of citrate, isocitrate and *cis*-aconitate (Koen and Goodman, 1969)
3. Malate dehydrogenase (NAD^+)	Mitochondria Cytosol	Conversion of malate to oxalacetate (Sophianopoulos and Vestling, 1960)
4. Malate dehydrogenase ($NADP^+$)	Mitochondria (heart, kidney) Cytosol (liver, brain, lung, skeletal muscle, kidney, adipose tissue)	Decarboxylation of malate to CO_2 and pyruvate (Henderson, 1966)

(*continued*)

Table 9 (continued)

Enzyme	Intracellular localization	Function/reference
5. Aspartate aminotransferase (glutamate–oxalacetate transaminase)	Mitochondria Cytosol	Transamination (Borst and Peeters, 1961)
6. Creatine kinase	Mitochondria Cytosol	Formation of phosphocreatine (Jacobs et al., 1964)
7. Isocitrate dehydrogenase (NADP$^+$)	Mitochondria Cytosol	Conversion of isocitrate to 2-oxoglutarate (Henderson, 1965)
8. Glucokinase (adaptive)	Cytosol	Formation of glucose 6-phosphate (Niemeyer et al., 1975)
Glucokinase (nonadaptive)	Golgi apparatus	(Berthillier et al., 1976)
9. α-D-Mannosidase	Cytosol	Hydrolysis of mannoside (Marsh and Gourlay, 1971; Shoup and Touster, 1976)
	Golgi apparatus	(Dewald and Touster, 1973)
	Lysosomes	(Carroll et al., 1972)
10. Protein–disulfide isomerase	Endoplasmic reticulum, cytoplasmic surface Endoplasmic reticulum, luminal surface	Folding of newly synthesized polypeptide chains (Ohba et al., 1977)

F. Nonenzymatic and enzymatic functions of certain proteins

Protein	Nonenzymatic function	Enzymatic function	References
1. Hemoglobin	O_2–CO_2 carrier	Peroxidase	Sasazuki et al. (1974); Polonovski and Jayle (1938)
		Hydroxylase	Mieyal and Blumer (1976)
2. Hemoglobin–haptoglobin complex	Hemoglobin binding	Enhanced peroxidase activity	Jayle (1951); Chung and Wood (1971)
3. Ceruloplasmin (ferroxidase)	Copper carrier	Oxidase activity for: serotonin, norepinephrine, adrenochrome, dihydroxynorepinephrine	Owen and Hazelrig (1968); Sankar (1959); Huber and Frieden (1970); Holmberg and Laurell (1951); Curzon and O'Reilly (1960)

(continued)

Table 9 (*continued*)

Protein	Nonenzymatic function	Enzymatic function	References
		Ferroxidase activity	Osaki et al. (1971)
4. Myoglobin	O_2 carrier	Oxidative decarboxylation	Milligan and Baldwin (1967)
5. Ligandin (Y protein)	Ligand binding of bilirubin, sulfobromophthalein, etc.	Glutathione-S-transferase	Litwack et al. (1971); Habig et al. (1974)

b. The Significance of the Special Functional Categories. We can speculate that the special functional categories of protein offer certain metabolic advantages, or represent intermediate states in the acquisition of metabolic function, or both. If the order of acquisition of metabolic function parallels the hierarchical organization of metabolic function (Table 8), then reversibly bound enzymes may represent a transitional phase between a completely soluble enzyme and a completely membrane-bound enzyme. We would expect to find certain enzymes that are reversibly bound. Such function would be retained in certain cases, since this would offer a mechanism of control. In the membrane-bound state, the enzyme would be inactive and would become active only in the soluble state. The polymeric states of enzyme subunits could give rise to new functions. With evolutionary alteration of subunit structure, sufficient similarity between the subunits would allow for aggregation, while each different subunit would retain its functional identity, giving rise to a multienzyme complex. As is known, the multienzyme complex forms a cyclic system allowing for repetitive action on the substrate or for a controlled reaction utilizing a substrate otherwise difficult to metabolize (e.g., α-ketoacids). Proteins that have both nonenzyme and enzyme function may represent a transitional phase between the evolutionary acquisition of substrate binding sites and the further acquisition of catalytic sites. That is, such proteins may represent primitive enzymes. A later developmental phase could then have been the acquisition of more than one catalytic site, for related substrates, on the same protein molecule.

The ability of proteins to bind prosthetic groups or cofactors and other proteins may have led to the development of enzymatic activity. Those proteins (e.g., hemoglobin, etc.) which have both nonenzymatic and enzymatic functions may represent one evolutionary pathway by which a protein may have acquired enzymatic activity. In the case of hemoglobin, enzymatic activity may have developed when the globin could bind a cofactor (heme) and may have acquired increased activity when the hemoglobin could bind a protein (haptoglobin) (see Tables 5 and 9). As enzyme activity evolved, a great variety of substances would be generated which could bind other proteins and lead to the appearance of new enzyme activities. Some enzymes bind to nonenzymatic proteins which alter their properties (see Table 5). The special functional categories of proteins may represent various evolutionary stages in the development of enzyme function. This is outlined in Fig. 4.

3. Protein Alterations and Modifications

Proteins and enzymes often are altered by covalent changes occurring directly on one or more constituent amino acids of the protein. These are referred to as posttranslational modifications and are discussed in Chapter 5.

C. Cellular Replacement and Replication

1. The Need for Cellular Replacement

The information necessary for protein synthesis resides in DNA and is transmitted to the ribosomal protein synthetic machinery via mRNA. While this system provides for the continual renewal of protein and enzymes (which are continually degraded), minimizing the effects of destabilizing events and degradation, the DNA also undergoes replication prior to mitosis. Cells become senescent and are dismantled and discarded. Many cellular systems possess a germ cell layer that undergoes mitosis, generating more cells. These cells undergo differentiation, forming an adult or mature form, function at this stage for a variable period of time, and then deteriorate. The cells are sloughed off or disposed of by an appropriate degradative system, with reutilization of certain components. Table 10 lists various cellular systems which undergo continual cellular renewal.

Continual cellular renewal would not be necessary if it were not for the senescense of the cell, which occurs by virtue of the unstable nature of its macromolecules. Thus, in certain cells, despite the biosynthetic renewal of proteins and enzymes and the presence of various stabilizing systems, the cell can only exist for a relatively short time, after which deterioration occurs. Since deterioration cannot be prevented in certain cell types, the nature of the life process dictates that cell renewal must occur, and DNA replication and mitosis are consequently neces-

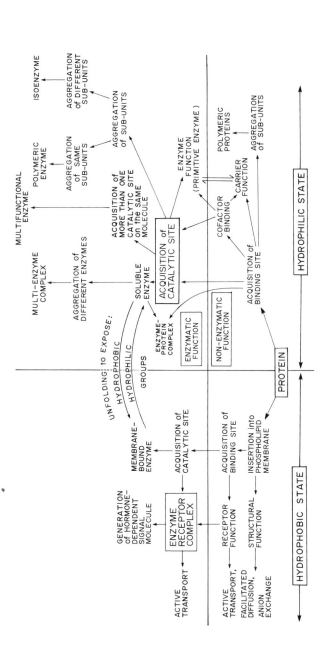

Figure 4. Speculative developmental relationships between different functional categories of proteins and enzymes. The figure depicts the distinction between hydrophobic and hydrophilic proteins. Sequential acquisition of function determines the nature of the protein. With the acquisition of catalytic function, some soluble enzymes (primitive enzymes) still retain the nonenzymatic functions. Other enzymes aggregate to form different complexes (multienzyme complexes, polymers, isoenzymes). Some enzymes acquire more than one catalytic site and become multifunctional enzymes. Certain soluble enzymes are able to alter their conformation and bind to membranes in a reversible fashion. Within membrane systems, hydrophobic proteins have receptor, transport, and structural functions. With the acquisition of catalytic sites, hydrophobic enzymes subserve a variety of functions but generally do not form aggregated complexes. Certain enzymes and nonenzymatic proteins have both hydrophobic and hydrophilic properties so that a portion of the polypeptide is embedded within a hydrophobic membrane while another portion exists within an aqueous milieu (not shown in the figure). The acquisition of control regions

Table 10. Cell Renewal Systems

A. Continuous mitotically-active cell populations

Cell system	Species	Turnover time[a] (days)	References
1. Bone marrow erythropoiesis	Rat	2.5	Widner et al. (1951)
2. Bone marrow myelopoiesis	Rat	1.4	Widner et al. (1951)
3. Small intestinal epithelium	Human	4–6	MacDonald et al. (1965); Shorter et al. (1964)
	Rat	24 h	Messier and Leblond (1960)
	Rat	48 h	Leblond and Stevens (1948); Imondi et al. (1969)
4. Epidermis	Human		
	Abdomen	100	Katzberg (1952)
	Thorax	45	Johnson et al. (1960)
	Forearm	13	Katzberg (1952)
5. Corneal epithelium	Human	7	Hanna et al. (1961)
6. Lymphatic cells	Rat		
	Thymus	7	Everett and Tyler
	Spleen	15	(Caffrey) (1967)
	Mesenteric lymph node	20	
7. Spermatogonia	Human	74	Heller and Clermont (1963)
8. Epithelial cells	Rat vagina	3.9	Bertalanffy and Lau (1963)
	Human cervix	5.7	Richart (1963)

B. Intermittent mitotically-active cell populations

Cell system	Species	Turnover time (days)	References
1. Hepatic parenchymal cells	Rat	400–450	MacDonald (1961)
2. Renal interstitial cells	Mouse	165	Cameron (1970)
3. Osteocytes	Mouse	Very slow, if at all	Cameron (1970)
4. Arterial endothelial	Rat	Very slow	Schwartz and Benditt (1976)

C. Non-mitotically-active (static) cell populations

Cell system	Species	References
1. Neurons	Mouse	Cameron (1970)
2. Cardiac muscle	Mouse	
3. Molar odontoblasts	Mouse	
4. Sertoli cells of testis	Mouse	

[a] Cell turnover includes the entire process of cell renewal: replication, migration, and extrusion for epithelial cell populations. Turnover time is the time required to replace the cells equal to that in the entire population.

sary. The exact molecular mechanisms leading to aging and loss of function are unknown. However, the process of continual cellular renewal is necessary because of cellular aging and loss of function.

It has been hypothesized that cells are genetically predetermined to have a given number of mitoses, after which senescence and death occur. This explanation has been proposed to account for the finite lifespan of human fibroblasts in tissue culture (Hayflick, 1973). Another suggestion has been that, in fibroblast tissue cultures of potentially ageless or immortal cells, some fibroblasts become irreversibly committed to senescence and death. During the time between commitment and senescense, the committed cells are assumed to maintain normal growth, so that uncommitted cells are diluted by the committed cells and may be lost on subculturing (Holliday et al., 1977). A third hypothesis has been advanced, attributing aging to errors in protein synthesis which accumulate and culminate in death (Orgel, 1963, 1970, 1973; Kirkwood, 1977). Aging of fibroblasts is accompanied by defects in chromosomes, genes, DNA replication and repair, and enzymes (Holliday et al., 1977). Defects in DNA may result from mutagenesis (Drake and Baltz, 1976; Wilson et al., 1977), while defects in enzymes may result from errors in gene transcription or translation (Kirkwood, 1977; Dreyfus et al., 1977) or from post translational changes (Houben and Remacle, 1978) (see concept of secondary isozymes in section I.B.2.a).

2. DNA Replication and Mitosis

DNA must replicate prior to mitosis. Because of the separate roles of DNA and RNA polymerases in these processes, it is not possible for DNA to direct RNA synthesis during DNA replication and mitosis. Hence, RNA synthesis and, consequently, protein synthesis cease. After mitosis is completed, protein synthesis can be reinitiated, and differentiation can proceed to bring the new cell to the adult stage. This elaborate intermeshing of these processes provides for the continuation of life and represents one of the most intensively studied and yet unsolved patterns of behavior in molecular biology. This subject is examined in some detail in Chapter 7.

II. The Second Fundamental Theorem of Theoretical Biology

The second fundamental theorem of theoretical biology states that the sequential acquisition of biological function by proteins and nucleic acids is a reflection of the evolutionary sequence of development. Simply stated this says that biological systems developed through the process of evolution (Wilson et al., 1977; Lahav et al., 1978; Schwartz and Dayhoff, 1978; Levine et al., 1978). Evolution is a natural process that occurs in incremental, quantized steps as a consequence of the interaction of matter and energy. We will not examine this theorem in detail here since it is discussed in Chapter 2 (evolutionary feedback). Our focus here, however, will be the relationship of this theorem to certain aspects of the principles of metabolic control. This theorem implies that there was a sequential evolutionary acquisition of metabolic control functions. The parallel development of function and control of function was obligatory, since the life process is not possible in the absence of control.

The details of the evolutionary process are largely unknown, but some of the mechanisms of mutation have come to light, and the rates of macromolecular evolution—the so-called evolutionary clock—have been calculated (Wilson et al., 1977). One of the molecular mechanisms involved in evolution has been ascribed to gene duplication (Wilson et al., 1977; Tang et al., 1978). With two copies of a gene, one may provide the original function while the other accumulates mutation. In this way, related proteins can be generated from duplicated genes. Many mammalian proteins can be clustered into superfamilies of related sequences, which are thought to represent the evolutionary products of gene duplication (Wilson et al., 1977; Dayhoff et al., 1978). Binding of the related proteins can give rise to protein polymers, isozymes, etc., as already discussed in the section on special functional categories of proteins.

The instability of DNA leads to mutations which may or may not be expressed in terms of alteration of protein structure which, in turn, may or may not result in alteration of function (Wilson et al., 1977). If the resulting alteration in function is deleterious in terms of survival, the result of the mutation is a genetic disease. If the resulting function is beneficial, then the result of the mutation is an evolutionary change. A single evolutionary change may allow an individual to survive in a hostile environment and reproduce, which might otherwise be more difficult, if not impossible. A now classic example is the mutation in the gene for the β-chain of hemoglobin, which gives rise to sickle cell hemoglobin (hemoglobin S) ($\beta6$ Glu $\rightarrow$ Val). In the heterozygous form, the red cells containing hemoglobins S and A are more resistant to the malaria parasite, *Plasmodium falciparum*. A heterozygous sickle cell (containing hemoglobins A and S) that is parasitized is more likely to sickle. It becomes rigid, is removed from the circulation, and is destroyed. Thus, the sickle cell trait allows an individual to survive in an area infested with *P. falciparum*. However, in the homozygous state, sickle cell disease occurs as a consequence of the instability of the β-chain of hemoglobin S in the deoxygenated state. Children homozygous for hemoglobin A (the normal hemoglobin) are at greater risk of death from the cerebral form of falciparum malaria (Power, 1975; Friedman, 1978).

A. The Order of Acquisition of Function

In Table 7, a hierarchical organization of metabolic processes is listed, progressing successively from hydrophobic and hydrophilic proteins to metabolic networks. Such a hierarchical organization is a logical and convenient arrangement but is not necessarily the order of evolutionary events that actually took place, since evolution proceeds through trial and error. However, such a hierarchical organization provides a convenient intellectual framework from which control mechanisms may be logically derived. But this logic may not be the logic of evolutionary events. Hence, we emphasize that the order of the evolutionary acquisition of function must be considered to be arbitrary. Likewise, the order of acquisition of metabolic control processes *must also be considered to be tentative*.

B. The Functional Specialization of DNA, RNA, and Protein

Why does DNA function only as a repository of information, i.e., why does the nucleotide sequence of DNA specify the order of amino acids in the corresponding protein? By contrast, protein has a great many functional categories: enzyme, hormone, structure, receptor, carrier, contraction, antibody, light refraction, storage, etc. Since DNA is composed of four nucleotide bases (adenine, guanine, cytosine, thymine), we can arrange these four in all possible combinations of sequence. All such combinations amount to 4^4 in number, or a total of 256 unique combinations. This number of combinations is too small to handle all of the biochemical reactions, structures, and other functions known to exist in living systems, even if DNA did have intrinsic enzymatic, hormonal, structural, or other functions. On the other hand, the genetic code need only specify 20 amino acids. In the case of the genetic code, 4 nucleotides taken 3 at a time (4^3) gives 64 combinations, which is more than enough to specify the 20 amino acids. Proteins can have as many as 20 amino acids arranged in all possible linear combinations (in the simplest arrangement). This can give rise to 20^{20} arrangements or approximately 1.27×10^{25} combinations. Although many of these sequences may be inert or biologically useless, there are still more sequences possible than have ever been used in living systems. Such a number of combinations could easily encompass the number of proteins generated by 50,000 genes (5×10^4) (Dayhoff et al., 1978), which is the estimated number of genes in man. Thus, proteins have a great number of functions as a direct result of their structure. The primary structure of a protein (the amino acid sequence) results in its secondary structure (coiling or helix formation) and tertiary structure (folding) (Guzzo, 1965; Sternberg and Thornton, 1978), with hydrophilic or hydrophobic groups on the surface, depending on whether the protein is in an aqueous or lipid environment (Crampin et al., 1978), and determines interactions between adjacent groups, charges on the protein surface, etc. And the structure of protein determines its function. The tertiary structure of a protein determines its oxido–reductive capacity (Stellwagen, 1978), thermostability (Stellwagen and Wilgus, 1978), binding characteristics, solubility (Hagler and Moult, 1978), enzymatic activity, etc.

Consider the versatility of the major human serum protein, albumin. It contains 585 amino acids, has a molecular weight of 66,248.3, occurs in 12 known genetic types, and binds more than 20 different physiological substances [bilirubin, uric acid, vitamin C, acetylcholine, cholinesterase, adenosine, fatty acids, progesterone, cortisol, corticosterone, aldosterone, testosterone, cholesterol, prostaglandins, zinc, and a variety of drugs and chemicals (Peters, 1975)].

The conformability of protein and its susceptibility to denaturation is determined by its tertiary structure. The very structure of protein which determines its function also renders it susceptible to unfolding and precipitation from solution (denaturation).

Nucleic acid is more rigid and stacked, while protein can assume a number of tertiary conformations so that many different surfaces can accommodate different molecular species. The various ribonucleotides that compose RNA can serve as cofactors for enzymatic action (e.g., ATP) and as substrates for polymer formation. For example, ATP can polymerize to poly(A) in the presence of a polynucleotide primer and poly(A) polymerase. Poly(A) polymerase can also function as a poly(A) hydrolase and liberate AMP, ADP, and ATP from poly(A) (Abraham and Jacob, 1978).

Once a pool of deoxyribonucleotides has formed, the individual deoxyribonucleotides cannot function effectively as enzyme cofactors but can only be utilized for DNA formation. The deoxyribose structure seems to preclude enzyme cofactor function. [In the evolution of macromolecules, it is conceivable that RNA served as a template for DNA formation in the presence of a reverse transcriptase (Verma, 1977).] Interestingly, it has been found that RNA serves as a primer for DNA replication (Kornberg, 1976).

C. Biochemical Implications of Evolution

Although we have, for purposes of convenience, divided the acquisition of function in biological systems between hydrophobic and hydrophilic capabilities, it is likely that, at a very early stage of evolution, these separate functionalities were combined in certain structures, e.g., membranes. Hence, the trial and error strategy of evolution has always been based on the ability of functional molecules and supramolecules to carry out biological activities essential for survival or reproduction.

As living forms became more complex, the very acquisition of metabolic function itself constituted the evolutionary process, and acquisition of metabolic control functions was a necessary consequence. One may postulate that the principles of metabolic control that developed were those which permitted the life process to adjust to environmental conditions that otherwise would have caused destabilization.

III. Other Fundamental Theorems of Theoretical Biology

A. Theorem 3

As is developed in Chapter 5 (Regulation of Protein Biosynthesis), this theorem states that the continued synthesis and degradation of proteins and enzymes constitutes the life process by enabling the metabolic network to achieve negative entropy and remain in a far-from-equilibrium condition.

B. Theorem 4

The entire chapter on the biochemical basis of disease (Chapter 15) is based on the thesis that all disease is a consequence of at least one abnormality of biochemical function. In other words, all disease results from the dysfunctional biology of intracellular events and therefore is biochemical in nature.

C. Theorem 5

As a consequence of theorem 4, it follows that psychiatric disorders and mental illness have a biochemical basis, however poorly understood. This is so because mental function is a consequence of the continued flow of nerve impulses through the enormous number of neuronal circuits of the nerve network of the central nervous system.

Mental function is dependent on a critical number of circuits that can be formed from all possible permutations and combinations of the individual neurons. (In man the number of circuits is estimated to be between factorial 10^{10} and factorial 10^{12}).

IV. Summary

1. The life process is the inevitable consequence of the chemical and physical reactions that occur in nature.
2. The characteristics of water permit the physical and chemical reactions of biological systems.
3. Biological systems are open systems which exist only in a far-from-equilibrium state where they can utilize the flow of energy and matter.
4. The hydrophobic properties of phospholipids and certain proteins leads to the development of cellular and subcellular membranes that constitute the mammalian cell and permit its subcellular organization.
5. The development of an increasing order of complexity of the hydrophobic and hydrophilic systems gives rise to the metabolic network, which operates as a steady-state system.
6. Because of the far-from-equilibrium nature of the life process, it is particularly susceptible to perturbations generated by the environment, the system itself, or both, and which destabilize the components of the living process.
7. The metabolic control mechanisms (which are part of the metabolic network) maintain the steady state of the metabolic network, thereby maintaining the constancy of the intracellular environment against internal and external perturbations, i.e., in a state of homeostasis.

From these various considerations we can now see that a finite number of regulatory principles are involved in the control of metabolic pathways. In theory, when all of the control principles are known, it should be possible to deduce the consequences of failure of these control mechanisms and to determine whether means of correction are possible and what these means should be.

With knowledge of what metabolic pathways exist and the nature of the regulatory controls, it should be possible, in principle, to calculate precisely the response of the metabolic pathways to all types of environmental conditions and to deduce the nature of the physiological processes that depend on these metabolic pathways.

References

Abraham, A. K., and Jacob, S. T., 1978, Hydrolysis of poly(A) to adenine nucleotides by purified poly(A) polymerase, *Proc. Natl. Acad. Sci. USA* **75**:2085.

Anderson, J. W., and King, P., 1975, Subcellular distribution of hexokinase activity in rat jejunal mucosa: Response to diabetes and dietary changes, *Biochem. Med.* **12**:1.

Arnold, H., and Pette, D., 1968, Binding of glycolytic enzymes to structure proteins of the muscle, *Eur. J. Biochem.* **6**:163.

Arnold, H., and Pette, D., 1970, Binding of aldolase and triosephosphate dehydrogenase to F-actin and modification of catalytic properties of aldolase, *Eur. J. Biochem.* **15**:360.

Arnone, A., 1974, Mechanism of action of hemoglobin, *Annu. Rev. Med.* **25**:123.

Babior, B. M., 1978, Oxygen-dependent microbial killing by phagocytes, *N. Engl. J. Med.* **298**:659.

Baker, B. L., and Hultquist, D. E., 1978, A copper-binding immunoglobulin from a myeloma patient, *J. Biol. Chem.* **253**:1195.

Bauer, C., and Kurtz, A., 1977, Oxygen-linked CO_2 binding to isolated β subunits of human hemoglobin, *J. Biol. Chem.* **252**:2952.

Beard, J. R., and Razzell, W. E., 1964, Purification of alkaline ribonuclease II from mitochondrial and soluble fractions of liver, *J. Biol. Chem.* **239**:4186.

Bell, J. L., and Baron, D. N., 1961, Isoenzyme of isocitric dehydrogenase, *Biochem. J.* **82**:5P.

Benesch, R., and Benesch, R. E., 1967, The effect of organic phosphates from the human erythrocyte on the allosteric properties of hemoglobin, *Biochem. Biophys. Res. Commun.* **26**:162.

Benesch, R., Benesch, R. E., and Yu, C. I., 1968, Reciprocal binding of oxygen and diphosphoglycerate by human hemoglobin, *Proc. Natl. Acad. Sci. USA* **59**:526.

Benesch, R. E., Benesch, R., and Yu, C. I., 1969, The oxygenation of hemoglobin in the presence of 2,3-diphosphoglycerate. Effect of temperature, pH, ionic strength, and hemoglobin concentration, *Biochemistry* **8**:2567.

Berglund, L., Ljungstrom, O., and Engstrom, L., 1977, Purification and characterization of pig kidney pyruvate kinase (type A), *J. Biol. Chem.* **252**:6108.

Bernard, C., 1878, *Leçons sur les phénomènes de la Vie Communs aux Animaux et aux Végétaux*, J. B. Baillière et Fils, Paris, France.

Bertalanffy, F. D., and Lau, C., 1963, Mitotic rates, renewal times, and cytodynamics of the female genital tract epithelia in the rat, *Acta Anat.* **54**:39.

Berthillier, G., Coleman, R., and Walker, D. G., 1976, The topographical location and unique nature of a glucokinase associated with the Golgi apparatus of rat liver, *Biochem. J.* **154**:193.

Black, H. S., and Lo, W. B., 1971, Formation of a carcinogen in human skin irradiated with ultraviolet light, *Nature* **234**:306.

Black, H. S., and Douglas, D. R., 1972, A model system for the evaluation of the role of cholesterol α-oxide in ultraviolet carcinogenesis, *Cancer Res.* **32**:2630.

Bokman, A. H., and Schweigert, B. S., 1951, 3-Hydroxyanthranilic acid metabolism. IV. Spectrophotometric evidence for the formation of an intermediate, *Arch. Biochem. Biophys.* **33**:270.

Bonaventura, J., Bonaventura, C., Amiconi, G., Tentori, L., Brunori, M., and Antonini, E., 1975, Allosteric interactions in non-α chains isolated from normal human hemoglobin, fetal hemoglobin, and hemoglobin Abruzzo (β 143(H21) His → Arg), *J. Biol. Chem.* **250**:6278.

Borst, P., and Peeters, E., 1961, The intracellular localization of glutamateoxaloacetate transaminases in heart, *Biochim. Biophys. Acta* **54**:188.

Brewer, G.J., 1969, Erythrocyte metabolism and function: Hexokinase inhibition by 2,3-diphosphoglycerate and interaction with ATP and Mg^{2+}, *Biochim. Biophys. Acta* **192**:157.

Brewer, G. J., 1974, 2,3-DPG and erythrocyte oxygen affinity, *Annu. Rev. Med.* **25**:29.

Brewer, G. J., and Eaton, J. W., 1971, Erythrocyte metabolism: Interaction with oxygen transport, *Science* **171**:1205.

Brissenden, J. E., and Cox, D. W., 1978, Electrophoretic and quantitative assessment of vitamin D-binding protein (group-specific component) in inherited rickets, *J. Lab. Clin. Med.* **91**:455.

Brok, F., Ramot, B., Zwang, E., and Danon, D., 1966, Enzyme activities in human red blood cells of different age groups, *Israel J. Med. Sci.* **2**:291.

Brumbaugh, P. F., and Haussler, M. P., 1974, 1α,25-Dihydroxycholecalciferol receptors in intestine. II. Temperature-dependent transfer of the hormone to chromatin via a specific cytosol receptor, *J. Biol. Chem.* **249**:1258.

Bunn, H. F. 1971, Differences in the interaction of 2,3-diphosphoglycerate with certain mammalian hemoglobins, *Science* **172**:1049.

Bunn, H. F., Forget, B. G., and Ranney, H. M. 1977, Unstable hemoglobin variants—congenital Heinz body hemolytic anemia, in: *Human Hemoglobins*, pp. 282–311, W. B. Saunders, Philadelphia.

Burger, R. L., Mehlman, C. S., and Allen, R. H., 1975, Human plasma R-type vitamin B_{12}-binding proteins. I. Isolation and characterization of transcobalamin I, transcobalamin III, and the normal granulocyte vitamin B_{12}-binding protein, *J. Biol. Chem.* **250**:7700.

Cameron, I. L., 1970, Cell renewal in the organs and tissues of nongrowing adult mouse, *Tex. Rep. Biol. Med.* **28**:203.

Cannon, W. B., 1929, Organization for physiological homeostatsis, *Physiol. Rev.* **9**:399.

Carroll, M., Dance, N., Masson, P. K., Robinson, D., and Winchester, B. G., 1972, Human mannosidosis—the enzymic defect, *Biochem. Biophys. Res. Commun.* **49**:579.

Castle, W. B., and Townsend, W. C. 1929, Observations on the etiologic relationship of achylia gastrica to pernicious anemia. II. The effect of the administration to patients with pernicious anemia of beef muscle after incubation with normal human gastric juice, *Am. J. Med. Sci.* **178**:764.

Catignani, G. L., 1975, An α-tocopherol binding protein in rat liver cytoplasm, *Biochem. Biophys. Res. Commun.* **67**:66.

Chanutin, A., and Curnish, R. R., 1967, Effect of organic and inorganic phosphates on the oxygen equilibrium of human erythrocytes, *Arch. Biochem. Biophys.* **121**:96.

Chapman, R. G., and Schaumburg, L., 1967, Glycolysis and glycolytic enzyme activity of aging red cells in man. Changes in hexokinase, aldolase, glyceraldehyde-3-phosphate dehydrogenase, pyruvate kinase and glutamic-oxalacetic transaminase, *Br. J. Haematol.* **13**:665.

Chen, S.-H., and Giblett, E. R., 1971, Polymorphism of soluble glutamicpyruvic transaminase: A new genetic marker in man, *Science* **173**:148.

Chian, L. T. Y., and Wilgram, G. F., 1967, Tyrosinase inhibition: Its role in suntanning and in albinism, *Science* **155**:198.

Chock, P. B., and Stadtman, E. R., 1977, Superiority of interconvertible enzyme cascades in metabolic regulation: Analysis of multicylic systems, *Proc. Natl. Acad. Sci. USA* **74**:2766.

Chou, P. Y., and Fasman, G. D., 1974, Prediction of protein conformation, *Biochemistry* **13**:222.

Chung, J., and Wood, J. L., 1971, Oxidation of thiocyanate to cyanide catalyzed by hemoglobin, *J. Biol. Chem.* **246**:555.

Clarke, S., 1976, A major polypeptide component of rat liver mitochondria: Carbamyl phosphate synthetase, *J. Biol. Chem.* **251**:950.

Cleaver, J. E., 1972, Excision repair: Our current knowledge based on human (xeroderma pigmentosum) and cattle cells, in: *Molecular and Cellular Repair Processes* (R. F. Beers, Jr., R. M. Herriott, and R. C. Tilghman, eds.), pp. 195–211, The Johns Hopkins University Press, Baltimore.

Coleman, P. F., Suttle, D. P., and Stark, G. R., 1977, Purification from hamster cells of the multifunctional protein that initiates *de novo* synthesis of pyrimidine nucleotides, *J. Biol. Chem.* **252**:6379.

Comper, W. D., and Laurent, T. C., 1978, Physiological function of connective tissue polysaccharides, *Physiol. Rev.* **58**:255.

Conklin, K. A., Yamashiro, K. M., and Gary, G. M., 1975, Human intestinal sucrase-isomaltase. Identification of free sucrase and isomaltase and cleavage of the hybrid into active distinct subunits, *J. Biol. Chem.* **250**:5735.

Conover, T. E., and Ernster, L., 1962, DT diaphorase. II. Relation to respiratory chain of intact mitochondria, *Biochim. Biophys. Acta* **58**:189.

Cook, J. S., 1972, Photoenzymatic repair in animal cells, in: *Molecular and Cellular Repair Processes* (R. F. Beers, Jr., R. M. Herriott, and R. C. Tilghman, eds.), pp. 79–94, The Johns Hopkins University Press, Baltimore,

Crampin, J., Nicholson, B. H., and Robson, B., 1978, Protein folding and heterogeneity inside globular proteins, *Nature* **272**:558.

Curzon, G., and O'Reilly, S., 1960, The effects of some ions and chelating agents on the oxidase activity of ceruloplasmin, *Biochem. J.* **77**:66.

Daddona, P. E., and Kelley, W. N., 1978, Human adenosine deaminase binding protein, *J. Biol. Chem.* **253**:4617.

Davis, N. R., and Anwar, R. A., 1970, On the mechanism of formation of desmosine and isodesmosine cross-links of elastin, *J. Am. Chem. Soc.* **92**:3778.

Dayhoff, M. O., Barker, W. C., and Schwartz, R. M., 1978, Evolution of the mammalian genome, a view based on protein and nucleic acid sequence data, *Fed. Proc.* **37**:1419.

Dean, R. T., and Barrett, A. J., 1976, Lysosomes, *Essays in Biochem.* **12**:1.

Demma, L. S., and Salhany, J. M., 1977, Direct generation of superoxide anions by flash photolysis of human oxyhemoglobin, *J. Biol. Chem.* **252**:1226.

Dewald, B., and Touster, O., 1973, A new α-D-mannosidase occurring in Golgi membranes, *J. Biol. Chem.* **248**:7223.

Dische, Z., 1941, Interdependence of various enzymes of the glycolytic system and the automatic regulation of their activity within the cells. I. Inhibition of the phosphorylation of glucose in red corpuscles by monophosphoglyceric and diphosphoglyceric acids; state of the diphosphoglyceric acid and the phosphorylation of glucose, *Trav. Membres Soc. Chim. Biol.* **23**:1140.

Dixon, J. W., and Sarkar, B., 1974, Isolation, amino acid sequence and copper(II)-binding properties of peptide (1–24) of dog serum albumin, *J. Biol. Chem.* **249**:5872.

Drake, J. W., and Baltz, R. H., 1976, The biochemistry of mutagenesis, *Annu. Rev. Biochem.* **45**:11.

Dreyfus, J. C., Rubinson, H., Schapira, F., Weber, A., Marie, J., and Kahn, A., 1977, Possible molecular mechanisms of aging, *Gerontology* **23**:211.

Drury, E. J., Bazar, L. S., and MacKenzie, R. E., 1975, Formiminotransferase-cyclodeaminase from porcine liver. Purification and physical properties of the enzyme complex, *Arch. Biochem. Biophys.* **169**:662.

Duley, J. A., and Holmes, R. S., 1976, L-α-Hydroxyacid oxidase isozymes. Purification and molecular properties, *Eur. J. Biochem.* **63**:163.

Ellenbogen, L., Burson, S. L., and Williams, W. L., 1958, Purification of intrinsic factor, *Proc. Soc. Exp. Biol. Med.* **97**:760.

Engasser, J.-M., Flamm, M., and Horvath, C., 1977, Hormone regulation of cellular metabolism: Interplay of membrane transport and consecutive enzymic reaction, *J. Theor. Biol.* **67**:433.

Enoch, H. G., Fleming, P. J., and Strittmatter, P., 1977, Cytochrome b_5 and cytochrome b_5 reductase-phospholipid vesicles, *J. Biol. Chem.* **252**:5656.

Evans, G. W., 1973, Copper homeostasis in the mammalian system, *Physiol. Rev.* **53**:535.

Evans, G. W., Majors, P. F., and Cornatzer, W. E., 1970, Mechanism of cadmium and zinc antagonism of copper metabolism, *Biochem. Biophys. Res. Commun.* **40**:1142.

Eventoff, W., Rossmann, M. G., Taylor, S. S., Torff, H-S., Meyer, H., Keil, W., and Kiltz, H-H., 1977, Structural adaptations of lactate dehydrogenase isozymes, *Proc. Natl. Acad. Sci. USA* **74**:2677.

Everett, N. B., and Tyler (Caffrey), R. W., 1967, lymphopoiesis in the thymus and other tissues: Functional implications. *Int. Rev. Cytol.* **22**:205.

Exton, J. H., and Park, C. R., 1967, Control of gluconeogenesis in liver. I. General features of gluconeogenesis in the perfused liver of rats, *J. Biol. Chem.* **242**:2622.

Exton, J. H., and Park, C. R., 1968, Control of gluconeogenesis in liver. II. Effects of glucagon, catecholamines, and adenosine 3', 5'-monophosphate on gluconeogenesis in the perfused rat liver, *J. Biol. Chem.* **243**:4189.

Fawcett, D. W., 1966, *The Cell. Its Organelles and Inclusions. An Atlas of Fine Structure*, W. B. Saunders, Philadelphia.

Fichera, G., Sneider, M. A., and Wyman, J., 1977a, On the existence of a steady state in a biological system, *Proc. Natl. Acad. Sci. USA* **74**:4182.

Fichera, G., Sneider, M. A., and Wyman, J., 1977b, On the existence of a steady state in a biological system, *Mem. Accad. Nazionale Lincei, Cl. Sc. Mat. Fis. Nat., Roma* **14**:1.

Florkin, M., and Stotz, E. H., 1962–1977, *Comprehensive Biochemistry*, Elsevier, Amsterdam.

Foemmel, R. S., Gray, R. H., and Bernstein, I. A., 1975, Intracellular localization of fructose 1,6-bisphosphate aldolase, *J. Biol. Chem.* **250**:1892.

Ford, H. C., and Engel, L. L., 1974, Purification and properties of the Δ^5-3 β-hydroxysteroid dehydrogenase-isomerase system of sheep adrenal cortical microsomes, *J. Biol. Chem.* **249**:1363.

Francis, S. H., Meriwether, B. P., and Park, J. H., 1973, Effects of photooxidation of histidine-38 on the various catalytic activities of glyceraldehyde-3-phosphate dehydrogenase, *Biochemistry* **12**:346.

Frey, W. H., II, and Utter, M. F., 1977, Binding of acetyl-CoA to chicken liver pyruvate carboxylase, *J. Biol. Chem.* **252**:51.

Fridovich, I., 1977, Oxygen is toxic! *BioScience* **27**:462.

Friedman, M. J., 1978, Erythrocytic mechanism of sickle cell resistance to malaria, *Proc. Natl. Acad. Sci. USA* **75**:1994.

Futterman, S., Saari, J. C., and Blair, S., 1977, Occurrence of a binding protein for 11-*cis*-retinal in retina, *J. Biol. Chem.* **252**:3267.

Gaertner, F. H., and Cole, K. W., 1977, A cluster-gene: Evidence for one gene, one polypeptide, five enzymes, *Biochem. Biophys. Res. Commun.* **75**:259.

Giroux, E. L., Durieux, M., and Schechter, P. J., 1976, A study of zinc distribution in human serum, *Bioinorg. Chem.* **5**:211.

Glass, G. B. J., 1963, Gastric intrinsic factor and its function in the metabolism of vitamin B_{12}, *Physiol. Rev.* **43**:529.

Glover, J., Jay, C., and White, G. H., 1974, Distribution of retinol-binding protein in tissues, *Vitamins Horm.* **32**:215.
Goedde, H. W., and Keller, W., 1967, Metabolic pathways in maple syrup urine disease, in: *Amino Acid Metabolism and Genetic Variation* (W. L. Nyhan, ed.), pp. 191–214, McGraw-Hill, New York.
Goldberg, B., and Stern, A., 1976, Production of superoxide anion during the oxidation of hemoglobin by menadione, *Biochim. Biophys. Acta* **437**:628.
Goldstein, S., and Moerman, E. J., 1975, Heat-labile enzymes in Werner's syndrome fibroblasts, *Nature* **255**:159.
Gonzalez, F., and Kemp, R. G., 1978, The A_2B_2 hybrid isozyme of phosphofructokinase, *J. Biol. Chem.* **253**:1493.
Goodman, DeW. S., 1974, Vitamin A transport and retinol-binding protein metabolism, *Vitamins Horm.* **32**:167.
Green, D. E., Murer, E., Hultin, H. O., Richardson, S. H., Salmon, B., Brierley, G. P., and Baum, H., 1965, Association of integrated metabolic pathways with membranes. I. Glycolytic enzymes of the red blood corpuscle and yeast, *Arch. Biochem. Biophys.* **112**:635.
Greene, M. L., Boyle, J. A., and Seegmiller, J. E., 1970, Substrate stabilization: genetically controlled reciprocal relationship of two human enzymes, *Science* **167**:887.
Guthöhrlein, G., and Knappe, J., 1968, Structure and function of carbamoylphosphate synthetase. On the mechanism of bicarbonate activation, *Eur. J. Biochem.* **7**:119.
Guzzo, A. V., 1965, The influence of amino-acid sequence on protein structure, *Biophys. J.* **5**:809.
Habig, W. H., Pabst, M. J., Fleischner, G., Gatmaitan, Z., Arias, I. M., and Jakoby, W. B., 1974, The identity of glutathione S-transferase B with ligandin, a major binding protein of liver, *Proc. Natl. Acad. Sci. USA* **71**:3879.
Hagler, A. T., and Moult, J., 1978, Computer simulation of the solvent structure around biological macromolecules, *Nature* **272**:222.
Hagler, L., Coppes, R. I., and Herman, R. H., 1976, Metmyoglobin reductase: Identification of reduced nicotinamide adenine dinucleotide (NADH)-dependent enzyme activity from bovine heart which reduces metmyoglobin, *Fed. Proc.* **35**:1423.
Haining, R. G., Hulse, T. E., and Labbe, R. F., 1969, Photohemolysis. The comparative behavior of erythrocytes from patients with different types of porphyria, *Proc. Soc. Exp. Biol. Med.* **132**:625.
Halestrap, A. P., and Denton, R. M., 1974, Hormonal regulation of adipose-tissue acetylcoenzyme A carboxylase by changes in the polymeric state of the enzyme, *Biochem. J.* **142**:365.
Hall, C. A., 1975, Transcobalamins I and II as natural transport proteins of vitamin B_{12}, *J. Clin. Invest.* **56**:1125.
Hanker, J. S., and Romanovicz, D. K., 1977, Phi bodies: Peroxidatic particles that produce crystalloidal cellular inclusions, *Science* **197**:895.
Hanna, C., Bicknell, D. S., and O'Brien, J. E., 1961, Cell turnover in the adult human eye, *Arch. Ophthalmol.* **64**:536.
Harris, H., 1969, Genes and isozymes, *Proc. R. Soc. London Ser. B* **174**:1.
Harrison, J. E., 1974, A proposed hydrogen transfer function for cytochrome *c*, *Proc. Natl. Acad. Sci. USA* **71**:2332.
Havir, E. A., Tamir, H., Ratner, S., and Warner, R. C., 1965, Biosynthesis of urea. XI. Preparation and properties of crystalline argininosuccinase, *J. Biol. Chem.* **240**:3079.
Hayflick, L., 1973, The biology of human aging, *Am. J. Med. Sci.* **265**:432.
Hekman, A., 1971, Association of lactoferrin with other proteins as demonstrated by changes in electrophoretic mobility, *Biochim. Biophys. Acta* **251**:380.

Heller, C. G., and Clermont, Y., 1963, Spermatogenesis in man: An estimate of its duration, *Science* **140**:184.
Henderson, N. S., 1965, Isozymes of isocitrate dehydrogenase: Subunit structure and intracellular location, *J. Exp. Zool.* **158**:263.
Henderson, N. S., 1966, Isozymes and genetic control of NADP-malate dehydrogenase in mice, *Arch. Biochem. Biophys.* **117**:28.
Hernandez, A., and Crane, R. K., 1966, Assocation of heart hexokinse with subcellular structure, *Arch. Biochem. Biophys.* **113**:223.
Holick, M. F., Frommer, J. E., McNeill, S. C., Richtand, N. M., Henley, J. W., and Potts, J. T., Jr., 1977, Photometabolism of 7-dehydrocholesterol to previtamin D_3 in skin, *Biochem. Biophys. Res. Commun.* **76**:107.
Holliday, R., and Tarrant, G. M., 1972, Altered enzymes in aging human fibroblasts, *Nature* **238**:26.
Holliday, R., Huschtscha, L. I., Tarrant, G. M., and Kirkwood, T. B. L., 1977, Testing the commitment theory of cellular aging, *Science* **198**:366.
Holmberg, C. G., and Laurell, C.-B., 1951, Investigations in serum copper. III. Ceruloplasmin as enzyme, *Acta Chem. Scand.* **5**:476.
Hooton, B. T., 1968, Creatine kinase isoenzymes and the role of thiol groups in the enzymic mechanism, *Biochemistry* **7**:2063.
Horecker, B. L., 1975, Biochemistry of isozymes, in: *Isozymes. Molecular Structure*, Vol. 1 (C. I. Markert, ed.), pp. 11–38, Academic Press, New York.
Houben, A., and Remacle, J., 1978, Lysosomal and mitochondrial heat labile enzymes in aging human fibroblasts, *Nature* **275**:59.
Hsieh, H.-S., and Jaffe, E. R., 1971, Electrophoretic and functional variants of NADH-methemoglobin reductase in hereditary methemoglobinemia, *J. Clin. Invest.* **50**:196.
Huang, W.-Y. Cohn, D. V., Hamilton, J. W., Fullmer, C., and Wasserman, R. H., 1975, Calcium-binding protein of bovine intestine. The complete amino acid sequence, *J. Biol. Chem.* **250**:7647.
Huber, C. T., and Frieden, E., 1970, Substrate activation and kinetics of ferroxidase, *J. Biol. Chem.* **245**:3973.
Hynes, R. O., 1976, Cell surface proteins and malignant transformation, *Biochim. Biophys. Acta* **458**:73.
Imondi, A. R., Balis, M. E., and Lipkin, M., 1969, Changes in enzyme levels accompanying differentiaion of intestinal epithelial cells, *Exp. Cell Res.* **58**:323.
Ingbar, S. H., 1963, Observations concerning the binding of thyroid hormones by human serum prealbumin, *J. Clin. Invest.* **42**:143.
Jackson, R. L., Morrisett, J. D., and Gotto, A. M. Jr., 1976, Lipoprotein structure and metabolism, *Physiol. Rev.* **56**:259.
Jacob, H. S., Brain, M. C., and Dacie, J. V., 1968a, Altered sulfhydryl reactivity of hemoglobins and red blood cell membranes in congenital Heinz body hemolytic anemia, *J. Clin. Invest.* **47**:2664.
Jacob, H. S., Brain, M. C., Dacie, J. V., Carrel, R. W., and Lehmann, H., 1968b, Abnormal haem binding and globin SH group blockade in unstable hemoglobins, *Nature* **218**:1214.
Jacobs, A., and Worwood, M., 1975, Ferritin in serum. Clinical and biochemical implications, *N. Engl. J. Med.* **292**:951.
Jacobs, H., Heldt, H. W., and Klingenberg, M., 1964, High activity of creatine kinase in mitochondria from muscle and brain and evidence for a separate mitochondrial isoenzyme of creatine kinase, *Biochem. Biophys. Res. Commun.* **16**:516.
Jakoby, W. B., and Bonner, D. M., 1956, Kynurenine transaminase from Neurospora, *J. Biol. Chem.* **221**:689.

Jayle, M. F., 1951, Méthode de dosage de l'haptoglobine sérique, *Bull Soc. Chim. Biol.* **33**:876.
Johnson, H. A., Haymaker, W. E., Rubini, J. R., Fliedner, T. M., Bond, V. P., Cronkite, E. P., and Hughes, W. L., 1960, A radioautographic study of a human brain and glioblastoma multiforme after the in vivo uptake of tritiated thymidine, *Cancer* **13**:636.
Katsunuma, T., Temma, M., and Katunuma, N., 1968, Allosteric nature of a glutaminase isozyme in rat liver, *Biochem. Biophys. Res. Commun.* **32**:433.
Katzberg, A. A., 1952, The influence of age on the rate of desquamation of the human epidermis, *Anat. Rec. (Suppl.)* **112**:418.
Kavipurapu, P. R., and Jones, M. E., 1976, Purification, size, and properties of the complex of orotate phosphoribosyltransferase:orotidylate decarboxylase from mouse Ehrlich ascites carcinoma, *J. Biol. Chem.* **251**:5589.
Kilmartin, J. V., and Rossi-Bernardi, L., 1973, Interaction of hemoglobin with hydrogen ions, carbon dioxide, and organic phosphates, *Physiol. Rev.* **53**:836.
Kirk, J. R., Brunner, J. R., Stine, C. M., and Schweigert, B. S., 1972, Effects of pH and electrodialysis on the bindings of vitamin B_{12} by β-lactoglobulin and associated peptides, *J. Nutr.* **102**:699.
Kirkwood, T. B. L., 1977, Evolution of aging, *Nature* **270**:301.
Kirschner, K., and Bisswanger, H., 1976, Multifunctional proteins, *Annu. Rev. Biochem.* **45**:143.
Kitamura, M., and Nishina, T., 1975, Hereditary deficiency of subunit B of lactate dehydrogenase, in: *Isozymes, Physiological Function*, Vol. 2 (C. I. Markert, ed.), pp. 97–111, Academic Press, New York.
Klatskin, G., and Bungards, L., 1956, Bilirubin-protein linkage in serum and their relationship to the Van den Berg reaction, *J. Clin. Invest.* **35**:537.
Koch-Weser, J., and Sellers, E. M., 1976a, Binding of drugs to serum albumin, *N. Engl. J. Med.* **294**:311.
Koch-Weser, J., and Sellers, E. M., 1976b, Binding of drugs to serum albumin, *N. Engl. J. Med.* **294**:526.
Koen, A. L., and Goodman, M., 1969, Aconitate hydratase isozymes: Subcellular location, tissue distribution and possible subunit structure, *Biochim. Biophys. Acta* **191**:698.
Kornberg, A., 1976, RNA priming of DNA replication, in: *RNA Polymerase* (R. Losich and M. Chamberlin, eds.), pp. 331–352, Cold Spring Harbor Laboratory, Cold Spring Harbor, New York.
Kosenow, W., and Treibs, A., 1953, Lichtüberempfindlichkeit und Porphyrinämie, *Z. Kinderheilk.* **73**:82.
Krebs, H. A., 1975, The role of chemical equilibria in organ function, *Adv. Enz. Regul.* **13**:449.
Lahav, N., White, D., and Chang, S., 1978, Peptide formation in the prebiotic era: Thermal condensation of glycine in fluctuating clay environments, *Science* **201**:67.
Lau, S.-J., Kruck, T. P. A., and Sarkar, B., 1974, A peptide molecule mimicking the copper (II) transport site of human serum albumin, *J. Biol. Chem.* **249**:5878.
Leblond, D. P., and Stevens, C. E., 1948, The constant renewal of the intestinal epithelium in the albino rat, *Anat. Rec.* **100**:357.
Lee, I.-Y., Strunk, R. C., and Coe, E.-L., 1967, Coordination among rate-limiting steps of glycolysis and respiration in intact ascites tumor cells, *J. Biol. Chem.* **242**:2021.
Leffell, M. S., and Spitznagel, J. K., 1972, Association of lactoferrin with lysozyme in granules in human polymorphonuclear leukocytes, *Infect. Immunol.* **6**:761.
Lentz, T. L., 1971, *Cell Fine Structure. An Atlas of Drawings of Whole-Cell Structure* W. B. Saunders, Philadelphia.

Levi, A. J., Gatmaitan, Z., and Arias, I. M., 1969, Two hepatic cytoplasmic protein fractions, Y and Z, and their possible role in the hepatic uptake of bilirubin, sulfobromophthalein and other anions, *J. Clin. Invest.* **48**:2156.
Levine, M., Muirhead, H., Stammers, D. K., and Stuart, D. I., 1978, Structure of pyruvate kinase and similarities with other enzymes: Possible implications for protein taxonomy and evolution, *Nature* **271**:626.
Lewis, C. M., and Tarrant, G. M., 1972, Error theory and aging in human diploid fibroblasts, *Nature* **239**:316.
Lien, E. L., Ellenbogen, L., Law, P. Y., and Wood, J. M., 1973, The mechanism of cobalamin binding to hog intrinsic factor, *Biochem. Biophys. Res. Commun.* **55**:730.
Lindahl, U., and Høøk, M., 1978, Glycosaminoglycans and their binding to biological macromolecules, *Annu. Rev. Biochem.* **47**:385.
Litwack, G., Ketterer, B., and Arias, I. M., 1971, Ligandin: A hepatic protein which binds steroids, bilirubin, carcinogens and a number of exogenous organic anions, *Nature* **234**:466.
Lo, H. H., and Schimmel, P. R., 1969, Interaction of human hemoglobin with adenine nucleotides, *J. Biol. Chem.* **244**:5084.
Lo, W.-B., and Black, H. S., 1972, Formation of cholesterol-derived photoproducts in human skin, *J. Invest. Dermatol.* **58**:278.
Long, C. L., Hill, H. N., Weinstock, I. M., and Henderson, L. M., 1954, Studies of the enzymatic transformation of 3-hydroxyanthranilate to quinolinate, *J. Biol. Chem.* **211**:405.
Longenecker, J. B., and Snell, E. E., 1957, Pyridoxal and metal ion catalysis of α,β elimination reactions of serine-3-phosphate and related compounds, *J. Biol. Chem.* **225**:409.
Lornitzo, F. A. Qureshi, A. A., and Porter, J. W., 1975, Subunits of fatty acid synthase complexes. Enzymatic activities and properties of the half-molecular weight nonidentical subunits of pigeon liver fatty acid synthase, *J. Biol. Chem.* **250**:4520.
Lynch, R. E., and Fridovich, I., 1978, Effects of superoxide on the erythrocyte membrane, *J. Biol. Chem.* **253**:1838.
MacDonald, R. A., 1961, "Lifespan" of liver cells, *Arch. Intern. Med.* **107**:335.
MacDonald, W. C., Trier, J. S., and Everett, N. B., 1965, Cell proliferation and migration in the stomach, duodenum and rectum of man: Radioautographic studies, *Gastroenterology* **46**:405.
Magnus, I. S., Jarrett, A., Prankerd, T. A. J., and Rimington, C., 1961, Erythropoietic protoporphyria: A new porphyria syndrome with solar urticaria due to protoporphyrinemia, *Lancet* **2**:448.
Mahler, H. R., and Cordes, E. H., 1971, *Biological Chemistry*, 2nd ed., Harper and Row, New York.
Marks, P. A., 1958, Red cell glucose-6-phosphate and 6-phosphogluconic dehydrogenases and nucleoside phosphorylase, *Science* **127**:1338.
Marsh, C. A., and Gourlay, G. C., 1971, Evidence for a non-lysosomal α-mannosidase in rat liver homogenates, *Biochim. Biophys. Acta* **235**:142.
Masson, P. L., and Heremans, J. F., 1968, Metal combining properties of human lactoferrin—the involvement of bicarbonate in the reaction, *Eur. J. Biochem.* **6**:579.
Masson, P. L., Heremans, J. F., and Schonne, E., 1969, Lactoferrin, an iron-binding protein in neutrophilic leukocytes, *J. Exp. Med.* **130**:643.
Masters, C., and Holmes, R., 1977, Peroxisomes: New aspects of cell physiology and biochemistry, *Physiol. Rev.* **57**:816.
McCord, J. M., and Fridovich, I., 1978, The biology and pathology of oxygen radicals, *Ann. Intern. Med.* **89**:122.

McDaniel, C. F., Kirtley, M. E., and Tanner, M. J. A., 1974, The interaction of glyceraldehyde 3-phosphate dehydrogenase with human erythrocyte membranes, *J. Biol. Chem.* **249**:6478.

McDonagh, A. F., 1976, Photochemistry and photometabolism of bilirubin IXα, *Birth Defects: Original Article Series* **12**:30.

McDonald, M. J., Shapiro, R., Bleichman, M., Solway, J., and Bunn, H. F., 1978, Glycosylated minor components of human adult hemoglobin, *J. Biol. Chem.* **253**:2327.

Mehler, A. H., 1956, Formation of picolinic and quinolinic acids following enzymatic oxidations of 3-hydroxyanthranilic acid, *J. Biol. Chem.* **218**:241.

Messier, B., and Leblond, C. P., 1960, Cell proliferation and migration as revealed by autoradiography after injection of thymidine-H^3 into male rats and mice. *Am. J. Anat.* **106**:247.

Metzger, B., Helmreich, E., and Glaser, L., 1967, The mechanism of activation of skeletal muscle phosphorylase A by glycogen, *Proc. Natl. Acad. Sci. USA* **57**:994.

Metzler, D. E., 1977, *Biochemistry. The Chemical Reactions of Living Cells,* Academic Press, New York.

Mieyal, J. J., and Blumer, J. L., 1976, Acceleration of the autooxidation of human oxyhemoglobin by aniline and its relation to hemoglobin-catalyzed aniline hydroxylation, *J. Biol. Chem.* **251**:3442.

Mikkelsen, R. B., Tang, D. H., and Triplett, E. L., 1975, Photochemical activation of *Rana pipiens* tyrosinase, *Biochem. Biophys. Res. Commun.* **63**:980.

Milligan, L. P., and Baldwin, R. L., 1967, The conversion of acetoacetate to pyruvaldehyde, *J. Biol. Chem.* **242**:1095.

Minato, S., and Werbin, H., 1971, Spectral properties of the chromophoric material associated with the deoxyribonucleic acid photoreactivating enzyme isolated from baker's yeast, *Biochemistry* **10**:4503.

Mittal, C. K., and Murad, F., 1977, Activation of guanylate cyclase by superoxide dismutase and hydroxy radical: A physiological regulator of guanosine 3',5'-monophosphate formation, *Proc. Natl. Acad. Sci. USA* **74**:4360.

Morimoto, H., Lehmann, H., and Perutz, M. F., 1971, Molecular pathology of human haemoglobin: Stereochemical interpretation of abnormal oxygen affinities, *Nature* **232**:408.

Muensing, R. A., Lornitzo, F. A., Kumar, S., and Porter, J. W., 1975, Factors affecting the reassociation and reactivation of the half-molecular weight nonidentical subunits of pigeon liver fatty acid synthetase, *J. Biol. Chem.* **250**:1814.

Muller-Eberhard, U., 1970, Hemopexin, *N. Engl. J. Med.* **283**:1090.

Musajo, L., Spada, A., and Bulgarelli, E., 1950, Synthesis of *o*-nitrobenzoyl-pyruvatic acid and its catalytic reduction, *Gazz. Chim. Ital.* **80**:161.

Nelson, D. P., Miller, W. D., and Kiesow, L. A., 1974, Calorimetric studies of hemoglobin function, the binding of 2,3-diphosphoglycerate and inositol hexaphosphate to human hemoglobin A, *J. Biol. Chem.* **249**:4770.

Neurath, H., and Walsh, K. A., 1976, Role of proteolytic enzymes in biological regulation (a review), *Proc. Natl. Acad. Sci. USA* **73**:3825.

Newman, S. A., Rossi, G., and Metzger, H., 1977, Molecular weight and valence of the cell-surface receptor for immunoglobulin E, *Proc. Natl. Acad. Sci. USA* **74**:869.

Niemeyer, H., Ureta, T., and Clark-Turri, L., 1975, Adaptive character of liver glucokinase, *Mol. Cell. Biochem.* **6**:109.

Nordlie, R. C., 1974, Metabolic regulation by multifunctional glucose-6-phosphatase, *Curr. Top. Cell. Regul.* **8**:33.

Ogasawara, N., Goto, H., and Watanabe, T., 1975, Isozymes of rat brain deaminase: Developmental changes and characterizations of five forms, *FEBS Lett.* **58**:245.

Ogasawara, N., Goto, H., Yamada, Y., and Yoshino, M., 1977, Subunit structures of AMP deaminase isozymes in rat, *Biochem. Biophys. Res. Commun.* **79**:671.
Ohba, H., Harano, T., and Omura, T., 1977, Presence of two different types of protein-disulfide isomerase on cytoplasmic and luminal surfaces of endoplasmic reticulum of rat liver cells, *Biochem. Biophus. Res. Commun.* **77**:830.
Oppenheimer, J. H., 1968, Role of plasma proteins in the binding, distribution and metabolism of the thyroid hormones, *N. Engl. J. Med.* **278**:1153.
Orgel, L. E., 1963, The maintenance of the accuracy of protein synthesis and its relevance to aging, *Proc. Natl. Acad. Sci. USA* **49**:517.
Orgel, L. E., 1970, The maintenance of the accuracy of protein synthesis and its revelance to aging: A correction, *Proc. Natl. Acad. Sci. USA* **67**:1476.
Orgel, L. E., 1973, Aging of clones of mammalian cells, *Nature* **243**:441.
Osaki, S., Johnson, D. A., and Frieden, E., 1971, The mobilization of iron from the perfused mammalian liver by a serum copper enzyme, ferroxidase I, *J. Biol. Chem.* **246**:3018.
Ostroy, S. E., 1977, Rhodopsin and the visual process, *Biochim. Biophys. Acta* **463**:91.
Ottaway, J. H., and Mowbray, 1977, The role of compartmentation in the control of glycolysis, *Curr. Top. Cell. Regul.* **12**:107.
Owen, C. A., Jr., and Hazelrig, J. B., 1968, Copper deficiency and copper toxicity in the rat, *Am. J. Physiol.* **215**:334.
Parisi, A. F., and Vallee, B. L., 1970, Isolation of a zinc α_2-macroglobulin from human serum, *Biochemistry* **9**:2421.
Park, D. J. M., 1974, An algorithm for detecting nonsteady state moieties in steady state subnetworks, *J. Theor. Biol.* **48**:125.
Park, D. J. M., 1975, SMISS, Stoichiometric matrix inversion for steady state metabolic networks, *Comp. Programs in Biomed.* **5**:46.
Parshad, R., Sanford, K. K., Jones, G. M., and Tarone, R. E., 1978, Fluorescent light-induced chromosome damage and its prevention in mouse cells in culture, *Proc. Natl. Acad. Sci. USA* **75**:1830.
Paukert, J. L., Straus, L. D'A., and Rabinowitz, J. C., 1976, Formylmethenyl-methylenetetrahydrofolate synthetase-(combined). An ovine protein with multiple catalytic activities, *J. Biol. Chem.* **251**:5104.
Peanasky, R. J., and Lardy, H. A., 1958, Bovine liver aldolase. I. Isolation, crystallization and some general properties, *J. Biol. Chem.* **233**:365.
Peat, R., and Soderwall, A. L., 1972, Estrogen stimulated pathway changes and cold-inactivated enzymes, *Physiol. Chem. Phys.* **4**:295.
Perutz, M. F., and Lehmann, H., 1968, Molecular pathology of human haemoglobin, *Nature* **219**:902.
Peterka, E. S., Fusaro, R. M., and Goltz, R. W., 1965, Erythropoietic protoporphyria. II. Histological and histochemical studies of cutaneous lesions, *Arch. Derm.* **92**:357.
Peters, T., Jr., 1975, Serum albumin in: *The Plasma Proteins, Structure, Function and Genetic Control*, Vol. 1 (F. W. Putnam, ed.), pp. 133–181, Academic Press, New York.
Phillips, A. T., and Wood, W. A., 1964, Basis for AMP activation of "biodegradative" threonine dehydrase from *Escherichia coli*, *Biochem. Biophys. Res. Commun.* **15**:530.
Phillips, D. R., Duley, J. A., Fennell, D. J., and Holmes, R. S., 1976, The self-association of L-α-hydroxyacid oxidase, *Biochim. Biophys. Acta* **427**:679.
Poillon, W. N., and Bearn, A. G., 1967, The molecular structure of human ceruloplasmin: Evidence for subunits, *Biochim. Biophys. Acta* **127**:407.
Polonovski, M., and Jayle, M. F., 1938, Existence dans le plasma sanguin d'une substance activant l'action peroxydasique de l'hémoglobine, *C. R. Seances Soc. Biol. Fil.* **129**: 457.

Power, H. W., 1975, A model of how the sickle-cell gene produces malaria resistance, *J. Theor. Biol.* **50**:121.
Purich, D. L., and Fromm, H. J., 1972, A possible role for kinetic reaction mechanism dependent substrate and product effects in enzyme regulation, *Curr. Top. Cell. Regul.* **6**:131.
Quevedo, W. C., Jr., and Smith, J. A., 1963, Studies on radiation-induced tanning of skin, *Ann. N. Y. Acad. Sci.* **100**:364.
Qureshi, A. A., Lornitzo, F. A., and Porter, J. W., 1974, The isolation of acyl carrier protein from the pigeon liver fatty acid synthetase complex II, *Biochem. Biophys. Res. Commun.* **60**:158.
Rachmilewitz, E. A., 1974, Denaturation of the normal and abnormal hemoglobin molecule, *Semin. Hematol.* **11**:441.
Raghavan, S. R. V., and Gonick, H. C., 1977, Isolation of low-molecular-weight lead-binding protein from human erythrocytes, *Proc. Soc. Exp. Biol. Med.* **155**:164.
Rapoport, T. A., Heinrich, R., and Rapoport, S. M., 1976, The regulatory principles of glycolysis in erythrocytes in vivo and in vitro. A minimal comprehensive model describing steady states, quasi-steady states and time-dependent processes, *Biochem. J.* **154**:449.
Ratner, S., 1973, Enzymes of arginine and urea synthesis, *Adv. Enzymol.* **39**:1.
Reed, L. J., and Cox, D. J., 1966, Macromolecular organization of enzyme systems, *Annu. Rev. Biochem.* **35**:57.
Rees, D. A., 1975, Stereochemistry and binding behaviour of carbohydrate chains, *MTP Int. Rev. Sci. Biochem. Ser. 1*, **5**:1.
Richards, M. P., and Cousins, R. J. 1977, Isolation of an intestinal metallothionein induced by parenteral zinc, *Biochem. Biophys. Res. Commun.* **75**:286.
Richart, R. M., 1963, A radioautographic analysis of cellular proliferation in dysplasia and carcinoma *in situ* of the uterine cervix, *Am. J. Obstet. Gynecol.* **86**:925.
Riggs, A., and Gibson, Q. H., 1973, Oxygen equilibrium and kinetics of isolated subunits from hemoglobin Kansas, *Proc. Natl. Acad. Sci. USA* **70**:1718.
Rose, I. R., and Warms, J. V. B., 1967, Mitochondrial hexokinase. Release, rebinding, and location, *J. Biol. Chem.* **242**:1635.
Rubin, C. S., Balis, M. E., Piomelli, S., Berman, P. H., and Dancis, J., 1969, Elevated AMP pyrophosphorylase activity in congenital IMP pyrophosphorylase deficiency (Lesch-Nyhan disease), *J. Lab. Clin. Med.* **74**:732.
Ruchti, J., and McLaren, A. D., 1965, Enzyme reactions in structurally restricted systems. VI. Activity of hexokinase on surfaces, *Enzymologia* **28**:201.
Rucker, R. B., and Murray, J., 1978, Cross-linking amino acids in collagen and elastin, *Am. J. Clin. Nutr.* **31**:1221.
Saari, J. C., and Futterman, S., 1976, Separable binding proteins for retinoic acid and retinol in bovine retina, *Biochim. Biophys, Acta* **444**:789.
Sankar, D. V. S., 1959, Enzymatic activity of ceruloplasmin, *Fed. Proc.* **18**:441.
Sasazuki, T., Tsunoo, H., Nakajima, H., and Imai, K., 1974, Interaction of human hemoglobin with haptoglobin or antihemoglobin antibody, *J. Biol. Chem.* **249**:2441.
Sass, M. D., Vorsanger, E., and Spear, P. W., 1964, Enzyme activity as an indicator of red cell age, *Clin. Chim. Acta* **10**:21.
Scallen, T. J., Seetharam, B., Srikantaiah, M. V., Hansbury, E., and Lewis, M. K., 1975, Sterol carrier hypothesis: Requirement for three substrate-specific soluble proteins in liver cholesterol biosynthesis, *Life Sci.* **16**:853.
Schade, A. L., and Caroline, L., 1944, Raw hen egg white and the role of iron in growth inhibition of *Shigella dysenteriae, Staphylococcus aureus, Escherichia coli* and *Saccharomyces cerevisiae, Science* **100**:14.

Schade, A. L., and Caroline, L., 1946, An iron-binding component in human blood plasma, *Science* **104**:340.
Scheiner, S., and Kern, C. W., 1978, Energies of polypeptides: Theoretical conformational study of polyglycine using quantum mechanical partitioning, *Proc. Natl. Acad. Sci. USA* **75**:2071.
Schimke, R. T., 1975, Protein synthesis and degradation in animal tissue, *MTP Int. Rev. Sci. Biochem. Ser. 1*, **9**:183.
Schneider, R. G., Ueda, S., Alperin, J. B., Brimhall, B., and Jones R. T., 1969, Hemoglobin Sabine β 91(F7) Leu $\rightarrow$ Pro. An unstable variant causing severe anemia with inclusion bodies, *N. Engl. J. Med.* **280**:739.
Schothorst, A. A., Van Steveninck, J., Went, L. N., and Suurmond, D., 1970, Protoporphyrin-induced photohemolysis in protoporphyria and in normal red blood cells, *Clin. Chim. Acta* **28**:41.
Schwartz, R. M., and Dayhoff, M. O., 1978, Origins of prokaryotes, eukaryotes, mitochondria, and chloroplasts, *Science* **199**:395.
Schwartz, S. M., and Benditt, E. P., 1976, Clustering of replicating cells in aortic endothelium, *Proc. Natl. Acad. Sci. USA* **73**:651.
Sel'kov, E. E., 1975, Stabilization of energy charge, generation of oscillations and multiple steady states in energy metabolism as a result of purely stoichiometric regulation, *Eur. J. Biochem.* **59**:151.
Setlow, R. B., 1968, The photochemistry, photobiology, and repair of polynucleotides, *Prog. Nucleic Acid Res. Mol. Biol.* **8**:257.
Shin, B. C., and Carraway, K. L., 1973, Association of glyceraldehyde 3-phosphate dehydrogenase with the human erythrocyte membrane, *J. Biol. Chem.* **248**:1436.
Shorter, R. G., Moertel, C. G., Titus, J. L., and Reitemeier, R. J., 1964, Cell kinetics in the jejunum and rectum of man, *Am. J. Dig. Dis.* **9**:760.
Shoup, V. A., and Touster, O., 1976, Purification and characterization of the α-D-mannosidase of rat liver cytosol, *J. Biol. Chem.* **251**:3845.
Sophianopoulos, A. J., and Vestling, C. S., 1960, Nature of the two forms of malic dehydrogenase from rat liver, *Biochim. Biophys. Acta* **45**:400.
Spector, A. A., John, K., and Fletcher, J. E., 1969, Binding of long-chain fatty acids to bovine serum albumin, *J. Lipid Res.* **10**:56.
Srikantaiah, M. V., Hansbury, E., Loughran, E. D., and Scallen, T. J., 1976, Purification and properties of sterol carrier protein, *J. Biol. Chem.* **251**:5496.
Stadtman, E. R., and Chock, P. B., 1977, Superiority of interconvertible enzyme cascades in metabolic regulation: Analysis of monocyclic systems, *Proc. Natl. Acad. Sci. USA* **74**:2761.
Stellwagen, E., 1978, Haem exposure as the determinate of oxidation-reduction potential of haem proteins, *Nature* **275**:73.
Stellwagen, E., and Cass, R. D., 1975, Complexation of iron hexacyanides by cytochrome *c*. Evidence for electron exchange at the exposed heme edge, *J. Biol. Chem.* **250**:2095.
Stellwagen, E., and Wilgus, H., 1978, Relationship of protein thermostability to accessible surface area, *Nature* **275**:342.
Sternberg, M. J. E., and Thornton, J. M., 1978, Prediction of protein structure from amino acid sequence, *Nature* **271**:15.
Stoops, J. K., Arslanian, M. J., Oh, Y. H., Aune, K. C., Vanaman, T. C., and Wakil, S. J., 1975, Presence of two polypeptide chains comprising fatty acid synthetase, *Proc. Natl. Acad. Sci. USA* **72**:1940.
Strittmatter, P., and Velick, S. F., 1956, A microsomal cytochrome reductase specific for diphosphopyridine nucleotide, *J. Biol. Chem.* **221**:277.
Sutherland, B. M., Chamberlin, M. J., and Sutherland, J. C., 1973, Deoxyribonucleic acid

photoreactivating enzyme from *Escherichia coli*. Purification and properties, *J. Biol. Chem.* **248**:4200.
Sutherland, B. M., Runge, P., and Sutherland, J. C., 1974, DNA photoreactivating enzyme from placental mammals. Origin and characteristics, *Biochemistry* **13**:4710.
Sutton, H. E., 1970, The haptoglobins, *Prog. Med. Genet.* **7**:163.
Svasti, J., and Bowman, B. H., 1978, Human group-specific component, *J. Biol. Chem.* **253**:4188.
Swartz, S. K., Soloff, M. S., and Suriano, J. B., 1974, Binding of estrogens by α-fetoprotein in rat amniotic fluid, *Biochim. Biophys. Acta* **388**:480.
Tanabe, T., Wada, K., Okazaki, T., and Numa, S., 1975, Acetyl-coenzyme-A carboxylase from rat liver, *Eur. J. Biochem.* **57**:15.
Tanford, C., 1973, *The Hydrophobic Effect: Formation of Micelles and Biological Membranes*, Wiley, New York.
Tanford, C., 1978, The hydrophobic effect and the organization of living matter, *Science* **200**:1012.
Tang, J., James, M. N. G., Hsu, I. N., Jenkins, J. A., and Blundell, T. L., 1978, Structural evidence for gene duplication in the evolution of the acid proteases, *Nature* **271**:618.
Taylor, B. L., Frey, W. H., II, Barder, R. E., Scrutton, M. C., and Utter, M. F., 1978, The use of the ultracentrifuge to determine the catalytically competent forms of enzymes with more than one oligomeric structure, *J. Biol. Chem.* **253**:3062.
Taylor, C., Cox, A. J., Kernohan, J. C., and Cohen, P., 1975, Debranching enzyme from rabbit skeletal muscle. Purification, properties and physiological role, *Eur. J. Biochem.* **51**:105.
Tomino, S., and Paigen, K., 1975, Egasyn, a protein complexed with microsomal β-glucuronidase, *J. Biol. Chem.* **250**:1146.
Tomoda, A., Matsukawa, S., Takeshita, M., and Yoneyama, Y., 1976, Effect of organic phosphates on methemoglobin reduction by ascorbic acid, *J. Biol. Chem.* **251**:1794.
Tsai, H. C., and Norman, A. W., 1973, Studies on calciferol metabolism. VIII. Evidence for a cytoplasmic receptor for 1,25-dihydroxy-vitamin D_3 in the intestinal mucosa, *J. Biol. Chem.* **248**:5967.
Tschudy, D. P., and Bonkowsky, H. L., 1973, A steady state model of sequential irreversible enzyme reactions. *Mol. Cell. Biochem.* **2**:55.
Turner, B. M., Fisher, R. A., and Harris, H., 1975, Post-translational alterations of human erythrocyte enzymes, in: *Isozymes. Molecular Structure*, Vol 1 (C. I. Markert, ed.), pp. 781–795, Academic Press, New York.
Ulevitch, R. J., Cochrane, C. G., Revak, S. D., Morrison, D. C., and Johnston, A. R., 1975, The structural and enzymatic properties of the components of the Hageman factor-activated pathways, in: *Proteases and Biological Control* (Reich, E., Rifkin, D. B., and Shaw, E., eds.), pp. 85–93, Cold Spring Harbor, New York.
Vaitukaitis, J. L., 1976, Peptide hormones as tumor markers, *Cancer* (Suppl. 1) **37**:567.
Varghese, A. J., 1972, Photochemistry of nucleic acids and their constituents, *Photophysiology* **7**:207.
Vendrely, R., and Vendrely, C., 1956, The results of cytophotometry in the study of deoxyribonucleic acid (DNA) content of the nucleus, *Int. Rev. Cytol.* **5**:171.
Verma, I. M., 1977, The reverse transcriptase, *Biochim. Biophys. Acta* **473**:1.
Verseé, V. C., and Barel, A. O., 1978, Spectroscopic evidence of binding properties of rat α-fetoprotein, *Fed. Proc.* **37**:1620.
Vestling, C. S., 1942, The reduction of methemoglobin by ascorbic acid, *J. Biol. Chem.* **143**:439.
Volpe, J. J., and Vagelos, P. R., 1973, Saturated fatty acid biosynthesis and its regulation, *Annu. Rev. Biochem.* **42**:21.

von Bertalanffy, L., 1950, The theory of open systems in physics and biology, *Science* **111**:23.
Wallach, D., Davies, P. J. A., and Pastan, I., 1978, Cyclic AMP-dependent phosphorylation of filamin in mammalian smooth muscle, *J. Biol. Chem.* **253**:4739.
Walser, M., 1961, Ion Association VI. Interactions between calcium, magnesium, inorganic phosphate, citrate and protein in normal human plasma, *J. Clin. Invest.* **40**:723.
Walter, H., and Selby, F. W., 1966, Counter-current distribution of red blood cells of slightly different ages, *Biochim. Biophys. Acta* **112**:146.
Wang, A.-C., and Sutton, H. E., 1965, Human transferrins C and D_1: Chemical difference in a peptide, *Science* **149**:435.
Waterman, M. R., and Cottam, G. L., 1976, Kinetics of the polymerization of hemoglobin S: Studies below normal erythrocyte hemoglobin concentration, *Biochem. Biophys. Res. Commun.* **73**:639.
Whanger, P. D., Phillips, A. T., Rabinowitz, K. W., Piperno, J. R., Shada, J. D., and Wood, W. A., 1968, The mechanism of action of 5'-adenylic acid-activated threonine dehydrase. II. Protomer-oligomer interconversions and related properties, *J. Biol. Chem.* **243**:167.
Wheeler, G. W., and Bitensky, M. W., 1977, A light-activated GTPase in vertebrate photoreceptors: Regulation of light-activated cyclic GMP phosphodiesterase, *Proc. Natl. Acad. Sci. USA* **74**:4238.
White R. C., and Nelson, T. E., 1975, Analytical gel chromatography of rabbit muscle amylo-1,6-glucosidase/4-α-gluconotransferase under denaturing and nondenaturing conditions, *Biochim. Biophys. Acta* **400**:154.
Widner, W. R., Storer, J. B., and Lushbaugh, C. C., 1951, The use of X-ray and nitrogen mustard to determine the mitotic and intermitotic time in normal and malignant rat tissues, *Cancer Res.* **11**:877.
Wilson, A. C., Carlson, S. S., and White, T. J., 1977, Biochemical evolution, *Annu. Rev. Biochem.* **46**:573.
Winterbourn, C. C., and Carrell, R. W., 1974, Studies of hemoglobin denaturation and Heinz body formation in the unstable hemoglobins, *J. Clin. Invest.* **54**:678.
Wittenberg, J. B., 1970, Myoglobin-facilitated oxygen diffusion: Role of myoglobin in oxygen entry into muscle, *Physiol. Rev.* **50**:559.
Yoshida, A., 1973, Hemolytic anemia and G6PD deficiency, *Science* **179**:532.
Zamierowski, M. M., and Wagner, C., 1977, Identification of folate binding proteins in rat liver, *J. Biol. Chem.* **252**:933.

2

Nonequilibrium Thermodynamics, Noncovalent Forces, and Water

Robert M. Cohn, Michael J. Palmieri, and
Pamela D. McNamara

I. Introduction

Living systems are characterized by a degree of complexity and associated order absent in the nonliving world. This complexity and order is maintained only at the expense of free energy from the environment, the sun being the ultimate source of this energy. Living systems are thus *open systems*—able to exchange matter and energy with their surroundings—and, as we shall see, when such systems are driven from equilibrium, there occurs the possibility for their increasing organization.

The hierarchy that characterizes life is maintained through a multiplicity of catabolic and anabolic reactions, governed by a complex set of mechanisms that control the rate and timing of these processes. Thus, while this book will focus primarily on intermediary metabolism and its control mechanisms, in this chapter we will consider metabolic control from the viewpoint of nonlinear, nonequilibrium thermodynamics, which deals with the issue of how molecular events can be *coupled* and *amplified* so that they are expressed on a *macroscopic level*. The concepts herein expressed may be difficult to grasp at first reading. With perseverance, the beauty of the concepts will become apparent and their great importance for biological systems will become evident.

Robert M. Cohn, Michael J. Palmieri, and Pamela D. McNamara • Department of Metabolic Research, Children's Hospital of Philadelphia, University of Pennsylvania, Philadelphia, Pennsylvania 19104.

We know that living systems are the apogees of coordinate organization. This biological organization has been termed *coherent behavior* (Weiss, 1968) and embraces the view that life is characterized both by a *structural order* (e.g., macromolecules, membranes, cells) and a *functional order* maintained by the myriad number of coupled biochemical pathways (intermediary metabolism).

While classical thermodynamics predicts that the evolution of natural processes leads to increasing *disorder,* the development of nonlinear, nonequilibrium thermodynamics has sought to explain the ubiquitous observation that biological evolution has led to increasing *order.* What the Brussels school (Glansdorff and Prigogine, 1971; Prigogine, 1978)—as it is sometimes called—has done is to develop a thermodynamic view of the hierarchy of life structures which relates the existence of sequential levels of organization to the development of a *succession of instabilities.* In this view, structures evolve through the continuous interaction of energy and matter with the outside world, a critical distance from equilibrium being required to maintain such structures.

The ability of living organisms to feed on "negative entropy" by maintaining the system at a distance from equilibrium, through metabolism, was first propounded by Schroedinger (1944), and expanded by Prigogine (1947) and Bertalanffy (1952). Bertalanffy's (1952) assertion that "living forms are not *in being,* they are *happening;* they are the expression of a perpetual stream of matter and energy which passes the organism and at the same time constitutes it" captures the essential interactions in life. This emphasis on the removal of an open system from equilibrium (irreversibility) represents the crucial conceptual advance permitting life's processes to obey the second law of thermodynamics while achieving greater order. Through the exchange of energy with the surroundings, a new supermolecular organization results from the amplification and stabilization of a preceding instability. This new principle of *"order through fluctuations"* is basic to understanding *dissipative structures,* which play an essential role in the living system.

In this chapter we will consider how organized patterns of structure and function may energe as a consequence of the coupling and amplification of molecular events occurring far from equilibrium and obeying nonlinear kinetics. The role of noncovalent forces in providing flexibility of response to the environment—a possibility not realizable with covalent forces only—will then be considered. Finally, we shall consider the role of water in living systems as it relates to the formation and stabilization of biopolymers and to energy relationships within the cell.

II. Stability, Thermodynamics, and Biological Organization

As already indicated, living systems are the paradigm of functioning, self-organizing, self-maintaining, self-replicating, and evolving systems. Certain phenomena, at present incompletely understood, are essential to a reductionist view of the hierarchy we understand as life. This view is a postulate providing certain arbitrary categories that are convenient to use and do not disagree with our first fundamental principle. These categories include:

1. A universal informational code (genetic) and functional molecules (proteins and polypeptides).
2. Employment of subunit construction, in which atoms form molecules, molecules form macromolecules, and so on, to form membranes, organelles, cells, tissues, and organs.
3. A thermodynamic principle which prescribes the emergence of order in systems driven far from equilibrium and which also possesses appropriate kinetic parameters to ensure their removal from equilibrium. This thermodynamic contribution involves the interplay of information, energy, and matter with the living system, allowing for the coherent behavior synonymous with life.

No single factor in the "reductionist view of life" is sufficient to explain the complexity of a biological system, yet each factor provides study areas which contribute significantly to the understanding of the whole. Enzymes may be considered the functional entities of life. They possess catalytic activity and conformational adaptability (implying plasticity or instability) and, when coupled with the protein synthetic apparatus that exists to replace denatured enzymes, they epitomize at the molecular level the semistable state known as "life."

Entropy is usually defined as the amount of disorder in a physical system, but, in a living system, entropy and its relationship to information and work requires redefinition. Brillouin (1962) has argued that entropy actually measures the lack of information about the true structure of the system. This lack of information introduces the possibility of a great variety of microscopically distinct structures which in practice we cannot distinguish from one another. This then raises the further question: Is entropy a property of the system, of the observer, or of the relationship between them?

Most experimental methods in biochemistry require significant disruption of the "native" state in order to yield quantitative information regarding the state. This insinuation into, or perturbation of, the "native" state so alters the system that it is no longer possible to determine the exact nature of the unperturbed system with any certainty. This corresponds to an application of the Heisenberg uncertainty principle to biological systems. In living organisms, however, the observer and the system may, in fact be one and the same (Elsasser, 1966), since phenomena occur on a molecular basis. As Morowitz (1968) emphasizes, a system which is itself the microstate may well have knowledge about its own state. Put in another way, the formation of an enzyme–substrate complex does not require that the enzyme perform a series of observations to see if substrate binding is stereospecific. The enzyme, existing in its own microstate, interacts specifically, without any necessity of an informational step (Comorosan, 1976). The important role that entropy plays in explaining the mechanism of enzyme action will be developed in Chapter 3.

In the newly developed general theory of thermodynamics (Glansdorff and Prigogine, 1971), entropy provides the basis of *global stability criteria*. Thermodynamic principles describe and can even predict the emergence of organized patterns and structures in biological systems. This emergence of order comes as a consequence of the regulated interactions between the myriad *enzymatic processes* occurring at the molecular level, and the *amplification processes* by which these molecular events manifest themselves collectively and coherently at the macroscopic level, i.e., within the province of physiology. Such coupling, amplification, and emergence of order depend upon organization in both time and space. Oscillations in metabolic sequences represent such emergent phenomena.

An apt analogy can be drawn between the functioning of the symphony orchestra and the life process. The modern symphony orchestra has developed over a long period, through an evolutionary process employing a trial and error strategy. The component parts of the modern orchestra, while possessing the ability to function in isolation, can only function as an orchestra when they are organized in time and space, so that the resultant is the *coherent* and *harmonious* performance of a musical work. And so it is—on an even grander scale—with a living organism. While a symphony orchestra has approximately 100 instruments and an equal number of performers, each living cell has several thousand enzymes, and each organ has millions to billions of cells. Thus it is clear that this analogy—while aiding our understanding of the synchronized interactions of cell components and ultimately of organisms—cannot begin to approach the complexity of the orchestrated processes comprising life.

But just as the coupling of the multiplicity of instruments in an orchestra leads to a result not achievable by any single instrument (i.e., harmony), so the coupling of molecular processes achieves a similar amplification, whose resultant is the emergence of function expressed at a macroscopic level (coherent behavior). Specifically, when the chemical reactions going on within cells are coordinated in space and time, they give rise to a functional order which provides the conditions for the emergence of more complex structures. In this chapter we shall discuss the relationship of these chemical reactions to nonequilibrium thermodynamics and the genesis of biologic order and, in Chapter 8, we shall further explore their relationship to metabolic control, oscillatory behavior or pathways, and morphogenesis.

A. The Development of a General Theory of Thermodynamics

The importance of thermodynamics lies in its ability to provide a description of complicated macroscopic processes in simplified language. A cursory review of certain concepts from equilibrium thermodynamics may be in order before embarking upon a consideration of far-from-equilibrium thermodynamics as applied to living systems. Classical thermodynamics is by no means a subject that has been neglected, even when applied to biochemistry, and several references are worthy of note (Ingraham and Pardee, 1967; Klotz, 1967; and Peusner, 1974).

The first law of thermodynamics states that energy can neither be created nor destroyed and hence that in any process the total quantity of energy of the system and its surroundings remains constant. The mathematical statement of the first law is given by the equation:

$$\Delta E = Q - W \tag{1}$$

where ΔE is the difference between the energy of the final and initial states of the system, Q is the heat absorbed by the body, and W is the quantity of work performed, defined as the product of a force by a displacement. For chemical reactions,

$$W = P\Delta V \tag{2}$$

where P equals the opposing pressure and ΔV is the infinitesimal increase in volume.

Since most chemical reactions are carried out at constant pressure (e.g., atmospheric pressure), the heat absorbed (or evolved) by the system is equivalent to another thermodynamic quantity, ΔH, which is known as the enthalpy, and which is sometimes called the heat content.

Thus, enthalpy is related to energy, E, by the following equation:

$$\Delta H = \Delta E + P\Delta V \tag{3}$$

In actual circumstances, volume changes may be negligible ($P\Delta V \cong 0$), in which case equation (3) reduces to $\Delta H = \Delta E$, i.e., the enthalpy describes the total energy of the system.

From the second law of thermodynamics, we know that all natural processes tend to approach equilibrium spontaneously, energy being required to drive the system away from equilibrium. The second law provides us with a criterion of spontaneity, i.e., whether a given process (e.g., biochemical or biophysical reactions) is feasible under a specified set of conditions. It points the direction for the evolution of physicochemical processes since it capsulizes the observation that, in all processes, some of the energy becomes unavailable to perform work, owing to the increased random motion of some of the component molecules of the system. In other words, the second law distinguishes between irreversible processes, which by their very nature are unidirectional, and reversible processes, which are bidirectional.

This relationship can be expressed classically as follows:

$$\Delta S = \frac{\Delta Q}{T} \geq 0 \text{ (for an isolated system)} \tag{4}$$

where Q represents that total energy which loses its ability to perform work, T is the absolute temperature at which the heat is absorbed, and S is the *entropy* and measures the randomness or disorder of the system at constant temperature. For systems at equilibrium, $\Delta S = 0$. Consequently, for a process to occur spontaneously, the entropy of the system must increase.*

Classical thermodynamics has concentrated on systems at equilibrium, and structures at equilibrium are formed and maintained by reversible processes which minimize deviations from equilibrium. Thus, the second law dictates that, in an equilibrium state, an increase in entropy prescribes the *destruction of structure* for a system. While examples of order-to-disorder transition are common in our experience, it is known that if an equilibrium structure is brought to sufficiently low temperature, the entropic component can be minimized. This is the Boltzmann order principle (Glansdorff and Prigogine, 1971), which

---

*Although entropy has been established as a criterion of spontaneity, there are practical difficulties in measuring entropy change. To circumvent these differences, Gibbs introduced a new thermodynamic quantity, the free energy, G, to define a criterion of spontaneity. The free energy is related to the entropy and enthalpy by the following: $\Delta G = \Delta H + T\Delta S \leq 0$. For a spontaneous change to occur at constant pressure and temperature, the free energy of the system must decrease. For a system at equilibrium, $\Delta G = 0$.

can explain the existence of crystals and phase transitions, two kinds of temperature-dependent ordering phenomena.

Crystallization is a classical example of a self-organizing, *equilibrium structure*, and one may inquire whether it serves as an adequate model for biological organization. Crystal organization is a state of minimum free energy, in which the geometric arrangement of the subunits is a repeating or periodic pattern that can be accomplished only under near-equilibrium conditions (on the thermodynamic branch). Biological subunits (macromolecules, membranes, and cells), while being composed of smaller units, are strikingly different from crystals in that they are characterized by the generation of complex and asymmetric structures, often from symmetric precursors (Rosen, 1972). Thus, crystal formation is not a valid prototype for the structural organization of biological systems which display a hierarchical order.

The extension of classical thermodynamics to cover irreversible processes whose flows or reaction rates are linear functions of thermodynamic forces (such as temperature, chemical affinities, etc.) was termed "linear nonequilibrium thermodynamics," and includes the Onsager reciprocity relations (1931) and the theorem of minimum entropy production (Glansdorff and Prigogine, 1971). The diffusion of two gases in a mixture subjected to a thermal gradient represents a situation occurring in the realm of linear nonequilibrium thermodynamics. The result of thermal diffusion under these conditions is a heterogenous situation, where the gas mixture at the hot wall is enriched in one component while that at the cold wall is enriched in the other. This nonequilibrium situation has a minimized entropy component, which the Boltzmann order principle associates with order; therefore, the situation provides an example where *nonequilibrium may be a source of order*. However, despite an increase in order, no new structures can be formed in the domain of linear nonequilibrium thermodynamics, because the subject deals with equilibrium structures in systems which are prevented from achieving equilibrium.

While Schroedinger (1944), Prigogine (1947), and Bertalanffy (1952) recognized that biological order requires that the system be removed from equilibrium, it was left to the Brussels school to develop a generalized theory of thermodynamics which would deal with the creation of new structures and in which the biological realities of nonlinear, nonequilibrium systems could be encompassed. Such systems are, by definition, *open* since they can exchange energy, matter, and information with their surroundings. According to the theory, as a system is forced further and further away from equilibrium, the stability criteria for the system on the thermodynamic branch (corresponding to equilibrium and linear nonequilibrium conditions) may not be met, resulting in an *instability of the thermodynamic branch.* Beyond this critical point, known as the *thermodynamic threshold,* the system may evolve through the formation of new structures or states of matter termed *dissipative structures,* which display *coherent behavior.* Dissipative structures, in contrast to equilibrium structures, are formed and maintained by the flow of energy and matter in a far-from-equilibrium situation. Thus elaborated, this general theory provides that near equilibrium there is destruction of order but that far from equilibrium there is creation of order, neither state violating the entropic constraints imposed by the second law of thermodynamics.

B. Order through Fluctuations

Complex systems have many degrees of freedom, which in a macroscopic system implies the possibility of spontaneous fluctuations. At equilibrium, near-equilibrium, and in the linear, nonequilibrium domain, a spontaneous fluctuation regresses in conformity with the Le Chatelier–Braun principle (Glansdorff and Prigogine, 1971). However, a fluctuation in a nonlinear, nonequilibrium situation does not always lead to a regression but may be amplified and lead to a new organization which originates from an instability generated by the fluctuation. This principle has been termed "order through fluctuations" and occurs through the formation of dissipative structures. In the far-from-equilibrium domain, order through fluctuations replaces the Boltzmann order principle. As noted above, dissipative structures are self-organized, ordered systems which appear as a response to a large deviation from thermodynamic equilibrium. As stated by Nicolis (1975, p. 31), "they are created and maintained by the dissipative, entropy-producing processes inside the system . . . they provide a sufficient condition that must be realized in order to have spontaneous emergent or ordered structures." Since nonlinearity is a general rule in living organisms, comprehension of ordering phenomena in such systems can be facilitated by the use of examples from less complex macroscopic systems which, when driven far from equilibrium, lead to increasing order. Examples may be found in hydrodynamics, chemical, and electrochemical phenomena.

Since the early 1900s, it has been known that when an ordinary fluid layer is heated from below, a critical point in the heating process is reached (the thermodynamic threshold) beyond which there emerges a macroscopic, hexagonal pattern reminiscent of a honeycomb. This is known as the *Bénard instability* and represents one of the most easily appreciated examples of a dissipative structure (Glansdorff and Prigogine, 1971). In this situation, initial heating of the horizontal fluid layer from below results only in fluctuations which are rapidly damped; but *beyond a critical value,* that energy of the system which exists as random thermal motion is redistributed into a convective flow which leads to the hexagonal pattern. However, only when the temperature gradient drives the system sufficiently *far from equilibrium* can the new order occur. This new order is the result of a cooperative phenomenon on the molecular level and occurs only in a system possessing nonlinear couplings.

A chemical example of the formation of a dissipative structure is the Belusov–Zhabotinski reaction. In this reaction, cerium ions catalyze the oxidation of malonic acid analogs by bromate, in a process that involves three separate reactions: the oxidation of malonic acid, the oxidation of cerium ions, and the transformation of malonic acid into bromomalonic acid. When this sequence is carried out in a medium which is continuously mixed, and for which certain values of products and temperature are maintained, the concentrations of the intermediates oscillate. If the reaction is carried out in a test tube and concentrations are changed slightly, the system evolves to a macroscopic pattern of dark layers separated by distinct light interfaces. When carried out instead in a shallow layer of unmixed solution, wavefronts are propagated from multiple centers, producing scroll-like patterns. The spatial patterns emerge from oscillations brought about by a short oxidation phase followed by a longer reduction phase. The patterning observed under both sets of conditions suggests that these are examples of dissipative structures arising beyond the thermodynamic branch because of perturbations which break the symmetry of the original state. For an excellent description and elegant pictorial presentation of this phenomenon, the reader is directed to the review by Winfree (1974).

It is in such situations that an explanation for the ordering phenomena characterizing biological organization is to be found. The new ordering is dependent upon the relative contribution of the entropy production within the entropy exchange (flow) with the surroundings. It is encountered in systems whose response to perturbations is nonlinear owing to autocatalysis or feedback phenomena in chemical reactions. When such systems are driven far from equilibrium, they become unstable and evolve to new structures which display coherent behavior. The importance of this formulation is that it demonstrates the existence of a class of systems which exhibit two kinds of behavior: the destruction of order in the neighborhood of thermodynamic equilibrium but the creation of order far from equilibrium. The most interesting and far-reaching result of the emphasis on far-from-equilibrium conditions has been that the stability of the macroscopic state of the system cannot be guaranteed. The system may thus depart from such a state abruptly, through an unstable transition, and enter into a domain of states not predictable from the domain of near-equilibrium states.

C. Stability Criteria, Instability, and Thermodynamic Theory

The emphasis of equilibrium thermodynamics has been on stability; thus, entropy provides a criterion for stability, where high entropy is synonymous with disorder according to the Boltzmann order principle. The Boltzmann order principle, however, cannot be applied to dissipative structures or nonlinear situations. Yet life is characterized by its nonlinearity, and just as nonlinearity in life is associated with multiple degrees of freedom and higher order, so nonlinearity in mathematics is described in higher order terms. Linear functions are relatively rigid regarding directional change (evolution), since the first differential (which assesses change) of a linear function is a constant. Thus, no fluctuation in the linear range can lead to evolution, since the rate of change (first differential) and the curvature (second differential) of the function will be a constant and zero, respectively. Thus the system will damp fluctuations unless they exceed the boundary conditions. In this case, the system becomes unstable in the linear range (on the thermodynamic branch) and passes a bifurcation point (the thermodynamic threshhold) into the nonlinear range, where a newly ordered system can result through the formation of dissipative structures. The development of a thermodynamic theory which would embrace equilibrium, linear nonequilibrium, and nonlinear nonequilibrium domains involved extensive use of the qualitative theory of partial differential equations as developed by Poincaré and Lyapounov (Andronov et al., 1966; Minorsky, 1962). A brief overview of the development of mathematical theory involved in the generalized theory of thermodynamics is presented in Appendix I.

Prigogine and his collaborators (Glansdorff and Prigogine, 1971; Prigogine et al., 1972; Prigogine and Lefever, 1975; Nicolis, 1975; Nicolis and Prigogine, 1977) have formulated an extended version of the second law of thermodynamics, focusing on the behavior and components of entropy. In this formulation, the variation of total entropy, S, of the system during a time interval, dt, is represented by the equation:

$$dS = d_eS + d_iS \qquad (5)$$

which divides entropy into two parts: d_eS, the flow of entropy due to exchanges with the surroundings; and d_iS, the entropy production due to irreversible processes within the system. The second law implies that $d_iS \geq 0$, the equality obtaining at equilibrium. For an isolated system, $d_eS = 0$, since by definition no interaction with the outside world exists, and the equation reduces to $dS = d_iS \geq 0$.

On the other hand, an open system possesses flow terms in the entropy variations (d_eS) which relate to exchanges of energy and matter with the outside world. These terms do not have any definite sign—in contrast to entropy production (d_iS) which, by the second law, cannot be negative. It is useful here to realize that the entropy production term, d_iS, corresponds to the entropy associated with molecular parameters within the system (i.e., chemical reactions, heat production, diffusion, etc.), while entropy flow, d_eS, represents the boundary conditions imposed on the system and its interactions with the outside world. Classical thermodynamics deals primarily with reversible processes at equilibrium. In such cases entropy production vanishes ($d_iS = 0$). In nonequilibrium thermodynamics, macroscopic states are studied on the basis of their entropy production. Since entropy flow terms have no definite sign, it is conceivable that a system may evolve to a more ordered state in which it attains a lower entropy than it had initially. This new state of order can be maintained only if a steady state is reached such that $dS = 0$, in which case $d_eS = -d_iS \leq 0$. In other words, only through negative entropy flow can the new ordered state be achieved and maintained. This can occur only in nonequilibrium conditions since, at the equilibrium steady state, d_iS, and consequently d_eS, would vanish. Thus, we begin to see the basis on which a new generalized theory of thermodynamics can be developed, where the second law ($d_iS \geq 0$) would be compatible with a decrease in overall entropy and where nonequilibrium can be a source of order.

The local equilibrium assumption was the basis on which the Brussels school developed a global thermodynamic theory. Use of this assumption makes possible the macroscopic evaluation of entropy production and entropy flow terms with macroscopic thermodynamic methods. The assumption states that "there exists within each small mass element of the medium a state of *local equilibrium* for which the local entropy, s, is the same function of the local macroscopic variables as at equilibrium state" (Glansdorff and Prigogine, 1971, p. 14). In other words, each small element of a system may be treated as a state near equilibrium but need not necessarily be at equilibrium. This does not mean that the system as a whole need be near equilibrium; thus, neighboring local elements may differ in parameters (temperatures, chemical affinities, etc.) which are reflected in the function describing their local entropy. The additional assumption is made that the sum of the criteria of local stability for each element corresponds to the global stability criterion for the whole system.

Thus, for each local equilibrium element, the entropy s is a function of local macroscopic variables, and the integration of the local entropy terms gives the entropy terms for the system. If we expand entropy about its reference value at steady state, s_0, taking the limit of *small* fluctuations (δ) we obtain:

$$s = s_0 + \delta s + 1/2 \delta^2 s + \ldots \tag{6}$$

Higher-order terms of the expansion are not retained, since we are dealing only with small fluctuations. Since s_0 is time independent, the variation of entropy with time is given by:

$$\partial_t[s] = \partial_t(\delta s) + 1/2\partial_t(\delta^2 s) \tag{7}$$

For systems both at equilibrium and removed from equilibrium, Einstein's theory of fluctuations indicates that the curvature of entropy ($\delta^2 s$) is a quantity which should be considered when seeking to describe the stability of a system. First-order terms, according to the stability theory of differential equations, are considered as general equilibrium conditions which reflect the irreversibility of the processes involved. Second-order terms, however, measure the ability to reestablish given boundary conditions when perturbations or fluctuations occur, and it is the sign of the curvature of entropy ($\delta^2 s$) which dictates stability for thermodynamic equilibrium. In the range for which the local equilibrium assumption remains valid,

$$\delta^2 s < 0 \tag{8}$$

where $\delta^2 s$ is termed the *"excess entropy."* The regression of fluctuations may be studied by considering the change in excess entropy with time, i.e., $\dfrac{\partial(\delta^2 s)}{\partial_t}$ which is called the *"excess entropy production."*

The minimum entropy production theorem dictates that, for a system near equilibrium to achieve a steady state, the entropy production must attain the least possible value compatible with the boundary conditions. Near equilibrium, if the steady state is perturbed by a small fluctuation (δ), the stability of the steady state is assured if the time derivative of entropy production (P) is less than or equal to zero. This may be expressed mathematically as $dP/dt \leq 0$. When this condition pertains, the system will develop a mechanism to damp the fluctuation and return to the initial state. The minimum entropy production theorem, however, may be viewed as providing an *evolution criterion* since it implies that a physical system open to fluxes will evolve until it reaches a steady state which is characterized by a minimal rate of dissipation of energy. Because a system on the thermodynamic branch is governed by the Onsager reciprocity relations and the theorem of minimum entropy production, it cannot evolve. Yet as a system is driven further away from equilibrium, an instability of the thermodynamic branch can occur and new structures can arise through the formation of *dissipative structures* which requires the constant dissipation of energy.

The new general thermodynamic theory extends the use of macroscopic thermodynamic methods to embrace the nonlinear couplings in life, exemplified by feedback and autocatalytic phenomena, which are most precisely described by kinetic arguments, as emphasized by its developers (Glansdorff and Prigogine, 1971). A clear analogy with the framing of new, powerful theories in physics, which have relied upon advances in mathematics and physics, can be made with the examples of the theory of relativity and quantum mechanics. In achieving this new general theory of thermodynamics, Prigogine and his colleagues have used a number of fertile areas of mathematical physics, specifically balance equations, classical thermodynamic stability theory, Lyapounov stability theory, and Einstein's fluctuation theory and, in doing so, have synthesized and extended these views to permit the development of what they term *"global stability criteria."* As elaborated below, these are

$$\delta^2 s < 0 \tag{8}$$

$$\frac{\partial(\delta^2 s)}{\partial_t} \geq 0 \tag{9}$$

At equilibrium, or in the linear nonequilibrium domain, equation (9) is a direct result of the second law of thermodynamics, and inequality (8) is derived as a sufficient criterion for stability. In the nonlinear range of nonequilibrium thermodynamics, as long as the local equilibrium assumption remains valid, inequality (8) is assumed and becomes the starting point from which equation (9) is derived as a sufficient stability criterion. Far from equilibrium, if

$$\frac{\partial(\delta^2 s)}{\partial_t} < 0 \quad \text{for all } t \geq t_0 \tag{9.1}$$

instability can occur; if

$$\frac{\partial(\delta^2 s)}{\partial_t} = 0 \quad \text{for all } t \geq t_0 \tag{9.2}$$

the system is marginally stable; and if

$$\frac{\partial(\delta^2 s)}{\partial_t} > 0 \quad \text{for all } t \geq t_0 \tag{9.3}$$

the system is asymptotically stable. The criteria above [(9.1) and (9.2)] provide only a general statement of classification of conditions compatible with instability, and derive from the qualitative theory of partial differential equations. Hence, only if the excess entropy production decreases with time can the system become *unstable* (9.1). When the excess entropy production vanishes (9.2), the system becomes *marginally stable* and is thought to be at the critical or transition point beyond which spatial differentiation may be possible. Thus, the condition of marginal stability represents the bifurcation point beyond which symmetry-breaking instabilities can occur. If excess entropy production increases with time, the system is said to be *asymptotically stable* and cannot evolve, owing to the fact that it responds to a fluctuation by damping it. Thus, the system becomes stabilized after a fluctuation by regressing to its initial state as time proceeds.

A more rigorous discussion of stability theory may be found in numerous sources (Minorsky, 1962; Andronov et al., 1966; Nicolis, 1971; Prigogine and Lefever, 1975) but may be rendered descriptively as follows: if an element is described as a function by a differential equation (see Appendix 1), the reference state of the element is said to be *asymptotically stable* if, for any perturbation (any change in the value of the variables of the function), the value of the perturbed state (solution of the differential equation of function) approaches its original value (the value of the function at the reference state) as time approaches infinity. For functions or equations which describe closed trajectories or *orbits,* the solution of the equation is *orbitally stable* if all solutions lie within a closely defined area of the solution at the reference state, thus, if a system (described by the equation) is perturbed, the new orbit resulting from the perturbation must be within a small, finite distance of the unperturbed orbit in order for the system to be "orbitally stable." If, as time proceeds, the orbit of the perturbed system approaches that of the unperturbed system, the system is said to be *asymptotically orbitally stable.* If, further, at every instant in time, a point or region on the orbit of the perturbed system lies within a small, finite distance of the position it would have occupied had the perturbation not occurred, it is said to be *stable in the sense of Lyapounov.* Thus, Lyapounov stability decrees that not only are the orbits of the perturbed and unperturbed systems within a given distance, but also that the periodicities of the two orbits are extremely similar. Oscillatory systems which are asymptotically orbitally stable were described by Poincaré as *limit cycles.* We shall encounter them in our consideration of oscillatory phenomena in chemical and biochemical systems and have already seen an example of them in the Belusov–Zhabotinski reaction. Their importance to the emergence of biological order is that they represent self-sustained oscillatory systems whose oscillations do not depend on initial conditions but are determined by the system itself. As such, they appear as a response to amplified fluctuations and represent a new order for the system, which is stabilized by the flow of energy. Thus, they are *dissipative structures* which present a coherent space–time behavior.

D. Model Dissipative Structures

In developing the theory of nonlinear, nonequilibrium thermodynamics, Prigogine and his co-workers have evaluated a set of nonlinear chemical networks in an attempt to present simpler, nonbiological models for the emergence of order encountered in living systems. One such model employing nonlinear feedback has been termed the Brusselator (Nicolis and Prigogine, 1977). It consists of the following chemical reaction chain, the third step being autocatalytic and responsible for the generation of nonlinearity into the system, which ultimately leads to the ordering phenomena observed:

$$\begin{align} A &\rightleftharpoons X \\ B + X &\rightleftharpoons Y + D \\ 2X + Y &\rightleftharpoons 3X \\ X &\rightleftharpoons E \end{align} \tag{10}$$

At or near equilibrium, this system is represented by a unique steady-state solution, which continues to be stable in the face of external perturbations. However, as the system is driven far from equilibrium, e.g., by making the concentration of E and D negligible, it enters into a domain where new types of behavior become possible. Maintaining time-independent boundary conditions, it is still possible to have a steady state which manifests equilibrium-like behavior. These states constitute the *thermodynamic branch*. By changing the values set for A and depending on the diffusion coefficients D_A, D_B, D_X, and D_Y, the excess entropy production may become negative and the system may leave the thermodynamic branch abruptly, resulting in an instability of this branch. The accompanying nonequilibrium phase diagram (Fig. 1)

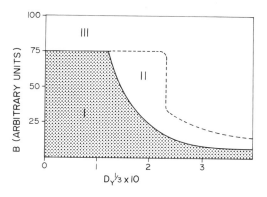

Figure 1. Nonequilibrium phase diagram showing stability of steady-state solutions after computer modeling of the Brusselator. For $\bar{Y} = B/\bar{A} = 1.86$, $\bar{X} = \bar{A} = 14.0$, $D_A = 197 \times 10^{-3}$, and $D_X = 1.05 \times 10^{-3}$, the steady state is stable with respect to fluctuations for the parameters of B and D_Y shown in domain I. This represents the solutions on the thermodynamic branch, where fluctuations are damped. Domain II represents an area in which the steady-state solution is unstable and where fluctuations increase monotonically. In domain III, the steady-state solution is also unstable; here, however, fluctuations are amplified and undergo oscillations. Taken with kind permission from Nicolis and Prigogine (1977), *Self-Organization in Nonequilibrium Systems*, p. 133, John Wiley and Sons.

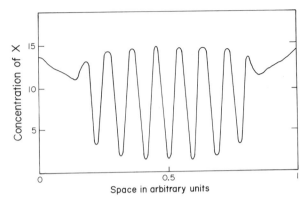

Figure 2. A dissipative structure in the form of a standing wave localized in space, arising from an instability in domain II of Fig. 1. Numerical parameters are the same as in Fig. 1. Taken with kind permission from Nicolis and Prigogine (1977) *Self-Organization in Nonequilibrium Systems*, p. 137, John Wiley and Sons.

demonstrates stability domains, achievable by manipulations of the internal parameters of the system, which were performed by the Brussels group (Prigogine and Nicolis, 1971).

Domain I represents the domain wherein the thermodynamic solution is stable. In domain II, with the parameters noted in Fig. 1, the thermodynamic branch has become unstable owing to fluctuations in the chemical composition of the system. Beyond the thermodynamic threshold (transition point), fluctuations increase uniformly in this domain (II), eventually resulting in a new steady state which corresponds to regular spatial distributions of X and Y (Fig. 2). This state represents a low entropy, dissipative structure localized in space and whose "natural" boundaries are determined by the system itself. The spatial localization of the resultant dissipative structure demonstrates the symmetry-breaking nature of the instability. It appears that the form which the dissipative structure takes depends on the type of initial perturbation; thus, the system possesses a primitive "memory effect" since the initial perturbation determines the form of the dissipative structure established.

When conditions are set such that $D_B \gg D_A \gg D_Y \cong D_X$, the thermodynamic branch may develop an instability which now drives the system to a new state periodic in both time and space (domain III). As shown in Fig. 3 in the course of one period, the system evolves with the appearance of nonstationary, concentrational wavefronts which propagate in the reaction volume. Temporal, quasi-discontinuous oscil-

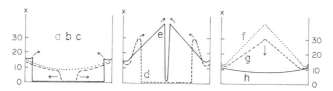

Figure 3. A dissipative structure, arising in domain III, taking the form of a propagating wave front. Here B is large and D_Y close to D_X, specifically, $B = 77.0$ and $D_Y = 0.66 \times 10^{-3}$, while $D_A = 195 \times 10^{-3}$, $D_X = 1.05 \times 10^{-3}$, $\bar{X} = \bar{A} = 14.0$, and $\bar{Y} = 5.5$. The propagation of a wave in time and space is depicted by the sequential patterns "a" through "h." After Nicolis (1975), *Adv. Chem. Phys.* **29**:38, with kind permission granted.

lations are characteristic of this system at each point in space and can produce significant recurring concentrations of chemical intermediates in restricted volumes. These recurring wavefronts propagate at a velocity much greater than that of wavefronts produced by simple diffusion and thus represent a potent means of transmitting information in a finite volume over both distance and time, i.e., a signal.

In terms of these models, if D_A is kept finite, the structure described by Figs. 2 and 3 are localized within the reaction volume, while for $D_A, D_B \to \infty$, the waves tend to occupy the entire reaction volume. Finally, if D_X, D_Y are maintained very large in comparison to the reaction rates, the spatial constraints disappear. The result is a system which, beyond the instability, oscillates with the same phase everywhere, the amplitude and periods being determined by the system itself. The periodic motion evolved is stable in that all fluctuations around this state are damped. Such periodic motion is a *limit cycle* (Fig. 4). As we have seen, limit cycles represent self-sustained oscillations in nonlinear, nonconservative systems, which depend only on the param-

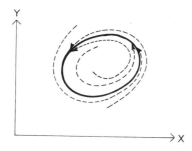

Figure 4. A stable limit cycle. The broken lines represent the perturbed path of an element approaching the stable orbit (solid line) asymptotically.

eters describing the system and are therefore independent of initial conditions.

These model dissipative structures demonstrate that the capacity of biological systems to store information, transmit information, and regulate chemical processes finds its primitive counterparts in such models so long as they are based upon nonlinearity and removal from thermodynamic equilibrium. In such systems, fluctuations play a crucial role near the point of instability. Thus, the intuitive concept that life is tenuous may be related, on physical–chemical grounds, to the importance of the generation of instabilities far from equilibrium.

In all of these examples, a new ordering mechanism, not predictable from the equilibrium principle, is encountered. Such *order through fluctuations* implies both chance and law and has been utilized by Prigogine and his colleagues (Glansdorff and Prigogine, 1971; Prigogine et al., 1972; Nicolis and Prigogine, 1977) to develop a theory for the evolution of living organisms.

E. Evolutionary Feedback

The principle of evolutionary feedback has been proposed by Prigogine and coworkers to account for evolution (Fig. 5), arguing that something in addition to the ability of macromolecules to self-replicate must play a role in the Darwinian behavior of evolution. This additional component is a thermodynamic parameter which provides that far from equilibrium, dissipation will increase, leading to a succession of instabilities, each with increasing dissipation. Each time that such an instability is followed by a higher level of energy dissipation, the driving force for the appearance of further instabilities has been

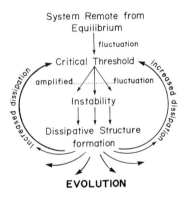

Figure 5. A scheme depicting the principle of evolutionary feedback.

increased. For example, studies with model systems show that upon switching from a system in which polymerization occurs by condensation of monomers to a system where monomers are assembled on a template, there is a distinct and decisive jump in the specific entropy production, an index of dissipation. Such an observation relates to Schroedinger's concept that life is built on discontinuous or quantized phenomena and that quantized increases in energy dissipation may thus be viewed as the driving force for evolution. Contrary to intuition, and as noted by deDuve (1974), as a living system evolves with an increase in organization and complexity, evolution does not proceed linearly, but is accelerated. Nicolis and Prigogine (1977) have reasoned that, each time instability leads to increased functional order, there exists a further possibility for the creation of other instabilities which will lead to evolution. The amplification that we encounter at the molecular level as a consequence of nonlinear couplings can now be seen to be operative at the macroscopic level of evolution, again because of nonlinear behavior of the system.

Utilizing increasing dissipation or specific entropy production as the parameter for following evolution, this theory provides not only for evolution but also for the acceleration of evolution noted by de Duve (1974). In this thermodynamic scheme, the creation of new order leads to an increase in entropy production, while maintenance steps appear to accord with the theorem of minimum entropy production.

III. Noncovalent Forces

Molecular interaction constitutes the central focus of chemistry. While covalent bonds represent the strongest molecular interactions that sustain components of living organisms, in this chapter we wish to confine our consideration to weaker or noncovalent interactions between molecules. Once the structure of living organisms is stabilized by covalent forces, it is these weaker forces that provide for the exceptional flexibility of living organisms to respond to environmental conditions. As noted earlier, it is also this reliance on noncovalent forces which introduces instability into protein structure and, in consequence, into protein function. Thus, proteins may be denatured by thermal agitation, heat being evolved in endogenous reactions and disrupting noncovalent bonds.

Both covalent and noncovalent forces are embraced under the umbrella of transferability (Parr, 1975), the fundamental concern of quantum chemistry. This problem deals with explaining how small changes can occur between atoms and molecules while the atoms, bonds, and functional groups maintain their identity. Because of the essential intactness of the parent atom or molecule, Parr emphasizes that a local property must be transferable; thus, for each individual interaction one must seek to isolate a local property contained in the Schroedinger equation of the entire system.

Kollman (1977a) has defined noncovalent interactions as those in which the electrons remain paired in both the reactants and the products and during which there is no change in the type of chemical bonding in either reactants or products. He notes that the intrinsic structures of noncovalent bonds are much harder to characterize than those of covalent bonds because of the exceptional sensitivity of the former to environmental conditions. It is this sensitivity which provides for the plasticity of response of living systems to their environment. Thus, noncovalent forces account for the stereospecific binding displayed by immunoglobulins, membrane receptors, and enzymes, as well as accounting in large measure for the stability of the native conformation of biopolymers. The major types of noncovalent forces which play a prominent role in biological systems include electrostatic interactions or salt bridges, van der Waals interactions, hydrogen bonding, and hydrophobic binding.

A. Electrostatic Interactions

Electrostatic interactions occur between fixed positive and negative charges, e.g., a positively charged amino group from the N-terminal residue of lysine or arginine which may attract or may be attracted by negatively charged groups, such as the carboxylate groups in the side chain of glutamic and aspartic acid residues of a protein.

B. Van der Waals Forces

These are weak forces, often termed nonbonded interactions, being neither the result of simple electrostatic attraction nor a gravitational attraction of matter. The special properties of these forces require an appeal to quantum mechanics. Ordinarily these forces arise from the interplay of electric charges which are polarizable, polarizability being a measure of the displacement of the positive and negative charges in an atom or molecule when placed in an external electric field. In this field, electrons are shifted relative to the nucleus, so that the electrical center of the negative charge no longer coincides with the positive center represented by the nucleus.

When two electrically neutral atoms approach one another, the instantaneous positions of the electrons determine the forces of interaction. While in the unperturbed atoms, the average position of the electrons coincides with the average position of the positive charge, with a resultant net charge of zero; when the two atoms approach one another, the electrons of one atom nearest those of the other will be subjected to a repulsive force. This causes a slight perturbation, resulting in the distorted motion of the electrons in their normal trajectory, which finally displaces the average position of negative charge relative to positive charge.

After establishing such a polarization, the atoms interact with one another as if they were dipoles. Thus, when the motions of the two sets of electrons are in phase, there is an attraction between the atoms; however, if the motions are independent, repulsion occurs. The van der Waals attraction between two atoms increases until their proximity is such that the electrostatic repulsion owing to the overlapping of the two electron clouds counterbalances the attraction between the instantaneous dipoles. This distance of nearest approach is called the *van der Waals radius* and is the radius at which the attraction is offset by the Coulomb repulsion. The attractive force depends on the number of electrons occupying the space between the two atoms, as well as the oscillations the electrons perform in space. Variability thus characterizes these forces, since they depend on the particular constitution of the individual atoms involved.

C. Hydrogen Bonding

When hydrogen is bonded to electronegative elements such as nitrogen or oxygen, the binding electrons are attracted toward the electronegative atom, with the hydrogen proton remaining at the outer end of the covalent bond, its charge unbalanced. The proton is thus capable of attracting an external, negatively charged group of another molecule. Such bonds possess a bond energy of approximately 3 to 5 kcal/mole. There are many opportunities for hydrogen bonding in proteins and nucleic acids, and groups with this ability usually complete their potential to do so. Morokuma et al. (1974), Morokuma (1977), and Kollman (1977a,b) have reviewed their studies on interactions between molecules involving noncovalent forces. These computer evaluations consider the total interactional energy to be comprised of the following components: electrostatic energy, electron exchange repulsion, polarization, and charge-transfer or delocalization energy. *Electrostatic interaction* involves interaction between undistorted electron distributions of two atoms or molecules, and such interaction may be either attractive or repulsive. *Polarization interaction* involves effects of distortion or polarization of the electron distribution of atom or molecule A or B and includes the interactions between all permanent charges or multipoles; these forces are always attractive. *Exchange replusion* involves interactions caused by exchange of electrons between A and B or, in other words, the short-range repulsion owing to overlap of the electron distribution of A with that of B. Such repulsion between closed electron shells is due mainly to the Pauli exclusion principle, since there is room for only one pair of electrons in a bonding orbital. Thus, a second pair must be assigned to an antibonding orbital which, in general, exerts a stronger repulsive effect than the attractive force of a bonding electron. Finally, *charge transfer* or *electron delocalization interaction* involves interactions caused by charge transfer from occupied molecular orbitals of A to vacant molecular orbitals of B, and vice versa. This transfer of electrons is accompanied by the appearance of a characteristic, intense absorption band in the visible or ultraviolet region of the spectrum.

The results of the studies reported in the above papers suggest that a hydrogen bond is a special case of an intermediate to weak electron donor–acceptor complex, in which linear bonding and appropriate directionality are encountered. These studies indicate that hydrogen bonds are strongly electrostatic (attractive) in nature, with a smaller but essential charge-transfer contribution as well as an exchange repulsion contribution. The computer analysis suggests that the three components are roughly of the same magnitude. Thus, a cancellation of two of the three components is achieved, with the electrostatic component providing, as a first approximation, an adequate explanation for the properties and directionality of normal hydrogen bonds.

In hydrogen bonds, the electron donor–acceptor complexes occur predominantly in the ground state. In contrast, in charge-transfer complexes, the system is in the excited state and is stabilized relative to the ground state by electrostatic interactions between the separate charges on the donor and acceptor. Again, as noted above, the charge-transfer complex is associated with the formation of a new absorption band. Indeed, on an even more general level, the results of these computer modeling studies suggest that hydrogen bonding, van der Waals complexes, charge-transfer interactions, and ionic interactions are all electrostatic interactions to a first approximation.

Hydrogen bonding is important in stabilizing the conformation of proteins and nucleic acids, accounting in the latter instance for the specificity of the interaction of base-pair formation in the double helix. Hydrogen bonding interactions within proteins include interpeptide hydrogen bonds, inter-side-chain hydrogen bonds, side-chain peptide hydrogen bonds, and proton-transfer complexes. The interpeptide hydrogen bond between the carbonyl group of one residue and the —NH group of another residue in a polypeptide chain appears to be one of the main interactions responsible for the maintenance of the alpha helix and pleated-sheet structures of proteins.

D. Hydrophobic Interactions

As mentioned above, an additional type of interaction, which is considered to be crucial for the integrity of biopolymers, is that which has been termed the hydrophobic or apolar interaction. The driving force for hydrophobic interactions appears to be largely entropic. By forming an oil-droplet-like structure, hydrophobic side chains avoid the increased order and unfavorable, decreased entropy that would occur were they to be incorporated into a water lattice. The genesis of these forces has been a subject of great interest, and several accounts are available (Tanford, 1973; Franks, 1975; Lewin, 1974).

IV. Water

At both the microscopic and macroscopic level, water seems to be uniquely equipped to serve a biological function. To begin with, life processes occur in aqueous systems in which hydrolytic processes, with water as the ubiquitous coreactant, are quite numerous. In addition, condensation reactions, often occurring during the course of the formation of biopolymers, involve the loss of water from the reactants. Furthermore, as we have noted, the hydrophobic interaction is clearly related to entropic effects associated with the ordering of water. At a more macroscopic level, the high heat capacity, substantial heat of vaporization, and wide range between freezing and boiling points, as well as the solubilizing ability, high dielectric constant, ionizing ability, and high surface tension of water (all of which are related particularly to the strong hydrogen bonding properties of water) are essential contributors to the ability of life to exist on this planet. Remarkably, while numerous theories abound in an attempt to define the structure of water, none has yet gained sufficient general acceptance to satisfy all concerned. As noted by Cooke and Kuntz (1974), while an adequate description of water may not yet be available, water–water interactions are of sufficient strength and directionality that there is available a short-time water structure which can be utilized to a first approximation in discussing the role of water in biological systems. Utilizing measures such as absorption isotherms, calorimetry, dielectric relaxation, and particularly NMR techniques, three classes of water within the cell may be defined. The bulk of the water within the cell, termed *bulk water*, is not appreciably altered by binding to macromolecules; the second type, *bound water*, shows some hindering of rotational motion by interactions with biopolymers; and finally, the third type of water, *irrotationally bound water*, appears to be bound to macromolecules for periods on the order of microseconds. However, it should be noted that even this last form of water within the cell undergoes movement regardless of binding to macromolecules. Irrotationally bound water is by far the most infrequently encountered.

As discussed by Eagland (1975) and Edelhoch and Osborne (1976), the integrity of the conformation of proteins and nucleic acids is, to a high degree, dependent upon the properties of water. Indeed, as emphasized by Lumry and Biltonen (1969), only when protein attains its three-dimensional conformation in the aqueous environment is it in its native state; in that sense, without water, a polypeptide (primary structure) cannot become a protein (tertiary structure). As noted before, the unique properties of water depend upon its propensity to hydrogen bond. Thus, the conformation which a macromolecule adopts in an aqueous environment is the ultimate energetic result of competing effects of the solvent for component peptides against other interactions in which peptides interact among themselves more strongly than with the solvent. Predictably, the ionic and polar groups of peptides engage in strong interactions with water, and thus the tendency to fold in an aqueous environment comes from the apolar groups. The polypeptide chain of the protein provides a backbone which prevents the hydrophobic tendency of apolar residues from turning the protein inward, forming a micelle-like structure; rather, because of the polypeptide chain, the three-dimensional array so characteristic of X-ray diffraction studies of proteins is encountered. However, it should be emphasized that the interior of the protein is not entirely apolar. In fact, approximately half of the polar groups of the protein which are being shielded from water are able to form hydrogen bonds, with a relatively high energy contribution, since they need not compete for hydrogen bond formation with the water. Thus, between the ordering effect of the polypeptide chain and the ability of internal polar groups to engage in hydrogen bonding, proteins exhibit characteristics of a densely packed crystal (Schultz, 1977; Page, 1977). The subject of the achievement of the native conformation of a protein has recently been reviewed (Richards, 1977; Nemethy and Scheraga, 1977).

V. Conclusions and Implications

We have seen that only in situations where dynamic systems are driven far from equilibrium can there be a development of higher-order structures which display coherent behavior. The complexity and flexibility of living systems depend at a molecular level upon noncovalent interactions, which are the essential ingredients of enzymic mechanisms. Energy must be utilized constantly to keep a system remote from equilibrium, and the utilization of energy in life requires enzymes, which are remarkable catalysts owing to their specificity and prodigious efficiency. Such efficiency at a molecular level relates to the ability of enzymes to act without the need for an information-seeking step during substrate binding. In Chapter 3, we shall explore the mode of action of enzymes and coenzymes.

As we have discussed in the section on thermodynamics, coupling a nonspontaneous reaction to one with a more negative free energy change is crucial in the functioning of intermediary metabolism. Spontaneous reactions are also accompanied by an increase in entropy.

We have emphasized the importance of irreversibility in the ability of structures to evolve to higher order. As regards the role of water in biochemical reactions, Lewin (1974) has proposed that the release of ordered water molecules during a chemical reaction serves as a potent driving force for successfully completing an otherwise energetically unfavorable reaction. Indeed, the conformation of biopolymers is maintained by the favorable entropy change occurring during hydrophobic interactions, when water molecules around apolar groups are displaced from their ordered pattern to a more probable, less ordered pattern.

The logical extension of this argument is, then, to view the displacement of water with its concomitant increase in entropy as one of the fundamental forces in the creation of biopolymers. Thus, during the early stages of polymer formation (peptides, nucleosides, and phosphohexoses), water is expelled during condensation, while in the final stages of achieving polymer formation, the extrusion of water as hydrophobic interactions develop provides the stabilizing force for these polymers. In this view, then, one of the major factors in keeping living systems far from equilibrium, and thereby in providing the conditions for increasing order, is the removal of water from precursor molecules, leading to polymer formation. These polymeric asymmetric structures are the functional and informational components of life, and it is the interaction between enzymes and genome which provides for the semipermanent information storage required both for reproduction and evolution. Living systems are responsive to weak fluctuations arising spontaneously as a consequence of thermal motion. Noncovalent forces confer plasticity on the system, allowing it to respond to fluctuations either by damping them (in a nonevolutionary situation) or by amplifying them. The amplification of fluctuations can, as we have seen, lead to the creation of new order. Thus, the power of this new general theory of thermodynamics to simplify and embrace the phenomena of biological order, which are, in fact, emergent expressions of underlying biochemistry, can now be seen.

Appendix I

In the sixteenth century, as a consequence of problems of motion, mathematics entered into the study of variable magnitudes and functions. In this instance, the abstract concepts of "variables" and "functions" have their real counterparts in the mutual dependencies of such parameters as time, distance, and velocity. From this concern with functions grew the branch of mathematics termed "analysis."

The development of differential and integral calculus by Newton and Leibnitz in the second half of the seventeenth century was a staggering advance in the mathematics of variable magnitudes in particular, and in the history of mathematics in general. The calculus was spawned from problems associated with the velocity of nonuniform motion and from those associated with the drawing of a tangent to a curve and of determining areas and volumes. The conceptual advance of Newton and Leibnitz was the discovery of the relationship between these two types of problems which, growing out of the method of coordinates, permitted one to construct a graphic representation of the dependence of one variable on another, i.e., of a function.

Differential calculus, which concerns us here, is fundamentally a study of infinitesimal changes and limiting processes contained essentially in the idea of the derivative of a function with respect to any independent variable. Specifically, it is a means of determining the rate of change of motion with respect to the stated variable at any point on the curve defined by the equation of that motion (i.e., the function). Differentiation, the means of solving such a problem, is identical to drawing a tangent to the curve. This tangent, then, represents the instataneous rate of change in the dependent variable (y) per unit change in the independent variable (x) at each value of x. Thus, if $y = f(x)$, the differential of this function is

$$dy = \frac{df}{dx} dx$$

the derivative of the function multiplied by the differential of the independent variable. A function z may contain two or more independent variables (i.e., x and y) in which case $z = f(x,y)$. Total differentiation of the function yields the total differential

$$dz = \frac{\partial f}{\partial x} dx + \frac{\partial f}{\partial y} dy$$

of the sum of each *partial derivative* with respect to one independent variable multiplied by the differential of that independent variable. With this in mind, let us remember that partial derivatives in the form $\partial/\partial t$ are included within an expression given as a total derivative, such as d_eS or d_iS, but may be focused upon, since that partial derivative may exemplify the crucial conditions being studied. The qualitative theory of partial differential equations, therefore, permits not only the determination of separate values for variable magnitudes but also the development of laws of dependence of certain variables on others. In differential equations, the unknown function may be found along with its derivatives of various orders. Since each differential equation has an infinite set of solutions, it defines a whole class of functions that satisfy it. The basic task of the qualitative theory of partial differential equations is to *investigate the functions* that satisfy the equation, not to solve the equation (Poincaré, 1952).

The fundamental problem of describing a phenomenon which changes with respect to a number of variables is to find the least number of quantities with sufficient exactness which describe the state of the phenomenon at a particular point in time (i.e., the reference state). It is often easier to state the problem than to find the exact equations, the latter often requiring a long empirical investigation. As noted by Andronov et al. (1966), theoretical investigation requires idealization of the true properties of the system, both to make the mathematics tractable and also to discern which parameters are not essential to an understanding of the experimental behavior of the system.

Appendix II: Glossary

Boltzmann order principle: As the entropy of a system is decreased, the probability that the system will become more ordered is increased. This statistical-mechanical principle is basic to the understanding of equilibrium structures.

coherent behavior: The integrated behavior of a number of interrelated pathways, best epitomized in the organization of biological systems. The term was orginally coined by Paul Weiss (1968).

dissipative structures: Structures which are formed and maintained by the flow of energy and matter under conditions which are remote from equilibrium. They are to be distinguished from equilibrium structures.

enthalpy: ΔH, also called the *heat content*, represents the heat absorbed or evolved by the system.

entropy: S, the randomness or disorder of a system.

equilibrium structures: Structures or systems which are formed and maintained through reversible transformations and consequently do not imply the removal from equilibrium.

excess entropy: $\delta^2 s$ or $\delta^2 S$, the curvature of the mathematical function describing the entropy of a system.

excess entropy production:

$$\frac{\partial(\delta^2 s)}{\partial_t}$$

or

$$\partial_t(\delta^2 s)$$

the partial derivative with respect to time of the excess entropy term.

free energy: G, the energy available to perform work, also known as Gibbs free energy.

"global stability criteria": Two thermodynamic statements of conditions which provide sufficient criteria for the stability of a system both near and far from equilibrium. If the conditions are not met, instability may occur.

Condition 1:

$$\delta^2 s < 0 \quad \text{Excess entropy} < 0$$

Condition 2:

$$\frac{\partial(\delta^2 s)}{\partial_t} \geq 0 \quad \text{Excess entropy production} \geq 0$$

Le Chatelier–Braun principle: When a system is in equilibrium, any stress applied to the system results in a change in direction such that the stress will be relieved.

limit cycle (stable): Self-sustained oscillations which are asymptotically orbitally stable and do not depend on initial conditions.

local equilibrium assumption: Within each small element of a system there exists a state of local equilibrium in which the local entropy, s, is the same function of the local macroscopic variables as at equilibrium (after Glansdorff and Prigogine, 1971).

minimum entropy production theorem: A system which is at the steady state is characterized as having achieved a state of least dissipation of free energy. Stated in another manner, if a system is at the steady state, internal irreversible processes will always operate to lower entropy production per unit time.

open system: One which can exchange energy and matter with its surroundings and is hence not isolated.

order through fluctuations: Under conditions remote from equilibrium, fluctuations in a system may be amplified and result in an instability which could, through the dissipation of energy and matter, lead to the formation of new, stable structures. This principle replaces the Boltzmann order principle in the domain of states remote from equilibrium.

steady state: A dynamic state in which the concentrations of components of the system remain fixed in the face of constant flux through the system.

thermodynamic branch: The branch of states composed of equilibrium structures which exist at equilibrium or in the linear nonequilibrium range (near equilibrium). All the states on this branch are governed by the Boltzmann order principle and the minimum entropy production theorem.

thermodynamic threshold: The critical point in the system at which the thermodynamic branch may develop an instability, owing to a fluctuation which is not damped.

References

Andronov, A. A., Vitt, A. A., and Khaikin, S. E., 1966, *Theory of Oscillators*, Pergamon Press, Oxford, England.
Bertalanffy, L., von, 1952, Problems of Life: *An Evaluation of Modern Biological and Scientific Thought*, Harper & Brothers, New York.
Brillouin, L., 1962, *Science and Information Theory*, 2nd ed., Academic Press, New York.
Comorosan, S., 1976, Biological observables, *Prog. Theor. Biol.* **4**:161.
Cooke, R., and Kuntz, I. D., 1974, The properties of water in biological systems, *Annu. Rev. Biophys. Bioeng.* **4**:95.
de Duve, C., 1974, La Biologie an XXème siècle, in: *Connaissance Scientifique et Philosophie*, Royal Academy of Belgium, Brussels.
Eagland, D., 1975, Nucleic acids, peptides, and proteins, in: *Water—A Comprehensive Treatise*, Vol. 4 (F. Franks, ed.) pp. 305–518, Plenum, New York.
Edelhoch, H., and Osborne, J. C., 1976, The thermodynamic basis of the stability of proteins, nucleic acids, and membranes, *Adv. Protein Chem.* **30**:183.
Elsasser, W. M., 1966, *Atom and Organism: A New Approach to Theoretical Biology*, Princeton University Press, Princeton, N.J.
Franks, F., 1975, The hydrophobic interaction, in: *Water—A Comprehensive Treatise*, Vol. 4 (F. Franks, ed.), pp. 1–94, Plenum, New York.
Glansdorff, P., and Prigogine, I., 1971, *Thermodynamic Theory of Structure, Stability, and Fluctuations*, John Wiley & Sons, London.
Ingraham, L. L., and Pardee, A. B., 1967, Free energy and entropy in metabolism, in: *Metabolic Pathways* (D. Greenberg, ed.), pp. 1–45, Academic Press, New York.
Klotz, I. M., 1967, *Energy Changes in Biochemical Reactions*, Academic Press, New York.
Kollman, P. A., 1977a, Noncovalent interactions, *Acc. Chem. Res.* **10**:365.

Kollman, P., 1977b, A general analysis of noncovalent intermolecular interactions, *J. Am. Chem. Soc.* **99**:4875.

Lewin, S., 1974, *Displacement of Water and Its Control of Biochemical Reactions*, Academic Press, New York.

Lumry, R., and Biltonen, R., 1969, Thermodynamic and kinetic aspects of protein conformations in relation to physiological function, in: *Structure and Stability of Biological Macromolecules* (S. N. Timasheff and G. D. Fasman, eds.), pp. 65–212, Marcel Dekker, New York.

Minorsky, N., 1962, *Nonlinear Oscillations*, D. Van Nostrand, Princeton, N.J.

Morokuma, K., 1977, Why do molecules interact? The origin of electron donor-acceptor complexes, hydrogen bonding, and proton affinity, *Acc. Chem. Res.* **10**:294.

Morokuma, K., Iwata, S., and Lathan, W. A., 1974, Molecular interactions in ground and exicted states, in: *The World of Quantum Mechnics* (R. Daudel and B. Pullman, eds.), pp. 277–316, D. Reidel Publishing, Dordrecht, Holland.

Morowitz, H. J., 1968, *Energy Flow in Biology*, Academic Press, New York.

Nemethy, G., and Scheraga, H. A., 1977, Protein folding, *Q. Rev. Biophys.* **10**:239.

Nicolis, G., 1971, Stability and dissipative structures in open systems far from equilibrium, *Adv. Chem. Phys.* **19**:209.

Nicolis, G., 1975, Dissipative instabilities, structure, and evolution, *Adv. Chem. Phys.* **29**:29.

Nicolis, G., and Prigogine, I., 1977, *Self-Organization in Nonequilibrium Systems: From Dissipative Structures to Order through Fluctuation*, John Wiley & Sons, New York.

Onsager, L., 1931, Reciprocal relations in irreversible processes, I. *Phys. Rev.* **37**:405; II. *Phys. Rev.* **38**:2265.

Page, M. I., 1977, Entropy, binding energy and enzymic catalysis, *Angew. Chem. (Engl.)* **16**:449.

Parr, R. G., 1975, The description of molecular structure, *Proc. Natl. Acad. Sci. USA* **72**:763.

Peusner, L., 1974, *Concepts in Bioenergetics, Concepts of Modern Biology Series*, Prentice Hall, Englewood Cliffs, N.J.

Poincaré, A., 1952, *Science and Method* (translated by F. Maitland), Dover Publishers, New York.

Prigogine, I., 1947, *Étude Thérmodynamique des Phénomènes Irréversibles*, Desoer, Liege.

Prigogine, I., 1978, Time, structure, and fluctuations, *Science* **201**:777.

Prigogine, I., and Lefever, R., 1975, Stability and self-organization in open systems, *Adv. Chem. Phys.* **29**:1.

Prigogine, I., and Nicolis, G., 1971, Biological order, structure, and instabilities, *Q. Rev. Biophys.* **4**:107.

Prigogine, I., Nicolis, G., and Babloyantz, A., 1972, Thermodynamics of evolution, *Physics Today* **25**:23, 39.

Richards, F. M., 1977, Areas, volumes, packing, and protein structures, *Annu. Rev. Biophys. Bioeng.* **6**:151.

Rosen, R., 1972, Morphogenesis, in: *Foundations of Mathematical Biology*, Vol. 2, *Cellular Systems* (R. Rosen, ed.), pp. 1–77, Academic Press, New York.

Schroedinger, E., 1944, *What Is Life?*, Cambridge University Press, Cambridge.

Schultz, G. E., 1977, Structural rules for globular proteins, *Angew. Chem. (Engl.)* **16**:23.

Tanford, E., 1973, *The Hydrophobic Effect: Formation of Micelles and Biological Membranes*, Wiley-Interscience, New York.

Weiss, P., 1968, *Dynamics of Development: Experiments and Inferences*, Academic Press, New York.

Winfree, A. T., 1974, Rotating chemical reactions, *Sci. Am.* **230 (6)**:82.

3

Enzymes and Coenzymes
A Mechanistic View

Robert M. Cohn

I. Introduction

Water is the common denominator in all biological systems. The nature and consequences of its interactions with the solutes of living organisms represent the central phenomena in explaining the existence of life. Proteins and enzymes require an aqueous environment for expression of their biologic activity—which is essential to the conduct of the life process—and the role of water, while only dimly understood, is the matrix upon which much of the present discussion of enzymic mechanisms is grounded. The reader wishing to pursue this topic further should consult Lewin (1974) and Franks (1975).

While enzymes, like organic catalysts, speed the approach to chemical equilibrium by providing an alternative pathway having a lower energy of activation, they can be distinguished by their enormous capabilities for stereospecific recognition and for rate acceleration. These faculties provide for the brisk conduct of the myriad chemical reactions, termed intermediary metabolism, that form the fabric of life's processes.

The specificity of enzymes for their substrates, conceptualized by Fischer as that of a lock for a key, while no longer entirely adequate, reflects the presence at the active site of an enzyme molecule, of electric charges or dipoles which complement those of the reactants. This active

---

Robert M. Cohn • Department of Metabolic Research, Children's Hospital of Philadelphia, University of Pennsylvania, Philadelphia, Pennsylvania 19104.

site, situated within the interior of the globular protein, is removed from the aqueous medium, permitting charged molecules to interact without hindrance from water, which may act as an insulator because of its high dielectric constant.

In this chapter, we will explore the organization of molecules, the mechanism of chemical reactions, and the diverse structural components which comprise enzymes, and we will account in some measure for their exceedingly efficient and specific catalytic activity. We shall also discuss theories of the molecular basis of catalysis, and will consider the role of coenzymes in enzymic catalysis.

II. Chemical Bonding

Understanding in the physical and biological sciences depends both upon the ability to frame theories which can explain experimental observations and upon the ability to develop models of those theories which permit evaluation and communication of the phenomena explained by the theories. In chemistry we are concerned with the number, energy, and spatial distribution of electrons. In biochemistry, covalent bonds abound, occurring when electrons are shared by two atomic nuclei. This model of a covalent bond was proposed by G. N. Lewis in 1916 and was the forerunner of the valence bond theory. The development of quantum mechanics was, however, a significant advance in providing not only for a qualitative picture of the bonding process but also for quantitative evaluation of bonding energies.

When two atoms approach each other, their electrons will experience increased attraction, as a consequence of the nuclei of the two atoms, as well as increased interelectronic repulsions. If the attractive force is greater than the repulsive, and if orbitals of suitable energy are available, then a new, more stable entity—a molecule—is formed through chemical bonding. Thus, bonding occurs when the ground state energy for a molecule is less than the sum of the ground state energies of the isolated atoms comprising the molecule. Since, as emphasized by Kutzelnigg (1973), electrons do not behave as classical particles, it is necessary to employ quantum mechanical arguments to explain the bonding process. Three components are involved: electrostatic forces, a decrease in kinetic energy (by the Heisenberg "uncertainty principle," the course of the electron over several atoms minimizes its kinetic energy since, if positional uncertainty is large, momentum is minimized), and deformation of molecular orbitals.

In the valence bond theory, a pair of electrons is considered to belong to the pair of atoms which is bonded together; in the molecular orbital theory—the other major theory of the chemical bond that emerges from quantum mechanics—the orbitals are viewed as serving the molecule as a whole. But while qualitative differences exist between these two theories—which are often employed to develop differing representations of a molecule—quantitative differences are obliterated when a complete set of wave functions are employed for computer calculations. Indeed, both theories share an emphasis on the importance of orbital overlap in leading to the strongest bonding conditions.

A model of carbon bonding established to account for the tetravalent nature of carbon involves two steps, in which the atoms are first raised to excited states (valence states), and are then permitted to interact in these states to form the molecule. In the case of carbon, one of the $2s$ orbitals is thus "promoted" to a p orbital to make available four unpaired electrons. The result is the "hybridization" of an s orbital with three p orbitals to form four equivalent and tetrahedrally oriented sp^3 orbitals. To devise molecular orbitals, one takes a linear combination of atomic orbitals. Following the valence bond theory, the atoms are in their excited state when hybridized, and then come together to form a molecule. In the molecular orbital approach, when the proper coefficients for the wave functions of the linear combination of atomic orbitals are chosen—i.e., those coefficients which will minimize the energy of the resultant molecule—the results will be the same as when hybrid orbitals are employed. Thus, both approaches lead to a minimization of energy and to stable bond formation.

Molecular orbital theory, in particular, has been successfully employed to develop computational techniques that permit calculation of the energy of each molecular orbital, the total electronic energy of the molecule versus that of its component atoms, and the total electronic energies of the orbitals occupied. This last value gives an estimate of molecular stability. Two main approaches have been developed to evaluate the energy of molecules (including macromolecules). The first are ab initio calculations which begin with first principles, utilizing an iterative procedure in which electron–electron repulsion is taken into account; the second are semiempirical methods which make approximations to simplify evaluation of electron repulsion. Both methods are based on self-consistent field (SCF) theory, which assumes that each electron travels in a field created by the nucleus and by an average diffuse charge distribution resulting from the other electrons. An excellent introduction to these methods can be found in Borden (1975).

A. Bond Energy

Bond energy denotes the chemical energy which is inherent in molecules by virtue of electrical forces, as well as an entropic contribution. These energies may be measured in the laboratory by measuring that energy required to break one specific bond within a molecule, the so-called "bond dissociation" energy. And, while an atom in isolation has zero potential energy, the approximation of two bodies, which can then interact through electrical forces, results in their potential energy becoming either positive or negative. As noted earlier, when a molecule forms from isolated atoms, stability is achieved, since the ground state energy of the molecule is less than that of the component atoms. Hence, bond energies are negative energies, with the strongest bonds possessing the lowest energies or the greatest negative values. Table 1 lists the bond energies for the more common bonds encountered in organic compounds.

B. Noncovalent Interactions

Covalent bonds, by virtue of their great strength, offer a paucity of opportunities for variation in bond strength, direction, or length. In particular, very little change in length occurs until covalent bonds are completely ruptured, a phenomenon requiring a sizable input of energy. Polymeric substances such as horn, chitin, and wood are almost entirely covalent in nature and are often thought of as nonliving. The limited repertoire of such covalent bonds cannot explain the immense variety of biochemical reactions and conformational changes which characterize the life process. Rather, it is through the study of noncovalent bonds, which are responsive to environmental fluctuations, that we obtain much of our understanding of the life process (Chapter 2). Enzymes and proteins act to a large degree by virtue of noncovalent forces, which account for the high degrees of organization of proteins and nucleic acids, as well as for the binding of compounds and effectors

Table 1. Bond Energies[a]

Bond	Mean value (kcal/mol at 25°C)
O—H	110–111
C—H	96–99
N—H	93
S—H	92
C—O	85–91
C—C	83–85
C—N	69–75
C≡C	199–200
C=C	146–151
C=O	173–181

[a] Taken from March (1977).

to proteins—be they enzymes, immunoglobulins, or membrane receptors.

Chotia and Janin (1975) have evaluated the role of noncovalent bonding forces in accounting for specificity and stabilization of structure and have concluded that hydrophobic forces, which are nondirectional, are crucial to stabilizing the side-chain interactions in proteins but are unlikely to contribute much to recognition and binding. Rather it is the van der Waals, electrostatic, and hydrogen bonds which, requiring complementary surfaces and proximity, are the decisive factors in binding by proteins.

III. Chemical Reactions

The essence of a chemical reaction is charge attraction and electron movement; with the breaking of sigma bonds in the electron movement, the molecules separate and form new molecules. Not uncommonly, a chemical reaction begins as a result of the complementary attraction of a site of negative charge in one molecule for a site of positive charge in another. Bond scission and formation may occur as separate, overlapping, or simultaneous events. The detailed course of the reaction is the *mechanism* of the reaction.

Three general types of mechanism are recognized, viz., substitutions, additions, and eliminations. In the first, an atom or group attached to the carbon atom involved in the reaction is displaced, being replaced by another atom or group, without change in the degree of unsaturation. With additions, the number of groups attached to the carbon atom increases, with the molecule becoming more nearly saturated. Finally, in an elimination reaction, the number of groups bound to the carbon atom decreases, with concomitant increase in unsaturation. *Fragmentation* reactions involve cleavage of carbon–carbon bonds and are a variant of elimination reactions. *Rearrangement* reactions involve internal reorganization of the carbon skeleton of the molecule and occur by a sequence of steps involving the above major classes of mechanisms.

A. Reaction Intermediates

Transient intermediates occur during the course of many organic reactions, and such intermediates include carbonium ions, carbanions, carbon radicals, and carbenes. *Carbonium ions*—or carbocations—occur when a group, with its pair of bonding electrons, is removed from a carbon atom. The positively charged carbon in the carbonium ion is

sp^2-hybridized and is therefore planar. As such, an unhybridized p orbital is left vacant, a situation that is energetically favorable, because the six bonding electrons fill the carbon $2s$ orbital, minimizing the energy of the system since $2s$-orbital electrons have lower energy than electrons in $2p$ orbitals.

Carbanions form by removal of one of the groups attached to a carbon atom, without removing the bonding pair of electrons and are thus negatively charged. *Carbon radicals* are formed from carbon compounds by removal of an attached group along with only one of the two bonding electrons and are thus intermediate in charge between carbanions and carbonium ions. Finally, *carbenes* are fragments of molecules in which two groups bound to carbon are removed with only one pair of bonding electrons. They are neutral, divalent carbon compounds, with only transient existence.

When a covalent bond is broken symmetrically in what is termed *homolytic cleavage,* one electron remains on each of the originally bonded atoms with the formation of a radical. In *heterolytic cleavage,* on the other hand, the bond is broken with the electron pair remaining with only one of the two originally bonded atoms, resulting in ion formation. Electrostatic forces represent the central forces initiating ionic reactions and determine the sequence of events and outcome. *Nucleophiles* are negative, electron-rich, or electron-donating molecules which seek a positively charged or electron-deficient site in another molecule, while *electrophiles* are positive, electron-deficient, or electron-accepting molecules and therefore seek electrons. A positive reactive site can be either an atom without a full outer electron shell, or an atom at the positive end of a dipole which may undergo additional polarization during the course of reaction. Usually, reactions are described as nucleophilic or electrophilic on the basis of the reagent involved. Carbanions are nucleophiles, while carbonium ions and carbenes are electrophiles owing to their unfilled valence shells. Not uncommonly, an initial proton transfer or removal readies a molecule to react further or to bring about the final collapse of intermediates to products. For example, when a nucleophile approaches an electrophile, the potential energy of the system rises as the molecules give up kinetic energy, coalescing to the strained, partially bonded state termed the transition state. This activated complex goes on to form new bonds between the original centers, leading to the production of either new molecules or another intermediate, which then forms a second or third transition state on the path to the final product. It is the energy barrier of the highest transition state that determines the reaction kinetics (Eyring et al., 1954).

The reactivity of organic molecules lies almost exclusively in their functional groups, and the primary distinction made among functional groups is between their reactivity as nucleophiles and their reactivity as electrophiles. In fact, much of the action of enzymes can be understood in terms of this principle. As we shall see, several amino acid side chains in proteins provide functional groups which account for the catalytic activity of certain enzymes. However, there are a number of functional groupings not provided by the amino acid side chains which are furnished by coenzymes and cofactors which augment or complete the catalytic potential. A list of common biochemical reaction types is found in Table 2.

IV. The Protein Nature of Enzymes

All enzymes are proteins and thus are composed of alpha-amino acid residues, arranged in a unique sequence for each protein, each sequence being determined genetically by a complex mechanism which will be considered elsewhere in this book. These amino acid residues are bonded by covalent peptide bonds, formed by a condensation pro-

Table 2. Common Biochemical Reaction Types

I. Nucleophilic substitution-saturated systems (occurs readily only if the carbon is attached to an atom with a positive charge)
   a. Transmethylases—sulfonium ions
   b. Glycosidases—acetals
II. Nucleophilic—reactions of carboxylic acid derivatives (attack of a base on an electrophilic carbon atom)
   a. Acyl-CoA reactions
   b. Acyl phosphates
   c. Esterases and peptidases—amides
III. Nucleophilic displacements on phosphorus and sulfur
   a. ATP and various nucleophiles—water, carboxylate ions
   b. Sulfate esters
IV. Addition and elimination involving carbon–carbon (enolate anions)
   a. Dehydration of alcohols—enoyl hydratase, fumarase, aconitase
V. Addition to carbonyl groups (usually as a step in a process resulting in hemiacetal, or carbanol-amine formation)
   a. Mutarotase
   b. Pyridoxal-catalyzed reactions—Schiff-base formation
   c. Tetrahydrofolic acid reactions
VI. Enols and enolate ions
   a. Isomerases
   b. Aldol condensations—aldolase
   c. Pyridoxal reactions—racemization, transamination
   d. Claisen condensations—Acetyl-CoA

cess in which, for each bond formed, one molecule of water is removed through loss of a hydrogen by the amino group and a hydroxyl by the carboxyl group that form the bond. Owing to the partial double-bond character of the carbon–nitrogen bond, the six atoms involved in a peptide linkage are planar, with adjacent peptide linkages oriented in different planes to minimize nonbonded linkages. It is this linear sequence of amino acids which is referred to as the *primary* structure of the protein.

Because of the condensation of amino acids into polypeptides by peptide bonding, the only distinguishing characteristics among the amino acids in a polypeptide are the side chains of each constituent amino acid. The multiplicity of interactions in which a particular amino acid is able to participate depends on the chemical properties of the side chains of the amino acids. Higher levels of organization of the protein include the *secondary* structure, which refers to the coiling or folding of the peptide chain as a consequence of hydrogen bond stabilization. Secondary structure is a direct and increasingly predictable function of the primary structure (Chou and Fasman, 1978). The *tertiary* structure results from further folding that is stabilized by the kinds of interactions possible between the various amino acid side chains remaining after peptide bond formation. There is an additional level of organization, *quaternary* structure, which refers to the association of separate peptide chains into larger aggregates, composed of dimers, trimers, and n-mers.

Maximization of intrapeptide hydrogen bonding is accomplished when the chain undergoes twists and foldings. As a result, two conformations frequently found in proteins are the α-helix and the β-pleated sheet. The cumulative nature of several hydrogen bonds contributes to maintaining the three-dimensional structure of a globular protein, but hydrogen bonds are probably more important in recognition processes. For further accounts of protein structure, as well as elegant illustrations, the reader is directed to Dickerson and Geis (1969).

A. The Amino Acid Side Chains

The general categories of side chains include those with acidic, basic, sulfur-containing, and neutral aliphatic residues, respectively. We shall discuss each in turn.

Acidic hydrophilic residues (aspartic acid and glutamic acid) bear single negative charges at physiologic pH, making involvement in ionic interactions and hydrogen bond formation likely. The un-ionized carboxyl group is a good hydrogen donor and the carboxylate a good acceptor. *Basic* hydrophilic residues are cationic in their protonated form and include lysine, arginine, and histidine, which are involved in charge interactions. The imidazole group of histidine is unusual in that it can accept or liberate a proton at physiologic pH. In its unprotonated form it contains two nitrogens, one electrophilic and one nucleophilic. Protonation causes a loss of the nucleophilic activity. In addition, imidazoles are also powerful ligands for the formation of coordination complexes of transition metal ions. For these reasons, as well as others we will discuss later, histidine is considered to function as a part of the active site of many enzymes.

The polarizability of *sulfur* in cysteine and methionine makes these amino acids effective both as entering groups and leaving groups in nucleophilic substitution. When the cysteine residue is positioned so that the sulfhydryl proton can hydrogen-bond with an acceptor, the sulfhydryl can function as a nucleophile. The *neutral* aliphatic residues include glycine, alanine, serine, threonine, valine, leucine, and isoleucine, as well as amino acids possessing aromatic hydrocarbon side chains (phenylalanine, tyrosine, and tryptophan). All of these neutral residues are capable of van der Waals interactions and are strongly hydrophobic.

B. The Active Site

X-ray diffraction studies demonstrate (Dickerson, 1972) that several functional groups, from different regions of the primary sequence, are brought together to form the active site of an enzyme. These neighboring groups comprise but a small portion of the amino acid side chains of the protein. The active site is made up of peptide bonds in direct contact with the substrate, as well as those which contribute to the catalytic action although not in direct contact with the substrate. This view of the active site envisions the amino acid side chains and cofactors involved in the chemical transformation as responsible for carrying out the chemical reaction, as well as for providing specificity for the reaction. The subject of the stereospecificity of enzymic reactions has been the subject of several reviews (Battersby and Staunton, 1974; Cornforth, 1974; Hanson, 1976). Model systems may provide an understanding of the structural aspects of recognition through complementary structures between enzyme and substrate. Studies in Cram's laboratory (Cram and Cram, 1974; Kyba et al., 1977; Timko et al., 1977) on "host–guest" chemistry are directed to developing multi-heteromacrocycles that can probe the nature of the stereospecific binding of small molecules by enzymes.

As noted at the beginning of this section, X-ray diffraction studies (Dickerson, 1972) have led to the accumulation of a large body of data regarding the mode of action of enzymes. In addition to the information furnished by such "static" studies, the effect of modification of enzyme structure by various chemical means has also contributed to greater understanding of enzymic action. Isolation and identification of covalent intermediates (Bell and Koshland, 1971), and chemical modification of the active site and residues which affect the conformation of the active site (Vallee and Riordan, 1969; Glazer, 1970; Sigman and Mooser, 1975) have further contributed to our understanding of enzymic mechanisms. More recently, investigation of crystalline enzymes by cryoenzymologic methods (at subzero temperatures) has allowed the isolation of hitherto short-lived intermediates, further clarifying our understanding of enzymic mechanisms (Makinen and Fink, 1977). Owing to the immense scope of this topic, the reader is encouraged to consult the above references, as well as Haschemeyer and Haschemeyer (1973).

V. Enzyme Mechanisms

Having cited techniques for identifying active site components of enzymes, as well as methods for elucidating the principal features of the pathway of enzyme-catalyzed reactions, we are now in a position to consider mechanisms which have been proposed for enzymic catalysis. As with all catalysts, enzymes alter the free energies of activation of the reactions which they catalyze. The enthalpic and entropic components of enzyme action are not always easy to separate, however. In carrying out cellular reactions, enzymes are assumed to follow the laws of physical chemistry, there being no unusual properties imputed to them. We wish to consider—one at a time—those factors that contribute to the ability to lower energy barriers. More extensive coverage of the topics considered in this section can be found in the following: Bruice and Benkovic (1966); Jencks (1969); Bender (1971); Zeffran and Hall (1973); and Fersht (1977).

A. Approximation and Orientation

From a purely intuitive point of view, both the approximation at the active site of reactants involved in an enzymatic reaction and the correct orientation of the groups destined to react have appealed to a number of researchers as representing the overriding factors in explaining both the binding specificity of enzymes and their enormous capacity to accelerate the rate of a chemical reaction (Bruice, 1970, 1976a; Jencks, 1975; Koshland, 1976). Furthermore, while some have considered it prudent to separate the approximation and orientation components, such a distinction often appears largely semantic, and thus a compromise which considers both factors together seems justifiable.

Before the chemical reaction catalyzed by the enzyme can occur, the substrate or substrates must bind to the enzyme, usually in a noncovalent complex. Binding of the substrate to the enzyme freezes the translational and rotational degrees of freedom available to the substrate and, in so doing, lowers the entropy of the enzyme–substrate complex, facilitating collision and subsequent reaction. This aspect of catalysis has been variously termed approximation, propinquity, proximity, and the like. The precise orientation of catalytic groups at the active site of the enzyme, which permits the performance of the catalytic act itself, has been variously referred to as orientation, stereo-population control, rotamer distribution, and orbital steering. A third component of the encounter complex embodies the view that binding of the substrate to the enzyme molecule forces the geometry into that of the transition state, this being termed a modern rack theory (Eyring and Johnson, 1971).

Experiments comparing the rate of intermolecular vs. intramolecular reactions have shown enormous enhancements in the rate of reaction for the latter. The reader is encouraged to consult Bruice (1976a), Jencks (1975), and Koshland (1976) for areas of disagreement and specific emphasis imposed by these particular authors. What they have shown is that placing the substrate or substrates at the active site in the appropriate position for the covalent changes necessary for the reaction constitutes a considerable kinetic advantage to the fruitful outcome of the reaction. Jencks (1975) has calculated that the overall entropy loss of going from two species in solution to one in solution represents approximately 30 to 35 entropy units, a factor equivalent to a rate enhancement of approximately 10^8. Hence, either by binding the substrate to the enzyme active site—and thereby freezing its translational and rotational degrees of freedom—or by converting an intermolecular to an intramolecular reaction, one is able to account for a rate enhancement of approximately 10^8. Koshland (1976) has combined the orientation and proximity terms into the term, "juxtaposition," emphasizing the importance of both the proximity and orientation components.

As regards the enzyme-catalyzed reaction, we should emphasize that when two reactants bind to the enzyme, in what is to become the transition state, they have already effectively lost their degrees of freedom, through immobilization on the enzyme. Thus, the value of the entropy change is much smaller, and the free energy requirements for the reaction are commensurately smaller; in other words, the unfavorable loss of entropy required for the formation of the transition state has occurred at the binding step, and it is no longer necessary for such an energetically unfavorable event to occur during the activation process itself.

B. The Transition State

As noted earlier, the transition state is a molecule in flux, containing partial bonds, i.e., bonds in the midst of formation and breakage. Long before experimental data were at hand, Pauling (1948) proposed that the enzyme would bond more strongly to the transition state than it would to substrates or products, owing to a complementarity in structure between the activated complex and the active site. In this view, the attraction of the enzyme for the activated complex would be a factor in decreasing the energy of activation of the reaction, thereby accounting to some degree for the rate enhancement. Evidence reviewed by Wolfeden (1974, 1976) and Lienhard (1973) provides strong support for this seminal contention by Pauling. Certain transition state analogs are bound more tightly to their enzymes than are the natural substrates or products of these reactions and, indeed, because of their tight binding, such analogs serve as the best inhibitors for these particular enzyme reactions.

The proposal that the energy of activation for an enzyme-catalyzed reaction could be lowered appreciably if some of the energy of substrate binding could be used to distort the substrate into a conformation closer to that of the transition state has been elaborated by Eyring et al. (1954), Eyring and Johnson (1971), Lumry (1974), and Jencks (1975). As presently viewed, it is suggested that, when a substrate binds, a portion of the intrinsic binding energy is utilized to strain the substrate complex toward the geometry of the transition state. This theory is, in a sense, an outgrowth of the induced-fit theory, but differs from it in that, in the induced-fit theory, the binding energy is used to change a catalytically inactive conformation of the enzyme to an active conformation. There is no change in the free energy of activation of the reaction. However, in the strain theory, a portion of the free energy of binding is used to reduce the free energy of activation, thereby enhancing the reaction rate. This strain theory has been discussed by Jencks (1975) and Bosshard (1976).

A point to emphasize is that the separation between specificity and catalytic ability is artificial, since it is clear that the binding process itself is responsible for a considerable portion of the rate enhancement brought about by the enzyme. In this modern strain theory, it is proposed that the intrinsic binding energy resulting from the noncovalent interaction of the substrate with the active site of the enzyme is larger than is generally perceived. It is this additional energy available from the binding of a substrate to an enzyme that represents a significant point of distinction between enzymes and nonenzymes. Owing to its large size, it is possible for an enzyme to utilize not only the binding energy consequent upon the binding of the substrate to the active site but also the energy available from the binding of nonreacting portions of the substrate to contiguous portions of the active site. Jencks and Page (1972) have reasoned that once the substrate or coenzyme binds to the active site—thus bringing about the required loss of translational and rotational entropy—any further binding interaction with the enzyme, as a consequence of interaction of additional substituents on the molecule bound to the enzyme, will provide the binding energy required to compensate for the entropy loss and for any substrate destabilization required to reach the transition state. This effect has been termed the "anchor principle" and has been put forward to explain how the large size of enzymes, coenzymes, and some substrates is utilized to permit specific binding between the enzymes and substrates and cofactors, such binding of nonessential groups providing the favorable binding energy that can be utilized to accelerate catalysis. Jencks (1975) proposes that the most important contribution to this strong binding between the substrate and the active site of the enzyme arises from van der Waals–London forces which favor an interaction between the substrate and enzyme rather than between the substrate and the aqueous solvent.

Electrostriction (Westley, 1969) is a term proposed for the decrease in volume of water resulting from the binding of ions and dipoles in proteins. Release of this bound and ordered water occurs when oppositely charged ions coalesce to form a neutral molecule. Water molecules which are bound to the active site will be displaced by the approaching substrate, thereby freeing this water. The entropy changes involved can be considerable, and calculations indicate that they may be large enough to facilitate substrate binding. In addition, as shown by Cohen et al. (1970) in studies on the active site of chymotrypsin, part of the energy of binding of the substrate to the enzyme is employed in desolvation of the reacting groups, thereby accelerating the reaction.

C. Other Factors in Catalysis

An important principle in catalysis appears to be minimization of charge separation during formation of the transition state (Knowles and Gutfreund, 1974). For example, the formation of a charge relay system, as occurs with serine proteases (Blackburn, 1976), may be important in that one group of the enzyme may be able to donate a proton from one side of the reacting pair of molecules, while another group accepts a proton from the opposite side during the passage through the transition state, thus stabilizing the electric charges generated.

Abundant evidence exists for the formation of *covalent* intermediates for a number of enzymes (Bell and Koshland, 1971). These compounds differ from the encounter complexes or the enzyme–substrate complexes discussed earlier, in that one part of the substrate, usually a transferable group, becomes covalently bound to the enzyme (Table 3). This type of covalent intermediate may serve one of a number of functions; in a double displacement (ping–pong mechanism), for example, it may be necessary to hold part of the substrate in place, as in a transamination reaction, while another part leaves as a product, so that the part remaining with the enzyme can then react with a second substrate, forming a second product. An additional role subserved by covalent intermediates may be freezing of the translational and rotational freedom of only one substrate molecule at the active site, rather than both reactant molecules. Moreover, as noted by Jencks (1975), since the enzyme requires but one binding site for both the entering and leaving groups, there need not evolve a second site for an additional step. Koshland and Neet (1968) have proposed, further, that such covalent intermediates may retain in the products the stereochemistry of the reactants as well as conserve the high-energy nature of a particular substrate.

1. General Acid–Base Catalysis

Acid–base catalysis occupies a venerable position in organic chemistry. The hydrogen ion is abundant, lacks bulk, and polarizes other groups easily, accounting for its particular importance. In the cell, however, high concentrations of hydrogen ion must be avoided. Therefore, general acid catalysis, where a weak acid functions by hydrogen bond-

Table 3. Covalent Enzyme–Substrate Intermediates

Enzyme class	Reacting group	Covalent Intermediate
Proteases: chymotrypsin trypsin, elastase, thrombin	OH (serine)	Acyl-serine
Papain, ficin, glyceraldehyde-3-phosphate dehydrogenase	SH (cysteine)	Acyl-cysteine
Alkaline phosphatase, phosphoglucomutase	OH (serine)	Phosphoserine
Aldolases, decarboxylases, pyridoxal phosphate requiring enzymes	C=O and C=N- (lysine and substrate amino)	Schiff base (imine)

ing-induced polarization, furnishes a means for achievement of a role analogous to that of the hydrogen ion. Since delocalization of electrons leads to increased stability, localization through general acid–base catalysis can serve to destabilize a molecule and prepare it for further reaction. Hence, general acid–base catalysis may often function as a priming reaction in a mechanistic sequence.

General acid catalysis denotes acceleration of a reaction as a consequence of partial transfer of a proton from any Brönsted acid to a reactant in the transition state, while general base catalysis occurs when any Brönsted base increases the reaction rate by engaging in partial transfer of a proton from a transition-state complex. Experimentally, reactions whose rates depend on the concentration of all acids and bases in solution, rather than on the concentration of hydronium or hydroxyl ions only, are susceptible to general acid or general base catalysis. The side chains of glutamic and aspartic acids, histidine, tyrosine, and cysteine have considerable acidic or basic character and are likely to function as general acid–base catalysts in the cell.

Reactions vulnerable to general acid–base catalysis include carbonyl addition reactions, hydrolysis of some carboxylic and phosphoric acid esters, and the aminolysis of esters. Efficiency of a general acid or general base catalyst depends upon the thermodynamic feasibility of hydrogen bonding between the catalyst and the substrate, the polarizability of the substrate bond resulting from complex formation, and the facility of proton transfer between the enzyme and the substrate. Since acid–base catalysis depends on increasing the electrophilic nature of the carbonyl carbon by partial donation of a proton to the carbonyl oxygen, it follows that formation of a coordinate covalent bond by a metal ion with an unshared electron pair accomplishes the same goal. The role of metals as "super acids" is thus easily appreciated. Extension of the idea of an acid to include such Lewis acids allows the range of reactions susceptible to general acid catalysis to embrace decarboxylations and many reactions with phosphate groups. Thus, either with certain amino acid side chains or with metals, the cell is able to achieve the catalytic advantage of an acid or base to drive proton transfer steps, without the deleterious consequences of accumulation of hydronium or hydroxyl ions. Excellent discussions of general acid–base catalysis may be found in Bruice and Benkovic (1966), Jencks (1969, 1972, 1976), Bender (1971), and Bell (1973).

2. Covalent Catalysis

This type of catalysis implies a role by a covalently bound substrate–enzyme intermediate. Evidence for such intermediates has been acquired by a number of means, and a listing of such intermediates has been prepared by Bell and Koshland (1971). Proof that a covalent intermediate is actually involved in the enzymatic reaction requires both the demonstration of the existence of the covalent intermediate and evidence that the rate of formation and decomposition of this intermediate accounts for the overall rate of the catalyzed reaction. Where these criteria have been satisfied, it has been shown that most enzyme reactions which involve such intermediates traverse a nucleophilic route.

In order for the covalent intermediate to serve a function in catalysis, the reactive amino acid on the enzyme must serve as a better attacking group than the final electron acceptor and as a better leaving group than the leaving group on the initial substrate donor. The nucleophiles that have been identified as giving rise to acyl intermediates in esterases and proteases are the side-chain oxygen of serine and the corresponding sulfur of cysteine. Thiols, in general, are substantially more nucleophilic than alcohols or nitrogen bases, so the cysteine $-SH$ group might be expected to perform well as a nucleophilic catalyst. Proteases and esterases catalyze hydrolysis and transfer of acyl groups, with the intermediate transfer of an acyl group to the hydroxyl or sulfhydryl group of a serine or cysteine residue on the enzyme.

In electrophilic covalent catalysis, the enzyme withdraws electrons from the reaction center, thereby activating the substrate. The distinction between electrophilic and nucleophilic catalysis is not always easy or profitable, since electrophilic attack is often preceded by a step in which the catalyst acts as a nucleophile to bind to the reactive center on the substrate. Further, electrophilic catalysis in one direction is likely to be nucleophilic in the other. Often, the definition depends upon whether the step presumed to be decisive is nucleophilic or electrophilic.

Electrophilic catalysis is often carried on by the proton which derives from the solvent, by a metal, or by combination with organic molecules which contain groups that can act as electron sinks. A principal mechanism of electrophilic catalysis embodies the combination of substrate and enzyme to produce a compound in which a positive nitrogen acts as an electron sink, or free nitrogen acts as the donor of the electron pair. Although nitrogen itself is not strongly electronegative, it can act as an electron sink since it is easily protonated to form cationic unsaturated adducts.

3. Catalysis by Protein Functional Groups

A number of enzymatic reactions depend upon the nucleophilic, basic, and acidic groups on an enzyme to carry out the chemical transformation. These are best exemplified by certain hydrolysis, elimination, and rearrangement reactions. In general, enzymes are able to catalyze these reactions at pH 7, although some of the esterases, such as pepsin (Fruton, 1976), operate at much lower pH. As noted before, an important property of enzymes is that they can have both acidic and basic groups. This allows the mechanism to be a combination of acid and base catalysis.

The groups that function in hydrolytic reactions include the carboxyl, thiol, and imidazole groups, as well as the serine hydroxyl. The carboxylate anion can act as a base to catalyze ester hydrolysis, the reaction appearing to proceed through an anhydride (Lowe and Ingraham, 1974). Imidazole, which occurs in the active site of many esterases, can serve as a nucleophilic catalyst. Indeed, imidazole is a very effective nucleophile even though it is neutral, because the positive charge that builds up in the transition state as a result of displacement can be dispersed to both nitrogens (Barnard and Stein, 1958). Thiol groups may also act as nucleophilic catalysts, as in the proteases ficin and papain (Lowe, 1976).

a. Elimination Reactions. Certain elimination reactions are catalyzed by enzymes not utilizing organic cofactors. The mechanism is usually an acid–base reaction, with loss of an anion producing a carbonium ion or loss of a proton producing a carbanion as the initial step. Carbonium ion mechanisms characterize fumarase and aconitase action, while reactions involving carbanion mechanisms are exemplified by crotonase and β-methyl-aspartase. Lowe and Ingraham (1974) have remarked upon the preponderance of carbanion elimination mechanisms, including pyridoxal phosphate-catalyzed reactions, suggesting that this may have come about because of the alkaline conditions under which life is presumed to have originated.

b. Hydrolytic Reactions: Serine Proteases. Chymotrypsin, trypsin, and elastase represent a group of enzymes termed serine proteases because of the role of this amino acid in the catalytic mechanism of these enzymes. Chymotrypsin is one of the most exhaustively studied of all enzymes; it and the other members of this class constitute a group of enzymes acting without additional cofactors. They have been the subject of numerous reviews, including those by Blackburn (1976), Blow (1976), Kraut (1977), and Stroud et al. (1977).

The serine proteases catalyze a nucleophilic displacement reaction of a group attached to a carbonyl carbon—such groups comprising esters, thioesters, and amides. The reaction carried out by this class of enzymes is essentially an acyl transfer in which the substrate binds to the enzyme, forming the noncovalent Michaelis complex, followed by the formation of a covalent bond between the carbonyl carbon of the substrate and the reactive serine γ-oxygen atom. A hydrogen ion is transferred from the reactive oxygen to the leaving group of the substrate in the tetrahedral intermediate, and, on decomposition of the intermediate, the acyl enzyme and leaving group are separated. The enzyme is then deacylated by reversing the formation of the acyl enzyme, with water serving as the nucleophile.

Transition-state stabilization is an important component of the mechanism of action of the serine proteases, with the multiple binding sites contributing favorable interactional energy that stabilizes the substrate in the form of the transition-state complex. The importance of interactional energy in catalytic mechanisms is demonstrated by the serine proteases, all of which possess multiple binding sites which are used as follows: (1) to facilitate hydrogen bonding between the substrate and enzyme; (2) for substrate side chains; (3) for the leaving group in this nucleophilic displacement; and (4) for the carbonyl oxygen of the bond to be cleaved—a site termed the "oxyanion hole." In addition, the proteases possess a serine side chain which binds covalently with the peptide bond to be cleaved and includes the so-called "charge relay system." It is this system that binds the proton which, in the transition state, is stabilized by the His–Arg couple of the active site. The Ser–His–Arg grouping indigenous to these enzymes has also been found in the bacterial enzyme subtilisin, suggesting that two widely divergent evolutionary pathways have led to the development of a unique grouping, which carries out a key chemical function in the catalytic action of the enzyme.

c. Lysozyme. This enzyme, brought to prominence through its rediscovery by Fleming in mucoid nasal discharge, lyses the bacterial cell wall by cleaving a polysaccharide chain containing alternating N-acetylglucosamine and N-acetylmuramic acid residues. Although its function in vivo is still not understood, it has been exceptionally useful in providing evidence for the roles of induced fit and of strain in distorting the substrate toward the conformation of the transition state (Imoto et al., 1972).

Lysozyme, ellipsoid in shape, binds six carbohydrate rings of the polymeric substrate to the active-site cleft, causing the pyranose ring at the fourth position of the six bound rings (subsite D) to be distorted toward the half-chair conformation. This distortion raises the free energy of the complex, bringing the substrate partially toward the conformation of the transition state . Recently, Pincus et al. (1976a,b) have employed theoretical models and conformational energy calculations to further probe the factors involved in the mechanism of binding of the substrate to the binding site. The reader is encouraged to refer to these papers for full details.

d. Pancreatic Ribonuclease. Ribonuclease represents one of the most intensively studied enzymes and, as its name implies, is involved in cleavage of members of the ribonucleic acid (RNA) chain, attacking only pyrimidine residues. Globular in structure, it possesses a crevice called the waist, which runs across the molecule. The form from bovine pancreas has a molecular weight of 13,600. Cleavage of the RNA occurs at the phosphorus–oxygen bond situated on the far side of a phosphorous connected to the 3'-carbon of the ribose bearing the pyrimidine. Abundant evidence, reviewed by Richards and Wykoff (1971) and Blackburn (1976), implicates histidine residues 12 and 119, and lysine residue 41 in the mechanism of action of pancreatic ribonuclease. The lysis of the phosphorous–oxygen bond proceeds in two steps, the first being a cyclization and the second a hydrolysis. It is proposed that, in the first step, one of the histidine imidazole residues removes a proton from the 2'-hydroxyl of the ribose ring while the acidic group from the other residue donates a proton to the departing 5'-oxygen—in other words, a push-pull mechanism. Originally, Meadows et al. (1967) assigned the lower pKa histidine to His 119 and the higher pKa to His 12 on the basis of NMR studies. However, Patel et al. (1975), Markley (1976), and Shindo et al. (1976) have reinvestigated these pKa assignments and have shown that the original designations were in error. Hence, the lower pKa histidine is His 12, and the higher pKa histidine is 119. The removal of the proton from the 2'-hydroxyl group leaves the oxygen of that group with a partial negative charge, permitting it to attack the phosphorus as a nucleophile. The donation of the proton to the oxygen atom on the 5'-CH_2OH group brings the substrate into the transition state. The cyclization step is concluded when the 5'-CH_2OH group is formed, with complete removal of the proton from the acidic imidazole at the same time that the basic imidazole binds the proton from the 2'-OH group. As a result, the cyclic phosphodiester has formed.

The hydrolysis step is more controversial, but one view (Roberts et al., 1969) is that it occurs by a reversal of the sequences followed during cyclization. Recently, Cozzone and Jardetzky (1977) studied the interaction of the single-stranded synthetic polynucleotide, poly(A), in the presence of RNase. They found that the enzyme can carry out the cyclization step on such polypurine nucleotides, but not the hydrolysis step. They interpreted their data to indicate that this nonoptimal substrate forms an enzyme–substrate complex which allows alignment with only one of the two active-site histidines. This emphasizes two important points for the mechanism of pancreatic RNase and other enzymes: (1) binding must be correct to have the energetically optimal interaction with the enzyme that will lead to the formation of the transition state;

and (2) incorrect binding will not permit the crucial active-site interactions for general acid–base catalysis in this particular reaction (which is essentially a two-step double-displacement in which a neighboring group in the substrate functions as a nucleophile).

e. Carboxypeptidase. Carboxypeptidase is a zinc metalloenzyme which acts on peptides containing a free carboxyl group, cleaving residues singly from the carboxyterminus. X-ray diffraction studies, reviewed by Lipscomb (1974), reveal that carboxypeptidase is ellipsoid in structure, with approximate dimensions of $50 \times 42 \times 38$ Å. The zinc is bound to the protein by histidines 69 and 196 and by glutamic acid 72.

The binding of substrate to the enzyme causes the displacement of several water molecules as the carboxyl side chain of the substrate binds to the pocket of the enzyme. Other interactions of the substrate with the enzyme, to form the Michaelis complex, include interactions with the guanido group of arginine 145 and with the carboxyl group of glutamic acid 270. In X-ray diffraction studies, several conformational changes are noted upon substrate binding, including movement of both the guanido group of Arg 145 and the carboxyl group of Glu 270. Additionally, the binding by Arg 145 to a carboxyl group of the substrate brings about a conformational change in Tyr 248, permitting it to act as a proton donor. As noted before, the binding of substrate to an enzyme, with conversion of the previously water-filled cavity to a hydrophobic region by displacement of water, provides a driving force for the catalytic reaction.

Zinc appears to act as a Lewis acid, polarizing the carbonyl group of the substrate and thereby making it more susceptible to nucleophilic attack, by reducing the double bond characteristic of the substrate carbonyl. It is suggested (Jencks, 1975) that the displacement of water by the substrate decreases the charge-separating effect caused by water and thereby enables the zinc to polarize the acyl group of the substrate, making it susceptible to nucleophilic attack. This form of "desolvation destabilization" is paid for by the favorable binding energy. Scheiner and Lipscomb (1977) have further enquired into the role of zinc, utilizing an approximate molecular orbital method.

It had been proposed that the Glu 270 of carboxypeptidase might function as a nucleophile by attacking the carbon of the susceptible carbonyl bond or by promoting general base catalysis of the oxygen atom of a water molecule at this carbon. Recently, Makinen et al. (1976), studying carboxypeptidase in ester hydrolysis, used cryo-enzymologic techniques to show that the intermediate formed in this reaction is a covalent, mixed anhydride, formed by acylation of the gamma carboxylate of Glu 270. A detailed review of carboxypeptidase activity may be found in Blackburn (1976).

VI. Coenzymes

As noted earlier, an enzyme may, through attachment of certain cofactors, dramatically increase its capability for substrate binding. Cofactors may be artifically categorized as: (1) prosthetic groups—those which complete their catalytic reaction cycles while attached to the same enzyme; and (2) coenzymes, which require two different enzyme forms in the course of their catalytic cycles. Both provide specific chemical properties not easily attainable with amino acid residues alone.

Pullman and Pullman (1963), in their seminal book on quantum biochemistry, noted the central role occupied in life processes by molecules possessing partly or completely conjugated systems. These molecules include the coenzymes, whose precursors are the water-soluble vitamins. Stabilization of transition states by electronic delocalization, transmission of electronic perturbations over several atoms, and facilitation of electron mobility were suggested as explaining, at least in part, the reaction capabilities of coenzymes. An additional role for coenzymes (Jencks, 1975) is the provision for optimal binding interactions with specific subsites on the enzyme, the so-called "anchor principle" discussed above. In this section, we shall consider the mechanism of action of the major coenzymes. A general description of coenzyme reaction mechanisms is available (Lowe and Ingraham, 1974).

A. Adenosine Triphosphate

In the living organism, the free energy derived from hydrolysis of the pyrophosphate bonds of ATP is available in a coupled enzymatic reaction to provide the energy necessary for driving otherwise unfavorable chemical reactions. Such otherwise unfavorable processes include biosynthetic reactions, muscle contraction, and transport across membranes. In serving thus as a source for the energy-requiring processes of life, ATP (Fig. 1) and its relatives can provide the negative entropy

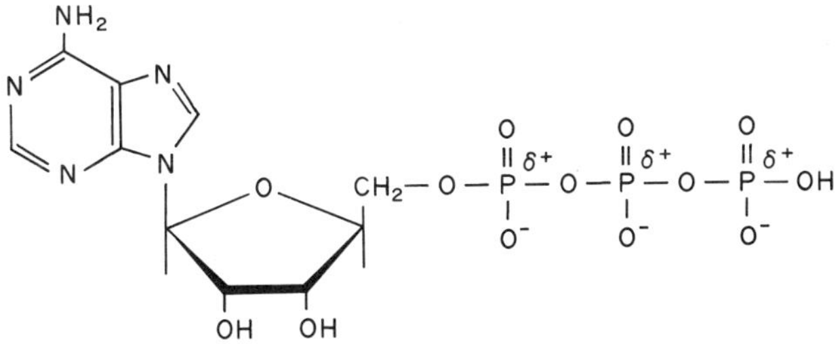

Figure 1. Adenosine triphosphate (ATP).

of which Schroedinger (1944) spoke regarding the need for a constant energy input to maintain the high degree of organization synonymous with life. Evolution has chosen ATP for these processes, evidently because it is capable of liberating a considerable amount of free energy on hydrolysis while, nonetheless, being stable kinetically in the presence of its ubiquitous coreactant, water. Indeed, as noted by Lehninger (1959), the kinetic stability of phosphoric acid derivatives is far greater than that of derivatives of sulfuric acid, hydrochloric acid, or common carboxylic acids. In addition, it is likely that the ability of the phosphorus atom to employ its d orbitals in forming the transition states of reactants constitutes an additional factor in understanding its biochemical importance. Thus, although the anhydrides of phosphates, and the mixed anhydrides of phosphates and carboxylic acids, may take advantage of the presence of water, they do not undergo hydrolysis rapidly except in the presence of a suitable enzyme.

ATP-dependent reactions may be divided into two broad categories: (1) the transfer of a portion of the nucleotide to a suitable acceptor molecule; and (2) the cleavage of ATP, providing the driving force for otherwise unfavorable reactions, e.g., ligases (synthesetases). All ATP-dependent reactions require, in addition, either Mg^{2+} or Mn^{2+} for activation, functioning principally to form chelates with the ATP. Such chelation increases the electrophilicity of the phosphate groups, thereby increasing their vulnerability to attack by nucleophilic reagents. This is important since, in essence, all reactions in which ATP is involved are nucleophilic in character. Phosphoryl transfer reactions have been reviewed by Benkovic and Schray (1973) and Douglas (1976).

The two factors which appear most important in explaining the high free energy of hydrolysis of the ATP molecule are electrostatic repulsion and competing or opposing resonance (Pullman and Pullman, 1963). At body pH, ATP exists mainly in the tetraanion form, allowing for a considerable degree of electrostatic repulsion between the negatively charged groups. The phosphorus–oxygen double bond is semipolar in character, owing to the greater electronegativity of oxygen as compared with phosphorus. Hence, the oxygen atom that is divorced from the main phosphorus–oxygen–phosphorus chain possesses a high negative charge. There are, at either side of the phosphorus–oxygen bonds, two groups that tend to attract electrons away from the central grouping, in a phenomenon termed competing resonance—because competing functions are attracting the electrons of a single oxygen.

Hydrolysis relieves the competition, and consequently the hydrolytic products are more stable than the parent compound. It is essential to note that the high-energy nature of ATP depends upon the relationship of ATP to water. A series of articles (Banks, 1969; Pauling, 1970;

Wilkie, 1970; Huxley, 1970) have dealt with the controversy in the scientific community concerning the role of ATP in serving as a source of energy for biological work. Ling (1977) has put forward a view of the role of ATP in supplying energy for biological work that involves an electronic interaction of ATP with proteins, mediated through conformational changes and not through coupled chemical reactions.

Reactions requiring ATP also require a divalent cation as cofactor, the cation bringing about partial neutralization of the excess negative charge. Thus, a nucleophile is able to attack the phosphorus, the nucleophile being either the oxygen of water or the hydroxyl ion. The high negative charge of the oxygens of ATP would strongly repel the nucleophile, thereby impeding hydrolysis. Metal-ion chelation increases the electrophilicity of the ATP molecule and enhances the ability of ATP to engage in nucleophilic reactions.

B. Nicotinamide Nucleotides

The nicotinamide (pyridine) nucleotides were the first coenzymes to be recognized; they function in a wide variety of oxidation–reduction reactions involving hydrogen transfer. Studies reviewed by Colowick et al. (1966) showed that reduction of the nicotinamide nucleotides occurs at C-4 of the pyridine nucleus as shown in Fig. 2. Hydrogen transfer at this position is stereospecific, as is abstraction of hydrogen from the substrate. Studies of alcohol dehydrogenase by Vennesland and Westheimer (1954), using deuterium isotopes, revealed that no hydrogen was supplied by the water. Instead, hydrogen transfer is direct from the substrate to the nucleotide. Dehydrogenases which utilize NAD and NADP abstract hydrogen from the pyridine ring in a stereospecific manner. The isomers of the reduced NAD are termed A (proR) and B (proS). A generalization which seems to have

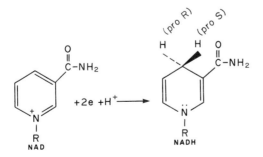

Figure 2. Nicotinamide adenine dinucleotide (NAD).

emerged from studies on the stereospecificity of enzymes acting with NAD indicates that enzymes acting on carbohydrates tend to utilize the B form of NAD, while enzymes acting on alpha-hydroxy acids utilize the A form. Thus lactate dehydrogenase, alcohol dehydrogenase, and malate dehydrogenase utilize the A form, whereas glycerophosphate dehydrogenase, glucose-6-phosphate dehydrogenase, and glutathione reductase utilize the B form. Cornforth et al. (1962) have established the absolute configuration corresponding to A and B stereospecificity. In addition, NMR studies have established the nonequivalence of the two hydrogen atoms at C-4 for NADH (Sarma and Kaplan, 1969), thus providing a rationale for the stereospecificity. This phenomenon is most likely a consequence of a stacked conformation that NADH adopts in aqueous solution. In such a conformation, the adenine and nicotinamide rings lie on top of one another in a parallel manner, much like the rungs of a ladder. In one instance, the adenine ring may lie on top of the nicotinamide ring, while, in the other, their positions are reversed, with the nicotinamide ring atop the adenine. Hence, the hydrogens at C-4 would then differ by virtue of their relationship to the adenine ring.

1. Mechanism of Nicotinamide-Linked Dehydrogenases

This subject forms the bulk of Volume 11 of *The Enzymes* (1975), and the interested reader should consult it for further details. High resolution X-ray crystallographic studies have shown that each of the dehydrogenases investigated folds into an NAD-binding domain, which is similar in all of these enzymes (Rossman and Argos, 1978). These studies have further demonstrated that the interaction between all of the dehydrogenases and the adenosine portion of NAD is very similar, with the difference in interaction at the A or B side of the coenzyme being caused by a rotation of 180° about the glycosidic bond of the pyridine ring.

Of particular importance from the point of view of mechanism is the finding that there is an obligatory binding order, in which the coenzyme precedes the substrate (Dalziel, 1975), binding causing a conformational change in the enzyme. Both NADH and NADPH are electron- and hydride-donating reagents, the donation of the hydride ion producing a pyridinium ring (Mildvan, 1974). Hamilton (1971) has proposed instead that the hydrogen is transferred as a proton and not a hydride ion.

Since most dehydrogenases, save for alcohol dehydrogenase, do not contain metals, the necessary polarization of the oxidized substrate must be carried out in another manner. For lactate dehydrogenase, X-ray analysis reveals that the imidazolium group of His 195 appears to accomplish this by hydrogen bonding, with the possible additional help of Arg 109 (Holbrook et al., 1975).

C. Coenzyme A

In acetyl coenzyme A, one finds a carbonyl group linked to sulfur, forming a thioester. Reactions involving this coenzyme proceed along polar mechanisms and may be divided into two types (Jaenicke and Lynen, 1960). The first type of reaction occurs at the methyl group, in which a hydrogen atom adjacent to the carbonyl group is activated, with the resulting carbanion uniting with an electrophile. This permits formation of new carbon–carbon bonds akin to Claisen-type condensations. In the other type of reaction, which occurs at the acyl carbon, the acetyl group is transferred to a new molecule, with the concomitant loss of the coenzyme A portion (Fig. 3). These are acetylation reactions, corresponding to those accomplished in the laboratory with acetyl chloride or acetic anhydride, and occur by nucleophilic displacement.

It is commonplace to find that the utilization of carboxylic acids involves the mediation of thioesters rather than oxygen esters. Thioesters differ from the common esters in a number of ways, the differences accounting for the greater reactivity of thioesters relative to oxygen esters. Since sulfur is substantially less electronegative than oxygen, being almost equal in electronegativity to carbon, it is larger and more polarizable than oxygen. This results in longer and weaker bond formation than with the corresponding oxygen esters. The higher atomic

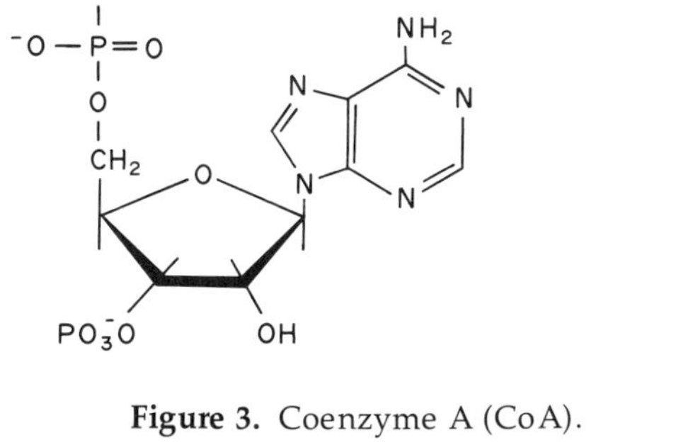

Figure 3. Coenzyme A (CoA).

charge density at sulfur relative to oxygen accounts for its reluctance to form a normal double bond. The resultant is greater charge localization in the carbonyl group, such that the thioester carbonyl carbon is decidedly more electrophilic than in the corresponding oxygen esters. The absence of resonance at the π double bond of the thioesters causes the carbonyl carbon to have a stronger positive charge than in oxygen ester carbonyls, imparting ketoacid character to it. It is therefore more accessible to nucleophilic attack, while the adjacent α carbon atom reacts with electrophilic reagents as a carbanion. These two areas of heightened activity in the thioesters account for their "head and tail" reactivity. A complete list of metabolic functions involving coenzyme A may be found in the review by Abiko (1975).

Two important reactions involving coenzyme A are worthy of mention: the first is the oxidative decarboxylation of pyruvate to acetyl-coenzyme A, which is catalyzed by the multienzyme complex, pyruvate dehydrogenase. This reaction involves a series of steps mediated by five coenzymes, during which a series of acid–base reactions and nucleophilic displacements take place, with pyruvate being oxidized and converted to acetyl-CoA. Another essential series of reactions involving acetyl-CoA is that of fatty acid synthesis. Here acetyl-CoA is converted to malonyl-CoA through condensation of the enzyme-bound thioester, with the occurrence of simultaneous decarboxylation.

D. Pyridoxal Phosphate

Pyridoxal phosphate is one of the most widely utilized of all coenzymes, being required for a multiplicity of reactions involving amino acids and amines. The general types of reactions involving pyridoxal phosphate include elimination, transamination, racemization, decarboxylation, and side chain cleavage. The underlying mechanism that accounts for most of the reactions involving pyridoxal phosphate is electron withdrawal toward the cationic nitrogen of the Schiff base (imine) that the enzyme forms with amino acids. This is shown in the accompanying diagram (Fig. 4). Such Schiff-base formation facilitates electron withdrawal from all three substituents on the α carbon, in a manner electronically akin to providing an adjacent carbonyl function, as is encountered in beta-keto acids. Hence, pyridoxal (like thiamine pyrophosphate) acts either to form a carbanion intermediate or to stabilize such an intermediate. As shown in Fig. 4, three general types of reactions are catalyzed by pyridoxal phosphate, depending on which of the three bonds, at reactive sites—a, b, or c (viz., the hydrogen, the carboxy, or the side chain)—are cleaved during the course of the reaction.

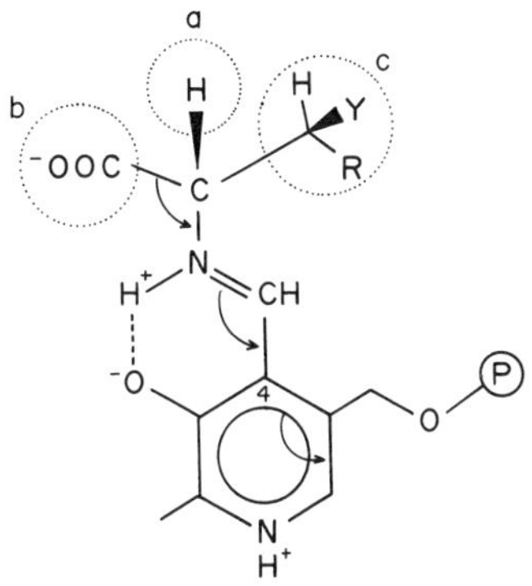

Figure 4. Pyridoxal phosphate + amino acid as a Schiff base. Ⓟ indicates a phosphate group. Bonds around the α-carbon can react as follows: a, removal of H⁺ to form semiquinoid intermediate which may then react in transamination (through ketimine formation), elimination, or racemization; b, decarboxylation; c, side chain cleavage.

When bond a is cleaved, resulting in the loss of the α hydrogen, a Schiff base forms, with the pyridoxal phosphate leading to what has been termed a semiquinoid intermediate (Fasella, 1967). Being similar to an enolate anion, this semiquinoid intermediate can react in one of several ways. In transamination—the most common reaction involving pyridoxal phosphate—the proton removed from a adds to the carbon attached to position 4 on the coenzyme ring, forming an additional Schiff base termed a ketimine. This intermediate may, through a double displacement reaction (ping–pong), undergo hydrolysis, forming a keto acid. In the reaction following, the coenzyme (pyridoxamine phosphate) transfers the amino group it has gained in the first step to another keto acid, forming a new amino acid.

Other paths may be traversed by the amino acid when the α hydrogen is removed, including elimination reactions, and racemization reactions whereby the proton is added back in a nonstereospecific manner. When bond b (that bearing the carboxyl function) is cleaved, a semiquinoid intermediate again forms, and completion of the reaction requires the addition of a proton to the site from which the carboxyl function had been lost. Finally, side chain cleavage involves breakage at the c bond, with the side chain being removed from the Schiff base, allowing an aldol cleavage to take place. Stereochemical aspects of pyridoxal action have been reviewed by Dunathan (1971).

The advantage gained by the formation of the enzyme-bound imine appears to be the greater reactivity of the imine as compared with aldehydes and ketones. This is because the more basic imine forms a stronger hydrogen bond and is more easily protonated than is oxygen. In addition, an imine carbon is more positive than a corresponding carbonyl carbon, thereby facilitating nucleophilic attack.

E. Lipoic Acid

Lipoic acid (1,2-dithiolane-3-valeric acid) carries out its biologic function by virtue of the presence of an intramolecular disulfide linkage in the molecule (Fig. 5). The lipoic acid of the pyruvate dehydrogenase multienzyme complex is bound covalently to the ϵ-amino group of lysine on the enzyme dihydrolipoyl transacetylase. In this complex series of reactions converting pyruvic acid to acetyl-CoA the "active aldehyde" bound to thiamine pyrophosphate is transferred to lipoic acid, causing the dithiolane ring to open, with the formation of the 6-acetyldihydrolipoic acid, which remains bound to the enzyme. In the next step, the acyl group is transferred to the next acceptor, CoA, and the dihydrolipoic acid can then be reoxidized to lipoic acid by a specific dihydrolipoic dehydrogenase (Koike and Koike, 1975).

F. Thiamine Pyrophosphate

Thiamine consists of a substituted pyrimidine linked by a methyl group to a substituted thiazole. As a coenzyme, it participates in the reactions involving the formation and scission of carbon–carbon bonds. Like cyanide ion in the laboratory, thiamine is employed where the bond broken is alpha to a carbonyl group. When thiamine acts as a cofactor in an enzymatic reaction, it does so as the pyrophosphate derivative (Fig. 6). Beta-type cleavage reactions at carbon–carbon bonds are accomplished readily, presumably as a consequence of the electron accepting properties of the carbonyl group. For example, decarboxylation of a beta-keto acid may occur through a cyclic transition state producing an enol, which avoids an anionic intermediate, with tautomerization to a methyl ketone. However, in the case of alpha-type cleavages, the electron-attracting properties of the adjacent carbonyl group are of little avail, since the group is unable to function via a resonance effect comparable to that encountered in beta-type cleavages. Thus alpha-keto acids decarboxylate with less facility than do the beta-keto acids.

The role of thiamine pyrophosphate appears to be the formation of compounds possessing structures analogous to those of the beta-keto

Figure 5. Lipoic acid (left) and dihydrolipoic acid (right).

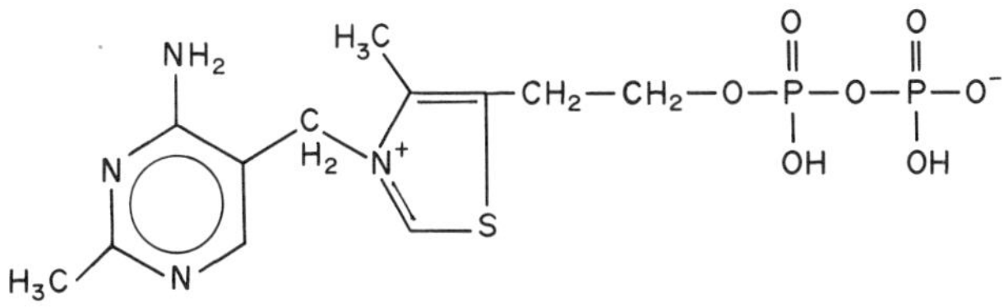

Figure 6. Thiamine pyrophosphate (TPP).

grouping, thereby facilitating the performance of carbon–carbon cleavage (Metzler, 1960; Krampitz, 1969). It is able to function in this manner because it possesses a cationic nitrogen, which is joined to the carbon atom at position 2 through a double bond. This nitrogen provides electrostatic stabilization of a -ylid (a dipolar compound containing a negative carbon and a positive nitrogen), easing removal of a proton, or a carbonyl or acyl group bonded at this position. The presence of sulfur in the ring imparts partial aromatic character, since sulfur does not conduct electrons as well as a normal π double bond, allowing the carbanion formed from the reactant to be stabilized without undergoing ketonization.

As shown in Fig. 7, the first step involved in the decarboxylation of an alpha-keto acid by thiamine pyrophosphate is the addition of the carbonyl by the carbanion nucleophile, followed by decarboxylation of the acid. As reviewed by Jencks (1975), investigations of the role of thiamine pyrophosphate analogs in the pyruvate dehydrogenase-catalyzed decarboxylation of pyruvate have provided evidence for rate enhancement by a desolvation effect. This effect comes about because of the removal of both the carboxylate group of the substrate and the cationic nitrogen of the coenzyme from an aqueous environment to the hydrophobic active site of the enzyme. The bulky and apparently chemically nonfunctional pyrophosphate and pyrimidine moieties of the

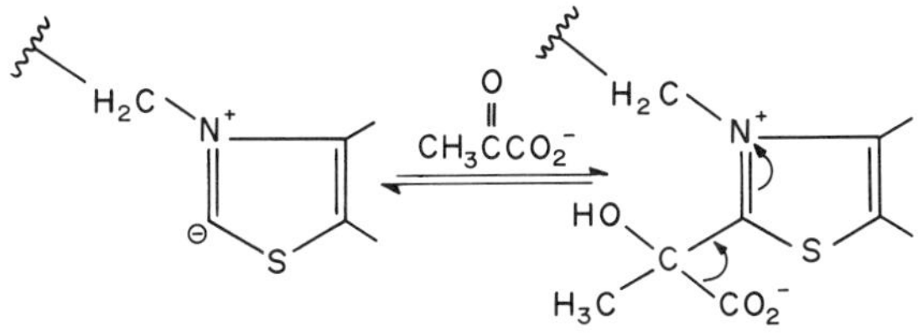

Figure 7. "Active-aldehyde" formation by TPP. Only the thiazole ring of TPP is shown to illustrate the carbanion nucleophilic addition which occurs during decarboxylation of an α-keto acid by TPP.

coenzyme provide the favorable binding energy to hold the charged substrate in this otherwise unfavorable environment and thus accelerate the reaction. Further discussion of the metabolic role of thiamine may be found in Leder (1975) and Gubler et al. (1976).

G. Flavins

Flavins can serve as either one-electron or two-electron transfer reagents, acting at the interface between the cytochromes and NADH. The vitamin, riboflavin, is converted by the body to the catalytically active coenzyme forms, FMN or FAD. Flavoproteins are involved in oxidation reactions with substrates which include the pyridine nucleotides, α-amino acids and α-hydroxy acids, and compounds containing saturated carbon–carbon bonds convertible into olefins. When oxidation occurs, the isoalloxazine ring of FMN (Fig. 8) or FAD becomes reduced.

The mechanism of action of the flavin coenzymes has been quite refractory of solution, but recently some advances in our knowledge have come about. The state of knowledge has been reviewed by Bruice (1976b), and the reader is directed there for areas which cannot be covered here.

Owing to the wide range of reactions in which flavin coenzymes are involved, a unitary mechanism of action appears unlikely. It seems clear that in enzymes where the fully reduced form is reacting with a one-electron acceptor (such as NADH-cytochrome b-reductase, NADPH-cytochrome c-reductase, and ferridoxin NADP reductase), the mechanism involves flavin semiquinones as intermediates. These are free-radical intermediates and can be detected by electron spin resonance (ESR). Flavin semiquinones can be stabilized because of multiple resonance forms, the odd electron being shared throughout the entire isoalloxazine ring. The free-radical intermediate may be further oxidized or reduced, one electron at a time. A number of enzymes requir-

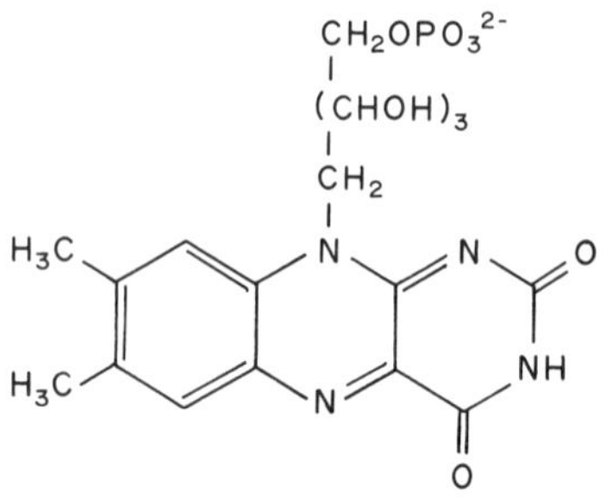

Figure 8. Flavin mononucleotide (FMN).

ing flavins also contain iron or molybdenum, e.g., xanthine oxidase. The function of the metal is not certain, but may involve chelation with the flavin. Such bonding would provide additional opportunity for delocalization of the odd electron, further stabilizing the semiquinone (Lowe and Ingraham, 1974).

Recently, Abramowitz and Massey (1976) have provided strong evidence for the formation of charge-transfer complexes in the interaction of phenols with the old-yellow enzyme (NADPH-oxidoreductase). Such complexes are identified by the demonstration of a new transfer absorption band. As reviewed by Massey and Ghisla (1974), the ability of the flavin to engage in charge-transfer complex formation derives from the fact that the ring system, as a p-quinoid molecule, is electron deficient when oxidized, thereby acting as a good electron acceptor in $\pi-\pi$ charge-transfer interactions with suitable electron donors, e.g., phenols, thiols, or amino compounds. These authors have explained the importance of such charge-transfer complexes in flavoenzymes as providing increased stability for the intermediates formed during such reactions. Recently, Bruice and Taulane (1976), Williams and Bruice (1976), and Williams et al. (1977) have provided strong evidence for a radical mechanism involving 1,5-dihydroflavin reduction of a series of carbonyl compounds including pyruvic acid, benzyl, and formaldehyde. Nonetheless, there are still a number of unanswered questions regarding the mechanism of action of flavins.

H. Biotin

Biotin is a bicyclic molecule involving the fusion of an imidazolidine ring with a tetrahydrothiophene ring bearing a valeric acid side chain. Since the molecule contains three asymmetric carbon atoms, eight stereoisomers are possible; however, only one is biologically active, such that the ureido and tetrahydrothiophene rings are fused *cis*, the aliphatic side chain having a *cis* conformation with respect to the ureido ring. For reviews of earlier work in this field, the reader is encouraged to consult Lynen (1967) and Knappe (1970). Eisenberg (1975) has considered the biosynthesis and degradation, as well as the biological role of biotin, and recently Wood and Barden (1977) have reviewed the subject of biotin-requiring enzymes.

Biotin-requiring enzymes include carboxylases, transcarboxylases, and decarboxylases. The function of biotin appears to be that of covalently bound CO_2 carrier. The mechanism of action of most biotin-containing enzymes appears to be a modified ping–pong mechanism, in which the biotin is first carboxylated at one site on the molecule and then migrates to another site, where the carboxybiotin transcarboxylates an acceptor. Thus, the first step is a synthetase reaction requiring an added divalent cation, while the second step is a transcarboxylation. Studies by Guchhait et al. (1974) have demonstrated that the biologically active form of biotin is the 1'-N-carboxy-D-biotin (Fig. 9).

The initial event in reactions involving biotin appears to be the formation of the carboxybiotin adduct through nucleophilic attack of the ureido ring on an electrophilic carrier of a carboxyl donor. Transfer of the carboxl group from the carboxybiotin requires electrophilic activation of the carboxyl carbon atom, as well as enzyme-mediated nucleophilic activation of the acceptor molecule. In decarboxylation reactions, electrophilic activation of carboxybiotin is necessitated. In studies reported by Guchhait et al. (1974), the 1'-N-carboxy-D-biotin was shown to act as a donor of CO_2 to acetyl-CoA in the presence of the carboxyl transferase subunit of acetyl-CoA carboxylase. Bruice (1976a) has questioned the assignment of the 1'-N position as the site of the covalently bound CO_2, indicating that CO_2 migration from the N to the O site cannot be ruled out by studies as presently performed.

Carboxylases which utilize bicarbonate as the CO_2 donor require ATP to provide energy for the formation of the new carbon–nitrogen bond, while in the case of transcarboxylases and decarboxylases, carboxyl-biotinyl enzyme formation occurs by a transcarboxylation pathway not requiring ATP. In the latter instance, either a malonyl-CoA derivative or a beta-keto acid serving as a carboxyl donor provide the active CO_2.

I. Folate Coenzymes

Folic acid (pteroylglutamic acid) is comprised of three units, a pterine ring, para-aminobenzoic acid, and one or more glutamic acid residues joined to the parent molecule through γ-glutamyl linkages. The structural formula of the coenzyme is shown in Fig. 10. Pterins differ from purines in the replacement of the imidazole ring by a pyrazine. When folate is converted through reduction of the pyrazine ring to

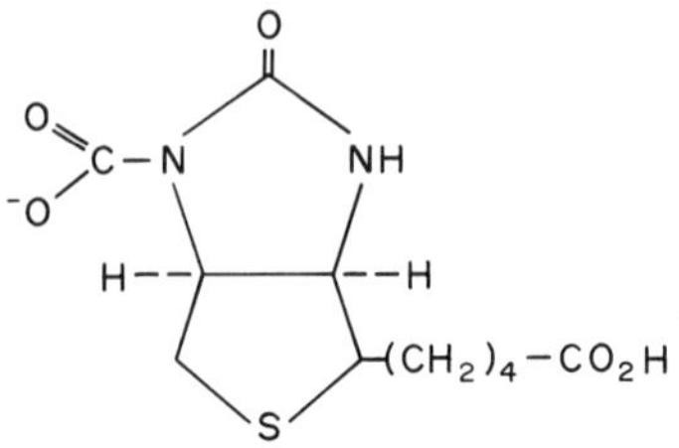

Figure 9. 1'-N-carboxy-D-biotin.

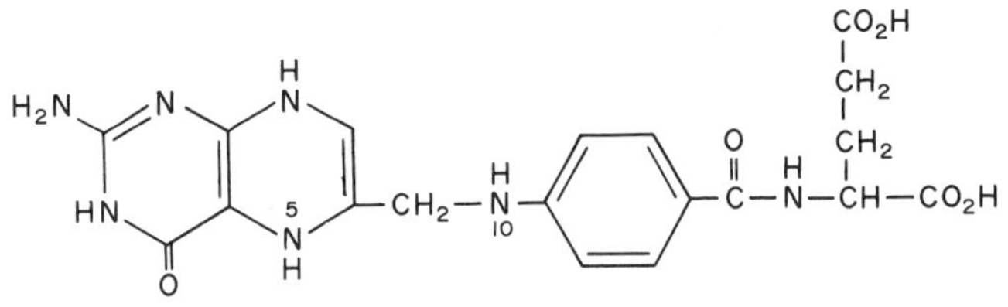

Figure 10. Folic acid.

tetrahydrofolate, the possibility then exists that C-1 groups can be attached by covalent linkage at the N-5 or N-10 position. Such C-1 groups include formyl (CHO), methenyl (=CH), methylene (CH_2), formimino (CHNH), and methyl (CH_3) groups. Moreover, N-5 acting with N-10 can allow a C-1 group to form a stable, five-membered ring with the coenzyme.

The N-5 position is considerably more basic than the N-10 position, and this basicity is one of several factors that control certain preferences in the course of reactions involving tetrahydrofolate. Thus, formylation occurs more readily at N-10 while alkylation occurs more readily at N-5. Benkovic and Bullard (1973) have reviewed evidence for an iminium cation at N-5 as the active donor in formaldehyde oxidation-level transfers. Recently, Barrows et al. (1976) have further studied such a mechanism for folic acid. The interconversion of these forms of folate coenzymes by enzymatic means has been reviewed by Stokstad and Koch (1967), and the reader is directed there for further details. Folate coenzymes are involved in a wide variety of biochemical reactions. These include purine and pyrimidine synthesis, conversion of glycine to serine, and utilization and generation of formate. In addition, the catabolism of histidine, with the formation of formiminoglutamic acid (FIGLU), is an important cellular reaction involving folate.

Molecular orbital calculations performed by Pullman and Pullman (1963) suggest that the ability of the folate coenzymes to relinquish one-carbon units is related to the di-positivity of the N—C bond which is ruptured during such transfers. This di-positivity denotes the presence of net positive charges on both the nitrogen and the carbon and has been found to be significantly greater in the active forms of the coenzyme than in the inactive forms. Owing to the wide range of reactions and the spectrum of intermediates involved in the utilization of folate coenzymes, it appears that no unitary mechanism of action of these coenzymes will be forthcoming. This subject has been considered by Benkovic and Bullard (1973) in a review which ought to be consulted by readers desiring a more detailed discussion of experimental data concerning folate action.

J. Metal Ions in Catalysis

Trace metals including Mg, Zn, Cu, Mn, Fe, and Co are employed as cofactors in a number of enzymatic reactions. Indeed, for many enzymes the only cofactor is a metal ion. Two broad classes of metal-utilizing enzymes are distinguishable by experimental means. These are the metalloenzymes where the metal is bound to the apoenzyme with sufficient strength to survive purification procedures and is present in a definite proportion to the protein, and the metal enzyme complexes where the metal, although easily dissociable, is required for full expression of enzyme activity.

The range of enzymatic reactions involving metals is large and includes hydrolyses, hydrogen transfers, hydroxylations, additions and eliminations, decarboxylations, and group transfer reactions. The metal in its association with the enzyme can participate in a number of different catalytic mechanisms. The metalloenzymes can engage in general acid–base catalysis, in nucleophilic catalysis through electron acceptance or donation, or, by chelation with the substrate and the enzyme, may enhance the propinquity of reactants as well as contributing to the development of the transition state. Mildvan (1970) has considered the gamut of metal involvement in enzymatic catalysis, and the reader is encouraged to consult that review, as well as one by Dixon et al. (1976).

K. Coenzyme B_{12}

Coenzyme B_{12} is composed of a corrin ring and a nucleotide, which contains 5,6-dimethylbenziminidazole instead of a purine or pyrimidine. A further distinction is the presence of ribose in α-glycosidic linkage, rather than in β linkage, as in the nucleic acids. The corrin ring contains trivalent cobalt linked in a coordination complex with the four nitrogen atoms of the ring (Fig. 11).

One bond between carbon and cobalt, while predominantly covalent, has some ionic character. The oxidation–reduction potentials of cobalt are modified by its incorporation into the corrin ring system, being able to exist in either a mono-, di-, or trivalent state. In the cobalt I derivatives (Schrauzer, 1976), the metal is in spin-paired d8 configuration. Two electrons occupy the weakly binding $3d_z^2$ orbital, and the resulting electron distribution is responsible for the high nucleophilic reactivity of cobalt I. Indeed, such cobalt derivatives have been termed *super nucleophiles*.

There are a number of reactions requiring adenosyl cobalamide, and, while most occur in bacteria, the conversion of methylmalonyl-CoA to succinyl-CoA is a mammalian enzyme reaction as well. All but one of these reactions can be understood as representing the migration

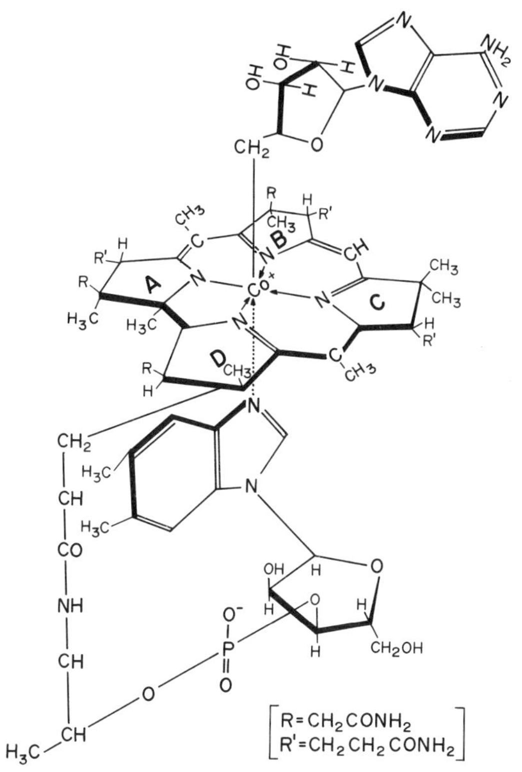

Figure 11. Coenzyme B_{12}.

of an H^+ from one carbon to an adjacent one, changing positions with the group attached to the adjacent carbon, which moves to the carbon formerly bearing hydrogen. Numerous studies have shown that, in this role, the coenzyme acts as an H^+ carrier, with the 5'-deoxyadenosine group serving as an intermediate in the reaction. H^+ migration occurs without exchange with solvent protons. This mechanism appears to involve cleavage of the cobalt–carbon bond, occurring by a homolytic route, with the production of a radical. Strong support of the radical formation has come from ESR studies. The enzyme-catalyzed homolytic cleavage of the cobalt–carbon bond is followed by removal of a substrate hydrogen by the 5'-deoxyadenos-5'-yl radical, leading to the production of a Cob(II)alamin-5'-deoxyadenosine and the substrate radical.

The mechanism of hydrogen transfer by coenzyme B_{12} seems to be clear. This is not the state of affairs as regards the nature of the other group which undergoes transfer. A number of different intermediates have been proposed, with elaborate mechanistic schemes. Epoxides, radical rearrangements, Faworskii rearrangements, allylcarbinyl ions, and reductive eliminations have all been proposed by different investigators. For further details the reader is directed to Abeles and Dolphin (1976), Golding and Radom (1976), Schrauzer (1977), and Krouwer and Babior (1977).

VII. Evolution of Enzyme Function

Koshland (1976) has considered the evolution of enzyme catalytic power from the point of view of the sequential acquisition of catalytic ability, specificity, regulability, and cooperativity. He contends that function is the driving force of evolution and must be subject to incremental improvements by small changes in structure, such incremental modifications occurring in a random manner.

A primordial catalytic function could develop in a protein, perhaps by the chance juxtaposition of serine and histidine (imidazole) moieties. Improvements in the spatial arrangement can be made by random mutations of amino acids removed from the catalytic site. Similarly, improvements in binding-site orientation with respect to proximity of the substrate molecules to one another and to enzyme functional groups could occur by small degrees, which would then be fixed by selection. Thus, orientation and proximity, the crucial factors for binding interaction and entropic effects, are ideally suited for evolutionary optimization.

Once a structure is developed that attracts substrates and catalytic groups into a fruitful orientation, the development of specificity becomes possible. Although specificity may not have been essential in early catalysts, complexity increased within primordial cells. It became necessary for each enzyme to catalyze a particular reaction, with a circumscribed spectrum of substrates. While a template-type enzyme would permit optimal orientation of specific substrates, the ever-present possibility existed that water, for example, might compete successfully for the active site of enzymes reacting with hydroxylic substrates present in much lower concentrations than water. The answer to this difficulty would be a flexible enzyme structure in which the substrate would have to possess certain minimal sites to bind to the flexible enzyme in order to induce the particular alignment of catalytic groups and to force the substrate into the conformation of the transition-state complex. As noted by Koshland (1976), it is not likely that flexibility was developed after creating a fully template-type enzyme. Flexibility provides the possibility for regulation by modifiers that are not substrates for the enzyme. Such regulatory behavior is the subject of the next chapter. Albery and Knowles (1976, 1977) have carried out a series of investigations relating to the optimization of catalytic ability, and the interested reader is directed there for further details of their experiments and a review of the literature.

References

Abeles, R. H., and Dolphin, D., 1976, The vitamin B_{12} coenzyme, *Accounts Chem. Res.* **9**:114.
Abiko, Y., 1975, Metabolism of Coenzyme A, in: *Metabolic Pathways*, 3rd ed., Vol. 7 (D. M. Greenberg, ed.), p. 1, Academic Press, New York.
Abramovitz, A. S., and Massey, V., 1976, Interaction of phenols with old yellow enzyme, *J. Biol. Chem.* **251**:5327.
Albery, W. J., and Knowles, J. R., 1976, Evolution of enzyme function and the development of catalytic efficiency, *Biochemistry* **15**:5631.
Albery, W. J., and Knowles, J. R., 1977, Efficiency and evolution of enzyme catalysis, *Angew. Chem. (Engl.)* **16**:285.
Banks, B., 1969, Thermodynamics and biology, *Chem. Br.* **5**:514.
Barnard, E. A., and Stein, W. D., 1958, The roles of imidazole in biological systems, *Adv. Enzymol.* **20**:51.
Barrows, T. H., Farina, P. R., Chrzanowski, R. L., Benkovic, P. A., and Benkovic, S. J., 1976, Studies on models for tetrahydrofolic acid. VII. Reactions and mechanisms of tetrahydroquinoxaline derivatives at the formaldehyde level of oxidation, *J. Am. Chem. Soc.* **98**:3678.
Battersby, A. R., and Staunton, J., 1974, Stereospecificity of some enzymic reactions, *Tetrahedron* **30**:1707.
Bell, R. M., and Koshland, D. E., Jr., 1971, Covalent enzyme-substrate intermediates, *Science* **172**:1253.
Bell, R. P., 1973, *The Proton in Chemistry*, Cornell University Press, Ithaca, N. Y.
Bender, M., 1971, *Mechanisms of Homogenous Catalysis from Protons to Proteins*, John Wiley & Sons, New York.
Benkovic, S. J., and Bullard, W. P., 1973, On the mechanism of action of folic acid cofactors, in: *Progress in Bioorganic Chemistry*, Vol. 2 (E. T. Kaiser and F. J. Kézdy, eds.), p. 133, John Wiley & Sons, N. Y.
Benkovic, S. J., and Schray, K. J., 1973, Chemical basis of biological phosphoryl transfer, in: *The Enzymes*, 3rd ed., Vol. 8 (P. D. Boyer, ed.), p. 201, Academic Press, New York.
Blackburn, S., 1976, *Enzyme Structure and Function*, M. Dekker, New York.
Blow, D. M., 1976, Structure and mechanism of chymotrypsin, *Accounts Chem. Res.* **9**:145.
Borden, W., 1975, *Modern Molecular Orbital Theory for Organic Chemistry*, Prentice-Hall, Englewood Cliffs, N. J.
Bosshard, H. R., 1976, Theories of enzyme specificity and their application to proteases and aminoacyl-transfer RNA synthetases, *Experientia* **32**:949.
Bruice, T. C., 1970, Proximity effects and enzyme catalysis, in: *The Enzymes*, 3rd ed., Vol. 2 (P. D. Boyer, ed.), p. 217, Academic Press, New York.
Bruice, T. C., 1976a, Some pertinent aspects of mechanism as determined with small molecules, *Ann. Rev. Biochem.* **45**:331.
Bruice, T. C., 1976b, Models and flavin catalysis, in: *Progress in Bioorganic Chemistry*, Vol. 4 (E. T. Kaiser and F. J. Kézdy, eds.), p. 1, John Wiley & Sons, New York.

Bruice, T. C., and Benkovic, S. J., 1966, *Bioorganic Mechanisms*, Benjamin, New York.
Bruice, T. C., and Taulane, J. P., 1976, The kinetics and mechanisms of 1,5-dihydroflavin reduction of carbonyl compounds and flavin oxidation of alcohols. 3. Oxidation of benzoin by flavin and reduction of benzil by 1,5-dihydroflavin, *J. Am. Chem. Soc.* **98**:7769.
Chotia, C., and Janin, J., 1975, Principles of protein–protein recognition, *Nature* **256**:705.
Chou, P. Y., and Fasman, G. D., 1978, Prediction of the secondary structure of proteins from their amino acid sequence, *Adv. Enzymol.* **47**:45.
Cohen, S. G., Vaidya, V. M., and Schultz, R. M., 1970, Active site of α-chymotrypsin activation by association-desolvation, *Proc. Natl. Acad. Sci. USA* **66**:249.
Colowick, S., Van Eys, J., and Park, J. H., 1966, Dehydrogenation, in: *Comprehensive Biochemistry* (M. Florkin and F. Stotz, eds.), p. 1, Elsevier, Amsterdam.
Cornforth, J. W., 1974, Enzymes and stereochemistry, *Tetrahedron* **30**:1515.
Cornforth, J. W., Ryback, G., Popják, G., Donninger, C., and Schroepfer, G., Jr., 1962, Stereochemistry of enzyme hydrogen transfer to pyridine nucleotides, *Biochem. Biophys. Res. Commun.* **9**:371.
Cozzone, P. J., and Jardetzky, O., 1977, The mechanism of purine polynucleotide hydrolysis by ribonuclease A, *FEBS Lett.* **73**:77.
Cram, D. J., and Cram, J. M., 1974, Host-guest chemistry, *Science* **183**:803.
Dalziel, K., 1975, Kinetics and mechanism of nicotinamide-nucleotide linked dehydrogenases in: *The Enzymes*, 3rd ed., Vol. 11 (P. D. Boyer, ed.), p. 1, Academic Press, New York.
Dickerson, R. E., 1972, X-ray studies of protein mechanisms, *Annu. Rev. Biochem.* **41**:815.
Dickerson, R. E., and Geis, I., 1969, *The Structure and Action of Proteins*, Harper & Row, New York.
Dixon, N. E., Gazzola, C., Blakely, R. L., and Zerner, B., 1976, Metal ions in enzymes using ammonia or amides, *Science* **191**:1144.
Douglas, K. T., 1976, A basis for biological phosphate and sulfate transfers-transition state properties of transfer substrates, in: *Progress in Bioorganic Chemistry*, Vol. 4 (E. T. Kaiser and F. J. Kézdy, eds.), p. 193, John Wiley & Sons, New York.
Dunathan, H. C., 1971, Stereochemical aspects of pyridoxal phosphate catalysis, *Adv. Enzymol.* **35**:79.
Eisenberg, M. A., 1975, Biotin, in: *Metabolic Pathways*, 3rd ed., Vol. 7 (D. M. Greenberg, ed.), p. 27, Academic Press, New York.
Eyring, H., and Johnson, F. H., 1971, The elastomeric rack in biology, *Proc. Natl. Acad. Sci. USA* **68**:2341.
Eyring, H., Lumry, R., and Spikes, J. D., 1954, Kinetic and thermodynamic aspects of enzyme-catalyzed reactions, in: *Mechanisms of Enzyme Action* (W. D. McElroy and B. Glass, ed.), p. 123, The Johns Hopkins Press, Baltimore.
Fasella, P., 1967, Pyridoxal phosphate, *Annu. Rev. Biochem.* **36**:185.
Fersht, A., 1977, *Enzyme Structure and Mechanism*, W. H. Freeman, San Francisco.
Franks, F., 1975, *Water—A Comprehensive Treatise* (5 volumes), Plenum, N. Y.
Fruton, J., 1976, The mechanism of the catalytic action of pepsin and related acid proteinases, *Adv. Enzymol.* **44**:1.
Glazer, A. N., 1970, Specific chemical modification of proteins, *Annu. Rev. Biochem.* **39**:101.
Golding, B. T., and Radom, L., 1976, On the mechanism of action of adenosylcabalamin, *J. Am. Chem. Soc.* **98**:6331.
Gubler, C. J., Fujiwara, M., and Dreyfus, P. M. (eds.), 1976, *Thiamin*, John Wiley & Sons, New York.
Guchhait, R. B., Polakis, S. E., Hollis, D., Fenselau, C., and Lane, M. D., 1974, Acetyl

coenzyme A carboxylase system of *E. coli* site of carboxylation of biotin and enzymatic reactivity of 1'-N-(ureido)-carboxybiotin derivatives, *J. Biol. Chem.* **249**:6646.

Hamilton, G. A., 1971, The proton in biological redox reactions, in: *Progress in Bioorganic Chemistry*, Vol. 1 (E. T. Kaiser and F. J. Kézdy, eds.), p. 83, John Wiley & Sons, New York.

Hanson, K. R., 1976, Concepts and perspectives in enzyme stereochemistry, *Annu. Rev. Biochem.* **45**:307.

Haschemeyer, R. H., and Haschemeyer, A. E., 1973, *Proteins—A Guide to Study by Physical and Chemical Methods*, John Wiley & Sons, New York.

Holbrook, J. J., Liljas, A., Steindel, S. J., and Rossman, M. G., 1975, Lactate dehydrogenase, in: *The Enzymes*, 3rd ed., Vol. 11 (P. D. Boyer, ed.), p. 61, Academic Press, New York.

Huxley, A. F., 1970, Energetics of muscle, *Chem. Br.* **6**:477.

Imoto, T., Johnson, L. N., North, A. C. T., Phillips, D. C., and Rupley, J. A., 1972, Vertebrate lysozymes, in: *The Enzymes*, 3rd ed., Vol. 7 (P. D. Boyer, ed.), p. 665, Academic Press, New York.

Jaenicke, L., and Lynen, F., 1960, Coenzyme A, in: *The Enzymes*, 2nd ed., Vol 3 (P. D. Boyer, H. Lardy, and K. Myrback, eds.), p. 3, Academic Press, New York.

Jencks, W. P., 1969, *Catalysis in Chemistry and Enzymology*, McGraw-Hill, New York.

Jencks, W. P., 1972, General acid–base catalysis of complex reactions in water, *Chem. Rev.* **72**:705.

Jencks, W. P., 1975, Binding energy, specificity and enzymatic catalysis: The Circe effect, *Adv. Enzymol.* **43**:219.

Jencks, W. P., 1976, Enforced general acid–base catalysis of complex reactions and its limitations. *Accounts Chem. Res.* **9**:425.

Jencks, W. P., and Page, M. I., 1972, On the importance of togetherness in enzymic catalysis, *Proc. Eighth FEBS Meeting, Amsterdam* **29**:45.

Knappe, J., 1970, Mechanism of biotin action, *Annu. Rev. Biochem.* **39**:757.

Knowles, J. R., and Gutfreund, H., 1974, The functions of proteins as devices, *MTP Intl. Rev. Sci.* **1**:375.

Koike, M., and Koike, K., 1975, Lipoic acid, in: *Metabolic Pathways*, 3rd ed., Vol. 7 (D. M. Greenberg, ed.), p. 87, Academic Press, New York.

Koshland, D. E., 1976, The evolution of function in enzymes, *Fed. Proc.* **35**:2104.

Koshland, D. E., Jr., and Neet, K. E., 1968, The catalytic and regulatory property of enzymes, *Annu. Rev. Biochem.* **37**:359.

Krampitz, L. O., 1969, Catalytic functions of thiamine diphosphate, *Annu. Rev. Biochem.* **38**:213.

Kraut, J., 1977, Serine proteases: Structure and mechanism of catalysis, *Annu. Rev. Biochem.* **46**:331.

Krouwer, J. S., and Babior, B. M., 1977, The mechanism of cobalamin-dependent rearrangements, *Molec. Cell. Biochem.* **15**:89.

Kutzelnigg, W., 1973, The physical meaning of the chemical bond, *Agnew. Chem. (Engl.)* **13**:456.

Kyba, E. P., Helgeson, R. C., Madan, K., Gokel, G. W., Tarnowski, T. L., Moore, S. S., and Cram, D. J., 1977, Host guest complexation. I. Concept and illustration. *J. Am. Chem. Soc.* **99**:2564.

Leder, I. G., 1975, Thiamine: Biosynthesis and function, in: *Metabolic Pathways*, 3rd ed., Vol. 7 (D. M. Greenberg, ed.), p. 57, Academic Press, New York.

Lehninger, A. L., 1959, Respiratory energy transformation, *Rev. Mod. Phys.* **31**:136.

Lewin, S., 1974, *Displacement of Water and Its Control of Biochemical Reactions*, Academic Press, N. Y.

Lewis, G. N., 1916, The atom and the molecule, *J. Am. Chem. Soc.* **38**:762.
Lienhard, G., 1973, Enzymatic catalysis and transition state theory, *Science* **180**:149.
Ling, G. N., 1977, The physical state of water and ions in living cells and a new theory of the energization of biological work performance of ATP, *Molec. Cell. Biochem.* **15**:159.
Lipscomb, W. N., 1974, Relationship of the three dimensional structure of carboxypeptidase A to catalysis, *Tetrahedron* **30**:1725.
Lowe, G., 1976, The cysteine proteinases, *Tetrahedron* **32**:291.
Lowe, J. N., and Ingraham, L. L., 1974, *An Introduction to Biochemical Reaction Mechanisms*, Prentice-Hall, Englewood Cliffs, N. J.
Lumry, R., 1974, Conformational mechanisms for free energy transduction in protein systems: Old ideas and new facts, *Ann. N. Y. Acad. Sci.* **227**:46.
Lynen, F., 1967, The role of biotin-dependent carboxylations in biosynthetic reactions, *Biochem. J.* **102**:381.
Makinen, M. W., and Fink, A. L., 1977, Reactivity and cryoenzymology of enzymes in the crystalline state, *Annu. Rev. Biophys. Bioeng.* **6**:301.
Makinen, M. W., Yamamura, K., and Kaiser, E. T., 1976, Mechanism of action of carboxypeptidase A in ester hydrolysis, *Proc. Natl. Acad. Sci. USA* **73**:3882.
March, J., 1977, *Advanced Organic Chemistry: Reactions, Mechanisms, and Structure*, 2nd ed., McGraw-Hill, New York.
Markley, J. L., 1976, Correlation proton magnetic resonance studies at 250 MHz of bovine pancreatic ribonuclease. I. Reinvestigation of histidine peak assignments, *Biochemistry* **14**:3546.
Massey, V., and Ghisla, S., 1974, Role of charge-transfer interactions in flavoprotein catalysis, *Ann. N. Y. Acad. Sci.* **227**:446.
Meadows, D. H., Markley, J. L., Cohen, J. S., and Jardetzky, O., 1967, Nuclear magnetic resonance studies of the structure and binding sites of enzymes. I. Histidine residues, *Proc. Natl. Acad. Sci. USA* **58**:1307.
Metzler, D. E., 1960, Thiamine coenzymes, in: *The Enzymes*, 2nd. ed., Vol. 2 (P. D. Boyer, H. Lardy, and K. Myrback, eds.), p. 295, Academic Press, New York.
Mildvan, A. S., 1970, Metals in enzyme catalysis, in: *The Enzymes*, 3rd ed., Vol. 2 (P. D. Boyer, ed.), p. 446, Academic Press, New York.
Mildvan, A. S., 1974, Mechanism of enzyme action, *Annu. Rev. Biochem.* **43**:357.
Patel, D. J., Canuel, L. L., and Bovey, F. A., 1975, Reassignment of the active site histidines in ribonuclease A by selective deuteration studies, *Biopolymers* **14**:987.
Pauling, L., 1948, Chemical achievement and hope for the future. *Am. Sci.* **36**:51.
Pauling, L., 1970, Structure of high energy molecules, *Chem. Br.* **6**:468.
Pincus, M. R., Burgess, A. W., and Scheraga, H. A., 1976a, Conformational energy calculations of enzyme substrate complexes of lysozyme. I. Energy minimization of monosaccharide and oligosaccharide inhibitors and substrates of lysozyme, *Biopolymers* **15**:2485.
Pincus, M. R., Zimmerman, S. S., and Scheraga, H. A., 1976b, Prediction of three-dimensional structures of enzyme–substrate and enzyme–inhibitor complexes of lysozyme, *Proc. Natl. Acad. Sci. USA* **73**:4261.
Pullman, B., and Pullman, A., 1963, *Quantum Biochemistry*, John Wiley & Sons, New York.
Richards, F. M., and Wykoff, H. W., 1971, Bovine pancreatic ribonuclease, in: *The Enzymes*, 3rd ed., Vol. 4, (P. D. Boyer, ed.), p. 647, Academic Press, N. Y.
Roberts, G. C. K., Dennis, E. A., Meadows, D. H., Cohen, J. S., and Jardetzky, O., 1969, The mechanism of action of ribonuclease, *Proc. Natl. Acad. Sci. USA* **62**:1151.
Rossmann, M. G., and Argos, P., 1978, The taxonomy of binding sites in proteins, *Molec. Cell. Biochem.* **21**:161.

Sarma, R. H., and Kaplan, N. O., 1969, 220 MH_z nuclear magnetic resonance spectra of oxidized and reduced pyridine dinucleotides, *J. Biol. Chem.* **244**:771.

Scheiner, S., and Lipscomb, W. N., 1977, Molecular orbital studies of enzyme activities. 4. Hydrolysis of peptides by carboxypeptidase A, *J. Am. Chem. Soc.* **99**:3466.

Schrauzer, G. N., 1976, New developments in the field of vitamin B_{12}: Reactions of the cobalt atom in corrins and in vitamin B_{12} model compounds, *Angew. Chem. (Engl.)* **15**:417.

Schrauzer, G. N., 1977, New developments in the field of vitamin B_{12}: Enzymatic reactions dependent upon corrins and coenzyme B_{12}, *Angew. Chem. (Engl.)* **16**:233.

Shindo, H., Hayes, M. B., and Cohen, J. S., 1976, Nuclear magnetic resonance titration curves of histidine ring protons. A direct assignment of the resonance of the active site histidine residues of ribonuclease, *J. Biol. Chem.* **251**:2644.

Sigman, D. S., and Mooser, G., 1975, Chemical studies of enzyme active sites, *Annu. Rev. Biochem.* **44**:889.

Stokstad, E. L. R., and Koch, J., 1967, Folic acid metabolism, *Physiol. Rev.* **47**:83.

Stroud, R. M., Kossiakoff, A. A., and Chambers, J. L., 1977, Mechanisms of zymogen activation, *Annu. Rev. Biophys. Bioeng.* **6**:177.

Timko, J. M., Moore, S. S., Walba, D. M., Hiberty, P. C., and Gram, D. J., 1977, Host guest complexation. II. Structural units that control association constants between polyethers and tert-butylammonium salts, *J. Am. Chem. Soc.* **99**:4207.

Vallee, B. L., and Riordan, J. F., 1969, Chemical approaches to the properties of active sites of enzymes, *Annu. Rev. Biochem.* **38**:733.

Vennesland, B., and Westheimer, F. H., 1954, Hydrogen transport and steric specificity in reactions catalyzed by pyridine nucleotide dehydrogenases, in: *Mechanisms of Enzyme Action* (W. D. McElroy and B. Glass, eds.), p. 357, The Johns Hopkins Press, Baltimore.

Westley, J., 1969, *Enzymic Catalysis,* Harper and Row, New York.

Wilkie, D., 1970, Thermodynamics and biology, *Chem. Br.* **6**:473.

Williams, R. F., and Bruice, T. C., 1976, The kinetics and mechanisms of 1,5-dihydroflavin reduction of carbonyl compounds and flavin oxidation of alcohols. 2. Ethyl pyruvate, pyruvamide, and pyruvic acid, *J. Am. Chem. Soc.* **98**:7752.

Williams, R. F., Shinkai, S. S., and Bruice, T. C., 1977, Kinetics and mechanisms of the 1,5-dihydroflavin reduction of carbonyl compounds and the flavin oxidation of alcohols. 4. Interconversion of formaldehyde and methanol, *J. Am. Chem. Soc.* **99**:921.

Wolfeden, R., 1974, Enzyme catalysis, conflicting requirements of substrate access and transition state affinity, *Molec. Cell. Biochem.* **3**:207.

Wolfeden, R., 1976, Transition state analog inhibitors and enzyme catalysis, *Annu. Rev. Biophys. Bioeng.* **5**:271.

Wood, H. G., and Barden, R. E., 1977, Biotin enzymes, *Annu. Rev. Biochem.* **46**:385.

Zeffran, E., and Hall, P., 1973, *The Study of Enzyme Mechanisms,* John Wiley & Sons, New York.

4

Modulation of Enzyme Activity

Robert M. Cohn and John R. Yandrasitz

I. Introduction

Control characterizes every level of biological function, often being most apparent in the prodigious ability of organisms to adapt to changes in their environment. But—while many studies have demonstrated the importance of hormones in affecting the direction and rate of simultaneously occurring reactions and thereby of providing for a response to an environmental perturbation—it is clear that the enzymes which catalyze the rate-controlling steps in a pathway exercise the most profound control over biological function. Such enzymes have been defined (Rolleston, 1972) as those whose kinetic properties respond to factors other than substrate concentration and which exert definitive control over the flux of the pathway.

Two functions of intermediary metabolism are to provide an adequate supply of high-energy compounds and to produce intermediates for the synthesis of other molecules and macromolecules. Regulation must then involve coordination of many independent chemical reactions and pathways so as to provide not only a given precursor, but also a proper balance of all precursors to ensure their uninterrupted incorporation into the more complex components of the cell's architecture.

In considering the modulation of enzyme activity, we are concerned with the various mechanisms by which such activity can be modified in vitro with the possible relevance of these mechanisms to the in vivo condition. Any attempt to describe quantitatively the inter-

---

Robert M. Cohn and John R. Yandrasitz • Department of Metabolic Research, Children's Hospital of Philadelphia, University of Pennsylvania, Philadelphia, Pennsylvania 19104.

locking reactions that comprise the whole of intermediary metabolism must include the chemical steps involved, the cellular activities of enzymes and concentrations of metabolic intermediates and modifers, as well as interconversion rates under a variety of circumstances. However, before attempting to present such a holistic view of metabolism, it is necessary to develop an appreciation for the multiplicity of regulatory capabilities available to the cell.

Regulatory mechanisms may be categorized according to the physical or chemical transformation which they accomplish and the time scale during which they occur. On this basis, we may conceptualize three levels of mechanism: *Noncovalent* interactions are rapid and freely reversible, occurring on a time scale of milliseconds. They involve direct modulation of an enzyme by substrate or modifier concentration, or by factors such as pH, temperature, ionic strength, or ionic composition. Polymerization and depolymerization are also noncovalent interactions, but occur on a more protracted time scale than the foregoing. *Covalent* modification of an enzyme by another enzyme results in the interconversion of forms differing in their catalytic or regulatory properties; these interconversions may be reversible or irreversible. Finally, there are regulatory mechanisms which involve alteration of the total enzyme concentration by influencing the rates of *synthesis or degradation* of enzymes. These regulatory mechanisms apparently function to maintain an optimal flux through the pathway. According to this last principle, there appears to be a reserve of catalytic activity inherent in pathways, since under normal circumstances the actual flux through a pathway is less than the potential maximal flux as determined by the least active enzyme in the pathway (Scrutton and Utter, 1968). Various aspects of enzymic regulation have been the subject of reviews (Stadtman, 1966, 1970) and of a monograph (Newsholme and Start, 1973).

II. Noncovalent Regulatory Mechanisms

A. Modulation by Substrate and Product Concentration in Enzymes Following Classical Michaelis–Menten Kinetics

Theoretically, any metabolic pathway could be regulated by the availability of substrate. A reduction of substrate concentration will decrease the activity of an enzyme, and this could result in a decreased flux stepwise through the entire pathway. However, studies demonstrating the relative constancy of substrate levels in vivo imply that such a mechanism, while perhaps important for single enzymes, is not operative for pathways in higher animals. Nonetheless, the plasma fatty acid concentration appears to play a fundamental role in the oxidation of these substances by various tissues, and the oxidation of fatty acids can, in turn, modify the rate of the animal's carbohydrate utilization.

Enzymes that respond to changes in substrate concentration by classical Michaelis–Menten kinetics are characterized by a hyperbolic substrate–velocity relationship; these enzymes are most responsive to changes in substrate concentration in the range from zero to the K_m value for the substrate. It appears that the K_m establishes an approximate value for the intracellular level of the substrate in the in vivo situation (Segel, 1975). At this level, the enzyme may still respond rapidly to a small change in substrate concentration. At substrate levels far below the K_m much of the catalytic potential of the enzyme would be wasted, and if substrate levels were maintained far in excess of the K_m, more enzyme would be required, so as to compensate for the decreased ability of the enzyme to respond to increases in substrate concentration near saturating levels. In order to maintain concentrations of metabolites within the small tolerances observed, reaction rates must be capable of changing very rapidly. For classical Michaelis–Menten enzymes, this is apparently achieved by maintaining the enzyme level that will maintain a substrate level near the K_m.

However, in the case of hepatic glucokinase, the K_m (10 mM) exceeds the usual glucose level, thus permitting a response to a large increase in the substrate load for the enzyme (Colowick, 1973; Weinhouse, 1976)—a condition expected after a carbohydrate-containing meal. This, coupled with the absence of inhibition of glucokinase by its product, glucose-6-phosphate, allows the enzyme to respond to the wide variation in portal vein glucose concentration, thus stimulating both glycolysis and glycogenesis.

In the case of multisubstrate enzymes, an extra degree of complexity is added to the possibility of catalytic regulation. For example, if the concentrations of more than one of the substrates are permitted to change simultaneously, instead of holding one constant, there occurs a different relative change in enzyme activity from that seen in the unisubstrate situation. As shown by Cleland (1963a), bi-substrate enzymes contain rate expressions with the factor V_{max} [A] [B]—a simultaneous change in both substrates will render the equation of a higher order than with a single substrate. The kinetics of multisubstrate enzymes have been considered by Cleland (1963a–c), Fromm (1975), Laidler and Bunting (1973), and Segel (1975) and will not be dealt with further here.

Product can also affect enzyme activity. If the product of an enzyme-catalyzed reaction is not removed, thermodynamic inhibition will be encountered as the reaction approaches equilibrium and the reverse reaction becomes significant. At equilibrium, the ratio of substrate to product becomes fixed according to the equilibrium constant. Maintenance of this ratio then serves to regulate the availability of reactants for further steps in the pathway.

If the concentration of product is too low to permit the reaction to proceed significantly in the reverse direction, it may nonetheless be sufficient to permit the formation of an end-product complex, thereby decreasing the actual concentration of the enzyme and resulting in kinetic inhibition. In a computer analysis of product inhibition of multisubstrate enzymes, Purich and Fromm (1972) have suggested that a number of possibly important regulatory effects can arise as a consequence of product inhibition in the absence of any special regulatory sites for modifiers. The authors suggest that the order of substrate addition and product release may be important as a regulatory mechanism for the enzyme in vivo. This conclusion came from the observation that enzymes having the same apparent kinetic constants gave, in some instances, very different responses to product concentration, depending on the reaction mechanism assumed. However, the tentative nature of these suggestions must be underscored.

While product inhibition in vitro is a common observation, there appear to be few examples which have physiologic meaning. Recently, Richman and Meister (1975) demonstrated the inhibition of glutamyl-

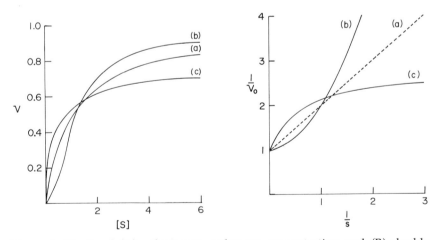

Figure 1. Idealized (A) velocity vs. substrate concentration and (B) double reciprocal plots (Lineweaver-Burk) for enzymes demonstrating: (a) hyperbolic (classical Michaelis–Menten) behavior; (b) positive cooperativity; (c) negative cooperativity.

cysteine synthetase by glutathione, the end product of this anabolic pathway.

L-glutamate + cysteine + ATP ⇌ L-γ-glutamyl-L-cysteine
  + ADP + P_i
L-γ-glutamyl-L-cysteine + glycine + ATP ⇌ glutathione + ADP + P_i (1)

Studies with the purified enzyme from rat kidney have demonstrated that glutathione is a competitive inhibitor with respect to glutamate in reaction (1), indicating that it occupies the product (γ glutamylcysteine) site. Since the K_i (2.3 mM) is in the range of physiological glutathione concentrations, it may be that this inhibition has relevance for the in vivo situation. Koshland and Neet (1968) have called this type of inhibition "autosteric." Kinetically, it resembles inhibition by immediate products, except that it permits modulation of an earlier step in the pathway. Hence this effect is analogous to the more common allosteric feedback inhibition to be considered shortly, except that the catalytic site is involved.

B. Cooperativity

Enzymes displaying classical Michaelis–Menten kinetics, as discussed above, are limited in their ability to maintain substrate levels within small tolerances, since the hyperbolic substrate–velocity relationship dictates that the responsiveness to change in substrate concentration decreases as the actual substrate level increases (Figure 1, curve a). However, many regulable enzymes are more flexible than this in their response to substrate, demonstrating the phenomenon of cooperativity (Citri, 1973). These enzymes have more than one catalytic site, generally on different polypeptides, and these sites communicate their status to each other.

In positive cooperativity, binding of substrates to one catalytic site increases the activity of other sites, either by increasing the affinity for substrate or by increasing catalytic efficiency. Enzymes showing positive cooperativity have sigmoidal velocity–substrate profiles (Figure 1, curve b). As the substrate concentration increases, additional sites are occupied. The remaining sites are converted to more active sites until, at high substrate concentration, all the sites are at their most active and velocity approaches a plateau. At the substrate concentration where the velocity is half maximal ($S_{0.5}$), the velocity curve is nearly vertical. Small changes in substrate influx at this concentration will produce very large changes in velocity, allowing for exquisite control of substrate concentrations.

Enzymes displaying negative cooperativity have velocity–substrate profiles less sensitive to change in substrate concentration than those for classical Michaelis–Menten enzymes (Figure 1, curve c). Numerous examples of negative cooperativity have now been recognized, and several have proved to be due to half-of-the-sites reactivity, an extreme case of negative cooperativity where binding at the initial half of the sites renders the remaining half of the sites virtually inactive. While positive cooperativity allows a rapid response to substrate influx—thus maintaining a constant substrate concentration—negative cooperativity may allow an enzyme to remain at a constant level of activity in the midst of environmental change, thus maintaining a constant rate of product formation. Additionally, negative cooperativity may allow an enzyme to effectively compete for a substrate when the substrate is in limited supply, while allowing for easy substrate turnover under ordinary conditions. Rabbit-muscle glyceraldehyde-3-phosphate dehydrogenase may represent an example of this (Levitski and Koshland, 1976). Since this enzyme is part of the glycolytic pathway, it is essential that it be able to compete for available NAD, to ensure the continual metabolism of glucose. Tenacious binding of NAD will accomplish this, but the enzyme may also bind NADH strongly, leading in some cases to kinetic inhibition due to slow dissociation of NADH. Negative cooperativity provides the flexibility required by the enzyme; while the first NAD site binds the coenzyme tightly, owing to negative cooperativity, the last NAD binds with low affinity.

The low affinity of the last coenzyme binding site (10^{-5} M) allows for easy turnover under normal, ambient levels of NAD, thereby avoiding kinetic inhibition. However, under circumstances of diminished NAD levels, the high affinity (10^{-11} M) of the first site ensures the presence at the active site of the coenzyme for carrying out the enzymic reaction. Thus the rate of oxidation of glyceraldehyde-3-phosphate is less subject to variation in NAD concentration than it would be if this enzyme displayed classical Michaelis–Menten kinetics for this substrate. Recently, negative cooperativity and half-of-the-site reactivity have been considered by Levitski and Koshland (1976). They have proposed that cooperativity is a device for enhancing or dampening responses of the enzyme.

C. Modulation of Enzyme Activity by Binding of Small Molecules to Regulatory Sites

Modulation of enzyme activity by substrate and product, as discussed above, involves interaction at the catalytic site of the enzyme. Inhibitors used in in vitro experiments also often act at the catalytic site, bearing a chemical resemblance to substrate or product and competing with substrate for the enzyme. In contrast, enzyme activity may be regulated by compounds which are chemically unrelated (allosteric) to the substrate and which are assumed to act at specific enzyme sites that are distinct from the catalytic site (Monod et al., 1963). This phenomenon has become known as allostery, and the regulatory site at which the ligand acts is called the allosteric site. The discrete nature of these sites may often be shown with agents which nullify the allosteric control, while leaving catalytic activity intact. In the case of *E. coli* asparate transcarbamylase (see below), the allosteric site has been shown to reside on a polypeptide distinct from that which contains the active site.

The ability of an enzyme to respond to concentrations of metabolites other than its substrate and product adds a new dimension to metabolic regulation. It allows the end product of the metabolic pathway to bring about feedback inhibition on earlier steps (Yates and Pardee, 1956). Such feedback control may be exerted by a metabolite several steps removed in a pathway or by metabolites from a different pathway which share a common intermediate with the first pathway. This regulability in strategically located enzymes can have profound effects on cellular metabolism. It allows certain key intermediates in one pathway to act as switches for another pathway; for example, citric acid can act as the switch for fatty acid metabolism. Regulation by central intermediates—e.g., adenine and pyridine nucleotides—may in fact determine the resultant direction of metabolism as anabolic or catabolic, depending on the energy reserves or redox state of the cell (see metabolite ratios below).

Allosteric modifiers may act as activators or inhibitors, originally being conceived as changing the enzyme's affinity for its substrate (K-type modifiers), or the maximal catalytic rate of the enzyme (V-type modifiers). Such a simplistic view no longer adequately accounts for the simultaneous effects on binding and catalytic efficiency observed with regulable enzymes (Sanwal et al., 1971). For cooperative enzymes, allosteric modifiers may profoundly affect the response of the enzyme to its substrate. An allosteric activator affects an enzyme showing positive cooperativity in the same manner as its substrate. At the maximal concentration of the activator, all of the catalytic sites on the enzyme may be converted to the most active form, and a hyperbolic substrate–velocity profile results. At an intermediate concentration of activators or inhibitors, the sigmoidal velocity profile is shifted to the left or right, respectively. Thus, the $S_{0.5}$, which is exquisitely controlled by cooperative enzymes, may be tuned to different metabolite levels by allosteric modifiers. Since the entire dynamic range of these enzymes lies in a small region about the $S_{0.5}$, an allosteric modifier may in fact switch the enzyme "on" or "off" by moving this point sufficiently below or above the existing substrate concentration. And if the affected enzyme occurs early in a pathway, the entire pathway may be controlled, determining the direction of substrate disposition. It is not surprising that many of the most effective regulable enzymes—e.g., phosphofructokinase—demonstrate both cooperativity and allosteric regulation (Mansour, 1972; Frieden et al., 1976).

D. Kinetics of Interacting Enzyme Sites

The capacities for cooperativity and allosteric regulation elevate enzymes from the level of simple catalysts to that of the regulators of metabolism. In fact, cooperativity and allosteric regulation share a common mechanism—the alteration of the properties of the catalytic site by binding of a ligand to a second site on the enzyme. Cooperativity may be thought of as the homotropic interaction of identical catalytic sites, and allostery as the heterotropic interaction of a catalytic site and a dissimilar site which binds an allosteric modifier.

The details of the mechanisms by which spatially distinct loci on a protein may interact are as yet unknown, but the process is undoubtedly initiated by a localized conformational change which accompanies the binding of ligands. This change may be thought of as an "induced fit" (Koshland, 1970), i.e., a rearrangement of the binding site into a conformation which does not exist in the absence of the ligand, or as the stabilization by the ligand of one particular conformation from many possible conformations. In any case, techniques such as X-ray crystallography (Dickerson, 1972), magnetic resonance (Metcalfe, 1970), and hydrogen exchange (Englander, 1975) have successfully documented that the conformations of enzyme–ligand complexes differ from those of the free enzyme.

Interactions between enzyme sites result in the appearance of kinetically different forms of the catalytic site. The first kinetic treatment of the effects of this phenomenon was not done for an enzyme but for the sigmoidal relationship between O_2 tension and its binding by hemoglobin. A.V. Hill (1913) suggested that Hb was an oligomeric protein and that binding of O_2 by one Hb subunit could increase the affinity of the remaining sites. If this process continued, the concentration of partially saturated molecules would be negligible, since the increased affinity at each step in saturation would ensure rapid conversion to the next, more saturated form. In effect, Hb with n binding sites would bind oxygen—n molecules at a time—with an overall equilibrium constant of

$$K' = \frac{[\text{Hb}(\text{O}_2)_n]}{[\text{Hb}][\text{O}_2]^n}$$

For the entire system, the fractional saturation

$$\overline{Y} = \frac{[\text{O}_2]^n}{K + [\text{O}_2]^n}$$

where K is the association constant ($1/K'$). This may be rearranged into

$$\log\left(\frac{\overline{Y}}{1-\overline{Y}}\right) = n \cdot \log[\text{O}_2] - \log K$$

which predicts that a log/log plot will give a straight line with a slope equal to the number of interacting sites (n). In fact, the data do not describe a straight line, rather another sigmoid curve. At very high and very low O_2 concentrations, Hb binds oxygen molecules one at a time. Further, the maximum slope is found to be 2.8 rather than 4, which is known from later structural and equilibrium binding studies.

While Hill's model has proved to be incorrect, the method of plotting remains a useful tool (Cornish-Bowden and Koshland, 1975). For binding of ligands, the plot may be used directly, and for enzymes where velocity is limited by the binding of substrate (rapid equilibrium assumption), the velocity as a fraction of V_{max} may be substituted for the fractional saturation. The maximal slope of the plot, which occurs at $S_{0.5}$, has become known as the Hill coefficient and is a measure of the number of binding sites and the degree of interaction among them. For positively cooperative enzymes, this coefficient is greater than one, and for classical Michaelis–Menten enzymes it is equal to one. Enzymes with negative cooperativity give slopes of less than one over some region of the curve.

The recognition of conformational changes in proteins has led to the development of theories of cooperativity and allostery grounded in this fundamental phenomenon. Regulable enzymes are often composed of subunits, with the conformational changes in one unit being communicated to another. However, the subunit structure is not an indispensable feature. For example, Ainslie et al. (1972) have described a slow transition between two or more different conformational forms, giving rise to nonclassical kinetic behavior in a monomeric enzyme. Since the two contending theories which attempt to explain cooperative phenomena have been well reviewed (Koshland, 1970; Kirschner, 1971; Wyman, 1972; Laidler and Bunting, 1973; Hammes and Wu, 1974), they will be discussed only briefly. The symmetry model of Monod et al. (1965) is based upon the postulate that at least two symmetrical conformational states of a protein exist (R and T) in equilibrium in the absence of ligand, the equilibrium favoring the T state. Positive cooperativity is explained by conversion of the T and the R state (to which the substrate binds preferentially or even exclusively), so that the equilibrium between the two states is altered in the direction of the preferred substrate binding state (i.e., the R state). The postulate embodies a conservation of symmetry in changes of protein structure, with the ligand having all or no effect on protein conformation. This model, while representing a major step in understanding protein cooperativity, is unable to reconcile negative cooperativity.

The simple sequential model of Koshland et al. (1966) requires the existence of only one state of protein in the absence of ligand. Symmetry is not conserved during ligand binding, since the site to which the ligand is bound is present in a different conformational state from the remaining sites. The ligand-saturated site interacts with the remaining binding sites in such a way that the microscopic binding constants are either increased (positive cooperativity) or decreased (negative cooperativity). The occurrence of negative cooperativity is consistent only with the assumption of several classes of independent binding sites and precludes the operation of the symmetry model (Paulus and DeRiel, 1975; Goldbeter, 1976). The distinction of the sequential model from the symmetry model is the existence in the sequential model of states in which some of the sites exist in conformations different from others. The substrate velocity profiles of some enzymes are best explained by a combination of positive and negative cooperativity, consonant only with the sequential model.

E. Modulation by Metabolite Ratios

Considerable investigative effort has been directed to the role of metabolite ratios and multiple metabolites in enzymic control. The complex nature of such control is exemplified by studies of oscillatory responses in the glycolytic pathway in both intact yeast cells and bovine heart cytosol. This subject has been considered by Higgins (1967), Hess and Boiteux (1971), Walter (1972, 1974), and Goldbeter and Caplan (1976) and is discussed in Chapter 8 of this volume.

Regulation by variation of the adenine nucleotide pool or the NAD/NADH ratio appears to be of particular physiologic import, since the adenylate nucleotides represent the principal storage form of chemical energy, while the pyridine nucleotides are the major source of reducing potential in the cell. With the two pools linked via electron transport and oxidative phosphorylation, any perturbation which affects either of the pools may have wide-ranging consequences on diverse reactions

Modulation of Enzyme Activity

in which they are involved. Other dissociable coenzymes, such as CoA, FAD, pyridoxal phosphate, folate, and coenzyme B_{12}, form common links in other metabolic processes. Consequently, the ratios of the free and bound forms of these coenzymes can affect the equilibrium of the reactions in which they are involved, as well as the metabolic pools which they interconnect.

1. Adenylate Energy Charge

The adenylate energy charge hypothesis championed by Atkinson (1970, 1971, 1976) is based on the premise that the ubiquitous utilization of adenine nucleotides as energy coupling agents permits them to direct metabolic sequences. The three components of the adenylate regulatory system are related by the following equation, which permits definition of the mole fraction available as ATP—the fully charged form of adenine nucleotide

$$\text{energy charge} = \frac{[ATP] + 1/2[ADP]}{[ATP] + [ADP] + [AMP]}$$

The energy charge thus represents the degree of pyrophosphorylation of the adenine nucleotides on a scale of 0 to 1, representing the total fraction of adenine nucleotides present as AMP or ATP, respectively. Under most circumstances the adenylate energy charge has been found to be close to 0.8–0.9, indicating that cells maintain their energy potential at a high level.

Atkinson proposes that it is this ratio, which denotes the extent of pyrophosphorylation of the adenine nucleotides, that plays a central role in regulating intermediary metabolism. Thus, with a low adenylate energy charge, catabolic enzymes will be activated, while biosynthetic pathways will be depressed. Alternatively, when the energy charge approaches 1, the activity of energy-generating pathways will decrease, with a concomitant increase in ATP-utilizing biosynthetic pathways. At the enzymic level, these relationships are embodied in two classes of enzyme: namely ATP-utilizing (U) type, and ATP regenerating (R) type. Enzymes of the (R) type—such as phosphofructokinase, pyruvate kinase, and other enzymes that lead to regeneration of ATP—decrease in activity with increased energy charge, while enzymes of the (U) type, e.g., aspartokinase, increase markedly in activity as the energy charge increases. The energy charge will influence enzyme activity if one or more of the adenine nucleotides is a substrate, product, or allosteric effector of that enzyme. For example, for phosphofructokinase, the energy charge affects the level of substrate (ATP), product (ADP), allosteric inhibitor (ATP), and allosteric activator (AMP). Such changes can alter the influence of other modifiers e.g., citrate on catalytic activity. Purich and Fromm (1973) have challenged this postulate for not considering other key physiologic factors, including pH, divalent cation concentration, and reaction product, as well as interaction of the adenine nucleotide system with other cellular nucleotides, e.g., the GTP/GDP/GMP system. Masters (1977) sounding a note of caution regarding the overambitious framing of unifying hypotheses of metabolic regulation, has proposed that in vivo conditions differ from in vitro, and that studies should attempt to reflect the microenvironment of the cell. In this context, he emphasizes the role of "median metabolites," i.e., substrates occurring in the midst of a pathway which have regulatory significance as well.

Despite these potential shortcomings, the adenylate energy-charge concept provides a useful frame of reference for evaluating the interaction of the adenine nucloetides with specific enzymes and entire metabolic sequences. Atkinson (1976) has argued that simultaneous analysis of the effect of the pool members on a metabolic sequence is more representative of the true in vivo control situation and that varying one parameter at a time is physiologically invalid. He points to the relative invariance of the ratios of the adenine nucleotide pool members as indicating that control resides in the ability of enzymes by binding at catalytic and regulatory sites to sense the ratio of the pool components, rather than pool size. In support of this, Chapman and Atkinson (1973) and Chapman et al. (1976) demonstrated that the activity of adenylate deaminase increased markedly as the size of the adenylate charge decreased from the normal physiological range of 0.9 to 0.5. This effect has been interpreted as a protective one, in which a large decrease in the adenylate charge is avoided, since removal of AMP results in the synthesis—by adenylate kinase—of ATP and AMP from ADP. The authors interpret these results as further confirmation of the primacy of the adenylate charge as a regulatory mechanism, its integrity being protected at the expense of the adenylate pool.

2. The NAD/NADH System

The NAD/NADH couple, as the major source of reducing potential, occupies a position of preeminence akin to that of the adenine nucleotides. Dehydrogenase, catalyzing freely reversible reactions, (e.g., alcohol and lactate dehydrogenases), are subject to rate and direction control by the NAD/NADH ratio. Williamson et al. (1967) proposed a means of determining the NAD/NADH ratio in various intracellular compartments by measuring substrate/product ratios in freeze-clamped tissues that have dehydrogenases that catalyze a reaction at or near equilibrium and are localized to only one subcellular compartment. Application of this methodology has revealed that hepatic mitochondria preserve the NAD/NADH ratio 50- to 100-fold in a more reduced state than that of the cytosol (Williamson et al., 1967; Garber et al., 1972). When animals are starved, the mitochondrial NAD/NADH ratio becomes even more reduced, without a change in the cytosolic ratio (Williamson et al., 1967).

In a liver perfusion system, addition of β-hydroxybutyrate results in a deviation in the mitochondrial NAD/NADH ratio, with this deviation accelerating the rate of gluconeogenesis from lactate or alanine (Garber et al., 1972). The authors propose that the shift in the mitochondrial NAD/NADH ratio should force the mitochondrial malate–oxaloacetate equilibrium in the direction of malate. This will facilitate malate transport into the cytosol, where its conversion to oxaloacetate is encouraged by the more oxidized NAD/NADH ratio of the cystosol. The reducing equivalents furnished should promote the reduction of the 1,3-diphosphoglycerate to glyceraldehyde-3-phosphate, an essential step in reversing the glycolytic flux, directing it to glucose synthesis.

F. Modulation by Protein–Protein Interaction

The intracellular environment may contain individual proteins at concentrations in the milligram-per-milliliter range (Srere, 1967, 1970; Segel, 1975). Considering that the number of different proteins within cells is in the thousands, the total protein concentration must be in the neighborhood of 100 mg/ml (Segel, 1975). Under such conditions, there may take place many protein–protein interactions which cannot be duplicated in in vitro studies. Such interactions undoubtedly play a role in membrane function, subcellular localization of enzyme activity, and membrane transport in cellular membranes. Multienzyme complexes, enzyme polymerization, and specific interactions that engender specialized catalytic functions not present in the isolated enzyme are other phenomena which depend upon protein–protein interactions.

Formation of large macromolecular aggregates through polymerization is a special instance of subunit interaction of enzymes. While the metabolic significance of polymerizing enzyme systems is uncertain because of difficulty in duplicating the in vivo environment of the enzyme, Frieden (1971) has postulated that polymerization evolved late, since it is characteristic of higher vertebrates. For example, acetyl-CoA carboxylase from chicken liver (Lane et al., 1974) undergoes polymerization caused by addition of citrate or phosphate and depolymerization caused by removal of these activators or by addition of malonyl-CoA or palmityl-CoA. Increased and decreased enzyme activity is correlated with these changes. Thus, in this instance, polymerization appears to have physiologic significance and provides an easily understood basis for control of acetyl-CoA carboxylase, which initiates fatty acid synthesis. Cytosolic citrate acts as both a source of acetyl-CoA and some of the NADH required for this biosynthesis, while palmityl-CoA is the major product.

1. Protein Modifiers

In this situation, two different polypeptide chains interact to form a new specific complex, whose biologic activity can be modified by ligands that bind to the precursor subunits. This type of regulation has been extensively studied for the aspartate transcarbamylase of *E. coli* (Gerhart, 1970). This enzyme catalyzed the initial step in the synthesis of cytidine nucleotides; it is allosterically inhibited by CTP and shows positive cooperativity for substrate. It may be dissociated by mercurials into catalytic subunits, which are insensitive to CTP and which exhibit hyperbolic kinetics for substrate, and into regulatory subunits which bind CTP.

Recently, studies in mammalian systems on the 3,'5'-cAMP-activated protein kinases (Rubin and Rosen, 1975; Swillens and Dumont, 1976) have disclosed a new variant of the subunit model. The cAMP-dependent protein kinase that converts phosphorylase b to a, and glycogen synthase from I to D is composed of two subunits, one catalytically active and the other inhibiting the first. Cyclic AMP binds to the inhibitory subunit, causing its dissociation from the catalytic subunit, thereby permitting expression of the enzyme activity.

2. Protein Specifiers

The protein kinase system exhibits characteristics of enzymes with interacting subunits—exemplified by aspartate transcarbamylase—and of systems where a protein itself modifies the biologic activity of another protein without the mediation of a metabolite or small ligand. The lactose synthetase system of the mammary gland is the best example of such a situation in which a "specifier" protein changes the enzymatic specificity of the precursor protein to which it binds. In this situation, N-acetyl glucosamine–galactosyl uridyltransferase, which is widely distributed, synthesizes the glycoprotein precursor N-acetyl lactosamine (Ebner, 1973) according to the following equation:

$$N\text{ - Acetyl glucosamine} + UDP-Gal \underset{}{\overset{Mn^{2+}}{\rightleftharpoons}} N\text{ - Acetyl lactosamine} + UDP$$

However, in the lactating mammary gland this enzyme associates with milk protein, α-lactalbumin, forming a complex, denoted lactose synthetase, which instead catalyzes the formation of lactose from glucose and UDP-galactose.

$$\text{glucose} + \text{UDP-Gal} \underset{}{\overset{Mn^{2+}}{\rightleftharpoons}} \text{Lac} + \text{UDP}$$

The components of the lactose synthetase complex are freely dissociable, the α-lactalbumin lowering the apparent K_m for glucose and other monosaccharides as compared with the K_m observed for the uridyltransferase alone. The function performed by lactalbumin—of enhancing the substrate binding for a reaction, (with the substrate in this instance being glucose)—is reminiscent of that of metals in certain enzyme systems (Hill and Brew, 1975). Recently, Powell and Brew (1976) have shown that prior attachment of Mn^{2+} to the enzyme is necessary for its binding of α-lactalbumin.

A modifier–specifier role for proteins, wherein a new biologic function or enhanced activity is achieved, may well be more common than is presently appreciated. For example, factors V and VIII in the blood coagulation process appear to act in this way (Davie et al., 1979), as do specific protein initiation factors which participate in protein synthesis (Weissbach and Ochoa, 1976).

3. Multienzyme Complexes

The specific aggregation, in constant relative proportion, of several enzymes which catalyze sequential or related reactions in a metabolic pathway to form a conglomerate of enzymes is termed a multienzyme complex. Examples of such complexes in mammalian tissues include pyruvate dehydrogenase, α-ketoglutarate dehydrogenase, fatty acid synthetase, and components of the early steps of pyrimidine biosynthesis, as well as the glycine cleavage system (Ginsburg and Stadtman, 1970; Reed, 1969,1974; Shoaf and Jones, 1973; Volpe and Vagelos, 1976; Kikuchi, 1973).

Many reactions carried out in intermediary metabolism require a temporal and spatial contiguity of substrates, products, and various enzymes. Atkinson (1969), considering the solvent capacity of the cell, has questioned the availability of sufficient cellular water to act as a solvent for these multifarious reactions. One solution to this problem is the existence of multienzyme complexes which permit localization of the intermediates within the complex. Under these circumstances, the concentration of reactants in the environment of the active site could be quite high, while the concentration of reactants in the particular compartment could simultaneously be maintained at very low levels. Within the multienzyme complex, transfer of the substrate from one active site to the next is markedly facilitated. In addition, multienzyme complexes, by sequestering reactants, obviate the possibility of a highly reactive product of one enzyme diffusing through the intracellular space and being consumed by abortive site reactions.

It has been proposed (Ackrell, 1974) that such complexes may be much more common than present data suggest, since available methods of subfractionation would permit only the most stable of such highly organized species to survive. The rapidity with which individual molecules traverse the glycolytic or gluconeogenic pathway suggests a high level of organization of the sequential steps in these pathways (Exton and Park, 1967). Indeed, Atkinson (1969) has suggested that a major function of the extensive membrane matrix permeating the cytoplasm may be to provide a surface for enzyme attachment, preventing their precipitation and the consequent loss of the reactive sites. In addition, such organization would explain the rapidity of substrate transformation which has been observed in a number of organized pathways.

4. Membrane-Bound Enzymes

In many cases, an enzyme protein appears to form an integral part of the structure of a membrane, requiring interactions with other membrane components for full biologic activity. As aspects of this topic are dealt with in two chapters in this monograph, we shall not treat it any further here.

G. Other Regulatory Phenomena

1. Regulation by Hydrogen-Ion Concentration

The H^+-ion concentration has myriad functions within the cell. It acts, for example, as a substrate or product of dehydrogenases and as a factor affecting the ionization of groups which comprise the active site or stabilize its conformation. Substrate ionization may also be responsive to changes in $[H^+]$. A discussion of the manner in which one may analyze the effect of pH on both the apparent K_m and the apparent V_{max} is available in Dixon and Webb (1964). Recently, Knowles (1976) has evaluated the nature of the experiments that have hitherto been used to assess the relationship of pH optima to putative sites involved in the catalytic activity of a particular enzyme. His review defines limits to overambitious interpretations and extrapolations regarding participation of particular groups in the reaction mechanism.

Intracellular pH has been measured by several methods. Data obtained for mammalian liver indicate that intracellular pH approximates 7.2 (Walker et al., 1969); however, as with other metabolites, the distribution of hydrogen ions is probably neither temporally nor spatially invariant. It is particularly noteworthy that relatively few enzymes evince pH optima that concur with intracellular pH as presently measured. This discrepancy may reside either in the response of the isolated enzyme to a different pH than that which characterizes the in vivo environment. On the other hand, the actual pH of the enzymic microenvironment may differ from that measured in the total incubation.

2. Temperature as a Possible Regulatory Mechanism

It is unlikely that under normal conditions, temperature will exercise any role in modulating enzymic activity. However, temperature effects on enzymes may achieve some importance in poikilothermic species and possibly also during the decrease in body temperature that occurs during hibernation. More importantly temperature may have been an important factor in determining the nature of binding interactions for specific enzymes which have evolved under different thermal environments, a fascinating possibility which has been elaborated by Somero and Low (1976).

3. Inorganic Cations

Divalent cations, univalent cations, or both are essential cofactors for a large number of enzymes. Kinases as a class, for example, share the requirement for such cations, while in other instances, other metal ions are inhibitors of metal-activated enzymes, e.g., activation by Mg^{2+} and inhibition by Ca^{2+} for many kinases and synthetases. Mildvan (1970) has reviewed the models that have been proposed to account for activation (or inhibition) of enzymes by metal ions. The "substrate bridge" and "metal bridge" models conceive of the metal ions either combining with the substrate to form a chelate or interacting with the enzyme to complete the required binding site. These complexes usually involve the active site, but Schramm (1974) has demonstrated activation of AMP nucleosidase by $MgATP^{2-}$ at a modifier site instead.

Of course, as with other ligands, metal ions may interact at sites distinct from those occupied by substrates or modifiers and may influence catalytic activity in a less direct manner. It is interesting in this regard that metal-ion activation does not usually display the kinetically higher-order initial rate profile considered characteristic of most allosteric modifiers (Barden and Scrutton, 1974).

Research efforts over the last few years have culminated in a growing awareness of the role of calcium in cell communication, a function which calcium shares with the cyclic nucleotides. This topic has been recently reviewed by Rasmussen and Goodman (1977) and the reader is referred to this review for an extensive discussion of systems shown to be responsive to calcium, as well as for an attempt to frame a set of general principles regarding informational transfer at the cell periphery.

4. Inorganic Anions

Little attention has been paid to modulation of enzymic activity by interaction with inorganic ions, with the possible exception of enzymes utilizing and producing PO_4. The selectivity for anions by biological systems has been recently reviewed (Wright and Diamond, 1977).

H. Summary of Noncovalent Regulation of Enzyme Activity

Of the mechanisms evolved to regulate cellular metabolism at the enzyme level, noncovalent interactions are perhaps the most useful. Owing to the easy reversibility of interactions, this mode of regulation is characterized by rapid onset and decay, with minimal expenditure of energy. We have only briefly considered substrate and product control on enzymes displaying classical Michaelis–Menten kinetics, since the changes in substrate or product levels needed to effect a significant change in enzyme velocity by this mechanism are generally not consonant with cellular ability to maintain metabolite concentrations within small tolerances. Furthermore, except in the case of "autosteric" inhibition, the regulation is exerted only by immediate reactants, precluding communication among interlocking pathways.

Because of the flexibility of the catalytic site, the conformational changes which accompany noncovalent interactions with ligands and other proteins permit finer control by a wider range of stimuli. Binding of ligands to an allosteric site on the enzyme permits regulation of the enzyme by metabolites several steps removed in a metabolic pathway, or by a different but interlocking pathway. Control may be specific, as in feedback inhibition of an early enzyme in a synthetic pathway, or general, as in enzyme response to cofactors used in many reactions central to cellular metabolism. A refinement of the latter phenomenon occurs when an enzyme responds in an opposite manner to two interconvertible forms of a cofactor. In such case, the enzyme is effectively responding to the metabolite ratio of different cofactor forms, which may more accurately reflect the cell's metabolic state than the level of any single cofactor form.

While the velocity of enzymic reactions ultimately depends upon substrate concentration, cooperativity modifies the sensitivity of an enzyme to this factor. Negative cooperativity decreases an enzyme's sensitivity to substrate; since the velocity of the enzymic reaction does not increase appreciably over a range of intermediate concentrations, its excess substrate is free to be shunted off into another cellular compartment or metabolic sequence. On the other hand, positive cooperativity ensures that an enzyme will respond rapidly to changes in substrate influx over a narrow range of concentrations about the $S_{0.5}$, yet be virtually inactive below this range. Because of this increased sensitivity, allosteric modification may switch such enzymes "on" or "off" at a given substrate level, again effecting a redirection of metabolism.

The kinetic form of an enzyme may also be changed by interaction with other proteins. Such interaction may modify the enzyme's affinity for substrate, its velocity, or its capacity for regulation by allosteric modifiers. In the case where a protein modifier lowers the K_m for a normally poor substrate to such an extent that it becomes the most favored substrate the modifier may be called a *specifier* for the new reaction. In polymerization, the interacting protein is an identical molecule of enzyme, with the polymer differing in kinetic properties from the monomer.

As the association and dissociation of proteins takes place over a longer time scale than that of the binding of small ligands, there is some hysteresis built into interactive regulation. However, allosteric modification of the enzyme or modifier, subsequent to ligand binding, often triggers these associations and disassociations. Thus, protein–protein interactions are ultimately governed by the products of intermediary metabolism.

III. Covalent Regulatory Mechanisms

As considered in this chapter, covalent regulatory mechanisms involve reversible or irreversible conversion between different forms of an enzyme. This interconversion results from covalent modification and is usually catalyzed by another enzyme. Such effects operate over a longer time scale than the noncovalent mechanisms described previously and provide a means of modulating the noncovalent regulatory mechanisms, allowing for additional levels of control. Thus, an enzyme may be stored in an active state, protecting the cell, or regulatory signals may be amplified and transmitted to other pathways. An example of the latter is the activation of adenylate cyclase by a variety of hormones, leading to a physiologic response (Chapter 14). The amplification mechanism operates via a cascade which utilizes a minimal initial stimulus, often emanating from binding to a receptor on the periphery of the plasma membrane. A further distinction of covalent modification is that little energy is required for chemical modification of a preexisting enzyme in contrast to the energy requirement for de novo synthesis of the same enzyme (Holzer and Duntze, 1971).

A. Covalent Modification by Irreversible Interconversions

Generation of an active enzyme from an inactive enzyme usually involves proteolytic modification of the precursor by excision of an amino acid or a small peptide. This type of mechanism is primarily important in the gastrointestinal tract, where it enables proteolytic enzymes to be stored and released in an inactive state, as zymogens, to be activated at the appropriate instant. Analogous reactions are involved in a great variety of biological processes, such as hormone production, fibrinolysis, coagulation of blood, and the complement reaction.

It appears that limited proteolysis is the last step in the synthesis of many biologically active proteins and is probably the first step in protein degradation. Limited proteolysis appears to be directed toward surface loops and random segments of polypeptide chains rather than toward internal-domain helices or pleated sheets (Neurath and Walsh, 1976).

Recent studies indicate that many zymogens possess some slight, inherent enzymic activity (Stroud et al., 1977). Investigation of the activation process, by kinetic and spectroanalysis of the zymogen, indicates that the catalytic apparatus of trypsin, chymotrypsin, and other serine proteases largely preexists in the zymogen forms, and that, during activation, the effectiveness of the binding site is improved more than 1000-fold. Regulation by proteases has recently been considered by Neurath and Walsh (1976), the reasons for inactivity of the zymogen by Stroud et al. (1977), and mechanisms in blood coagulation by Davie et al. (1979).

B. Modification by Reversible Covalent Action

While the examples considered above of enzymic interconversions in extracellular fluids are irreversible, covalent modification of intracellular enzymes is usually reversible. Two major mechanisms of reversible covalent modification are known: phosphorylation–dephosphorylation, which we will consider further, and adenylation–deadenylation, which is a microbial system. Holtzer (1969) and Segal (1973) have reviewed this topic.

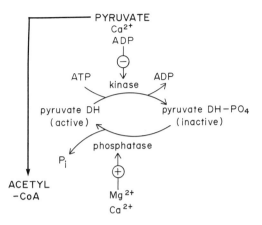

Figure 2. Pyruvate dehydrogenase multienzyme complex. The component of the pyruvate dehydrogenase multienzyme complex which undergoes reversible phosphorylation/dephosphorylation is E_1, pyruvate decarboxylase. In this manner the activity of the entire complex is regulated.

1. Modification by Phosphorylation

Four examples of this mechanism have been described, and all involve the formation of specific seryl-O-phosphate moieties on the protein. The systems involved are pyruvate dehydrogenase, triglyceride lipase, glycogen phosphorylase–glycogen synthetase, and acetyl-CoA carboxylase. The persistence of the regulatory effects, even after the triggering effector has disappeared, represents an important difference between regulation by chemical modification and physical or noncovalent modification, the latter mechanisms causing an immediate response of the controlled enzyme to changes in the concentration of specific effectors.

a. Pyruvate Dehydrogenase. This system is an example of a multienzyme complex catalyzing the conversion of pyruvate to acetyl-CoA, one of the metabolites most widely employed by the cell. The complex is composed of three groups of enzymes: pyruvate decarboxylase (E_1), lipoyl reductase–transacetylase (E_2), and dihydrolipoyl dehydrogenase (E_3). The complex is characterized by an association of the component enzymes in a fixed stoichiometry, by noncovalent binding forces (Hucho, 1975).

Besides the three enzymes described, the mammaliam complex contains a kinase and a phosphatase (Fig. 2), which are responsible for phosphorylation and dephosphorylation of the decarboxylase (E_1) (Reed, 1969, 1974). In adipose tissue, the active dehydrogenase content is increased after treatment with insulin and is conversely decreased by incubation with epinephrine (Denton et al., 1975), the activity increase resulting from a larger proportion of the enzyme complex being in the active nonphosphorylated form. Several factors, including the ATP/ADP ratio, and Ca^{2+} and pyruvate concentrations have been described as controlling the activities of the pyrvute dehydrogenase kinase and phosphatase. The kinase is inhibited by pyruvate, ADP, Ca^{2+}, Mg^{2+} and CoA and is activated by NADH and acetyl-CoA. Kerbey et al. (1977) and Hansford (1977) have shown that a decrease in activity of the pyruvate dehydrogenase complex is brought about by phosphorylation, stimulated by increased ratios of either acetyl CoA/CoA or NADH/NAD. Thus, the suggestion by Garland and Randle (1964) that fatty acid inhibition of pyruvate oxidation is brought about by an increase in the acetyl-CoA/CoA ratio has been amply confirmed by these more recent studies.

b. Triglyceride Lipase in Adipose Tissue. Hormonal and nervous stimulation exercise decisive control over the rate of lipolysis in adipose tissue. Activation of triglyceride lipase by protein kinase-catalyzed phosphorylation has been shown by Huttunen and Steinberg (1971), and inactivation of the lipase, by phosphoprotein phosphatase from a variety of tissues and species, has been demonstrated by Severson et al. (1977). Some doubt exists regarding the physiologic significance of the cycle (Fig. 3) in the regulation of lipolysis, since the level of epinephrine necessary to maximally stimulate lipolysis engenders only a meager increase in the cAMP level and in the protein kinase activity expressed in the absence of added cAMP (Soderling et al., 1973). Moreover, while dibucaine blocks epinephrine-induced lipolysis in isolated adipocytes, it nonetheless causes an increase in cellular cAMP, whether or not epinephrine is present (Siddle and Hales, 1974). The role of tri-

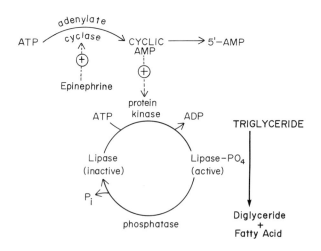

Figure 3. Adipose tissue triglyceride lipase. + denotes activation.

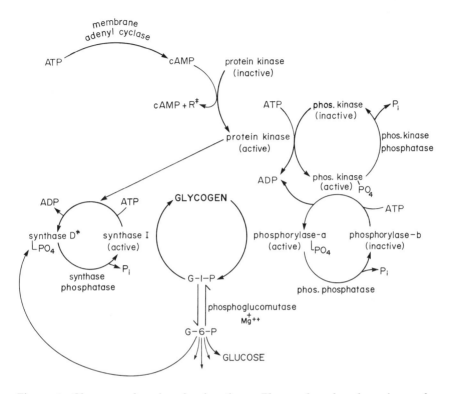

Figure 4. Glycogen phosphorylase/synthase. Phos., phosphorylase; ‡, regulatory subunit of protein kinase (cf, protein modifiers, this chapter, Section II.F.1.); * denotes that degree of phosphorylation is variable, providing for a continuum of activity in the presence of G-6-P; L_{PO_4} indicates this form of the enzyme is phosphorylated.

glyceride lipase in fat transport and metabolism has been recently reviewed (Steinberg and Khoo, 1977).

c. *Glycogen Phosphorylase/Glycogen Synthetase.* Cori et al. (1943) demonstrated interconvertible forms of muscle glycogen phosphorylase which were subsequently found to differ in phosphate content, response to various metabolic effectors (specifically AMP), and the position of the dimer–tetramer equilibrium under various conditions (Fischer et al., 1971).

The interconversion of phosphorylated (active "*a*" form) and dephosphorylated (less active "*b*" form) forms of the enzyme is mediated by a kinase and a phosphatase, respectively (Fig. 4). An additional level of control encountered here is that the kinase and phosphatase are also interconvertible between active and inactive species.

The protein kinase (initially called phosphorylase kinase kinase) (Krebs, 1972; Rosen et al., 1977), which phosphorylates the phosphorylase kinase, initiates glycogenolysis through its activation by cAMP. Thus, stimulation of adenylate cyclase by epinephrine or glucagon (in liver), which causes elevation of intracellular cAMP, results in an increased rate of glycogen breakdown via sequential phosphorylations of phosphorylase kinase and phosphorylase. This enhancement of glycogenolysis is coordinated with a decreased rate of glycogen synthesis, since glycogen synthetase also exists in interconvertible phospho- and dephospho- forms. Predictably, it is the dephosphorylated form which is the active species under physiological conditions.

While liver phosphorylase b has little activity under usual conditions, skeletal muscle phosphorylase b possesses considerable activity with adequate concentrations of AMP. In resting aerobic muscles, phosphorylase b is virtually without activity in the presence of normal ambient levels of ATP and G-6-P, which inhibit the enzyme. Anoxia stimulated glycogenolysis by enhancing phosphorylase b activity rather than conversion to phosphorylase a (Morgan and Parmeggiani, 1964). It seems that the essential ingredient is the decrease in the ATP/ADP ratio, which relieves the ATP inhibition of both the dephosphoenzyme and phosphofructokinase. However, stimulation of muscle glycogenolysis by epinephrine or electrical stimulation is brought about by the conversion of b to a, the active form of the phosphorylase. Hormonal and nervous control thus override metabolic control, thereby permitting a response of the organism to a situation requiring rapid energy utilization

Glycogen synthase has been conceived of as existing in two forms, distinguishable by their degree of phosphorylation and requirement of G-6-P for activity. The dephosphorylated, or "I" form, is independent of G-6-P, while the maximally phosphorylated "D" form is completely dependent on G-6-P. Roach et al. (1976) and Roach and Larner (1976) have provided evidence for the existence of glycogen synthase in forms varying in their degree of phosphorylation and which likewise vary in their dependence on G-6-P for expression of catalytic activity. Hormonal stimulation by epinephrine causes further phosphorylation of the glycogen synthase, converting it to a more G-6-P-dependent form, while insulin causes dephosphorylation, making the enzyme less dependent. Thus the effectiveness of the control of the enzyme by metabolites may be modified in a continuum by hormonal influences. Such a spectrum of forms permits finer control of enzyme activity than the previous, limited view of "I" and "D" forms. Saugmann (1977) has also found, in human leukocytes, a glycogen synthase whose properties differ from those of the "I" and "D" forms. He has called the form "R" (rheostatic potential) and it is formed from the "D" form by partial dephosphorylation.

The enormous flexibility and amplification achievable by enzymes involved in a cascade of activation—and which are also sensitive to allosteric regulation—has been the focus of a theoretical analysis by Stadtman and Chock (1977) and Chock and Stadtman (1977). They emphasized the dynamic aspect of the interconversion process, providing for a spectrum of specific activity of modifiable enzymes rather than an all-or-none response. Their analysis is amply borne out by the recent data on glycogen synthase.

d. Acetyl-CoA Carboxylase. This enzyme, which is the first committed step in fatty acid synthesis, catalyzes carboxylation of acetyl-CoA to form malonyl-CoA. It is known to be regulated by a number of metabolites and long-chain fatty-acyl-CoA (See Chapter 8). Carlson and Kim (1973,1974a,b) presented evidence for covalent modification of this enzyme through a phosphorylation–dephosphorylation cycle, which they proposed as an additional form of regulation. However, this contention has not been generally accepted (Lane et al., 1974). Recently Lee and Kim (1977), using antibodies prepared to a highly purified rat liver carboxylase, demonstrated the occurrence of such a cycle in association with activation–inactivation of the enzyme. Confirmation of this study will be further evidence for the multiple layers of control associated with enzymes catalyzing reactions at the crossroads of many different pathways or cellular functions.

C. Substrate Cycles

In liver and kidney cortex there are four enzymes, unique to gluconeogenesis, which are required to reverse the sequence of reactions in glycolysis. These enzymes constitute a set of energy requiring reactions that permit circumvention of the energy barrier of the glycolytic pathway.

Newsholme and Gevers (1967) have proposed that an additional form of amplification and control of a pathway may be obtained if a substrate cycle is operative in the pathway, e.g., a cycle, between fructose-6-phosphate and fructose-1,6-diphosphate, which may participate in the regulation of glycolytic flux in muscle. A cycle between triglycerides and fatty acids, which appears to control the rate of fatty acid release from adipose tissue, may also occur (Newsholme and Crabtree, 1976).

To be an effective form of regulation, substrate cycling requires that both enzymes respond to the same metabolites in opposite ways but that neither do so in an all-or-none fashion. Under circumstances of simultaneous activity of both enzymes, a countinuing cycle between the substrates of the enzymes obtains. In such a situation, sensitivity of control of the step depends on the rate of cycling being greater than the flux rate through the pathway. For example, AMP is an activator of phosphofructokinase (PFK), which is an inhibitor of of FDPase, while ATP inhibits PFK. Increased control of this step may be accomplished by utilizing the intracellular concentration of AMP to regulate this couple. Hence, low ATP and high AMP concentrations in the enzymic milieu will inhibit FDPase and activate PFK, thereby stimulating glycolysis, which would elevate the ATP concentration. When the ATP level rises and the AMP level falls, the FDPase at some point becomes more active than PFK. Then, gluconeogenesis will preponderate, again utilizing ATP.

The amount of energy lost as heat from such cycles may be minimized. The maximum possible amount of cycling is determined by the enzyme possessing the lowest maximal catalytic activity in the cycle. Since the substrate cycle ensures that either enzyme can be regulated in both the catabolic and anabolic directions, an energetic advantage accrues to the cell. By means of the cycle, the efficiency by which regulatory enzymes respond is enhanced beyond the limit imposed by equilibrium binding of small modifiers (Newsholme and Start, 1972). In addition, an important role in heat production has been shown in insect flight muscle (Clark et al., 1973), and the role of substrate cycles in heat production has been elaborated by Newsholme and Crabtree (1976), Katz and Rognstad (1976), and Stein and Blum (1978).

While substrate cycles consume energy, control implies a decrease in entropy, and this requires expenditure of energy. The synthesis and turnover of large and complex allosteric proteins with regulatory functions require energy, both at the level of RNA and protein synthesis. The functioning of a substrate cycle may serve as a sensitive regulatory mechanism whose energy expenditure may be considerably less than that required by de novo enzyme synthesis.

IV. Enzyme Synthesis and Degradation

Since the mechanisms which govern the rate of enzyme synthesis and degradation in mammalian tissue will be considered elsewhere in this volime, we will have little to remark here, save to note that, both from the point of view of energy expenditure and of time, such mechanisms are more costly forms of regulation than those we have thus far considered. Synthesis, in particular, must operate over a longer time frame than other forms of enzyme modulation and thus cannot be responsible for the rapid or fine tuning response required by intermediary metabolism. For some pathways, hormones provide long-term regulation by increasing the rate of synthesis of specific enzymes (Pitot and Yatvin, 1973). In fact, changes in pathway flux consequent upon dietary or hormonal manipulation have been shown to occur prior to increases in enzyme levels for a given pathway (Foster et al., 1966), further emphasizing the importance of rapid responses.

V. Evaluation of the Physiologic Importance of Regulatory Mechanisms

Up to this point we have considered the nature of the mechanisms which may be responsible for modulating enzymic activity, having chosen examples of mechanisms of presumed physiologic relevance. In this section we inquire into methods by which the site(s) of rate control in a given metabolic pathway can be identified.

A. Identification of Potentially Rate-Controlling Steps

1. Measurement of Relative Maximal Capacities

Krebs (1963) proposed that probable regulatory sites in a pathway could be identified by determining which enzymes exhibit a relatively low maximal capacity-to-flux ratio. This approach (Newsholme and Gevers, 1967) assumes that rate control requires a lesser degree of inhibition at such steps than at those for which a much greater excess of catalytic capacity exists in the tissue. While enzymes exhibiting a low capacity-to-flux ratio frequently are the sites of action of metabolic effectors, this is not sufficient reason to exclude an enzyme from a regulatory role because of lack of conformity to that criterion. Further, while the low excess-capacity presumption is useful, it rarely leads to unequivocal assignment of the site of regulation in the pathway. Thus hexokinase, phosphofructokinase, and pyruvate kinase fall into this category for glycolysis, while all four enzymes unique to gluconeogenesis, i.e., pyruvate carboxylase, phosphoenolpyruvate (PEP) carboxykinase, fructose-1,6-diphosphatase, and glucose-6-phosphatase, also fulfill this criterion.

2. Identification of Equilibrium and Nonequilibrium Reactions

Another approach to the identification of sites of possible metabolic control is to compare the apparent equilibrium constant (K')—determined in vitro under certain conditions (temperature, pH, and ionic strength) which approximate the in vivo environment—with the mass action ([substrate]/[product]) ratio obtained from tissue metabolite measurements (Rolleston, 1972). While no reaction in a living system can be at equilibrium, it is the extent of deviation of the mass-action ratio from K' that provides an indication of a step likely to be involved in regulation. The enzyme involved in such a step will be found to have an inequality in mass-action ratio and K', owing to its limited catalytic capacity relative to the contiguous enzyme in the pathway. Thus it should function as a bottleneck. In practice, a continuous spectrum of this relationship is encountered, ranging from the equilibrium ($K' \cong$ mass-action ratio) to the nonequilibrium ($K' >>$ mass-action ratio) situation. Although Rolleston (1972) has proposed that a difference of 20-fold or more between K' and the mass-action ratio may characterize a nonequilibrium reaction, this relationship is purely empirical.

The experimental aspects involved in collecting data to carry out the above calculations—reviewed by Rolleston (1972) and Newsholme and Start (1973)—may pose some serious difficulties, however. Often, K' may be obtained directly by measurement of the equilibrium concentration of the substrates and products of the reaction. However, if K' is very large or if one of the reactants is unstable, microcalorimetric analysis (Kitzinger and Benzinger, 1960) or calculation from kinetic data using the appropriate Haldane relationship (Cleland, 1963a; Segel, 1975) provide alternate methods for estimating this parameter. The mass-action ratio is usually obtained from measurement of metabolites in "freeze-clamped" tissue specimens. Owing to disruption in intracellular structure by "freeze-clamping," however, the value obtained does not take compartmentation—which is discussed elsewhere in this book—into account.

Despite these reservations, use of either the relative maximal capacity or the equilibrium/nonequilibrium reaction approach identifies the same enzymes as catalyzing potentially rate-controlling reactions. For example, in most mammalian tissues both methods suggest that the potential exists for control of glycolytic flux at the hexokinase, phosphofructokinase, and pyruvate kinase reactions (cf. Newsholme and Start, 1973).

3. The Crossover Theorem

The above approaches identify reactions in a pathway which might exercise a regulatory function. Proof of the actual site of rate control under specified conditions is, however, only possible when measurements are made on a pathway whose flux has been perturbed by an activator or inhibitor. Such an approach characterizes the crossover theorem of Chance et al. (1958) and was originally applied to studies of electron-transport chain. According to this theorem, one may define the locus of action of an inhibitor in a reaction sequence by finding out which step exhibits a decrease in product and an increase in substrate concentration resulting from the reduction in flux. Alternatively, an activator would bring about the opposite changes to those observed with the inhibitor. This theorem was quite fruitful when used to evaluate the components of the electron-transport chain, which do not vary in concentration over the time of the experimental observation. However, when applied to the analysis of metabolic pathways which did not share the property of conservation of intermediates, such as glycolysis and gluconeogenesis, it proved less well grounded (Rolleston, 1972). In this case, changes discerned in the substrate and product concentrations of the rate controlling step need not be reciprocal, being dependent not only on the properties of the reaction itself but also on those properties which would remove intermediates to another pathway.

VI. Conclusion: An Overview of Regulation

Any consideration of metabolic regulation must focus on those enzymes which exercise control over the flux of the pathway and the experimental means for identifying such enzymes. We have reviewed such experimental approaches, but we must now attempt to put the disparate pieces of information into some holistic approach. Kacser and Burns (1973) have questioned the general applicability of varying one factor at a time in trying to arrive at the importance of a particular factor in control and have developed instead an approach which attempts to interrelate the multiplicity of factors known to impinge on the flux of a pathway. The reader is directed to their paper for a full description of the theorem, including the mathematical manipulations involved in its application.

The point of departure for Kacser and Burns is that the flux through a pathway (flow into and out of the cellular pools) is a property of the system as a whole and cannot be examined one step at a time. Proper analysis requires investigation into the quantitative interrelationships of the component parts. The authors have defined three factors crucial to control, viz.: controllability, sensitivity, and elasticity. The first, controllability, defines the response of the enzyme in isolation to an inhibitor or perturbation. Sensitivity measures whether a small change in enzyme concentration brings about any change in the overall flux of the pathway. Finally, elasticity is the percentage change in rate brought about by a change in concentration of substrate, product, or effector. Elasticities may be evaluated by classical kinetic approaches. With the aid of a computer, the information gained from evaluating these three factors and the components that regulate them can permit development of a quantitative model of the factors which regulate a pathway under a given set of environmental circumstances, an example of which is provided by Kohn et al. (1979).

In this chapter we have sought to define levels of control within the cell and to demonstrate by suitable examples how such regulatory mechanisms may function in the context of a particular enzyme reaction or metabolic sequence. However, regulation of a single enzymic reaction can only be understood in the context of the sequence in which the reaction occurs and the overall function of that pathway. Since metabolic reactions may serve multiple functions, modulation of these reactions by a single parameter would not provide the organism with the flexibility it requires for the brisk interrelationship of simultaneously occurring reactions. A regulatory enzyme, then, is one which is capable of sensing the concentration of several metabolites and modifying its kinetic behavior accordingly. Interactions between sequences that compete for a common intermediate or supply precursors for an anabolic pathway are of particular interest in the context of metabolic regulation.

References

Ackrell, B., 1974, Metabolic functions of oxaloacetate, in: *Horizons in Biochemistry*, Vol. 1 (E. Quagliariello, ed.), Addison-Wesley, New York.

Ainslie, G.R., Shill, J.P., and Neet, K. E., 1972, Transients and cooperativity: A slow transition model for relating transients and cooperative kinetics of enzymes, *J. Biol. Chem.* **247**:7088.

Atkinson, D. E., 1969, Limitation of metabolite concentrations and the conservation of solvent capacity in the living cell, in: *Current Topics in Cellular Regulation*, Vol. 1 (B. L. Horecker and E. R. Stadtman, eds.), p. 29, Academic Press, New York.

Atkinson, D. E., 1970, Enzymes as control elements in metabolic regulation, in: *The Enzymes*, 3rd ed., Vol. 1 (P. D. Boyer, ed.), p. 461, Academic Press, New York.

Atkinson, D. E., 1971, Adenine nucleotides as universal stoichiometric metabolic coupling agents, *Adv. Enzyme Regul.* **9**:207.

Atkinson, D. E., 1976, Adaptations of enzymes for regulation of catalytic function, *Biochem. Soc. Sympos.* **41**:205.

Barden, R. E., and Scrutton, M. C., 1974, Pyruvate carboxylase from chicken liver, *J. Biol. Chem.* **249**:4829.

Carlson, C. A., and Kim, K. H., 1973, Regulation of hepatic acetyl coenzyme A carboxylase by phosphorylation and dephosphorylation, *J. Biol. Chem.* **248**:378.

Carlson, C. A., and Kim, K. H., 1974a, Regulation of hepatic acetyl CoA carboxylase by phosphorylation-dephosphorylation, *Arch. Biochem. Biophys.* **164**:478.

Carlson, C. A., and Kim, K. H., 1974b, Differential effects of metabolites on active and inactive forms of hepatic ac-CoA carboxylase, *Arch. Biochem. Biophys.* **164**:490.

Chance, B., Holmes, W., Higgins, J. J., and Connelly, C. M., 1958, Localization of interaction sites in multicomponent transfer systems: Theorems derived from analogues, *Nature* **182**:1190.

Chapman, A. G., and Atkinson, D. E., 1973, Stabilization of adenylate energy charge by the adenylate deaminase reaction, *J. Biol. Chem.* **248**:8309.

Chapman, A. G., Miller, A. L., and Atkinson, D. E., 1976, Role of adenylate deaminase reaction in regulation of nucleotide metabolism in Ehrlich ascites tumor cells, *Cancer Res.* **36**:1144.

Chock, P. G., and Stadtman, E. R., 1977, Superiority of interconvertible enzyme cascades in metabolic regulation: Analyses of multicyclic systems, *Proc. Natl. Acad. Sci. USA* **74**:2766.

Citri, N., 1973, Conformational adaptability in enzymes, *Adv. Enzymol.* **37**:397.

Clark, M. G., Bloxham, D.P., Holland, P. C., and Lardy, H. A., 1973, Estimation of the fructose diphosphatase–phosphofructokinase substrate cycle in the flight muscle of *Bombus affinis*, *Biochem. J.* **134**:589.

Cleland, W. W., 1963a, The kinetics of enzyme catalyzed reactions with two or more substrates or products. I. Nomenclature and rate equations, *Biochim. Biophys. Acta* **67**:104.

Cleland, W. W., 1963b, The kinetics of enzyme catalyzed reactions with two or more substrates or products. II. Inhibition: Nomenclature and theory, *Biochim. Biophys. Acta* **67**:173.

Cleland, W. W., 1963c, The kinetics of enzyme catalyzed reactions with two or more substrates or products. III. Prediction of initial velocity and inhibition patterns by inspection, *Biochim. Biophys. Acta.* **67**:188.

Colowick, S.P., 1973, The hexokinases, in: *The Enzymes*, 3rd ed., Vol. 9 (P. D. Boyer, ed.), p. 1, Academic Press, New York.

Cori, C. F., Cori, G. T., and Green, A. A., 1943, Crystalline muscle phosphorylase, *J. Biol. Chem.* **151**:39.

Cornish-Bowden, A., and Koshland, D. E., 1975, Diagnostic uses of the Hill (Logit and Nernst) plots, *J. Mol. Biol.* **95**:201.

Davie, E. W., Fujikawa, K., Kurachi, K., and Kisiel, W., 1979, The role of serine proteases in the blood coagulation cascade, *Adv. Enzymol.* **48**:277.

Denton, R. M., Randle, P. J., Bridges, B. J., Cooper, R. H., Kerbey, A. L., Pask, H. T., Severson, D. L., Stansbie, D., and Whitehouse, S., 1975, Regulation of mammalian pyruvate dehydrogenase, *Molec. Cell. Biochem.* **9**:27.

Dickerson, R. E., 1972, X-ray studies of protein mechanisms, *Annu. Rev. Biochem.* **41**:815.

Dixon, M., and Webb, E. C., 1964, *Enzymes*, Academic Press, New York.

Ebner, K. E., 1973, Lactose synthetase, in: *The Enzymes*, 3rd ed., Vol. 9 (P. D. Boyer, ed.), p. 363, Academic Press, New York.

Englander, S. W., 1975, Measurement of structural and free energy changes in hemoglobin by hydrogen exchange methods, *Ann. N. Y. Acad. Sci.* **244**:10.

Exton, J. H., and Park, C. R., 1967, Control of gluconeogenesis in liver. I. General features of gluconeogenesis in the perfused livers of rats, *J. Biol. Chem.* **247**:2622.

Fischer, E. H., Heilmeyer, L., and Haschke, R., 1971, Phosphorylase and the control of glycogen degradation, in: *Current Topics in Cellular Regulation*, Vol. 4 (B. L. Horecker and E. R. Stadtman, eds.), p. 211, Academic Press, New York.

Foster, D. O., Ray, P. D., and Lardy, H. A., 1966, Studies on the mechanisms underlying adaptive changes in rat liver phosphoenolpyruvate carboxykinase, *Biochemistry* **5**:555.

Frieden, C., 1971, Protein–protein interaction and enzymatic activity, *Annu. Rev. Biochem.* **40**:653.

Frieden, C., Gilbert, H.R., and Bock, P. E., 1976, Phosphofructokinase. III. Correlation of the regulatory, kinetic and molecular properties of the rabbit muscle enzymes, *J. Biol. Chem.* **251**:5644.

Fromm, H. J., 1975, *Initial Rate Enzyme Kinetics*, Springer-Verlag, Berlin.

Garber, A. J., Ballard, F. J., and Hanson, R. W., 1972, Significance of mitochondrial phosphoenolpyruvate formation in the regulation of gluconeogenesis in guinea pig liver, in: *Energy Metabolism and the Regulation of Metabolic Processes* (R. W. Hanson and M. A. Mehlman, eds.), p. 109, Academic Press, New York.

Garland, P. B., and Randle, P. J., 1964, Regulation of glucose uptake by muscle, *Biochem. J.* **93**:678.

Gerhart, J. C., 1970, A discussion of the regulatory properties of aspartate transcarbamylase from *Escherichia coli*, in: *Current Topics in Cellular Regulation*, Vol. 2 (B. L. Horecker and E. R. Stadtman, eds.), p. 276, Academic Press, New York.

Ginsburg, A., and Stadtman, E. R., 1970, Multienzyme systems, *Annu. Rev. Biochem.* **39**:429.

Goldbeter, A., 1976, Kinetic cooperativity in the concerted model for allosteric enzymes, *Biophys. Chem.* **4**:159.

Goldbeter, A., and Caplan, S. R., 1976, Oscillatory enzymes, *Annu. Rev. Biophys. Bioeng.* **5**:449.

Hammes, G., and Wu, C., 1974, Kinetics of allosteric enzymes, *Annu. Rev. Biophys. Bioeng.* **43**:1.

Hansford, R. G., 1977, Studies on inactivation of pyruvate dehydrogenase by palmitoylcarnitine oxidation in isolated rat heart mitochondria, *J. Biol. Chem.* **252**:1552.

Hess, B., and Boiteux, A., 1971, Oscillatory phenomena in biochemistry, *Annu. Rev. Biochem.* **40**:237.

Higgins, J., 1967, The theory of oscillating reactions, *Ind. Eng. Chem.* **59**:18.

Hill, A.V., 1913, The combinations of haemoglobins with oxygen and with carbon monoxide, *Biochemical J.* **7**:471.

Hill, R. L., and Brew, K., 1975, Lactose synthetase, *Adv. Enzymol.* **43**:441.

Holzer, H., 1969, Regulation of enzymes by enzyme-catalyzed chemical modification, *Adv. Enzymol.* **32**:297.

Holzer, H., and Duntze, W., 1971, Metabolic regulation by chemical modification of enzymes, *Annu. Rev. Biochem.* **40**:345.

Hucho, F., 1975, The pyruvate dehydrogenase multienzyme complex, *Angew. Chemie (Engl.)* **14**:591.

Huttunen, J. K., and Steinberg, D., 1971, Activation and phosphorylation of purified adipose tissue hormone-sensitive lipase by cyclic AMP-dependent protein kinase, *Biochim. Biophys. Acta* **239**:411.

Kacser, H., and Burns, J. A., 1973, The control of flux, *Symp. Soc. Exp. Biol.* **27**:65.

Katz, J., and Rognstad, R., 1976, Futile cycles in the metabolism of glucose, in: *Current Topics in Cellular Regulation*, Vol. 10 (B. L. Horecker and E. R. Stadtman, eds.), p. 237, Academic Press, New York.

Kerbey, A. L., Radcliffe, P. M., and Randle, P. J., 1977, Diabetes and the control of pyruvate dehydrogenase in rat heart mitochondria by concentration ratios of adenosine triphosphate/adenosine diphosphate, of reduced/oxidized nicotinamide-adenine dinucleotide, and of acetyl-Coenzyme A/Coenzyme A, *Biochem. J.* **164**:509.

Kikuchi, G., 1973, The glycine cleavage system: Composition, reaction mechanism, and physiological significance, *Molec. Cell. Biochem.* **1**:169.

Kirschner, K., 1971, Kinetic analysis of allosteric enzymes, in: *Current Topics in Cellular Regulation*, Vol. 4 (B. L. Horecker and E. R. Stadtman, eds.), p. 167, Academic Press, New York.

Kitzinger, C., and Benzinger, T. H., 1960, Principle and method of heat-burst microcalorimetry and the determination of free energy, enthalpy and entropy changes, *Methods Biochem. Anal.* **8**:309.

Knowles, J. F., 1976, The intrinsic pKa values of functional groups in enzymes: Improper deductions from the pH dependence of steady-state parameters, *Crit. Rev. Biochem.* **4**:165.

Kohn, M.C., Whitley, L.M., and Garfinkel, D., 1979, Instantaneous flux control analysis for biochemical systems, *J. Theor. Biol.* **76**:437.

Koshland, D. E., Jr., 1970, The molecular basis for enzyme regulation, in: *The Enzymes*, 3rd ed., Vol. 1 (P. D. Boyer, ed.), p. 342, Academic Press, New York.

Koshland, D. E., and Neet, K.E., 1968, The catalytic and regulatory properties of enzymes, *Annu. Rev. Biochem.* **37**:359.
Koshland, D. E., Némethy, G., and Filmer, D., 1966, Comparison of experimental binding data and theoretical models in proteins containing subunits, *Biochemistry* **5**:365.
Krebs, E. G., 1972, Protein kinases, in: *Current Topics in Cellular Regulation, Vol. 5* (B. L. Horecker and E. R. Stadtman, eds.), p. 99, Academic Press, New York.
Krebs, H. A., 1963, Renal gluconeogenesis, *Adv. Enzyme Regul.* **1**:385.
Laidler, K., and Bunting, P., 1973, *The Chemical Kinetics of Enzyme Action*, Oxford Press, London.
Lane, M. D., Moss, J., and Polakis, S. E., 1974, Acetyl Coenzyme A carboxylase, in: *Current Topics in Cellular Regulation*, Vol. 8 (B. L. Horecker and E. R. Stadtman, eds.), p. 139, Academic Press, New York.
Lee, K. H., and Kim, K. H., 1977, Regulation of rat liver acetyl Coenzyme A carboxylase, *J. Biol. Chem.* **252**:1748.
Levitzki, A., and Koshland D. E., Jr., 1976, Role of negative cooperativity and half of the sites reactivity in enzyme regulation, in: *Current Topics in Cellular Regulation*, Vol. 10 (B. L. Horecker and E. R. Stadtman, eds.), p. 1, Academic Press, New York.
Mansour, T. E., 1972, Phosphofructokinase, in: *Current Topics in Cellular Regulation*, Vol. 5 (B. L. Horecker and E. R. Stadtman, eds.), p. 1, Academic Press, New York.
Masters, C. J., 1977, Metabolic control and the microenvironment, in: *Current Topics in Cellular Regulation*, Vol. 12 (B. L. Horecker and E. R. Stadtman, eds.), p. 75, Academic Press, New York.
Metcalfe, J. C., 1970, Nuclear magnetic resonance spectroscopy, in: *Physical Principles and Techniques of Protein Chemistry* (Part B) (J. Leach, ed.), p. 275, Academic Press, New York.
Mildvan, A. S., 1970, Metals in enzyme catalysis, in: *The Enzymes*, 3rd ed., Vol. 2 (P. D. Boyer, ed.), p. 446, Academic Press, New York
Monod, J., Changeux, J. P., and Jacob, F., 1963, Allosteric proteins and cellular control systems, *J. Mol. Biol.* **6**:306.
Monod, J., Wyman, J., and Changeux, J. P., 1965, On the nature of allosteric transitions: A plausible model, *J. Mol. Biol.* **12**:88.
Morgan, H. E., and Parmeggiani, A., 1964, Regulation of glycogenolysis in muscle, in: *Control of Glycogen Metabolism* (W. J. Whelan, ed.), p. 254, J. & A. Churchill, London.
Neurath, H., and Walsh, K. A., 1976, Role of proteolytic enzymes in biological regulation, *Proc. Natl. Acad. Sci. USA* **73**:3825.
Newsholme, E. A., and Crabtree, B., 1976, Substrate cycles in metabolic regulation and in heat generation, *Biochem. Soc. Symp.* **41**:61.
Newsholme, E. A., and Gevers, W., 1967, Control of glycolysis and gluconeogenesis in liver and kidney cortex, *Vitam. Horm.* **25**:1.
Newsholme, E. A., and Start, C., 1972, General aspects of the regulation of enzyme activity and the effects of hormones, in: *Handbook of Physiology-Endocrinology I* (S. R. Geiger, ed.), p. 369, American Physiological Society, Washington, D. C.
Newsholme, E. A., and Start, C., 1973, *Regulation in Metabolism*, p. 97, John Wiley and Sons, London.
Paulus, H., and DeRiel, K., 1975, Absence of kinetic negative cooperativity in the allosteric model of Monod, Wyman, and Changeux, *J. Mol. Biol.* **97**:667.
Pitot, H. C., and Yatvin, M. B., 1973, Interrelationships of mammalian hormones and enzyme levels in vivo, *Physiol. Rev.* **53**:228.
Powell, J. T., and Brew, K., 1976, A comparison of the interactions of galactosyltransferase with a glycoprotein substrate (ovalbumin) and with d-lactalbumin, *J. Biol. Chem.* **251**:3653.
Purich, D. L., and Fromm, H. J., 1972, A possible role for kinetic reaction mechanism

dependent substrate and product effects in enzyme regulation, in: *Current Topics in Cellular Regulation*, Vol. 6 (B. L. Horecker and E. R. Stadtman, eds.), p. 131, Academic Press, New York.
Purich, D. L., and Fromm, H. J., 1973, Additional factors influencing enzyme responses to the adenylate energy charge, *J. Biol. Chem.* **248**:461.
Rasmussen, H., and Goodman, D. B. P., 1977, Relationships between calcium and cyclic nucleotides in cell activation, *Physiol. Rev.* **57**:421.
Reed, L. J., 1969, Pyruvate dehydrogenase complex, in: *Current Topics in Cellular Regulation*, Vol. 1 (B. L. Horecker and E. R. Stadtman, eds.), p. 233, Academic Press, New York.
Reed, L. J., 1974, Multienzyme complexes, *Accounts Chem. Res.* **7**:40.
Richman, P., and Meister, A., 1975, Regulation of γ-glutamyl-cysteine synthetase by non-allosteric feedback inhibition by glutathione, *J. Biol. Chem.* **250**:1422.
Roach, P. J., and Larner, J., 1976, Rabbit skeletal muscle glycogen synthase. II. Enzyme phosphorylation state and effector concentration as interacting control parameters, *J. Biol. Chem.* **251**:1920.
Roach, P. J., Takeda, Y., and Larner, J., 1976, Rabbit skeletal muscle glycogen synthase. I. Relationship between phosphorylation state and kinetic properties, *J. Biol. Chem.* **251**:1913.
Rolleston, F. S., 1972, A theoretical background to the use of measured concentrations of intermediates in study of the control of intermediary metabolism, in: *Current Topics in Cellular Regulation*, Vol. 5 (B. L. Horecker and E. R. Stadtman, eds.), p. 47, Academic Press, New York.
Rosen, O. M., Rangel-Aldao, R., and Erlichman, J., 1977, Soluble cyclic AMP-dependent protein kinases: Review of the enzyme isolated from bovine cardiac muscle, in: *Current Topics in Cellular Regulation*, Vol. 12 (B. L. Horecker and E. R. Stadtman, eds.), p. 39, Academic Press, New York.
Rubin, C. S., and Rosen, O. M., 1975, Protein phosphorylation, *Annu. Rev. Biochem.* **44**:831.
Sanwal, B. D., Kapoor, M., and Duckworth, H., 1971, The regulation of branched and converging pathways, in: *Current Topics in Cellular Regulation*, Vol. 3 (B. L. Horecker and E. R. Stadtman, eds.), p. 1, Academic Press, New York.
Saugmann, P., 1977, Glycogen synthase-"R". The occurrence and significance of a previously unknown form of GS, found in metabolically active leucocytes, *Biochem. Biophys. Res. Commun.* **74**:1511.
Schramm, V. L., 1974, Kinetic properties of allosteric adenosine monophosphate nucleosidase from *Azotobacter vinelandii*, *J. Biol. Chem.* **249**:1729.
Scrutton, M. C., and Utter, M. F., 1968, The regulation of glycolysis and gluconeogenesis in animal tissues, *Annu. Rev. Biochem.* **37**:245.
Segal, H. L., 1973, Enzymatic interconversion of active and inactive forms of enzymes, *Science* **180**:25.
Segel, I. H., 1975, *Enzyme Kinetics*, John Wiley & Sons, New York.
Severson, D. L., Khoo, J. C., and Steinberg, D., 1977, Role of phosphoprotein phosphatases in reversible deactivation of chicken adipose tissue hormone-sensitive lipase, *J. Biol. Chem.* **252**:1484.
Shoaf, W. T., and Jones, M. E., 1973, Uridylic acid synthesis in Ehrlich ascites carcinoma: Properties, subcellular distribution, and the nature of the enzyme complexes of the six biosynthetic enzymes, *Biochemistry* **12**:4039.
Siddle, K., and Hales, C. N., 1974, The action of local anesthetics on lipolysis and on adenosine 3',5'-cyclic monophosphate content in isolated rat fat-cells, *Biochem. J.* **142**:345.
Soderling, T. R., Corbin, J. D., and Park, C. R., 1973, Regulation of adenosine 3',5'-mono-

phosphate dependent protein kinase II. Hormonal regulation of the adipose tissue enzyme, *J. Biol. Chem.* **248**:1822.
Somero, G. N., and Low, P. S., 1976, Temperature: A "shaping force" in protein evolution, *Biochem. Soc. Symp.* **41**:33.
Srere, P., 1967, Enzyme concentrations in tissues, *Science* **158**:936.
Srere, P., 1970, Enzyme concentrations in tissue. II. An additional list, *Biochem. Med.* **4**:43.
Stadtman, E. R., 1966, Allosteric regulation of enzyme activity, *Adv. Enzymol.* **28**:41.
Stadtman, E. R., 1970, Mechanisms of enzyme regulation in metabolism, in: *The Enzymes*, 3rd ed., Vol. 1 (P. D. Boyer, ed.), p. 397, Academic Press, New York.
Stadtman, E. R., and Chock, P. B., 1977, Superiority of interconvertible enzyme cascades in metabolic regulation: Analysis of monocyclic systems, *Proc. Natl. Acad. Sci. USA* **74**:2761.
Stein, R. B., and Blum, J.J., 1978, On the analysis of futile cycles in metabolism, *J. Theor. Biol.* **72**:487.
Steinberg, D., and Khoo, J. C., 1977, Hormone-sensitive lipase of adipose tissue, *Fed. Proc.* **36**:1986.
Stroud, R. M., Kossiakoff, A. A., and Chambers, J. L., 1977, Mechanisms of zymogen activation, *Annu. Rev. Biophys. Bioeng.* **6**:177.
Swillens, S., and Dumont, J. E., 1976, Analysis of the control of protein kinase activity, *J. Molec. Med.* **1**:273.
Volpe, J. J., and Vagelos, P. R., 1976, Mechanisms and regulation of biosynthesis of saturated fatty acids, *Physiol. Rev.* **56**:339.
Walker, W. D., Goodwin, F. J., and Cohen, R. D., 1969, Mean intracellular hydrogen ion activity in the whole body, liver, heart, and skeletal muscle of the rat, *Clin. Sci.* **36**:409.
Walter, C. F., 1972, Kinetic and thermodynamic aspects of biological and biochemical control mechanisms, in: *Biochemical Regulatory Mechanisms in Eucaryotes* (E. Kun and S. Gresolia, eds.), p. 355, John Wiley & Sons, New York.
Walter, C. F., 1974, Some dynamic properties of linear, hyperbolic and sigmoidal multienzyme systems with feedback control, *J. Theor. Biol.* **44**:219.
Weinhouse, S., 1976, Regulation of glucokinase in liver, in: *Current Topics in Cellular Regulation*, Vol. 11 (B. L. Horecker and E. R. Stadtman, eds.), p. 1, Academic Press, New York.
Weissbach, H., and Ochoa, S., 1976, Soluble factors required for eukaryotic protein synthesis, *Annu. Rev. Biochem.* **45**:191.
Williamson, D. H., Lund, P., and Krebs, H. A., 1967, The redox state of free nicotinamide-adenine dinucleotide in the cytoplasm and mitochondria of rat liver, *Biochem. J.* **103**:514.
Wright, E. M., and Diamond, J. M., 1977, Anion selectivity in biological systems, *Physiol. Rev.* **57**:109.
Wyman, J., 1972, On allosteric models, in: *Current Topics in Cellular Regulation*, Vol. 6 (B. L. Horecker and E. R. Stadtman, eds.), p. 211, Academic Press, New York.
Yates, R. A., and Pardee, A. B., 1956, Control of pyrimidine biosynthesis in *Escherichia coli* by a feed-back mechanism, *J. Biol. Chem.* **221**:757.

5

Regulation of Protein Biosynthesis

Robert M. Cohn, Pamela D. McNamara, and
Robert H. Herman

I. Introduction

A. The Reason for Protein Biosynthesis

Before embarking on a consideration of the regulation of protein biosynthesis, it may be useful to examine the need for protein biosynthesis. Although the life process depends absolutely on structural and functional proteins, it seems reasonable to inquire why, once proteins and enzymes are synthesized, the system does not remain stable, requiring no further protein synthesis. The answer has been dealt with in Chapters 1, 2, and 3. Suffice it to say here that enzymes function by virtue of their conformability, using noncovalent forces that permit substrate binding and product release in addition to the catalytic action of making and breaking covalent bonds in the substrate. The details of this process are treated in Chapter 3. Because of this conformability, which depends on noncovalent forces, enzymes are subject to denaturation and consequent loss of function; hence, *enzymes are semistable*. The relatively delicate balance between form, functionality, and denaturation is maintained, on one hand, by the orderly process of enzyme biosynthesis, and, on the other, by degradation of denatured enzymes. In this way—by regulation of the overall rate of enzyme synthesis and degradation—a relatively constant enzyme concentration can be main-

---

Robert M. Cohn and Pamela D. McNamara • Department of Metabolic Research, Children's Hospital of Philadelphia, University of Pennsylvania, Philadelphia, Pennsylvania 19104. Robert H. Herman • Endocrine-Metabolic Service, Letterman Army Medical Center, Presidio of San Francisco, California 94129.

tained despite the rate of denaturation of enzymes. This represents a steady-state system (see Chapter 1).

Protein biosynthesis is one method by which the cell can respond to changing conditions, synthesizing more or less of a given set of enzymes, which in turn regulate the rate and direction of particular metabolic pathways. In order to determine how protein biosynthesis can be influenced to produce more or less of a given enzyme, it is necessary to examine the details of the protein biosynthetic mechanism. In general, if the concentration of an enzyme increases, it is because there is increased synthesis or decreased degradation of the enzyme or both. Conversely, if the enzyme concentration decreases, there is decreased synthesis or increased degradation. Thus, the very nature of enzyme behavior necessitates an orderly process of biosynthesis and degradation, implying intrinsic regulatory controls in order to maintain a relatively constant concentration of enzyme under given environment conditions.

B. The Complexity of Protein Biosynthesis

The protein biosynthetic mechanism and its attendant regulatory controls are immensely complex, and the scientific literature about them is voluminous and intricate. It is not our intent—nor is it possible in one chapter—to cover all details of the biosynthetic process. Rather, in this chapter we shall emphasize the regulatory controls of protein biosynthesis. However, such a consideration requires review of some details of the biosynthetic pathways, so that the nature of the regulatory controls can be appreciated. For more detailed information, the reader is referred to the various reviews and books on the subject (e.g., Weissbach and Pestka, 1977; Vogel, 1977; Nienhuis and Benz, 1977; Chan, et al., 1977; Busch, et al., 1976; Goldberger, 1979).

The mechanisms that regulate protein biosynthesis must be such that appropriate concentrations and types of enzymes can be synthesized in response to extracellular conditions. Furthermore, the protein biosynthetic mechanism is of such paramount importance to life, and so complex, that we would expect multiple controls to exist. This expectation is completely justified—such controls being found at all levels of information transfer required in the protein biosynthetic mechanism. In order to maintain a balance between the component parts, the regulatory mechanisms must control the rates of protein synthesis, utilization, and degradation. Of course, it need hardly be pointed out that, while nucleic acids provide the information for protein synthesis, enzymes mediate the biosynthetic processes themselves.

C. The Third Fundamental Theorem of Theoretical Biology

The protein synthetic mechanism is the fundamental project of life and is thus the embodiment of the life process. The continued synthesis and degradation of proteins and enzymes maintains the metabolic network in a state of negative entropy, so that all reactions occur under far-from-equilibrium conditions. This nonequilibrium state, in a sense, constitutes the life process. Enzymes are the functional entities of the life process, and, in accordance with the principle of nonequilibrium thermodynamics (see Chapter 2), semistable enzymes constitute the functional basis of life. A totally stable enzyme (denatured enzyme) is inert, having no catalytic function, and is incapable of interacting with its substrates and products. An unstable enzyme has too transient an existence to carry out a catalytic reaction in a steady-state network. Yet for catalytic action to persist for a sufficient time, there must be a certain degree of stability, and the catalytic function of an enzyme requires flexibility or conformity. It is in this sense that enzymes can be considered as semistable.

II. The Mechanism of Protein Biosynthesis

A. A General Description of Protein Biosynthesis

The information contained in genes is expressed, in part, by the proteins that are within a cell. The genes contain information for their own replication, for the synthesis of other nucleic acids and proteins including enzymes, and for the various control mechanisms of protein biosynthesis.

The genes are represented by polymers of purine and pyrimidine nucleoside triphosphates arranged to form deoxyribonucleic acid (DNA) (Kornberg, 1974). DNA is double-stranded, and each strand is joined with its partner through hydrogen bonding between complementary purines and pyrimidines, forming a double helix. The arrangement of DNA, histones, and nonhistone proteins in the chromatin of the cell nucleus is described in Chapter 7.

DNA carries out two functions: (1) it provides for its own exact replication, thus permitting genetic continuity; and (2) it directs the linear sequence of amino acids in proteins on the basis of the linear sequence of nucleotides in the DNA. This latter function is carried out by the process of transcription, during which the genetic message in DNA is expressed in the various classes of RNA (viz., messenger, transfer, and ribosomal RNA).

Protein biosynthesis begins with the action of a DNA-dependent RNA polymerase which uses one of the strands of DNA as a template on which to polymerize purine and pyrimidine nucleoside triphosphates into a messenger RNA (mRNA) or into other nucleic acids. In eukaryotes, it is believed that a large amount of precursor RNA is synthesized (heterogeneous nuclear RNA, hnRNA) and is then processed into mRNA. The mRNA is further modified by the addition of poly(A) at the 3' end and a so-called "cap," i.e., $m^7G(5')ppp(5')X$, at the 5' end. The purine and pyrimidine bases in mRNA are ordered in a sequence corresponding to the purine and pyrimidine sequence in a single strand of the unwound DNA template. Thus, the sequence of bases in the mRNA represents the sequence of bases of the gene. In an increasing number of cases, it has been found that the mRNA contains only a portion of the gene sequence, and that the mRNA sequences are thus not colinear with the DNA sequences. The gene may contain DNA sequences which interrupt structurally translated sequences. The RNA sequences complementary to the interrupting DNA sequences are excised from the precursor RNA, giving rise to the mRNA by a ligation or splicing process, so that the sequences in the mature mRNA do not reflect the exact sequence of purines and pyrimidines within the gene. The physiological significance of these interrupting sequences is unknown, but will be considered subsequently.

In like manner, ribosomal ribonucleic acid (rRNA) and transfer ribonucleic acid (tRNA) are synthesized from their specific genes through the action of DNA-dependent RNA polymerase. And as in the case of mRNA, each of the precursor forms of these nucleotides are also processed until they reach their final forms. The rRNAs and a large variety of ribosomal proteins are assembled to form the ribosome. Ribosomes, the supramolecular structures in which protein synthesis takes place and which functionally resemble multienzyme complexes, are elaborate structures. They are attached, by a binding site on the smaller subribosomal particle, to the endoplasmic reticulum. The tRNAs that are synthesized finally become charged with specific amino acids.

Translation from the linear sequences of mRNA into proteins is an extremely complex process, as befits the requirement for fidelity at all stages of the process. The translational keys to the process are the tRNA molecules, which are akin to the Rosetta stone. These specific adaptor molecules (tRNA) are responsible for aligning each amino acid with its specific mRNA codon, and tRNA molecules thus read the information in both mRNA and in amino acids.

The various mRNAs become attached to the assembled ribosomes and order the sequence of amino acids that are attached to the tRNAs.

Each tRNA anticodon corresponds to a particular codon of the mRNA, so that each specific amino acid attached to its tRNA becomes ordered according to the sequence of codons in the mRNA. The information of the amino acid nucleotide code found in mRNA is translated into polypeptides, the messenger RNA being read in the 5' to 3' direction, with the polypeptides being synthesized from the amino-terminal to the carboxy-terminal amino acid. Upon completion of synthesis of a polypeptide chain, the chain leaves the ribosome and assumes its particular conformation. It is able to combine with other polypeptide chains, and subsequent covalent modifications of the protein molecule can take place.

Specific codons in mRNA signal the initiation and termination of synthesis of each protein. And there exist within the ribosome a number of factors which permit the stepwise synthesis of the polypeptide. In addition, soluble factors are required for initiation, elongation, translocation, and termination. In prokaryotes, translation is closely linked temporally to transcription, since both processes take place in the same milieu; in eukaryotes, the nucleus compartmentalizes these two processes, permitting further controls and delaying the coupling of translation to transcription.

At each level of the protein biosynthetic pathway, control mechanisms exist to regulate the transcription of the gene and the translation of the mRNA into protein. On the whole, the abundant use of electrostatic interactions, including protein–RNA as well as protein–protein interactions, and the relative paucity of covalent bond involvement during polypeptide synthesis, presents an interesting aspect of the process. It may be that the stringent fidelity necessary for genetic continuity is most easily regulated through the use of noncovalent forces which provide specificity as well as flexibility.

B. Gene Structure and Protein–Nucleic Acid Interactions

1. Gene Structure

It is now clear that the eukaryotic genome is not a macro variation on a prokaryotic theme (Van Holde and Isenberg, 1975; Kornberg, 1977; Felsenfeld, 1978; Chambon, 1978). In contrast to the relatively unadorned DNA of prokaryotes, eukaryotic DNA is associated with a variety of basic (histone) and acidic (nonhistone) proteins, as well as polymerases, forming the chromatin. While the amount of histones (in weight) equals that of DNA in the chromatin, there are only five types present. In contrast, the nonhistone proteins, which are present in much smaller amounts, are striking for their heterogeneity. Interestingly, while the polycationic histones would be expected to bind to the polyanionic phosphate moieties in the DNA, Van Holde and Isenberg (1975) have noted that almost 50% of the phosphate moieties are not involved in a histone interaction. Histone–histone interaction may also play an important role in chromatin structure.

Brief digestion of chromatin with micrococcal nuclease results in the production of discrete units termed nucleosomes (nu bodies). On electron microscopy, these protein–DNA particles are approximately 10–13 nm long, linked by shorter (2–4 nm) DNA sequences. Further details of chromatin structure may be found in Chapter 7.

The amount of DNA in the eukaryotic cell is approximately 500 times that found in prokaryotes (Watts and Watts, 1975). Since there has not been a commensurate increase in the number of proteins manufactured in the eukaryotic cell, this potential information explosion in the size of the genome must be accounted for in ways other than by structural genes. In fact, only 2–15% of the total information in the genome is expressed in differentiated cells. The interaction of the genome with its cellular environment, and the change in that environment introduced upon fertilization must eventually be incorporated into any theory of differentiation.

2. Protein–Nucleic Acid Interactions

Proteins which interact with DNA do so either by recognizing specific nucleotide sequences or by disregarding the nucleotide sequence. Such binding must involve several points of attachment so as to provide sufficient discrimination and free energy for the fruitful interaction to take place. In general, proteins binding to DNA carry out one of two functions: (1) they mediate or facilitate information expression in the DNA sequence; or (2) they inhibit or stop such information expression.

A number of enzymes which bind polyanions such as DNA may be separated broadly into those which recognize special sequences with higher affinity than "bulk DNA" and those which bind nonspecifically. The first group includes RNA polymerases, repressors, cAMP receptor protein (CRP), the *rho* factor, restriction endonucleases, and methylases; the second group includes histones and DNA-unwinding proteins.

The key to all biochemistry is the understanding of the relationship between form and function. Thus, before protein–nucleic acid interactions can be understood, we will require a detailed knowledge of the *static* three-dimensional, structural details of such interactions, as well as of the *dynamic* interactions of these macromolecules (Rich, 1977). This latter data must ultimately predict stability as well as reactivity.

This is a fertile area for investigation that will require X-ray crystallographic analysis of these macromolecules in their interactional state. Because of the axial roles played by both nucleic acids and proteins in the life process, it is reasonable to assume that the origins of such interactions date back as long as 3.5–4.5 billion years, when life began to emerge on earth.

As we have noted before (Chapters 1 and 2), while nucleic acids are especially suited for the storage and transmission of genetic information, they do not possess the requisite functional groups and higher-order structure for binding and carrying out the catalytic chemical reactions essential to the metabolism of living cells or to forming the varied structural elements of cells. Proteins, however, have a vast array of chemical environments, as the result of amino acid side chains which can create hydrophobic, hydrophilic, or charged environments (Chapter 3). Such variety makes proteins uniquely suited to serve as catalysts and supermolecular structural elements. Hence, the informational content of nucleic acids, present in the four nucleotide bases, is amplified as it is expressed in the 20 amino acids. This is accomplished by the nucleotide triplet in mRNA that is complementary to the triplet codon preserved in DNA which specifies one amino acid. The key to the translation of mRNA into protein is the tRNA molecule which reads the information in mRNA using codon–anticodon interactions at one site of the molecule, while supplying the appropriate amino acid, which is attached to the other end of that tRNA molecule.

Since DNA is never involved directly in protein synthesis, it has been proposed that RNA arose before DNA and that DNA represents a newer molecule whose role is solely information storage and replication (Rich, 1962). For carrying out this biologic function, the proteins that interact with DNA are those primarily concerned with information storage or with replication. Proteins involved in DNA replication and repair include polymerases, unwinding proteins, and ligases. The ability to recognize individual deoxyribonucleotides is not highly developed in these proteins, except possibly during the early steps of replication.

The RNA polymerases evince more sequence specificity than the repair enzymes since they transcribe the information in DNA into specific classes of RNA. In prokaryotes, transcription can be regulated by repressor and related proteins, which are very sensitive to the base sequence in DNA. Similar regulatory requirements undoubtedly exist for the eukaroytic genome but are not as simple. These are discussed subsequently. Regulating proteins combine with certain DNA molecules through recognition of a particular nucleotide base sequence.

Other classes of proteins interact with DNA and include a variety of nucleases (Table 1). An interesting subclass are the restriction endonucleases. These appear to be specific for double-stranded DNA and recognize specific base sequences which have a particular twofold rotational symmetry, often called *palindromes* (see below).

In contrast to its deoxy template, RNA exhibits a wider variety of interactions with proteins, because the three RNA species have a wider range of functions, involving various stages of protein synthesis. Table 2 lists some of the proteins we wish to consider. Proteins interacting with RNA include those that trim RNA before ribosome formation. Since in eukaryotes there are approximately 70 ribosomal structural proteins, it is expected, and realized that these structural proteins undergo multiple interactions with the species of RNA. For example, evidence suggests that the interaction of 5 S RNA with the L5, L18, and L25 proteins situated in the large ribosomal subunit is involved in translocation of the peptidyl tRNA.

Messenger RNA interacts with several maturation enzymes, including those which cleave it to its mature forms, those which cap the 5' end and those which add polyadenylate groups to the 3' end. During protein synthesis, mRNA movement inside the ribosome is mediated by ribosomal proteins.

Table 1. Proteins That Interact with DNA[a]

---

1. Replication and repair
    a. DNA polymerases of different types
    b. DNA-unwinding proteins and others in the replication complex
    c. Ligases and repair proteins
    d. Nucleases and excision enzymes of the repair process
2. DNA packaging proteins
    a. Chromatin: Nucleosomes and histones
    b. Proteins of the sperm, protamines, and other proteins
    c. Virus condensation proteins: internal and coat
3. Transcription
    a. RNA polymerase and its various subunits
    b. Repressors and regulation of the initiation of transcription
    c. Cyclic AMP receptor proteins (CRP)
    d. Rho factors in the termination of transcription
4. Nucleases
    a. Restriction and endonucleases
    b. Exo- and endonucleases
    c. Nucleases of hydrolysis
5. DNA-modifying proteins
    a. Methylases

---

[a]This table is reproduced with the kind permission of Dr. A. Rich and Academic Press, from An overview of protein–nucleic acid interactions, in: *Nucleic Acid–Protein Recognition* (H. J. Vogel, ed.), 1977, p. 6.

Table 2. Proteins That Interact with RNA [a]

1. Proteins interacting with ribosomal RNA
   a. Proteins cleaving rRNA precursors
   b. rRNA modifying proteins
   c. Structural proteins of the ribosome
   d. Special complex of 5 S RNA with L5, L18, L25 proteins
   e. Peptidyltransferase
   f. Initiation factors and elongation factors
2. Proteins interacting with messenger RNA
   a. Maturation enzymes
   b. Polyadenylation enzymes
   c. Capping enzymes
   d. mRNA movement proteins in the ribosome
   e. mRNA hydrolytic enzymes
3. Proteins interacting with transfer RNA
   a. Maturation enzymes including RNAse-P
   b. Modification enzymes, such as methylases; ψ, isopentynyl and acetyl adding enzymes
   c. Nucleotidyltransferase, the CCA adding enzyme
   d. Aminoacyl–tRNA synthetases
   e. Elongation factors and initiation factors
   f. G factor for translocation
   g. Ribosomal proteins involved in translocation, guanosine tetraphosphate synthesis
   h. Aminoacyl-tRNA transferases which transfer amino acids to preformed proteins, to phosphatidyl glycerol in cell membranes or to the peptidoglycan components of bacterial cell walls
   i. Reverse transcriptase
   j. RNA polymerase interacts with initiator $tRNA_f^{met}$
   k. Regulatory proteins controlling transcription
4. Polymerizing and repair enzymes
   a. RNA replicases with single-stranded substrates, e.g., Qβ replicase
   b. RNA replicases with double-stranded substrates
   c. RNA polymerase
   d. Polynucleotide phosphorylase-nontemplate polymerases
   e. Reverse transcriptase and viral RNA
   f. RNA ligases
5. Nucleases
   a. Various processing enzymes
   b. Exo- and endonucleases
   c. Hydrolases

[a] This table is reproduced with the kind permission of Dr. A. Rich and Academic Press, from An overview of protein–nucleic acid interactions, in: *Nucleic Acid–Protein Recognition* (H. J. Vogel, ed.), 1977, p. 9.

Transfer RNA also undergoes a maturation process, involving cleavage as well as covalent modification by methylases and nucleotidyl transferases, the latter adding the CCA sequence to the 3'-hydroxyl end of all tRNA molecules. Other enzymes modify the purine on the 3' end of the anticodon and form pseudouracil and dihydrouracil groupings.

Interactions between each tRNA species and its own aminoacyl synthetase are among the most specific of all such interactions, since the synthetase distinguishes both the tRNA and its amino acid. *The fidelity of protein synthesis depends upon this step, since the action of tRNA-aminoacyl synthetase results in translation of the nucleotide code into the amino acid sequence.* Movement of the tRNAs within the ribosomes is mediated by certain ribosomal proteins.

It may be instructive to contrast the interactions between the ribosome and the nucleosomal particle which contains DNA. As noted above, the eukaryotic ribosome contains approximately 70 different proteins, and there is experimental evidence that the proteins interact with specific sequences of RNA. The nucleosome particles, which are the subunits comprising the chromatin, contain five major histone proteins, most represented twice, which interact nonspecifically with 140–170 DNA base pairs.

C. Genetic Code

Jukes (1977) has remarked that "the continuity of both DNA and life is sustained through the aeons by the amino acid code." Indeed, many believe that the origin of life and, most assuredly, the continuation of life depend upon the code. Since both DNA and proteins are linear polymers, it follows logically that some linear sequence of the bases in DNA might code for the sequence of amino acids in proteins. However, owing to the fact that there are only four bases in DNA, which must code for at least 20 different amino acids in proteins, specification of each amino acid requires more than one base. Since 16 pairs of bases are too few to specify 20 different amino acids, the minimal number of bases required to accomplish this specification would be a group of at least three nucleotides. A triplet code utilizing four nucleotide bases can be expressed in 64 permutations. Experiments carried out by Nirenberg and Matthaei (1961) and Ochoa and his colleagues (Lengyel et al., 1961) demonstrated that the blueprint for the primary structure of proteins consisted of consecutive, nonoverlapping nucleotide triplets, shown in Table 3. These investigations answered the question of whether all of the 64 possible codons were used and, further, whether any served purposes other than specifying amino acids. The investigators found that several codons were used to specify a single

Table 3. The Amino Acid Code and the Identified Anticodons in Sequenced Transfer RNAs[a]

Identified anticodons[b]	Codons		Identified anticodons[b]	Codons	
	UUU	Phenylalanine		UAU	Tyrosine
GAA	UUC	Phenylalanine	GUA	UAC	Tyrosine
UAA	UUA	Leucine	UUA	UAA	Chain Term.
CAA	UUG	Leucine	CUA	UAG	Chain Term.
	CUU	Leucine		CAU	Histidine
GAG	CUC	Leucine	GUG	CAC	Histidine
UAG	CUA	Leucine	UUG	CAA	Glutamine
CAG	CUG	Leucine	CUG	CAG	Glutamine
IAU	AUU	Isoleucine	GUU	AAU	Asparagine
GAU	AUC	Isoleucine		AAC	Asparagine
UAU	AUA	Isoleucine	UUU	AAA	Lysine
CAU	AUG	Methionine	CUU	AAG	Lysine
IAC	GUU	Valine		GAU	Aspartic acid
GAC	GUC	Valine	GUC	GAC	Aspartic acid
UAC	GUA	Valine	UUU	GAA	Glutamic acid
CAC	GUG	Valine		GAG	Glutamic acid
IGA	UCU	Serine		UGU	Cysteine
	UCC	Serine	GCA	UGC	Cysteine
UGA	UCA	Serine		UGA	Chain Term.
	UCG	Serine	CCA	UGG	Tryptophan
	CCU	Proline	ICG	CGU	Arginine
	CCC	Proline	GCG	CGC	Arginine
UGG	CCA	Proline		CGA	Arginine
	CCG	Proline		CGG	Arginine
IGU	ACU	Threonine		AGU	Serine
GGU	ACC	Threonine	GCU	AGC	Serine
	ACA	Threonine	UCU	AGA	Arginine
	ACG	Threonine		AGG	Arginine
IGC	GCU	Alanine		GGU	Glycine
	GCC	Alanine	GCC	GGC	Glycine
UGC	GCA	Alanine	UCC	GGA	Glycine
	GCG	Alanine	CCC	GGG	Glycine

[a]This table is reproduced with the kind permission of Dr. T. H. Jukes and Elsevier Scientific Publishers, from *Comprehensive Biochemistry*, Vol. 24, 1977, Chapter 5, p. 240, Elsevier North-Holland Biomedical Press.
[b]The bases in anticodons are shown in unmodified form, except for I (inosine) which is a modification of A. The unidentified anticodons are as follows: IAG: leucine; GGA, CGA: serine; IGG, GGG, CGG: proline; UGU, CGU: threonine; GGC, CGC: alanine; CUC: glutamic acid; UCG, CCG: arginine; CCU: arginine; ICC: glycine.

amino acid and that the code was universal, being employed by all organisms. Khorana (1968) developed methods for synthesizing oligonucleotides of known sequence, including regular alternating polymers of known sequence, which were used as templates for the in vitro synthesis of polypeptide chains. Evaluation of the amino acids encoded by these particular oligonucleotides led to the deciphering of the code. Indeed, certain codons did not code for any amino acid but rather directed chain termination. These terminating codons include UAA, UAG, and UGA and are frequently referred to as nonsense codons. The codon AUG (for methionine) was shown to serve as the initiation signal for protein synthesis (Jukes, 1977).

The genetic code includes the information for amino acid ordering, as well as information contained in other sequences of DNA molecules which serve ends other than translation into polypeptide chains. Examples of these latter types of regions in DNA are sequences which act as binding sites for endonucleases and RNA polymerase. The informational capability of the four-base system is enormous.

The amino acid codons represent a prescription for pairing between molecules of tRNA—with their attached amino acids—and molecules of mRNA. Functionally, the code works via antiparallel pairing of codon and anticodon in an RNA species. Translation into polypeptides is directed only by those triplet nucleotides which lie between the start and stop signals in an mRNA molecule. The corresponding complementary regions in DNA are the structural genes which are transcribed into mRNA by RNA polymerase.

However, the majority of base sequences in DNA serve functions other than coding for structural genes. As mentioned above, one important function is the binding of certain enzymes—for example, of endonucleases that cut the double strand of DNA. Other specific sequences are recognized by enzymes that bind and either initiate or terminate transcription. Still other sequences bind proteins that inhibit transcription. Translation into a polypeptide begins when an initiator codon in a strand of mRNA becomes bound to a ribosome, and continues consecutively, at the behest of the amino acid code, until a termination codon is reached. Translation from nucleotide to amino acid occurs at the level of the tRNA molecule, since on the one hand it pairs with mRNA by codon–anticodon pairing, while on the other it binds a specific amino acid.

What accounts for the great increase of nucleotide sequences in eukaryotic DNA as opposed to bacterial DNA? Addressing this subject, Watts and Watts (1975) suggested one possible distribution of information in the genome. We have modified their table in light of the most recent information on the eukaryotic genome (Table 4). Apparently,

Table 4. Types of Information Contained in the Genetic Material[a]

Information content in the DNA	Role(s)
1. None apparent	a. "Spacer" DNA
	b. Evolutionary stock
	1. Conserved
	2. Potential new genes
	3. Discarded
	c. Possible coordinate control sequences for noncoding, homologous genes
2. Nuclear housekeeping	a. Maintenance of chromosome structure
	b. In nuclear division
3. Nuclear control systems	a. Interaction with regulatory molecules controlling transcription
4. RNA structure	Encodes structure of
	a. Working rRNA and tRNA molecules
	b. Extra RNA required for stability, transport, or maturation of rRNA, tRNA, or mRNA
	c. RNA involved in the translation mechanism but not itself translated
5. Protein structure	Encodes structure of
	a. Working protein molecules for use anywhere in the cell or for export
	b. Parts of polypeptides discarded in formation of the working protein conformation

[a]Modified from Chapter 6 of the MTP International Review of Science, Series One, *Biochemistry*, Vol. 7 (H.R. Arnstein, ed.), 1975, p. 281, with kind permission from the authors, R.L. Watts and D.C. Watts, and University Park Press, Baltimore, Md.

gene duplication accounts for the enormous eukaryotic informational proliferation, the reiterated sequences permitting preservation of material in the face of mutation, as well as providing the possibility of new mutations with greater adaptive value. Recently, evidence for *gene amplification* (Alt et al., 1978) and *discontinuous sequences* (Darnell, 1978) has provided a partial answer to the question. Certain persistent patterns are emerging, but, in the latter instance, the function is not well understood. Some possibilities are discussed below.

More than 250 amino acids have been isolated from living systems, but only 20 appear to be encoded by the triplet nucleotide sequence in mRNA. Posttranslation modification accounts for the panoply of amino acids found in contemporary proteins. Hence, the limited number of amino acids encoded by the genome must be a consequence of evolutionary selection. Rohlfing and Saunders (1978) have proposed that it is the specific tripartite interaction among activating enzymes, tRNAs, and amino acid which explains why the amino acid code is limited to the 20 commonly found amino acids.

Various hypotheses concerning the origin and evolution of the genetic code have been proposed. Woese (1967) has proposed that stereospecific interactions between codons at all nucleotide levels (DNA, tRNA, mRNA, rRNA) and amino acids determine the makeup of the code. Evidence supporting this direct interaction model is meager. Most other models are stochastic, concerning themselves with the nature and evolution of the progenitor code without attempting to delineate how it arose. Crick (1968) has proposed the frozen accident theory, which posits the immutability of the code, since any fundamental change in the amino acid code would so profoundly change the sequence of amino acids as to destroy the functional capabilities of the original protein. In this view, the code is frozen because the consequence otherwise would be lethal. Two other theories, the vocabulary expansion theory (Jukes, 1967) and the coevolution theory (Wong, 1975, 1976), propose that the original code was limited to fewer amino acids than are provided for in the present code. This coheres with the fact that ferredoxins, primitive bacterial proteins, have fewer than 20 amino acids represented in their composition (Malkin, 1973; Palmer, 1975).

D. DNA-Dependent RNA Polymerase

The product of DNA replication is a *complete* copy of each parental strand of the DNA duplex. In contrast, the products of transcription are *limited* copies of one or the other strand of the DNA duplex. As noted above, transcription is initiated and terminated as a result of recognition of specific codons, and the entire process is subject to regulation. The initial products undergo processing or maturation via modification of the base or ribose moieties, or both, cleavage by exoribonucleases and endoribonucleases, and addition of one or more nucleotide residues at the 5' or 3' ends, or both, of the transcription products.

The enzymes responsible for this transcription process from the DNA template are RNA polymerases. Base pairing is involved in this process as in DNA replication, but, since only one strand of RNA is produced, it appears that the double-stranded DNA unwinds and that only one strand (the plus strand) acts as a template for the RNA polymerase.

Both prokaryotic and eukaryotic RNA polymerases are composed of several subunits. Although only one form of RNA polymerase predominates in bacteria and blue–green algae, in eukaroytes there are multiple forms of RNA polymerase which have different nuclear and organelle compartmentalization, function, and subunit composition. Polymerase I (or A) is located in the nucleolus and transcribes 18 S and 28 S ribosomal RNA genes. By contrast, polymerase II (or B) and III (or

C) are situated in the nucleoplasm, with type II transcribing hnRNA and mRNA and type III forming transfer RNAs and 5 S RNA (Chambon, 1975; Roeder, 1976).

1. Prokaryotic RNA Polymerase

In prokaryotes, transcription is far better understood than in eukaryotes. In bacteria, it seems that RNA polymerase rapidly and reversibly binds to random DNA sites until a promoter region is located. Calculations predict that a minimum of 12 base pairs are necessary for interaction to be stereospecific (Chamberlin, 1974, 1976a).

It is believed that when the initial specific polymerase-promoter complex forms, the bases in the DNA chains are still paired. This is termed the "closed complex" and is believed to be in equilibrium through a conformational change with an open complex. The latter serves as the template to initiate mRNA synthesis, since interbase hydrogen bonds have been broken, thus uncovering the bases of the template chain so that they can pair with ribonucleoside triphosphates.

In broad review, the steps involved in bacterial RNA synthesis include location of the promoter by the polymerase, formation of an enzyme–substrate complex, initiation of RNA formation, elongation of the RNA strand by the addition of nucleotides complementary to one or both of the DNA strands, and finally termination, this last being directed by a specific sequence in DNA (Chamberlin, 1976b). As described earlier, RNA polymerase belongs to the class of proteins which, while capable of binding polyanions such as DNA and glycosaminoglycans, are able to bind to certain specific sequences in DNA with greater affinity than to other, nonspecific sequences. To carry out transcription, the RNA polymerase holoenzyme must disregard general sequences until the promoter sequence is located. Nonspecific binding appears to be electrostatic in nature (Chamberlin, 1976b). Considering that nonspecific DNA sequences far outnumber promoter sequences, it is necessary for RNA polymerase to dissociate rapidly from, or disregard entirely, all sequences other than the promoter. The nonspecific sequences do in fact have low affinity for the enzyme. On the other hand, the enzyme binds to the promoters with an affinity 1000 times that with which it binds to general DNA sequences.

One of the most intriguing aspects of polymerase–promoter interaction is that initial recognition of promoter sequences occurs with native DNA, i.e., closed, double-stranded DNA. Such interaction must then involve the functional groups comprising the enzyme binding site and a promoter sequence which can be detected along the major or minor groove of the DNA double helix. Apparently, as soon as the closed promoter is recognized, tight binding occurs, and a starting complex of enzyme and promoter is formed. Further details of this process have been reviewed by Chamberlin (1974, 1976a,b). As currently viewed, the polymerase must interact with about 55 base pairs in DNA when it binds initially to the promoter. Modulating factors which facilitate initiation, such as the cAMP receptor protein (CRP), apparently function by causing a conformational change in the promoter sequence, thereby promoting formation of the open-polymerase–promoter complex.

RNA synthesis is mediated by RNA polymerase under the direction of a locally denatured region of a double-stranded DNA template. Much data on this has been accumulated in bacterial systems and has been summarized by Krakow et al. (1976), which the interested reader may consult for details. RNA polymerase has been described as "processive" (Krakow and Kumar, 1977) since, with the beginning of the chain initiation reactions directed by the native (double-stranded) DNA template, dissociation of growing RNA polymers does not take place until the termination signal is reached.

In prokaryotes, termination of RNA chain transcription may be considered from a number of viewpoints, including cessation of RNA elongation, release of the newly synthesized RNA polymer, and dissociation of the RNA polymerase from the DNA template, making it free to again engage in RNA transcription (Chamberlin, 1976b). It seems reasonable, then, to expect specific termination signals in the DNA template, which could interact with the RNA polymerase and which would result in release of both the product RNA and the polymerase. The discovery in bacteria of the *rho* factor, a protein of some 200,000 molecular weight, provided a possible mechanism by which such RNA chain termination can be accomplished. The *rho* factor has been shown to contain both RNA-dependent ATPase activity and transcription termination activity which can be easily uncoupled. As yet, the process of terminating RNA transcription has not been fully elucidated, and details of RNA chain termination in eukaryotes remain unknown.

2. Eukaryotic RNA Polymerases

As noted at the beginning of this section, the mechanism of interaction of RNA polymerases with DNA in bacteria is well on the way to being worked out. The situation with eukaryotes is not nearly as clear. In the first place, the eukaryotic genome, ensconced in its chromatin blanket, is not easy to isolate intact; in the second place, there is little evidence for a promoter sequence as is found in bacteria. Finally, there are three RNA polymerases in eukaryotes with distinctive subcellular localization and function.

Nonetheless, evidence is mounting to support the view that modification of the chromatin structure, through covalent or conformational changes, controls accessibility of individual genes to RNA polymerase (O'Malley et al., 1977; Chambon, 1975). Since eukaryotic cells possess several forms of RNA polymerase, with specific compartmentalization and function (in contrast to prokaryotic cells), this puts regulation of transcription partially under the aegis of these separate polymerases, since they transcribe only certain genes. And while promoter sequences as such do not exist in eukaryotic cells, specific protein–nucleic acid interactions are almost certainly involved in regulating transcription of the genome.

In studying the factors governing transcription specificity in eukaryotes, there are at least three essential ingredients; viz., purified polymerases free of nucleases and protease activity, an intact DNA template free from artifactual nicks and denaturation, and a means to accurately measure selective transcription. Having listed these requirements, we must recognize two impediments that must be overcome. First, owing to the complexity of the eukaryotic genome, DNA sequences encoding a specific mRNA are not likely to account for more than 1×10^{-6} of the total DNA sequence. And owing to breakage of the DNA during isolation, nonspecific starts are quite likely. Two alternative approaches—the use of chromatin, since it is likely to be the in vivo state of the genome, or DNA from animal viruses—have met with some success. The other major impediment in these studies is the difficulty of determining whether the RNA synthesized represents an RNA from a particular cistron or whether many RNAs were synthesized collectively. Selectivity of transcription has been evaluated by DNA–RNA hybridization. Several reviewers (Chambon, 1975; Biswas et al., 1975; Roeder, 1976) have concluded that as-yet-uncharacterized proteins, in addition to chromatin proteins, probably play a role in controlling transcription in eukaryotes.

E. Posttranscriptional Modification of Ribonucleic Acids

Modification of a precursor molecule, such as the zymogens (Chapter 4), by covalent changes presents the opportunity to make a biologically functional molecule available at the proper site for action within the cell. The various RNAs have also been shown to undergo such modification or processing, being converted from essentially linear molecules into mature forms which assume higher-order structures, thereby enabling them to interact in a specific manner with other nucleic acids and proteins by virtue of weak forces (Perry, 1976).

1. Heterogeneous Nuclear DNA and Messenger RNA

In contrast to the short-lived (order of minutes) mRNA molecules in prokaryotes, these message units in eukaryotes have a half life of 7–12 h. We have already mentioned the recent discovery of the discontinuous informational content of the genome which codes for mRNA, and the phenomenon of splicing which appears to be so widespread. With all of these complexities, it is noteworthy that all mRNA molecules thus far studied in eukaryotes are monocistronic, i.e., they code for only one protein, as opposed to a series of proteins as seen in prokaryotes.

After transcription in eukaryotic cells, several modifying reactions are carried out on the initial RNA species, including 3' polyadenylation, 5' end capping, cleavage of large RNA precursors, and splicing together of different segments of pre-mRNA into one mRNA chain.

a. Posttranscriptional Modification of Pre-mRNA. As a consequence of posttranscriptional polyadenylation of the 3' end of pre-mRNA molecules, cytoplasmic mRNA has a poly(A) tail of 50–200 residues. From a temporal point of view, addition of the polyadenyl groups can occur either before or after sizing of hnRNA in the nucleus as well as in the cytoplasm. Removal of the poly(A) end from globin mRNA does not affect the initial rate of globin synthesis in the Krebs ascites cell-free system, but, over longer periods of time, the deadenylated mRNA becomes less active. Thus, this poly(A) end appears to serve a role in stabilizing the message but is not involved in facilitating translation. There is evidence that certain proteins may bind to the poly(A) segment to inhibit nuclease activity (Revel and Groner, 1978; Perry, 1976). The process of polyadenylation and mRNA formation is depicted in Fig. 1.

b. "Capping" and Methylation in mRNA Translation. The 5' terminus of pre-mRNA has been shown to have a cap designated $m^7G^{5'}ppp^{5'}XmpYm$, in which a 5',5'-pyrophosphate group binds the terminal 7-methyl guanosine to the methylated nucleotide. The 5' cap plays at least two roles in translation; the m^7G moiety acts as a recognition site during initiation, while the cap protects the mRNA from exonuclease hydrolysis (Filipowicz, 1978; Revel and Groner, 1978). Removal of the cap inhibits translation, permitting degradation of nontranslated mRNAs, and, thus, enzymatic removal of the cap may act as one control of translation. Internal methylation of adenosine residues has been reported in eukaryotic and viral mRNAs. The role of the meth-

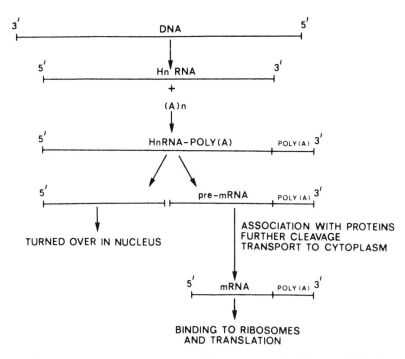

Figure 1. Schematic representation of biogenesis of globin mRNA. Reproduced, by kind permission of the authors and Grune and Stratton, from "The biosynthesis of hemoglobin" by Benz and Forget (1974), *Semin. Hematol.* 11:463–523.

ylated adenyl residues is not yet clear, but it has been speculated that they may represent recognition sites for enzymes which cleave or splice RNA. Another possibility is that these methylated bases could, in conjunction with specific signals, act as terminators of translation.

c. Speculation of the Biological Function of Intervening DNA Sequences. As a consequence of the recognition of "discontinuous genes" in the eukaryotic genome, the heterogeneous nuclear RNA (hnRNA) is now seen to be the unabridged transcriptional unit which, through the processes of excision of intervening sequences and splicing of the remaining RNA segments, becomes the final mRNA (Fig. 2). Several investigators have proposed that these intervening sequences accelerate evolution by providing the possibility of quantal-type jumps in functional modifications of proteins (Tonegawa et al., 1978; Doolittle,

1978; Darnell, 1978). With the presence of "silent" intervening DNA sequences in the genome, the mutational consequence of a single base change can extend beyond the traditional single amino acid substitution. Thus, the theoretical result of a base change occurring at the boundary of an intervening segment, which can alter the manner of reconstitution of the messenger, could thereby affect an entire amino acid sequence. The longer stretches of DNA, with intervening sequences imposed among functional genes, could allow for an increased rate of mutation as a result of recombinational events (Tonegawa et al., 1978). The splicing process itself and a variable degree of efficiency associated with it provide an added site at which evolutionary changes could occur. A variation in splicing efficiency can produce a "mutated" final product from a nonmutated precursor; thus, duplication of a gene need not be a necessary step in the production of a mutant expression. Further discussions on the evolutionary and functional role of intervening sequences can be found in Gilbert (1978), Doolittle (1978), Darnell (1978), Tonegawa et al. (1978), and Leder (1978).

2. Transfer Ribonucleic Acid (tRNA)

These adaptor molecules serve a multiplicity of functions in protein synthesis, as well as certain other functions in RNA metabolism (Rich

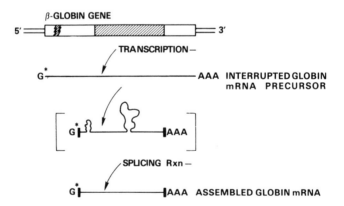

Figure 2. Diagram showing elimination of the intervening sequence that interrupts the globin gene and the final assembly of mRNA that will be translated. Reprinted, by kind permission, from *The New England Journal of Medicine*, **298**:1080, 1978.

and RajBhandary, 1976; Clark, 1977). A summary of their functions in the biosynthesis of proteins is impressive and illuminating. Transfer RNAs interact both with specific aminoacyl-tRNA synthetases and elongation factor EF-1 and with ribosomes. The tRNA molecule binds initially to the A site of the ribosome where, by virtue of its anticodon, it reads the codon in mRNA. Hence, tRNA molecules function at the interface between the nucleotide sequence of mRNA on the one hand and the amino acid sequence of proteins on the other.

Base pairing between nucleotides is the manner by which information is transmitted between nucleic acids. However, only tRNAs demonstrate specific base pairing interactions with amino acids. Of course, there is an additional component which is the key in the recognition process involving tRNA, that being the aminoacyl-tRNA synthetases. These cytoplasmic enzymes possess specificity both for particular amino acids and for specific tRNAs. The presence of a tertiary structure in the tRNA and the variety of modified bases in the different loops probably contributes significantly to the specificity of these interactions with synthetases on the one hand and nucleic acids on the other.

Each amino acid has at least one specific tRNA with which it becomes esterified and through which it is carried to the mRNA codon, although, in general, several tRNA molecules differing in primary structure are able to accept the same amino acid. These related tRNAs are termed isoaccepting species. Transfer RNAs have a molecular weight of approximately 26,000, being composed of about 75 nucleotides and sedimenting at 4 S. There is considerable similarity in size and basic structure between prokaryotic and eukaryotic tRNA molecules, since all exhibit a cloverleaf-type structure (Fig. 3). By X-ray crystallography, the four-part cloverleaf folds onto itself, producing a tertiary structure with an L shape (Fig. 4). As shown in Fig. 3, there are four hydrogen-bonded stems. One stem terminates in the sequence CCA and serves as the site to which the amino acid is esterified. Modified bases are common in the loops of the other three stems (I, II, IV). The dihydrouracil loop (I) contains 5,6-dihydrouridine to a variable degree and in diverse locations. The anticodon loop (II) is shown directly opposite to the amino acid acceptor end in the cloverleaf. This loop is responsible for the complementary interaction with the codon in mRNA. The TψC loop (IV) contains the pseudouridine base in its nucleotide sequence. Recognition sites for aminoacyl-tRNA synthetases, as well as one for ribosome attachment, must also exist on each tRNA molecule. It is presumed that higher-order structure is involved in such recognition.

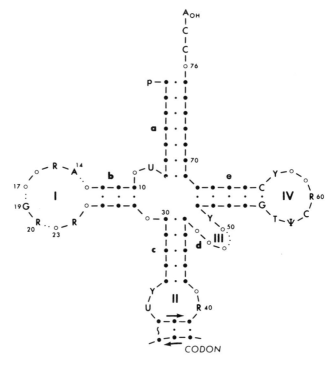

Figure 3. Generalized two dimensional cloverleaf formula for tRNA. Reprinted with the kind permission of T. H. Jukes and Elsevier Scientific Publishers, Elsevier North-Holland Biomedical Press, from *Comprehensive Biochemistry*, Vol. 24, Chapter 5, p. 255, 1977.

Abundant evidence exists for a precursor–product relationship between a metabolically unstable and heterogeneous (4–5 S) pre-tRNA fraction and mature 4 S tRNA (Perry, 1976; Krakow and Kumar, 1977). Cytoplasmic nucleases process the pre-tRNA, converting it to the mature form. The most significant posttranscriptional modification involves certain bases which undergo methylation, isopentenylation, and formation of pseudouridine, in which the N-glycosyl linkage is at position 5 rather than at position 1 as in uracil. The maturation process for tRNA is schematized in Fig. 5. All tRNAs are characterized by a common base sequence at the 3′ end, namely CCA_{OH}. As a result of an enzyme-catalyzed reaction, amino acids specified by the tRNA anticodon become covalently linked to the 3′ terminal adenosine of tRNAs. Coalescence of amino acids with their specific tRNAs occurs through a two-step reaction mediated by cytoplasmic enzymes called aminoacyl tRNA synthetases, which have specificity both for their amino acid as well as their tRNA (Holler, 1978). The two-step reactions catalyzed by these enzymes are (1) nucleophilic attack of the anhydride bond between the α and β phosphate groups in ATP, with the formation of aminoacyl–AMP

$$AA + ATP + E \rightleftharpoons AA\text{-}AMP\text{-}E + PP_i$$

and (2) transfer of the amino acid from the adenylate to the tRNA

$$AA\text{-}AMP\text{-}E + tRNA \rightleftharpoons AA\text{-}tRNA + AMP + E$$

While the first step is not very discriminating, the second step exhibits considerable specificity for both the amino acid and tRNA species. The

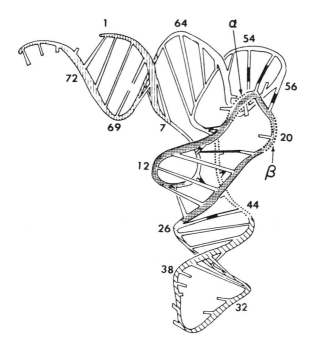

Figure 4. A schematic diagam showing the folding of the yeast tRNA[phe] molecule. The anticodon is at the bottom of the figure while the amino acid acceptor is at the upper left. Reprinted with the kind permission of T. H. Jukes and Elsevier Scientific Publishers, Elsevier North-Holland Biomedical Press, from *Comprehensive Biochemistry*, Vol. 24, Chapter 5, p. 280, 1977.

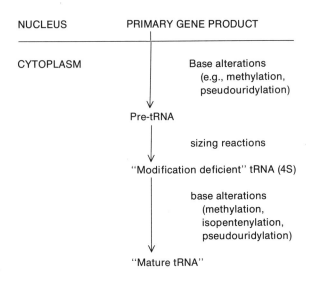

Figure 5. Maturation process for transfer RNA. Reprinted with the kind permission of J. S. Krakow and S. Anand Kumar, and Elsevier Scientific Publishers, Elsevier North-Holland Biomedical Press, *Comprehensive Biochemistry*, Vol. 24, Chapter 3, p. 165, 1977.

first step is analogous to the initial step in the activation of fatty acids prior to β-oxidation. In the present case, an enzyme-bound aminoacyl adenylic acid is formed, in which the carboxyl group of the amino acid is linked by an acid anhydride bond with the 5'-phosphate of AMP, forming a mixed anhydride. In the second step, the aminoacyl group is transferred to the 3' end of tRNA, forming a single ester linkage in which the activated state of the carboxyl group persists.

An important feature of this reaction highlights a significant fact about biosyntheses generally. The elimination of pyrophosphate as a reaction product, which at first glance would appear to be wasteful, assures the completion of the activating reactions because of the ubiquitous availability of pyrophosphatases. These enzymes hydrolyze pyrophosphates into orthophosphates, thus ensuring the irreversibility of the acylation reaction by removing pyrophosphate as a reaction product (cf. Chapter 2, the role of water).

Selection of the correct amino acid by an aminoacyl-tRNA synthetase is extremely important, since once an amino acid is covalently linked to its tRNA, it is the tRNA anticodon which specifies pairing with its cognate codon in mRNA. Thus, a misacylated amino acid could be inserted into the growing peptide. A moment's reflection will underscore the problem confronting a particular aminoacyl-tRNA synthetase that must discriminate accurately between two amino acids with similar structures, such as isoleucine and valine. Apparently this specificity problem has been solved by making the loading of an amino acid on its cognate tRNA a multistep process, where aminoacyl rejection is possible at any one of the steps. While controversy exists as to the exact nature of the process (Hopfield, 1974; Fersht, 1977a,b), the process involves an "editing" or proofreading mechanism, whereby a covalent intermediate in the reaction sequence undergoes hydrolysis before it can be brought to the mRNA codon. Thus, a lag time is introduced which permits an "information seeking step" and can result in hydrolytic cleavage of the wrongly charged amino acid from its linkage with tRNA.

3. Ribosomal Ribonucleic Acid (rRNA) and Ribosomes

The ribosome, the most complex multicomponent apparatus of the protein-synthesizing machinery, is made up of two unequal subparticles. In eukaryotes, there are over 70 proteins and at least three different RNA species which comprise the ribosome. An analogy with multienzyme complexes is suggested by the peptidyl transferase and translocation activities of the ribosome and by the multiple subunits comprising this complex. With so many proteins involved, it is likely that some serve structural roles while others serve either recognition functions or catalytic functions. The peptide bond formation requires the approximation at the ribosome of mRNA, aminoacyl-tRNA, and peptidyl-tRNA, along with necessary initiation, elongation, and termination factors.

a. Structural Features. Eukaryotic cytoplasmic ribosomes are large particles with a molecular weight in the range of 4–4.6 million. Ribosomes from mitochondria and chloroplasts differ from their cytoplasmic counterparts, resembling the prokaryotic ribosomes in structure. As noted earlier, there is a substantial protein component to the ribosome, of which many proteins possess basic lysine and arginine residues that undoubtedly facilitate interaction with the nucleic acid. In both prokaryotes and eukaryotes, the ribosome is composed of two nonidentical subunits, called the small and large subunits. In prokaryotes, the subunits are 30 S and 50 S, respectively. The former contains 16 S RNA along with 21 distinct proteins, while the latter contains 23 S and 5 S RNA as well as 34 different proteins. One protein is apparently shared by both ribosomal subunits. Both ribonucleoprotein subunits have been reconstituted in vitro, suggesting that ribosomes undergo self assembly.

While the basic features of the cytoplasmic eukaryotic ribosome resemble those of the prokaryotic type, the eukaryotic ribosome is an 80 S particle composed of 40 S and 60 S subunits. The smaller subunit is constructed from 18 S RNA and about 30 proteins, while the larger subunit is composed of 28 S and 5 S RNA as well as about 40 proteins. Perhaps the most striking difference exhibited on electron microscopy is the attachment of the 40S subunit to the endoplasmic reticulum in the cytoplasm.

b. Processing of Ribosomal RNA. In contrast to mRNA and tRNA, transcription of rRNA precursors by polymerase I (A), and subsequent processing of the precursor rRNAs are carried out in the nucleolus (Krakow and Kumar, 1977). In mammals, rRNA is synthesized as a large precursor molecule (45 S) containing about 14,000 nucleotides. It is composed of the sequences of one molecule of 28 S RNA (the rRNA in the 60 S ribosomal subunit), one molecule of 18 S RNA (the component of the small 40 S ribosomal subunit) and a nonribosomal portion which does not appear in any ribosomal subunit. Precursor–product labeling experiments have shown that cleavage of 45 S RNA results in formation of 18 S and 32 S RNA. The 32 S RNA is subsequently trimmed to 28 S.

c. Ribosome Function. As with the other nucleic acids in protein biosynthesis, rRNA molecules engage in specific interactions essential to translation. The 40 S subunit carries the decoding site, while the 60 S subunit bears the peptidyl transferase activity. The latter particle, on binding to the 40 S unit, also provides an A (aminoacyl) site for the aminoacyl-tRNA carrying the next amino acid to be linked to the nascent polypeptide, which is bound to the D (donor) or P (peptidyl) site through a tRNA molecule. These two binding sites (A and D) are juxtaposed so that peptide bond formation occurs readily. With formation of the new peptide link, the growing polypeptide becomes attached to the aminoacyl-tRNA on the A site. Before translation can proceed further, this newly formed peptidyl-tRNA must be translocated to the D site. This occurs after expulsion of the deacylated tRNA from the D site, as the mRNA shifts one codon closer toward the 3' end, carrying the tRNA attached to the growing polypeptide chain from the A to the D site. This now renders the A site free to bind the next amino acid to be incorporated into the chain. This chain of events recurs (elongation) until the termination codon is reached (see Figs. 6–8). In addition to the mRNA codon–tRNA anticodon interaction, it may be worthwhile to consider the stereospecific binding requirements for the A and D sites on the complete ribosomal unit. Kurland (1977) has summarized these needs, noting that the A site, binding the specific aminoacyl-tRNA complex, must possess the ability to distinguish the tertiary structure of the tRNAs, while the donor site (D or P) has less stringent recognitive requirements. This latter site apparently must recognize only the fact that the tRNA is esterified to an amino acid via the 3' terminal adenylate residue of the sequence -CCA$_{OH}$. The two elongation factors, EF-1 and EF-2 (see below), aid in these interactions.

A single mRNA may accommodate several ribosomes interspersed along the length of the message at any one time. These elaborate functional entities are called "polysomes." The ribosomes which comprise the polysomes are heterogeneous in the sense that the growing polypeptide chain attached to each ribosome differs in length. The farther an individual ribosome has travelled from the 5' end of the mRNA, the longer its nascent polypeptide chain. Thus, a polysome provides a means by which a single genetic message can be translated at a rapid rate.

F. Protein Biosynthesis

1. Protein Factors

Specific protein factors are required for fidelity of mRNA translation. Others ensure a rate of synthesis consistent with the needs of the metabolic economy of a particular cell type. Protein synthesis may be divided into formation of the initiation complex, initiation, elongation, and termination. This subject is quite broad in scope and has been reviewed recently by Mazumder and Szer (1977), Weissbach and Pestka (1977), and Safer and Anderson (1978). The interested reader may wish to turn to these articles for further details, as well as for extensive reference lists.

2. Initiation and Initiator-Complex Formation

Owing to the complexity of the eukaryotic ribosome and the variability of the initiation nucleotide sequences in mRNA to which ribosomes bind, it is likely that both RNA–RNA and protein–RNA interactions mediate recognition of tRNA by the ribosome (Revel and Groner, 1978).

We recall that mRNA is read from the 5' to the 3' end, while consecutive additions to the growing polypeptide chain occur from the amino terminus toward the carboxy terminus.

The codon AUG, for methionine, serves as the initiation codon. The nucleotide sequence preceding AUG probably also plays a role in location of the start signal. As in prokaryotes, a special methionyl-tRNA (Met-tRNA$_f$) functions as the initiator tRNA in eukaryotes. This tRNA is capable of being formylated by a bacterial enzyme (thus the subscript f), although it possesses no formyl group in the eukaryotic cell. In both prokaryotes and eukaryotes, internal methionine positions in a polypeptide appear to be supplied by another species of tRNA, designated Met-tRNA$_m$, although some investigators believe that Met-tRNA$_f$ also supplies internal methionine residues (Revel, 1977).

The role of the cap structure in this recognition process has been reviewed by Revel and Groner (1978) and Filipowicz (1978). For example, ribosome binding experiments have shown preferential binding of mRNA molecules containing a 5' terminal mG to 40 S subunits from wheat germ. In addition, the use of cap analogs, which inhibit in vitro binding and translation, has shown them to interfere with the formation of the initiation complex through interaction with eIF-4B, the soluble factor that seems to function in binding to the methyl cap. This factor thus plays a key role in mRNA binding (Safer and Anderson, 1978).

In 1965, with a return to studies using natural mRNAs instead of synthetic polyribonucleotides, several investigators reported the requirement of a number of specialized protein factors for initiation of translation. They were found to be associated, in both prokaryotes and eukaryotes, with the small ribosomal subunit. Table 5 lists the eukaryotic initiation factors and their functions in this process.

The formation of the initiation complex necessary to translate the amino acid sequences specified in mRNA is, in reality, a series of reactions involving ribosomal subunits, protein factors, initiator tRNA, mRNA, and GTP. This process has been reviewed by Revel (1977), Mazumder and Szer (1977), and Safer and Anderson (1978) and is presented in Fig. 6. The first step in the formation of the complete initiator

Table 5. Eukaryotic Initiation Factors

Factor	Function
eIF-1	40 S complex formation, 80 S ribosome dissociation
eIF-2	Met-tRNA$_f$ binding, GTP binding
eIF-2A	tRNA binding to 40 S subunit
eIF-3	Prevents reassociation of 40 S and 60 S subunit, mRNA binding
eIF-4A	Natural mRNA binding
eIF-4B	mRNA (cap) recognition, mRNA binding
eIF-4C	Stabilization
eIF-4D	Subunit joining, stabilization
eIF-5	GTPase, initiation factor release, subunit joining to form 80 S unit

Regulation of Protein Biosynthesis

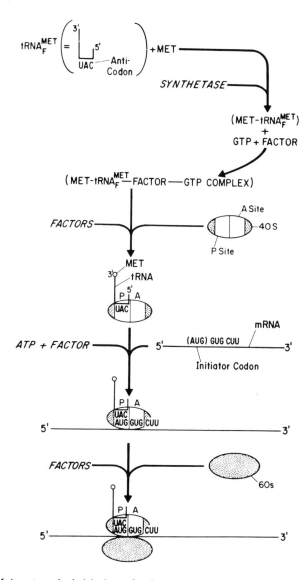

Figure 6. Major steps in initiation of polypeptide chain synthesis. Reproduced, by kind permission of the authors and Grune and Stratton, from "The biosynthesis of hemoglobin" by Benz and Forget (1974), *Semin. Hematol.* **11**:463–523.

complex is the binding of eIF-3, a multimeric protein (mol. wt. 700,000), to the 40 S subunit, in an energy-independent process. When this factor binds to the 40 S subunit, the 60 S subunit is prevented from reassociating with the 40 S subunit until Met-tRNA$_f$ binds to the 40 S subunit on the donor or peptidyl site. Binding of Met-tRNA$_f$, the second step, requires the prior formation of a stable tertiary complex with eIF-2, GTP, and Met-tRNA$_f$, in that ordered sequence. Binding of this complex to the 40 S subunit does not require the AUG initiation codon or mRNA. In the third step, mRNA binds to the 40 S preinitiation complex in an interaction that requires factors eIF-4A and eIF-4B, as well as ATP. In addition, the 5'-m^{7}G(5')ppp(5')X cap appears to be essential for binding and translation. In the final step, the 60 S ribosomal subunit joins the 40 S preinitiation complex to form the 80 S initiation complex; this interaction requires eIF-5 and the GTP which was bound in the second step above. In this final step, eIF-5 makes possible the expression of ribosome-dependent GTPase activity, which mediates dissociation of factors eIF-2 and eIF-3 prior to coalescence of the 60 S and 40 S subunits.

3. Elongation and Translocation

When the 60 S ribosomal subunit joins the 40 S, the acceptor or A site is actually formed for the first time. Two proteins, EF-1 and EF-2, as well as GTP, are required for this process. EF-1 recognizes and binds all aminoacyl-tRNAs, save for the initiator tRNA, to the A site. EF-2 is involved in translocation of peptidyl-tRNA from the A site to the D or P site, in association with the coordinated movement of mRNA and the ejection of deacylated tRNA from the D or P site on the ribosome. Diphtheria toxin catalyzes an ADP-ribosylation of EF-2 in a reaction where NAD acts as the donor of the ADP-ribosyl moiety. As a consequence of this reaction, the EF-2 is rendered ineffective in the translocation process. The process of elongation has been reviewed by Miller and Weissbach (1977), Mazumder and Szer (1977), and Safer and Anderson (1978), and that of translocation by Brot (1977) and Spirin (1978). Both processes are depicted in Figs. 7 and 8.

Recalling the Met-tRNA$_f$ occupies the D or P site, the codon next to AUG (toward the 3' end of the mRNA) directs binding of the specified aminoacyl-tRNA to the A site. In the next step, catalyzed by the peptidyl transferase activity associated with the 60 S subunit, the first peptide bond is formed between the carboxyl group of methionine and the free α-amino group of the incoming aminoacyl-tRNA. The reaction is formally a nucleophilic attack of the amino group of the aminoacyl-tRNA on the esterified carboxyl carbon of Met-tRNA$_f$ or, in subsequent steps, the peptidyl-tRNA (Harris and Pestka, 1977). Several proteins associated with the 60 S subunit are involved in this activity and in binding, but details of the process are unknown.

Subsequent to peptide bond formation, both the D or P and the A sites are occupied, the former with the deacylated tRNA and the latter by a peptidyl-tRNA. This is the "so-called" pretranslocational state (Fig. 7). Further elongation of the peptide chain requires departure of the deacylated tRNA from the P (or D) site. In this process, the ribosome moves one codon toward the 3' end of the mRNA. As a conse-

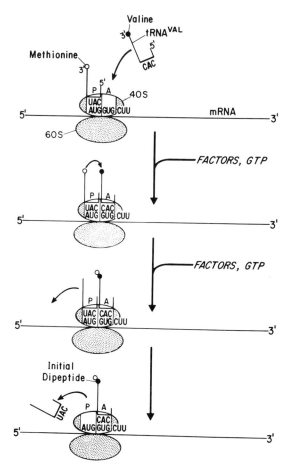

Figure 7. Initial dipeptide formation. Reproduced, by kind permission of the authors and Grune and Stratton, from "The biosynthesis of hemoglobin" by Benz and Forget (1974), *Semin. Hematol.* **11**: 463–523.

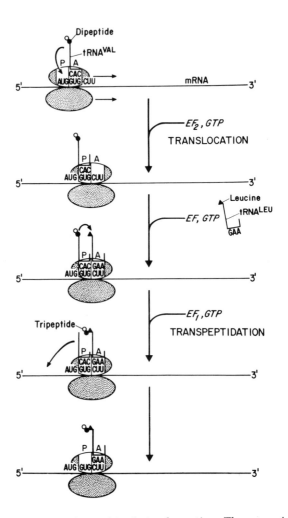

Figure 8. Major steps of polypeptide chain elongation. The arrow in the uppermost panel shows the shift of the peptidyl tRNA^{val} from the A to P (D) site as the ribosome advances one codon. This advance puts the CUU codon for leucine in the A site, and the GUG in the P site as the ribosome moves toward the 3' end of the mRNA. Reproduced, by kind permission of the authors and Grune and Stratton, from "The biosynthesis of hemoglobin" by Benz and Forget (1974), *Semin. Hematol.* **11**:463–523.

quence of this translocation process, the new codon at the A site instructs the binding of the newly specified aminoacyl-tRNA (Fig. 8).

The energetics of the translocation process and the role of GTP therein have fascinated investigators for many years. Brot (1977) and Spirin (1978) have reviewed this process and the proposed means by which translocation is carried out. Spirin has suggested that, while the ribosome binds peptidyl-tRNA on the acceptor site (i.e., prior to translocation), the whole system is akin to a high-energy intermediate in which eventual translocation is thermodynamically ensured. This provocative hypothesis merits careful reading in view of its similarity to the role of the membrane in the chemiosmotic hypothesis for oxidative phosphorylation (Chapter 9).

4. Termination and Release of Proteins

Termination of growth of the peptide chain is signaled by specific codons (UAA, UAG, and UGA) in the mRNA at the A site of the ribosome, and release requires a soluble protein factor, RF. In the process, the peptidyl-tRNA occupying the P or D site is hydrolyzed, liberating the polypeptide. These termination codons are the so-called nonsense codons.

Caskey (1977) has proposed the following model for termination. With addition of the final amino acid to the nascent polypeptide, the chain moves from the A site to the P or D site, with the resultant placement of a terminator codon in the A site. Since this codon does not specify an amino acid, RF and GTP bind to the A site. It is proposed that RF activates the peptidyl transferase activity on the 60 S subunit which mediates hydrolysis of the ester bond of peptidyl-tRNA. In this model, GTP hydrolysis permits dissociation of the RF. This process has been considered in greater detail by Mazumder and Szer (1977), as well as by Caskey (1977). The role of GTP in this process appears to be analogous to that in initiation and elongation; it is not required for binding of soluble factors to the ribosome but rather is utilized to mediate dissociation of these factors.

Posttranslational modification of proteins is a ubiquitous phenomenon since approximately 140 amino acids and derivatized amino acids have been found in various proteins, while, as we have discussed before, the amino acid code provides information for only 20 amino acids (Uy and Wold, 1977; Dreyfus et al., 1978). Cleavage of peptide bonds or addition of substituents to individual amino acids are the means of posttranslational processing of proteins. Cleavage of peptide bonds is the strategy utilized with the zymogens, pepsin, trypsin, and chymotrypsin, coagulation proteins, immunoglobulins, complement fixation, collagen synthesis, and hormone activation, to name several. Derivatization of individual amino acids includes the formation of disulfide bonds, and glycosyl, acyl, phospho-, and methyl substituents.

While there is evidence that some modification reactions take place while the growing polypeptide chain is still attached to the polysomes, most modifications occur after release of the chain from the ribosome. Examples of processing of chains still on the ribosome include formation of hydroxylysine and hydroxyproline in collagen elaboration.

Postrelease modification is carried out by enzymes in several loci: the cytosol, the endoplasmic reticulum, the Golgi complex, storage granules, and the extracellular space. While protein kinases and methylases are found in the cytosol, it appears that the preponderance of glycosylation, disulfide bond formation, halogenation, and some acetylation occurs in the Golgi complex. Precursor–product peptide bond cleavage takes place in secretory granules, and cross-link formation in collagen occurs extracellularly.

Since posttranslational modification reactions involve only a few types of amino acids—and then only some of those in a protein—certain sequences must provide recognitive specificity for the processing enzymes. As we have considered in Chapter 3, such specificity is the hallmark of enzymatic action. Even after a protein folds to a higher-order structure, the amino acid sequence must be the key to determining the sites for processing, with the added proviso that the sequences must be accessible to the processing enzymes. This specificity has been reviewed by Uy and Wold (1977). As an example, glycosylation of asparagine residue requires the sequence Asn-X-Thr(Ser), while phosphorylation of serine or threonine requires the sequence Arg-X-Y-(Z)-Ser(Thr). Similar sequential information is required for hydroxylation of proline (X-Pro-Gly), or lysine (X-Lys-Gly).

As yet unsolved is the observation that only a portion of a particular protein undergoes processing. For example, bovine pancreatic ribonuclease is secreted in both nonglycosylated and glycosylated forms. Undoubtedly, as in collagen, some derivatives provide the opportunity to further stabilize a protein for its ultimate biological function. Thyroxin, a posttranslationally modified amino acid stored in the macromolecular glycoprotein thyroglobulin (mol. wt. 670,000), is released to the cell by degradation of the protein; this may represent a means of ensuring a rapid response to changing demands for this hormone. We have considered phosphorylation–dephosphorylation reactions in the context of enzyme modulation (Chapter 4), and it seems that certain acetylation and methylation reactions also play a role in controlling the activity of enzymes.

III. The Regulation of Protein Biosynthesis

A. Regulation at the Gene Level

Since the genome carries in it the information that dictates the formation of all cellular regulatory mechanisms, the control of genetic expression (protein synthesis) at all levels becomes of primary interest when examining control mechanisms. Since protein is the product of gene expression, regulation of protein synthesis at the gene level could be the most energetically economical method for suppressing information contained in the genome. The more complex the organism, the more important it becomes to have a genetic "light-switch" mechanism to turn portions of the genome on or off and provide for the differentiation of multitissued organisms.

1. The Prokaryotic System

Protein biosynthesis is the ultimate gauge of gene expression. In bacteria, some genes are permanently "turned on" in that they produce enzymes *constitutively*, i.e., regardless of environmental conditions. These enzymes are generally involved in universally occurring, energy-yielding pathways and are sometimes referred to as "housekeeping"enzymes. Other genes are *inducible*, often producing only trace amounts of protein, except in response to given environmental stimuli, when enzyme production is greatly increased.

The most well studied inducible genes are those in *E. coli* that encode the enzyme required for lactose degradation. Jacob and Monod (1961) first proposed the existence of the *operon* to explain the phenomenon of enzyme induction. This model has become one of the most powerful concepts of regulation in the field of biochemical genetics. The operon is a cluster of functionally related genes including a control region. In *E. coli*, the genes for the enzymes that degrade lactose, and their controls, are clustered in the *lac* operon. This much-investigated entity has been amply reviewed elsewhere (Chamberlin, 1974), but serves as an excellent example of the operon model. It consists of a control region comprised sequentially, from the 3' end of the template, of a promoter (the site for initial RNA polymerase binding) and an operator [(which binds a *repressor* (R)], followed by the structural genes for three proteins involved in lactose catabolism. When the promoter is free, transcription can begin as the RNA polymerase binds to the promoter sequence in the DNA. Certain proteins, such as the cAMP receptor protein which also binds to the promoter, aid in this process, evidently by causing a conformational change. If a repressor is bound to the operator, transcription is blocked and the operon is repressed. Genes which code for repressor proteins may be located at any point in the genome, and are usually transcribed at a slow but constant rate. Inducer substances can bind to repressor proteins and, by virtue of the conformational change they induce in the repressor, block its ability to bind to the operator. This results in continued transcription of the operon and, consequently, continued protein synthesis of the inducible enzyme encoded in the operon.

Mutations in the regulatory genes usually produce dramatic results. A mutant repressor gene may produce a defective protein which cannot bind to the operator and would result in continued protein synthesis. Enzymes which, in normal cells, are inducible thus become constitutive. Constitutive operator mutants also occur where an altered base sequence in the DNA of the operator region results in loss of the ability to bind the normal repressor protein.

2. Eukaryotic Gene Expression

The segregation of genetic information into a compact organelle, the nucleus, must raise the question of the functional or survival value, or both, of such an organization. The persistence of the nucleus and the remarkable proliferation of organisms employing this mode of information storage strongly support an ability to compete successfully in the evolutionary process. What advantages accrue to the cell from such an organization? Busch (1977) has proposed that such segregation may protect the genome from degradative cytoplasmic enzymes. The advantage for control derives from the compartmentalization of metabolic processes (Chapter 13).

Owing to the greater complexity of the eukaryotic genome, one would anticipate finding regulatory factors charged with aiding in the selective transcription of specific DNA sequences. Additionally, the phenomenon of differentiation requires permanent repression of certain genes within the complete genome possessed by a particular cell type and the ability of other genes to be transcribed with the appropriate signal, with the best understood signals being certain hormones. The fraction of the total eukaryotic genome susceptible to reversible induction or depression is quite small.

While control of gene expression can involve mRNA processing, translational control, and postribosomal covalent modification, it appears that, in eukaryotes as in prokaryotes, transcriptional control is the cardinal means for regulating gene expression.

a. The Discontinuous Eukaryotic Genome. Joining of distant nucleotide sequences in pre-mRNA by a "splicing" mechanism to form the final messenger appears to be a widespread phenomenon. Indeed, an ever-increasing body of evidence demonstrates that long stretches of noncoding, nontranslatable intervening sequences interrupt the eukaryotic genome (Flavell et al., 1978; Kourilsky and Chamber, 1978; Revel and Groner, 1978; Darnell, 1978; Crick, 1979). Hence, the colinear relationship between DNA and its functional RNA transcripts characteristic of bacteria does not exist in eukaryotes, since the information encoded in noncontiguous DNA sequences is ultimately transcribed into a monocistronic functional mRNA. Thus, the "one gene–one polypeptide" dogma for prokaryotes must be modified for the eukaryotic system along the lines expressed by Blake (1978)—"genes-in-pieces → a protein (-in-pieces?)."

Two wide-reaching experimental approaches employed to study the eukaryotic genome have led to the above discoveries, which add to a growing list of structural or functional features separating the eukaryotic genome from the prokaryotic. These are (1) gene mapping, a field that has exploded because of the availability of a wide range of restriction endonucleases with sequence specificity; and (2) cloning of mammalian DNA sequences in bacteria.

1. Restriction endonucleases. These bacterial enzymes are endodeoxyribonucleases which cleave the phosphodiester bonds of both strands of double-stranded DNA at specific nucleotide sequences (Nathans and Smith, 1975; Arber, 1978). These nucleotide sequences always exhibit twofold rotational symmetry about the cleavage points such that the identical sequence may be read on each of the two DNA strands when read with the same polarity. The points at which the complementary DNA strands are cleaved may be staggered, thus facilitating recombination by producing "sticky," single-stranded regions of complementary sequences at the ends of double-stranded DNA segments (Linn, 1978). Armed with a group of such enzymes, exhibiting a spectrum of sequence specificities, it is possible, as with the proteolytic enzymes, to prepare a variety of hydrolytic digests of overlapping DNA sequences. Thus, DNA mapping by fingerprint techniques analogous to those used in peptide sequencing can permit delineation of the linear array of deoxyribonucleotide bases in the genome. Through use of several restriction enzymes, the β-globin structural gene in the rabbit (Tilghman et al., 1978) has been shown to contain an intervening sequence of approximately 600 nucleotide bases, located somewhere between the sequences coding for amino acid 100 and amino acid 120. Additionally, similar studies in the mouse have demonstrated an inter-

vening sequence after the triplet coding for amino acid 104 (Jeffreys and Flavell, 1977).

2. *Gene cloning.* The other method for exploring the eukaryotic gene is one of great discriminative power and depends upon the codon–anticodon recognition and binding of complementary nucleotide strands. Thus, in a manner analogous to binding a specific antibody to a membrane protein in order to isolate and identify it, an mRNA strand can be transcribed into its complementary DNA (cDNA) by a reverse transcriptase, which then provides specific DNA for further replication by DNA polymerase. The product, double-stranded DNA (ds-cDNA), is then recombined in an appropriate bacterial plasmid, which provides the essential functions for autonomous replication of the plasmid and associated nonbacterial DNA sequences. Thus, large quantities of specific segments of the mammalian genome can be obtained and purified from the modified plasmid DNA using specific endonucleases (Nathans and Smith, 1975).

b. Sequence Frequencies and Homologies. Prior to the discovery of intervening sequences in the eukaryotic genome, evidence gained by studies of the reassociation of complementary strands of purified DNA supported a division of DNA sequences into three classes, on the basis of frequency of repetition (Britten and Kohne, 1968). These classes are (1) single copies, or the unique sequence class which represents most structural and enzymic proteins; (2) moderately repetitive sequences, which include the genes for histones and rRNA as well as sequences which occur adjacent to structural genes; and (3) highly repetitive sequences which occur in thousands of copies and are localized in chromomeres and other chromosomal constrictions. The unique sequences account for approximately 60% of the DNA, while the moderately repetitive sequences represent 30%, with the highly repetitive comprising the remainder (Schmid and Deininger, 1975; Davidson et al., 1977). Several possible roles in regulating gene expression on the same and separate chromosomes have been proposed, and are discussed below in the section called "An Operon for Eukaryotes?"

c. Gene Amplification. Recently, evidence has been presented that genes themselves may be multiplied or amplified in some types of cultured cells during treatment with selection agents. Such a phenomenon points to an exciting possibility for an additional means of regulation of protein biosynthesis and for a process contributing to adaption of differentiation of cells. For the first time, Alt et al. (1978) have demonstrated that selective multiplication of dihydrofolate reductase genes can account for the increased production of this enzyme in methotrexate (MTX)-treated murine sarcoma and lymphoma cells. While the evidence for overprotection of enzymes by gene amplification in such systems is limited to cells which are not highly differentiated, these studies clearly indicate that the potential for gene amplification does exist. Identification of MTX-resistant cells is accomplished by culturing the cells in sequentially increasing concentrations of the folic acid analog. Among the MTX-resistant cells, some lose their highly resistant characteristic after growth in the absence of MTX, while others appear to be retained as a resistant line. In all cases, the degree of resistance correlates with the degree of enzyme overproduction, a proportional increase in the mRNA coding for the enzyme, and a corresponding primary increase in the number of gene copies per cell coding for the enzyme mRNA. Using in vivo hybridization studies with DNA complementary to the enzyme mRNA, the excess gene copies have been localized to a single chromosome and, indeed, to a specific "homogeneously staining region" (HSR), which appears to be expanded in MTX-resistant cells (Nunberg et al., 1978). Loss of MTX resistance is correlated with a diminution of the HSR region. These elegant experiments of Schimke's group cannot answer the question of whether such gene amplification is present in cells other than those infected by viruses or grown under selection pressure. The possibilities exist, however, for an exciting regulatory mechanism of an evolutionary nature, which could be present during development or differentiation.

d. Irreversible Repression of Genes. We have exphasized that differentiation of multicellular organisms requires selective repression of the potentially enormous amount of information encoded in the genome of the fertilized ovum, so that, in the mature cells of a particular organ, only certain structural and functional proteins are synthesized (Rutter et al., 1973; Caplan and Ordahl, 1978). While all cell types possess enzymes required for essential life processes (protein biosynthesis and glycolysis, for example), specialized cells produce specific proteins which are their biochemical fingerprint and, in fact, account for the properties of the phenotype. O'Malley et al. (1977) and Caplan and Ordahl (1978) have proposed that the expression of phenotype-specific genes can exist in a number of categories of repression and accessibility to transcription, at different stages in the cell's existence. The proposed categories span the following possibilities: (1) genes that are never repressed and that encode proteins which are the biochemical hallmark of a particular cell; (2) genes whose transcription is reversibly repressed and are therefore able to respond to a signal; (3) genes whose transcription ceases, either coincidentally with repression—and which are therefore not inducible—or those where there is a hiatus between cessation of transcription and irreversible repression, so that the genes are inducible during that period.

e. Hormones. Induction of specific enzymes and, indeed, differentiation itself must involve regulation of the genome in response to signals from either the nucleus, transmitted through the cytoplasm, or from the plasma membrane. The latter signals are often amplified by an additional response of the cell membrane. The organization of nucleoproteins into the chromatin and the accessibility of actively transcribing genes to pancreatic DNAse supports the view that an important means of transcriptional control involves alterations of these proteins which surround the DNA. Both histones and nonhistones influence binding to the DNA template, as well as activity of RNA polymerases (Jungmann and Russell, 1977).

Both polypeptide hormones and steroid hormones have been shown to act at the level of transcription (Krakow and Kumar, 1977). Certain polypeptide hormones act by binding to membrane receptors of target cells, rather than by penetrating the cell, and thereby control the activity of protein kinases by causing the dissociation of a regulatory subunit from the holoenzyme. This results in full expression of the catalytic activity of the remaining subunit. There is evidence that phosphorylation may play a role in gene regulation. Indeed, both histones and nonhistones undergo phosphorylation. Studies with mutant cells, devoid of cAMP-dependent protein kinase, showed the inability of these cells to induce phosphodiesterase; cells with the enzyme were able to induce phosphodiesterase (Jungmann and Russell, 1977). It has been proposed that nuclear cAMP binding activity might function in a manner analogous to that of the cAMP receptor protein of *E. coli*, causing a conformational change in a control protein, which then stimulates transcription (Krakow and Kumar, 1977).

Activity of eukaryotic RNA polymerase may also be modulated by a phosphorylation–dephosphorylation modification. For example, in vitro studies with ovarian nuclear RNA polymerase and a cAMP-dependent protein kinase led to marked stimulation of the polymerase activity. A growing body of data supports the intriguing contention that cAMP mediates the translocation of the cytoplasmic protein kinases into the nucleus. This would then be the means by which the information first received at the cell membrane is communicated to the genome (Jungmann and Russell, 1977; Jungmann and Kranias, 1977).

The mechanism of action of insulin and its role in protein synthesis is not well understood. It appears that, on a physiologic level, insulin functions as a growth hormone in the human fetus (Villee, 1975); but studies on a biochemical level have failed to show the level at which insulin acts. Recently, Peavy et al. (1978) investigated the role of alloxan-induced diabetes and subsequent insulin administration on albumin synthesis by rat liver. They found diminution of albumin synthesis in experimental diabetes, with resumption of normal synthesis when insulin was administered. Evaluation of albumin mRNA in a cell-free protein synthesizing system, or by hybridization to cDNA, revealed that the messenger itself was decreased in these animals, and that when insulin was given to these animals, mRNA levels for albumin increased. These intriguing results must await confirmation and elucidation of the precise site at which insulin is acting.

Thyroid hormone, like insulin, continues to present a puzzle in terms of its mechanism of action. Since thyroid hormones increase the activity of a number of enzymes, as well as protein synthesis in general, an effect on protein synthesis either at the level of transcription or translation has been sought (Rall, 1978). Nuclear binding of T_4 has been demonstrated, apparently to nonhistone proteins. It appears that it is free T_4 which binds to the nucleus and not the T_4 already bound to the cytoplasmic receptor. In fact, it appears that cytoplasmic receptors compete for T_4 bound to nuclear sites. Nonetheless, there is strong inferential evidence for a role of T_4 on transcription of mRNA (Rall,1978), but this does not represent the only means of explaining the pleiotropic effects of the hormone (Sterling, 1979).

A role for steroid hormones in enzyme induction has been known for a long time, but the discrete steps involved in their mode of action have only recently become clear. The sequence of events in steroid binding to the nucleus begins with penetration of the lipid-soluble steroid into the cell without prior binding to a receptor, followed by entry into the cytosol where binding to a specific receptor occurs. Thereafter, this hormone–receptor complex moves to the nucleus, where it binds to the DNA or chromatin (Chan and O'Malley, 1976; O'Malley et al., 1977). In some yet undefined manner, the binding of this complex to DNA causes activation of specific genes and their transcription by RNA polymerase, forming multiple copies of phenotype-specific proteins. There is evidence that chromatin proteins, particularly nonhistones, may be involved in defining sites for selective transcription.

Work on progesterone action has become a prototype for the sex steroids (Chan and O'Malley, 1976). The progesterone receptor is a dimer composed of two subunits, both of which bind the hormone. On binding progesterone, the receptor translocates to the nucleus; what mediates this process is not known. The binding affinities of the subunits differ greatly; the B subunit binds to the nonhistone protein–

DNA complex, while the A subunit binds to purified DNA. As a consequence of these divergent specificities, it has been proposed that the B subunit locates the specific site at which to bind to chromatin, while A may cause a conformational change in a localized region of chromatin–DNA, thereby providing sites for RNA polymerase to initiate mRNA synthesis. An alternate view, proposed by Yamamoto and Alberts (1976), is that binding of the receptor causes a cascade of histone acetylation, which makes the genome accessible to RNA polymerase. Perhaps the receptor, in a manner analogous to the cAMP receptor protein of prokaryotes, influences formation of an open-promoter complex with RNA polymerase (Krakow and Kumar, 1977).

Experimental evidence for phenotype-specific induction is impressive. While the immature chick oviduct has no ovalbumin mRNA, chronic estrogen administration causes a striking increase of these messages, to almost 50,000 copies per tubule cell. Cessation of estrogen administration causes a diminution of ovalbumin mRNA to 60 copies per cell. A further source of support for this relationship is the demonstration that chromatin isolated from estrogen-stimulated oviduct directs synthesis of ovalbumin mRNA, whereas that from unstimulated oviduct does not. Finally, Jungmann and Kranias (1977) have suggested that steroids may stimulate nuclear *cAMP-independent* kinases, which phosphorylate nuclear proteins, thus influencing transcription. The role of steroid hormones in enzyme induction has been reviewed by Gelehrter (1976) and by Feigelson and Kurtz (1978). These extensive discussions merit the attention of the reader.

f. An Operon for Eukaryotes? The processes of morphogenesis and differentiation—the hallmarks of the multicellular organism—represent such an enormous increase in organizational complexity when compared to that of the unicellular organism that we would expect the genome of the multicellular organism to be far more complex than that of the prokaryote. This expectation is justified and has been documented in the foregoing discussion. We must then consider whether the operon hypothesis of Jacob and Monod (1961)—so successful in explaining the strategies of genetic regulation in the unicellular organism—can be taken over, without substantial modification, to explain the genetic process in multicellular organisms. Abundant evidence, reviewed above, would support the contention that uncritical use of the operon-type model in the eukaryote situation is unwarranted. That is not to say that none of its predicates are relevant to the eukaryotic system, but rather to emphasize certain distinctive differences between these two classes of organisms that must be reconciled in any theory of genetic regulation which would explain differentiation, and morphogenesis.

As we emphasized at the beginning of this chapter, protein–nucleic acid interactions form the keystone of both the specific and nonspecific interactions so necessary for selective transcription and organization of the translational process, thereby ensuring the fidelity of the entire protein biosynthetic process. In consequence of this truism, it is to be expected that nucleotide sequences will play an important role in binding by proteins that recognize particular sequences. However, as emphasized, the DNA in the eukaryotic genome is not in a naked state, as it exists in the prokaryote, but is clothed for substantial sequences by histones and other proteins. It is, therefore, to be expected that some control of transcription must reside in the selective covalent modification or conformational changes of some of these proteins, thereby permitting certain areas of the hitherto inaccessible eukaryotic genome to become accessible to transcription. Indeed, the studies by Weintraub and Groudine (1976) and others have demonstrated that genes undergoing active transcription are more accessible to pancreatic DNAse that those not being transcribed.

Inferential evidence supports a role for nonhistone proteins in regulating eukaryotic gene expression (O'Malley et al., 1977). The nonhistone proteins increase in amount in tissues carrying out RNA transcription, in the face of unchanged histone levels. The nonhistone proteins also exhibit a greater heterogeneity than the five histone proteins and possess DNA binding specificity. Nonetheless, a demonstration of an unequivocal role for these proteins will depend upon more definitive evidence than is presently available.

While the operon hypothesis is based on the view that the unicellular bacterium always makes certain enzymes—the constitutive enzymes—while most other enzymes are produced only in response to environmental changes, as the consequence of production of an inducer which blocks a repressor, evidence reviewed by Caplan and Ordahl (1978) leads to the inescapable conclusion that, in higher organisms, the vast majority of the information encoded in the genome is reversibly or irreversibly repressed for the bulk of the life of that particular organism. Surely this postulate is a necessary consequence of the differentiated state of many tissues within the multicellular organism; for example, kidney and muscle, while sharing certain housekeeping enzymes, are distinctively different in their biochemical phenotypes. Hence, any theory regarding expression of the genome in eukaryotes must address itself to this fact. In addition, there is now a large body of data demonstrating conclusively that genes encoding enzymes in the same metabolic pathway, or, for example, the α- and β-globin genes, are present on different chromosomes. Thus, coordinate regulation of these different chromosomes, which does occur, cannot be explained by sequential transcription of contiguous genes present on the same chromosome, as explained in the operon hypothesis. In addition, the long stretches of intervening sequences so characteristic of the eukaryotic genome have not yet been described in any prokaryotic organism. We have considered the possibility that these intervening sequences provide for the acceleration of evolution. The moderately repetitive and highly repetitive sequences probably serve roles essential to chromosomal organization and as binding sites for different proteins which interact specifically and nonspecifically with DNA. Since moderately repeated sequences have been shown to occur next to certain structural genes, it is quite likely that they are involved in orchestrating coordinated gene expression, since a signal molecule can bind to these homologous sequences which occur on separate chromosomes, thereby coordinating the synthesis of related proteins, as in the α- and β-globin genes (Davidson et al., 1977; Leder et al., 1978). This would then permit coordinate or so-called "trans" regulation of these spatially separate genes.

The nature of these inducers remains a matter of considerable speculation; RNA molecules, proteins, and as yet unknown molecules, in addition to hormones, have been proposed for this role (Davidson et al., 1977). We list the following references for the reader who wishes to delve more deeply into these hypotheses, with the caveat that our present level of ignorance regarding the nature of the eukaryotic genome suggests that, as with the discovery of intervening sequences, as yet unknown characteristics of this more complex genome must be described before a coherent theory of regulation can be proposed. Nonetheless, present theories are useful, since they point toward possible experimental approaches which will reconcile some of these issues (Rutter et al., 1973; Schmid and Deininger, 1975; Davidson and Britten, 1979; as well as the references included in this section).

B. Translational Control of Protein Synthesis

The frenetic pace of research in protein biosynthesis has led to a reasonable understanding of the transcriptional process and its attendant controls. However, the situation is not so clear regarding translational controls; but one can expect an improved understanding of that process to emerge during the next few years.

Here we wish to consider the role of hemin in translational control of globin synthesis. The subject of hemin control has been considered in further detail by Lodish (1976), Nienhaus and Benz (1977), Revel and Groner (1978), and Safer and Anderson (1978). The main phenotypic protein of reticulocytes is hemoglobin, which accounts for about 90%

of the protein made by these nucleated cells. The hemoglobin tetramer contains equal amounts of α- and β-globin chain. As the reticulocyte matures, the nucleus disappears and, with it, mRNA and ribosomes, which are degraded so that the mature erythrocyte cannot make protein.

Evidence reviewed by Lodish (1976) indicates that both human and rat β-globin mRNAs are able to initiate protein synthesis more efficiently than α-globin mRNA. These differences may result from a combination of intrinsic differences in the primary and secondary structure of these mRNAs, as well as from differences in the affinity for binding either to the 40 S–Met–tRNA$_f$ initiation complex or to eIF-2.

Many studies have demonstrated a role for hemin in controlling the initiation of hemoglobin synthesis. Indeed, hemin exerts this stimulatory effect even when iron chelating agents are added to the reticulocyte cell-free lysates. In the presence of hemin, synthesis of the α and β chains are equal, but, when hemin is deficient, α-chain synthesis is decreased to a greater extent than β-chain synthesis. If eIF-2 is added to this in vitro system, synthesis of the two chains equalizes once again. Hemin will stimulate globin synthesis in intact reticulocytes. In iron deficiency and lead poisoning, two disorders which primarily affect heme synthesis, α and β-globin synthesis again becomes unequal.

It appears that a protein inhibitor termed "hemin controlled repressor" (HCR) forms in hemin-deficient cells from a latent "prorepressor." This inhibitor appears to function as a protein kinase, adding one to two phosphorus atoms to the small subunit of eIF-2. Whether this activity is cAMP dependent (Datta et al., 1978) or not (Gross, 1978; Safer and Anderson, 1978) has not been resolved. The effect of the HCR appears to be to decrease the amount of initiator Met–tRNA$_f$ which binds to the 40 S subunit. The ability of eIF-2 to counteract the effect of HCR, in conjunction with the data on phosphorylation, strongly supports a role for HCR as inhibitor of the action of eIF-2. However, the actual means by which phosphorylation of the initiator factor leads to inhibition, or how hemin works, is not clear. Clarification of this phenomenon should prove illuminating not only for hemoglobin synthesis but also for nonerythroid cells, since the effect of hemin is seen with extracts from other cells as well.

Coordination of heme and globin synthesis is the rule during erythropoiesis. While heme is synthesized in mitochondria, protein synthesis occurs in the endoplasmic reticulum. Control between compartments may have more far-reaching effects than merely coordinated hemoglobin manufacture. Indeed, there is evidence from studies with deprivation of amino acids, glucose, or oxygen that protein synthesis is coupled in some manner to the metabolic economy. So conceived, the role of hemin in the cytochromes may serve as the nexus between the energy state of the cell and the translation of mRNA (Revel and Groner, 1978).

References

Alt, F. W., Kellems, R. E., Bertino, J. R., and Schimke, R. T., 1978, Selective multiplication of dihydrofolate reductase genes in methotrexate-resistant variants of cultured murine cells, *J. Biol. Chem.* **253**:1357.

Arber, W., 1978, Restriction endonucleases, *Angew. Chem. (Engl.)* **17**:73.

Benz, E. J., Jr., and Forget, B. G., 1974, The biosynthesis of hemoglobin, *Semin. Hemat.* **11**:463.

Biswas, B. B., Ganguly, A., and Das, A., 1975, Eukaryotic RNA polymerases and the factors that control them, *Prog. Nucleic Acid Res. Mol. Biol.* **15**:145.

Blake, C. C. F., 1978, Do genes-in-pieces imply proteins-in-pieces? *Nature* **273**:267.

Britten, R. J., and Kohne, D. E., 1968, Repeated sequences in DNA, *Science* **161**:529.

Brot, N., 1977, Translocation, in: *Molecular Mechanisms of Protein Biosynthesis* (H. Weissbach and S. Pestka, eds.), pp. 375–411, Academic Press, New York.

Busch, H., 1977, The eukaryotic nucleus, in: *Receptors and Hormone Action*, Vol. 1 (B. W. O'Malley and L. Birnbaumer, eds.), pp. 32–103, Academic Press, New York.

Busch, H., Choi, Y. C., Daskal, Y., Liarakos, C. D., Rao, M. R. S., RoChoi, T. S., and Wu, B. C., 1976, Methods for studies on messenger RNA, *Methods Cancer Res.* **13**:101.

Caplan, A. I., and Ordahl, C. P., 1978, Irreversible gene repression model for control of development, *Science* **201**:120.

Caskey, C. T., 1977, Peptide chain termination, in: *Molecular Mechanisms of Protein Biosynthesis* (H. Weissbach and S. Pestka, eds.), pp. 443–465, Academic Press, New York.

Chamberlin, M. J., 1974, The selectivity of transcription, *Annu. Rev. Biochem.* **43**:721.

Chamberlin, M. J., 1976a, RNA polymerase — an overview, in: *RNA Polymerase* (R. Losick and M. Chamberlin, eds.), pp. 17–68, Cold Spring Harbor Laboratory Press, Cold Spring Harbor, L.I., New York.

Chamberlin, M. J., 1976b, Interaction of RNA polymerase with the DNA template, in: *RNA Polymerase* (R. Losick and M. Chamberlin, eds.), pp. 159–192, Cold Spring Harbor Laboratory Press, Cold Spring Harbor, L.I., New York.

Chambon, P., 1975, Eukaryotic nuclear RNA polymerases, *Annu. Rev. Biochem.* **44**:613.

Chambon, P., 1978, Summary: The molecular biology of the eukaryotic genome is coming of age, *Cold Spring Harbor Symp. Quant. Biol.* **42**:1209.

Chan, L., and O'Malley, B. W., 1976, Mechanism of action of the sex steroid hormones, *N. Engl. J. Med.* **294**:1322, 1372, 1430.

Chan, L., Harris, S. E., Rosen, J. M., Means, A. R., and O'Malley, B. W., 1977, Processing of nuclear heterogeneous RNA: Recent developments, *Life Sci.* **20**:1.

Clark, B. F. C., 1977, Correlation of biological activities with structural features of transfer RNA, *Prog. Nucleic Acid Res.* **20**:1.

Crick, F., 1968, The origin of the genetic code, *J. Mol. Biol.* **38**:367.

Crick, F., 1979, Split genes and RNA splicing, *Science* **204**:264.

Darnell, J. E., Jr., 1978, Implications of RNA•RNA splicing in evolution of eukaryotic cells, *Science* **202**:1257.

Datta, A., deHaro, C., and Ochoa, S., 1978, Translational control by hemin is due to binding to cyclic AMP-dependent protein kinase, *Proc. Natl. Acad. Sci. USA* **75**:1148.

Davidson, E. H., and Britten, R. J., 1979, Regulation of gene expression: Possible role of repetitive sequences, *Science* **204**:1052.

Davidson, E. H., Klein, W. H., and Britten, R. J., 1977, Sequence organization in animal DNA and a speculation on hnRNA as a coordinate regulatory transcript, *Dev. Biol.* **55**:69.
Doolittle, W. F., 1978, Genes in pieces: Were they ever together? *Nature* **272**:581.
Dreyfus, J. C., Kahn, A., and Schapira, F., 1978, Post-translational modifications of enzymes, *Curr. Top. Cell. Regul.* **14**:243.
Feigelson, P., and Kurtz, D. T., 1978, Hormonal modulation of specific messenger RNA species in normal and neoplastic rat liver, *Adv. Enzymol.* **47**:275.
Felsenfeld, G., 1978, Chromatin, *Nature* **271**:115.
Fersht, A. R., 1977a, Editing mechanisms in protein synthesis. Rejection of valine by the isoleucyl-tRNA synthetase, *Biochemistry* **16**:1025.
Fersht, A. R., 1977b, *Enzyme Structure and Mechanism*, W. H. Freeman, San Francisco.
Filipowicz, W., 1978, Functions of the 5'-terminal m^{7}G cap in eukaryotic mRNA, *FEBS Lett.* **96**:1.
Flavell, R. A., Glover, D. M., and Jeffreys, A. J., 1978, Discontinuous genes, *Trends Biochem. Sci.* **3**:241.
Gelehrter, T. D., 1976, Enzyme induction, *N. Engl. J. Med.* **294**:522, 589, 646.
Gilbert, W., 1978, Why genes in pieces? *Nature* **271**:501.
Goldberger, R. F., ed.,1979, *Biological Regulation and Development*, Vol. 1, *Gene Expression*, Plenum, New York.
Gross, M., 1978, Regulation of protein synthesis by hemin: Effect of dithiothreitol on the formation and activity of the hemin-controlled translational repressor, *Biochim. Biophys. Acta* **520**:642.
Harris, R. J., and Pestka, S., 1977, Peptide bond formation, in: *Molecular Mechanisms of Protein Biosynthesis* (H. Weissbach and S. Pestka, eds.), pp. 413-442, Academic Press, New York.
Holler, E., 1978, Protein biosynthesis: The codon-specific activation of amino acids, *Angew. Chem. (Engl.)* **17**:648.
Hopfield, J. J., 1974, Kinetic proofreading: A new mechanism for reducing errors in biosynthetic processes requiring high specificity, *Proc. Natl. Acad. Sci. USA* **71**:4135.
Jacob, F., and Monod, J., 1961, Genetic regulatory mechanisms in the synthesis of proteins, *J. Mol. Biol.* **3**:318.
Jeffreys, A. J., and Flavell, R.A., 1977, The rabbit of β-globin gene contains a large insert in the coding sequence, *Cell* **12**:1097.
Jukes, T. H., 1967, Indications of an evolutionary pathway in the amino acid code, *Biochem. Biophys. Res. Commun.* **27**:573.
Jukes, T. H., 1977, The amino acid code, in: *Comprehensive Biochemistry*, Vol. 24, *Biological Information Transfer* (M. Florkin, A. Neuberger, and L. L. M. VanDeenen, eds.), pp. 235–293, Elsevier, Amsterdam.
Jungmann, R. A., and Kranias, E. G., 1977, Minireview: Nuclear phosphoprotein kinases and the regulation of gene transcription, *Int. J. Biochem.* **8**:819.
Jungmann, R. A.,and Russell, D. H., 1977, Minireview: Cyclic AMP, cyclic AMP-dependent protein kinase, and the regulation of gene expression, *Life Sci.* **20**:1787.
Khorana, H. G., 1968, Synthesis in the study of nucleic acids, *Biochem. J.* **109**:709.
Kornberg, A., 1974, *DNA Synthesis*, W. H. Freeman, San Francisco.
Kornberg, R. D., 1977, Structure of chromatin, *Annu. Rev. Biochem.* **46**:931.
Kourilsky, P., and Chambon, P., 1978, The ovalbumin gene: An amazing gene in eight pieces, *Trends Biochem. Sci.* **3**:244.
Krakow, J. S., and Kumar, S. A., 1977, Biosynthesis of ribonucleic acid, in: *Comprehensive Biochemistry*, Vol. 24, *Biological Information Transfer* (M. Florkin, A. Neuberger, and L. L. M. VanDeenen, eds.), pp. 105–184, Elsevier, Amsterdam.

Krakow, J. S., Rhodes, G., and Jovin, T. M., 1976, RNA polymerase: Catalytic mechanisms and inhibitors, in: *RNA Polymerase* (R. Losick and M. Chamberlin, eds.), pp. 127-158, Cold Spring Harbor Laboratory Press, Cold Spring Harbor, L.I., New York.

Kurland, C. G., 1977, Aspects of ribosome structure and function, in: *Molecular Mechanisms of Protein Biosynthesis* (H. Weissbach and S. Pestka, eds.), pp. 81–116, Academic Press, New York.

Leder, A., Miller, H. I., Hamer, D. H., Seidman, J. G., Norman, B., Sullivan, M., and Leder, P., 1978, Comparison of cloned mouse α- and β-globin genes: Conservation of intervening sequence locations and extragenic homology, *Proc. Natl. Acad. Sci. USA* **75**:6187.

Leder, P., 1978, Discontinuous genes, *N. Engl. J. Med.* **298**:1079.

Lengyle, P., Speyer, J. F., and Ochoa, S., 1961, Synthetic polynucleotides and the amino acid code, *Proc. Natl. Acad. Sci. USA* **47**:1936.

Linn, S., 1978, The 1978 Nobel prize in physiology or medicine, *Science* **202**:1069.

Lodish, H., 1976, Translational control of protein synthesis, *Annu. Rev. Biochem.* **45**:39.

Malkin, R., 1973, The chemical properties of ferredoxins, in: *Iron-Sulfur Proteins*, Vol. 11 (W. Lovenberg, ed.), pp. 1–28, Academic Press, New York.

Mazumder, R., and Szer, W., 1977, Protein biosynthesis, in: *Comprehensive Biochemistry*, Vol. 24, *Biological Information Transfer* (M. Florkin, A. Neuberger, and L. L. M. VanDeenen, eds.), pp. 186–233, Elsevier, Amsterdam.

Miller, D. L., and Weissbach, H., 1977, Factors involved in the transfer of aminoacyl-tRNA to the ribosome, in: *Molecular Mechanisms of Protein Biosynthesis* (H. Weissbach and S. Pestka, eds.), pp. 323–373, Academic Press, New York.

Nathans, D., and Smith, H. O., 1975, Restriction endonucleases in the analysis and restructuring of DNA molecules, *Annu. Rev. Biochem.* **44**:273.

Nienhuis, A. W., and Benz, E. J., Jr., 1977, Regulation of hemoglobin synthesis during the development of the red cell, *N. Engl. J. Med.* **297**:1318,1371,1430.

Nirenberg, M. W., and Matthaei, J. H., 1961, The dependence of cell-free protein synthesis in *E. coli* upon naturally occurring or synthetic polyribonucleotides, *Proc. Natl. Acad. Sci. USA* **47**:1588.

Nunberg, J. H., Kaufman, R. J., Schmike, R. T., Urlaub, G., and Chasin, L. A., 1978, Amplified dihydrofolate reductase genes are localized to a homogeneously staining region of a single chromosome in a methotrexate-resistant Chinese hamster ovary cell line, *Proc. Natl. Acad. Sci. USA* **75**:5553.

O'Malley, B. W., Towle, H. C., and Schwartz, R. J., 1977, Regulation of gene expression in eukaryotes, *Annu. Rev. Genet.* **11**:239.

Palmer, G., 1975, Iron-sulfur proteins, in: *The Enzymes*, Vol. XII (P.D. Boyer, ed.), pp. 2–56, Academic Press, New York.

Peavy, D. E., Taylor, J. M., and Jefferson, L. S., 1978, Correlation of albumin production rates and albumin mRNA levels in livers of normal diabetic, and insulin-treated diabetic rats, *Proc. Natl. Acad. Sci. USA* **75**:5879.

Perry, R. P., 1976, Processing of RNA, *Annu. Rev. Biochem.* **45**:605.

Rall, J. E., 1978, Chemistry and actions of thyroid hormone: Mechanism of action of T_4, in: *The Thyroid*, 4th ed. (S. C. Werner and S.H. Ingbar, eds.), pp. 138–148, Harper & Row, Hagerstown, Md.

Revel, M., 1977, Initiation of messenger RNA translation into protein and some aspects of its regulation, in: *Molecular Mechanisms of Protein Biosynthesis* (H. Weissbach and S. Pestka, eds.), pp. 246–321, Academic Press, New York.

Revel, M., and Groner, Y., 1978, Post-transcriptional and translational controls of gene expression in eukaryotes, *Annu. Rev. Biochem.* **47**:1079.

Rich, A., 1962, On the problems of evolution and biochemical information transfer, in: *Horizons in Biochemistry* (M. Kasha and B. Pullman, eds.), pp. 103–126, Academic Press, New York.

Rich, A., 1977, An overview of protein-nucleic acid interactions, in: *Nucleic Acid–Protein Recognition* (H. J. Vogel, ed.), pp. 3–11, Academic Press, New York.

Rich, A., and RajBhandary, U. L., 1976, Transfer RNA: Molecular structure, sequence, and properties, *Annu. Rev. Biochem.* **45**:805.

Roeder, R. G., 1976, Eukaryotic nuclear RNA polymerases, in: *RNA Polymerase* (R. Losick and M. Chamberlin, eds.), pp. 285–330, Cold Spring Harbor Laboratory Press, Cold Spring Harbor, L.I., New York.

Rohlfing, D. L., and Saunders, M. A., 1978, Evolutionary processes possibly limiting the kinds of amino acids in proteins to twenty: A review, *J. Theoret. Biol.* **71**:487.

Rutter, W. J., Pictet, R. L., and Morris, P. E., 1973, Toward molecular mechanisms of developmental processes, *Annu. Rev. Biochem.* **42**:601.

Safer, B., and Anderson, W. F., 1978, The molecular mechanism of hemoglobin synthesis and its regulation in the reticulocyte, *CRC Crit. Rev. Biochem.* **5**:261.

Schmid, C. W., and Deininger, P. L., 1975, Sequence organization of the human genome, *Cell* **6**:345.

Spirin, A. S., 1978, Energetics of the ribosome, *Prog. Nucleic Acid Res.* **21**:39.

Sterling, K., 1979, Thyroid hormone action at the cell level, *N. Engl. J. Med.* **300**:117.

Tilghman, S., Tiemier, D. C., Seidman, J. G., Peterlin, B. M., Sullivan, M., Maizel, J. V., and Leder, P., 1978, Intervening sequence of DNA identified in the structural portion of a mouse β-globin gene, *Proc. Natl. Acad. Sci. USA* **75**:725.

Tonegawa, S., Mazam, A. M., Tizard, R., Bernard, O., and Gilbert, W., 1978, Sequence of a mouse germ-line gene for a variable region of an immunoglobulin light chain, *Proc. Natl. Acad. Sci. USA* **75**:1485.

Uy, R., and Wold, F., 1977, Post-translational covalent modification of proteins, *Science* **198**:890.

Van Holde, K. E., and Isenberg, I., 1975, Histone interactions and chromatin structure, *Accounts Chem. Res.* **8**:327.

Villee, D. B., 1975, *Human Endocrinology, A Developmental Approach*, W. B. Saunders, Philadelphia.

Vogel, H. J., ed., 1977, *Nucleic Acid–Protein Recognition*, Academic Press, New York.

Watts, R. L., and Watts, D. C., 1975, The genetic code, in: *MTP International Review of Science: Biochemistry*, Series One, Vol. 7, *Synthesis of Amino Acids and Proteins* (H. L. Kornberg and D.C. Phillips, consultant eds., H. R. V. Arnstein, volume ed.), pp. 255–294, Unniversity Park Press, Baltimore, Md.

Weintraub, H., and Groudine, M., 1976, Chromosomal subunits in active genes have an altered conformation, *Science* **193**:848.

Weissbach, H., and Pestka, S., eds., 1977, *Molecular Mechansims of Protein Biosynthesis*, Academic Press, New York.

Woese, C. R., 1967, *The Genetic Code*, Harper & Row, New York.

Wong, J. T.-F., 1975, A co-evolution theory of the genetic code, *Proc. Natl. Acad. Sci. USA* **72**:1909.

Wong, J. T.-F., 1976, The evolution of a universal genetic code, *Proc. Natl. Acad. Sci. USA* **73**:2336.

Yamamoto, K. R., and Alberts, B. M., 1976, Steroid receptors: Elements for modulation of eukaryotic transcription, *Annu. Rev. Biochem.* **45**:721.

6

Degradation of Enzymes

F. John Ballard

I. Introduction

A. Why Enzyme Degradation?

The capacity to adapt in response to a variety of changes in environment confers a substantial evolutionary advantage on an organism. Although this capacity is often considered in terms of the whole organism, it is equally important in multicellular animals, where there are groups of cells, each with a specialized role to play in the overall functioning of the organism. For example, the parenchymal cells in mammalian liver regulate the supply of nutrients to other tissues, receive metabolic end products from those tissues and process them for reuse or excretion, and remove toxic compounds that have been ingested by the mammal and absorbed into the portal blood. These major functions of the liver are in addition to the normal maintenance or growth of liver tissue and are of no immediate benefit to the liver cell itself.

Examination of the three groups of liver functions shows that each must change quantitatively and qualitatively in response to a change in the nutrient intake of the animal. Thus a rat eating a high-carbohydrate–low-fat meal once a day will deliver large amounts of glucose to the liver, where much of it will be converted to lipids for storage in extrahepatic tissues. If blood concentrations of glucose and amino acids are to be maintained within fairly narrow limits—seemingly essential for mammalian well-being—then an alteration of the diet to one poor

---

F. John Ballard • CSIRO Division of Human Nutrition, Adelaide, S.A. 5000, Australia.

in carbohydrate and rich in fat, or even to a diet whose composition is not changed but in which small frequent meals are taken, must result in a series of adaptation responses.

Replacing one metabolic pathway with another more suitable to a new nutritional state involves both increases and decreases of enzyme activities. How does this occur?

In general terms, a cell has three ways by which its complement of enzymes can respond:

1. New enzymes can be synthesized and less appropriate enzymes degraded.
2. Inactive enzymes can be reactivated and those not required can be converted to inactive or less active forms.
3. The enzymes may have kinetic properties so designed that changes in activity are produced simply by alterations in the amounts of substrates or products.

Replacement of one group of enzymes by another requires a large amount of energy, which is dissipated when the enzymes are later degraded to amino acids. For this mechanism of adaptation to be effective, control should be on both synthesis and degradation and should show a degree of overlapping specificity. Thus an excess of dietary protein over that needed to maintain the protein synthesis required for the growth of the animal should not only result in the synthesis of limiting enzymes in the pathways of amino acid degradation and the utilization of the deaminated compounds, but the spectrum of enzymes synthesized should ideally reflect, quantitatively, the relative excesses of each amino acid in the diet. Since these excesses will depend on the amino acid proportions in the average protein eaten, situations may arise where one amino acid is present in adequate amounts while most other amino acids are in oversupply.

Adaptation by enzyme replacement is more important in nondividing cells than in a continuously growing population. In the latter case, a less appropriate enzyme complement can be diluted out by daughter cells containing enzymes more suitable for the new nutritional state. A limited life for the cells would also facilitate such adaptation. However, adaptation via the production of new cells is only relevant to those mammalian cell types where the cells are not physically confined to a discrete area or to those cells which have a short life span. Examples are sperm, erythrocytes, other blood cells, and gut mucosal cells. Yet these cells are typically highly specialized, of relatively constant composition, and rarely show adaptation.

Enzyme synthesis and degradation could be avoided if an enzyme no longer needed could be reversibly inhibited in response to the nutritional or hormonal change. And, when conditions revert to those where the enzyme is again required, it could be reactivated. Yet the large number of physiologic conditions to which some cell types (e.g., liver) could be exposed would require a situation in which many enzymes would be held in the inactive form for the majority of cell life. Accordingly, constraints on cell size in relation to nutrient supply would restrict the development of this type of regulation during evolution of the organism. Moreover, it seems unlikely, from a teleological point of view, to have such a system together with one in which enzymes are encoded by a single pair of genes. The genetic code is surely more efficient, both on the basis of concentration of information and on flexibility to respond to an adaptation stimulus.

The logic in favor of "storing" enzymes indirectly as DNA sequences does not exclude inactivation–reactivation as an important regulatory event. Rather, this event must be restricted to a limited number of enzymes and is most appropriate as a response to brief environmental changes. The processes of inactivation and reactivation include allosteric control of enzyme activities and have been treated in detail in Chapter 4.

Certain enzymes can "adapt" without changes in enzyme amount and without ligand-induced change in specific enzymic activity. Hepatic glucokinase is an example of this type of enzyme (Niemeyer et al., 1975). Glucose, the substrate for glucokinase, varies in concentration between 3 and 20 mM in most mammals. Since the Michaelis–Menten constant for glucokinase is in that range, changes in blood glucose will automatically regulate glucose metabolism via glucokinase without any modification to the enzyme. In assessing the biological relevance of this type of "adaptation," it is interesting to note that those species in which blood glucose does not vary over a wide range do not have hepatic glucokinase. Included in this group are fetal mammals, who receive a constant supply of glucose from the mother, and ruminants, who obtain only trivial amounts of dietary glucose. The liver in these cases is never in a situation where rapid glucose uptake is facilitated. Of course, this regulation is only appropriate to a few enzymes at initial steps in pathways, and its widespread occurrence would be limited according to cell size and efficiency constraints put forward in relation to allosteric enzymes.

Some mammalian enzymes are present outside the limits of the cell plasma membrane. These include digestive enzymes secreted into the intestinal lumen, enzymes involved in blood clotting, and also enzymes which have "leaked" from cells. This last group includes most of the blood enzymes used as tissue-specific markers of cell damage. In this chapter I shall not discuss the breakdown of extracellular enzymes except where comparisons are being made with normal intracellular enzymes. However, I shall consider briefly the molecular modifications introduced into the structures of secreted extracellular enzymes which permit their functioning in harsh environments such as the extremes of pH in the gut. The avoidance of self-hydrolysis by proteolytic enzymes also comes into the category of enzymes functioning in harsh situations.

From the above discussion, it is apparent that enzyme degradation is an essential process for efficient adaptation. A second need for enzyme breakdown is as a means of removing incorrect enzymes. Such molecules may be formed as a result of errors in transcription or translation or may be produced by posttranslational modifications to a normal enzyme molecule. In either case, the enzyme differs from that encoded by the gene and may have altered catalytic properties which may be harmful to the cell. Although the well-being of the cell dictates a high specificity of the protein-synthesizing sequence in order to prevent the formation of error proteins, recent experiments have shown the presence of such mistakes. A protein breakdown pathway that is capable of recognizing modified enzymes and other proteins seems to be present as a protective system in all cell types (Goldberg and St. John, 1976; Ballard, 1977).

There are thus two major roles of enzyme degradation in mammalian cells: (1) to permit adaptation by removal of inappropriate enzymes; and (2) to protect cells from enzymes of altered catalytic specificity.

B. Partial or Complete Degradation?

Enzymes, like other cellular proteins, are completely degraded to their constituent amino acids via the proteolytic sequence. There is no evidence that peptides which might occur as breakdown products can be utilized for the synthesis of the same or different proteins—another way of stating that protein degradation and protein synthesis do not share any intermediates. However, this does not mean that proteolytic cleavage necessarily leads to loss of function, nor the converse: that activity is always retained unless proteolytic hydrolysis has occurred.

With some enzymes, the cleavage of one or more internal peptide bonds does not cause protein disruption and can in fact lead to an increase in enzyme activity. This occurs not only during the processing of proenzymes, such as the conversion of trypsinogen to trypsin or chymotrypsinogen to chymotrypsin, but also with intracellular enzymes such as hepatic fructose bisphosphatase (Pontremoli et al., 1973). This enzyme does not fall apart upon limited proteolytic cleavage, because peptide chains and different regions on a single chain are held in appropriate conformation by disulfide linkages, hydrogen bonds, or other types of intramolecular forces. So long as the enzyme does not unfold as a result of normal state transitions or "breathing," and, so long as the bond cleaved does not modify the catalytic site, it is quite plausible for proteolytic nicking to occur without any detectable change in enzyme function. However, the process is irreversible since there are no indications of repair processes such as seen for damaged DNA molecules.

The nicking of hepatic fructose bisphosphatase in response to nutritional changes is readily detected when peptide chains are separated by molecular sieving procedures, especially sodium dodecyl sulfate (SDS)-acrylamide gel electrophoresis (Pontremoli et al., 1973). The modified enzymes seen in aged cells may also reflect partial proteolysis (Rothstein, 1975). However, cautious interpretation of all these results as being representative of changes in vivo is needed, because the nicking might occur between excision of tissue from the animal and analysis.

In general, the initiation of proteolysis can be expected to lead rapidly to complete degradation of the enzyme to amino acids. Nicking is perhaps best considered as an irreversible posttranslational modification that bears some similarities to methylation or phosphorylation of amino acids in proteins. It is doubtful whether it occurs with many enzymes.

II. Kinetics of Enzyme Degradation

The complexity of enzyme or protein degradation is readily apparent, since not only is the susceptibility of a peptide bond dependent on which two amino acids are linked, but folding of the primary peptide chain will create a molecule in which some potential sites for hydrolysis are buried in the molecule and thus relatively unavailable to proteinases. Accordingly, enzyme degradation includes reactions necessary to unfold the molecule, separate individual chains, and hydrolyze disulfide linkages, as well as to break peptide bonds. We must assume that any of these reactions can be rate-limiting in enzyme degradation, and thus a range of breakdown kinetics can be expected.

A. Kinetic Order

A large number of observations on the decay of enzyme activity after prevention of enzyme synthesis suggests simple first-order kinet-

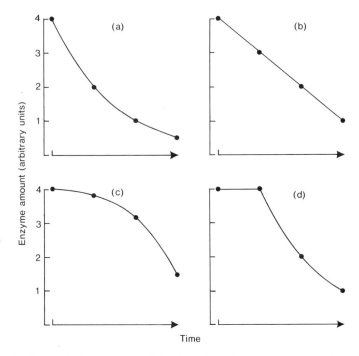

Figure 1. Theoretical examples of the kinetics of enzyme degradation. Loss of enzyme has been plotted on the basis that the process: (a) follows first-order kinetics; (b) follows zero-order kinetics; (c) describes a gradual destabilizing of the enzyme; and (d) shows lifetime kinetics.

ics; that is, the log of activity declines as a linear function of time (Fig. 1a). Similar results have been found when enzyme content rather than activity is followed. In those experiments where steady-state conditions are maintained under conditions in which the enzyme pool is labeled briefly by a radioactive precursor, the decay of enzyme radioactivity also follows first-order kinetics (Schimke and Doyle, 1970; Schimke, 1970). It is appropriate to mention at this stage that the finding of a single first-order degradation rate constant does not imply a single pathway of degradation for the enzyme. Breakdown by two or more pathways, each with different rate constants, will give a plot indistinguishable from the one shown in Fig. 1a.

Although experiments showing enzyme degradation to be a first-order process are based on data from many different enzymes measured under a variety of conditions, it is not entirely clear why first-order kinetics are found. Thus, if an enzyme acts as a high-affinity substrate for a degrading system responsible for breakdown of the enzyme, a low activity of the breakdown system could lead to a situation in which the breakdown system is rate limiting. This condition would be analogous to a normal enzyme assay performed in the presence of a saturating substrate concentration. In such a case, substrate loss (in our example the substrate is an enzyme) will be linear with time and consistent with zero-order kinetics (Fig. 1b). The lack of observations showing zero-order kinetics of protein catabolism argues against the conditions of a rate-limiting degradation system having a high affinity for an enzyme as a substrate.

A further type of kinetic behavior is depicted in Fig. 1c. Here the enzyme is relatively stable for an initial period and then loses activity. These kinetics are frequently found when enzyme inactivation is followed in broken-cell preparations and have been reported for glucose 6-phosphate dehydrogenase (G-6-PD), phosphoenolpyruvate carboxykinase, and tyrosine aminotransferase in liver homogenates (Hopgood and Ballard, 1974). Whereas such kinetics of enzyme loss can probably be explained by a gradual activation of the degradation pathway, perhaps by disruption of a cellular organelle, they could also occur if the enzyme was metabolized through one or more slow steps without loss of activity prior to its becoming susceptible to a crucial inactivation reaction.

The pattern of activity loss shown in Fig. 1d is really a special case of the previous example. Here an enzyme is fully active for a period and is then rapidly degraded. These "lifetime" kinetics will be found in all cases where the intracellular enzyme is stable while the cell itself has a fixed life. The best example of this type of kinetic behavior occurs with erythrocytes—cells which have a definite life in the blood. In erythrocytes, all enzymes are required to be stable because the cell does not have the capacity to replenish them.

A similar process is likely to occur with other cells as a response to autophagy or gross tissue damage. Thus, all the proteins within a cell might be degraded at the same time in an analogous way to the "clearing" of old erythrocytes from the blood. Also, a mitochondrion may be destroyed by an autophagic stimulus that is not severe enough to lead to destruction of the whole cell. Glucagon or vinblastine injection produces this effect in mammalian liver (Arstila et al., 1972, 1974). These examples of lifetime kinetics are characterized by an increase in susceptibility of the cell or organelle toward the degradation process. They can be considered as two periods of normal first-order degradation, the first with a very long half-life and the second with a short half-life. It is important to note that these examples are real events and that non-first-order kinetics of degradation should not be assumed to be incorrect until reconsidered under all models of protein breakdown.

B. First-Order Kinetics

Only a brief account of the kinetics of degradation will be given in this section. The reader is referred to reviews by Schimke (1970) and Schimke and Doyle (1970) for more extensive treatment and for the derivation of equations. As mentioned in the previous section, the large majority of experiments are consistent with enzyme degradation being a first-order process. Enzyme synthesis, on the other hand, is regulated by the rate at which the system produces a particular enzyme or other protein and follows zero-order kinetics. Accordingly, the change in enzyme content can be described by:

$$\frac{dE}{dt} = k_s - k_d E \qquad (1)$$

where E is the enzyme content, k_s is the zero-order rate constant for synthesis, and k_d the first-order rate constant for degradation.

In any steady state, the rate of enzyme change per unit time is zero, so that

$$k_s = k_d E$$

or

$$E = \frac{k_s}{k_d} \qquad (2)$$

which expresses, in mathematical terms, the observation that enzyme content at a steady state is proportional to the rate of synthesis and inversely proportional to the rate constant of degradation. Thus an alteration to either constant will affect the amount of enzyme present and lead to a new steady state unless a compensatory change is made to the other constant.

The rate constant for degradation is related to the half-life of an enzyme ($t_{1/2}$) by the equation:

$$k_d = \frac{\ln 2}{t_{1/2}} \qquad (3)$$

The half-life is readily obtained from a plot of time against enzyme content using semilog graph paper.

It should be stressed that a plot of the log of enzyme content against time will only lead to a valid measurement of $t_{1/2}$ or k_d if the synthesis rate of the enzyme is zero. Otherwise the plot will underestimate k_d. Whereas k_s is near zero when synthesis is prevented by inhibitors, or when enzyme radioactivity is measured after a brief pulse of radioactive amino acid followed by an effective chase, the k_s frequently cannot be determined when changes in enzyme amount are followed. Even when k_s is markedly reduced by a deinduction stimulus, it will still lead to a steady state at an enzyme level greater than zero.

For changes between two steady states, k_s and $t_{1/2}$ can be obtained if the log of the decrement ($E_2 - E_1$) is plotted against time (Haining, 1971).

A corollary of degradation being a first-order process (Fig. 1a) is that the measured rate of enzyme degradation increases with an increase in enzyme content. For example, if an enzyme with a half-life of 1 h increases in content from 2 nmol/g tissue to 10 nmol/g tissue, the rate of degradation will increase from 1 nmol to 5 nmol in the first hour after synthesis has been stopped. Although this point is well recognized, it is confusing if the terms "degradation rate" and "degradation rate constant" are used interchangeably. Rate is continuously variable, while the rate constant is independent of the amount of enzyme present.

II. Techniques for the Measurement of Enzyme Degradation

Rate constants of enzyme degradation can be obtained either directly, by measuring loss of enzyme or enzyme radioactivity, or indirectly from the enzyme content and enzyme synthesis rate. With each method, the most satisfactory results are obtained under steady-state conditions where the enzyme content does not change, although nonsteady-state conditions can also be used. The assumptions and limitations differ between each method and will be listed below. More lengthy discussions of the various methods available have been given in the reviews of Schimke (1970), Schimke and Doyle (1970), Goldberg and Dice (1974), and Goldberg and St. John (1976).

A. Steady-State Methods for the Determination of Degradation Rate Constants

Degradation rate constants may be determined under steady-state conditions either by following the conversion of a labeled precursor into the enzyme or by measuring loss of label from the enzyme pool after a brief pulse and effective dose. For both methods, it is essential that the enzyme can be isolated in a pure form from the tissue concerned. This is most conveniently accomplished by addition to the tissue extract of antibody specific for the enzyme. Label in the resultant antibody–enzyme precipitate can then be measured and expressed as total enzyme radioactivity in the enzyme pool or per unit weight of tissue (see, for example, Hopgood et al., 1973). Meaningful results are only obtained when the antiserum is completely specific for the enzyme, otherwise the precipitate will be contaminated with other proteins. This point is often given only cursory attention, and researchers rarely provide information on the specificity of the antibody–enzyme reaction or even on whether the enzyme has been quantitatively precipitated by the antibody.

1. Synthesis Method (k_d Calculated from k_s and E)

The most convenient method of measuring the conversion of precursor to enzyme is by the constant infusion of radioactive amino acid. So long as enzyme is isolated at a time after the specific radioactivity of precursor has attained a constant value, multiple sampling of tissue is not needed. The fractional rate of enzyme synthesis (k'_s), given in units of percent per day, can be obtained from a solution of the equation

$$\frac{SA_E}{SA_p} = \frac{\lambda_p}{(\lambda_p - k'_s)} \cdot \frac{(1 - e^{-k'_s t})}{(1 - e^{-\lambda p t})} - \frac{k'_s}{(\lambda_p - k'_s)} \quad (4)$$

as described by Swick (1958) and Garlick et al. (1973). In this equation SA_E is the specific radioactivity of precursor amino acid in the enzyme at time t, SA_p is the specific radioactivity of the precursor pool of labeled amino acid, and λp is the rate constant for the approach to constant specific radioactivity in the precursor amino acid. The major problems in the application of this technique for the measurement of enzyme synthesis rate are the uncertainty as to the actual precursor pool of radioactive amino acid, and the difficulty associated with a specific radioactivity measurement of amino acid in the enzyme. Garlick et al. (1973) and Fern and Garlick (1974) have measured the specific radioactivities of infused amino acids in plasma and tissue water and conclude that plasma free amino acids are not the immediate precursors of protein. Accordingly, values of SA_p and λp in equation (4) should apply not to the plasma amino acid but to the amino acid in tissue water. However, if the rate constant for approach to constant specific radioactivity of tissue amino acids is similar to one obtained from serial sampling of plasma, then SA_p can be readily obtained for the tissue (Garlick et al., 1973).

A measurement of the specific activity of the infused amino acid in the enzyme being studied is not immediately applicable to an antibody–enzyme precipitate because most of the protein in the precipitate is derived from the antibody rather than from the enzyme. However, enzyme can be separated from the immunoglobulin chains by SDS-acrylamide gel electrophoresis prior to the specific radioactivity determination. Use of a minimally metabolized amino acid is also a necessity, unless the enzyme protein is hydrolyzed and the specific radioactivity of the amino acid itself measured.

2. Direct Determination of k_d

More extensive use has been made of the decay of enzyme radioactivity as a measure of enzyme catabolism (Schimke and Doyle, 1970; Goldberg and St. John, 1976). In this method, proteins are first labeled by injection of a radioactive amino acid or other precursor, and the decline in total radioactivity in the enzyme is then followed. The technic requires multiple sampling times, either by serial biopsies or by using several animals, but is easier than measuring synthesis rate because neither a continuous infusion nor a measure of specific radioactivity is needed (see Fig. 2a). A major disadvantage with the method is the difficulty in preventing reincorporation of radioactive amino acids released during degradation back into the enzyme pool. Reincorporation occurs because mixing between intracellular and extracellular amino acids is neither rapid nor complete (Fern and Garlick, 1974), and will lead to an overestimation of the enzyme half-life. Reincorporation is also a potential problem with measurements of enzyme synthesis

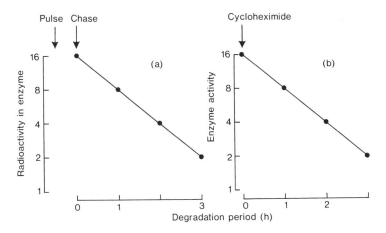

Figure 2. Measurement of enzyme degradation. The degradation of an hypothetical enzyme having a half-life of 1 h has been measured by: (a) loss of enzyme radioactivity after an initial brief pulse of precursor has been followed by an effective chase; (b) loss of catalytic activity after protein synthesis has been completely inhibited with cycloheximide.

rate, because a gradually increasing precursor specific radioactivity will be found instead of a plateau.

Reincorporation may be reduced by:

1. Selection of an amino acid or precursor that mixes rapidly with a large body pool. Examples suitable for liver are guanidino-labeled arginine, since this is rapidly converted to urea; and $NaH^{14}CO_3$, which is quickly converted to $^{14}CO_2$ (Swick and Ip, 1974).
2. Addition of a massive dose of the unlabeled amino acid at the conclusion of the pulse. Unfortunately, the amount needed to reduce the specific radioactivity of the intracellular amino acid may be so large that secondary changes will be produced and the steady state destroyed.
3. Inhibition of protein synthesis at the conclusion of the pulse. Not only does this disrupt the steady state, but the inhibitors often reduce protein degradation as well as protein synthesis (Goldberg and St. John, 1976; Ballard, 1977).

Each of the methods used to reduce reincorporation of radioactive amino acids has disadvantages, but combinations may be the most acceptable. For the special case of enzyme breakdown in isolated mammalian cells, the pulse can be conveniently removed by washing and a chase given by addition of a new medium containing high concentrations of the precursor amino acid (Hopgood et al., 1977; Gunn et al., 1976). Use of rapidly metabolized amino acids has the disadvantage that more label must be injected in order to adequately label the intracellular amino acid pool.

Modifications of the decay method have been used in special cases. Thus radioactive δ-aminolevulinate is a precursor of heme and will thus label enzymes with heme prosthetic groups. Provided the heme does not dissociate from the enzyme until the enzyme is degraded, isotope will then be lost by the rapid removal of heme as bilirubin, and reincorporation will be prevented (Poole et al., 1969).

Although the isotope infusion method determines k_s and the decay method gives k_d, equation (2) can be used to calculate the other rate constant so long as the enzyme content is known and a steady state holds. Thus the respective changes in each rate constant can be assessed at different steady states, although more complex analysis is needed if reliable information on k_s and k_d *between* steady states is sought.

B. Non-Steady-State Methods

Enzyme degradation can be measured after the removal of an inducing stimulus by following the time course of enzyme loss. If the assumption is made that enzyme activity is a valid measure of enzyme content, the method offers an extremely simple technique for determining k_d, and has been used in numerous studies (Schimke and Doyle, 1970). Thus a degradation rate constant can be obtained between steady states using equation (1) by plotting against time the log of the difference in enzyme activities at the two steady states. Further assumptions inherent in the method limit its usefulness for the calculation of accurate values of k_d. Thus, the degradation rate constant must not change during the alteration in enzyme content, and the synthesis rate of the enzyme must decrease immediately from that occurring at the high steady state to that present at the lower steady state. This latter restriction can only occur if the mRNA coding for the enzyme has an extremely short half-life. Although direct measurements of enzyme mRNA turnover rates have not been performed, there are indications that the stability of mRNA templates is similar to the enzymes for which they code (Bohley et al., 1971). However, the mRNA for phosphoenolpyruvate carboxykinase during deinduction is much less stable than the enzyme itself (Tilghman et al., 1974).

Degradation rate constants may also be obtained from the decrease in enzyme activity after protein synthesis has been inhibited (Fig. 2b). This simple method should give reliable measurements of k_d, provided enzyme activity is equivalent to enzyme content, the protein synthesis inhibitor is totally effective, and side effects of the inhibitor are not seen over the time course of the experiment. Of these assumptions, the first is the easiest to satisfy since the only examples reported in which an inactive enzyme protein is present in vivo are those in which an apoenzyme accumulates during a deficiency of the required coenzyme (see Section VII.C).

Of the protein synthesis inhibitors used in experiments to determine k_d, cycloheximide has the advantage of being readily available and a highly effective inhibitor of nonmitochondrial protein synthesis. Unfortunately, cycloheximide and other inhibitors of protein synthesis also reduce the degradation rate of bulk cell proteins and some enzymes (Goldberg and St. John, 1976). Although this problem must be considered in all experiments where cycloheximide is used to inhibit protein synthesis and to permit the calculation of k_d, it may be less of a difficulty in measurements on enzymes of short half-life, because the partial inhibition of degradation takes some time to take effect (Gunn et al., 1976). Thus the rapid degradation of abnormal proteins is not inhibited by cycloheximide (Ballard, 1977). It is probably safe to consider measurements of k_d by this method as giving a minimum value.

The change in enzyme activity or content from a low to a high level can also be used to calculate the degradation rate constant (Schimke and Doyle, 1970), but, as mentioned above for measurements on the reduction in enzyme content, the method assumes no change in k_d between the two conditions. Similarly, an instantaneous increase in k_s from the value at the lower steady state must also me assumed.

In most cases, measurements made at steady-state concentrations of enzymes are far more precise than those dependent on changes in steady states. Yet the simplicity of detecting activity changes has made the methods very attractive, and even the rough information so obtained is useful.

IV. Variability of Enzyme Half-Lives

Although half-lives have been measured for many different enzymes, the results are not always comparable since the variety of technics used have different degrees of precision. Yet it is clear from the available data that enzymes in the liver cytosol have a wide range of half-lives (Table 1). The less complete information from other tissues or for different cellular compartments also shows a diversity of enzyme stabilities (Schimke and Doyle, 1970; Goldberg and St. John, 1976; Dice and Goldberg, 1975). While information on half-lives or degradation rate constants is available for only a few enzymes—even in rat liver—some general concepts are apparent with regard to the types of reactions catalyzed by unstable enzymes and by stable enzymes.

A. Enzymes with Short Half-Lives

Enzymes present in the liver cytosol with short half-lives include ornithine decarboxylase, thymidine kinase, tyrosine aminotransferase, tryptophan oxygenase, hydroxymethylglutaryl-CoA reductase, serine dehydratase, and phosphoenolpyruvate carboxykinase. All of these enzymes have degradation rate constants greater than 0.1/h—more than 10 times more rapid than the average k_d for liver cytosol proteins (Schimke, 1970). Perhaps a scrutiny of the group can provide information on the enzyme properties as well as the nature of reactions catalyzed by enzymes with rapid turnover rates.

Without exception, the enzymes are adaptable, often showing 20-fold induction or more to an appropriate stimulus. Indeed, the degree of adaptation shown by these enzymes far exceeds that shown by any enzymes of long half-life listed in Table 1.

Tyrosine aminotransferase and phosphoenolpyruvate carboxykinase are the only enzymes of the group which catalyze readily reversible reactions, and, even with these enzymes, reversibility would not occur to any extent in vivo. Accordingly, the enzymes may be consid-

Table 1. Enzymes in Rat Liver Cytosol with Short and Long Half-Lives

Enzyme	$t_{1/2}$ (days)	Method of measurement[a]	Reference
Ornithine decarboxylase	0.007	2	Russell and Snyder (1969)
Tyrosine aminotransferase	0.06	1	Cihak et al. (1973)
Tryptophan oxygenase	0.12	1	Schimke et al. (1965a)
Thymidine kinase	0.11	2	Bresnick et al. (1967)
Hydroxymethylglutaryl-CoA reductase	0.12	3	Shapiro and Rodwell (1971)
Serine dehydratase	0.18	1	Jost et al. (1968)
Phosphoenolpyruvate carboxykinase	0.25	1	Hopgood et al. (1973)
Acetyl-CoA carboxylase	2.1	1	Majerus and Kilburn (1969)
Fatty acid synthetase	2.8	1	Volpe et al. (1973)
Malate dehydrogenase	4.0	2	Tarentino et al. (1966)
Arginase	4.0	1	Schimke (1964)
Aldolase	4.9	1	Kuehl and Sumsion (1970)
Glyceraldehyde-3-phosphate dehydrogenase	5.4	1	Kuehl and Sumsion (1970)
Lactate dehydrogenase, isoenzyme 5	6.0	1	Kuehl and Sumsion (1970)
Phosphofructokinase	7.0	1	Dunaway and Weber (1974)

[a]The methods used for these calculations were: (1) loss of enzyme radioactivity; (2) loss of activity after protein synthesis was inhibited by cycloheximide; (3) loss of activity after a deinduction stimulus.

ered to catalyze irreversible reactions in which there are large negative changes in free energy.

As is especially appropriate, for regulatory enzymes, all either catalyze the initial reaction in a metabolic pathway or the first reaction committed to a particular pathway (e.g., phosphoenolpyruvate carboxykinase and hydroxymethylglutaryl-CoA reductase). The enzymes are at points of regulatory control where an increase in activity should result in an increase in substrate throughput. In the basal state, each enzyme is present at a low activity.

What of the pathways in which the enzymes are involved?

Ornithine decarboxylase and thymidine kinase are associated with cell division and DNA synthesis, i.e., growth. Although the exact biological function of ornithine decarboxylase is unknown, we can assume from its rapid appearance and disappearance that it is only required at a certain time during cell division. The cell cycle usually lasts about 1 day in growing mammalian cells, and, accordingly, an enzyme functioning at only one stage may require a short half-life.

Tyrosine aminotransferase, tryptophan oxygenase, and serine dehydratase catalyze reactions in the degradative pathways of tyrosine, tryptophan, and serine, respectively. Since an amino acid excess is a transient event following a high-protein meal, it is not surprising that the degradative pathways are adaptive, with the rate-limiting enzymes degraded when the amino acids are not present at high concentrations. The appearance of these enzymes in response to their substrates seems analogous to substrate-mediated enzyme induction in bacteria, but with the added advantage in mammalian liver that the enzyme is degraded and thus removed at a rapid rate.

Phosphoenolpyruvate carboxykinase and hydroxymethylglutaryl-CoA reductase are limiting enzymes in gluconeogenesis and cholesterogenesis, respectively. Each enzyme participates in the synthesis of compounds that are both essential to mammals and normally available in the diet. From the tightly regulated blood concentrations of glucose and cholesterol, we can infer a need for the rapid production of these compounds once dietary sources are cut off. Accordingly, both enzymes are adaptable.

A short half-life is not sufficient to produce rapid changes in enzyme content unless the synthesis rate of the enzyme is also responsive. In fact, the half-lives of translatable mRNA for the three enzymes in the group for which measurements have been made—phosphoenolpyruvate carboxykinase (Tilghman et al., 1974), tryptophan oxygenase (Killewich et al., 1975), and tyrosine aminotransferase (Steinberg et al., 1975)—are extremely short. Values of less than 1 h were found in each case when the activities were falling. Since measurements for phosphoenolpyruvate carboxykinase were made during deinduction caused by refeeding of glucose to fasted animals, a lengthy sequence of events must have occurred prior to any effect on the enzyme template in liver. Perhaps the half-life of the enzyme mRNA is only a few minutes (Tilghman et al., 1974).

With the exception of the two enzymes involved in cell division, the others catalyze reactions that are predominantly liver specific. Few of the enzymes of long half-life listed in Table 1 are confined to liver.

A summary of the properties of unstable enzymes and their reactions is given in Table 2.

B. Stable Enzymes

The enzymes with the long half-lives listed in Table 1 include: the glycolytic enzymes aldolase, glyceraldehyde-phosphate dehydroge-

Table 2. Comparison of Biological Properties of Hepatic Enzymes with Short and Long Half-Lives[a]

	Adaptable	Reversible in vivo	High basal activity	Initial reaction in a pathway	Potentially rate-limiting	Relatively liver-specific	Allosteric control
Short half-lives							
Ornithine decarboxylase	Yes	No	No	Yes	Yes	No	No
Tyrosine aminotransferase	Yes	No	No	Yes	Yes	Yes	No
Thymidine kinase	Yes	No	No	Yes	Yes	No	No
Tryptophan oxygenase	Yes	No	No	Yes	Yes	Yes	No
Hydroxymethylglutaryl-CoA reductase	Yes	No	No	Yes	Yes	Yes	No
Serine dehydratase	Yes	No	No	Yes	Yes	Yes	No
Phosphoenolpyruvate carboxykinase	Yes	?	No	Yes	Yes	Yes	No
Long half-lives							
Acetyl-CoA carboxylase	Yes	No	No	Yes	Yes	No	Yes
Fatty acid synthetase	Yes	No	No	No	Yes	No	Yes
Arginase	Yes	No	Yes	No	No	Yes	?
Phosphofructokinase	?	No	No	No	Yes	No	Yes
Malate dehydrogenase	No	Yes	Yes	No	No	No	No
Aldolase	No	Yes	No	No	No	No	No
Glyceraldehyde 3-phosphate dehydrogenase	No	Yes	Yes	No	No	No	No
Lactate dehydrogenase	No	Yes	Yes	No	No	No	No

[a] The yes/no rating is based on a large number of experiments reported in the literature. The decisions should not be taken as unequivocal but on the best information presently available. Doubtful decisions indicated by a "?".

nase, phosphofructokinase, and lactate dehydrogenase; two enzymes (acetyl CoA carboxylase and fatty acid synthetase) catalyzing reactions in fatty acid synthesis; and arginase and malate dehydrogenase.

The properties of these enzymes fall into two groups. Of the four enzymes with a regulatory function in liver, three—fatty acid synthetase, acetyl-CoA carboxylase, and phosphofructokinase—are tightly regulated by allosteric ligands. Accordingly, there is less need for these enzymes to show rapid adaptation to the dietary supply of nutrients. However, in response to long-term dietary modifications, all show adaptive increases, which can be considered as changing baseline activities for the short-term allosteric control. Arginase is also inducible and possibly regulates urea synthesis, but has no well-characterized allosteric effectors. The effective activity of arginase in liver cells is largely limited by the input of arginine, as indicated by the extremely low hepatic concentration of this amino acid. In this sense, control of arginase activity is somewhat similar to control of hepatic glucokinase, an enzyme which responds to changing glucose concentrations as a result of its high Michaelis–Menten constant for glucose.

The second group of stable enzymes consists of noninducible enzymes present at very high activities. The only partial exception to this generalization is aldolase, and even this enzyme is present at a sufficiently high activity to maintain equilibrium between substrate and products. All the enzymes in the group catalyze reversible reactions in vivo, a property essential to their roles in both glycolysis and gluconeogenesis. The enzymes are present in most mammalian tissues and are devoid of allosteric control.

A summary of the properties of both types of stable enzymes is given in Table 2.

C. Abnormal Enzymes

Protein synthesis is a series of reactions, each with a high degree of fidelity. Yet mistakes can occur as a result of mistranslation, incorporation of unusual amino acids that are substrates of tRNA acylation reactions, premature termination of nascent peptides, and point mutations where amino acids are substituted (Ballard, 1977). Errors in protein synthesis are especially important if they are introduced into an enzyme, since the catalytic properties of the enzyme may be modified. If such errors occurred in ribosomal proteins or in other enzymes associated with protein synthesis, RNA synthesis, or DNA replication, a cell could be faced with a cataclysmic series of mistakes. Indeed, this concept forms the basis of a theory of aging, the "error catastrophe" theory of Orgel (1963, 1973). It is hardly surprising, therefore, that cells

Table 3. Half-Lives of Abnormal Enzymes

Enzyme	Source	Modification	$t_{1/2}$(h)
Phosphoenolpyruvate carboxykinase[a]	(H35 hepatoma)	Normal enzyme	4.7
		Canavanine replaces Arg	2.1
		6-Fluorotryptophan replaces Trp	2.8
Tyrosine aminotransferase[b]	(H35 hepatoma)	Normal enzyme	2.2
		6-Fluorotryptophan replaces Trp	2.1
Hypoxanthine-guanine phosphoribosyltransferase[c]	(L cells)	Normal enzyme	22.0
		Mutant 345	5.7
		Mutant 343	2.2
		Mutant 398	1.3
		Mutant 463	1.0

[a]Knowles et al. (1975).
[b]Johnson and Kenney (1973).
[c]Capecchi et al. (1974).

are capable of recognizing and degrading error proteins at a rapid rate (Goldberg and St. John, 1976; Ballard, 1977).

The majority of measurements on the degradation of abnormal proteins in mammals have concentrated on bulk cellular protein rather than individual enzymes (Ballard, 1977). However, incorporation of canavanine in place of arginine, or 6-fluorotryptophan in place of tryptophan, into phosphoenolpyruvate carboxykinase gives rise to enzyme forms that are considerably less stable than the normal enzyme (Table 3). On the other hand, Johnson and Kenney (1973) altered the heat stability and antibody reactivity of tyrosine aminotransferase in hepatoma cells without producing an increase in k_d. Since normal tyrosine aminotransferase is very labile in vivo, perhaps its k_d is already close to a maximum limited by the activity of the relevant degradation system. If this interpretation is correct, stable enzymes which have unusual amino acids inserted into their chains may show much larger changes in k_d than unstable enzymes. Indeed, some mutants of the relatively stable enzyme hypoxanthine–guanine phosphoribosyltransferase are very unstable (Table 3).

In addition to these findings with cancer cells, abnormal enzymes are also produced in erythrocytes, where the response seems to be precipitation of hemoglobin within the cell and eventual removal of the distorted cell from the blood. This result also occurs when "unstable" hemoglobins are encoded by mutant genes (White, 1976)—probably a reflection of the absence of proteinases in erythrocytes.

D. Lysosomal Enzymes

I have selected this group of enzymes for discussion since they are present inside an organelle which contains a wide range of proteolytic and other hydrolytic activities. Turnover studies of the proteins in whole lysosomes, lysosomal membranes, and lysosomal soluble fractions show an average half-life of 4 days, while β-glucuronidase has an even longer half-life (Wang and Touster, 1974). Since these proteins are continually exposed to lysosomal proteinases, it is pertinent to ask whether there are any properties of lysosomal enzymes and proteins which can explain their long half-lives.

Goldstone and Koenig (1974) have shown that lysosomal hydrolases are glycoprotein enzymes containing N-acetylglucosamine, mannose, glucose, neuraminic acid, galactose, and fucose. The enzymes are resistant to autolytic inactivation, particularly at neutral pH, while glycosidase inhibitors further reduce the autolytic cleavage reactions. The authors interpret these findings as protection of the polypeptide chains by the carbohydrate moiety. The carbohydrate might protect the polypeptide chains of the enzymes by linking them to lysosomal membranes. Such attachment does indeed occur (Baccino et al., 1971). The binding of lysosomal hydrolases to plasma membranes is similarly dependent on the integrity of the carbohydrate side chains (Stahl et al., 1976).

E. Mitochondrial Enzymes

The mean half-life of mitochondrial proteins in rat liver is about 7 days (Schimke, 1970). Like enzymes in other subcellular fractions, the half-lives of mitochondrial enzymes cover a wide range, extending from 1 h for δ-aminolevulinate synthetase to several days for cytochromes b, c, and oxidase (Schimke, 1970; Schimke and Doyle, 1970). This diversity is good evidence that mitochondria are not entirely degraded as units, despite morphological studies which show the engulfment of mitochondria within autophagic vacuoles (Arstila et al., 1972). Certainly mitochondria can be totally degraded in such vacuoles, but some additional process is required to explain the rapid turnover of enzymes such as δ-aminolevulinate synthetase and ornithine aminotransferase (see Section VI.B).

F. Protein Properties Correlating with Half-Lives

Attempts to explain the diversity of protein half-lives and relate them to biological function have resulted in a number of highly significant correlations being discovered (Bohley, 1968; Goldberg and St. John, 1976; Ballard, 1977). In general, these relationships apply to intracellular proteins from several tissues, but most information has been obtained from rat liver. Thus proteins with the shortest half-lives in vivo have the following characteristics:

1. The largest subunit molecular weights. This also applies to individual enzymes, at least in rat liver.
2. More acid isoelectric points. A similar correlation was found when the isoelectric points of liver enzymes were compared to their half-lives.
3. The highest proportion of surface hydrophobic regions. Proteins which had short half-lives in vivo showed selective movement to the organic phase in partition experiments between aqueous solution and CCl_4, isobutanol, xylol, oleic acid, benzene, or liver lipids.
4. A greater susceptibility to purified proteinases. This correlation is also valid for liver cathepsins and for autolytic degradation of liver proteins.

In addition to these effects applying to proteins in general, enzymes which have short half-lives in vivo are

5. Most sensitive to nonproteolytic inactivation reactions catalyzed by membranes in the presence of disulfides (Ballard, 1977).
6. More heat labile. I have discussed previously the correlation between stability in vivo and various enzyme functions or regulatory responses (see Section IV.A and B).

G. Half-Lives of the Same Enzymes in Different Tissues

There are several examples where half-lives for the same enzyme show wide variations between tissues. Thus phosphoenolpyruvate carboxykinase, which has a half life of about 6 h in rat liver and in cultured hepatoma cells, has a half-life of 20 h or more in adipose tissue and kidney (Hopgood et al., 1973; Iynedjian et al., 1975). Also, lactate dehydrogenase isoenzyme 5 has been reported to have half-lives of 1.6, 16, and 31 days in heart, liver, and skeletal muscle, respectively (Fritz et al., 1969), while the half-life of ornithine aminotransferase is 1 day in liver and 3.5 days in the Morris hepatoma 44 (Kobayashi et al., 1976).

These findings can possibly result from different activities of rate-limiting degradation systems in each tissue, but in that case many other enzymes should be affected than have been reported to date. A second alternative would be differential stabilization of the enzymes, perhaps by ligands (see Section V.A). The findings might really be artifacts produced by different degrees of isotope reincorporation in the measurements of k_d, which could somehow be specific to the enzymes concerned or could more likely result from variations in the extent of amino acid equilibration between plasma and the different tissues. Extensive investigations are needed before unequivocal proof can be given for tissue-specific degradation rate constants for a particular enzyme. Such studies are warranted because they could help resolve the extent to which enzyme breakdown is regulated by the activity of the degradation system in any one tissue.

V. Changes to Degradation Rate Constants

In this section I shall give selected examples of instances in which the degradation rate constant of an enzyme has been shown to change as a response to some stimulus. The examples are grouped into ligand-induced changes in stability, alterations caused by nutritional adaptations, and alterations during the initial synthesis of an enzyme during development. For each finding one must always question whether the observed change is more apparent than real. This somewhat cynical reaction to changes in k_d is warranted, because it is virtually impossible to eliminate changes in reincorporation as being the true cause of an apparent change in k_d, unless protein synthesis is completely suppressed by an agent which does not alter the k_d of the enzyme. The possibility of reincorporation is an especially serious problem if a stimulus influences the average half-life of cell proteins as well as that of the enzyme under study.

The examples given below exclude changes in k_d that can be attributed to modifications of the enzyme structure. Accordingly, mutant enzymes, enzymes in which certain residues are replaced by unusual amino acids, or enzymes which have been altered by posttranslational modifications are not considered. This restriction in scope has been made because any meaningful explanation of changes in k_d must assume that the enzyme being studied under diverse conditions has in fact the same primary structure.

A. Effects of Ligands

It is generally accepted that enzymes are flexible molecules which can have their conformations altered upon binding of substrates or other ligands. Thus it is not surprising that conformational changes alter the susceptibility of enzymes to proteolytic attack.

Perhaps the best documented example of ligand-induced enzyme stabilization occurs with tryptophan oxygenase. The activity of this enzyme can be induced in liver by its substrate, tryptophan, an effect which occurs via a decrease in k_d, with k_s not affected (Schimke et al., 1965a,b). Stabilization of tryptophan oxygenase by tryptophan or tryptophan analogs can also be demonstrated in vitro against inactivation produced by heat, denaturation, or proteolytic enzymes. These effects are best explained by the binding of tryptophan to the enzyme and the resultant fixing of the enzyme in a stable conformation (Schimke et al., 1965b).

Tryptophan administration to rats results in the stabilization of a number of enzymes for which tryptophan is not a substrate. These include tyrosine aminotransferase (Cihak et al., 1973), phosphoenolpyruvate carboxykinase (Ballard and Hopgood, 1973), threonine dehydratase (Peraino et al., 1965), and ornithine aminotransferase (Chee and Swick, 1976). With these enzymes, the decreased lability cannot be shown in vitro, so it is doubtful whether tryptophan itself is the active compound. Perhaps a tryptophan metabolite either binds to the enzymes or affects the dissociation of substrates or cofactors. This latter possibility could be important in the case of the aminotransferases, which have pyridoxal phosphate cofactors, since Litwack and Rosenfield (1973) demonstrated a correlation between the degradation rate constants of a number of enzymes and the rate of coenzyme dissociation from the molecules.

Further indications that coenzyme binding results in increased enzyme stability come from work on group-specific proteinases (Katunuma et al., 1971a,b; Kominami and Katunuma, 1976). Two proteinases were described, one of which was active against NAD-dependent dehydrogenases and the other against pyridoxal enzymes. In both cases, the proteinases were specific to the enzyme group, with no activity against a wide range of other enzymes. Each proteinase attacked the apoenzymes and had little or no activity against holoenzymes. It is not yet possible to decide whether the group-specific proteinases are involved in the intracellular degradation of NAD-linked or pyridoxal enzymes, but the experiments strongly support the concept that ligands (in this case cofactors) reduce the degradation rate constants.

There are many examples where substrates, products, and cofactors have been shown to protect enzymes against inactivation in vitro (Goldberg and Dice, 1974), but few of the agents have been tested in intact animals or in isolated tissues. However, it is likely that ligand-induced stabilization has a widespread occurrence and is an important factor in the alteration of degradation rate constants.

Although no ligands have, to my knowledge, been shown to increase the degradation rate constant of an enzyme, their existence can be expected simply on statistical grounds. Perhaps end products of a reaction sequence, or even direct products of an enzyme reaction, might bind to the enzyme and increase its catabolism. Although this seems a rather crude means of reducing the substrate throughput of a pathway, in comparison to product inhibition or feedback inhibition, it would offer an increased flexibility to regulatory processes within a cell.

B. Effects of Hormones and Nutrients

Protein degradation in cultured mammalian cells is increased by removal of amino acids or by other types of nutritional step-down, and decreased by the addition of enriched media. In addition, insulin, serum, and other growth-promoting compounds have been shown to inhibit protein catabolism in many cell types (Ballard, 1977; Goldberg and St. John, 1976). Although the mechanisms responsible for these effects are not understood, there is evidence that most if not all such agents have a common site of action. Thus combinations of different nutrient or hormonal inhibitors of degradation do not give additive effects (Knowles and Ballard, 1976).

The effect of nutrients in inhibiting protein breakdown is most noticeable on proteins of long half-life (Poole and Wibo, 1973; Knowles and Ballard, 1976) and is not seen at all with very unstable proteins. Whereas comparable studies have not been performed using enzymes with a range of half-lives, it is reasonable to expect that *general* effects of nutrients would follow such a pattern and preferentially affect enzymes with long half-lives.

Specific effects of nutrients and hormones on enzyme degradation may be superimposed on top of the general effects. Four examples where nutrients have been shown to markedly alter degradation rate constants are given in Table 4. Acetyl-CoA carboxylase and the fatty acid synthetase complex are present in liver at high activity in animals fed low-fat diets. Starvation or fat-feeding reduces the activities of these enzymes, in part by increasing the respective k_d. Serine dehydratase is induced by the administration of an amino acid mixture to rats on a protein-free diet, a change which is produced by a four- to five-fold increase in the enzyme half-life. With these examples, the direction of change in enzyme stability is the same as that found for the stability of total protein. Thus starvation increases general protein degradation, as does a protein-free diet. Accordingly, it is pertinent to ask to what extent the changes are specific to the three enzymes and what can be attributed to changes in overall protein breakdown. This question also introduces a methodological problem, because the decrease in half-lives may simply reflect decreases in reincorporation of radioactive amino acids into enzyme proteins. If this is true, the longer half-lives measured under the nutritional step-up conditions may be incorrect. It is not possible to eliminate, unequivocally, such an interpretation, but the large changes in half-life would argue against it. Nevertheless, it is a problem that should always be considered when alterations in half-life are seen.

Serum addition to hepatoma cells maintained in serum-free medium initiates cell growth and division and produces a transient but

Table 4. Effect of Nutrients on Enzyme Degradation

Enzyme	Tissue	Stimulus	Half-life (h)
Acetyl-CoA carboxylase[a]	Rat liver, in vivo	Normal diet	50
		Fat-free diet	48
		Starved	18
Fatty acid synthetase[b]	Rat liver, in vivo	Normal diet	67
		Fat-free diet	65
		Starved	18
Serine dehydratase[c]	Rat liver, in vivo	Protein-free diet	4.5
		Amino acids added	20
Ornithine decarboxylase[d]	Hepatoma cells	Serum-free medium	0.08
		Serum added	0.6

[a]Majerus and Kilburn (1969).
[b]Volpe et al. (1973).
[c]Jost et al. (1968).
[d]Hogan et al. (1976).

large increase in ornithine decarboxylase activity (Hogan et al., 1974). Accompanying the activity increase is a seven- to eight-fold increase in the enzyme half-life (Table 4). In this case, the half-life increase cannot be an artifact caused by a change in the extent of reincorporation, because protein synthesis was blocked by cycloheximide.

Although insulin decreases protein breakdown in cultured cells, it has no effect on the degradation rate constants of two enzymes with short half-lives, phosphoenolpyruvate carboxykinase (Gunn et al., 1976) and tyrosine aminotransferase (Reel et al., 1970). This negative result is in keeping with the concept that insulin acts by reducing autophagy, a process that seems trivial in the breakdown of proteins with short half-lives (Knowles and Ballard, 1976).

While I have chosen to separate ligand-induced and nutrient-induced changes in k_d under two headings, a distinction between them may be vague. Some of the effects attributed to nutrients or hormones may be caused by the binding of unknown metabolic products to the enzymes. This point can only be resolved by further experimentation and is probably the most likely explanation of all effects except those which are caused by changes in the activity of the degrading system, i.e., by changes in the extent of autophagy.

C. Effects of Growth and Development

Growth of a tissue implies an excess of protein synthesis over degradation and can be produced by changes in the rate constants for either pathway. As a result of the technical difficulties in measuring protein synthesis and degradation in vivo, most reliable information on changes in k_d during growth has come from studies with growing cells in culture. For example, insulin and serum induce growth by increasing protein synthesis in fibroblasts and hepatoma cells, as well as by reducing the degradation rate constant of the average cellular protein (Hershko et al., 1971; Gunn et al., 1976). On the other hand, compensatory liver growth after partial hepatectomy can be entirely explained by a reduction in protein degradation. But whereas this decrease in k_d should be questioned on the basis of a likely increase in amino acid retention and, thus, reincorporation, Scornik and Botbol (1976) appear to have avoided such problems by the use of minimally reutilized $NaH^{14}CO_3$ in their experiments.

It would seem likely from a teleological standpoint that protein degradation should be low during growth, because it would lead to a high efficiency of protein accretion in relation to amino acid supply. However, muscle growth in young rats is accompanied by a larger rate constant of protein degradation than found in later life (Millward et al., 1975), and a similar conclusion has been obtained for total protein breakdown in man (Winterer et al., 1976, Tomas and Ballard, unpublished experiments).

There are a number of examples in which the initial appearance of an enzyme during development is accompanied by a complete lack of its degradation. The experiments reported in Fig. 3 relate the appearance of phosphoenolpyruvate carboxykinase in rat liver at birth to changes in the relative synthesis rate and degradation of this enzyme (Philippidis et al., 1972). The increase in enzyme content accompanies a dramatic increase in the rate of enzyme synthesis, from approximately 0.1% of cytosol protein synthesis in the term fetus to 1% by 2 h after birth to 2.4% in the 1-day-old rat. This relative synthesis rate remains high for the next 2 days (Fig. 3) and, indeed, up to 2 weeks after birth. Enzyme degradation could not be detected during the period of rapid enzyme accumulation. By 3 days after birth, the half-life of phosphoenolpyruvate carboxykinase had decreased to 10 h, still somewhat longer than found in the adult liver.

Analogous results showing a lack of degradation during the initial appearance of enzymes have been reported for chick liver malic enzyme (Silpananta and Goodridge, 1971), xanthine dehydrogenase (Epstein and Newburgh, 1972, 1973), and acetyl-CoA carboxylase (Teraoka and Numa, 1975), as well as for fatty acid synthetase in the differentiating mammary gland (Speake et al., 1975). With the seemingly perennial proviso that the results may be explained in part by a high reincorporation rate of labeled amino acids, an extremely low degradation rate constant during development would greatly facilitate a rapid increase in enzyme activity. Moreover, this pattern of activity increase is very

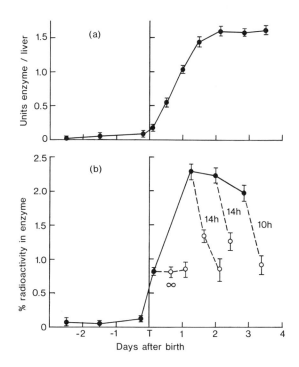

Figure 3. Appearance of phosphoenolpyruvate carboxykinase in developing rat liver. (a) The activity increase is expressed as units of enzyme per total liver. (b) Enzyme synthesis (●) is expressed as the percent of radioactivity in the enzyme pool as compared to radioactivity in cytosol protein after injections of radioactive leucine. Degradation (o), in the same terms, is shown at various times after a leucine chase was given. The half-lives at each age are indicated. Details are given by Philippidis et al. (1972). Values are means ±SEM. T, term.

similar to substrate-induced enzyme induction in bacteria, a process which is not accompanied by enzyme degradation (Mandelstam, 1963).

D. Relative Contribution of Changes in k_d to Alterations in Enzyme Content

The change in enzyme content from one steady state to another can be analyzed in terms of the individual changes in k_s and k_d at the two steady states. This is accomplished according to equation (2) from measurements of any two of E, k_d, and k_s.

As mentioned in the discussion on enzyme changes during development, the most efficient and rapid way to alter enzyme content would be to change both k_s and k_d. This is seen with the changes in hepatic

Table 5. Relative Contributions of k_s and k_d to Changes in Enzyme Content in Rat Liver

Enzyme	Treatment	E (mg/liver)	k_d (h^{-1})	k_s (mg/h)
Fatty acid synthetase[a]	Chow diet	4.71	0.010	0.047
	Starved	0.28	0.039	0.011
	Fat-free diet	14.6	0.011	0.156
Phosphoenolpyruvate carboxykinase[b,c]	Fed	1.20	0.113	0.135
	Starved	2.56	0.104	0.264
	Starved, triamcinolone	1.35	0.094	0.127
	Starved, refed	0.93	0.118	0.110
Ornithine aminotransferase[d]	60% casein diet	36.0	0.017	0.60
	12% casein diet	11.0	0.016	0.17

[a]Volpe et al. (1973).
[b]Hopgood et al. (1973).
[c]Gunn et al. (1975).
[d]Chee and Swick (1976).

fatty acid synthetase between fed and starved animals, where a fourfold increase in k_d and a comparable decrease in k_s are found (Table 5). On the other hand, only k_s is affected by the increase in enzyme content after the switch from a chow diet to a fat-free diet. A series of measurements on rat liver phosphoenolpyruvate carboxykinase showed large alterations in enzyme content. However, the degradation rate constant for this enzyme remained at about 0.11/h, and all changes in content were caused by alterations in k_s. Similar results were found with ornithine aminotransferase (Table 5). The opposite result was found with tryptophan oxygenase, since the observed increase in enzyme activity following tryptophan administration to rats could be attributed entirely to a reduction in the degradation rate constant of the enzyme (Schimke et al., 1965a).

The examples listed represent the three possible results: content increases caused by changes in k_s and k_d, k_s alone, or k_d alone. An examination of data on a number of enzymes suggests that although the three responses are often seen, changes in k_s are more frequent than changes in k_d. Of course, k_s must alter when the translation rate of an enzyme template is altered, either by a translation-specific event or as a result of an altered transcription rate. Accordingly, alterations in k_d may be limited to either conformational changes in enzymes or changes in the activity of a rate-limiting proteolytic system.

The values in Table 5 are at steady states. During the transition from one steady state to another there can be transient but dramatic changes in either k_s or k_d. For example, refeeding starved rats decreases the synthesis rate for hepatic phosphoenolpyruvate carboxykinase to about 5% of the initial rate, even though the rate of enzyme synthesis is much higher by the time a new steady state is attained (Hopgood et al., 1973). Similarly, the degradation rate constants for arginase (Schimke et al., 1964) and for ornithine aminotransferase (Chee and Swick, 1976) increased two- to threefold soon after rats were transferred from high-casein to low-casein diets, but returned to the initial value when the lower steady states were established.

VI. Intracellular Localization of Degradative Pathways

Although there have been numerous and comprehensive studies on intracellular proteinases, they have been carried out predominantly with the purified enzymes. This work has provided information on catalytic mechanisms, susceptibility of proteinases to different classes of inhibitors, specificities of the enzymes toward a variety of artificial and natural substrates, and details on optimal conditions for each proteinase (Barrett, 1969). Unfortunately, these experiments provide little if any direct evidence about whether a particular enzyme is involved in the breakdown of intracellular enzymes or other proteins.

Proteinases are found in all subcellular fractions of liver, but most of the activity associated with an acid pH optimum is localized in lysosomes. Although neutral proteinases occur in various cell membranes, it is not yet clear whether any selective enrichment occurs in a particular membrane type. The total neutral proteinase activity found in the microsomal fraction, which includes the Golgi and plasma membranes as well as the smooth and rough endoplasmic reticulum, is adequate to explain the observed breakdown of proteins in these fractions (Bohley, 1968; Bohley et al., 1971). Of course this does not imply self-proteolysis, only the *capacity* to degrade the membrane proteins.

Whereas proteolytic activities in mitochondria, nuclei, and the microsomal fraction are relatively high, very little proteolytic capacity is found in the cell cytosol. In part this may reflect technical difficulties, because the highly efficient proteolytic inhibitors found in serum, especially α_2-macroglobulin and α_1-antitrypsin, would occur in the cytosol fraction of an homogenized tissue. Nevertheless, separation of cytosol proteins from serum proteolytic inhibitors by Sephadex chromatography does not give rise to any fractions with substantial proteolytic capacity (Bohley et al., 1971).

A. Lysosomes and Autophagy

Lysosomes are organelles of varying sizes that contain an entire range of hydrolytic enzymes. Normally, lysosomes have a single membrane surrounding a matrix in which are often seen partially digested cell constituents, some recognizable as mitochondria or glycogen granules. Lysosomes occur in all tissues that have been examined and comprise about 1% of the cell volume in liver parenchymal cells. Much higher contents are found in macrophages and other cells involved in digesting extracellular material, while muscle cells and reticulocytes have lower proteolytic activities than hepatocytes (Ericsson, 1969).

Lysosomes contain several proteinases which have been historically called cathepsins. The proteinases are glycoproteins, have acid pH optima, and are possibly attached to the inner side of the lysosomal membrane. Lysosomal involvement in the breakdown of intracellular proteins has been inferred from electron microscopic data showing regions of membrane engulfing cell constituents, as well as whole and partially degraded mitochondria enclosed in organelles which are rich in lysosomal marker enzymes (Ericsson, 1969; Barrett, 1969).

Further indications of the lysosomal engulfing process come from studies of human genetic defects. In patients with type II glycogenosis (Pompe's disease), lysosomes are seen to be crammed full of glycogen granules, compatible with the concept that lysosomes engulf the cytoplasm. Glycogen accumulates because it cannot be degraded, due to the absence of α-glucosidase (Hers and van Hoof, 1969), and cannot be released from the lysosomes because the lysosomal membrane is not permeable to large-molecular-weight compounds. No genetic defects have been found in which lysosomal proteinases are absent (Hers and van Hoof, 1969; Neufeld et al., 1975). While the absence of a defect cannot be used as evidence for the essentiality of proteinases for cell viability, it does present a strong case for such essentiality, especially since some 30 other types of lysosomal deficiency diseases have been observed.

The concept of a lysosomal cycle has arisen largely on the basis of morphological experiments (Ericsson, 1969; Dean and Barrett, 1976). An area of cytoplasm newly surrounded with membrane is considered to fuse with a primary lysosome, a smaller vesicle containing the hydrolytic activities. Digestion then occurs in this autophagic vacuole, with eventual reduction in organelle size until a secondary lysosome is formed (Fig. 4). This lysosome is indistinguishable from a primary lysosome, and is able to fuse again and repeat the cycle. The autophagic lysosomes may recycle many times and will also mix with other lysosomes involved in the uptake and digestion of extracellular material (heterophagic vacuoles). A detailed analysis of lysosomal properties and function has been given by Dean and Barrett (1976).

Circumstantial evidence for equating autophagy with the normal degradation of intracellular proteins comes from studies in which the number of autophagic vacuoles have been seen to decrease after treatment with agents known to depress protein breakdown, such as insulin

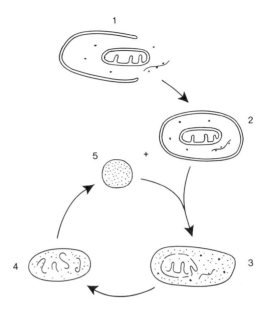

Figure 4. A schematic representation of lysosomal autophagy. A double membrane is shown enclosing a portion of cytoplasm (1) to form a vacuole (2). This fuses with a primary or secondary lysosome (5) to produce an autophagic vacuole (3). Digestion of intravacuolar contents continues (4) until complete, with the breakdown products passing out of the organelle, which is then reduced in size and becomes a secondary lysosome.

or cycloheximide. Moreover, glucagon increases protein degradation and also increases the number of autophagic vacuoles in liver (Woodside et al., 1974; Arstila et al., 1972). The requirement for protein synthesis in order to have maximal degradation of intracellular proteins can be interpreted as a need for new proteins in the vacuole membranes. Moreover, the often observed energy requirement for protein degradation—often observed in the form of an inhibition of catabolism in the presence of any one of a large number of compounds which interfere with glycolysis or oxidative phosphorylation (Goldberg and Dice, 1974; Goldberg and St. John, 1976)—can be explained as a requirement for energy either in the engulfing process or in the establishment of low intralysosomal pH. But it should be emphasized that the energy requirement for protein degradation is not complete.

It is difficult to imagine how an exclusively lysosomal breakdown of intracellular proteins could account for the wide range of enzyme half-lives. Perhaps some selective entrapment or micropinocytosis of labile proteins might occur, but this is hard to envisage.

As an explanation for these findings, we have recently hypothesized that lysosomal autophagy accounts for the part of protein degradation that is sensitive to nutritional step-up and step-down (Knowles and Ballard, 1976). We envisage autophagy as a nonselective process in which all cytoplasmic constituents are degraded at a relatively slow rate. Variations in enzyme half-lives are explained by a second degradative system as yet uncharacterized, but possibly associated with microsomes or other membranes. In this second system the rate constant of degradation for an enzyme is not set by the amount of proteinase present but is defined by the primary sequence of each enzyme as substrate, perhaps modifed by the binding of ligands to the enzyme or by other processes that alter enzyme conformation. The second pathway thus offers an explanation for: (1) the diversity of degradation rate constants; (2) decreases in k_d by ligand binding; (3) increases in k_d associated with enzyme errors; (4) the relative insensitivity of unstable enzymes toward inhibition of their k_d by nutrients or cycloheximide; (5) the rapid inactivation of some enzymes by cell membranes and the correlation of inactivation rates with degradation rate constants determined on intact tissues.

As yet there is no direct evidence for this second pathway and no information as to whether lysosomes may be involved in later non-rate-limiting steps in enzyme degradation. However, lysosome involvement is not obligatory, since proteinases do occur elsewhere in the cell.

Direct evidence for a lysosomal role in the breakdown of intracellular proteins has recently been provided by Dean (1975). In his experiment, pepstatin, a potent inhibitor of cathepsin D, was incorporated into liposomes by sonication of the inhibitor in the presence of a lipid mixture. The relatively stable liposomes fused with the cells and introduced pepstatin into the cell interior. Although the means by which pepstatin became incorporated into lysosomes are not clear, degradation in the treated cells was rapidly inhibited to the extent of about 50%. Dean concluded that he had observed a direct lysosomal effect, because pepstatin did not inhibit any proteinase in cells other than cathepsin D, and cathepsin D is totally localized in lysosomes.

B. Degradation of Proteins within Organelles

In general, the efforts aimed at elucidating the degradation sequence for intracellular proteins have centered on the breakdown of cytosol proteins rather than on those present in cell organelles. I mentioned previously that organelles, including mitochondria, are degraded by the autophagic vacuole system (Ericsson, 1969). Presumably this degradation results in the catabolism of all mitochondrial proteins at approximately equal rates, determined not by the sensitivity of individual proteins to proteinases, but by the rate and extent of autophagy. Although the turnover of several mitochondrial enzymes approximates the half-life of average mitochondrial proteins (about 7 days, Schimke, 1970) and can be considered to occur via autophagy, difficulties occur upon consideration of mitochondrial enzymes with short half-lives. Two such enzymes, δ-aminolevulinate synthetase and ornithine aminotransferase, have half-lives of 1 h and 17 h respectively (Tschudy et al., 1965; Chee and Swick, 1976). The most reasonable explanation for these differences would be a selective degradation of the enzymes within mitochondria. Certainly proteinases have been reported in mitochondria (Bohley et al., 1971; Lovaas, 1974), but I am not aware whether purified mitochondrial preparations have been shown to degrade their complement of δ-aminolevulinate synthetase. Otherwise, the enzyme would need to be extruded from the organelle for degradation in the cytosol. The nature of a "force" responsible for extruding the enzyme is difficult to envisage.

C. Possible Experimental Approaches for Defining the Intracellular Localization of Protein Breakdown

There are two major difficulties in assigning a role for a particular proteinase in the degradation of an enzyme or other protein:

1. In order to resolve the different catabolic steps, degradation products derived from cleavage of the enzyme need to be distinguished from degradation products of other proteins.
2. Some means must be possible for investigating one proteinase or degradation system without interference from other catabolic enzymes.

The inability to satisfy either of these conditions means that the degradation sequence has not been even partly resolved for any single intracellular protein. Theoretically at least, the first point can be satisfied by introducing a pure, labeled protein into cells or by selective labeling of a single protein within cells. Insertion of proteins into cells is possible with the liposome-fusion technique, exactly as carried out by Dean with pepstatin (Dean, 1975). Unfortunately, it is not completely clear whether material in liposomes is added to the cytosol fraction or whether liposome fusion induces the formation of heterophagic vacuoles via pinocytosis. If the latter is true, the proteins would be immediately inserted into lysosomes, and the technique would be irrelevant to a solution of our problem. An alternate but related technique is to enclose the labeled protein within resealed erythrocyte membranes and to fuse these to the cell membrane, perhaps using Sendai virus (Loyter et al., 1975). Although this approach has not yet been used for an analysis of protein breakdown, morphological studies confirm a mixing of the entrapped material into the cytoplasm of the host cell (Loyter et al., 1975).

Once the labeled protein is inside the cell, it should be simple to compare its k_d with the k_d of the same but endogenous protein, using a double-label procedure. The finding of equal rate constants would be a minimum requirement for justifying the approach. An advantage of the method is that it avoids reincorporation problems, by labeling the inserted proteins in such a manner that subsequent degradation releases a modified amino acid that is not a substrate for the protein-synthesizing system. Protein tyrosine residues labeled with ^{125}I would be suitable for this purpose.

The second difficulty listed above can be at least partly resolved. Specific inhibitors or additional proteinases can be inserted into cells by the same liposome method and the effects on the degradation assessed. Even this would not be necessary if the compounds were of sufficiently low molecular weight to permeate the cell membrane. Alternatively, conditions can be sought whereby the activity of a degrading enzyme is altered by nutritional or hormonal changes or perhaps in genetically deficient cell lines. However, it is possible, even likely, that degradation is not limited by proteinases until their activities have been very markedly reduced.

The emphasis I have given in this section to *potential* methods and the difficulties they are designed to circumvent illustrates how little data exist on the degradation process and its intracellular localization. Indeed, even the methods outlined above will be applicable only to cultured cells, and other techniques must be developed before useful information can be obtained on the mechanism and site of degradation in whole tissues or intact animals.

VII. Initial Reactions in Enzyme Degradation

Attempts to measure intermediates in the in vivo degradation of proteins have been restricted by the technical limitations outlined in Section VI. One approach, the use of specific antibodies to search for breakdown products of phosphoenolpyruvate carboxykinase or ornithine aminotransferase, gave no indications of an accumulation of intermediates in vivo, even though the antibody preparations could be expected to react and precipitate fragments of the enzymes (Ballard et al., 1974; Kominami and Katunuma, 1976). However, the method is rather crude, since an intermediate would need to make up a few percent of the starting enzyme in order to be detected. More satisfactory results can be expected when pure labeled proteins are inserted into cells and the accumulation of all radioactive products is followed.

A. Inactivation of Enzymes in Vitro

It is a simple matter to label a pure enzyme, add it to a homogenate or other tissue preparation, and measure loss of activity and enzyme protein as well as the accumulation of products. Unfortunately, the approach has two major drawbacks: (1) it is virtually impossible to establish whether the reaction sequence in vitro bears any resemblance to enzyme turnover in the intact cell; and (2) cooperation between more than one subcellular fraction will be difficult to attain in vitro. Notwithstanding these problems, the technique has been applied to several enzymes, with the results showing a consistent pattern. Firstly, comparisons between the inactivation rates of a group of enzymes with known degradation rate constants in vivo show good correlations, with the order of inactivation rates being similar to the order of degradation rate constants. This is especially true when enzymes without cofactors or allosteric ligands are compared (Hopgood and Ballard, 1974), as is shown in Table 6. Although only suggestive, comparisons of this type are consistent with inactivation being related to degradation and perhaps being rate limiting.

Inactivation experiments with pure, labeled phosphoenolpyruvate carboxykinase in vitro suggest the following sequence of events (Ballard et al., 1974; Ballard and Hopgood, 1976; Ballard, 1977).

1. Native enzyme is inactivated initially without loss of antibody reactivity. The inactivation reaction is catalyzed by a membrane protein present in all liver membranes but at highest specific activity in plasma membranes and at lowest activity in lysosomal membranes. Inactivation is greatly accelerated in the presence of disulfides such as oxidized glutathione or cystine and retarded by thiols. Disulfides on the membrane protein are implicated because treatment of membranes with dithiothreitol in the presence of iodoacetamide destroys the capacity to inactivate phosphoenolpyruvate carboxykinase. This treatment would reduce and fix protein disulfides. Inactivation requires a membrane protein that shows some tissue specificity, since plasma membranes from reticulocytes or erythrocytes are not active, nor are liposomes prepared from the lipids of liver microsomes.

2. Inactivation of enzyme is followed by loss of antibody reactivity.

3. The inactive, "denatured" enzyme has a conformation which

Table 6. Relative Rates of Enzyme Inactivation in Vitro[a]

Enzyme	$t_{1/2}$ in vivo (days)	Relative rate of inactivation	
		Liver slice	Liver homogenate
Lactate dehydrogenase	6.0	−	−
Aldolase	4.9	+	−
Glucose-6-phosphate dehydrogenase	1	+	+
Glucokinase	1	++	++
Phosphoenolpyruvate carboxykinase	0.25	+	++
Thymidine kinase	0.1	+++	++

[a] Data are from Hopgood and Ballard (1974).

makes it bind to cell membranes. This binding reaction has not been investigated in any detail, but the bound enzyme still has a molecular weight identical to that of native enzyme (a single-chain protein), while proteolytic inhibitors do not affect the reaction. Accordingly, there is no evidence for proteolysis at any step prior to membrane binding.

4. Proteolytic cleavage occurs subsequently, with the appearance of lower molecular weight fragments.

All of the steps in the breakdown of phosphoenolpyruvate carboxykinase in vitro occur at neutral pH and are not accelerated either by intact or by disrupted lysosomes. Yet a role for lysosomes cannot be eliminated, especially in the latter stages of enzyme catabolism, since the "denatured" enzyme might be selectively translocated into lysosomes after being attached to their outer surfaces. Nevertheless, it is most unlikely that lysosomes *initiate* the degradation process. Initiation is reduced by a combination of the ability of enzyme to maintain a stable conformation and by the high ratio of reduced to oxidized glutathione in cells.

Although I have chosen to describe in some detail our experiments on the inactivation of phosphoenolpyruvate carboxykinase, the conclusions are compatible with inactivation measurements on tryptophan oxygenase (Li and Knox, 1972) and, with the exception of an early proteolytic cleavage, on ornithine aminotransferase (Kominami and Katunuma, 1976).

B. Sulfhydryl Reactions and Protein Catabolism

Whereas disulfide links within peptide chains are thought to play an important role in stabilizing enzymes (Anfinsen and Scheraga, 1975), free sulfhydryl groups are often required for enzyme activity and stability. Are these statements as conflicting as they first appear?

Most experiments on protein folding or on the formation and reformation of disulfide links have been carried out with enzymes that are normally secreted from cells or packaged into lysosomes (Anfinsen and Scheraga, 1975). Indeed, lysosomal enzymes are formed by the same mechanism, with the same sequence of steps, as occurs with secreted proteins, except for the secretion process itself, and they can be considered as a common group (Palade, 1975). Exported proteins are frequently glycoproteins, as are lysosomal enzymes and often have stable disulfide linkages. Is the stability "fixed" as a result of proteolytic cleavage reactions in which the proenzymes or prohormones are converted to forms which are active in the extracellular milieu? Studies on the degradation of these extracellular proteins have argued for *reduction* of disulfide bonds as an early, perhaps initial, event in the breakdown sequence (Ansorge et al., 1973; Varandani, 1973). However, as mentioned in the previous section, *oxidation* of sulfhydryl groups in intracellular enzymes facilitates their catabolism. It is possible that enzymes have evolved in such a way that extracellular molecules are largely maintained in the oxidized form and intracellular proteins in the reduced state. Further research on the two classes of protein should clarify the situation.

C. Coenzyme Dissociation

I have already mentioned that enzymes with prosthetic groups are more stable in the holoenzyme form than after dissociation of coenzyme. In addition, Litwack and Rosenfield (1974) reported a correlation between the rate of coenzyme dissociation of a number of liver enzymes and their degradation rate constants in vivo. Of course these experiments were carried out in vitro, often with purified enzymes, and it is not established whether apoenzymes are especially unstable in the intact animal. In fact, three biotin-dependent enzymes, acetyl-CoA carboxylase, propionyl-CoA carboxylase, and pyruvate carboxylase, are present in normal amounts during biotin deficiency but with very low catalytic activities. Catalytic activity is rapidly restored when biotin is administered to the enzymes (Chiang and Mistry, 1974; Jacobs et al., 1970).

D. Specific Proteolytic Enzymes

Katunuma and coworkers have described proteolytic enzymes that specifically cleave the apo forms of pyridoxal enzymes and NAD enzymes (Katunuma et al., 1971a,b; Kominami and Katunuma, 1976). The pyridoxal enzyme-specific proteinase reacts with most pyridoxal enzymes tested, with greatest activity against ornithine aminotransferase and phosphorylase, and with moderate activity against homoserine deaminase and serine dehydratase. Aspartate aminotransferase and tyrosine aminotransferase are not affected (Katunuma et al., 1975). In searching for a role for these enzymes, it is apparent that they cannot be responsible for the initial degradation of all pyridoxal enzymes, because the order of their effectiveness is not consistent with the rapid degradation in vivo of serine dehydratase and tyrosine aminotransferase (see Table 1). However, the pyridoxal proteinase in liver is localized in mitochondria, where it would be active owing to its neutral pH optimum. Perhaps the liver enzyme catalyzes the breakdown of ornithine aminotransferase in that organelle. Such a role would explain why the cytosol pyridoxal enzymes, tyrosine aminotransferase and serine dehydratase, are barely affected, since neither enzyme would be exposed to an intramitochondrial proteinase.

Homogenates of the muscle layer of rat small intestine inactivate ornithine aminotransferase, with the catalytic activity being lost more rapidly than reactivity toward an antiornithine aminotransferase antibody (Kominami and Katunuma, 1976). Examination of antibody–antigen precipitates taken at different stages during the inactivation process showed the accumulation of enzyme breakdown products, consistent with hydrolysis by the group-specific proteinase. Moreover, the inactivation was prevented by keeping ornithine aminotransferase in the holo form, and also by proteinase inhibitors known to act on the group-specific proteinase. Yet no breakdown products were seen if the experiment was carried out with intact animals rather than with tissue preparations. The authors consider that this could be explained by a rate-limiting conversion of holoenzyme to apoenzyme in vivo (Kominami and Katunuma, 1976)—an interpretation consistent with the concept of coenzyme dissociability limiting enzyme degradation (Litwack and Rosenfield, 1974).

Whatever the true function of group-specific proteinases, it seems reasonable that they play a significant role in intracellular enzyme catabolism. Further investigations should clarify this role, as well as establish whether a wide range of such proteinases occurs in tissues.

VIII. Conclusions

All intracellular enzymes in mammalian cells are degraded by a pathway which produces amino acids. To date no intermediates in the sequence have been described, and information on the intracellular localization of the pathway and on the interplay between various subcellular fractions is very sparse. Nevertheless, measurement of half-lives of many enzymes has been made, especially those present in the liver cytosol, and a general picture is appearing as to why some enzymes are stable in vivo and others extremely unstable.

The contribution of enzyme degradation to the various adaptation responses seen in mammalian tissues is becoming clear. Although changes in enzyme content are usually slower than those caused by allosteric ligands, they are extremely important for the long-term modifications which enable a cell to carry out its major functions in a changing environment.

References

Anfinsen, C. B., and Scheraga, H. A., 1975, Experimental and theoretical aspects of protein folding, *Adv. Protein Chem.* **29**:205.

Ansorge, S., Bohley, P., Kirschke, H., Langner, J., Marquardt, I., Wiederanders, B., and Hanson, H., 1973, The identity of the insulin-degrading thiol-protein disulphide

oxidoreductase (glutathione–insulin transhydrogenase) with the sulfydryl–disulfide interchange enzyme, *FEBS Lett.* **37**:238.
Arstila, A. U., Shelburne, J. D., and Trump, B. F., 1972, Studies on cellular autophagocytosis. A histochemical study on sequential alterations of mitochondria in the glucagon-induced autophagic vacuoles of rat liver, *Lab. Invest.* **27**:317.
Arstila, A. W., Nuuja, I. J. M., and Trump, B. F., 1974, Studies on cellular autophagocytosis. Vinblastine-induced autophagy in the rat liver, *Exp. Cell Res.* **87**:249.
Baccino, F. M., Rita, G. A., and Zuretti, M.NF., 1971, Studies on the structure-bound sedimentability of some rat liver lysosome hydrolases, *Biochem. J.* **122**:363.
Ballard, F. J., 1977, Intracellular protein degradation, *Essays Biochem.* **13**:1.
Ballard, F. J., and Hopgood, M. F., 1973, Phosphopyruvate carboxylase induction by L-tryptophan. Effects on synthesis and degradation of the enzyme, *Biochem. J.* **136**:259.
Ballard, F. J., and Hopgood, M. F., 1976, Inactivation of phosphoenolpyruvate carboxykinase (GTP) by liver extracts, *Biochem. J.* **154**:717.
Ballard, F. J., Hopgood, M. F., Reshef, L., and Hanson, R. W., 1974, Degradation of phosphoenolpyruvate carboxykinase (GTP) in vivo and in vitro, *Biochem. J.* **140**:531.
Barrett, A. J., 1969, Properties of lysosomal enzymes, in: *Lysosomes in Biology and Pathology*, Vol. 2 (J. T. Dingle and H. B. Fell, eds.), pp. 245–312, North-Holland, Amsterdam.
Bohley, P., 1968, Intrazelluläre proteolyse, *Naturwissenschaften* **55**:211.
Bohley, P., Kirschke, H., Langner, J., Ansorge, S., Wiederanders, B., and Hanson, H., 1971, Intracellular protein breakdown, in: *Tissue Proteinases* (A. J. Barrett and J. T. Dingle, eds.), pp. 187–219, North Holland, Amsterdam.
Bresnick, E., Williams, S. S., and Mosse, H., 1967, Rates of turnover of deoxythymidine kinase and of its template RNA in regenerating and control liver, *Cancer Res.* **27**(Part 1):469.
Capecchi, M. R., Capecchi, N. E., Hughes, S. H., and Wahl, G. M., 1974, Selective degradation of abnormal proteins in mammalian tissue culture cells, *Proc. Natl. Acad. Sci. USA* **71**:4732.
Chee, P. Y., and Swick, R. W., 1976, Effect of dietary protein and tryptophan on the turnover of rat liver ornithine aminotransferase, *J. Biol. Chem.* **251**:1029.
Chiang, G. S., and Mistry, S. P., 1974, Activities of pyruvate carboxylase and propionyl-CoA carboxylase in rat tissues during biotin deficiency and restoration of the activities after biotin administration, *Proc. Soc. Exp. Biol. Med.* **146**:21.
Cihak, A., Lamar, C., Jr., and Pitot, H. C., 1973, L-Tryptophan inhibition of tyrosine aminotransferase degradation in rat liver in vivo, *Arch. Biochem. Biophys.* **156**:188.
Dean, R. T., 1975, Direct evidence of importance of lysosomes in degradation of intracellular proteins, *Nature* **257**:414.
Dean, R. T., and Barrett, A. J., 1976, Lysosomes, *Essays Biochem.* **12**:1.
Dice, J. F., and Goldberg, A. L., 1975, A statistical analysis of the relationship between degradative rates and molecular weights of proteins, *Arch. Biochem. Biophys.* **170**:213.
Dunaway, G. A., Jr., and Weber, G., 1974, Effects of hormonal and nutritional changes on rates of synthesis and degradation of hepatic phosphofructokinase isozymes, *Arch. Biochem. Biophys.* **162**:629.
Epstein, A., and Newburgh, R. W., 1972–1973, Synthesis and degradation of chicken liver xanthine dehydrogenase during development, *Mech. Ageing and Dev.* **1**:431.
Ericsson, J. L. E., 1969, Mechanism of cellular authophagy, in: *Lysosomes in Biology and Pathology*, Vol. 2 (J. T. Dingle and H. B. Fell, eds.), pp. 345–394, North-Holland, Amsterdam.
Fern, E. B., and Garlick, P. J., 1974, The specific radioactivity of the tissue free amino acid pool as a basis for measuring the rate of protein synthesis in the rat in vivo, *Biochem. J.* **142**:413.

Fritz, P. J., Vesell, E. S., White, E. L., and Pruitt, K. M., 1969, The roles of synthesis and degradation in determining tissue concentrations of lactate dehydrogenase-5, *Proc. Natl. Acad. Sci., USA* **62:**558.

Garlick, P. J., Millward, D. J., and James, W. P. T., 1973, The diurnal response of muscle and liver protein synthesis in vivo in meal-fed rats, *Biochem. J.* **136:**935.

Goldberg, A. L., and Dice, J. F., 1974, Intracellular protein degradation in mammalian and bacterial cells, *Annu. Rev. Biochem.* **43:**835.

Goldberg, A. L., and St. John, A. C., 1976, Intracellular protein degradation in mammalian and bacterial cells. II. *Annu. Rev. Biochem.* **45:**747.

Goldstone, A., and Koenig, H., 1974, Autolysis of glycoproteins in rat kidney lysosomes in vitro. Effects on the isoelectric focusing behaviour of glycoproteins, arylsulphatase and β-glucuronidase, *Biochem. J.* **141:**527.

Gunn, J. M., Hanson, R. W., Meyuhas, O., Reshef, L., and Ballard, F. J., 1975, Glucocorticoids and the regulation of phosphoenolpyruvate carboxykinase (GTP) in the rat, *Biochem. J.* **150:**195.

Gunn, J. M., Ballard, F. J., and Hanson, R. W., 1976, Influence of hormones and medium composition on the degradation of phosphoenolpyruvate carboxykinase (GTP) and total protein in Reuber H35 cells, *J. Biol. Chem.* **251:**3586.

Haining, J. L., 1971, On the kinetics of liver enzyme regression following induction, *Arch. Biochem. Biophys.* **144:**204.

Hers, H. G., and van Hoof, F., 1969, Genetic abnormalities of lysosomes, in: *Lysosomes in Biology and Pathology*, Vol. 3 (J. T. Dingle and H. B. Fell, eds.), pp. 19–40, North-Holland, Amsterdam.

Herschko, A., Mamont, P., Shields, R., and Tomkins, G. M., 1971, Pleiotypic response, *Nature (New Biol.)* **232:**206.

Hogan, B. L. M., McIlhinney, A., and Murden, S., 1974, Effect of growth conditions on the activity of ornithine decarboxylase in cultured hepatoma cells. II. Effect of serum and insulin, *J. Cell. Physiol.* **83:**353.

Hopgood, M. F., and Ballard, F. J., 1974, The relative stability of liver cytosol enzymes incubated in vitro, *Biochem. J.* **144:**371.

Hopgood, M. F., Ballard, F. J., Reshef, L., and Hanson, R. W., 1973, Synthesis and degradation of phosphoenolpyruvate carboxylase in rat liver and adipose tissue. Changes during a starvation–refeeding cycle, *Biochem. J.* **134:**445.

Hopgood, M. F., Clark, M. G., and Ballard, F. J., 1977, Inhibition of protein degradation in isolated rat hepatocytes, *Biochem. J.* **164:**399.

Iynedjian, P. B., Ballard, F. J., and Hanson, R. W., 1975, The regulation of phosphoenolpyruvate carboxykinase (GTP) synthesis in rat kidney cortex. The role of acid-base balance and glucocorticoids, *J. Biol. Chem.* **250:**5596.

Jacobs, R., Kilburn, E., and Majerus, P. W., 1970, Acetyl coenzyme A carboxylase. The effects of biotin deficiency on enzyme in rat liver and adipose tissue, *J. Biol. Chem.* **245:**6462.

Johnson, R. W., and Kenney, F. T., 1973, Regulation of tyrosine aminotransferase in rat liver. XI. Studies on the relationship of enzyme stability to enzyme turnover in cultured hepatoma cells, *J. Biol. Chem.* **248:**4528.

Jost, J. P., Khairallah, E. A., and Pitot, H. C., 1968, Studies on the induction and repression of enzymes in rat liver. V. Regulation of the rate of synthesis and degradation of serine dehydratase by dietary amino acids and glucose, *J. Biol. Chem.* **243:**3057.

Katunuma, N., Kito, K., and Kominami, E., 1971a, A new enzyme that specifically inactivates apo-protein of NAD-dependent dehydrogenases, *Biochem. Biophys. Res. Commun.* **45:**76.

Katunuma, N., Kominami, E., and Kominami, S., 1971b, A new enzyme that specifically inactivates apo-protein of pyridoxal enzymes, *Biochem. Biophys. Res. Commun.* **45:**70.

Katunuma, N., Kominami, E., Kobayashi, K., Banno, Y., Susuki, K., Chichibu, K., Hamaguchi, Y., and Katsunuma, T., 1975, Studies on new intracellular proteases in various organs of rat. 1. Purification and comparison of their properties, *Eur. J. Biochem.* **52**:37.
Killewich, L., Schutz, G., and Feigelson, P., 1975, Functional level of rat liver tryptophan 2,3-diogenase messenger RNA during superinduction of enzyme with actinomycin D, *Proc. Natl. Acad. Sci. USA* **72**:4285.
Knowles, S. E., and Ballard, F. J., 1976, Selective control of the degradation of normal and aberrant proteins in Reuber H35 hepatoma cells, *Biochem. J.* **156**:609.
Knowles, S. E., Gunn, J. M., Reshef, L., Hanson, R. W., and Ballard, F. J., 1975, Properties of phosphoenolpyruvate carboxykinase (GTP) synthesized in hepatoma cells in the presence of amino acid analogues, *Biochem. J.* **146**:585.
Kobayashi, K., Morris, H. P., and Katunuma, N., 1976, Studies on the turnover rates of ornithine aminotransferase in Morris hepatoma 44 and host liver, *J. Biochem.* **80**:1085.
Kominami, E., and Katunuma, N., 1976, Studies on new intracellular proteases in various organs of rats. Participation of proteases in degradation of ornithine aminotransferase in vitro and in vivo, *Eur. J. Biochem.* **62**:425.
Kuehl, L., and Sumsion, E. N., 1970, Turnover of several glycolytic enzymes in rat liver, *J. Biol. Chem.* **245**:6616.
Li, J. B., and Knox, W. E., 1972, Inactivation of tryptophan oxygenase in vivo and in vitro, *J. Biol. Chem.* **247**:7550.
Litwack, G., and Rosenfield, S., 1973, Coenzyme dissociation, a possible determinant of short half-life of inducible enzymes in mammalian liver, *Biochem. Biophys. Res. Commun.* **52**:181.
Lovaas, E., 1974, Evidence for a proteolytic system in rat liver mitochondria, *FEBS Lett.* **45**:244.
Loyter, A., Zakai, N., and Kulka, R. G., 1975, Ultramicroinjection of macromolecules or small particles into animal cells. A new technique based on virus-induced cell fusion, *J. Cell. Biol.* **66**:292.
Majerus, P. W., and Kilburn, E., 1969, Acetyl coenzyme A carboxylase. The roles of synthesis and degradation in regulation of enzyme levels in rat liver, *J. Biol. Chem.* **244**:6254.
Mandelstam, J., 1963, Protein turnover and its function in the economy of the cell, *Ann. N.Y. Acad. Sci.* **102**:621.
Millward, D. J., Garlick, P. J., Stewart, R. J. C., Nnanyelugo, D. O., and Waterlow, J. C., 1975, Skeletal-muscle growth and protein turnover, *Biochem. J.* **150**:235.
Neufeld, E. F., Lim, T. W., and Shapiro, L. J., 1975, Inherited disorders of lysosomal metabolism, *Annu. Rev. Biochem.* **44**:357.
Niemeyer, H., Ureta, T., and Clark-Turri, L., 1975, Adaptive character of liver glucokinase, *Molec. Cell. Biochem.* **6**:109.
Orgel, L. E., 1963, The maintenance of the accuracy of protein synthesis and its relevance to ageing, *Proc. Natl. Acad. Sci. USA* **49**:517.
Orgel, L. E., 1973, Ageing of clones of mammalian cells, *Nature* **243**:441.
Palade, G. E., 1975, Intracellular aspects of the process of protein synthesis, *Science* **189**:347.
Peraino, C., Blake, R. L., and Pitot, H. C., 1965, Studies on the induction and repression of enzymes in rat liver. III. Induction of ornithine δ-transaminase and threonine dehydrase by oral intubation of free amino acids, *J. Biol. Chem.* **240**:3039.
Philippidis, H., Hanson, R. W., Reshef, L., Hopgood, M. F., and Ballard, F. J., 1972, The initial synthesis of proteins during development. Phosphoenolpyruvate carboxylase in rat liver at birth, *Biochem. J.* **126**:1127.

Pontremoli, S., Melloni, E., Balestrero, F., Franzi, A. T., De Fiora, A., and Horecker, B. L., 1973, Fructose 1,6-bisphosphatase: The role of lysosomal enzymes in the modification of catalytic and structural properties, *Proc. Natl. Acad. Sci. USA* **70**:303.

Poole, B., and Wibo, M., 1973, Protein degradation in cultured cells. The effect of fresh medium, fluoride and iodoacetate on the digestion of cellular protein in rat fibroblasts, *J. Biol. Chem.* **248**:6221.

Poole, B., Leighton, F., and De Duve, C., 1969, The synthesis and turnover of rat liver peroxisomes. II. Turnover of peroxisome proteins, *J. Cell Biol.* **41**:536.

Reel, J. R., Lee, K.-L., and Kenney, F. J., 1970, Regulation of tyrosine α-ketoglutarate transaminase in rat liver. VIII. Inductions by hydrocortisone and insulin in cultured hepatoma cells, *J. Biol. Chem.* **245**:5800.

Rothstein, M., 1975, Aging and the alteration of enzymes: A review, *Mech. Ageing and Devel.* **4**:325.

Russell, D. H., and Snyder, S. H., 1969, Amine synthesis in regenerating rat liver: Extremely rapid turnover of ornithine decarboxylase, *Mol. Pharmacol.* **5**:293.

Schimke, R. T., 1964, The importance of both synthesis and degradation in control of arginase levels in rat liver, *J. Biol. Chem.* **239**:3808.

Schimke, R. T., 1970, Regulation of protein degradation in mammalian tissues, in: *Mammalian Protein Metabolism*, Vol. 4 (H. N. Munro, ed.), pp. 177–228, Academic Press, New York.

Schimke, R. T., and Doyle, D., 1970, Control of enzyme levels in animal tissues, *Annu. Rev. Biochem.* **39**:929.

Schimke, R. T., Sweeney, E. W., and Berlin, C. M., 1965a, The roles of synthesis and degradation in the control of rat liver tryptophan pyrolase, *J. Biol. Chem.* **240**:322.

Schimke, R. T., Sweeney, E. W., and Berlin, C. M., 1965b, Studies of the stability in vivo and in vitro of rat liver tryptophan pyrolase, *J. Biol. Chem.* **240**:4609.

Scornik, O. A., and Botbol, V., 1976, Role of changes in protein degradation in the growth of regenerating livers, *J. Biol. Chem.* **251**:2891.

Shapiro, D. J., and Rodwell, V. W., 1971, Regulation of hepatic 3-hydroxy-3-methylglutaryl coenzyme A reductase and cholesterol synthesis, *J. Biol. Chem.* **246**:3210.

Silpananta, P., and Goodridge, A. G., 1971, Synthesis and degradation of malic enzyme in chick liver, *J. Biol. Chem.* **246**:5754.

Speake, B. K., Dils, R., and Mayer, R. J., 1975, Regulation of enzyme turnover during tissue differentiation. Studies on the effects of hormones on the turnover of fatty acid synthetase in rabbit mammary gland in organ culture, *Biochem. J.* **148**:309.

Stahl, P., Six, H., Rodman, J. S., Schlesinger, P., Tulsiani, D. R. P., and Touster, O., 1976, Evidence for specific recognition sites mediating clearance of lysosomal enzymes in vivo, *Proc. Natl. Acad. Sci. USA* **73**:4045.

Steinberg, R. A., Levinson, B. B., and Tomkins, G. M., 1975, Kinetics of steroid induction and deinduction of tyrosine aminotransferase synthesis in cultured hepatoma cells, *Proc. Natl. Acad. Sci. USA* **72**:2007.

Swick, R. W., 1958, Measurement of protein turnover in rat liver, *J. Biol. Chem.* **231**:751.

Swick, R. W., and Ip, M. M., 1974, Measurement of protein turnover in rat liver with [^{14}C]-carbonate, *J. Biol. Chem.* **249**:6836.

Tarentino, A. L., Richert, D. A., and Westerfeld, W. W., 1966, The concurrent induction of hepatic α-glycerophosphate dehydrogenase and malate dehydrogenase by thyroid hormone, *Biochim. Biophys. Acta* **124**:295.

Teraoka, H., and Numa, S., 1975, Content, synthesis and degradation of acetyl coenzyme A carboxylase in the liver of growing chicks, *Eur. J. Biochem.* **53**:465.

Tilghman, S. M., Hanson, R. W., Reshef, L., Hopgood, M. F., and Ballard, F. J., 1974, Rapid loss of translatable messenger RNA of phosphoenolpyruvate carboxykinase during glucose repression in liver, *Proc. Natl. Acad. Sci. USA* **71**:1304.

Tschudy, D. P., Marver, H. S., and Collins, A., 1965, A model for calculating messenger RNA half life: Short lived messenger RNA in the induction of mammalian δ-aminolevulinic acid synthetase, *Biochem. Biophys. Res. Commun.* **21**:480.

Varandani, P. T., 1973, Insulin degradation, V. Unmasking of glutathione-insulin transhydrogenase in rat liver microsomal membrane, *Biochim. Biophys. Acta* **304**:642.

Volpe, J. J., Lyles, T. O., Roucari, D. A. K., and Vagelos, P. R., 1973, Fatty acid synthetase of developing brain and liver. Content, synthesis and degradation during development, *J. Biol. Chem.* **248**:2502.

Wang, C.-C., and Touster, O., 1974, Turnover studies on proteins of rat liver lysosomes, *J. Biol. Chem.* **250**:4896.

White, J. M., 1976, The unstable haemoglobins, *Br. Med. Bull.* **32**:219.

Winterer, J. C., Steffee, W. P., Davy, W., Perera, A., Uauy, R., Scrimshaw, N. S., and Young, V. R., 1976, Whole body turnover in aging man, *Exp. Gerontol.* **11**:79.

Woodside, K. H., Ward, S. P., and Mortimore, G. E., 1974, Effects of glucagon on general protein degradation and synthesis in perfused rat liver, *J. Biol. Chem.* **249**:5458.

7

DNA Replication and the Cell Cycle

Robert H. Herman

I. Introduction

A recurring theme in this text has been the importance of weak forces in providing flexibility in life. This carries with it the proclivity for destabilization. While myriad systems exist for the repair and renewal of the macromolecular and supramolecular entities which comprise and confine the life process, none furnish the possibility of biochemical immortality. Therefore, cell replication is necessary for the complete transmission of genetic information to the progeny and for the rejuvenation of the cell. Since DNA acts as a repository for the cellular information, it is crucial that replication of the genetic information be accurate. In this chapter we will review, briefly, the current view of how this most pivotal of life functions is carried out.

II. Chromatin Structure

Compartmentalization of the genetic material (chromatin) into the eukaryotic nucleus is so widespread that its survival value cannot be doubted. As discussed in Chapter 13, such compartmentalization confers profound control capabilities on the cell while, in this instance, protecting the precious information contained in the DNA from gratuitous destruction by cytoplasmic hydrolases (Chapter 5). The chromatin of the nucleus is composed of deoxyribonucleic acid (DNA), histones

---

Robert H. Herman • Endocrine-Metabolic Service, Letterman Army Medical Center, Presidio of San Francisco, California 94129.

(basic proteins), nuclear acidic protein, and a variety of enzymes including RNA polymerase, poly(A) polymerase, DNA polymerase, DNA endonuclease, DNA ligase, deoxyribonuclease, terminal DNA-nucleotidyltransferase, poly(adenosine diphosphateribose) polymerase, poly(ADP-ribose) glycohydrolase, histone acetyltransferase, histone methylase, histone kinase (acid-labile phosphatase), histone protease, and nonhistone chromosomal (NHC) protein kinases (Elgin and Weintraub, 1975). There are NHC proteins (Walker et al., 1976; Peterson and McConkey, 1976) which partly, but not entirely, include the nuclear acidic proteins and nuclear phosphoproteins. There are other proteins contained in RNA transport particles (informosomes) which are involved in the processing of mRNA.

The fundamental structural unit of chromatin in eukaryotic nuclei is the nucleosome or nu body, which consists of about 200 base pairs (with a range of 140 to 240 base pairs) of DNA associated with histones (Kornberg, 1974b; Olins and Olins, 1974; Woodcock, 1973; Thomas and Kornberg, 1975; Hewish and Burgoyne, 1973). Approximately 140–160 DNA base pairs are associated with two molecules each of histones H2A, H2B, H3, and H4, forming an octomer or nu body (Thomas and Kornberg, 1975), with the remaining base pairs being associated with an adjacent H1 histone in the internucleosome region (spacer or linkage region) (Whitlock and Simpson, 1976) which is not part of the histone octomer itself. The DNA is wound around a core of the nu body (Baldwin et al., 1975). The DNA's extended length of 68 nm (140 base pairs) is wound around a 10-nm nu body with two turns (Finch et al., 1977). A short length of DNA (spacer or linker region) connects adjacent nu bodies. The nucleosome core is a flat disc about 110Å in diameter and 57Å in height (Finch et al.,1977).

The DNA is bound to the histone by electrostatic forces, which are large at physiological ionic strength. Assuming that the DNA forms a supercoil with the dimensions of the nucleosome by continuous deformation of the DNA double helix, it can be calculated that 20–28 kcal/mol of nucleosome are required for formation of the supercoil (Finch et al., 1977). Each octomer of core histones induces super-coiling in closed circular DNA (Fuller, 1971; Crick, 1976). There is conversion between relaxed and nucleosomal DNA. The arginine-rich histones H3 and H4 are necessary and sufficient for the formation of nucleosomes (Camerini-Otero et al., 1976; Sollner-Webb et al., 1976) and induce supercoiling in closed circular DNA (Camerini-Otero and Felsenfeld, 1977; Bina-Stein and Simpson, 1977).

A protein with a molecular weight of 29,000 has been isolated from the eggs of *Xenopus laevis* which is capable of producing nucleosome structure by incubation with histones followed by the addition of DNA. In the absence of this protein, the mixture of histones and DNA rapidly form a precipitate (Laskey et al., 1978).

DNA occurs in a double helical structure, with the complementary strands held together by hydrogen bonds between the purine of one DNA strand and the pyrimidine of the complementary strand (adenine with thymine; guanine with cytosine). The details of the structure of DNA may be found in an extensive and voluminous literature and need not be discussed here (see, for example, Kornberg, 1974a).

The nucleosome forms a thin chromatin filament, with a diameter of 100 Å, which is a linear array of nucleosome cores in contact with one another (Felsenfeld, 1978). Coiling of the thin fiber generates a thicker fiber with a diameter of 200–300 Å, with the nucleosomes arranged in a solenoid with a 100A hole down the center (Carpenter et al., 1976). It is not entirely clear how the chromatin is packed into the nucleus in the intact state, but a number of tentative models have been proposed (Felsenfeld, 1978).

As emphasized in Chapter 5, nucleic acid–protein interactions provide the discriminative and functional mechanics of the entire protein synthesizing apparatus, from DNA replication through protein biosynthesis. On the one hand, many proteins bind with specific nucleotide sequences and thus serve a highly specific role in the transmission of biological information. On the other hand, proteins such as the histones bind to wide stretches of nucleotides in an apparently nonspecific manner. The absence of histones from the relatively simple prokaryotic genome argues for a structural or regulatory function, or both, for these ubiquitous basic proteins. While histone H1 may have a cross-linking function, it may also act as a general repressor, keeping the chromatin tightly folded so as to prevent transcription. Histone H1 is phosphorylated by a specific protein kinase during the initiation of mitosis (Bradbury et al., 1974a,b; Lake and Salzman, 1972; Balhorn et al., 1972; Goldstein, 1976). H1 histone stabilizes SV-40 DNA superhelical turns and prevents the action of DNA-relaxing enzyme (Bina-Stein and Singer, 1977; Elgin and Weintraub, 1975). SV-40 chromatin is highly condensed in its native state, this condensed state being converted to the extended "beaded-string" structure of nucleosomes upon the removal of histone H1. Addition of purified histone H1 restores the condensed state of the nucleoprotein complex (Muller et al., 1978). H1 histone is known to interact preferentially with superhelical DNA as compared to relaxed DNA (Vogel and Singer, 1975), the globular region of histone H1 (amino acids 72–106) being involved in this recognition of superhelical DNA (Singer and Singer, 1976).

DNA-relaxing enzyme, which can be inhibited by histone H1, is present in chicken erythrocyte nuclei (Bina-Stein et al., 1976). Calf thymus unwinding protein 1 (UP-1) has been purified and has been shown to have many of the same properties as prokaryote DNA-unwinding proteins (Herrick and Alberts, 1976a). UP-1 facilitates DNA denaturation by strong, selective binding to single-stranded DNA, forming complexes in which the otherwise folded strands are extended and completely covered by protein, and stimulates the rate of in vitro DNA synthesis by a homologous DNA polymerase by a factor of five to tenfold when tested with appropriate DNA templates. Calf thymus UP-1 destabilizes the DNA helix, especially "hairpin" helical regions (Herrick and Alberts, 1976b). After denaturation of DNA by UP-1, renaturation could not be effected despite the presence of helix-stabilizing conditions. Calf UP-1 forms a complex with single-stranded DNA and holds this DNA in a rigid extended conformation (Herrick et al., 1976a). UP-1 stimulates calf thymus DNA polymerase α.

Transcriptionally active sequences of DNA in chromatin are quite sensitive to digestion by pancreatic DNAse; these include those DNA sequences corresponding to the genes for globin in chick erythrocytes (Weintraub and Groudine, 1976) and ovalbumin in hen oviduct (Garel and Axel, 1977). The structural differences between transcriptionally active and inactive DNA that account for the increased sensitivity to DNAse-1 appear to be due to the particular interactions between DNA and histones. The temporal relationship to stages of the transcriptional process which confers this sensitivity is as yet unknown (Felsenfeld, 1978).

Covalent modification of enzymes and proteins is a definitive form of metabolic control utilized in many metabolic sequences in the body (Chapter 4). The effectiveness of this form of control and its ability to be further modulated has been capitalized upon in the controls of genome expression and cell division. Histones are modified by acetylation, methylation, and phosphorylation (Elgin and Weintraub, 1975). These posttranslational modifications are thought to be important for the changes in chromatin structure during the cell cycle (Gurley et al., 1975). Histones H2A, H2B, H3, and H4 are extensively acetylated in vivo but histone H1 is not (Candido and Dixon, 1972). Histones may be modified by poly(ADP–ribosyl)ation (Ueda et al., 1975). The histone H1 of trout testis nuclei is poly(ADP–ribosyl)ated at three sites at the γ-carboxyl group of glutamic acid (Dixon et al., 1976).

The protein ubiquitin, which has a structure identical to that of thymic hormone, is covalently linked to histone H2A (Goldknopf and Busch, 1977). The function of ubiquitin is unknown.

Certain nonhistone proteins of calf thymus chromatin have been designated as high-mobility-group (HMG) proteins on the basis of their electrophoretic mobilities on gel (Johns et al., 1977). The HMG proteins include the following four: HMG_1, HMG_2, HMG_{14}, and HMG_{17}. It has been shown recently that HMG_1 and HMG_2 alter the configuration of the DNA double helix either by unwinding the double helix or by inducing a supercoiling of the DNA (Javaherian et al., 1978).

III. The Cell Cycle

The events of cell replication are embodied in the cell cycle, which has been divided into four successive time intervals: G_1, S, G_2, and D (or M) (Howard and Pelc, 1953) (Fig. 1). G_1 is the time between the end of cell division and the start of DNA replication. S is the period of DNA replication, and G_2 the time between the end of DNA replication and the start of mitosis. D (or M) is the period of cell division or mitosis. G_0 is postulated to be the state that a cell enters when cell replication is arrested at a critical point during G_1 (Lajtha, 1963). Although the time periods for each state vary depending on species, cell type, and environmental conditions, approximate times can be given. Thus, G_1 may last for 5 h, S for 7 h, G_2 for 3 h, and D for 1 h, with a total generation time of 16 h. The G_1 period is the most variable of the four periods (Sisken and Kinosita, 1961).

There may be a range of different times for each period even in a homogeneous population of cells. Cessation of cell reproduction in cell cultures occurs in the G_1 period. Arrest of cell reproduction by density-dependent inhibition (see Chapter 11) also occurs in the G_1 period (Grove, 1974) and can be reversed with fresh serum (Tsuboi and Baserga, 1972) or by treatment with pronase (Noonan and Burger, 1973). Density-dependent regulation of the growth of BSC-1 cells is mediated by lactic acid, ammonia, and what appears to be an unstable protein. Additional factors which seem to be responsible for density-dependent inhibition of cell growth are decreased availability of receptor sites for serum growth factors and limiting concentrations of low-molecular-weight nutrients in the medium. In 3T3 mouse embryo fibroblasts, density-dependent inhibition of growth appears to be entirely a consequence of the inactivation of serum factors (Holley et al., 1978). Other factors that arrest cell reproduction in the G_1 phase are liver extracts (Aujard et al., 1973), amino acid deprivation, and increased intracellular cAMP (Pardee, 1974). The site of arrest in the G_1 phase is termed the R point or restriction point (Pardee, 1974; Pardee et al., 1978). The more

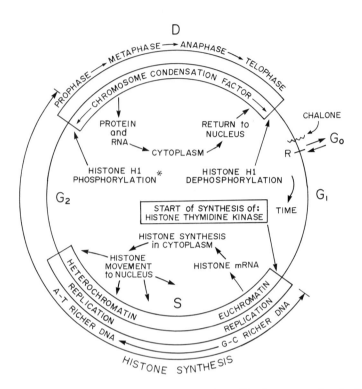

Figure 1. The cell cycle. S is the time for DNA replication or synthesis. D is the time for cell division or mitosis. G_1 is the time period between the end of D and the start of S. G_2 is the time period between the end of S and the start of D. G_0 is a postulated state that occurs when progression through G_1 toward S is arrested. R refers to a restriction point where arrest of the cell cycle occurs in G_1. The symbol ∿ denotes inhibition. The interphase is composed of periods G_1, S, and G_2.

time a cell remains in the G_1 phase, the longer the delay before it resumes the cell cycle. Such observations support the concept of the G_0 state (Prescott, 1957). The differences in generation time of various cells can be accounted for by differences in the length of the G_1 period (Cameron and Greulich, 1963). Chalones (tissue-specific local hormones) inhibit cell reproduction by acting primarily in the G_1 period (Houck and Hennings, 1973). Chalones are tissue-specific, so that epidermal chalone inhibits epidermal cell proliferation but has no effect on other tissues (Bullough and Laurence, 1964) (see Chapter 14).

At least two mammalian cells in tissue culture lack the G_1 period: Syrian hamster fibroblasts (Bürk, 1970) and the V79 line of Chinese hamster cells (Robbins and Scharff, 1967). The significance of this finding is not clear.

Inhibition of protein synthesis (Highfield and Dewey, 1972) with puromycin or cycloheximide arrests the cell cycle in the G_1 phase. Actinomycin D, which inhibits RNA synthesis, also arrests the cell cycle in the G_1 phase (Frankfurt, 1968). Thymidine kinase activity is low during the G_1 period in cultured mammalian cells but begins to rise as the S period begins, reaches a maximum during the S period and declines at the end of mitosis (Stubblefield and Murphree, 1967). Cycloheximide or actinomycin D administration during the G_1 period inhibits the increase in thymidine kinase. Similar changes have been described for thymidylate synthetase (Conrad, 1971), ribonucleotide reductase (Lowdon and Vitols, 1973), deoxycytidine deaminase (Gelbard et al., 1969), deoxycytidine kinase (Howard et al., 1974), and deoxycytidylate synthesis (Adams et al., 1966).

Dibutyryl-cAMP arrests mouse lymphoma cells in the G_1 phase, but a mutant cell line which is deficient in cAMP-dependent protein kinase is not affected by exogenous dibutyryl-cAMP (Coffino et al., 1975). After the release of density-dependent inhibition, cGMP increases transiently and cAMP decreases within minutes (Seifert and Rudland, 1974a,b). After phytohemagglutinin stimulation of mitosis in lymphocytes, cGMP rises immediately and reaches levels 17 times greater than control levels in 20 min, while cAMP does not change for 30 min (Hadden et al., 1972). It has been suggested that the cGMP/cAMP ratio governs the release of cells from G_1 arrest and the reversal of the G_0 state. Exogenously administered cGMP stimulates DNA synthesis in splenic lymphocytes (Weinstein et al., 1974). Serum may somehow reduce the level of cAMP, enabling the cell to pass the point of cAMP sensitivity and continue through the G_1 phase even though the cAMP level may increase once again (Millis et al., 1972). Cyclic AMP may also mediate the inhibitory effect of epinephrine on cell proliferation in epidermal tissue (Bullough and Laurence, 1968).

A cytoplasmic factor is postulated to be responsible for the initiation of the S phase, and various investigators have demonstrated that DNA synthesis can be initiated by extracts of HeLa cells (Friedman and Mueller, 1968; Kumar and Friedman, 1972), L cells and ascites cells (Thompson and McCarthy, 1968), and embryos and eggs of *Xenopus* (Benbow and Ford, 1975). Although these factors are thought to be proteins, their exact nature and mechanisms of action are unknown. Recently it was suggested that diadenosine-tetra-phosphate (Ap_4A) may act as an initiator of DNA replication in animal cells (Grummt, 1978).

During the cell cycle, plasma membrane changes are observed, but their significance is not known. During the G_1 phase, the cell membrane has projections which disappear during the S phase (Hale et al., 1975), so that the cell membrane becomes smooth. In the G_2 phase, microvilli increase in number and the cells thicken. In the late G_2 phase, long filapodia appear and these become abundant during mitosis.

IV. DNA Synthesis

In the S phase, DNA replication takes place. In eukaryotic cells, there are a great number of replicons, a replicon being the replicating unit of genetic material (Taylor, 1960). In Chinese hamster and HeLa cells replicons have been found to be between 7 and 100 μm long (Huberman and Riggs, 1968). Using the assumption that an average replicon is 30 μm long (6×10^7 daltons) and that a haploid set of mammalian chromosomes measures about 90 cm (3 pg or 1.8×10^{12} daltons), the number of replicons per haploid genome is calculated to be about 30,-000. Since the Chinese hamster cells have 21 chromosomes, the average number of replicons would be about 2800 per chromosome. Since replication is bidirectional (occurs simultaneously on both strands of DNA) (Callan, 1972; Hand and Tamm, 1973) the time necessary for replication of a 30-μm replicon is about 17 min (0.9 μm/min/fork) (Prescott, 1976). Certain replicons initiate replication at the same time and are termed banks of replicons. The members of a bank are distributed throughout the various chromosomes. Thus, if the average duration of the mammalian S period is 7 h, a minimum of 25 banks of over 1000 replicons each is required (Prescott, 1976) for the replication of the entire genome.

Early in the S period, DNA has an average GC (guanine–cytosine) content of 43.6%, while in the late S period the GC content is 38.7%. This transition of synthesis from a relatively rich early GC DNA to a relatively rich late AT (adenine–thymine) DNA appears to be a general property of mammalian cells (Tobia et al., 1970), but its signficiance is unknown.

Actinomycin D, added early in the S phase, prevents DNA replication in the late S phase (Mueller and Kajiwara, 1966). If added two or more hours after the initiation of DNA replication, actinomycin D does not inhibit DNA replication. These observations are consistent with the view that actinomycin D inhibits synthesis of the RNA needed for the priming of DNA synthesis (Kornberg, 1976). The addition of cycloheximide and puromycin to a growing cell causes a prompt decrease in the rate of DNA synthesis (Hand and Tamm, 1973; Seki and Mueller, 1975). This may represent inhibition of histone synthesis in the cytoplasm, which may be necessary to stabilize the newly formed DNA in the nucleus in nucleosome structures (Weintraub, 1972). Interestingly, histone synthesis begins during DNA synthesis, and the inhibition of DNA synthesis leads to the cessation of histone synthesis.

Histone synthesis occurs during the S phase and ceases at the end of the G_2 phase or early in mitosis (Breindl and Gallwitz, 1973; Pederson and Robbins, 1970), and histone mRNA is degraded during mitosis and the beginning of the G_1 period (Prescott, 1976). Histones are encoded by repetitious DNA sequences, which has the advantage that a large amount of mRNA can be produced rapidly. Histone mRNA is processed and transported within several minutes of its synthesis to the cytoplasm (Schochetman and Perry, 1972), where it is found in cytoplasmic ribosomes (Zauderer et al., 1973). Histone synthesis is not inhibited by cordycepin (Breindl and Gallwitz, 1974), which is consistent with the finding that histone mRNA does not contain the 3'-terminal poly(A) sequence characteristic of other mRNAs (Adesnik and Darnell, 1972).

Histone synthesis is initiated with the universal initiator methionyl-tRNA (Jacobs-Lorena and Baglioni, 1973). As the histone polypeptide is synthesized, the methionyl group is removed and the N-terminal serine is acetylated (Pestana and Pitot, 1974). The newly synthesized histone is transported within 10 s into the nucleus (Weintraub, 1972).

The nuclear enzyme, poly(ADP-ribose)synthetase, catalyzes the ADP-ribosylation of histones H1, H2a, H2b, and H3 (Ueda et al., 1975; Dixon et al., 1976). It also has been reported that a Ca^{2+}, Mg^{2+}-dependent endonuclease is ADP-ribosylated and consequently inactivated (Yoshihara et al., 1975). Poly(ADP-ribose) synthetase requires DNA for activity and is stimulated severalfold by histone (Okazaki et al., 1976). Two types of glycohydrolases degrade poly(ADP-ribose) and both are found within cell nuclei (Miyakawa et al., 1972). PolyADP-ribosylation may inhibit (regenerating rat liver, Ehrlich carcinoma cells, HeLa S3 cells), stimulate (HeLa cells), or have no effect (lymphoid cells, hepatoma cells, human fibroblasts) on DNA replication in in vitro systems (Hayaishi and Ueda, 1977). It has been suggested that ADP-ribosylation inhibition of Ca^{2+}, Mg^{2+}-dependent alkaline endonuclease is the reason for NAD^+-induced inhibition of DNA synthesis (Hayaishi and Ueda, 1977). It has been proposed that polyADP-ribosylation may regulate the initiation of successive sets of replicons (Shall, 1972). An alternate idea is that polyADP-ribosylation functions in the G_2 phase to sustain continuous cell proliferation (Hayaishi and Ueda, 1977). The role of polyADP-ribosylation is not clear, despite the fact that various nuclear proteins are ADP-ribosylated, that ADP-ribosylation patterns vary among different nuclear proteins during the cell cycle, and that it is mono-ADP-ribosylation that is mainly affected when growth rates change (Hayaishi and Ueda, 1977). Another postulate is that poly(ADP-ribose) functions as a bridge in the reversible condensation of chromatin, via a linkage of neighboring H1 molecules (Lorimer et al., 1976).

DNA synthesis requires a complex set of factors (see Kornberg, 1976; Sheinin et al., 1978), including DNA polymerase, DNA ligating enzyme, unwinding enzyme, untwisting enzyme, and probably a number of other proteins and enzymes that have been identifed, such as single-stranded DNA-dependent ATPase, DNA binding protein, endodeoxyribonuclease, exodeoxyribonuclease, RNase H, RNA polymerase, and various structural proteins of chromatin. Although a great deal is known about prokaryotic DNA synthesis (Dressler, 1975), much less is known about eukaryotic and mammalian DNA synthesis (Sheinin et al., 1978; Pardee et al., 1978). The following discussion will be concerned primarily with mammalian DNA synthesis and will refer to non-mammalian DNA synthesis only for purposes of comparison.

DNA synthesis proceeds along both strands of DNA, organizing its substrates (dATP, dGTP, dTTP, dCTP) into new strands of DNA in a semiconservative fashion. The resultant daughter helices carry one parental DNA strand and the one that has just been synthesized (Edenberg and Huberman, 1975).

From a number of studies in noneukaryotic systems (Sheinin et al., 1978), it appears that the initiation of DNA synthesis occurs at sequences within larger DNA segments that contain palindromes or tandem repeats of unique DNA segments. The term palindrome is applied to double-stranded DNA regions which have an axis of twofold rotational symmetry. Such regions transcribe in the same way from both ends of the region (Engberg et al., 1976). A palindromic region has the properties of a hairpin loop (Soeda et al., 1977). An example of a palindromic region is shown below:

$$A\text{-}A\text{-}T\text{-}T\text{-}G | C\text{-}A\text{-}A\text{-}T\text{-}T$$
$$T\text{-}T\text{-}A\text{-}A\text{-}C | G\text{-}T\text{-}T\text{-}A\text{-}A$$

The mechanism of termination of DNA synthesis is not clear. Whether these same mechanisms apply to eukaryotic DNA synthesis is not yet known.

DNA polymerases function only in the 5' to 3' direction (Weissbach, 1977), suggesting that the discontinuous synthesis of DNA must occur on at least one strand of the DNA double helix. The discontinuous fragments (Okazaki fragments) (Okazaki and Okazaki, 1969) are joined by the action of ligases to form replicon-sized pieces, which polymerize to form chromosome-sized DNA strands. There may be formation of small pieces sedimenting at 2–4 S, then conversion to intermediate (20–100 S) pieces, and finally conversion to fragments with molecular weights of greater than 10^6 (Tseng and Goulian, 1975; Reinhard et al., 1977). One DNA strand may be replicated continuously in the 5' to 3' direction without the formation of the intermediate Okazaki fragments (Brewer, 1975). In HeLa cells it has been found that the 4-S piece may be attached to a larger DNA segment (14–30 S) rather than to other 4-S pieces (Krokan et al., 1975).

Some recent studies have shown that discontinuous DNA synthesis occurs on both template strands in certain mammalian cells (Tseng and Goulian, 1975, 1977; Krokan et al., 1975; Gautschi et al., 1977). At present, in vivo studies suggest that the first stage of DNA synthesis involves initiation of the replicons with the formation of single-strand DNA intermediates with a size up to 4–7 S (6.6–7.4 × 10^4 daltons; 220–280 nucleotide residues) (Sheinin et al., 1978). In the second stage, elongation of the DNA results in fragments of 6–26 S and those of 20–100 S. In the third stage, the various-sized DNA fragments are united. If DNA replication within chromatin is regulated by the nucleosome organization, it might be expected that DNA intermediates that are multiples of the 140 base-pair DNA unit associated with nucleosomes would be detected. A DNA strand corresponding to the 140 base pairs would have a sedimentation value of 4–5 S, with a molecular weight of 6.6 × 10^4 daltons, which is what has been observed.

Inhibition of protein synthesis inhibits DNA biosynthesis (Thompson et al., 1973; Haralson and Roufa, 1975). A variety of studies have suggested that DNA synthesis is a component of chromatin replication. The chromatin containing the DNA is destabilized at the replication fork during the process of replication. Inhibition of protein synthesis may prevent the formation of histones and nonhistone proteins necessary to stabilize the DNA after its synthesis.

RNA primers are involved in eukaryotic DNA synthesis (Kornberg, 1976; Edenberg and Huberman, 1975). DNA polymerases are unable to initiate DNA synthesis directly on a DNA template (Weissbach, 1977). In prokaryotes it has been shown that an RNA segment consisting of 40–50 nucleotide residues, copied from a single-stranded DNA template, serves as the primer for attachment of deoxyribonucleotides at the 3' end (Dressler, 1975). The primer RNA is later cleaved by a specific ribonuclease.

Labeled uridine is found in the 4-S Okazaki pieces after very brief pulses and, in CsCl gradients, there is evidence of both RNA and DNA synthesis (Taylor et al., 1975). In Ehrlich ascites cells, no evidence of

RNA primers has been found, however (Probst et al., 1974). It is not yet clear whether an RNA primer mechanism generally occurs in mammalian cell systems (Sheinin et al., 1978).

DNA unwinding proteins have been identified in calf thymus (Herrick and Alberts, 1976a), mouse ascites cells (Otto et al., 1977), and HeLa cells (Enomoto and Yamata, 1976). These unwinding proteins bind to single-stranded DNA and have a low affinity for double-stranded DNA. It is possible that the unwinding proteins act as initiators of DNA synthesis by unwinding the DNA double helix, exposing the unwound strands to RNA primers and DNA polymerases. Calf thymus unwinding proteins stimulate DNA synthesis in vitro (Herrick and Alberts, 1976b). Mouse ascites cell unwinding protein stimulates polymerase I when bound to a single-stranded template. A chromatin-associated protein kinase can phosphorylate the mouse ascites cell unwinding protein, thereby inactivating it.

It has been postulated that during replication of the DNA double helix, a torque develops which must be relieved by a so-called swivel or untwisting enzyme. Such enzymes have been found in rat liver (Champoux and McConaughy, 1976; Champoux, 1976; Thomas and Patel, 1976), normal human lymphocytes (Rosenberg et al., 1976), mature chick erythrocytes (Bina-Stein et al., 1976), and human fibroblasts (Keller, 1975). Unwinding proteins are required in stoichiometric amounts so that, in the HeLa cell, they constitute 3% of the cell protein (Herrick and Alberts, 1976a). Swivel enzymes are catalytic and produce a nick in one strand, allowing free rotation and relaxation of the supercoiling (Kornberg, 1974a), and then reseal the nick. The rat liver swivel enzyme has a molecular weight of 65,000. It produces a nicked intermediate (Champoux, 1976). The activity of the lymphocyte enzyme is greatest during the S period (Rosenberg et al., 1976).

Although the exact DNA replicase of animal cells has not been conclusively identified, it is believed that DNA polymerase I (or α) is most likely the DNA replicase. In phytohemagglutinin-stimulated human lymphocytes, polymerase I activity increased with DNA synthesis (Bertazzoni et al., 1976). In HeLa cells, DNA polymerase I increased tenfold during the G_1 period and decreased after the S period (Chiu and Baril, 1975). In mice, polymerase I increased during liver regeneration (Hecht, 1975).

Polymerase I is a high-molecular-weight molecule (6–8 S) (Weissbach, 1977) that is found in the nucleus (Herrick et al., 1976b; Foster and Gurney, 1976). DNA polymerase I is template dependent and is able to extend DNA chains on 3'-hydroxyl primers (Weissbach, 1977).

Ligases which convert the 4-S pieces to larger DNA fragments have been found in liver regeneration and in rapidly replicating tissues (Söderhäll and Lindahl, 1975; Söderhäll, 1976). Two ligases have been identified which require ATP and are found in the nucleus and cytoplasm, respectively. DNA ligase I is a high-molecular-weight protein which appears to be active during DNA replication, while DNA ligase II is a low-molecular-weight protein whose activity remains constant.

V. Mitosis

At the end of the S period, the cell enters in the G_2 period. It is believed that condensation of chromosomes occurs as a result of synthesis of a number of uncharacterized factors as the cell enters the D phase from the G_2 phase. This is the beginning of the prophase of mitosis. The HeLa cell in mitosis can cause premature chromosome condensation (chromosomal pulverization) of an interphase nucleus that enters the cell by cell fusion (Johnson and Rao, 1970; Matsui et al., 1972). The chromosomal condensation factor is not species specific (Johnson et al., 1970). Arrest of the G_2 phase may occur but usually involves only a few percent of the cells. It has been studied best in the mouse ear epidermis (Gelfant, 1963). Cyclic AMP may mediate the reversible arrest of cells in the G_2 phase (Nose and Katsuta, 1975; Willingham et al., 1972). Fusion of a HeLa cell in the G_2 phase with a cell in an earlier phase can arrest the G_2 phase and prevent progression into the D phase (Rao et al., 1975; Rao and Johnson, 1970). Protein synthesis is necessary until 10 min before prophase in order for cells to enter into mitosis (Tobey et al., 1966). RNA synthesis is also required but is completed before the protein synthesis requirement (Tobey et al., 1966).

The phosphorylation of H1 histone occurs at the start of the D phase (Bradbury et al., 1974 a,b; Lake and Salzman, 1972; Balhorn et al., 1972; Goldstein, 1976), and H1 is dephosphorylated in the G_1 phase (Lake et al., 1972; Marks et al., 1973). Phosphorylation of H1 histone also has been reported to take place in the S and G_2 phases, with histone remaining phosphorylated through mitosis (Gurley et al., 1973a,b; Lake, 1973; Lake et al., 1972). Chinese hamster cells have two phosphokinases, KI and KII, which are specific for histone H1 (Lake, 1973). KI is cAMP dependent, but KII is cAMP independent. Histone H1 is not methylated (Borun et al., 1972). Phytohemagglutinin and other mitogens stimulate the proliferation of lymphocytes and phosphorylation of nuclear proteins (Johnson and Hadden, 1975) and cause a tenfold rise in cGMP (Schumm et al., 1974); cGMP also stimulates phosphorylation of nuclear acidic proteins (NAP) while cAMP inhibits their phosphorylation (Johnson and Hadden, 1975).

As the prophase progresses, RNA synthesis decreases and finally stops before metaphase is reached (Prescott, 1964), except for the synthesis of 4-S and 5-S RNA (Zylber and Penman, 1971). RNA synthesis resumes in the late telophase. The cessation of RNA synthesis occurs because of the lack of DNA for DNA-dependent RNA synthesis, since the DNA is condensed within the chromosomes (Farber et al., 1972). When RNA synthesis ceases, protein moves from the nucleus into the cytoplasm. At the same time protein synthesis declines to 25% of its initial level by late telophase (Stein and Baserga, 1970). Metaphase ribosomes appear to be inhibited by a protein which may have been released from the nucleus (Salb and Marcus, 1965). At the same time the nuclear membrane loses its structural integrity, the nucleolus breaks down, and nuclear material enters the cytoplasm. In late telophase it appears that many of the proteins return to the nucleus (Sims, 1965). Most of the RNA of the nucleus is lost during mitosis, although precursor ribosomal RNA is found associated with metaphase chromosomes (Fan and Penman, 1971), since ribosome processing is inhibited during mitosis. Heterogeneous nuclear RNA (hnRNA) is found in the cytoplasm and is not degraded (Abramova and Neyfakh, 1973). As precursor rRNA returns to the nucleus, the nucleolus reforms (Phillips and Phillips, 1973). During mitosis, cAMP concentrations decrease in a variety of cell lines (Burger et al., 1972; Sheppard and Prescott, 1972) and are increased in the G_1 period. Growth of cells (3T3 cells) is inhibited by cAMP (Papahadjopoulos et al., 1974; Froehlich and Rachmeler, 1974).

The stages of mitosis are well known, although much remains to be learned about the mechanisms and the control of the mitotic events. Classically, mitosis is divided into four stages: prophase, metaphase, anaphase, and telophase (see, for example, Ham, 1957; McIntosh et al., 1969).

In the prophase, the chromosomes condense and the cell membrane disappears. A body, termed the centrosome, which contains the centrioles (Fawcett, 1966) is located near the apical pole of the nucleus, where it is partially surrounded by the Golgi complex. The centrosome contains a pair of centrioles which are in the form of short rods. The centrioles divide in the prophase so that each daughter cell receives two. After division of the centrioles, they move apart and come to occupy opposite positions in the cell, forming poles toward which half of the chromosome complement will ultimately move.

In megakaryocytes there may be as many as 40 pairs of centrioles (Fawcett, 1966). In the differentiation of ciliated epithelial cells the centrioles reduplicate to form several hundred basal bodies, giving rise to cilia and serving as their kinetic centers. Centrioles are hollow cylinders 300–500 nm in length and 150 nm in diameter. Their walls are composed of nine evenly-spaced, triplet, hollow fibrils or tubules embedded in a dense amorphous matrix which in transverse section gives the the appearance of a pinwheel. One end of the cylinder is closed while the other is open. The two centrioles may be arranged at right angles to one another. Various satellite structures are associated with or attached to the centrioles.

Radiating from dense satellite bodies in front of the centrioles are fibers with a diameter of 15 nm which connect to the satellite bodies of the opposite centriole and to each of the chromosomes at a specialized region, the centromere or kinetochore, which is a constricted attachment point for each replicated chromosome formed during the S phase. In this state, connected chromosomes are referred to as chromatids. The fibers radiating from the region of the centrioles form the spindle and are composed of microtubules which appear to have contractile properties resembling those of actin. Microtubule-associated protein may be able to hydrolyze ATP, from which energy is obtained for microtubule contraction which is postulated to be involved in the movement of the chromosomes in the anaphase (McIntosh et al., 1969).

The prophase ends when the nuclear membrane breaks down. During prophase the chromosomes condense and the kinetochores differentiate into a pair of disc-shaped plaques at the centromere of each chromosome. The microtubules become attached to the kinetochores. It has been proposed that metaphase develops as a consequence of the opposite force generated by the kinetochore tubules. When the chromatids break apart, the chromosomes are moved to the opposite poles because some of the tubules anchored to chromatids slide past other tubules (using cross bridges) and pull the chromatids which are no longer attached to one another (McIntosh et al., 1969). An alternative idea is that polymerization of tubulin pushes the chromosomes to form the metaphase plate and that subsequent depolymerization causes shortening of the microtubules which results in the movement of the chromosomes to each pole (Inoué and Sato, 1967). In this regard, a Ca^{2+}-dependent ATPase, specific for the mitotic spindle, has recently been shown to exist (Petzelt, 1978). It may well be that poleward-sliding of antiparallel microtubules and an opposite-end assembly/disassembly of all microtubules occur constantly throughout mitosis and act coordinately to produce mitotic movement, as has recently been suggested (Margolis et al., 1978).

At metaphase, the centrioles are in place and the spindle is fully formed. The chromosomes are lined up midway between the centrioles, forming a metaphase plate. There are two centrioles at each pole, lying at right angles to one another. The chromatids are attached to one another in the regions of the centromeres. The microtubular fibers appear to contract while the point of attachment of the chromatids breaks apart. The anaphase has now begun.

Each set of chromosomes moves to their respective poles. As has been mentioned, the exact mechanism whereby the spindle causes the chromosomal movement is unknown, but inhibition of spindle function with various inhibitors will prevent the anaphase movement of the chromosomes. A calcium-binding protein termed calcium-dependent regulator (CDR) protein has been specifically localized to the mitotic spindle. CDR is structually similar to troponin-C, which is the calcium-binding subunit of troponin that regulates skeletal muscle ATPase activity and muscle contraction. Since there is evidence that myosin and actin localize and appear to function in the mitotic apparatus, it is suggested that CDR might serve a troponin-C-like function in the spindle (Welsh et al., 1978). Another possibility is that CDR could mediate microtubule depolymerization in a Ca^{2+}-dependent fashion. Based on data which show that H1 histone induces the formation of small nuclei of actin at low ionic strength and increases the rate of polymerization of G-ATP actin, it has been suggested that histone H1 may trigger the formation of actin fibers which, at metaphase, connect the kinetochore and the pole of the mitotic spindle (Magri et al., 1978).

In the telophase, the chromosomes begin to leave the condensed state, a nuclear membrane is reestablished, and the cell membrane dividing the cell is reformed; the actual cell division is termed cytokinesis and can be inhibited by cytochalasin-B, yielding binucleate cells (Rao and Smith, 1976). Cytokinesis rapidly follows mitosis. A portion of the chromatin remains condensed and is termed heterochromatin. The inactive X chromosome of females is an example of heterochromatin and is the so-called Barr body (Barr and Bertram, 1949). Heterochromatin is distinguished from unwound or dispersed chromatin, which is the euchromatin.

Microtubules are fibrous aggregates of the globular protein tubulin (Snyder and McIntosh, 1976; Olmsted and Borisy, 1973). They have an outer diameter of 25 nm, a wall 5 nm thick, and an inner hollow core 15 nm across. Microtubules may have a length from a fraction of a micrometer to several micrometers. Colchicine binds to tubulin and prevents its polymerization into microtubules.

Tubulin is a dimer composed of similar but not identical polypeptides, α and β, one of which (α) is phosphorylated. The molecular weight of tubulin is 100,000, with each of the subunits having a molecular weight of 55,000. Tubulin polymerizes, forming a polar, helical surface lattice. Microtubules are stabilized by Mg^{2+} and GTP, which also promote the polymerization. Ca^{2+} inhibits polymerization and may promote depolymerization (Weisenberg, 1972; Staprans et al., 1975). Brain tubulin is phosphorylated, possibly by cAMP-stimulated protein kinase, which is one of many microtubule-associated proteins. Colchicine blocks polymerization by binding to the growing end of a microtubule. Podophyllotoxin also binds to tubulin, inhibits polymerization, and is a competitive inhibitor of colchicine binding. *Vinca* alkaloids (vinblastine, vincristine, and desacetylvinblastine) form small paracrystals with tubulin. Spindle growth results from tubulin polymerization into microtubules.

The amount of tubulin increases during the late S and early G_2 phases. In HeLa cells, there is a two- to three-fold increase in the phosphate content of tubulin in the S and D phase as compared with the G_1 and G_2 phases (Piras and Piras, 1975).

Dibutyryl cAMP applied to Chinese hamster ovary cells in culture causes an increase in the number of microtubules per unit volume of cytoplasm (Porter et al., 1974). Microtubules end on microtubule organizing centers. A microtubule-associated protein, tau, will facilitate the initiation of tubulin polymerization and its subsequent elongation (Witman et al., 1976). The α chain of tubulin can be acted on by the enzyme tyrosyltubulin ligase, so as to add tyrosine, phenylalanine, or 3,4-dihydroxyphenylalanine to the C-terminus (Deanin and Gordon, 1976).

It has been shown (Adolph et al., 1977a) that when histones and nonhistone proteins are removed from HeLa cell metaphase chromosomes, the chromosomal DNA remains in an organized structure. Each chromatid remains paired with its sister chromatid. About 30 proteins (nonhistone) still remain with the DNA. Further studies have shown (Paulson and Laemmli, 1977) that the core of condensed material retains the original chromosomal morphology. This core is composed of a protein scaffold and is surrounded by many loops of DNA. The protein scaffold can be isolated (Adolph et al., 1977b). In its isolated form, the protein scaffold resembles the shape of the original metaphase chromosomes and contains small pieces of tightly bound DNA. These studies suggest that nonhistone proteins are responsible for the chromosomal shape.

VI. Gene Activation and Inactivation

Although it is generally considered that all adult cells have the same gene content, it is obvious that different cell types produce different proteins. The concept has developed that certain genes are actively functioning while others are in a dormant or nonfunctioning state. The mechanism whereby certain genes remain permanently active and others permanently inactive is one of the major unsolved problems of molecular biology. Other genes can be activated or inactivated by a number of factors including substrates, products, hormones, and genetic preprogramming or predetermination. The control of genes in the synthesis of proteins is considered in Chapter 5. There are examples of gene activation and inactivation which represent relatively permanent transitions from a state of active function to one of inactivity.

In early human fetal life, at 8 weeks or less of gestation, the various hemoglobin chains appear, probably in the order $\epsilon, \alpha, \zeta, \gamma, \beta$, although this is not entirely certain. At this time there is the appearance of hemoglobins F ($\alpha_2\gamma_2$), Gower 1 ($\zeta_2\epsilon_2$), Gower 2 ($\alpha_2\epsilon_2$), and Portland ($\gamma_2\zeta_2$) (Kazazian, 1974; Weatherall, 1976). The synthesis of the α hemoglobin chain continues throughout life. However, ϵ-chain synthesis stops before the eighth week. The mechanism of the termination is unknown. By the eighth week the synthesis of the γ-chain is established. Thus, hemoglobin F is the major hemoglobin in the fetus. Synthesis of the β-chain starts at a low level in the eighth week, so that hemoglobin A ($\alpha_2\beta_2$) is present as 8–10% of the total hemoglobin from the eighth to the thirty-fourth week of gestation. After the thirty-fourth week, β-chain synthesis increases so that, at birth, equal amounts of β and γ chains are synthesized. Hemopoiesis stops at this time for 3–4 weeks. Following this, β-chain synthesis is fully active, while γ-chain production persists only in a small population of erythropoietic cells. The nature of the γ- to β-chain switch is unknown. The γ-chain synthesis continues to decrease so that, at 1 year of age, hemoglobin F is less than 5% of the total hemoglobin. Hemoglobin F reaches the adult level of 1% at a variable age within the first few years of life. There is a minor adult hemoglobin called hemoglobin A_2 ($\alpha_2\delta_2$). The output of the δ gene is low at all times, so that hemoglobin A_2 is generally less than 3.5% of the total hemoglobin. The physiological significance of hemoglobin A_2 is unknown.

In some cases, hemoglobin F may remain elevated (Weatherall et al., 1976). Thus, in sickle cell anemia, hemoglobin F is elevated (10–30% in Saudi Arabians, but generally 5% in all other individuals with sickle cell anemia), as it is also in β-thalassemia (where it may represent from 10 to 100% of the total hemoglobin). It has been postulated that some of this increase may result from an increase in the total mass of erythropoietic tissue.

Hemoglobin F production may increase in pregnancy, leukemia (especially juvenile chronic myeloid leukemia), aplastic anemia, and in a few other disorders (Weatherall, 1976).

Neoplastic tissues are capable of producing a variety of proteins, enzymes, and hormones that are not produced by the tissues from which the neoplasm arose. It is believed that the neoplastic process results in the activation of certain genes that otherwise would have remained inactive. Thus, in juvenile chronic myeloid leukemia, hemoglobin F gradually increases in the red blood cells, which also assume a fetal pattern with respect to hemoglobin A_2 and carbonic anhydrase, which disappear as the illness evolves (Weatherall, 1976). Somehow, the nonneoplastic erythropoietic cells are influenced by the neoplastic granulocytes to produce the γ-chains of hemoglobin. Tumor-associated fetal proteins have been found in a variety of neoplastic diseases. Carcinoembryonic antigen (CEA), which has a molecular weight of 200,000, has been found in carcinoma of the colon, pancreas and liver (Kraft, 1972). The protein α-fetoprotein, with a molecular weight of 70,000, is produced by primary liver cancer (Kraft, 1972). Elevated plasma fibrinogen occurs in tumor-bearing rats (Hilgard and Hiemeyer, 1970). A variety of neoplasias produce different types of hormones and hormonal-like substances which include, among others, adrenocorticotropin (lung, thymus, pancreas, etc.), parathormone (lung, kidney, ovary, etc.), intact human chorionic gonadotropin or its α or β subunit (choriocarcinoma, lung, teratoma), erythropoietin (kidney, cerebellar hemangioblastoma, benign renal lesions, uterine fibromata, etc.), thyroid-stimulating hormone-like substance (choriocarcinoma), and growth hormone (lung) (Vaitukaitis, 1976; Gomez-Uria and Pazianos, 1975).

A deficiency of lactase is quite common and is prevalent in many ethnic groups. Lactase deficiency results from abnormally low activity of the small intestinal brush border enzyme, lactase (Dahlqvist et al., 1963), and may be an example of gene inactivation. It has been estimated that 10-20% of individuals of Northern European ethnic origin may be lactase deficient (Welsh et al., 1967). In other ethnic groups the percentage may be as high as 70% or more (Paige et al., 1972). After weaning, the activity of lactase declines in animals (Welsh et al., 1974; Ekstrom et al., 1975). In some ethnic groups the decline occurs relatively early, so that individuals become lactase deficient in childhood, while in other individuals, the decline occurs late in life so that lactase deficiency may appear in the fourth to sixth decade or later. In the human fetus, lactase activity appears in the middle of gestation and reaches a maximum level at birth (Auricchio et al., 1965; Doell and Kretchmer, 1962). In adults, lactase is not adaptable to lactose (Rosensweig and Herman, 1968), so that the decrease in lactase activity with age is not due to the absence of lactose in the diet.

Another example of control of genetic expression is that of histone production during the cell cycle in replicating cells. During the S phase of the cell cycle, when DNA synthesis is occurring, there is concomitant histone synthesis. During the preceding G_1 period, there is no detectable mRNA for histones. The transcription of histone genes begins near the boundary between G_1 and S. Hence, the histone genes are activated at the start of the S period. Histone synthesis continues during DNA replication and ceases at the end of the G_2 period just before or during mitosis. During mitosis or at the beginning of the G_1 period all histone mRNA is degraded (Prescott, 1976; Pederson and Robbins, 1970).

An analysis of the activation–inactivation of the gene for tyrosine aminotransferase has been made in rat hepatoma cell cultures (Martin and Tomkins, 1970). It is known that the synthesis of tyrosine aminotransferase can be induced by glucocorticoids during the latter half of the G_1 and S periods of the cell cycle but not during the G_2 or early part of the G_1 periods. Although the mechanism that is postulated is complex, it involves, in part, the activation and inactivation of genes regulating the synthesis of tyrosine aminotransferase.

In cultured mammalian cells, thymidine kinase activity is low during the G_1 period and increases at the start of the S period of the cell cycle (Stubblefield and Murphree, 1967; Loeb et al., 1970), reaches a maximum during the S period, remains elevated through the G_2 period, and decreases abruptly at the end of mitosis (Stubblefield and Murphree, 1967; Thilly et al., 1975; Howard et al., 1974; Bello, 1974). This would appear to represent activation–inactivation of the gene for thymidine kinase.

VII. Summary

The nucleus of the cell is filled with a substance called chromatin, which consists of deoxyribonucleic acid (DNA), histones (basic proteins), and nuclear acidic proteins. The nucleus also contains ribosomes and undoubtedly carries out some protein synthesis. DNA is able to synthesize or replicate itself utilizing a DNA-polymerase and deoxyribonucleotide diphosphates. Thus, DNA is a long-chain polymer which employs itself to organize nucleotides into a copy of itself by means of an enzymatic reaction. From one point of view, this is a unique enzymatic reaction, in which a substance provides the framework for making more of itself without itself necessarily being altered. Of course, as is now well known, DNA contains the information for the proteins and events that will occur in the cell. DNA is the repository of hereditary information, thus serving as a blueprint for the resulting cell architecture and function.

The chromatin of the nucleus can condense into discrete bodies called chromosomes, which separate into two groups, forming two cells from the original cell. This process is called mitosis. There is an intricate relationship between DNA replication and the process of mitosis, both evincing a number of control mechanisms. A number of investigators believe that the DNA-dependent formation of RNA precludes DNA replication. The sequences of DNA replication and mitosis may be drastically altered if some injury to DNA occurs. Interestingly enough, there exist mechanisms for the detection, excision, and repair of certain types of damage to DNA. Failure of the repair mechanism may also allow injury to interfere with DNA function, DNA replication or the mitotic process, or both.

Mitosis results in daughter cells bearing the same genetic material as the parent cell. Thus it would be expected that the cells in a mammalian organism would have the same genetic complement in all tissues. The differing expressions of this genetic complement may be explained by the concept of gene activation–inactivation. Thus, certain genes are permanently active (constitutive genes), permanently inactive, inducible by signal molecules (hormones, metabolites), time-dependent (activation–deactivation occurring during embryogenesis, birth, growth, and aging), cell cycle-dependent, and derepressed by neoplastic processes.

References

Abramova, N. B., and Neyfakh, A. A., 1973, Migration of newly synthesized RNA during mitosis. III. Nuclear RNA in the cytoplasm of metaphase cells, *Exp. Cell Res.* **77**:136.

Adams, R. L. P., Abrams, R., and Lieberman, I., 1966, Deoxycytidylate synthesis and entry into the period of deoxyribonucleic acid replication in rabbit kidney cells, *J. Biol. Chem.* **241**:903.

Adesnik, M., and Darnell, J. E., 1972, Biogenesis and characterization of histone messenger RNA in HeLa cells, *J. Mol. Biol.* **67**:397.

Adolph, K. W., Cheng, S. M., and Laemmli, U. K., 1977a, Role of nonhistone proteins in metaphase chromosome structure, *Cell* **12**:805.

Adolph, K. W., Cheng, S. M., Paulson, J. R., and Laemmli, U. K., 1977b, Isolation of a protein scaffold from mitotic HeLa cell chromosomes, *Proc. Natl. Acad. Sci. USA* **74**:4937.

Aujard, C., Chany, E., and Frayssinet, C., 1973, Inhibition of DNA synthesis of synchronized cells by liver extracts acting in G_1 phase, *Exp. Cell. Res.* **78**:476.

Auricchio, A., Rubino, A., and Muiset, G., 1965, Intestinal glycosidase activities in the human embryo, fetus and newborn, *Pediatrics* **35**:944.

Baldwin, J. P., Boseley, P. G., Bradbury, E. M., and Ibel, K., 1975, The subunit structure of the eukaryotic chromosome, *Nature* **253**:245.

Balhorn, R., Chalkley, R., and Granner, D., 1972, Lysine-rich histone phosphorylation. A positive correlation with cell replication, *Biochemistry* **11**:1094.

Barr, M. L., and Bertram, E. G., 1949, A morphological distinction between neurons of the male and female, and the behaviour of the nucleolar satellite during accelerated nucleoprotein synthesis, *Nature* **163**:676.

Bello, L. J., 1974, Regulation of thymidine kinase synthesis in human cells, *Exp. Cell Res.* **89**:263.

Benbow, R. M., and Ford, C. C., 1975, Cytoplasmic control of nuclear DNA synthesis during early development of *Xenopus laevis*: A cell-free assay, *Proc. Natl. Acad. Sci. USA* **72**:2437.

Bertazzoni, U., Stefanini, M., Noy, G. P., Giulotto, E., Nuzzo, F., Falaschi, A., and Spa-

dari, S., 1976, Variations of DNA polymerases-α and -β during prolonged stimulation of human lymphocytes, *Proc. Natl. Acad. Sci. USA* **73**:785.
Bina-Stein, M., and Simpson, R. T., 1977, Specific folding and contraction of DNA by histones H3 and H4, *Cell* **11**:609.
Bina-Stein, M., and Singer, M. F., 1977, The effect of H1 histone on the action of DNA-relaxing enzyme, *Nucleic Acids Res.* **4**:117.
Bina-Stein, M., Vogel, T., Singer, D. S., and Singer, M. F., 1976, H5 histone and DNA-relaxing enzyme of chicken erythrocytes, *J. Biol. Chem.* **251**:7363.
Borun, T. W., Pearson, D., and Paik, W. K., 1972, Studies of histone methylation during the HeLa S-3 cell cycle, *J. Biol. Chem.* **247**:4288.
Bradbury, E. M., Inglis, R. J., Matthews, H. R., and Langan, T. A., 1974a, Molecular basis of control of mitotic cell division in eukaryotes, *Nature* **249**:553.
Bradbury, E. M., Inglis, R. J., and Matthews, H. R., 1974b, Control of cell division by very lysine-rich histone (F1) phosphorylation, *Nature* **247**:257.
Breindl, M., and Gallwitz, D., 1973, Identification of histone messenger RNA from HeLa cells, *Eur. J. Biochem.* **32**:381.
Breindl, M., and Gallwitz, D., 1974, Effects of cordycepin, hydroxyurea, and cycloheximide on histone mRNA synthesis in synchronized HeLa cells, *Mol. Biol. Rep.* **1**:263.
Brewer, E. N., 1975, DNA replication by a possible continuous-discontinuous mechanism in homogenates of *Physarum polycephalum* containing dextran, *Biochim. Biophys. Acta* **402**:363.
Bullough, W. S., and Laurence, E. B., 1964, Mitotic control by internal secretion: the role of the chalone-adrenalin complex, *Exp. Cell Res.* **33**:176.
Bullough, W. S., and Laurence, E. B., 1968, The role of glucocorticoid hormones in the control of epidermal mitosis, *Cell Tissue Kinet.* **1**:5.
Burger, M. M., Bombik, B. M., Breckenridge, B. McL., and Sheppard, J. R., 1972, Growth control and cyclic alterations of cyclic AMP in the cell cycle, *Nature (New Biol.)* **239**:161.
Bürk, R. R., 1970, One-step growth cycle for BHK21/13 hamster fibroblasts, *Exp. Cell Res.* **63**:309.
Callan, H. G., 1972, Replication of DNA in the chromosomes of eukaryotes, *Proc. R. Soc. London (Biol.)* **181**:19.
Camerini-Otero, R. D., and Felsenfeld, G., 1977, Supercoiling energy and nucleosome formation: The role of the arginine-rich histone kernel, *Nucleic Acids Res.* **4**:1159.
Camerini-Otero, R. D., Sollner-Webb, B., and Felsenfeld, G., 1976, The organization of histones and DNA in chromatin: Evidence for an arginine-rich histone kernel, *Cell* **8**:333.
Cameron, I. L., and Greulich, R. C., 1963, Evidence for an essentially constant duration of DNA synthesis in renewing epithelia of the adult mouse, *J. Cell Biol.* **18**:31.
Candido, E. P. M., and Dixon, G. H., 1972, Acetylation of trout testis histones in vivo. Site of modification in histone IIb$_1$, *J. Biol. Chem.* **247**:3868.
Carpenter, B. G., Baldwin, J. P., Bradbury, E. M., and Ibel, K., 1976, Organization of subunits in chromatin, *Nucleic Acids Res.* **3**:1739.
Champoux, J. J., 1976, Evidence for an intermediate with a single-strand break in the reaction catalyzed by the DNA untwisting enzyme, *Proc. Natl. Acad Sci. USA* **73**:3488.
Champoux, J. J., and McConaughy, B. L., 1976, Purification and characterization of the DNA untwisting enzyme from rat liver, *Biochemistry* **15**:4638.
Chiu, R. W., and Baril, E. F., 1975, Nuclear DNA polymerases and the HeLa cell cycle, *J. Biol. Chem.* **250**:7951.
Coffino, P., Gray, J. W., and Tomkins, G. M., 1975, Cyclic AMP, a nonessential regulator of the cell cycle, *Proc. Natl. Acad. Sci. USA* **72**:878.

Conrad, A. H., 1971, Thymidylate synthetase activity in cultured mammalian cells, *J. Biol. Chem.* **246**:1318.
Crick, F. H. C., 1976, Linking numbers and nucleosomes, *Proc. Natl. Acad. Sci. USA* **73**:2639.
Dahlqvist, A., Hammond, J. B., Crane, R. K., Dunphy, J. V., and Littman, A., 1963, Intestinal lactase deficiency and lactose intolerance in adults, *Gastroenterology* **45**:488.
Deanin, G. G., and Gordon, M. W., 1976, The distribution of tyrosyltubulin ligase in brain and other tissues, *Biochem. Biophys. Res. Commun.* **71**:676.
Dixon, G. H., Wong, N., and Poirier, G. G., 1976, Adenosine diphospho-ribosylation of basic chromosomal proteins in trout testis nuclei, *Fed. Proc.* **35**:1623.
Doell, R. G., and Kretchmer, N., 1962, Studies of small intestine during development. I. Distribution and activity of β-galactosidase, *Biochim. Biophys. Acta* **62**:353.
Dressler, D., 1975, The recent excitement in the DNA growing point problem, *Annu. Rev. Microbiol.* **29**:525.
Edenberg, H. J., and Huberman, J. A., 1975, Eukaryotic chromosome replication, *Annu. Rev. Genet.* **9**:245.
Ekstrom, K. E., Benevenga, N. J., and Grummer, R. H., 1975, Changes in the intestinal lactase activity in the small intestine of two breeds of swine from birth to 6 weeks of age, *J. Nutr.* **105**:1032.
Elgin, S. C. R., and Weintraub, H., 1975, Chromosomal proteins and chromatin structure, *Annu. Rev. Biochem.* **44**:725.
Engberg, J., Andersson, P., Leick, V., and Collins, J., 1976, Free ribosomal DNA molecules from *Tetrahymena pyriformis* GL are giant palindromes, *J. Mol. Biol.* **104**:455.
Enomoto, T., and Yamata, M., 1976, Binding specificity of a HeLa DNA-binding protein to DNA and homopolymers, *Biochem. Biophys. Res. Commun.* **71**:122.
Fan, H., and Penman, S., 1971, Regulation of synthesis and processing of nucleolar components in metaphase-arrested cells, *J. Mol. Biol.* **59**:27.
Farber, J., Stein, G., and Baserga, R., 1972, The regulation of RNA synthesis during mitosis, *Biochem. Biophys. Res. Commun.* **47**:790.
Fawcett, D. W., 1966, *The Cell. Its Organelles and Inclusions. An Atlas of Fine Structure*, W. B. Saunders, Philadelphia.
Felsenfeld, G., 1978, Chromatin, *Nature* **271**:115.
Finch, J. T., Lutter, L. C., Rhodes, D., Brown, R. S., Rushton, B., Levitt, M., and Klug, A., 1977, Structure of nucleosome core particles of chromatin, *Nature* **269**:29.
Foster, D. N., and Gurney, T., Jr., 1976, Nuclear location of mammalian DNA polymerase activities, *J. Biol. Chem.* **251**:7893.
Frankfurt, O. S., 1968, Effect of hydrocortisone, adrenalin and actinomycin D on transition of cells to the DNA synthesis phase, *Exp. Cell Res.* **52**:220.
Friedman, D. L., and Mueller, G. C., 1968, A nuclear system for DNA replication from synchronized HeLa cells, *Biochim. Biophys. Acta* **161**:455.
Froehlich, J. E., and Rachmeler, M., 1974, Inhibition of cell growth in the G_1 phase by adenosine 3',5'-cyclic monophosphate, *J. Cell Biol.* **60**:249.
Fuller, F. B., 1971, The writhing number of a space curve, *Proc. Natl. Acad. Sci. USA* **68**:815.
Garel, A., and Axel, R., 1977, Selective digestion of transcriptionally active ovalbumin genes from oviduct nuclei, *Proc. Natl. Acad. Sci. USA* **73**:3966.
Gautschi, J. R., Burkhalter, M., and Reinhard, P., 1977, Semiconservative DNA replication in vitro. II. Replicative intermediates of mouse P-815 cells, *Biochim. Biophys. Acta* **474**:512.
Gelbard, A. S., Kim, J. H., and Perez, A. G., 1969, Fluctuations in deoxycytidine monophosphate deaminase activity during the cell cycle in synchronous populations of HeLa cells, *Biochim. Biophys. Acta* **182**:564.

Gelfant, S., 1963, A new theory on the mechanism of cell division, *Symp. Int. Soc. Cell Biol.* **2**:229.
Goldknopf, I. L., and Busch, H., 1977, Isopeptide linkage between nonhistone and histone 2A polypeptides of chromosomal conjugate-protein A24, *Proc. Natl. Acad. Sci. USA* **74**:864.
Goldstein, L., 1976, Role for small nuclear RNAs in "programming" chromosomal information? *Nature* **261**:519.
Gomez-Uria, A., and Pazianos, A. G., 1975, Syndromes resulting from ectopic hormone-producing tumors, *Med. Clin. North. Am.*, **59**:431.
Grove, G. L., 1974, A cytophotometric analysis of nuclear DNA contents of cultured human diploid cells in log and in plateau phases of growth, *Exp. Cell Res.* **87**:386.
Grummt, F., 1978, How cells cycle, *Nature* **273**:594.
Gurley, L. R., Walters, R. A., and Tobey, R. A., 1973a, Histone phosphorylation in late interphase and mitosis, *Biochem. Biophys. Res. Commun.* **50**:744.
Gurley, L. R., Walters, R. A., and Tobey, R. A., 1973b, The metabolism of histone fractions. VI. Differences in the phosphorylation of histone fractions during the cell cycle, *Arch. Biochem. Biophys.* **154**:212.
Gurley, L. R., Walters, R. A., and Tobey, R. A., 1975, Sequential phosphorylation of histone subfractions in the Chinese hamster cell cycle, *J. Biol. Chem.* **250**:3936.
Hadden, J. W., Hadden, E. M., Haddox, M. K., and Goldberg, N. D., 1972, Guanosine 3′,5′-cyclic monophosphate: A possible intracellular mediator of mitogenic influences in lymphocytes, *Proc. Natl. Acad. Sci. USA* **69**:3024.
Hale, A. H., Winkelhake, J. L., and Weber, M. J., 1975, Cell surface changes and rous sarcoma virus gene expression in synchronized cells, *J. Cell Biol.* **64**:398.
Ham, A. W., 1957, *Histology*, 3rd ed., pp. 53–72, J. B. Lippincott, Philadelphia.
Hand, R., and Tamm, I., 1973, DNA replication: Direction and rate of chain growth in mammalian cells, *J. Cell Biol.* **58**:410.
Haralson, M. A., and Roufa, D. J., 1975, A temperature-sensitive mutation affecting the mammalian 60S ribosome, *J. Biol. Chem.* **250**:8618.
Hayaishi, O., and Ueda, K., 1977, Poly(ADP-ribose) and ADP-ribosylation of proteins, *Annu. Rev. Biochem.* **46**:95.
Hecht, N. B., 1975, The relationship between two murine DNA-dependent DNA polymerases from the cytosol and the low molecular weight DNA polymerase, *Biochim. Biophys. Acta* **383**:388.
Herrick, G., and Alberts, B., 1976a, Purification and physical characterization of nucleic acid helix-unwinding proteins from calf thymus, *J. Biol. Chem.* **251**:2124.
Herrick, G., and Alberts, B., 1976b, Nucleic acid helix-coil transitions mediated by helix-unwinding proteins from calf thymus, *J. Biol. Chem.* **251**:2133.
Herrick, G., Delius, H., and Alberts, B., 1976a, Single-stranded DNA structure and DNA polymerase activity in the presence of nucleic acid helix-unwinding proteins from calf thymus, *J. Biol. Chem.* **251**:2142.
Herrick, G., Spear, B. B., and Veomett, G., 1976b, Intracellular localization of mouse DNA polymerase-α, *Proc. Natl. Acad. Sci. USA* **73**:1136.
Hewish, D. R., and Burgoyne, L. A., 1973, Chromatin substructure. The digestion of chromatin DNA at regularly spaced sites by a nuclear deoxyribonuclease, *Biochem. Biophys. Res. Commun.* **52**:504.
Highfield, D. P., and Dewey, W. C., 1972, Inhibition of DNA synthesis in synchronized Chinese hamster cells treated in G1 or early S phase with cycloheximide or puromycin, *Exp. Cell Res.* **75**:314.
Hilgard, P., and Hiemeyer, V., 1970, Increased plasma fibrinogen and the release of a fibrinogen enhancing factor in tumour-bearing rats, *Experientia* **20**:182.
Holley, R. W., Armour, R., and Baldwin, J. H., 1978, Density-dependent regulation of

growth of BSC-1 cells in cell culture: Growth inhibitors formed by the cells, *Proc. Natl. Acad. Sci. USA* **75**:1864.
Houck, J. C., and Hennings, H., 1973, Chalones: Specific endogenous mitotic inhibitors, *FEBS Lett.* **32**:1.
Howard, A., and Pelc, S. R., 1953, Synthesis of deoxyribonucleic acid in normal and irradiated cells and its relation to chromosome breakage, *Heredity (Suppl.)* **6**:261.
Howard, D. K., Hay, J., Melvin, W. T., and Durham, J. P., 1974, Changes in DNA and RNA synthesis and associated enzyme activities after the stimulation of serum-depleted BHK21/C13 cells by the addition of serum, *Exp. Cell Res.* **86**:31.
Huberman, J. A., and Riggs, A. D., 1968, On the mechanism of DNA replication in mammalian chromosomes, *J. Mol. Biol.* **32**:327.
Inoué, S., and Sato, H., 1967, Cell motility by labile association of molecules. The nature of mitotic spindle fibers and their role in chromosome movement, *J. Gen. Physiol. (Suppl.)* **50**:259.
Jacobs-Lorena, M., and Baglioni, C., 1973, Initiation of histone synthesis by f-met-tRNA$_f$ in an ascites cell-free system, *Mol. Biol. Rep.* **1**:113.
Javaherian, K., Liu, L. F., and Wang, J. C., 1978, Nonhistone proteins HMG$_1$ and HMG$_2$ change the DNA helical structure, *Science* **199**:1345.
Johns, E. W., Goodwin, G. H., Hasting, J. R. B., and Walker, J. M., 1977, The histones and some histone-like chromosomal proteins, in: *The Organization and Expression of the Eukaryotic Genome* (E. M. Bradbury and K. Javaherian, eds.), pp. 3–19, Academic Press, London.
Johnson, E. M., and Hadden, J. W., 1975, Phosphorylation of lymphocyte nuclear acidic proteins: regulation by cyclic nucleotides, *Science* **187**:1198.
Johnson, R. T., and Rao, P. N., 1970, Mammalian cell fusion: Induction of premature chromosome condensation in interphase nuclei, *Nature* **226**:717.
Johnson, R. T., Rao, P. N., and Hughes, D. S., 1970, Mammalian cell fusion. III. A HeLa cell inducer of premature chromosome condensation active in cells from a variety of animal species, *J. Cell. Physiol.* **76**:151.
Kazazian, H. H., Jr., 1974, Regulation of fetal hemoglobin production, *Semin. Hematol.* **11**:525.
Keller, W., 1975, Characterization of purified DNA-relaxing enzyme from human tissue culture cells, *Proc. Natl. Acad. Sci. USA* **72**:2550.
Kornberg, A., 1974a, *DNA Synthesis*, W. H. Freeman, San Francisco.
Kornberg, A., 1976, RNA priming of DNA replication, in: *RNA Polymerase* (R. Losick and M. Chamberlin, eds.), pp. 331–352, Cold Spring Harbor Laboratory, Cold Spring Harbor, New York.
Kornberg, R., 1974b, Chromatin structure: A repeating unit of histones and DNA, *Science* **184**:868.
Kraft, S. C., 1972, "Humors from tumors": Carcinoembryonic antigen, alpha-fetoprotein, and digestive system cancer, *Ann. Intern. Med.* **76**:502.
Krokan, H., Cooke, L., and Prydz, H., 1975, DNA synthesis in isolated HeLa cell nuclei. Evidence for in vitro initiation of synthesis of small pieces of DNA and their subsequent ligation, *Biochemistry* **14**:4233.
Kumar, K. V., and Friedman, D. L., 1972, Initiation of DNA synthesis in HeLa cell-free system, *Nature (New Biol.)* **239**:74.
Lajtha, L. G., 1963, On the concept of the cell cycle, *J. Cell. Comp. Physiol.* **62**:143.
Lake, R. S., 1973, F1-histone phosphorylation in metaphase chromosomes of cultured Chinese hamster cells, *Nature (New Biol.)* **242**:145.
Lake, R. S., and Salzman, N. P., 1972, Occurrence and properties of a chromatin-associated F1-histone phosphokinase in mitotic Chinese hamster cells, *Biochemistry* **11**:4817.

Lake, R.S., Goidl, J. A., and Salzman, N. P., 1972, F1-histone modification at metaphase in Chinese hamster cells, *Exp. Cell Res.* **73**:113.
Laskey, R. A., Honda, B. M., Mills, A. D., and Finch, J. T., 1978, Nucleosomes are assembled by an acidic protein which binds histones and transfers them to DNA, *Nature* **275**:416.
Loeb, L. A., Ewald, J. L., and Agarwal, S. S., 1970, DNA polymerase and DNA replication during lymphocyte transformation, *Cancer Res.* **30**:2514.
Lorimer, W. S., III, Stone, P. R., and Kidwell, W. R., 1976, Adenosine diphosphate ribosylation of basic nuclear proteins, *Fed. Proc.* **35**:1624.
Lowdon, M., and Vitols, E., 1973, Ribonucleotide reductase activity during the cell cycle of *Saccharomyces cerevisiae*, *Arch. Biochem. Biophys.* **158**:177.
Magri, E., Zaccarini, M., and Grazi, E., 1978, The interaction of histone and protamine with actin. Possible involvement in the formation of the mitotic spindle, *Biochem. Biophys. Res. Commun.* **82**:1207.
Margolis, R. L., Wilson, L., and Kiefer, B. I., 1978, Mitotic mechanism based on intrinsic microtubule behaviour, *Nature* **272**:450.
Marks, D. B., Paik, W. K., and Borun, T. W., 1973, The relationship of histone phosphorylation to deoxyribonucleic acid replication and mitosis during the HeLa S-3 cell cycle, *J. Biol. Chem.* **248**:5660.
Martin, D. W., Jr., and Tomkins, G. M., 1970, The appearance and disappearance of the post-transcriptional repressor of tyrosine amino-transferase synthesis during the HTC cell cycle, *Proc. Natl. Acad. Sci USA* **65**:1064.
Matsui, S.-I., Yoshida, H., Weinfeld, H., and Sandberg, A. A., 1972, Induction of prophase in interphase nuclei by fusion with metaphase cells, *J. Cell Biol.* **54**:120.
McIntosh, J. R., Hepler, P. K., and van Wie, D. G., 1969, Model for mitosis, *Nature* **224**:659.
Millis, A. J. T., Forrest, G., and Pious, D. A., 1972, Cyclic AMP in cultured human lymphoid cells: Relationship to mitosis, *Biochem. Biophys. Res. Commun.* **49**:1645.
Miyakawa, N., Ueda, K., and Hayaishi, O., 1972, Association of poly ADP-ribose glycohydrolase with liver chromatin, *Biochem. Biophys. Res. Commun.* **49**:239.
Mueller, G. C., and Kajiwara, K., 1966, Actinomycin D and *p*-fluorophenylalanine, inhibitors of nuclear replication in HeLa cells, *Biochim. Biophys. Acta* **119**:557.
Müller, U., Zentgraf, H., Eicken, I., and Keller, W., 1978, Higher order structure of simian virus 40 chromatin, *Science* **201**:406.
Noonan, K. D., and Burger, M. M., 1973, Induction of 3T3 cell division at the monolayer stage. Early changes in macronuclear processes, *Exp. Cell Res.* **80**:405.
Nose, K., and Katsuta, H., 1975, Arrest of cultured rat liver cells in G_2 phase by the treatment with dibutyryl cAMP, *Biochem. Biophys. Res. Commun.* **64**:983.
Okazaki, H., Niedergang, C., and Mandel, P., 1976, Purification and properties of calf thymus polyadenosine diphosphate ribose polymer, *FEBS Lett.* **62**:255.
Okazaki, T., and Okazaki, R., 1969, Mechanism of DNA chain growth. IV. Direction of synthesis of T4 short DNA chains as revealed by exonucleolytic degradation, *Proc. Natl. Acad. Sci. USA* **64**:1242.
Olins, A. L., and Olins, D. E., 1974, Spheroid chromatin units (*v* bodies), *Science* **183**:330.
Olmsted, J. B., and Borisy, G. G., 1973, Microtubules, *Annu. Rev. Biochem.* **42**:507.
Otto, B., Baynes, M., and Knippers, R., 1977, A single-strand-specific DNA-binding protein from mouse cells that stimulates DNA polymerase. Its modification by phosphorylation, *Eur. J. Biochem.* **73**:17.
Paige, D. M., Leonardo, E., Cordano, A., Nakashima, J., Adrianzen, B., and Graham, G. G., 1972, Lactose intolerance in Peruvian children: Effect of age and early nutrition, *Am. J. Clin. Nutr.* **25**:297.

Papahadjopoulos, D., Poste, G., and Mayhew, E., 1974, Cellular uptake of cyclic AMP captured within phospholipid vesicles and effect on cell-growth behavior, *Biochim. Biophys. Acta* **363**:404.

Pardee, A. B., 1974, A restriction point for control of normal animal cell proliferation, *Proc. Natl. Acad. Sci. USA* **71**:1286.

Pardee, A. B., Dubrow, R., Hamlin, J. L., and Kletzien, R. F., 1978, Animal cell cycle, *Annu. Rev. Biochem.* **47**:715.

Paulson, J. R., and Laemmli, U. K., 1977, The structure of histone-depleted metaphase chromosomes, *Cell* **12**:817.

Pederson, T., and Robbins, E., 1970, Absence of translational control of histone synthesis during the HeLa cell life cycle, *J. Cell Biol.* **45**:509.

Pestana, A., and Pitot, H. C., 1974, N-terminal acetylation of histone-like nascent peptides on rat liver polyribosomes in vitro, *Nature* **247**:200.

Peterson, J. L., and McConkey, E. H., 1976, Non-histone chromosomal proteins from HeLa cells, *J. Biol. Chem.* **251**:548.

Petzelt, C., 1978, How cells cycle, *Nature* **273**:594.

Phillips, D. M., and Phillips, S. G., 1973, Repopulation of postmitotic nucleoli by preformed RNA. II. Ultrastructure, *J. Cell Biol.* **58**:54.

Piras, R., and Piras, M. M., 1975, Changes in microtubule phosphorylation during cell cycle of HeLa cells, *Proc. Natl. Acad. Sci. USA* **72**:1161.

Porter, K. R., Puck, T. T., Hsie, A. W., and Kelley, D., 1974, An electron microscopy study of the effects of dibutyryl cyclic AMP on Chinese hamster ovary cells, *Cell* **2**:145.

Prescott, D. M., 1957, Change in the physiological state of a cell population as a function of culture growth and age *(Tetrahymena geleii)*, *Exp. Cell Res.* **12**:126.

Prescott, D. M., 1964, Cellular sites of RNA synthesis, *Progr. Nucleic Acid Res. Mol. Biol.* **3**:33.

Prescott, D. M., 1976, *Reproduction of Eukaryotic Cells*, Academic Press, New York.

Probst, H. Gentner, P. R., Hofstätter, T., and Jenke, S., 1974, Newly synthesized mammalian cell DNA: Evidence for effects simulating the presence of RNA in the nascent DNA fraction isolated by nitrocellulose column chromatography, *Biochim. Biophys. Acta* **340**:361.

Rao, P. N., and Johnson, R. T., 1970, Mammalian cell fusion, I. Studies on the regulation of DNA synthesis and mitosis, *Nature* **225**:159.

Rao, P. N., and Smith, M. L., 1976, Regulation of DNA synthesis in cytochalasin B(CB)-induced binucleate HeLa cells, *Exp. Cell Res.* **103**:213.

Rao, P. N., Hittelman, W. N., and Wilson, B. A., 1975, Mammalian cell fusion. VI. Regulation of mitosis in binucleate HeLa cells, *Exp. Cell Res.* **90**:40.

Reinhard, P., Burkhalter, M., and Gautschi, J. R., 1977, Semiconservative DNA replication in vitro. I. Properties of two systems derived from mouse P-815 cells by permeabilization or lysis with Brij-58, *Biochim. Biophys. Acta* **474**:500.

Robbins, E., and Scharff, M. D., 1967, The absence of a detectable G_1 phase in a cultured strain of Chinese hamster lung cell, *J. Cell Biol.* **34**:684.

Rosenberg, B. H., Ungers, G., and Deutsch, J. F., 1976, Variation in DNA swivel enzyme activity during the mammalian cell cycle, *Nucleic Acids Res.* **3**:3305.

Rosensweig, N. S., and Herman, R. H., 1968, Control of jejunal sucrase and maltase activity by dietary sucrose or fructose in man, *J. Clin. Invest.* **47**:2253.

Salb, J. M., and Marcus, P. I., 1965, Translational inhibition in mitotic HeLa cells, *Proc. Natl. Acad. Sci. USA* **54**:1353.

Schochetman, G., and Perry, R. P., 1972, Characterization of the messenger RNA released from L cell polyribosomes as a result of temperature shock, *J. Mol. Biol.* **63**:577.

Schumm, D. E., Morris, H. P., and Webb, T. E., 1974, Early biochemical changes in phytohemagglutinin-stimulated peripheral blood lymphocytes from normal and tumor-bearing rats, *Eur. J. Cancer* **10**:107.

Seifert, W. E., and Rudland, P. S., 1974a, Cyclic nucleotides and growth control in cultured mouse cells: Correlation of changes in intracellular 3',5'-cGMP concentration with a specific phase of the cell cycle, *Proc. Natl. Acad. Sci. USA* **71**:4920.

Seifert, W. E., and Rudland, P. S., 1974b, Possible involvement of cyclic GMP in growth control of cultured mouse cells, *Nature* **248**:138.

Seki, S., and Mueller, G. G., 1975, A requirement for RNA, protein, and DNA synthesis in the establishment of DNA replicase activity in synchronized HeLa cells, *Biochim. Biophys. Acta* **378**:354.

Shall, S., 1972, Poly (ADP-ribose), *FEBS Lett.* **24**:1.

Sheinin, R., Humbert, J., and Pearlman, R. E., 1978, Some aspects of eukaryotic DNA replication, *Annu. Rev. Biochem.* **47**:277.

Sheppard, J. R., and Prescott, D. M., 1972, Cyclic AMP levels in synchronized mammalian cells, *Exp. Cell Res.* **75**:293.

Sims, R. T., 1965, The synthesis and migration of nuclear proteins during mitosis and differentiation of cells in rats, *Q. J. Microsc. Sci. (N. S.)* **106**:229.

Singer, D. S., and Singer, M. F., 1976, Studies on the interaction of H1 histone with superhelical DNA: Characterization of the recognition and binding regions of H1 histone, *Nucleic Acids Res.* **3**:2531.

Sisken, J. E., and Kinosita, R., 1961, Timing of DNA synthesis in the mitotic cycle in vitro, *J. Biophys. Biochem. Cytol.* **9**:509.

Snyder, J. A., and McIntosh, J. R., 1976, Biochemistry and physiology of microtubules, *Annu. Rev. Biochem.* **45**:699.

Söderhäll, S., 1976, DNA ligases during rat liver regeneration, *Nature* **260**:640.

Söderhäll, S., and Lindahl, T., 1975, Mammalian DNA ligases. Serological evidence for two separate enzymes, *J. Biol. Chem.* **250**:8438.

Soeda, E., Miura, K., Nakaso, A., and Kimura, G., 1977, Nucleotide sequence around the replication origin of polyoma virus DNA, *FEBS Lett.* **79**:383.

Sollner-Webb, B., Camerini-Otero, R. D., and Felsenfeld, G., 1976, Chromatin structure as probed by nucleases and proteases: Evidence for the control role of histones H3 and H4, *Cell* **9**:179.

Staprans, I., Kenney, W. C., and Dirksen, E. R., 1975, Calcium affinity of chick brain tubulin, *Biochem. Biophys. Res. Commun.* **62**:92.

Stein, G., and Baserga, R., 1970, Continued synthesis of non-histone chromosomal proteins during mitosis, *Biochem. Biophys. Res. Commun.* **41**:715.

Stubblefield, E., and Murphree, S., 1967, Synchronized mammalian cell cultures. II. Thymidine kinase activity in colcemid synchronized fibroblasts, *Exp. Cell Res.* **48**:652.

Taylor, J. H., 1960, Asynchronous duplication of chromosomes in cultured cells of Chinese hamster, *J. Biophys. Biochem. Cytol.* **7**:455.

Taylor, J. H., Wu, M., Erickson, L. C., and Kurek, M. P., 1975, Replication of DNA in mammalian chromosomes. III. Size and RNA content of Okazaki fragments, *Chromosoma* **53**:175.

Thilly, W. G., Arkin, D. I., Nowak, T. S., Jr., and Wogan, G. N., 1975, Behavior of subcellular marker enzymes during the HeLa cell cycle, *Biotechnol. Bioeng.* **17**:695.

Thomas, J. O., and Kornberg, R. D., 1975, An octamer of histones in chromatin and free in solution, *Proc. Natl. Acad. Sci. USA* **72**:2626.

Thomas, T. L., and Patel, G. L., 1976, DNA unwinding component of the non-histone proteins, *Proc. Natl. Acad. Sci. USA* **73**:4364.

Thompson, L. H., Harkins, J. L., and Stanners, C. P., 1973, A mammalian cell mutant with a temperature-sensitive leucyl-transfer RNA synthetase, *Proc. Natl. Acad. Sci. USA* **70**:3094.

Thompson, L. R., and McCarthy, B. J., 1968, Stimulation of nuclear DNA and RNA synthesis by cytoplasmic extracts in vitro, *Biochem. Biophys. Res. Commun.* **30**:166.

Tobey, R. A., Petersen, D. F., Anderson, E. C., and Puck, T. T., 1966, Life cycle analysis of mammalian cells. III. The inhibition of division in Chinese hamster cells by puromycin and actinomycin, *Biophys. J.* **6**:567.

Tobia, A. M., Schildkraut, C. L., and Maio, J. J., 1970, Deoxyribonucleic acid replication in synchronized cultured mammalian cells. I. Time of synthesis of molecules of different average guanine + cytosine content, *J. Mol. Biol.* **54**:499.

Tseng, B. Y., and Goulian, M., 1975, DNA synthesis in human lymphocytes: Intermediates in DNA synthesis, in vitro and in vivo, *J. Mol. Biol.* **99**:317.

Tseng, B. Y., and Goulian, M., 1977, Initiator RNA of discontinuous DNA synthesis in human lymphocytes, *Cell* **12**:483.

Tsuboi, A., and Baserga, R., 1972, Synthesis of nuclear acidic proteins in density-inhibited fibroblasts stimulated to proliferate, *J. Cell Physiol.* **80**:107.

Ueda, K., Omachi, A., Kawaichi, M., and Hayaishi, O., 1975, Natural occurrence of poly (ADP-ribosyl) histones in rat liver, *Proc. Natl. Acad. Sci. USA* **72**:205.

Vaitukaitis, J. L., 1976, Peptide hormones as tumor markers, *Cancer (Suppl. 1)* **37**:567.

Vogel, T., and Singer, M., 1975, The interaction of histones with simian virus 40 supercoiled circular deoxyribonucleic acid in vitro, *J. Biol. Chem.* **250**:796.

Walker, J. M., Shooter, K. V., Goodwin, G. H., and Johns, E. W., 1976, The isolation of two peptides from a nonhistone chromosomal protein showing irregular charge distribution within the molecule, *Biochem. Biophys. Res. Commun.* **70**:88.

Weatherall, D. J., 1976, Fetal haemoglobin synthesis, in: *Congenital Disorders of Erythropoiesis*, Ciba Foundation Symp. 37 (new series), pp. 307–328, Elsevier, Amsterdam.

Weatherall, D. J., Clegg, J. B., and Wood, W. G., 1976, A model for the persistence or reactivation of fetal hemoglobin production, *Lancet* **2**:660.

Weinstein, Y., Chambers, D. A., Bourne, H. R., and Melmon, K. L., 1974, Cyclic GMP stimulates lymphocyte nucleic acid synthesis, *Nature* **251**:352.

Weintraub, H., 1972, A possible role for histone in the synthesis of DNA, *Nature* **240**:449.

Weintraub, H., and Groudine, M., 1976, Chromosomal subunits in active genes have an altered conformation, *Science* **193**:848.

Weisenberg, R. C., 1972, Microtubule formation in vitro in solutions containing low calcium concentrations, *Science* **177**:1104.

Weissbach, A., 1977, Eukaryotic DNA polymerases, *Annu. Rev. Biochem.* **46**:25.

Welsh, J. D., Rohrer, V., Knudsen, K. B., and Paustian, F. F., 1967, Isolated lactase deficiency: Correlation of laboratory studies and clinical data, *Arch. Intern. Med.* **120**:261.

Welsh, J. D., Russell, L. C., and Walker, A. W., 1974, Changes in intestinal lactase and alkaline phosphatase activity levels with age in the baboon *(Papio papio)*, *Gastroenterology* **66**:993.

Welsh, M. J., Dedman, J. R., Brinkley, B. R., and Means, A. R., 1978, Calcium-dependent regulator protein: Localization in mitotic apparatus of eukaryotic cells, *Proc. Natl. Acad. Sci. USA* **75**:1867.

Whitlock, J. P., Jr., and Simpson, R. T., 1976, Removal of histone H1 exposes a fifty base pair DNA segment between nucleosomes, *Biochemistry* **15**:3307.

Willingham, M. C., Johnson, G. S., and Pastan, I., 1972, Control of DNA synthesis and mitosis in 3T3 cells by cyclic AMP, *Biochem. Biophys. Res. Commun.* **48**:743.

Witman, G. B., Cleveland, D. W., Weingarten, M. D., and Kirschner, M. W., 1976, Tubulin requires tau for growth onto microtubule initiating sites, *Proc. Natl. Acad. Sci. USA* **73**:4070.

Woodcock, C. L. F., 1973, Ultrastructure of inactive chromatin, *J. Cell Biol.* **59**:368a.

Yoshihara, K., Tanigawa, Y., Burzio, L., and Koide, S. S., 1975, Evidence for adenosine diphosphate ribosylation of Ca^{2+}, Mg^{2+}-dependent endonuclease, *Proc. Natl. Acad. Sci. USA* **72**:289.

Zauderer, M., Liberti, P., and Baglioni, C., 1973, Distribution of histone messenger RNA among free and membrane-associated polyribosomes of a mouse myeloma cell line, *J. Mol. Biol.* **79**:577.

Zylber, E. A., and Penman, S., 1971, Synthesis of 5S and 4S RNA in metaphase-arrested HeLa cells, *Science* **172**:947.

8

Servomechanisms and Oscillatory Phenomena

Robert M. Cohn, Marc Yudkoff, and Pamela D. McNamara

I. Introduction

The activity of some enzymes is modulated by binding of a ligand to the enzyme at a so-called allosteric site, which is distinct from the catalytic site. Such ligands often are dissimilar to the substrate. They induce a conformational change of enzyme structure which increases or decreases enzyme activity. Thus, the intracellular concentration of a particular metabolite may regulate the rate of a metabolic pathway in accordance with the organism's needs. Usually, the final product of the pathway modifies (with a variable degree of cooperativity) the activity of the enzyme catalyzing the first committed step of the pathway. In many instances, however, the product of one metabolic sequence modulates the flux through a separate but related sequence. In both instances, the metabolic regulation encountered is termed *feedback inhibition*. It is an essential mechanism for maintaining the biochemical steady-state, since it prevents—or rapidly corrects—the undue accumulation of a given metabolite. The converse of feedback inhibition is *feedforward activation*—the stimulation of enzyme activity by a metabolite. While negative feedback may correct for an overshoot in the response of a pathway to a perturbation, positive feedback may result in a propagation phenomenon. It, too, represents a regulatory pattern

---

Robert M. Cohn, Marc Yudkoff, and Pamela D. McNamara • Department of Metabolic Research, Children's Hospital of Philadelphia, University of Pennsylvania, Philadelphia, Pennsylvania 19104.

which ensures constancy of the biochemical environment but through a more complicated mechanism which involves an initial destabilization of the system.

The terminology of electromechanical systems has been appropriated to describe both feedback inhibition and feedforward activation as *servomechanisms*, or automatic devices which control the flow of large amounts of energy with the expenditure of only a minute amount of energy. Implicit in the servomechanism concept is the capacity to sense and correct errors, such capability being automatic, constant, and immediate.

The existence of *oscillatory phenomena* in complex biochemical pathways is a function of the existence of servomechanisms. Oscillatory phenomena are expressions of nonlinear kinetics in a system displaced far from equilibrim and are thus examples of the dissipative structures necessary for the maintenance and evolution of biological systems (Chapter 2). Such structures are termed "dissipative" because they require the dissipation of energy through the system both for creation and stabilization of the new structure.

In this chapter, we will review a variety of metabolic pathways regulated by servomechanisms. We will also discuss oscillatory phenomena and speculate on their relationship to chemical signals, higher-order structures, morphogenesis, and hypercycles, the last having been proposed as the essential functional organization for evolution (Eigen, 1971; Eigen and Schuster, 1978).

II. Feedback and Feedforward Phenomena

Although feedback inhibition of enzyme activity was first described in 1940 by Dische (Krebs, 1972), its significance was not appreciated for many years. Dische observed that phosphoglyceric acid specifically inhibited hexokinase-mediated glucose phosphorylation in red cell hemolysates. The mechanism of this inhibition was not immediately apparent, since phosphoglycerate is neither a substrate nor a product of the hexokinase reaction. Dische proposed that the phosphoglycerate effect represented a novel and significant type of metabolic regulation.

A fuller understanding of the phosphoglycerate effect awaited the development of our understanding of hemoglobin function. Red cell 2,3-diphosphoglycerate (2,3-DPG) may either exist freely in the cell or may be bound to hemoglobin. Only free 2,3-DPG inhibits hexokinase, the first step in glycolysis. Hypoxia favors the binding of 2,3-DPG to hemoglobin. This has two immediate consequences. First, as 2,3-DPG binds to hemoglobin, it induces a conformational change of the hemoglobin molecule such that hemoglobin binding of oxygen is decreased, thereby promoting increased oxygen delivery to tissues. In addition, as the concentration of free 2,3-DPG is reduced, the inhibition of hexokinase is relaxed, with an attendant acceleration of glycolysis until the free 2,3-DPG level is restored to a concentration at which it again inhibits glycolysis. Hence, modulation of 2,3-DPG formation by feedback at the hexokinase step could serve an important role in regulating the level of this intermediate which accounts, in the red cell, for the majority of organic phosphates (Beutler, 1971).

The role of the end products of a metabolic pathway in regulating their own biosynthesis was first demonstrated by Roberts et al. (1955). Working with *E. coli*, they showed that amino acid synthesis from glucose is inhibited by the addition of amino acids to the incubation medium. Umbarger (1956) demonstrated that end products may inhibit the activity of enzymes mediating end-product synthesis. Often this inhibition is exerted on the first enzyme of the metabolic sequence. End products may also inhibit enzyme synthesis itself, as is frequently observed in anabolic pathways for amino acids, purines, and pyrimidines. This latter mode of metabolic regulation is termed *repression* and may occur independently of feedback inhibition. Both mechanisms may be involved in regulation of the same biosynthetic pathway. However, unlike feedback inhibition, which provides very rapid control, repression is a relatively slow process which permits adjustment of metabolism over an extended period of time.

The demonstration that the end product of a linear biosynthetic pathway exerts feedback inhibition of the first committed step of the pathway provided the requisite intellectual stimulus for further research into this pattern of metabolic regulation. A central question was the molecular basis of feedback inhibition. One theoretical solution to this problem was the allosteric model of Monod et al. (1965), in which the binding of a modifier to an enzyme at the allosteric site induces a conformational change of the enzyme that alters enzyme kinetics, usually making the kinetics higher than first order. This informational exchange occurs with such rapidity that the attachment of the modifier to the allosteric site must be mediated through equilibrium binding. Evidence for distinct catalytic and allosteric binding sites is abundant (Citri, 1973), including the demonstration of catalytic and allosteric sites on separate subunits of oligomeric enzymes, the failure of substrate to inhibit effector binding, and the selective inhibition of the allosteric site.

Feedback inhibition has been studied most closely in microbial systems and has been reviewed by Stadtman (1966, 1970). Servomechanisms are also well suited to the more complex regulatory needs of eukaryotic cells. As argued by Atkinson (1976), this kind of metabolic regulation, affecting the affinity for substrate binding, permits greater sensitivity in discrimination than can be achieved by variation of the V_{max}. Atkinson further notes that control can be made more sensitive, permitting fine tuning, when the enzyme involved exhibits cooperativity following higher than first-order kinetics for both its substrate and modifiers. The higher-order kinetics of modifier binding then permits ligands to exert the same degree of control as substrates exert. Masters (1977) cautions against the gratuitous extrapolation of in vitro data to the in vivo situation since the concentration of many enzymes in vivo is comparable to the concentration of substrates and effectors, a circumstance seldom duplicated experimentally. Furthermore, adsorption of enzymes to membranes may be an important factor in modifying the catalytic activity of the enzyme as well as its specificity for its substrate. Finally, it is likely that a metabolic pathway is regulated by many different compounds, a factor almost never reproduced with in vitro experiments.

In the following pages we will describe the role of servomechanisms in modulating the activity of three key pathways of mammalian metabolism: glycolysis, fatty acid synthesis, and cholesterol synthesis. The regulation of each sequence depends upon a constant flow of information to key enzymes and receptors. A defect of function of any single component could be a cause of disease.

A. Glycolysis: The Pasteur Effect

Most cells derive energy from the enzymatic breakdown of glucose. When this metabolism involves oxygen, the process is termed respiration. When it occurs without oxygen, it is called fermentation. Although both pathways are present in most mammalian cells, the energy yield from respiration is greater than that from fermentation. Furthermore, the end products of respiration are carbon dioxide and water, while the end products of fermentation are lactic acid and other organic acids which may cause an acidosis. Thus, from a teleological perspective, it is advantageous to an organism to suppress fermentation and to favor respiration when oxygen is present. The biochemical correlate of this view is the Pasteur effect, so termed after an 1861 observation by the French scientist that anoxia stimulates glycolysis while oxygen inhibits glucose utilization. The cause of the Pasteur effect is a multifocal regulation of the glycolytic pathway at the levels of hexokinase, phosphofructokinase (PFK), and glyceraldehyde-3-phosphate dehydrogenase (G-3-PD) (Krebs, 1972; Racker, 1975).

While glycolysis is controlled at three distinct steps in the pathway, perhaps the most crucial control occurs at the PFK step (Mansour, 1972). PFK is subject to allosteric inhibition by ATP. The inhibition is relieved by a variety of compounds, including AMP, inorganic phosphate, fructose diphosphate, and fructose-6-phosphate. Studies with bovine heart PFK have shown the presence of three ATP binding sites and one fructose-6-phosphate binding site per monomer. Desensitization studies have demonstrated that the allosteric binding sites for ATP are distinct from the catalytic site. A sigmoid curve is obtained if PFK activity is plotted against fructose-6-phosphate activity at an inhibitory concentration of ATP. The release of ATP inhibition by physiologic concentrations of AMP is important, since PFK activity may be controlled by the ATP/AMP ratio rather than by the absolute concentrations of either nucleotide (Newsholme and Start, 1973; Chapter 4, this volume).

Experiments with perfused rat heart have demonstrated that physiologic concentrations of citrate inhibit glycolysis by potentiating ATP inhibition of phosphofructokinase. This factor might permit additional modulation of glycolysis by the oxidation of pyruvate, fatty acids, and ketone bodies, each of which could increase citrate concentration. Of course, the actual diminution of glucose utilization relates to an inhibition of hexokinase in response to the elevated glucose-6-phosphate concentration which occurs consequent to the inhibition of PFK.

Another important inhibitor of PFK activity is 3-phosphoglyceric acid (3-PGA). As the citrate concentration reflects the availability of fuels in the tricarboxylic acid cycle, so does the 3-PGA concentration reflect the level of metabolites at the terminal sequence of the glycolytic pathway. In this fashion, additional information is available to the cell to modulate the rate of glycolysis in accordance with its ever-changing needs.

In addition to the postulated roles of the modulation of PFK and hexokinase activity in mediating the Pasteur effect, Racker (1975) has proposed a role for G-3-PD. In this instance, the effector molecule would be NADH, which exerts an allosteric inhibition of G-3-PD activity. Since NADH is one immediate product of the reaction, its modulating effect is not a classic example of feedback inhibition. Nonetheless, it may play a role akin to that of other servomechanisms in regulating flux through the glycolytic pathway.

Clearly, the biochemical controls of glucose utilization are coupled tightly to the organism's need to derive energy from glucose oxidation. In the Pasteur effect, the biochemical correlate of the organism's energy needs is reflected in the level of ATP or in the ATP/AMP ratio. Thus, the entry of glycosyl residues into the glycolytic pathway is determined, in large measure, by the ATP generated from the metabolism of glucose residues which have already traversed the glycolytic sequence and have entered the tricarboxylic acid cycle. Since hypoxia would diminish the intracellular ATP concentration, the entry of new glucose molecules into the glycolytic sequence would be enhanced. The Pasteur effect, then, is an expression of a servomechanistic biochemical control.

B. Fatty Acid Synthesis

On oxidation, 1 g of fat liberates 9.3 kcal—more than twice the energy yield obtained by the oxidation of either carbohydrate or protein. Fat makes a better energy repository than glycogen, since the storage of fat, unlike the storage of glycogen, does not involve the concomitant storage of water. For these reasons, fat is a very efficient form of stored energy, a fact amply exploited by evolutionary forces: the average 70-kg man has fat stores adequate to satisfy energy needs for approximately a 40-day period of starvation.

Consistent with the central role of fats as the prime storage form of energy are the number of mechanisms which ensure the synthesis of fatty acids whenever caloric intake exceeds caloric requirement. Many of these strategies affect the activity of acetyl-CoA carboxylase. This enzyme catalyzes the conversion of acetyl-CoA to malonyl-CoA in a biotin-requiring reaction which is the first committed step of fatty acid synthesis. Acetyl-CoA carboxylase may exist in an active polymeric form or in an inactive protomeric form. A number of metabolic signals, including acetyl-CoA, a low pH (6.5–7.0), high protein concentration, and citrate, favor formation of the active form of the enzyme. Activation involves polymerization of the inactive protomer and a conformational change of enzyme structure which increases the V_{max} without affecting the K_m. In the case of citrate, the conformational change involves the biotin prosthetic group, rendering it relatively inaccessible to avidin, a biotin inhibitor. The stimulatory effect of citrate is competitively inhibited by palmityl-CoA and other long-chain fatty acid derivatives. The physiologic significance of the citrate and palmityl-CoA effects is open to question, however, since insulin does not alter the concentration of citrate in epididymal fat, and the actual rate of fatty acid synthesis is not well correlated with the concentration of either citrate or acetyl-CoA (Denton and Halperin, 1968). Thus, while ample evidence exists for a citrate role in promoting the in vitro activation of the carboxylase (Lane and Moss, 1971; Volpe and Vagelos, 1976), documentation of such a role in vivo must await measurement of citrate levels in different subcellular compartments. As Greenbaum et al. (1971) have shown, sizeable changes of citrate in these compartments may occur without a perceptible shift in total tissue citrate concentration.

Attention has been directed also at fatty acid synthetase as a control site for the regulation of fatty acid synthesis. This multienzyme complex mediates the synthesis of saturated fatty acids from malonyl-CoA in a reaction requiring acetyl-CoA and NADPH. Details of the enzyme complex and reaction sequence have been reviewed recently (Volpe and Vagelos, 1976; Katiyar and Porter, 1977; Block and Vance, 1977). All available data suggest that while modulation of acetyl-CoA carboxylase activity represents the essential site of short-term control of fatty acid synthesis, long-term control of fatty acid synthesis may rest with the rate of de novo synthesis of the fatty acid synthetase complex.

C. Cholesterol Synthesis

Cholesterol is distributed ubiquitously throughout the animal world, serving as a component of the cell membrane, of plasma lipoproteins, and as a precursor to bile acids and steroid hormones. Active growth is associated with increased cholesterol synthesis. While significant cholesterol synthesis occurs in the developing brain, in the mature animal the liver and intestine account for over 80% of all cholesterol synthesized. The liver is also the primary source of plasma cholesterol. Physiologic factors associated with increased cholesterol synthesis include bile duct fistula or obstruction, high fat diet, some hormones (growth hormone, ACTH, epinephrine, glucagon), and stress. Factors leading to diminished cholesterol synthesis are bile acid feeding, a low fat diet, cholesterol feeding, fasting, and hypoendocrine conditions. Obviously, cholesterol synthesis is related directly to the flow of fat and bile acids to the liver. Indeed, Krumdieck and Ho (1977) have proposed that a factor produced in the intestine stimulates hepatic cholesterogenesis whenever the need for bile acids and bile salts is increased.

The rate-limiting step of hepatic cholesterol synthesis is the conversion of hydroxymethylglutaryl-CoA to mevalonic acid in a reaction catalyzed by hydroxymethylglutaryl-CoA reductase. Both in vivo and in vitro studies have shown that exogenous cholesterol inhibits this reaction by interfering with synthesis of the reductase (Higgins, 1976). Although classical feedback inhibition by exogenous cholesterol had been anticipated, there is no evidence that exogenous cholesterol serves as an allosteric effector (Sabine and James, 1976). Evidence does exist, however, that the fluidity of the microsomal membrane, to which the hydroxymethylglutaryl-CoA reductase is bound, may play a decisive role in determining the activity of this enzyme in response to changes of the exogenous cholesterol concentration. Sabine and James (1976) have termed such control "viscotropic." An inhibitory role for endogenous cholesterol is suggested by recent work of Edwards et al. (1977), who demonstrated that hydroxymethylglutaryl-CoA reductase activity in rat liver is decreased both in vivo and in vitro in rat hepatocytes upon administration of mevalonolactone, enhancing cholesterol synthesis.

Understanding of nonhepatic cholesterogenesis and its relationship to atherosclerosis has increased rapidly in recent years. A unified theory of the control of nonhepatic cholesterogenesis is now possible (Brown and Goldstein, 1976; Goldstein and Brown, 1977). According to this theory, the liver secretes very-low-density lipoproteins (VLDL) containing cholesterol and triglycerides. The triglycerides are then removed by adipose tissue, converting the VLDL into low-density lipoproteins (LDL). The low-density lipoproteins constitute the primary vehicle for the transport of plasma cholesterol. A specific high-affinity receptor for the LDL is present on the plasma membrane of many tissues, including human fibroblasts, lymphocytes, and arterial smooth muscle cells. The LDL become bound to this receptor and then enter the cell by absorptive endocytosis. The relatively small contribution of nonhepatic tissue to cholesterol synthesis—only 20% of the total—reflects feedback inhibition of cholesterol synthesis mediated by the binding of low-density lipoproteins to the cell membrane. When the lipoproteins are not bound to the membranes of isolated cells, cholesterol synthesis is stimulated. Indeed, under such circumstances, cholesterol synthesis increases sharply (Brown and Goldstein, 1976). The emergent concept, then, is that the cholesterol utilized for membranogenesis is derived from an exogenous source when the low-density lipoproteins are abundant. Recent experimental proof for this arrangement is the work of Balasubramaniam et al. (1976), who have shown that administration to rats of 4-aminopyrazolopyrimidine, an adenine analog that inhibits hepatic lipoprotein synthesis, will cause increased activity of hydroxymethylglutaryl-CoA reductase in nonhepatic tissues. An interesting application of this experimental data to a clinical setting is the demonstration that patients with familial hypercholesterolemia suffer a marked inability to bind low-density lipoproteins to cell membranes because of an absence or defect of the LDL receptor.

D. Other Examples of Feedback Control

Servomechanisms play important roles in the regulation of several other pathways of mammalian metabolism, including the synthesis of purines (Wyngaarden and Kelley, 1978), pyrimidines (Levine et al., 1974), porphyrins (Meyer and Schmid, 1978), and sialic acids (Kornfeld et al., 1964). A detailed discussion of each instance is beyond the scope of this chapter. Some examples of feedback control are cited in Table 1.

Table 1. Examples of Feedback Inhibition of Enzymes Involved in Biosynthetic Pathways

Reaction	Inhibitor	Tissue	References
1. Phosphoserine $\longrightarrow$ serine	Serine	Rat liver	Borkenhagen and Kennedy (1958), Neuhaus and Byrne (1958)
2. Phosphoribosyl pyrophosphate + glutamate $\longrightarrow$ phosphoribosylamine	ATP (purine ribotides)	Pigeon liver	Wyngaarden and Ashton (1959)
3. Beta-hydroxy-beta'-methyl glutarate $\longrightarrow$ mevalonic acid	Cholesterol	Human liver	Siperstein and Fagan (1964)
4. Fructose-6-P + ATP $\longrightarrow$ Fructose-1,6-diP + ADP	Citrate	Rabbit muscle	Passonneau and Lowry (1963)
5. Phosphoenolpyruvate + ADP $\longrightarrow$ Pyruvate + ATP	Fatty acyl-CoA	Rat heart	Tsutsumi and Takenaka (1969)
6. Acetyl-CoA + CO_2 + ATP $\longrightarrow$ malonyl-CoA + ADP + P_i	Palmityl-CoA	Chick liver	Goodridge (1973)
7. Aspartate + carbamylphosphate $\longrightarrow$ ureidosuccinate	Uridine, cytidine derivatives (especially UMP and CMP)	Rat liver	Bourget and Tremblay (1972)
8. ATP + CO_2 + Glutamine $\longrightarrow$ carbamyl-P + ADP + P_i	UTP	Mouse spleen	Tatibana and Ito (1967)
9. ATP + CO_2 + "NH_3" $\longrightarrow$ carbamyl-P + ADP + P_i	CTP	Rat liver	Kerson and Appel (1968)
10. Glutamine + phosphoribosyl pyrophosphate $\longrightarrow$ glutamate + phosphoribosylamine	AMP, IMP, GMP	Rat liver	Katunuma et al. (1970)
11. Glucose + ATP $\longrightarrow$ Glucose-6-P + ADP	Free fatty acids (and CoA derivatives)	Rat liver	Weber et al. (1968)
12. Acetyl-CoA + oxaloacetate $\longrightarrow$ citrate	Long chain acyl-CoAs	Rat liver	Tubbs (1963)
		Pig heart	Wieland and Weiss (1963)
13. Carbamylaspartic acid $\longrightarrow$ dihydroorotic acid	CMP, d-CMP	Ehrlich ascites cells	Bresnick and Blatchford (1964)

III. Oscillatory Phenomena

Vibrations or oscillations occur in nature, and some commonplace examples can be easily recognized, among which are the motion of a child on a swing, the flow pattern of electrons in a radio tube, and the vibrations of a plucked guitar string. Understanding vibrations is essential to the atomic theory of matter since vibrations occur in atoms and are represented by wave patterns. Waves are propagated by oscillations and can transport energy, a common example being represented by electromagnetic radiation. Thus, an oscillation can propagate an energetic signal. Remarkably, several wave patterns may exist in the same medium without interfering with one another, just as radio signals from several transmitters may be received simultaneously at a given point, each signal arriving intact and distinguishable from the others. Thus, the existence of several energetic signals within a given medium is not precluded. Waves may be superpositioned such that the additive effect of two wave displacements results in a pulse. Resonance is a phenomenon which occurs when energy of a certain frequency stimulates the development of an oscillation of the same or similar frequency. Resonant phenomena occur in both atomic and molecular vibrations.

A. Oscillations in Open Systems

Life can be viewed as an open system, i.e., one which can exchange energy and matter with the environment. When such systems attain a steady state, they may behave in several ways (Fig. 1). Normally, the concentration of a substance in an open system will approach a steady state without evincing periodicity. The steady state may also be reached through rhythmically changing concentrations of the substance as it approaches the steady state asymptotically. Finally, there is the possibility of some type of oscillatory behavior around the steady state. The latter type of behavior presents one of the most interesting areas for investigation and speculation regarding the propagation of informational signals. While a number of examples of such phenomena exist, the biochemical examples are of specific interest here. These include oscillations in the glycolytic pathway, oscillations in mitochondrial volume, and cAMP oscillations in the slime mold.

Cooperativity in enzyme action and servomechanism-type control of enzyme activity are important principles of metabolic control which we have already encountered. Separately or in concert, they introduce nonlinearity into the kinetic behavior of a pathway. It is noteworthy that nonlinearity is at the heart of the genesis of oscillations as well.

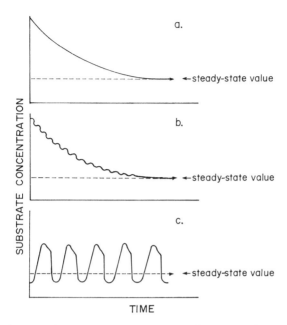

Figure 1. Behavior of an open system as it reaches a steady state. (a) Normal asymptotic approach. (b) Rhythmically changing concentration as the steady state is approached asymptotically. (c) Oscillations around the steady state. Adapted from Franck (1978), p. 1, with kind permission from Verlag Chemie International Inc.

This is a result of autocatalytic or feedback phenomena in systems where components are extensively coupled, so that collective behavior becomes possible. Servomechanisms in biological systems act as closed loops in which the output feeds back upon the input, modifying the kinetics of the process. Far-from-equilibrium conditions are a prerequuisite for any state in which we will encounter oscillatory behavior. The implications seem clear that oscillatory phenomena can propagate signals which can be amplified and can stimulate other, similar systems, which can then act collectively. The occurrence of oscillatory phenomena in biochemical systems has been documented (Chance et al., 1973; Nicolis and Portnow, 1973); however, as yet, there is no conclusive proof that such phenomena actually generate signals responsible for such hierarchical behavior.

As discussed in Chapter 2, the ability of living organisms to utilize energy and matter to form dissipative structures beyond the thermodynamic branch is crucial to the ordering of the life process. Oscillatory behavior around a steady state of a biochemical pathway is an example of a dissipative structure. Such behavior exhibits periods in the order of a minute. These oscillations are of the limit-cycle type (Chapter 2) since they possess unique amplitude and frequency regardless of the starting conditions. Such limit-cycle behavior is encountered only in open systems operating far from equilibrium under conditions of nonlinear kinetics.

B. Biological Examples

Studies with yeast, heart, and brain have shown that concentrations of intermediates within the glycolytic pathway often follow an oscillating function. Continuous spectrophotometric recording techniques for determining the $NAD^+/NADH$ ratio in cell-free extracts first revealed oscillations of the NADH level in these systems. These studies then led to the discovery of glycolytic oscillations in yeast cell and cell-free extracts, beef heart extracts, rat skeletal muscle extracts, and in ascites tumor cells, with concentrations of intermediates varying in the range between 10^{-5} and 10^{-3} M (Chance et al., 1973).

Studies in all of the systems thus far investigated have confirmed the primary role of PFK as the progenitor of the oscillations. Support for this conclusion comes from the observation that in the glycolytic pathway, fructose-6-phosphate is the last substrate that is able to generate oscillatory behavior, and the addition of fructose diphosphate, the product of the PFK step, does not generate periodic oscillations in the systems investigated. The oscillations of the glycolytic pathway appear to relate to the allosteric nature of PFK and to the autocatalytic control of PFK by AMP and ADP, the product of the reaction. It is noteworthy that in yeast, at least, FDP is not involved in generating the periodicity; in beef heart extracts, oscillations of the glycolytic pathway with about 20 min periods have been demonstrated.

Models proposed for glycolytic oscillations rely upon positive feedback and cooperativity of PFK, both contributing to the generation of periodicity (Goldbeter, 1977). Modeling has shown that when the sigmoid kinetics associated with PFK are converted to hyperbolic kinetics, the oscillatory behavior ceases. Hence, positive effectors, such as NH_4^+ ions, and negative effectors, such as citrate, obliterate the oscillations by decreasing the cooperative behavior of the enzyme.

Oscillatory behavior has also been demonstrated in mitochondrial preparations and appears to be mediated by a combination of the mitochondrial membrane and ion transport processes. Such oscillations involve volume changes, ion movements, and oxidation–reduction states of the respiratory chain proteins. Experiments have demonstrated that these oscillations require the simultaneous presence of a monovalent cation, a permeant anion, and substrate, or ATP. Membrane-bound ATPase appears to be involved, and the fundamental factor in the development of the oscillation appears to be ion transport and associated water flow, with consequent swelling of the mitochondria.

Another example of an oscillatory phenomenon is provided by the behavior of the slime mold, *Dictyostelium discoideum*. In this organism, nutritional deprivation results in the formation of a central core of cells which emit pulses of cAMP in a regular, wavelike manner with a period of several minutes. Ameboid cells aggregate around these central cells, as a result of a chemotactic response to the cAMP gradient produced, and relay the attraction signal outward, resulting in wavelike patterns of aggregation and the development of multicellular fruiting bodies (Gerisch, 1968). Goldbeter and Segel (1977) have presented an allosteric model for adenyl cyclase which can account for both the relay of cAMP pulses and the oscillations in cAMP concentration and can also provide an explanation for the formation of aggregation centers.

C. Involvement of Oscillatory Behavior in Collective Phenomena

Collective phenomena on a biochemical level are assumed to involve the coupling of myriad enzymatic processes with the emergence of behavior observable at a macroscopic level. Models to explain dynamic processes such as evolution and morphogenesis have involved the use of oscillatory phenomena to achieve collective behavior.

1. Hypercycles

Hypercycles represent systems in which self-replicative (informational) units and functional units are coupled in a cyclic linkage (Fig. 2). Here the informational intermediates (nucleic acids) are able both to direct their own reproduction and to provide the functional support (enzymes) for the reproduction of subsequent intermediates, an energy requiring process. The importance of a hypercyclic organization is that, while noncoupled self-replicative units ensure the preservation of a lower limit of heritable information, catalytic hypercycles possess integrating characteristics which allow for cooperation among otherwise competitive units.

Eigen (1971) investigated the information content of the prokaryotic genome and found that the amount of information capable of being stored in a single replicative unit is small. He concluded that, in order to amass sufficient information for the development of a translation apparatus, it is necessary to integrate several replicative units through functional linkages. Such integration leads to a new level of organiza-

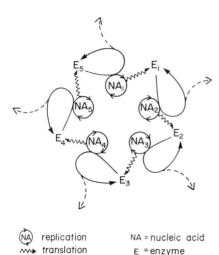

Figure 2. A hypercycle. A catalytic hypercycle consists of both nucleic acids and proteins which together function as self-instructive units. Each replicative unit, NA_n, is able to instruct both its own reproduction as well as the information for the enzyme, E, which carries out the formation of a subsequent component of the cycle. Where the enzyme possesses catalytic capabilities beyond those required for the reproduction of more nucleic acid (i.e., translation of metabolism) additional cooperativity is achieved. Adapted from Eigen (1971), p. 504, with kind permission from Springer-Verlag.

tion which augments the capacity to store information. In Eigen's view, the molecular evolution that preceded biological evolution must have been governed by an optimization principle based on the relative immunity of the genetic code to single point mutations. A randomly produced advantageous mutant must be able to proliferate in the presence of less adaptable competitors. Thus, the property which permits the mutant to gain ascendancy is represented by inherent autocatalysis, i.e., self-reproductive behavior. This ability to self-replicate is essential if the information already accumulated by the system is to be conserved for continued transmission. Mutability is required for evolution, and copying errors represent the main source of this new heritable information. Thus, Eigen concludes, molecular evolution was based on the universal genetic code which became fixed, not as a consequence of a lack of viable alternatives, but because of what he terms a *"once-forever"* selection mechanism which could begin by any random process. The crux of the argument is, then, that "once-forever" selection is a direct consequence of the behavior of hypercycles.

Computer modeling of hypercycles (Küppers, 1975; Eigen and Schuster, 1977, 1978) demonstrates the development of oscillatory behavior around a steady state which, through feedback and coupling of the members of the hypercycle, minimizes the effect of mistakes in reproduction and thus allows for selection and optimization of the dynamic structures. So viewed, hypercycles represent the minimum structural organization for a system to accumulate, maintain, and process the information in the genome.

2. Morphogenesis

The problem of differentiation and morphogenesis (i.e., the origin of spatial patterns in developing systems) constitutes one of the most enigmatic of all problems in biology. It involves the formation of dynamic patterns (i.e., patterns created and maintained by dissipation of energy), as opposed to patterns which may be termed static (e.g., immiscible liquids, a crystal, or a viral capsid), whose form is determined by specific affinities between each of the other components.

Turing (1952) confronted this enigma and developed a mathematical model of morphogenesis which was dependent upon interaction between reactions and diffusions such that, as kinetic and transport coefficients traverse critical (bifurcation) values, the initial uniform state becomes unstable and a spatially nonuniform state emerges.

A number of workers have suggested that the "morphogens" suggested by Turing and others may be inducers or repressors of a particular gene. Formation of a gradient in the concentration of such a morphogen in a population of cells which are chemically and genetically identical will cause derepression of a previously repressed gene or repression of a previously activated gene in a localized region of the aggregate. In either circumstance, the initial inhomogeneity in the morphogen can be amplified in the group of cells. Such a pattern has been studied by Babloyantz and Hiernaux (1974) and Martinez (1972). Their studies emphasize the importance of a succession of instabilities in the generation of spatial patterns through the formation of dissipative structures. This accords, then, with the principle of Prigogine et al. of evolutionary feedback (see Chapter 2). A number of elaborate models for morphogenesis based on the original Turing model have now appeared (Meinhardt, 1978; Gierer, 1977; Kauffman et al., 1978; Lacalli and Harrison, 1979).

IV. Proposed Physiological Significance of Oscillatory Phenomena

Oscillations appear at several levels of biological function, and Nicolis and Portnow (1973) have proposed a number of situations in which they may play a regulatory or progenitor role. At a physiological level, supercellular oscillations provide the basic property of *entrainment* (ability to synchronize to an external periodic action), characteristic of circadian rhythms. This periodic behavior could give the organism the flexibility it requires to adapt to fluctuations in the environment while maintaining a suitable phase relation, by providing an identical sequence of essential metabolic processes during a 24-h cycle. At the genetic level, oscillations may provide positional information, and when coupled with diffusion processes, they could provide the signals propagated in time and space during morphogenesis. At the metabolic level, enhanced sensitivity to environmental perturbations may be imputed to oscillations. Indeed, periodicity in a pathway may occur as a response to changing microenvironmental conditions, thus providing constant evaluation of hydrogen ion concentration, ionic strength, and electrolyte flux, to name but a few conditions.

Finally, as we have discussed, oscillatory behavior, in association with feedback phenomena and autocatalysis, has been utilized by Eigen to develop a theory of evolution which, in conjuction with the evolutionary feedback principle of Prigogine (see Chapter 2) involving increasing dissipation, may represent a model for biological evolution.

References

Atkinson, D. E., 1976, Adaptations of enzymes for regulation of catalytic function, *Biochem. Soc. Symp.* **41**:205.

Babloyantz, A., and Hiernaux, J., 1974, Models for positional information and positional differentiation, *Proc. Natl. Acad. Sci. USA* **71**:1530.

Balasubramaniam, S., Goldstein, J. L., Faust, J. R., and Brown, M. S., 1976, Evidence for regulation of 3-hydroxy-3-methylglutaryl Coenzyme A reductase activity and cholesterol synthesis in nonhepatic tissues of rat, *Proc. Natl. Acad. Sci. USA* **73**:2564.

Beutler, E., 1971, 2,3-Diphosphoglycerate affects enzymes of glucose metabolism in red blood cells, *Nature (New Biol.)* **232**:20.

Bloch, K., and Vance, D., 1977, Control mechanisms in the synthesis of saturated fatty acids, *Annu. Rev. Biochem.* **46**:263.

Borkenhagen, L. F., and Kennedy, E. P., 1958, Enzymic equilibration of L-serine with O-phospho-L-serine, *Biochim. Biophys. Acta* **28**:222.

Bourget, P. A., and Tremblay, G. C., 1972, Control of pyrimidine biosynthesis in rat liver by feedback inhibition of aspartate carbamyltransferase, *Biochem. Biophys. Res. Commun.* **46**:752.

Bresnick, E., and Blatchford, K., 1964, Inhibition of dihydroorotase by purines and pyrimidines, *Biochim. Biophys. Acta* **81**:150.

Brown, M. S., and Goldstein, J. L., 1976, Receptor-mediated control of cholesterol metabolism, *Science* **191**:150.

Chance, B., Pye, E. K., Ghosh, A. K., and Hess, B., eds., 1973, *Biological and Biochemical Oscillators*, Academic Press, New York.

Citri, N., 1973, Conformational adaptability in enzymes, *Adv. Enzymol.* **37**:397.

Denton, R. M., and Halperin, M. L., 1968, The control of fatty acid and triglyceride synthesis in rat epididymal adipose tissue, *Biochem. J.* **110**:27.

Edwards, P. E., Popják, G., Fogelman, A. M., and Edmond, J., 1977, Control of 3-hydroxy-3-methylglutaryl Coenzyme A reductase by endogenously synthesized sterols in vitro and in vivo, *J. Biol. Chem.* **252**:1057.

Eigen, M., 1971, Self organization of matter and the evolution of biological macromolecules, *Naturwissenschaften* **58**:465.

Eigen, M., and Schuster, P., 1977, The hypercycle, a principle of natural self-organization. Part A. Emergence of the hypercycle, *Naturwissenschaften* **64**:541.

Eigen, M., and Schuster, P., 1978, The hypercycle: A principle of natural self-organization, Part B: The abstract hypercycle, *Naturwissenschaften* **65**:7, and Part C. The realistic hypercycle, *Naturwissenschaften* **65**:341.
Franck, U. F., 1978, Chemical oscillations, *Angew. Chem. (Engl.)* **17**:1.
Gerisch, G., 1968, Cell aggregation and differentiation in *Dictyostelium*, *Curr. Top. Dev. Biol.* **3**:157.
Gierer, A., 1977, Physical aspects of tissue evagination and biological form, *Q. Rev. Biophys.* **10**:529.
Goldbeter, A., and Segel, L. A., 1977, Unified mechanism for relay and oscillation of cyclic AMP in *Dictyostelium discoideum*, *Proc. Natl. Acad. Sci. USA* **74**:1543.
Goldstein, J. L., and Brown, M. S., 1977, Atherosclerosis: The low-density lipoprotein hypothesis, *Metabolism* **26**:1257.
Goodridge, A. B., 1973, Regulation of the activity of acetyl CoA carboxylase by palmitoyl CoA and citrate, *J. Biol. Chem.* **247**:6946.
Greenbaum, A. L., Gumaa, K. A., and McLean, P., 1971, The distribution of hepatic metabolites and the control of the pathways of carbohydrate metabolism in animals of different dietary and hormonal status, *Arch. Biochem. Biophys.* **143**:617.
Higgins, M. J. P., 1976, The regulation of cholesterol metabolism, *Biochem. Soc. Trans.* **4**:572.
Katiyar, S. S., and Porter, J. W., 1977, Minireview: Mechanism of fatty acid synthesis, *Life Sci.* **20**:737.
Katunuma, N., Matsuda, Y., and Kuroda, Y., 1970, Phylogenic aspects of different regulatory mechanisms of glutamine metabolism, *Adv. Enzyme Regulation* **8**:73.
Kauffman, S. A., Shymuko, R. M., and Trubert, K., 1978, Control of sequential compartment formation in *Drosophilia*, *Science* **199**:259.
Kerson, L. A., and Appel, S. H., 1968, Kinetic studies on rat liver carbamylphosphate synthetase, *J. Biol. Chem.* **243**:4279.
Kornfeld, S., Kornfeld, R., Neufeld, E. F., and O'Brien, P. J., 1964, The feedback control of sugar nucleotide biosynthesis in liver, *Proc. Natl. Acad. Sci. USA* **52**:371.
Krebs, H. A., 1972, The Pasteur effect and the relations between respiration and fermentation, *Essays Biochem.* **8**:1.
Krumdieck, C. L., and Ho, K-J., 1977, Intestinal regulation of hepatic cholesterol synthesis: An hypothesis, *Am. J. Clin. Nutr.* **30**:255.
Küppers, B., 1975, The general principles of selection and evolution at the molecular level, *Prog. Biophys. Molec. Biol.* **30**:1.
Lacalli, T. C., and Harrison, L. G., 1979, Turing's conditions and the analysis of morphogenetic models, *J. Theor. Biol.* **76**:419.
Lane, M. D., and Moss, J., 1971, Regulation of fatty acid synthesis in animal tissues, in: *Metabolic Pathways*, Vol. V (D. Greenberg, ed.), p. 23, Academic Press, New York.
Levine, R. L., Hoogenraad, N. J., and Kretchmer, N., 1974, A review: Biological and clinical aspects of pyrimidine metabolism, *Pediat. Res.* **8**:724.
Mansour, T. E., 1972, Phosphofructokinase, in: *Current Topics in Cellular Regulation*, Vol. 5 (B. L. Horecker and E. R. Stadtman, eds.), p. 1, Academic Press, New York.
Martinez, H. M., 1972, Morphogenesis and chemical dissipative structures, a computer simulated case study, *J. Theor. Biol.* **36**:479.
Masters, C. J., 1977, Metabolic control and the microenvironment, *Curr. Top. Cell. Regul.* **12**:75.
Meinhardt, H., 1978, Models for the ontogenetic development of higher organisms, *Rev. Physiol. Biochem. Pharmacol.* **80**:48.
Meyer, U. A., and Schmid, R., 1978, The porphyrias, in: *The Metabolic Basis of Inherited Disease*, 4th ed. (J. B. Stanbury, J. B. Wyngaarden, and D. S. Fredrickson, eds.), p. 1166, McGraw-Hill, New York.

Monod, J., Wyman, J., and Changeux, J-P., 1965, On the nature of allosteric transitions: A plausible model, *J. Mol. Biol.* **12**:88.
Neuhaus, F. C., and Byrne, W. L., 1958, O-Phosphoserine phosphatase, *Biochim. Biophys. Acta* **28**:223.
Newsholme, E. A., and Start, C., 1973, *Regulation in Metabolism*, John Wiley and Sons, New York.
Nicolis, G., and Portnow, J., 1973, Chemical oscillations, *Chem. Rev.* **73**:365.
Passonneau, J. V., and Lowry, O. H., 1963, P-fructokinase and the control of the citric acid cycle, *Biochem. Biophys. Res. Commun.* **13**:372.
Racker, E., 1975, Control of energy transducing systems, *MTP Intl. Rev. Sci. Biochem.* **3**:163.
Roberts, R. B., Abelson, P. H., Cowie, D. V., Bolton, E. T., and Britten, R. J., 1955, *Studies on biosynthesis in E. coli*, Carnegie Institute, Washington, D.C., Publ. No. 607.
Sabine, J. R., and James M. J., 1976, Minireview: The intracellular mechanism responsible for dietary feedback control of cholesterol synthesis, *Life Sci.* **18**:1185.
Siperstein, M. D., and Fagan, V. M., 1964, Studies on the feedback regulation of cholesterol synthesis, *Adv. Enzyme Regul.* **2**:249.
Stadtman, E. R., 1966, Allosteric regulation of enzyme activity, *Adv. Enzymol.* **28**:41.
Stadtman, E. R., 1970, Mechanisms of enzyme regulation in metabolism, in: *The Enzymes*, 3rd ed., Vol. 1 (P. D. Boyer, ed.), p. 397, Academic Press, New York.
Tatibana, M., and Ito, K., 1967, Carbamylphosphate synthetase of the hematopoietic mouse spleen and the control of pyrimidine biosynthesis, *Biochem. Biophys. Res. Commun.* **26**:221.
Tsutsumi, E., and Takenaka, F., 1969, Inhibition of pyruvate kinase by free fatty acids in rat heart muscle, *Biochim. Biophys. Acta* **171**:355.
Tubbs, P. K., 1963, Inhibition of citrate formation of long chain acyl thioesters of Coenzyme A as a possible control mechanism in fatty acid biosynthesis, *Biochim. Biophys. Acta* **70**:608.
Turing, A. M., 1952, The chemical basic of morphogenesis, *Philos. Trans. Roy. Soc. London B* **237**:37.
Umbarger, H. E., 1956, Evidence for a negative-feedback mechanism in the biosynthesis of isoleucine, *Science* **123**:848.
Volpe, J. J., and Vagelos, P. R., 1976, Mechanisms and regulation of biosynthesis of saturated fatty acids, *Physiol. Rev.* **56**:339.
Weber, G., Lea, M. A., and Stamm, N. B., 1968, Sequential feedback inhibition and regulation of liver carbohydrate metabolism through control of enzyme activity, *Adv. Enzyme Regul.* **6**:101.
Wieland, O., and Weiss, L., 1963, Inhibition of citrate-synthase by palmityl coenzyme A, *Biochem. Biophys. Res. Commun.* **13**:26.
Wyngaarden, J. B., and Ashton, D. M., 1959, The regulation of activity of phosphoribosylpyrophosphate amidotransferase by purine ribonucleotides: A potential feedback control of purine biosynthesis. *J. Biol. Chem.* **234**:1492.
Wyngaarden, J. B., and Kelley, W. N., 1978, Gout, in: *Metabolic Basis of Inherited Disease*, 4th ed. (J. B. Stanbury, J. B. Wyngaarden, and D. S. Fredrickson, eds.), p. 916, McGraw-Hill, New York.

9

Membrane-Bound Enzymes

Robert M. Cohn and Pamela D. McNamara

I. Introduction

Converging lines of inquiry in various areas of biochemistry have provided unequivocal evidence for the close, if not indivisible, relationship between structure and function. This is perhaps nowhere more striking—yet elusive—than in the case of membrane-bound enzymes. In this chapter we shall explore the relationship of enzymes to membrane structural elements in order to attempt to arrive at an understanding of the relationship of these membrane-bound enzymes to the extraordinarily complex chemistry that goes on within the cell.

Biological membranes are supramolecular aggregates of proteins and complex lipids to which carbohydrate moieties are bound in covalent linkage. Membranes are integral parts of cellular organelles including the mitochondria, endoplasmic reticulum, Golgi apparatus, and nucleus, while the plasma membrane of the cell is itself considered to be an organelle. In association with cellular structures, membranes provide for retention of material within and regulation of transport into and out of the cell or organelle, as well as for a multiplicity of anabolic and catabolic functions essential to the life of the cell and the organism. The limiting membrane, of course, provides for compartmentation of function, an essential ingredient for control of the simultaneous metabolic events that take place within the cell. Several reviews covering aspects of this subject are Coleman (1973), DePierre and Ernster (1977), Gennis and Jonas (1977), and Sandermann (1978).

Robert M. Cohn and Pamela D. McNamara • Department of Metabolic Research, Children's Hospital of Philadelphia, University of Pennsylvania, Philadelphia, Pennsylvania 19104.

II. Membrane Composition and Structure

At present, the fluid mosaic model of membrane structure (Singer, 1974, 1976) enjoys the widest acceptance. It conceptualizes the membrane as being predominantly a phospholipid bilayer interrupted to an as yet undetermined extent by proteins. These proteins are either extrinsic (peripheral) or intrinsic (integral). According to this view the extrinsic proteins are bound weakly to the membrane through noncovalent forces, and have little if any interaction with the membrane lipids. On the other hand, intrinsic proteins require hydrophobic bond-breaking agents, including detergents and chaotropic agents, to disrupt their binding to the membrane. Hence, intrinsic proteins are bound more tightly to the membrane and appear to interact with the membrane lipids in a manner that influences the biological activity of the proteins. This particular point will be a major focus of our consideration of membrane-bound enzymes.

A further aspect of the fluid mosaic model for which there is abundant experimental evidence is the globular and amphipathic nature of the integral proteins. The hydrophobic ends of these proteins are viewed as passing through the transverse plane of the membrane, engaging in sizable areas of interaction with the fatty acid moieties of the membrane lipids. For example, studies with nonionic detergents (Helenius and Simons, 1972) have shown that intrinsic proteins may bind to and be solubilized by nonionic detergents without loss of enzymatic activity, whereas use of ionic detergents to solubilize the enzymes often inactivates them. In this instance, it appears that the hydrophilic end of such an enzyme corresponds to the catalytic end and that its interaction with the ionic detergent leads to its inactivation.

Perhaps the most striking physical property of membranes is the asymmetric distribution of both their proteins and phospholipids. Deriving from this asymmetric orientation is the vectorial nature of many of the metabolic processes involving membranes, including transport, binding of hormones and other informational molecules, and formation of electrochemical gradients, as well as energy transduction. Experimental methods to document and elucidate this anisotropic orientation include chemical labeling, nuclear magnetic resonance (NMR) spectroscopy using shift and broadening reagents, intermembrane phospholipid exchange using transfer proteins, immunochemical studies, lectin binding, and enzymatic treatment with proteases, phospholipases, or neuraminidase, as well as enzyme-catalyzed iodination studies (Rothman and Lenard, 1977; Bergelson and Barsukov, 1977; Oseroff et al., 1973).

A. Isolation and Solubilization of Membrane-Bound Enzymes

A recent review (Maddy and Dunn, 1976) deals with the problems of obtaining a membrane preparation from a particular cell organelle which is at once enriched in enzymes known to be indigenous to that membrane while evincing no evidence of contamination with an enzyme from another cellular subfraction. Practically speaking, this problem is sufficiently formidable that specific preparative procedures must be consulted, making all-embracing generalizations therefore clearly unwarranted. The investigator having obtained a preparation enriched in certain organelle-specific enzymes may then set out to solubilize them. Again, such methods require careful detail to particulars, but useful procedures have been reviewed (Maddy and Dunn, 1976) and are summarized in Table 1.

From an operational viewpoint, there is a spectrum of proteins associated with membranes. At one end of the spectrum are those proteins such as erythrocyte glyceraldehyde-3-phosphate dehydrogenase (G-3-PD) that appear to be bound to the membrane through coulombic forces and which are released under conditions utilizing changes in ionic strength. Beyond this, one encounters the vast majority of membrane-bound proteins which interact with both the hydrophilic and hydrophobic domains of the membrane, while at the other end of the spectrum lie those proteins located entirely within the hydrocarbon core of the membrane. This wide spectrum dictates careful selection

Table 1. Methods of Solubilization of Enzymes from Membranes

Method	Function	Reference
Low ionic strength, very dilute buffers, metal chelating agents	Removes adsorbed proteins-such enzymes are believed to carry out their functions in association with their membranes; e.g., glycolytic enzymes in red cells, hexokinase in mitochondria	Maddy and Dunn (1976)
High ionic strength	Extracts 30–40% of red cell protein acetylcholinesterase	Maddy and Dunn (1976)
Chaotropic agents (SCN^- and ClO_4^-)	Extracts complex I from mitochondria	Hatefi and Hanstein (1974)
Organic solvents	Succinate DH and coupling factor F_2 solubilized	Beechey and Cattell (1973)
Detergents	Most widely used agents for preparation of membrane-bound enzymes	Helenius and Simons (1975); Tanford and Reynolds (1976)
Phospholipases	May be used to determine transverse topology	Zwaal et al. (1973)

and application of procedures for protein solubilization, with particular attention to the purpose of the isolation procedure. Thus, if retention of enzymatic activity is necessary in the procedure, some interaction with lipids will be essential—either through milder extraction or reconstitution. If, on the other hand, constitutional analysis of the isolated material is to be carried out, then full dissociation of the membrane component of interest is essential for further study. This may well necessitate disruption of the conformation of the protein.

Detergents have been the most widely utilized agents for solubilizing enzymes from membranes. Their mode of action and the nature of their interaction with lipids and proteins have been subjects of numerous studies (Helenius and Simons, 1975; Tanford and Reynolds, 1976). Formation of mixed micelles of membrane lipids and proteins is involved in the solubilization process with detergents.

III. Endoplasmic Reticulum

In many cells the cytoplasm is traversed by a complex membrane network, canalicular in nature and demonstrating irregular dilatations, cisterns, and vesicles. This system is called the endoplasmic reticulum. In the rat hepatocyte, 19% of total cellular proteins, 48% of the total phospholipid, and 58% of the total RNA are found in association with this network. Metabolic functions associated with the endoplasmic reticulum include protein synthesis, detoxification of foreign chemicals, synthesis of cholesterol, phospholipids, and triglycerides, and involvement of the reticulum in glycoprotein synthesis, transport, and glycogen catabolism (DePierre and Dallner, 1975). The granular or rough endoplasmic reticulum, comprising 60% of this organelle, is involved in protein synthesis.

There is a recent, detailed account of methods for preparing and characterizing the components of the endoplasmic reticulum (DePierre and Dallner, 1976). Table 2 lists a number of enzymes associated with the endoplasmic reticulum. Several of these enzymes have been localized to the cytoplasmic surface of the reticular membranes, including cytochrome b_5, NADH-cytochrome b_5-reductase, Mg^{2+}-ATPase, 5'-nucleotidase, nucleoside pyrophosphorylase, and GDP-mannosyl transferase. A number of others, including nucleoside diphosphatase, glucose-6-phosphatase, acetanilide-hydrolyzing esterase, and β-glucuronidase have been localized to the luminal surface (DePierre and Ernster, 1977). Cytochrome P-450 appears to be present at both the cytoplasmic and luminal surfaces.

Secretory processes carried out by the endoplasmic reticulum have been compartmentalized to the lumen of this organelle. This compart-

Table 2. Selected Enzymes Associated with the Endoplasmic Reticulum

Enzyme	Reference
Alkenyl-glycerophosphinicholine hydrolase	Wykle and Snyder (1976)
Glucose-6-phosphatase and UDP-glucuronyltransferase	Zakim and Vessey, Chapter 10, this volume
Acyl-CoA desaturase, cytochrome b_5, cytochrome b_5 reductase	Enoch et al. (1976); Strittmatter and Rogers (1975)
Cytochrome P-450 system	Lu (1976)
Hydroxymethylglutaryl-CoA reductase	Goldstein and Brown (1977)
Steroid dehydrogenases, steroid synthetic enzymes	Dugan and Porter (1976)
Phospholipid synthesis	Lands and Crawford (1976)

mentation permits the proteins and glycoproteins to be concentrated without interference from cytosolic enzymes. Glycoprotein synthesis is an example of one such process and has been well studied (Jentoft et al., 1976; Molnar, 1976).

A. Microsomal Acyl-CoA Desaturation System

There are three proteins required for the formation of unsaturated fatty acids from precursor saturated acids by the microsomal electron-transport chain. They are NADH-cytochrome b_5 reductase, cytochrome b_5, and acyl-CoA desaturase. The reaction sequence (Fig. 1A) requires molecular oxygen as a proton acceptor, and electrons from NADH. The electron-transport sequence commences with the reductase, passes through cytochrome b_5, and terminates with the cyanide-sensitive factor, acyl-CoA desaturase (Enoch et al., 1976). Both cytochrome b_5 reductase and cytochrome b_5 are amphipathic proteins in which the redox center is located at the hydrophilic end while the hydrophobic portion binds to lipid moieties (DePierre and Ernster, 1977). Several studies have demonstrated the ability of cytochrome b_5 and cytochrome b_5 reductase to undergo lateral diffusion both in microsomes and in liposomes during the course of electron transfer (Hackenbrock, 1976).

Cytochrome b_5 can be solubilized either by detergents or with trypsin in a form which retains biologic activity (DePierre and Ernster, 1977). However, the proteins thus obtained differ in a number of properties. The protein extracted by trypsin treatment is soluble in aqueous solutions, does not bind phospholipids, and has a molecular weight of 11,000. The detergent-solubilized protein forms aggregates in aqueous

A

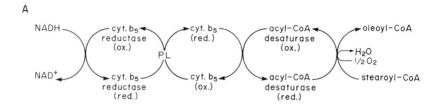

B

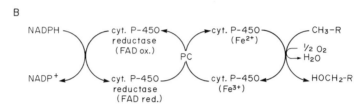

Figure 1. Electron transport chains of the endoplasmic reticulum. (A) Microsomal acyl-CoA desaturation system composed of NADH-cytochrome b_5 reductase, cytochrome b_5 (a flavoprotein), and fatty acyl-CoA desaturase. (B) Microsomal hydroxylase system depicting participation of the NADPH-cytochrome P-450 reductase (a flavoprotein), cytochrome P-450, and phosphatidylcholine. The role of the phospholipid appears to be in enhancing interaction of the proteins. The reduced form of the hemoprotein cytochrome P-450, on addition of carbon monoxide, envinces a Soret maximum at 450 nm, accounting for its designation. There is evidence that these two systems (A and B) interact in the membrane.

solutions, binds phospholipids, and, in contrast to the protease-prepared cytochrome, can be incorporated into microsomal vesicles in a biologically active state. The detergent cytochrome, called d-cyt b_5, has a molecular weight of 16,700 and possesses 44 amino acid residues which compromise the hydrophobic tail believed to be responsible for interaction with the hydrocarbon core of the membrane.

The desaturase is more nonpolar than the other components of the microsomal acyl-CoA desaturation system, possessing 62% hydrophobic amino acid residues and one atom of nonheme iron and having a molecular weight of 53,000 (Strittmatter et al., 1974).

The phospholipid requirement for this system has recently been studied by Enoch et al. (1976), who concluded that the hydrocarbon is necessary both as an initial attachment site for the stearoyl-CoA prior to binding to the enzyme and as an anchor for the hydrophobic regions of the desaturase system components. The study by Enoch et al. supports the view that the fluidity of the hydrocarbon chains in the membrane allows for lateral diffusion of the three enzymes of the system. In the partially purified systems studied, phosphatidylcholine and dimyristyl lecithin have fulfilled the lipid requirements of the system (Enoch et al., 1976).

B. Microsomal Hydroxylation System

The other electron-transport chain situated in the endoplasmic reticulum of liver is the system consisting of cytochrome P-450 and NADPH cytochrome P-450 reductase (Fig. 1B). This system utilizes molecular oxygen as a proton acceptor and NADPH as the electron carrier. Both endogenous (bile acids, fatty acids, heme, and steroids) and exogenous (carcinogens, drugs, and insecticides) compounds are used as substrates by the system in association with a number of accessory enzymes for carrying out hydroxylation, dealkylation, epoxidation, and desulfuration reactions, among others. In catalyzing such a wide spectrum of reactions, the microsomal hydroxylation system is undoubtedly one of the most versatile of enzyme sequences. The properties of this system have been the subject of reviews (Lu and Levin, 1974; Lu, 1976) and a recent monograph (Cooper et al., 1975).

Phosphatidylcholine was demonstrated to play an essential role in the biological activity of the microsomal hydroxylation system (Lu et al., 1969), functioning in some poorly understood manner in electron transfer from NADPH to cytochrome P-450. Reconstitution studies of the microsomal hydroxylation system from its isolated components have demonstrated the ability of a number of nonionic detergents in low concentration to substitute for phospholipid in the process of benzphetamine N-demethylation (Lu et al., 1974). While restoration of catalytic activity was not complete, it is likely that the natural lipid and the detergents act in a similar manner, perhaps enhancing the interaction of the two proteins of this system.

C. Sarcoplasmic Reticulum

In muscle there is an extensive endoplasmic reticulum called the sarcoplasmic reticulum. It seems to be primarily concerned with regulating Ca^{2+} ion fluxes during the contraction–relaxation cycle. The components of the sarcoplasmic reticulum are in contact with invaginations of the cell membrane which conduct the wave of depolarization into the interior of the muscle cell and to the myofibrils. Relaxation of muscle is brought about by accumulation of calcium within the sarcoplasmic reticulum, whereas contraction occurs as a consequence of an increase in calcium released from the sarcoplasmic reticulum secondary to an action potential. During relaxation, calcium ion is transported into the lumen of the sarcoplasmic reticulum in a tightly coupled process whose stoichiometry is two moles of calcium for each mole of ATP (MacLennan and Holland, 1975).

There is abundant evidence for the view that calcium translocation, as well as ATP hydrolysis, is mediated by a Ca^{2+}-ATPase situated in the sarcoplasmic reticulum membranes (MacLennan and Holland, 1975). Lipid dependency of both of these functions has also been known for some time. Treatment of membrane preparations or of the Ca^{2+}-ATPase with phospholipases results in parallel loss of both functions, whereas the restoration of both follows phospholipid addition. In early studies, lecithin alone has been effective in restoring ATPase activity of the delipidated enzyme while, for calcium translocation, phosphatidylethanolamine was required as well (MacLennan and Holland, 1976).

Recently, a number of investigators have set out to reinvestigate the nature and presumed indispensability of phospholipids for biological activity of this ATPase (Hidalgo et al., 1976; Knowles et al., 1976). Knowles et al. (1976) were able to restore approximately 50% of the original ATPase activity to a delipidated enzyme preparation using phosphatidylcholine alone. As better and better methods for isolation of Ca^{2+}-ATPase were developed, the bound lipid content of the isolated enzyme, which was necessary to allow reconstitution of some catalytic activity, was drastically decreased. Most recently, Dean and Tanford (1977) have reported a method for the delipidation of this enzyme to the level of 4 mol phospholipid/mol ATPase (far below the natural 90 mol phospholipid/mol ATPase). The addition of phosphatidylcholine to this preparation resulted in almost total restoration of catalytic activity (90%). Strikingly, Dean and Tanford achieved the same results with the nonionic detergent dodecyl octaoxyethylene glycol monoether, furnishing a model system to investigate the manner in which the phospholipid provides for full expression of biological activity.

IV. Golgi Apparatus

Electron microscopy of the cell shows the Golgi apparatus to be composed of several membrane-bounded, flattened saccules or cisternae arranged in parallel array, approximately 300 Å apart. These saccules are disc-like and often slightly curved, so that they appear to be stacked. Functions believed to be accomplished by the Golgi apparatus include glycosylation and transport of molecules designed for secretion, formation, and differentiation of membranous organelles such as the plasma membrane and lysosomes, and sulfation.

Attempts at preparing and isolating different components of the Golgi apparatus are in their infancy. The membranes composing the apparatus are particularly susceptible to breakage during mechanical shearing, and care must be exercised during the homogenization step (Whaley, 1975).

The Golgi apparatus functions as part of a complex system in which carbohydrate moieties are added in a sequential manner to a protein moiety which has been previously synthesized in the rough endoplasmic reticulum. The manner in which the protein leaves the endoplasmic reticulum and the nature of the putative recognition sites necessary for carbohydrate uptake by the Golgi apparatus are areas of intensive investigation (Whaley et al., 1972). Nonetheless, the bulk of the addition of glycosyl moieties to proteins to form glycoproteins undoubtedly is carried out in the Golgi apparatus. Thus, this organelle performs an important function in determining the characteristics of certain macromolecules associated with the cell surface, many of which carry out an informational function. Many of the substances secreted by the Golgi apparatus are secreted through the mediation of vesicles formed in the apparatus, through which transport to the cell surface occurs. The glycosyltransferases that carry out these synthetic processes have been the subject of intense investigation, and have been reviewed by a number of authors (Molnar, 1975; Jentoft et al., 1976). The addition of glycosyl moieties to the nascent (glyco)protein in a faultless manner, unaided by the genome, suggests a high degree of specificity and organization of the individual glycosyltransferases. Roseman (1970) has defined this specificity as cooperative sequential specificity, indicating the important role played by the previously incorporated glycosyl moieties in determining the absolute specificity for the next glycosyl moiety to be incorporated by these enzymes located within the Golgi apparatus.

V. Mitochondria

Mitochondria are the focus of one of the most intensive research efforts in all of biochemistry, owing to the essential metabolic processes they carry on (the citric acid cycle, fatty acid oxidation, and electron transport are but a few) and most particularly to the nature of the process of oxidative phosphorylation. Table 3 lists the major metabolic pathways and processes that occur within the confines of the mitochondrion. A complete list of enzymes associated with mitochondria is available (Altman and Katz, 1976), as is a review on methods of preparation (Sottocasa, 1976).

Table 3. Selected Membrane-Associated Pathways in the Mitochondria

Pathway	References
Citric acid cycle	Lowenstein (1969)
Branched-chain amino acid metabolism	Bender (1975)
Fatty acid oxidation and activation	Stumpf (1969)
Ketone-body metabolism	Krebs et al. (1971)
Electron-transport chain and phosphate-transfer enzymes	Baltscheffsky and Baltscheffsky (1974)
Several enzymes of heme biosynthesis	Gidari and Levere (1977)
Part of urea cycle	Grisolia et al. (1976)

The inner membrane of the mitochondrion accounts for 80–95% of the protein found in mitochondrial membranes and over 90% of the lipid. It is the site of the respiratory chain and the synthesis of ATP. It is this membrane, in conjunction with studies on transport through the plasma membrane, that has contributed most forcefully both to the viewpoint of the anisotropic organization of membrane structural elements and of biochemical events carried out by or in membranes. As regards mitochondria, the interaction of the inner membrane components in carrying out electron transport and oxidative phosphorylation is the focal investigative question both for mitochondrial function and for the organization of vectorial events in general.

A. Respiratory Chain and Electron Transport

NADH and $FADH_2$ are produced as a result of substrate level dehydrogenations. Oxidation of these reduced coenzymes by oxygen is accomplished by the intervention of a series of electron carriers between the primary reductant and the terminal oxidant (Fig. 2). The electron-transport components represent redox couples of increasing redox potential and are therefore favored thermodynamically. The respiratory chain can be separated into four multienzyme complexes: NADH-Q reductase (complex I), succinate-Q reductase (complex II), QH_2-cytochrome c reductase (complex III), and cytochrome c oxidase (complex IV). At each of these successive oxidation–reduction steps, a certain amount of free energy is available, the amount being determined by the difference in the oxidation–reduction potential of the two sequential components. The difference in the redox potential between

NADH and oxygen is approximately 1.13 V, representing an energy availability of 51 kcal/mol of NADH oxidized. At steps I, III, and IV of the respiratory chain, the free energy is coupled to the formation of ATP, a process which requires approximately 25 kcal to form 3 mol of ATP (Baltscheffsky and Baltscheffsky, 1974).

1. Components of the Respiratory Chain

The cytochromes, which have a prosthetic heme group at the active site, possess characteristic absorption maxima originating from the porphyrin rings as modified by the iron atom held in coordinate linkage. It is this iron that accounts for the electron acceptor and donor properties of the cytochromes.

Some of the electron-transport carriers contain iron that is not found in a porphyrin prosthetic group. This is termed *nonheme iron*, the identification of iron depending on instrumental methods of electron paramagnetic resonance (EPR). Nonheme-iron carriers have an asymmetric EPR signal at $g = 1.94$, which is generated when they are reduced. In addition, they are usually found to be bound to sulfur as part of the active site.

The *ubiquinones* (coenzyme Q) are a group of lipid-soluble benzoquinones found in most aerobic organisms. For a long while their function was unclear, but studies with reconstituted electron-transport systems have shown that coenzyme Q is a component which occurs at a

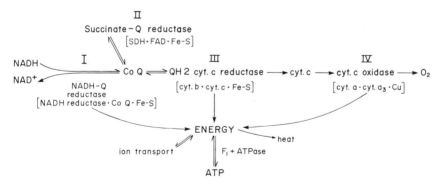

Figure 2. Multienzyme complexes of the electron-transport chain. The four multienzyme complexes comprising the electron-transport chain are shown in order of increasing reduction potential. The properties of each component are described in the text. Free energy available at each of these oxidation–reduction steps can be utilized to produce ATP, generate heat, or drive membrane ion translocation. Sites I, III, and IV are the probable points of energy transduction along the respiratory chain.

junction between three complexes involving NADH and succinate dehydrogenase and cytochrome b (Fig. 2).

NADH-Q reductase (complex I) has a molecular weight exceeding 850,000. In addition to NADH reductase, which is a flavoprotein, this complex contains coenzyme Q and 16 to 24 iron–sulfur centers. The NADH oxidizing site in complex I has been shown to be on the M (matrix) side of the mitochondrial membrane (Ragan, 1976).

Succinate-Q reductase (complex II) contains succinate dehydrogenase, FAD, two iron–sulfur centers, and an additional iron–sulfur protein. As with complex I, the substrate binding site of succinate reductase is also on the M side of the membrane (Hatefi and Stiggall, 1976).

QH2-cytochrome c reductase (complex III), also termed the cytochrome $b-c_1$ complex, has a molecular weight of approximately 300,000, with the protein components accounting for 240,000 of this. It is comprised of 6–8 polypeptides including cytochrome b, cytochrome c_1, an iron–sulfur protein, an antimycin-binding protein, and a core protein (Rieske, 1976).

All of these complexes are integral components of the mitochondrial membrane. In contrast, cytochrome c is an extrinsic or peripheral protein and appears to be associated with the C (cristae) side of the inner mitochondrial membrane. Indeed, Nicholls (1974) showed that the interaction of cytochrome c with cytochrome c oxidase accounts for the binding of cytochrome c to the mitochondrial membrane.

Cytochrome c oxidase (complex IV), the final mitochondrial complex, is composed of cytochromes a and a_3 and contains two atoms of Cu^{2+}. As with other complexes of the inner mitochondrial membrane, it is an integral protein with a molecular weight of approximately 200,000, is composed of 6 to 7 subunits, and contains two hemes and two atoms of Cu^{2+} (Caughey et al., 1976). Studies to define its topology indicate that cytochrome a occurs on the C side of the membrane and cytochrome a_3 on the M side. Hackenbrock (1976) has shown that cytochrome c oxidase, like a number of other integral membrane proteins, manifests lateral mobility within the plane of the inner mitochondrial membrane. He suggested that such mobility might be of material importance in terms of the actual energy transduction processes occurring during oxidative phosphorylation, suggesting that this mobility might be involved in a conformational interaction in the formation of ATP.

B. H^+–ATPase

The mitochondrial ATPase system functions as a proton translocating, reversible ATPase and is thus the proximate cause for ATP synthesis as a consequence of the energy transduced through electron transport. The enzyme consists of three parts: (1) F_1, the catalytically active part of the enzyme, occurs on the M side of the membrane and is cold labile; (2) a hydrophobic protein located within the membrane lipid bilayer and thought to contain the proton translocating portion of the enzyme; and (3) a protein component which connects the two (Senior, 1973).

Studies employing mitochondrial ATPase reconstituted into proteoliposomes and evaluating its ability to create a membrane potential with the penetrating ion method have shown clearly that the ATPase can generate a membrane potential. Drachev et al. (1976) showed that when ATP was added to the reconstituted ATPase lipsome in the presence of Ca^{2+}, a transmembranous electric potential difference was detected.

By appropriate treatment of the ATPase and its reconstitution into liposomes, it is possible to show that the H^+-ion translocation and ATP-hydrolyzing activities of this complex are separable. Thus, coupling factor F_1, devoid of hydrophobic proteins, is able to carry out hydrolysis of ATP, whereas the addition of hydrophobic proteins to the ATPase increases H^+ transport, such transport being sensitive to inhibition by oligomycin. These studies have been reviewed by Skulachev (1975).

Electron transport and oxidative phosphorylation represent the most complex membrane processes yet uncovered in living mammalian cells. The broad outlines of this pathway are not much in question. Thus, the oxidation of one molecule of NADH by the respiratory chain results in the formation of three molecules of ATP. Three complexes along this chain—NADH-Q reductase, QH_2-cytochrome c reductase, and cytochrome c oxidase—contain the sites where energy is transduced and enabled to interact with the H^+–ATPase to generate ATP (Fig. 2). Abundant evidence exists, as well, for the transfer of energy among these three complexes without the involvement of ATP synthesis; for instance in ion translocation by the mitochondrion (Ernster, 1977).

The manner in which the energy liberated during the movement of electrons down the respiratory chain is coupled to the formation of ATP has been the subject of elaborate theorizing and investigation, often not without its acrimony. The early history of this field has been captured in a collection of papers edited by Kalckar (1969). Recently, the *Annual Review of Biochemistry* (Boyer et al., 1977) published six thoughtful essays written by leaders in this field which were put together in a spirit of compromise and further experimentation.

To better comprehend the differences between the contending hypotheses of the mechanism of oxidative phosphorylation, it is useful to break that process into its phenomenologically separate events. The first major event must be the capture of energy derived from oxidation. The captured energy must then be transmitted into some other form for the final step, the transformation of this energy by the H⁺–ATPase. All participants in the *Annual Review of Biochemistry* symposium appear in general agreement that the first two steps operate by a chemiosmotic mechanism, i.e., by translocation of protons across a membrane in a process that generates an electrochemical potential gradient for protons, demonstrable by a change in pH and a membrane potential. However, it is with the third event, i.e., the transformation of the energy resulting in the formation of ATP, that disagreement is encountered and experimental evidence is inconclusive. The conformational hypotheses (which accept the proton gradient in the initial coupling) maintain that some intermediary is required and, therefore, an indirect change in protein conformation serves as the driving force for ATP formation.

In the chemiosmotic hypothesis, the proton flux serves to generate the ATP. Recently, Mitchell (1976, 1977a, 1979) has reviewed the postulates of the chemiosmotic hypothesis and has attempted to point to areas of future investigation which are at present only dimly understood even in the most expectant of schemes. The sum and substance of such postulates is that various energy transducing units located within the mitochondrial membrane can generate either a proton gradient or membrane potential, and that mitochondrial uncoupling agents which abolish the synthesis of ATP cause the abolition of the proton gradient. An important area of support for these postulates has emerged from work with electron-transport complexes reconstituted in liposomes. Such reconstituted complexes are capable of translocating protons across the liposome membrane, and when ATPase is reconstituted along with the respiratory chain complex, oxidative phosphorylation is observed (Racker, 1977).

In the Mitchell hypothesis, electron transfer is coupled to transmembrane movement of protons, and this transport process results in the creation of an electrochemical potential (proton motive force) at the outside of the membrane. When protons return, they do so through a proton channel in the membrane that leads to the H⁺–ATPase, where synthesis is accomplished. Mitchell (1976) has elaborated on the concept of *proticity*, i.e., proton flow. One key feature of the chemiosmotic theory is the expectation that $H^+/e^- = H^+/ATP$, the value of 2 for both ratios being determined experimentally. Reevaluation of experimental data led Brand and Lehninger (1977) to propose a modification of the chemiosmotic theory which accommodated H^+/ATP ratios greater than 2 and $H^+/e^- \neq H^+/ATP$. Stoichiometric considerations of proton translocation have been reviewed by Papa (1976).

While the chemiosmotic hypothesis does not embrace the need for the mediation of a chemical coupling compound(s) or for conformational coupling interaction(s) between the redox system and the ATP-synthesizing system, Mitchell notes that there is every reason to believe that conformational interactions may be involved within the translocation system itself, in a manner consistent with the role of conformational changes in enzymic catalysis, as elaborated in Chapters 3 and 4 of this book.

Boyer (1977) and Slater (1977) have proposed a mechanism of conformational coupling which is brought about by the proton flux. In so doing, they have attempted to reconcile what for them have been data not embraced by the chemiosmotic hypothesis. In their hypothesis, the essence is an "indirectly linked function" (Lumry, 1974) in which the binding of substrate or ligand at one site initiates a transformation at another, distant site on the macromolecule. Specifically, they propose that electron transfer brings about a conformational change in protein(s), resulting in a change in the dissociation constant of a side chain in the protein. This liberates what are termed Bohr protons (Chance, 1977), which are delivered to the outside of the membrane. Thus, the conformational change caused by electron transfer brings about a conformational change in the ATPase which is associated with the creation of a (Bohr) proton gradient.

One piece of experimental data embraced by the conformational hypothesis is the finding that the major energy requirement in the formation of ATP is for dissociating the ATP bound to the enzyme active site, and not for actually forming the covalent bond between ADP and P_i (Boyer, 1977, Slater, 1977).

Specifics concerning the chemical reactions involved in the generation of ATP are quite involved and not within the direct purview of this chapter. The reader is encouraged to pursue articles by Mitchell (1977b), Boyer (1975), and Kozlov and Skulachev (1977) to gain the full flavor of this exciting area of inquiry and theory.

In summary, the direct chemiosmotic hypothesis visualizes an overlap of the domains involving proton transport and ATP synthesis, while the conformational couplng hypothesis proposes an indirect interaction of these two domains mediated through an intervening macromolecule.

In addition to the references included in this section, another paper worthy of study is that of Skulachev (1977), in which the author has proposed that the formation of transmembrane electrochemical H^+ potentials along with ATP constitute the energy currency for life processes. This particular view is fortified by studies Skulachev details in which organisms devoid of ATPase were able to carry out energy-requiring processes utilizing the transmembrane electrochemical potential of hydrogen ions. Further support for Skulachev's view comes from studies of the energy transduction system of *Halobacterium halobium* (Henderson, 1977).

Thus, as we have seen in this section on mitochondrial membrane-bound enzymes, not only is the membrane essential for carrying out the energy transducing functions associated with oxidative phosphorylation, but the ability of a membrane to generate a transmembrane hydrogen gradient may well be a universal form of energy, quite possibly antedating ATP in evolution.

Equally universal in its application is the vectorial nature of membrane-mediated processes, herein epitomized by oxidative phosphorylation. Evidence reviewed earlier supports the asymmetric organization of membrane structural elements, and evidence presented here supports the view that such organization determines the directional nature of certain membrane-associated functions. To date no convincing demonstration of an ATP-generating system devoid of a membranous structure has been shown, indicating the essential role subserved by the membrane in this process (Racker, 1977). Thus, the high-energy compound sought for so many years turns out, instead, to be the dynamic membrane and its ability to generate an electrochemical gradient (Harold, 1978).

VI. Plasma Membrane

The plasma membrane is the outermost boundary of the mammalian cell and represents the organelle which ultimately controls accessibility of substrates to the cell. Table 4 lists a number of selected enzymes associated with the plasma membrane. A detailed discussion of several of these has been covered elsewhere in this volume (Chapter 11).

Table 4. Selected Enzymes Associated with Plasma Membranes

Enzyme	Reference
Acetylcholinesterase	Potter, 1970
Alkaline phosphatase	Kinne et al. (1971); Fernley (1971)
5'-Nucleotidase	Neville (1976)
Adenylate cyclase	Herman and Taunton, Chapter 14, this volume
Aminopeptidase	Pattus et al. (1976)
(Na^+, K^+)-ATPase	Dahl and Hokin (1974); Kinne et al. (1971)
Glycosyltransferases	Shur and Roth (1975)

VII. Temperature Effects

A. Lipid Liquid Crystals

Increasing the temperature of certain solid organic compounds, including lipids, brings about changes in state through the formation of one or more intermediate phases, rather than direct conversion into a liquid. Liquid crystals constitute the intermediate phases that exist between true crystalline solid and liquid phases. These intermediate phases (mesophases) possess characteristics both of a crystal and of a liquid. Asymmetrical compounds tend to form liquid crystals, the molecular shape tending to parallel molecular alignment in both the crystalline state and the mesophases. The smectic mesophase (*smectic* coming from soap-like) has a stratified structure, with the molecules arranged in layers, such as might be found on stacked beds of nails. The *nematic* mesophase is more fluid than the smectic mesophase and the asymmetric molecules align themselves in a near-parallel orientation without forming a layer as seen in the smectic mesophase. These mesophases are brought about through thermotrophic reactions, i.e., heat changes. However, solvent may also cause disruptions of the crystal lattice, resulting in the formation of intermediate phases, such behavior being termed lyotrophic mesomorphism.

Phospholipids and glycolipids exhibit both thermotrophic and lyotropic mesomorphism. The temperature at which such phase transitions are observed depends on the head group, the hydrocarbon chain length, and the degree and type of unsaturation present. Unsaturation tends to lower the transition temperature, as do shorter chains. This subject has been reviewed by Saupe (1973).

B. Lipid–Protein Interactions

Evidence presented throughout the course of this review has emphasized the interaction of proteins and lipids in carrying out membrane-associated function. Thus, it is no longer admissible to conceive of lipids as acting merely as a matrix in which proteins exist; rather, there is evidence that proteins can immobilize the lipids which bind them and that lipids can influence the temperature dependence of a number of membrane-protein functions. This mutual interaction of lipids and proteins is often viewed in the context of solvent effects. To a significant degree, this point of view derives from studies based on thermotropic phase transitions occurring in biomembranes. This subject has recently been discussed by Melchior and Steim (1976), Papahadjopoulos et al. (1975), and Lee (1977). They cite abundant evidence for a reversible, thermally-induced transition of lipid bilayers from the ordered crystal-like state encountered at low temperature to a disordered fluid-like state at raised temperature. This change of state is not accompanied by a rearrangement of the bilayer to a different lipid fluid-crystalline mesophase. Rather what occurs is a melting of the bilayers, with conservation of the lamellar conformation throughout. While the underlying physiological and biochemical significance of these effects is not at all clear, the solvent-like nature of these observed effects suggests, at least, a role in the control of membrane organization. Not only is the topology of the membrane proteins susceptible to control by thermal factors, but the actual biological activities of the proteins appear to be similarly affected. Data presented as Arrhenius plots demonstrate discontinuities in the plot at different temperatures for enzymes found in membranes. This behavior is believed to be evidence that the physical state of the lipid in relationship to the protein is materially involved in the activity of that membrane-bound protein.

It is important to understand that the underlying biochemical cause for the discontinuities in the Arrhenius plots is not well understood. Moreover, the abrupt change seen in these plots should not be taken as an indication of a drastic change in behavior of a system as a consequence of addition of thermal energy. Rather, it is more likely that the discontinuity represents the initiation of a new phenomenon at that temperature. Hence, the actual physiologic effect at that temperature may be minimal.

In addition to changes in the lipid–protein interaction as temperatures increase, another cause for the discontinuities in the Arrhenius plot may be changes in the conformation of the protein itself. Morrisett et al. (1975), examining thermotropic phase transitions in *Escherichia coli*, proposed that the membrane lipids are organized heterogeneously. This heterogeneity embraces the view that lipids within the membrane exist in physically separate domains, with particular emphasis being placed on boundary lipid, which is immediately associated with membrane-bound proteins. The bulk of the membrane lipid, then, is in a solid or semisolid form, imparting substance to the membrane. The annulus or boundary lipid associated with proteins is that which undergoes fluid transformation, thereby modulating the activity of the membrane-bound enzymes.

VIII. Conclusion: Effects of Lipids on Enzymatic Activity

One important way in which a lipid may be involved in the expression of enzymatic activity for a particular protein would be to serve as an anchor point for the protein to the membrane. In the case of cytochrome b_5, and cytochrome b_5 reductase, this is apparently how the lipid acts with regard to these proteins. Another role for lipid may be to affect the specificity or affinity of the protein for a particular ligand. This particular effect has been well shown with β-hydroxybutyrate dehydrogenase from beef heart mitochondria, in which the lipid phosphatidylcholine affects the ability of the enzyme to bind the cofactor, NADH (Gazzotti et al., 1974). The delipidated enzyme is without biological activity, activity being restored by a variety of lecithins of chain lengths varying from C-4 to C-18 (Grover et al., 1975). Several detergents were without effect in activating β-hydroxybutyrate dehydrogenase.

While the exact mode of action of the phosphatidylcholines is not known, the presence of the choline head group is evidently essential, since phosphatidylethanolamine and cardiolipin are unable to restore activity to the purified enzyme (Gazzotti et al., 1975; Isaacson et al., 1979). It should be remembered, however, that the interesting finding of Dean and Tanford (1977) of the ability of a nonionic detergent to reactivate the highly purified, 96% delipidated calcium ATPase suggests that the role of the naturally occurring lipids in membranes may occasionally be subserved by a synthetic chemical. This phenomenon observed in vitro should be very useful in elucidating those aspects of a lipid which are essential in providing a specific protein with the necessary interactions to allow it to exercise or manifest its biological activity. It should not, however, be taken to mean that the naturally occurring lipids can be supplanted in terms of their biological activity in providing for full expression of an inherent protein-related function. In addition, phospholipids may provide a nonpolar environment in which certain reactions may be facilitated.

References

Altman, P. L., and Katz, D. D., eds., 1976, *Biological Handbooks I: Cell Biology,* Federation of American Societies of Experimental Biology, Bethesda, Md.

Baltscheffsky, H., and Baltscheffsky, M., 1974, Electron transport phosphorylation, *Annu. Rev. Biochem.* **43:**871.

Beechey, R. B., and Cattell, K. J., 1973, Mitochondrial coupling factors, *Curr. Top. Bioenergetics* **5:**305.

Bender, D. M., 1975, *Amino Acid Metabolism,* John Wiley and Sons, New York.

Bergelson, L. D., and Barsukov, L. I., 1977, Topological asymmetry of phospholipids in membranes, *Science* **197:**224.

Boyer, P. D., 1975, A model for conformational coupling of membrane potential and proton translocation to ATP synthesis and to active transport, *FEBS Lett.* **58:**1.

Boyer, P. D., 1977, Coupling mechanisms in capture, transmission and use of energy, *Annu. Rev. Biochem.* **46:**957.

Boyer, P. D., Chance, B., Ernster, L., Mitchell, P., Racker, E., and Slater, E. C., 1977,

Oxidative phosphorylation and photophosphorylation, *Annu. Rev. Biochem.* **46:** 955.
Brand, M. D., and Lehninger, A. L., 1977, H$^+$/ATP ratio during ATP hydrolysis by mitochondria: Modification of the chemiosmotic theory, *Proc. Natl. Acad. Sci. USA* **74:**1955.
Caughey, W. S., Wallace, W. J., Volpe, J. A., and Yoshikawa, S., 1976, Cytochrome *c* oxidase, in: *The Enzymes,* Vol. 13 (P. Boyer, ed.), Academic Press, New York.
Chance, B., 1977, Electron transfer: Pathways, mechanisms and controls, *Annu. Rev. Biochem.* **46:**967.
Coleman, R., 1973, Membrane-bound enzymes and membrane ultrastructure, *Biochim. Biophys. Acta* **300:**1.
Cooper, D., Rosenthal, O., Snyder, R., and Witmer, C., eds., 1975, Cytochrome P-450 and cytochrome b_5, in: *Advances in Experimental Medicine and Biology,* Vol. 58, Plenum, New York.
Dahl, J. L., and Hokin, L. E., 1974, The sodium-potassium adenosinetriphosphatase, *Annu. Rev. Biochem.* **43:**327.
Dean, W. L., and Tanford, C., 1977, Reactivation of lipid-depleted Ca^{2+}−ATPase by a nonionic detergent, *J. Biol. Chem.* **252:**3551.
DePierre, J. W., and Dallner, G., 1975, Structural aspects of the membrane of the endoplasmic reticulum, *Biochim. Biophys. Acta* **415:**411.
DePierre, J., and Dallner, G., 1976, Isolation, sub-fractionation and characterization of the endoplasmic reticulum, in: *Biochemical Analysis of Membranes* (A. H. Maddy, ed.), John Wiley and Sons, New York.
DePierre, J. W., and Ernster, L., 1977, Enzyme topology of intracellular membranes, *Annu. Rev. Biochem.* **46:**201.
Drachev, L. A., Jasuitis, A. A., Mikelsaar, H., Nemeček, I. B., Semenov, A. Y., Semenova, E. G., Severina, I. I., and Skulachev, V. P., 1976, Reconstitution of biological molecular generators of electric current, *J. Biol. Chem.* **251:**7077.
Dugan, R. E., and Porter, J. W., 1976, Membrane-bound enzymes of sterol metabolism, in: *The Enzymes of Biological Membranes,* Vol. 2 (A. Martonosi, ed.), Plenum, New York.
Enoch, H. G., Catalá, A., and Strittmatter, P., 1976, Mechanism of rat liver microsomal stearyl-CoA desaturase: Studies of the substrate specificity, enzyme–substrate interactions, and the function of lipid, *J. Biol. Chem.* **251:**5095.
Ernster, L., 1977, Chemical and chemiosmotic aspects of electron transport-linked phosphorylation, *Annu. Rev. Biochem.* **46:**981.
Fernely, H. N., 1971, Mammalian alkaline phosphatases, in: *The Enzymes,* Vol. IV (P. D. Boyer, ed.), pp. 417–447, Academic Press, New York.
Gazzotti, P., Bock, H. G., and Fleischer, S., 1974, Role of lecithin in D-β-hydroxybutyrate dehydrogenase function, *Biochem. Biophys. Res. Commun.* **58:**309.
Gazzotti, P., Bock, H. G., and Fleischer, S., 1975, Interaction of D-beta-hydroxybutyrate apodehydrogenase with phospholipids, *J. Biol. Chem.* **250:**5782.
Gennis, R. B., and Jonas, A., 1977, Protein–lipid interactions, *Annu. Rev. Biophys. Bioeng.* **6:**195.
Gidari, A. S., and Levere, R. D., 1977, Enzymatic formation and cellular regulation of heme synthesis, *Semin. Hematol.* **14:**145.
Goldstein, J. L., and Brown, M. S., 1977, The low-density lipoprotein pathway and its relation to atherosclerosis, *Annu. Rev. Biochem.* **46:**897.
Grisolia, S., Baguena, R., and Mayor, F., eds., 1976, *The Urea Cycle,* John Wiley and Sons, New York.
Grover, A. K., Slotboom, A. J., DeHaas, G. H., and Hammes, G. G., 1975, Lipid specificity of beta-hydroxybutyrate dehydrogenase activation, *J. Biol. Chem.* **250:**31.

Hackenbrock, C. R., 1976, Molecular organization and the fluid nature of the mitochondrial energy transducing membrane, in: *The Structure of Biological Membranes* (S. Abrahamsson and I. Pascher, eds.), Nobel Foundation Symposium, Vol. 34, p. 199, Plenum, New York.

Harold, F. M., 1978, The 1978 Nobel prize in chemistry, *Science* **202**:1174.

Hatefi, Y., and Hanstein, W. G., 1974, Destabilization of membranes with chaotropic ions, in: *Methods in Enzymology*, Vol. 31 (S. Fleischer and L. Packer, eds.), p. 770, Academic Press, New York.

Hatefi, Y., and Stiggal, D. L., 1976, Metal containing flavoprotein dehydrogenases, *The Enzymes* **13**:175.

Helenius, A., and Simons, K., 1972, The binding of detergents to lipophilic and hydrophilic proteins, *J. Biol. Chem.* **247**:3656.

Helenius, A., and Simons, K., 1975, Solubilization of membranes by detergents, *Biochim. Biophys. Acta* **415**:29.

Henderson, R., 1977, The purple membrane from *Halobacterium halobium*, *Annu. Rev. Biophys. Bioeng.* **6**:87.

Hidalgo, C., Ikemoto, N., and Gergely, J., 1976, Role of phospholipids in the calcium-dependent ATPase of the sarcoplasma reticulum. Enzymatic and ESR studies with phospholipid-replaced membranes, *J. Biol. Chem.* **251**:4224.

Isaacson, Y. A., Deroo, P. W., Rosenthal, A. F., Bittman, R., McIntyre, J. O., Bock, H. G., Gazzotti, P., and Fleischer, S., 1979, The structural specificity of lecithin for activation of purified D-β-hydroxybutyrate apodehydrogenase, *J. Biol. Chem.* **254**:117.

Jentoft, N., Cheng, P. W., and Carlson, D. M., 1976, Glycosyl transferases and glycoprotein biosynthesis, in: *The Enzymes of Biological Membranes* (A. Martonosi, ed.), p. 343, Plenum, New York.

Kalckar, H., 1969, *Biological Phosphorylations: Development of Concepts*, Prentice Hall, Englewood Cliffs, N.J.

Kinne, R., Schmitz, J. E., and Kinne-Saffran, E., 1971, The localization of the Na^+-K^+-ATPase in the cells of rat kidney cortex, *Pflügers Arch.* **329**:191.

Knowles, A. F., Eytan, E., and Racker, E., 1976, Phospholipid-protein interactions in the Ca^{2+}-adenosine triphosphatase of sarcoplasmic reticulum, *J. Biol. Chem.* **251**:5161.

Kozlov, I. A., and Skulachev, V. P., 1977, The H^+-adenosine triphosphatase and membrane coupling, *Biochim. Biophys. Acta* **463**:29.

Krebs, H. A., Williamson, D. H., Bates, M. W., Page, M. A., and Hawkins, R. A., 1971, The role of ketone bodies in caloric homeostasis, *Adv. Enzyme Regul.* **9**:387.

Lands, W. E. M., and Crawford, C. G., 1976, Enzymes of membrane phospholipid metabolism in animals, in: *The Enzymes of Biological Membranes*, Vol. 2 (A. Martonosi, ed.), p. 3, Plenum, New York.

Lee, A. G., 1977, Lipid phase transitions and phase diagrams. I. Lipid phase transitions, *Biochim. Biophys. Acta* **472**:237.

Lowenstein, J. M., ed., 1969, *Citric Acid Cycle*, Academic Press, New York.

Lu, A. Y. H., 1976, Liver microsomal drug-metabolizing enzyme system: Functional components and their properties, *Fed. Proc.* **35**:2460.

Lu, A. Y. H., and Levin, W., 1974, The resolution and reconstitution of the liver microsomal hydroxylation system, *Biochim. Biophys. Acta.* **344**:205.

Lu, A. Y. H., Junk, K. W., and Coon, M. J., 1969, Resolution of the cytochrome P-450 containing ω-hydroxylation system of liver microsomes into three components, *J. Biol. Chem.* **244**:3714.

Lu, A. Y. H., Levin, W., and Kuntzman, R., 1974, Reconstituted liver microsomal enzyme system that hydroxylates drugs, other foreign compounds and endogenous substrates. VII. Stimulation of benzphetamine N-demethylation by lipid and detergent, *Biochem. Biophys. Res. Commun.* **60**:266.

Lumry, R., 1974, Conformational mechanisms for free energy transduction in protein systems: Old ideas and new facts, *Ann. N.Y. Acad. Sci.* **227**:46.
MacLennan, D. H., and Holland, P. C., 1975, Calcium transport in sarcoplasmic reticulum, *Annu. Rev. Biophys. Bioeng.* **4**:377.
MacLennan, D. H., and Holland, P. C., 1976, The calcium transport ATPase of sarcoplasmic reticulum, in: *The Enzymes of Biological Membranes*, Vol. 3 (A. Martonosi, ed.), p. 221, Plenum, New York.
Maddy, A. H., and Dunn, M. J., 1976, The solubilization of membranes, in: *Biochemical Analysis of Membranes* (A. H. Maddy, ed.), p. 177, John Wiley and Sons, New York.
Melchior, D. L., and Steim, J. M., 1976, Thermotropic transitions in biomembranes, *Annu. Rev. Biophys. Bioeng.* **5**:205.
Mitchell, P., 1976, Vectorial chemistry and the molecular mechanics of chemiosmotic coupling: Power transmission and proticity, *Biochem. Soc. Trans.* **4**:399.
Mitchell, P., 1977a, Vectorial chemiosmotic processes, *Annu. Rev. Biochem.* **46**:996.
Mitchell, P., 1977b, A commentary on alternative hypotheses of protonic coupling in the membrane systems catalysing oxidative and photosynthetic phosphorylation, *FEBS Lett.* **78**:1.
Mitchell, P., 1979, Compartmentation and communication in living systems. Ligand conduction: A general catalytic principle in chemical, osmotic and chemiosmotic reaction systems, *Eur. J. Biochem.* **95**:1.
Molnar, J., 1975, A proposed pathway of plasma glycoprotein synthesis, *Molec. Cell. Biochem.* **6**:3.
Molnar, J., 1976, Role of endoplasmic reticulum and Golgi apparatus in the biosyntheses of plasma glycoproteins, in: *The Enzymes of Biological Membranes*, Vol. 2 (A. Martonosi, ed.), p. 385, Plenum, New York.
Morrisett, J. D., Pownall, H. J., Plumlee, R. T., Smith, L. C., Zehner, Z. E., Esfahani, M., and Wakil, S. J., 1975, Multiple thermotropic phase transitions in *Escherichia coli* membranes and membrane lipids, *J. Biol. Chem.* **250**:6969.
Neville, D. M., Jr., 1976, The preparation of cell surface membrane enriched fractions, in: *Biochemical Analysis of Membranes* (A. H. Maddy, ed.), p. 27, John Wiley and Sons, New York.
Nicholls, P., 1974, Cytochrome *c* binding to enzymes and membranes, *Biochim. Biophys. Acta* **346**:261.
Oseroff, A. R., Robbins, P. W., and Burger, M. M., 1973, The cell surface membrane: Biochemical aspects and biophysical probes, *Annu. Rev.Biochem.* **42**:647.
Papa, S., 1976, Proton translocation reactions in the respiratory chains, *Biochim. Biophys. Acta* **456**:39.
Papahadjopoulos, D., Moscarello, M., Eylar, E. H., and Isac, T., 1975, Effects of proteins on thermotropic phase transitions of phospholipid membranes, *Biochim. Biophys. Acta* **401**:317.
Pattus, F., Verger, R., and Desnuelle, P., 1976, Comparative study of the interactions of the trypsin and detergent form of the intestinal aminopeptidase with liposomes, *Biochem. Biophys. Res. Commun.* **69**:718.
Potter, L. T., 1970, Acetylcholine, choline acetyltransferase and acetylcholinesterase, in: *Handbook of Neurochemistry*, Vol. 4 (A. Lajtha, ed.), Plenum, New York.
Racker, E., 1977, Mechanisms of energy transformation, *Annu. Rev. Biochem.* **46**:1006.
Ragan, C. I., 1976, NADH-ubiquinone oxidoreductase, *Biochim. Biophys. Acta* **456**:249.
Rieske, J. S., 1976, Composition, structure and function of complex III of the respiratory chain, *Biochim. Biophys. Acta* **456**:195.
Roseman, S., 1970, The synthesis of complex carbohydrates by multiglycosyltransferase systems and their potential function in intercellular adhesion, *Chem. Phys. Lipids* **5**:270.

Rothman, J. E., and Lenard, J., 1977, Membrane asymmetry, *Science* **195**:743.
Saupe, A., 1973, Liquid crystals, *Annu. Rev. Physical Chem.* **24**:441.
Sandermann, H., Jr., 1978, Regulation of membrane enzymes by lipids, *Biochim. Biophys. Acta* **515**:209.
Senior, A. E., 1973, The structure of the mitochondrial ATPase, *Biochim. Biophys. Acta* **301**:249.
Shur, B. D., and Roth S., 1975, Cell surface glycosyltransferases, *Biochim. Biophys. Acta* **415**:473.
Singer, S. J., 1974, The molecular organization of membranes, *Annu. Rev. Biochem.* **43**:805.
Singer, S. J., 1976, The fluid mosaic model of membrane structure, in: *The Structure of Biological Membranes* (S. Abrahamsson and I. Pascher, eds.), p. 443, Plenum, New York.
Skulachev, V. P., 1975, Energy coupling in biological membranes: Current state and perspectives, in: *MTP International Review of Science*, Vol. 3 (E. Racker, ed.), p. 31, University Park Press, Baltimore.
Skulachev, V. P., 1977, Transmembrane electrochemical H^+ potential as a convertible energy source for the living cell, *FEBS Lett.* **74**:1.
Slater, E. E., 1977, Mechanisms of oxidative phosphorylation, *Annu. Rev. Biochem.* **46**:1015.
Sottocasa, G. L., 1976, The isolation of mitochondria and their membranes, in: *Biochemical Analysis of Membranes* (A. H. Maddy, ed.), p. 55, John Wiley and Sons, New York.
Strittmatter, P., Spatz, L., Corcoran, D., Rogers, M. J., Setlow, B., and Redline, R., 1974, Purification and properties of rat liver microsomal stearyl coenzyme A desaturase, *Proc. Natl. Acad. Sci. USA* **71**:4565.
Strittmatter, P., and Rogers, M. J., 1975, Apparent dependence of interactions between cytochrome b_5 and cytochrome b_5 reductase upon translational diffusion in dimyristoyl lecithin liposomes, *Proc. Natl. Acad. Sci. USA* **72**:2658.
Stumpf, P. K., 1969, Metabolism of fatty acids, *Annu. Rev. Biochem.* **38**:159.
Tanford, C., and Reynolds, J. A., 1976, Characterization of membrane proteins in detergent solutions, *Biochim. Biophys. Acta* **457**:133.
Whaley, W. G., 1975, *The Golgi Apparatus*, Springer-Verlag, Vienna.
Whaley, W. G., Dauwalder, M., and Kephart, J. E., 1972, Golgi apparatus: Influence on cell surface, *Science* **175**:596.
Wykle, R. L., and Snyder, F., 1976, Microsomal enzymes involved in the metabolism of ether-linked glycerolipids and their precursors in mammals, in: *The Enzymes of Biological Membranes*, Vol. 2 (A. Martonosi, ed.), p. 87, Plenum, New York.
Zwaal, R. F. A., Roelofsen, B., and Colley, C. M., 1973, Localization of red cell membrane components, *Biochim. Biophys. Acta* **300**:159.

10

The Importance of Phospholipid–Protein Interactions for Regulation of the Activities of Membrane-Bound Enzymes

David Zakim and Donald A. Vessey

I. Introduction

Soluble enzymes have been shown to undergo changes in conformation which result in modification of their kinetic properties. These changes can result from alterations in the enzyme's environment or from the binding of soluble ligands. Both of these well-known phenomena are useful for the dynamic regulation of rates of substrate flux in various metabolic pathways (Atkinson, 1966; Stadtman, 1966; Koshland, 1970). Interest in this type of regulatory mechanism has focused on the properties of soluble proteins and enzymes, but the same general principles are likely to apply also to proteins which are bound to membrane structures. In fact, the number of parameters which potentially can alter protein structure—and thereby protein function—is greater for membrane-bound enzymes than for soluble enzymes, since particulate enzymes are likely to be in contact with or partially embedded in the

---

David Zakim and Donald A. Vessey • Departments of Medicine and Pharmacology, University of California, San Francisco, California 94122; Liver Study Unit, Department of Medicine, Division of Molecular Biology, Veterans Administration Hospital, San Francisco, California 94121.

hydrophobic portion of the membrane bilayer and exposed to an aqueous environment (Singer and Nicolsen, 1972). Furthermore, besides the binding of soluble ligands, changes in the gross composition or structure of the lipid bilayer could alter the properties of proteins which interact with it.

As early as 1954, Beaufay and deDuve (1954) observed that the treatment of liver microsomes with phospholipases A and C inactivated microsomal glucose-6-phosphatase (G-6-Pase). Subsequently, other microsomal enzyme systems have been reported to be inactivated by treatment with phospholipases (Emmelot and Bos, 1967; Martonosi et al., 1968; Abou-Issa and Cleland, 1969; Atwood et al., 1971) or by extraction with organic solvents (Jones et al., 1969); and in some cases, the inactivated enzymes appear to be reactivated by treatment with phospholipid micelles (Martonosi et al., 1968; Abou-Issa and Cleland, 1969; Atwood et al., 1971; Jones et al., 1969; Duttera et al., 1968). On the basis of such results it has been suggested that several tightly-bound membrane enzymes have an absolute requirement for the phospholipid environment of the native membrane in order to be active. Besides appearing to explain the effects of phospholipases and organic solvents on the behavior of membrane-bound enzymes, an additionally attractive feature of this hypothesis is in offering a rationale for the difficulties encountered in preparing "soluble," membrane-free forms of several membrane-bound enzymes. It is becoming apparent, however, that previous interpretations of the effects of phospholipases, phospholipids, and organic solvents on the activities of tightly-bound enzymes are not completely correct (Zakim, 1970; Vessey and Zakim, 1971), since modification of membrane lipids can also inactivate or alter the stabilities of some membrane-bound enzymes. It is clear that there is a dynamic rather than a static interaction between lipid and protein components in the membrane.

A variety of physical techniques such as NMR, electron spin resonance (ESR), X-ray diffraction, fluorescence spectroscopy, and differential thermal analysis have been applied to the study of natural and synthetic membranes. The usefulness of these methods has been discussed in several reviews (Chapman, 1968; Luzzati, 1968; Luzzati et al., 1969; Chapman and Dodd, 1971; Gaffney, 1974; Azzi, 1974; Lee et al., 1974a) which indicate that considerable progress has been made in understanding the structure and properties of membranes. Nevertheless, the manner in which the lipid and protein portions of membranes interact is unresolved. For this reason it is not possible to discuss in detail the mechanisms which control the function of membrane-bound enzymes. This chapter, therefore, presents evidence useful in establishing a chemical basis for postulating that the properties of membrane-bound proteins depend on lipid–protein interactions and that dynamic regulation of membrane function can be affected by the perturbation of membrane lipids. The early sections of the chapter emphasize the general problem of those physical properties of the lipid phase of membranes that are subject either to short- or to long-term modification, and on phospholipid–protein interactions within membranes. Data pertaining to specific soluble and particulate enzymes whose activities can be modified by treatment with phospholipids or by perturbation of membrane lipids are considered later. These data are included to indicate the ways in which phospholipids affect the properties of enzymes, rather than as an encyclopedic review of enzymes and proteins which have been shown to interact with phospholipids. For this reason, we have stressed enzymes with which we have an extensive experience in this laboratory. In addition, we have considered only the problem of enzymes and proteins for which phospholipids have no presumed functions as cofactors or substrates.

II. Effect of Lipid Composition on the Properties of Membranes and Membrane-Bound Proteins

The structure of the lipid portion of membranes is potentially subject to regulation via variation of a number of independent parameters not related to interactions between lipid and protein phases. These parameters are the hydrocarbon chain-lengths of fatty acids and their degree of unsaturation, the nature of phospholipid headgroups, the cholesterol content of the membrane, and the presence or absence of certain ions. Further, it is becoming clear that subtle changes in the physical properties of membrane lipids correlate with changes in the behavior of membrane proteins.

The extent to which the lipid composition of a membrane can influence the behavior of the membrane and the properties of membrane-bound proteins is discussed in this section by considering, separately, the effects of the acyl chains, the phospholipid headgroups, and cholesterol content.

A. Influence of Chain Length and Unsaturation of Phospholipid Fatty Acids on Membrane Structure and Function

It is clear that the hydrophobic acyl chains of the phospholipids contact each other and that the energy which stabilizes this contact is derived from the unfavorable entropy changes associated with the entry of hydrophobic groups into water (Kauzman, 1959). Van der Waals interactions are less important in maintaining this structure but are probably important for restricting the motion of the hydrocarbon chains in the hydrophobic portion of the membrane. Studies of the relationship between the pressure and area of a phospholipid monolayer (Demel et al., 1972a), the electron paramagnetic resonance (EPR) spectra of suitable probes (Keith et al., 1968; Eletr and Keith, 1972), and differential thermal analysis (Chapman, 1968; Chapman and Collin, 1965) indicate that this motion depends, in part, on the nature of the acyl chains. Thus, monolayers of phospholipids containing primarily saturated fatty acids are packed more tightly (less expanded) than similar structures containing unsaturated fatty acids. The angular distortion resulting from a *cis* configuration about a double bond is especially effective in expanding the lipid phase. Also, membranes containing short- as compared to long-chain acyl groups have more expanded structures. The relationship between fatty acid length and pressure–area curves for membranes presumably results from the number and strength of Van der Waals contacts between hydrocarbon chains. In association with the looser packing of membranes containing unsaturated as compared to saturated fatty acids, EPR studies with nitroxide-labeled fatty acids reveal greater mobility for the hydrocarbon chains. The extent of motion of these acyl chains and the role of chain length and unsaturation in restricting such motion is reflected also in the temperature at which the hydrophobic portions of a phospholipid bilayer appear to "melt." For example, differential thermal analysis (Ladbrooke and Chapman, 1969) of lipid bilayers reveals endothermic transitions associated with marked changes in the motion of the alkyl hydrocarbon groups reflected by X-ray, EPR, and NMR. Below the transition temperature (or melting point), the molecular motion of the hydrocarbon chain is restricted. Above the transition temperature there is a considerable increase in the mobility of individual fatty acid chains. Transition temperatures are lower for membranes containing short, branched, or unsaturated acyl groups than for those with long-chain, saturated fatty acids.

That the fatty acid composition of membranes influences their properties has been demonstrated in several laboratories. In liposomes, which are prepared by gentle but extensive sonication of phospholipid mixtures, permeability to electrolytes and nonelectrolytes is modified by varying the fatty acids of the phospholipids from which the liposomes are made (DeGier et al., 1970; Chen et al., 1971; Scarpa and DeGier, 1971; Demel et al., 1972a,c; DeKruyff et al., 1972; Hsia and

Boggs, 1972). For example, membranes containing relatively large amounts of unsaturated fatty acids are more permeable to glucose (Chen et al., 1971), glycol, glycerol, and erythritol (Demel et al., 1972a) than similar membranes containing saturated fatty acids (McElhaney et al., 1973). The rate of leakage of Rb^+ from liposomes is also greater in those containing unsaturated as compared to saturated fatty acids. Further, fatty acid composition is important not only for passive diffusion of Rb^+ through the membrane, but for its facilitated diffusion as well. In the presence of valinomycin, Rb^+ and K^+ transport are increased in membranes. There is selectivity in valinomycin–lipid interactions, however, in that valinomycin-facilitated diffusion of K^+ from liposomes is relatively greater for those liposomes containing saturated fatty acids (Scarpa and DeGier, 1971), whereas for Rb^+ the valinomycin effect is greater in membranes containing unsaturated fatty acids (DeGier et al., 1970). Although liposomes composed of phospholipids containing primarily unsaturated fatty acids are, generally, more permeable than those composed of phospholipids containing saturated fatty acids, this does not apply to all ions. Data reported by Scarpa and DeGier (1971) are consistent with the idea that the extent of unsaturation has relatively little effect on Na^+ and H^+ permeability. Thus, variation in the fatty acid composition of liposomes can lead to the production of membranes which discriminate between different cations.

The chain length and unsaturation of phospholipid fatty acids also influence the behavior of protein-catalyzed membrane functions in biological membranes. By using mutant strains of *Escherichia coli* which cannot synthesize unsaturated fatty acids, it is possible to grow *E. coli* with specific phospholipid fatty acid compositions. Wilson and Fox (1971) have examined galactose and glucose transport in these organisms as a function of the degree of saturation of fatty acids added to the growth media. The rates of glucose and galactose transport were measured at various temperatures, and the data were plotted according to the Arrhenius equation. Discontinuities in these plots occurred at transition temperatures characteristic for the fatty acid compositions of the membrane specified by the growth conditions of the organism. Since the slope of an Arrhenius plot is a measure of the activation energy for a reaction, discontinuities reflect a change in behavior of the proteins catalyzing the process. Although it is not possible to isolate the transport process in order to demonstrate the linearity of Arrhenius plots in the absence of the membrane, such plots are linear over temperatures of 3–40°C for most reactions catalyzed by soluble enzymes, and it is reasonable to conclude from these data that changes in the mobility or phase equilibria (see below) of the *E. coli* phospholipids, resulting from changes in the fatty acid composition of the membrane, lead to modifications of the properties of the glucose and galactose transport proteins. Subsequently, a large number of similar studies have been conducted (cf. Machtiger and Fox, 1973).

The effects of variations in the fatty acid composition of membranes on the function of membrane-bound enzymes have been investigated in limited instances, under physiological conditions. For example, with mutant Saccharomyces cerevisiae species which cannot synthesize unsaturated fatty acids, depletion of unsaturated fatty acids inhibits mitochondrial oxidative phosphorylation (Haslam, 1971). This does not result from decreased synthesis or activity of any mitochondrial enzyme but represents an uncoupling between substrate and NAD/NADH metabolism and the ability to synthesize ATP. It is also of interest that the morphology of Mycoplasma laidlawii is susceptible to variation as a function of the fatty acid composition of its membranes (Steim et al., 1969).

B. Influence of Phospholipid Headgroups on Membrane Structure and Function

In addition to acyl chains, the phospholipid headgroups should contribute to the stability of phospholipid bilayers as reflected by the considerably higher melting points (200°C) of pure phospholipids than of fatty acids (60°C). Also, the type of headgroup is important for the properties of a phospholipid membrane. For example, monolayers of phosphatidylethanolamine are less expanded than those derived from phosphatidylcholine; and the melting point of the solid, pure phospholipid is lower for the former than for the latter, irrespective of the length or unsaturation of the fatty acids (Williams and Chapman, 1970).

Because phospholipids have ionizable headgroups, the membrane structure should be destabilized by repulsion between similar charges under certain conditions. Over the pH range of 3.5 to 10.0, however, phosphatidylcholine is a zwitterion and thus bears no net charge (Williams and Chapman, 1970). In fact, evidence suggests that phosphatidylcholine has an internal salt link between its quaternary ammonium group and its negatively charged phosphate group (Shah and Schulman, 1967a,b). On the other hand—probably because of charge repulsion—increasing the pH from 4.0 to 7.0 expands monolayers of phosphatidylethanolamine (Hsia and Boggs, 1972); indeed, large changes in the fluidity of phosphatidylethanolamine phases in water are seen when the pH is raised to levels higher than 8.0. And conversely, as expected from the properties of the ionizable groups of phosphatidylcholine, variation of pH in this range has no effect on the packing of membranes composed of phosphatidylcholine (Van Deenen et al., 1962). In microsomal membranes from liver, which contain large amounts of both phosphatidylcholine and phosphatidylethanolamine, treatment with alkaline pH increases the activity of G-6-Pase (Stetten and Burnett, 1966) and UDP-glucuronyltransferase (Vessey and Zakim, 1971). The shapes of pH activity curves for these enzymes over the pH range 8.5–12.0 are identical, suggesting that activation is not mediated via a direct action on these two enzymes, but indirectly, via modification of the lipid portion of the membrane.

An important observation indicating the interdependence of the properties of the hydrophobic and hydrophilic portions of the phospholipids is that the extent of unsaturation of its fatty acids modifies the reactivity of phosphatidylcholine with Ca^{2+}. Thus, saturated acyl chains, presumably by allowing for closer packing in the membrane, destabilize the internal salt structure and lead to a greater interaction between Ca^{2+} and the phospate ion of the phosphatidylcholine headgroup (Shah and Shulman, 1967a; Abramson and Pisetsky, 1972). Whatever the exact chemical basis for these results, they have important implications for protein–lipid interactions, since they indicate that the nature of the acyl groups of the membrane phospholipids could modify the function of a membrane protein, even if the interaction of the ions is with the phospholipid headgroups.

Treatment of nerve membranes with phospholipase C, producing hydrolysis of 70–80% of the membranes' phospholipid headgroups, leaves the membranes apparently intact according to evidence from electron micrographs. It therefore appears that membrane structure is reasonably well-preserved even in the absence of the charged headgroup; but treatment with phospholipase C leads to swelling of the membrane, and studies with EPR labels indicate a considerable increase in fluidity of the hydrocarbon portions of nerve membranes after treatment with phospholipase C (Simpkins et al., 1971). Anisotropy of EPR signals (i.e., the difference between signals from membranes aligned perpendicular or parallel to the applied magnetic field), which can be taken as evidence for bilayer character, decreases after treatment of membranes with phospholipase C. Thus, treatment with phospholipase C does destroy extensive regions of bilayer in the native membrance, indicating again the importance of both hydrophobic and hydrophilic portions for native membrane structure.

C. Influence of Cholesterol on the Properties of Phospholipid Membranes

Cholesterol has a relatively rigid structure, with less random motion than long-chain acyl groups in the liquid-crystalline state. As a result, the insertion of cholesterol into phospholipid membranes can condense these membranes (Leathes, 1925) and can restrict the motion of their alkyl side chains (Chapman and Penkett, 1966; Oldfield and Chapman, 1971; Dark et al., 1972; Keough et al., 1973). On the other hand, cholesterol increases the mobility of the hydrocarbon chains of lipids in the gel phase (Dark et al., 1972). In a sense, addition of cholesterol tends to obscure the differences between membranes in the gel and liquid-crystalline states in a functional as well as structural manner. For example, insertion of cholesterol into liposomes containing unsaturated (fluid) fatty acids leads to a reduction in their permeability (DeKruyff et al., 1972; Demel et al., 1972a,c). Cholesterol generally enhances the permeability of liposomes containing saturated lipids, although high concentrations of cholesterol gave anomalous results (Inoue, 1974).

Cholesterol also decreases the heat absorbed in the transition from the gel to the liquid-crystalline state (melting) of a phospholipid:cholesterol membrane, such that by 33 mole% cholesterol, no transition is apparent (Oldfield and Chapman, 1971; Ladbrooke and Chapman, 1969). Apparently, a cholesterol molecule is able to complex with a phospholipid molecule (Dark et al., 1972) to yield a complex which is more fluid than the gel state and less fluid than the liquid-crystalline state.

The effect of cholesterol on the properties of mono- or bilayers is mediated only in part by interactions between the hydrophobic portions of the cholesterol molecule and the hydrocarbon chains of phospholipid fatty acids. The evidence for this is that, for membrane activity, a sterol must have the 3-hydroxyl group in the β orientation (Demel et al., 1972a,b,c). Thus, epicholesterol (3-α-hydroxyl cholesterol) has no effect on the mean molecular area of a phospholipid membrane and only slightly reduces the temperature for its endothermic transitions, even though the epicholesterol is incorporated into the membrane. Also, some mycoplasma species have an absolute growth requirement for cholesterol which is not satisfied by the 3-α-hydroxy analog (DeKruyff et al., 1972). Data from X-ray and infrared experiments indicate that the 3-β-hydroxyl group of cholesterol is inserted into the polar region of the lecithin molecule (Zull et al., 1968; Rand and Luzzati, 1968). Hydrogen bonding between the 3-β-hydroxyl and the oxygen of the phosphate headgroups, or water, could account for the importance of this hydroxyl for membrane structure (Long et al., 1970). Of interest in this regard is that cholesterol restricts the increase in the molecular motion of the phospholipid fatty acids and the area of a phosphatidylethanolamine monolayer which is normally produced by increasing the pH from 8 to 10 (Hsia and Boggs, 1972).

D. Inhomogeneous Nature of the Lipid Phase of Biological Membranes

In the previous sections, we have primarily discussed pure phospholipid systems in order to illustrate general concepts. Pure lipids exist in either a solid (gel) phase or a melted (liquid-crystalline) phase. More complex phase behavior is encountered in heterogeneous-lipid systems. For example, a simple vesicle containing two types of phospholipids, differing in their acyl chain compositions by an unsaturated bond or by two methylene groups, has two critical temperatures. The first is the temperature below which all lipids are in the solid phase; just above this temperature, the lowest-melting component becomes liquid-like. The second critical temperature is the temperature above which the most solid species is also completely melted, thereby yielding a relatively homogeneous, liquid-crystalline vesicle. The bilayer is inhomogeneous in the region between these two temperatures, containing both a solid and a liquid region with differing chemical composition (Ladbrooke and Chapman, 1969; Shimshick and McConnell, 1973).

Biological membranes typically contain 100–200 chemically distinct lipid molecules. As a result of this compositional heterogeneity, biological membranes do not display the distinct phase separations observed with homogeneous vesicles. Rather, the phase transition from gel to liquid crystalline state is extremely complex in biological membranes and occurs over a broad temperature range (Steim et al., 1969; Melchoir et al., 1970; Esfahani et al., 1971). Nevertheless, phase transitions in biological membranes are characterized by temperature ranges in which a liquid-crystalline region coexists with regions of gel-phase lipids (Esfahani et al., 1971; Shechter et al., 1974). Such an inhomogenous lipid phase appears not to be physiologically essential, at least in microorganisms. Their membranes can be entirely in the liquid crystalline state at the growth temperature without apparent effect (McElhaney, 1974). However, microorganisms do adjust their lipid composition in response to changes in growth temperature in order to maintain the fluidity of their membranes in some constant range. Complete solidification of the membrane does not permit growth of the microorganism (McElhaney, 1974). LM cells in culture show a similar tendency to maintain a constant value for membrane fluidity (Glaser et al., 1974). Finally, it should be pointed out that phase transitions can occur in biological membranes that are in the liquid state (melted) owing to the formation of transient clusters of specific lipids (Lee et al., 1974).

E. Sensitivity of Membrane-Bound Proteins to Temperature-Induced Changes in Membrane Lipids

In Section II.A, it was pointed out that the Arrhenius plots of sugar transport rates have discontinuities at the same temperature at which spin-label probes of the bulk lipid phase undergo discontinuous changes. This implies a sensitivity on the part of the protein to changes in the mobility (fluidity) of membrane lipids. In this section, we will examine this point in greater detail.

Arrhenius plots of log v vs. $1/T$ for succinic dehydrogenase in rat liver mitochondria are not linear but have a discontinuity at approximately 25°C. If, however, the mitochondria are treated with detergent in order to increase the fluidity of their membrane lipids, the Arrhenius plot gives a single straight line over the temperature range of 5–35°C (Raison et al., 1971b). Similar results apply for the oxidation of succinate by mitochondria from sweet potato and for a variety of other oxidizable substrates. On the other hand, with mitochondria from potato or fish liver, Arrhenius plots for succinic dehydrogenase are linear (Raison et al., 1971a). Known differences in fatty acid composition between rat and fish liver mitochondria, or between mitochondria from chilling-sensitive (sweet potato) and chilling-insensitive plants (potato), suggest that the lipid phases from these preparations would have different physical properties, since mitochondria from fish liver and chilling-insensitive plants contain more unsaturated fatty acids than mitochondria from rat liver and chilling-sensitive plants. This may account for the variability in behavior of succinic dehydrogenase in different species. This postulate was confirmed, in fact, in studies comparing the temperatures at which discontinuous changes occur in spectral parameters for ESR probes of the lipid bilayer, with Arrhenius plots for the activity of succinic dehydrogenase (Raison et al., 1971a). The critical temperatures obtained from the molecular motion parameter (τ_0) of a 12-nitroxide stearate spin label (12 NS) in rat liver and sweet potato mitochondrial membranes are the same as the temperature for which discontinuities occur in Arrhenius plots for the activity of succinic dehydrogenase. In contrast to the results with mitochondria from rat liver and sweet potato, Arrhenius plots for the τ_0 of 12 NS are linear in mitochondria from fish liver or potato.

Correlations between critical temperatures for membrane lipid phases determined with EPR techniques and discontinuities in Arrhenius plots have also been shown for the ATPase from sarcoplasmic reticulum (Eletr and Inesi, 1972), for UDP-glucuronyltransferase, and for G-6-Pase (Eletr et al., 1973) from liver microsomes. In the case of the microsomal membranes, perturbation of the membrane lipids by treatment with detergents or phospholipase A leads to linear Arrhenius plots for both enzyme activities and τ_0 between 5° and 30°C. For UDP-glucuronyltransferase, the phase change in the lipids also results in a loss of substrate specificity and a loss of sensitivity to an allosteric effector (Vessey and Zakim, 1974).

Using X-ray diffraction, Wakil and his coworkers (Esfahani et al., 1971) observed the temperature-induced melt in membranes from *E. coli* and sought to associate this with discontinuities in Arrhenius plots for proline uptake and the activity of succinic dehydrogenase. The melt occurs over a broad temperature range, and the breaks in the Arrhenius plots for proline uptake and succinic dehydrogenase did not agree with the onset or completion of the melt. Since the *E. coli* used in these experiments were not capable of synthesizing unsaturated fatty acids, the fatty acid composition of the membranes was varied by controlling the fatty acid composition of the growth medium, with the result that the transition temperature for the uptake of proline could be raised or lowered in the expected way. Thus, as fatty acids of increasing unsaturation were added to the growth media, the temperature range for the transition decreased. This was not the case, however, for succinic dehydrogenase activity. The anomalous results with succinic dehydrogenase remain unexplained, but the data are consistent with the hypothesis of an unequal distribution of phospholipids of differing compositions in the membrane, and of selective effects of certain regions of a membrane on specific proteins. Further evidence for a heterogeneous distribution of membrane components and selectivity in protein–lipid interactions is available also in the work of Mavis and Vagelos (1972), who determined the effect of membrane fatty acid composition on the linearity of Arrhenius plots for glycerol-3-P acyltransferase, 1-acylglycerol-3-P acyltransferase, and glycerol-3-P dehydrogenase. The first enzyme had a linear Arrhenius plot and was independent of the membrane fatty acid composition. The second enzyme had a linear Arrhenius plot but its slope was dependent on the fatty acid composition. The third enzyme had a nonlinear Arrhenius plot, the slope of which was sensitive to the fatty acid composition. Recent studies of model membrane systems also provide evidence for heterogeneity in the distribution of phospholipids in synthetic membranes (see Section II.D).

Thus, there appear to be temperature-induced changes within the lipid bilayer of some membranes that can cause certain membrane-bound proteins to alter their conformation. What is striking is that these changes are very subtle. The nature of the changes detected by the spin label probes remains unknown, but these changes are often not associated directly with a melting of the membrane lipids. Differential scanning calorimetry of mitochondrial and microsomal membranes does not reveal any endothermic processes between approximately 5° and 50° C (Blazyk and Steim, 1972). Lee et al. (1974b) have presented evidence that in microsomal and mitochondrial membranes, the spin label probes and membrane-bound proteins are sensing a temperature-induced clustering of specific lipids. Regardless of the mechanism, the end result is that the spin labels inserted into the lipid bilayer are experiencing a discontinuous change in the motion of the lipid molecules in their immediate environment at certain specific temperatures. This abrupt change in the "fluidity" of certain microenvironments in the lipid bilayer correlates well, in some cases, with changes in the properties of certain membrane-bound proteins and thus appears to be the cause of the presumed conformational change in the protein. A model for the translation of fluidity changes in the lipids into a conformational change in a membrane-bound protein is presented in Section VII.

In most mammals, the normal body temperature is far above that at which membrane lipids "melt." Motions within the membrane appear to be important for membrane function, however, and there are adaptive mechanisms for preserving this parameter. For example, the average chain length of the phospholipid fatty acids of rat liver decreased when essential fatty acids were removed from the diet; as a result of the substitution of short-chain, saturated fatty acids for long-chain, polyunsaturated fatty acids, the loss of membrane fluidity associated with removal of polyunsaturated fatty acids from the diet is minimized (see Vessey and Zakim, 1974). Another type of adaptive response in the fluidity of membranes is seen in hibernating squirrels. In liver mitochondria from animals living at 23°C, Arrhenius plots for succinic dehydrogenase show a discontinuity at approximately 25°C, but in hibernating squirrels (living at 1°C) Arrhenius plots for succinic dehydrogenase are linear over the temperature range of 5° to 35°C. As squirrels hibernate, there is a change in the mitochondrial membrane lipids which eliminates the transition (Raison and Lyons, 1971; Keith et al., 1975).

III. The Effect of Proteins on the Properties of Membrane Lipids

In view of the effect of lipids on the properties of membrane-bound proteins, the latter also might alter the behavior of the lipid portion of the membrane. Studies aimed at elucidating how proteins interact with pure lipids are limited in scope, partly because of the technical difficulties in isolating appropriate membrane-bound proteins. Consistent with the idea of interactions between apolar portions of both the lipid and protein components of the membrane, however, is the observation that two different forms of cytochrome b_5 can be prepared from liver microsomes, depending on the method used for "solubilizing" this protein. After treatment of microsomes with lipase or trypsin, a cytochrome b_5 with molecular weight of approximately 11,000 can be purified (Strittmatter, 1967). If, on the other hand, microsomes are first treated with detergent, a cytochrome b_5 with a molecular weight of 16,700 is isolated (Spatz and Strittmatter, 1971). The difference between these two forms of cytochrome b_5 is a "hydrophobic appendage of 40 amino acids" in the detergent-extracted form of the enzyme. The property of having a high percentage of apolar amino acid side chains does not appear to be limited to cytochrome b_5, since the polarity of several membrane proteins—as calculated by summing the mole fractions of polar amino acids—is less than for soluble proteins (Capaldi and Vanderkooi, 1972).

The insertion of the hydrophobic core of cytochrome b_5 into a lipid bilayer was found to result in an immobilization of 2–4 molecules of lipid per molecule of cytochrome b_5 (Dehlinger et al., 1974). This so-called "boundary lipid" was also seen when cytochrome oxidase was added to lipid vesicles (Jost et al., 1973).

The addition of a variety of proteins to lipid monolayers increases their pressure at constant surface area (Eley and Hedge, 1956, 1957; Snart and Sanyal, 1968). This increase in surface pressure is presumably due to the insertion of nonpolar amino acid side chains into the hydrophobic portions of the membrane, but this is not yet proven. More direct evidence that proteins alter the properties of the membrane lipids is apparent in studies of the fluorescence of 8-anilino-1-napthalene-sulfonic acid (ANS) in mitochondrial membrane, as modified by the addition of substrates such as ATP and O_2 (Azzi et al., 1969; Chance and Lee, 1969). Although it seems that the binding of substrates to discrete enzymes influences the structure of the entire membrane in this case, care must be taken in the interpretation of studies of ANS fluorescence since the exact location of the bound ANS is not known.

Another approach to defining the extent of the effect of proteins on the lipid phase is by comparing the physical properties of pure extracted lipids with those of the intact membrane. In *E. coli*, for example, the transition temperatures for dispersions of pure lipids extracted from the membrane are not identical to those for the intact membrane (Esfahani et al., 1971). In addition, Eletr and Inesi (1972), using EPR techniques, have found transitions in the fluidity of the lipid phase of sarcoplasmic reticulum at 22° and 40°C. The melt at 22°C is present also in dispersions of lipids extracted from sarcoplasmic reticulum, but the transition at 40°C is absent. On the other hand, studies with spin-labeled proteins and the observation that thermal denaturation removes the 40°C transition in membranes indicate that the phase change in membrane lipids at 40°C results from a conformational change in the proteins. An identical effect is seen in spin label studies with microsomes (Eletr et al., 1973). These results suggest that proteins can affect the bulk properties of the lipids and thereby provide an explanation as to how protein–protein cooperativity might be mediated by the lipid phase. This role in cooperativity is an additional way in which lipids could regulate the activity of membrane-bound enzymes.

IV. The Effect of Phospholipids on the Activities of Soluble Enzymes and Proteins

Reversible structural changes in soluble proteins, as a result of the addition of nonpolar compounds, have been demonstrated for a variety of proteins (Featherstone et al., 1961; Wishnia, 1962; Wetlaufer and Lovrien, 1963). Based on current knowledge of protein structure and structure–function interrelations, this is not surprising, nor should it be surprising to find that nonpolar substances enhance the activities of some soluble enzymes which appear to function in vivo in the cytoplasm.

An interesting aspect of the problem of fatty acid synthesis in liver-cell supernatant is stimulation of activity by microsomes. Explanations of this phenomenon have been based on the effect of microsomes on transacylase reactions but have never been completely convincing because of the inability to demonstrate expected effects on the concentration of long-chain acyl-CoA derivatives (Zakim, 1973). On the other hand, recent studies of the effect of microsomes on fatty acid synthesis have shown that they contain a heat-stable factor, extractable with lipid solvents, which increases the activity of acetyl-CoA carboxylase severalfold (Foster and McWhorter, 1969). This factor is a phospholipid, and its stimulation of acetyl-CoA carboxylase is prevented by prior treatment with phospholipase C. Purified phospholipids, not derived from microsomes, are also effective in stimulating the activity of acetyl-CoA carboxylase. The exact physiological significance of changes in the functional status of acetyl-CoA carboxylase in vitro associated with changes in its physical properties are uncertain, but the activity of this enzyme is regulated in vitro by a polymer–monomer equilibrium, the polymeric form being the high-activity form of the enzyme (Moss and Lane, 1971). Whether phospholipids can alter the equilibrium in the polymer–monomer distribution of acetyl-CoA carboxylase was not examined, but the activity of the enzyme after incubation with phospholipids was equal to that after incubation with citrate, a known promoter of polymerization. Unfortunately, the combined effect of citrate plus phospholipids was not studied.

A situation analogous to that for acetyl-CoA carboxylase is seen with pyruvate oxidase from *E. coli* (Cunningham and Hager, 1971). The activity of this soluble flavoprotein, when measured with the unphysiologic electron acceptor ferricyanide, is quite low in the absence of a cellular particulate fraction. Activity is increased markedly, however, by treatment of the enzyme with such surface-active agents as phospholipids, fatty acids, and detergents. The source of the phospholipids can be cell envelopes, but this is not an absolute requirement. What is important, however, is the requirement for a micellar form of phospholipid for full activation of the oxidase. The degree of unsaturation of the phospholipid fatty acids may be important for activation of the enzyme, since the conversion to micelles of phospholipids containing saturated fatty acids is less extensive than for those containing unsaturated fatty acids (Cunningham and Hager, 1971). Although fatty acids of varying chain length and unsaturation had different effects on the activity of pyruvate oxidase, these differences have not been correlated as yet with variable solubilities and critical micelle concentrations for the fatty acids used.

A more interesting interaction between an enzyme and phospholipids, because of the complexity of the kinetics, is seen with phenylalanine hydroxylase from rat liver. With its naturally occurring substrates, phenylalanine and tetrahydrobiopterin, plots of v vs. [phenylalanine] are sigmoidal (Kaufman, 1971); propanol converts these kinetics to typical Michaelis–Menten form and also increases the activity of the enzyme at V_{max} (Fisher and Kaufman, 1972). Low concentrations of lysolecithin and lysophosphatidylserine have a similar effect, as do high concentrations of phosphatidylserine and sphingomyelin. Phosphatidyl- and lysophosphatidylethanolamine are without effect. Lecithin inhibits the enzyme, and the lecithin-induced inhibition is prevented by addition of lysolecithin. Detergents also activate the enzyme, but more sodium dodecylsulfate than lysophosphatide is needed for maximum activation. High concentrations of SDS (0.5 mM) inhibit phenylalanine hydroxylase, an effect not seen with high concentrations of lysophosphatides.

V. Reconstituted Systems

β-Hydroxybutyrate dehydrogenase can be extracted from rat liver mitochondria and prepared as a homogeneous protein. In this form, it is without enzymatic activity, but the holoenzyme is reconstituted by the addition of phospholipid mixtures (Jurtshuk et al., 1963; Gotterer, 1967a,b) in the presence of a thiol reagent. A more active form of the reconstituted enzyme is prepared if treatment with phospholipids and a thiol is carried out in the presence of NAD^+ or NADH. The holoenzyme can be fractionated reversibly, by acetone extraction, into the apoenzyme and the phospholipid component.

Little is known about the physical basis for the effects of phospholipids on the catalytic parameters of β-hydroxybutyrate dehydrogenase, but the phospholipids needed to regenerate the activity of this enzyme have been examined in some detail. Vesicular bilayers of phosphatidylcholine are the most effective in increasing the activity of β-hydroxybutyrate dehydrogenase, although there is variability in the extent of activation achieved with different preparations of phosphatidylcholine (Grover et al., 1975). For example, unsaturated lecithins stabilize the activity of β-hydroxybutyrate dehydrogenase to a greater extent than saturated lecithins. This difference is not related to variability in the critical micelle concentrations of the lipid phases or to related properties. A phosphatidylcholine with short fatty acid chains is a less effective activator of the apo-dehydrogenase than a long-chain phosphatidylcholine. This difference was related to the binding of the different phospholipids by the enzyme. Interestingly, the addition of cholesterol to phospholipid micelles decreases the maximum activity of the lipid-activated enzyme system (Gotterer, 1967a,b).

That phosphatidylcholine is the only phospholipid that activates the apoenzyme of β-hydroxybutyrate dehydrogenase has been interpreted as evidence for interactions between the apoenzyme and the phospholipid headgroup. This is difficult to reconcile, however, with the observations that sphingomyelin is ineffective in reconstituting dehydrogenase activity, that the source of phosphatidylcholine is important for the catalytic properties of the reconstituted dehydrogenase, and that the degree of unsaturation of the phospholipid fatty acids is important for reconstitution (Jurtshuk et al., 1963). The data suggest that activation of the apo-β-hydroxybutyrate dehydrogenase by phospholipids is also related to interactions between the hydrophobic portions of the lipid and protein phases. Unfortunately, the effect of salt on activation has not been studied.

Like β-hydroxybutyrate dehydrogenase, malate dehydrogenase can be isolated from mitochondria, but in aqueous phase the purified enzyme is extremely unstable. Analysis of the amino acid composition of the "soluble" enzyme reveals a large number of nonpolar side chains (Criddle et al., 1961). Since those procedures which lead to instability also remove phospholipids, it was proposed that interactions between the apolar portions of malate dehydrogenase and phospholipids stabilize the enzyme (Criddle et al., 1961; Callahan and Kosicki, 1967). This notion was confirmed by Callahan and Kosicki (1967), who also found that phospholipids enhanced the activity of the enzyme under a variety of treatment conditions.

VI. Alteration of the Properties of Tightly-Bound Membrane Enzymes by Perturbation of Their Membrane Lipid Environment

The major difficulty in examining the effects of phospholipids on the properties of most membrane-bound enzymes is their firm attachment to the membrane, which precludes studies with highly purified systems. On the other hand, it is likely that if the technology were available for purification of these enzymes in homogeneous form, some enzymes would have properties completely different from those of the enzyme in intact membranes. It is impossible, therefore, to test conclusions and hypotheses derived from experiments with intact membranes in reconstitution experiments. For this reason, emphasis in the study of tightly-bound membrane enzymes should be directed to intact systems. In this regard, the basic design of experiments is to perturb the lipid environment and determine the effects of this perturbation on the properties of the enzyme of interest. This has been done with a variety of proteins, and details are given below for a few of these studies.

Perhaps one of the most interesting aspects of the function of microsomal enzymes is that their catalytic properties may not be constant during the life of an animal. For example, $K_{\text{G-6-P}}$ for G-6-Pase can be changed by modification of the dietary (Nordlie et al., 1968) or hormonal status of animals (Segal and Washko, 1959); and K_{aniline} for aniline hydroxylase varies in animals of different ages (Gram et al., 1969). Another interesting problem is that, for many enzymes, the activity at $V_{\max}$ is less in the intact microsome than in microsomes treated with lipid-active agents. The data reviewed below for G-6-Pase and UDP-glucuronyltransferase indicate the manner in which the peculiarities in function of membrane-bound enzymes depend on interrelations between enzymes and their lipid environments.

A. Glucose-6-Phosphatase

As mentioned in the Introduction, G-6-Pase was one of the first enzymes shown to be inactivated by treatment with phospholipases (Beaufay and DeDuve, 1954). Studies in this laboratory have demonstrated, however, that treatment of microsomes with phospholipase A does not directly inactivate G-6-Pase (Zakim, 1970). Instead, treatment of microsomes for short periods of time with small amounts of purified phospholipase A from *Naja naja* venom produces an unstable form of G-6-Pase. The phospholipase-A-treated form of G-6-Pase can be stabilized by albumin and phospholipid micelles. In fact, with proper selection of the conditions for treatment of microsomes with phospholipase A, and subsequently with albumin, the activity of G-6-Pase can be increased to levels greater than those in untreated microsomes. The exact role of albumin in activating G-6-Pase that has been treated with phospholipase A is not understood completely, but it is not related to binding of fatty acids released by the actions of phospholipase A.

Since G-6-Pase is not inactivated directly by treatment with phospholipase A, these experiments indicate that this enzyme does not have an absolute dependence for activity on the phospholipid configuration present in the native microsome. More importantly, the activity of G-6-Pase in untreated microsomes is less than its maximum potential activity at $V_{\max}$.

Studies of the action of phospholipase C on the activity of G-6-Pase yield further insights into the mechanism of the effects of phospholipase A on activity. The effects of phospholipase C (*Clostridium welchii*) on G-6-Pase differ from those of phospholipase A in that the action of phospholipase C per se decreases activity, seemingly without inducing instability. Also, the decrease in the activity of G-6-Pase produced by phospholipase C and be reversed by incubation of the phospholipase-C-treated form of the enzyme with phospholipid micelles. However, the combined treatment of microsomes with phospholipase C followed by addition of phospholipid micelles leads to an activity greater than that in untreated microsomes (Zakim, 1970). It seems, therefore, that just as constraint on the activity of G-6-Pase is relieved by alteration of the phospholipid environment by treatment with phospholipase A, such constraint is increased by treatment with phospholipase C. Furthermore, considerable evidence indicates that reversal of the effects of phospholipase C, upon subsequent treatment of G-6-Pase with phospholipid micelles, does not reflect a specific action of phospholipids, in the sense that the micellular phospholipids replaced, functionally and morphologically, those removed by the action of phospholipase. Rather, in this instance, the effects of phospholipids are largely based on their ability to nonspecifically sequester the end products of phospholipase C digestion.

In addition to altering the maximum potential activity of glucose-6-phosphatase, treatment with phospholipase A leads to a reversible increase in the affinity of this enzyme for glucose-6-phosphate (G-6-P), and to differential stabilities in the PP_i-glucose phosphotransferase activity and in the PP_i- and glucose-6-P-phosphohydrolase activities of the enzyme. All of these activities (equations 1a, 1b, 2, and 3, below) have been shown to be common to one enzyme (cf., Nordlie and Arion, 1964).

$$\text{G-6-P} \rightleftharpoons \text{E-P} + \text{glucose} \tag{1a}$$
$$PP_i + E \rightleftharpoons \text{E-P} + P_i \tag{1b}$$
$$\text{E-}P_i + H_2O \rightleftharpoons E + P_i \tag{2}$$
$$\text{E-}P_i + \text{glucose} \rightleftharpoons E + \text{G-6-P} \tag{3}$$

After treatment with phospholipase A, the phosphotransferase activity of G-6-Pase is unstable at pH 5.75 but stable at pH 8, whereas the phosphohydrolase activity is unstable at pH 8.0 but may be stabilized at pH 5.75 by G-6-P or PP_i. As a result, with proper selection of experimental conditions, preparations with variable phosphotransferase and phosphohydrolase activities can be made (Zakim, 1970). As alluded to above, treatment with phospholipase C also has differential actions on these phosphotransferase and phosphohydrolase activities of the enzyme.

Since reactions (1a) and (1b) are common for either the phosphotransferase or phosphohydrolase reactions, and since glucose does not affect K_{PP_i} (Stetten, 1964), the differential effects must be on reactions

(2) and (3). We have not measured K_{H_2O}, but $K_{glucose}$ is unaltered by treatment of the enzyme with phospholipase A. Hence, phospholipid-induced modulations of specificity are primarily directed toward effects on reaction rates rather than toward changes in binding constants. The data indicate that enzyme–phospholipid interactions are important not only for modulation of activity at V_{max} but also for defining the substrate specificity of G-6-Pase.

The effects of perturbation of microsomal lipids on the catalytic properties of G-6-Pase can be interpreted in two different ways. By modifying the microsomal membrane, treatment with phospholipases could "unmask" G-6-Pase to which substrates were not accessible previously. Alternatively, the structure of all the G-6-Pase could have been altered by modification of the environment. The first possibility provides a simple explanation of the "latency" of enzyme activity but does not clarify the nature of selective changes in binding constants or substrate specificity. Also, there is no direct evidence to support it. On the other hand, conformational changes of G-6-Pase in association with changes in kinetic parameters could explain all of the results. The data reviewed in Section II, indicating that lipids can influence protein structure and function, are also compatible with this hypothesis. Perhaps most important, an Arrhenius plot of the phosphohydrolase activity of G-6-Pase has a discontinuity at the transition temperature of the microsomal phospholipids (Eletr et al., 1973), indicating that subtle changes in lipid structure do in fact alter the kinetic parameters of G-6-Pase. Thus, it seems reasonable to conclude that G-6-Pase can exist in several different conformational states with variable kinetic properties. The stabilities of these conformational isomers, relative to each other, depend on their interactions with the lipid environment within the microsomal membrane.

Although structure–function relationships of G-6-Pase are of interest in themselves, it is important to point out that the alterations of activity seen in vitro as a result of treatment with phospholipases have functional counterparts in vivo. Already mentioned above is the K_{G-6-P} change in fasted or alloxan-diabetic animals (Segal and Washko, 1959; Nordlie et al., 1968). In some settings it is also possible to show that apparent increases in the activity of G-6-Pase at V_{max} are not due to the synthesis of new enzyme. In rats treated with the corticosteroid triamcinolone, for example, G-6-Pase activity increases, but this increase is not inhibited by prior administration of actinomycin D (Arion and Nordlie, 1967). On the other hand, whereas in vitro treatment of microsomes with deoxycholate increases the activity of G-6-Pase in microsomes from control animals by about 50 percent, the increase on deoxycholate treatment of microsomes from animals treated with actinomycin D plus triamcinolone is not significant (Arion and Nordlie, 1967). The interpretation to be drawn from these data is that triamcinolone-induced increases in the activity of G-6-Pase are not associated with increased synthesis of enzyme, but result from relief of the constraint imposed on the activity of the enzyme by the microsomal environment.

A contrasting situation prevails in fasted or diabetic rats. In these animals, the baseline activity level of G-6-Pase increases in untreated microsomes as compared with control rats, but the activity after treatment of microsomes with deoxycholate increases to a greater extent. Depending on experimental conditions, it is possible, therefore, to increase or relieve the extent of constraint on the activity of G-6-Pase in vivo by manipulation, with drugs or diet, of the animal's environment. An additional important aspect of the function of G-6-Pase is that in fasted rats the phosphohydrolase activity of G-6-Pase (G-6-P $\rightarrow$ glucose + P_i) increases, while the phosphotransferase (G-6-P + sugar $\rightarrow$ glucose + sugar–P_i) activity does not (Nordlie et al., 1968). It is also possible to produce differential changes in these independent activities of G-6-Pase via perturbation of the microsomal lipids in vitro (Zakim, 1970).

The striking similarities between the catalytic-property changes of G-6-Pase as a result of treatment of whole animals or by perturbation of the structure and composition of microsomal phospholipids in vitro suggest that phospholipid–protein interactions are important for the regulation of activity of this enzyme. We recognize, however, that as cited in the earlier parts of this section, interpretations of experiments conducted with intact microsomes must be drawn cautiously. Many of the problems arising in this regard have been discussed at length in a previous publication (Vessey and Zakim, 1972a), and need not be repeated here.

B. UDP-Glucuronyltransferase

UDP-glucuronyltransferase from hepatic microsomes has several properties in common with G-6-Pase (Zakim, 1970; Vessey and Zakim, 1971; Zakim and Vessey, 1971). Because of its apparent similarity to G-6-Pase, the effects of treatment with phospholipases A and C on the properties of UDP-glucuronyltransferase, assayed with p-nitrophenol as acceptor substrate, have been examined in this laboratory, in order to determine, among other things, whether the results for G-6-Pase are a general property of tightly-bound microsomal enzymes, and, if so, to determine the extent to which there is selectivity in the effects of phospholipases on microsomal enzyme properties.

Treatment of beef liver microsomes with phospholipase A leads to the production of three new forms of UDP-glucuronyltransferase, which are distinguishable on the basis of their kinetic properties and stabilities (Vessey and Zakim, 1971). Each form exhibits greater activity than the enzyme in untreated microsomes, but to varying degrees. Phospholipase A digestion leads first to form I, which is sixfold activated and stable at 37°C. Form II, which is produced by continued treatment of form I with phospholipase A, has approximately the same activity as form I but is unstable at 37°C. In addition, K_{p-NP} increases 10-fold on conversion of form I to form II. Form III is produced spontaneously at 37°C from form II, without the need for further action of phospholipase A, and is the stable decay product of form II. Treatment with phospholipase C also activates UDP-glucuronyltransferase, but to a lesser extent than treatment with phospholipase A. In contrast to the results with phospholipase A, activation by phospholipase C is reversed on subsequent incubation of microsomes with phospholipid micelles. The activity of UDP-glucuronyltransferase is also increased by treatment of microsomes with small amounts of Triton X-100, alkaline pH, and other agents capable of influencing protein–lipid interactions.

Obviously, as with G-6-Pase, the native phospholipid environment is not essential for the full activity in vitro of UDP-glucuronyltransferase, but acts to constrain it. In contrast to G-6-Pase, however, the exact relation between lipids and enzyme activity is more complex for UDP-glucuronyltransferase, since there are more forms of this enzyme than a constrained and a fully active form. Another important point is that despite the basic similarity in the protein–lipid interactions for UDP-glucuronyltransferase and G-6-Pase, careful comparison of the detailed effects of phospholipases on the activities of each enzyme reveal significant differences in their behavior (Vessey and Zakim, 1971). Thus, although phospholipid-induced constraint and its relaxation appear to be general characteristics of some microsomal enzymes in vitro, the details of the interactions between enzymes and microsomal phospholipids vary for different enzymes.

Because of the functional properties of UDP-glucuronyltransferase, it is possible to probe more deeply into the complexities of enzyme–protein interactions with this enzyme than with G-6-Pase. One such property, other than activity, which is susceptible to modification by perturbation of membrane lipids is the reactivity of the –SH groups of UDP-glucuronyltransferase, of which there are three types (Zakim and Vessey, 1972, 1973a; Vessey and Zakim, 1972b). Prior treatment of microsomes with detergents of phospholipases alters the reactivities of the types 2 and 3 –SH. In fact, in some species the type 3 –SH can be titrated only in phospholipase-A-treated microsomes (Vessey and Zakim, 1972b).

UDP-glucuronyltransferase has allosteric regulatory properties, and lipid–protein interactions are important determinants of this aspect of enzyme function. For example, with the p-nitrophenol-metabolizing form of the enzyme from guinea pig liver microsomes, plots of 1/v vs. 1/[UDPGA] are nonlinear, and detailed analysis of this phenomenon indicates that it results from negative cooperativity in the binding of UDP-glucuronic acid (Vessey et al., 1973). The subunit interactions which promote this negative cooperativity, as evidenced by the binding of UDP-glucuronic acid, are mediated, in part, by the microsomal lipids, since after treatment of microsomes with phospholipase A or detergents, the binding of UDP-glucuronic acid follows typical Michaelis–Menten kinetics (Zakim and Vessey, unpublished observations).

The affinity of UDP-glucuronyltransferase for UDP-glucuronic acid is enhanced in the presence of UDP-N-acetylglucosamine, which binds to the enzyme at a site separate from the active site (Vessey et al., 1973). The action of this effector is so great, in fact, that the rate of synthesis of p-nitrophenylglucuronide is nearly independent of the concentration of UDP-glucuronic acid at levels below 5 mM if UDP-N-acetylglucosamine is added to assays. The binding of UDP-N-acetylglucosamine, like that for UDP-glucuronic acid, reveals negative cooperativity, but, after treatment with detergents or phospholipase A, UDP-N-acetylglucosamine no longer alters the properties of the p-nitrophenol-metabolizing form of UDP-glucuronyltransferase. This is not a trivial point, since the rate of synthesis of p-nitrophenylglucuronide at low concentrations of UDP-glucuronic acid is greater for the native enzyme assayed with UDP-N-acetylglucosamine than it is for the phospholipase-A- or detergent-treated forms of UDP-glucuronyltransferase. Further, the detergent and phospholipase-A-treated forms of the enzyme show far greater sensitivity to inhibition by UMP and UDP, which are prevalent in vivo. Also, the phospholipase-A-treated enzyme is inhibited by UDP-glucose and a number of other sugar nucleotides (Zakim and Vessey, 1974). It appears that, although the native phospholipid environment acts to constrain the activity of the p-nitrophenol-metabolizing form of UDP-glucuronyltransferase, it also allows for sensitivity to the allosteric effector UDP-N-acetylglucosamine, limits inhibition of the enzyme in vivo by nucleotides, and preserves substrate specificity at the UDPGA binding site. For this enzyme, therefore, constraint actually provides for maximum catalytic efficiency under in vivo conditions (Zakim and Vessey, 1974).

Selective defects in the rate of glucuronidation of aglycones occur in livers from the Gunn strain of rat (Schmid, 1972). Studies of the properties of UDP-glucuronyltransferase in these rats afford an opportunity, therefore, for examining enzyme–phospholipid interactions for a presumably defective enzyme and also for exploring whether phospholipid–protein interactions might contribute to the pathogenesis of the genetically determined aberrant glucuronidation. With regard to this possibility, in vitro diethylnitrosamine (DEN) raises the abnormally low rates of o-aminophenylglucuronide synthesis in Gunn liver microsomes to normal levels (Stevenson et al., 1968), suggesting that the defect in these animals is regulatory rather than being in the primary structure of the active site of the enzyme. Studies of the effect of perturbation of microsomal phospholipids on the properties of UDP-glucuronyltransferases from Gunn rat liver microsomes demonstrate, in fact, that there are abnormalities in the way in which these enzyme proteins interact with their environments (Zakim et al., 1973b). For example, treatment of liver microsomes from normal Wistar rats with Triton X-100 increases the rate of synthesis of p-nitrophenylglucuronide approximately 20-fold, but has no effect on microsomes from Gunn rats. Also, although relatively large concentrations of Triton X-100 inactivate UDP-glucuronyltransferase from normal rats, the enzyme is unaffected in homozygous Gunn rats even at high concentrations of detergent. Further, treatment of microsomes with phospholipase A stimulates enzyme activity 10-fold in normal rats but is without effect in microsomes from homozygous Gunn rats. Qualitatively similar results are seen for assays with o-aminophenol and o-aminobenzoate—aglycones for which the rates of glucuronidation are low in Gunn rats and which appear to be metabolized by separate forms of UDP-glucuronyltransferase (Zakim and Vessey, 1973a; Zakim et al., 1973a).

It could be argued that the p-nitrophenol form of UDP-glucuronyltransferase is fully active and not constrained in microsomes from Gunn rats, and for this reason treatment with phospholipase A or detergent does not stimulate activity. However, since constraint on the maximum potential activity of the enzyme in microsomes from normal rats can be demonstrated, the argument of full activity in the Gunn rat still implies an alteration in interactions between UDP-glucuronyltransferase and its lipid environment. Moreover, with o-aminophenol or o-aminobenzoate as glucuronyl acceptors, it is clear that the rates of glucuronidation in the Gunn rat are less than their potential maxima, since activity is increased eight- and fourfold respectively in the presence of DEN (Zakim et al., 1973b). That treatment with phospholipase A or detergent did not relieve this constraint in the Gunn rat is additional evidence for abnormal interrelations between UDP-glucuronyltransferases and their environments in the membrane. In an attempt to study the question of a generalized membrane defect, detergent activation of G-6-Pase was examined in microsomes from Gunn rats and was found to be normal (Zakim et al., 1973b).

VII. Model for Lipid–Protein Interactions

As pointed out in the Introduction, the arrangement of lipids and proteins in biological membranes is uncertain, and there are reasons to expect differences among various types of membranes. We have stressed that the data for the effects of phospholipids on the properties of membranes can be interpreted without reference to specific membrane structures. It is obvious, however, that any membrane model must account for the observed interrelations between lipid structure and enzyme function. In view of the extensive evidence for interaction between the hydrophobic portions of membrane-bound proteins and the alkyl hydrocarbon chains of the membrane, it is a reasonable presumption that the changes in structure of the alkyl hydrocarbon regions are transmitted to the proteins via alterations in hydrophobic interactions. The problem is in visualizing how modification of these relationships can alter the overall conformation of a protein. (There should be little variability in the energy of different hydrophobic contacts.) The commonly accepted model for membrane structure (Singer and Nicolson, 1972) does not account for the fact that modification of the properties of membrane lipids leads to conformational change in membrane-bound enzymes as evidenced by changes in their kinetic parameters. A model which relates membrane lipids to the stability of different conformational isomers of tightly-bound microsomal enzymes is presented in Fig. 1. The essential feature of this model is that it becomes increasingly difficult to immerse a protein in the lipid phase of the membrane as the rigidity of the alkyl chains increases (decreasing "fluidity"). As a result, the protein tends to be extruded from the lipid phase as the

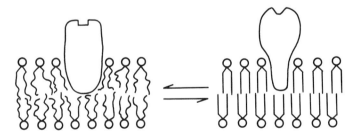

Figure 1. A model which relates a change in the fluidity of the membrane lipid phase to a change in the conformation of a tightly-bound protein. On the left is shown a liquid-crystalline lipid phase which will accommodate a relatively large portion of the protein molecule. On the right is represented a gel phase lipid bilayer from which the protein has been partially extruded, thereby exposing hydrophobic residues on the protein to water; the protein will thus rearrange its conformation to internalize these hydrophobic residues.

"fluidity" of the lipid portion of the membrane decreases. Because of the need to internalize hydrophobic groups previously in contact with the hydrocarbon chains, and also because of the opportunity for contacts between newly exposed polar groups on the enzyme and water, the partially extruded protein will assume a conformation different from that of the more fully immersed protein. Experimental support for this model is provided by the finding that nitroxide-labeled methylstearate cannot be inserted fully into sarcoplasmic reticulum or microsomal membranes below the transition temperature for the lipid phase, but, as the fluidity of these membranes is increased by raising the temperature, the extent of insertion of the methyl stearate into the membranes increases. Similar behavior is seen with certain partitioning spin labels in which the partitioning of the label between the internal lipid phase and the membrane–aqueous interface can be discerned directly from the spectra (Eletr et al., 1973). In these systems, the fluidity of the lipid phases determines the extent of hydrophobic interactions between the lipid phase and exogenous compounds. The notion that primary changes in the conformation of membrane-bound proteins can be transmitted to the lipid phase is also encompassed by the model in Fig. 1, since the insertion of a molecule into an ordered lipid region perturbs the matrix rigidity, thereby creating a local melt.

VIII. Consideration of Factors Regulating the Activities of Membrane-Bound Enzymes in Vivo

Although there is ample evidence for phospholipid-induced modification in the behavior of proteins and a chemical foundation for interpreting these effects, the manner in which these interactions may be modified in vivo remains unclear. Obviously, variation in the lipid components of a membrane, especially in chain length and unsaturation of phospholipid fatty acids, might alter the properties of the enzymes contained within the membrane. Unfortunately, however, little specific work has been done on this in higher organisms. Except for the transporting proteins of microorganisms (Machtiger and Fox, 1973), modification of the fatty acid composition of membranes does not appear to influence the properties of most membrane-bound enzyme systems. The answer to why this should be so, despite the evidence that perturbation of the lipid phase alters the properties of membrane-bound proteins, may lie in the fact that "animal tissues are equipped with systems attempting to maintain the physicochemical properties of the phospholipids within certain limits" (Van Deenen, 1966). For example, the distribution of different fatty acids in liver lecithins changes considerably when the type and amount of dietary fat is altered, yet the physical properties of the whole lecithin fraction derived from livers of rats fed different diets show only limited differences (Van Deenen, 1966).

The need to maintain constant membrane properties with respect to the physical behavior of the lipid phases may reflect requirements for the normal function and regulation of several membrane proteins. On the other hand, that changes in the dietary composition have relatively little influence on the physical properties of lipid extracts of membrane does not necessarily mean that such changes do not influence the function of selected enzymes. An observation with broad implications is that the minimum temperature at which chilling-sensitive plants grow can be lowered by perturbation of their membrane lipids with compounds which increase the fluidity of the membrane (Eletr, unpublished observations). There is indirect evidence also that some long-term control of membrane function may be effected by variation in the fatty acid composition of membranes in pathological states. In patients with Refsum disease, in whom there is increased storage of phytanic acid (3,7,11,15-tetramethylhexadecanoic acid) (Kahlke, 1964), the major clinical manifestations are related to abnormal rates of nerve conduction. Since branched-chain fatty acids perturb membrane structure by expanding the membrane, thereby decreasing Van der Waals interactions, and also result in a change in state for the protein-bound phase (Arvidsson et al., 1971), the manifestations of Refsum disease may be due simply to the effect of branched-chain hydrocarbons on the physical properties of the membrane. Partial reversal of the disease may be achieved by careful control of the intake of phytanic acid. As yet, however, the specific effects of phytanic acid on the properties of membrane-bound enzymes remain uninvestigated. A similar mechanism may explain the neuropathy in vitamin B_{12} deficiency (Frenkel, 1973; Kishimoto et al., 1973).

It is exceedingly interesting that modification of the diet alters the rate of catabolism of cholesterol, which takes place almost exclusively in a series of reactions in the endoplasmic reticulum of liver. The feeding of diets high in polyunsaturated fatty acids increases the degradation of cholesterol to bile acids, whereas diets containing primarily saturated fatty acids interfere with the rate of these degradative reactions. No explanations have been offered to explain these data, and though speculative, it is possible that they reflect modifications in the rate of specific microsomal enzyme reactions as a result of changes in the lipid composition of the microsomal membrane.

To a limited extent, it has been shown that phospholipid headgroups also influence the properties of membrane-bound enzymes.

Studies in choline-deficient animals indicate that the activity of UDP-glycosyltransferases, important for the synthesis of glycoproteins in liver microsomes, are abnormally low. After addition of phosphorylcholine plus CTP to a suspension of liver microsomes, there is a rapid repletion of UDP-glycosyltransferase activities (Mookerjea, 1971). CDP-choline had a similar effect. These results suggest that the functional status of UDP-glycosyltransferases is altered by manipulation of the compositions of phospholipid headgroups.

Alteration of the fatty acid composition of membranes is a long-term process, and rapid changes in the properties of membrane-bound enzymes would have to take place independently of this kind of change. In this regard, it is important that microsomal, mitochondrial, and plasma membranes contain endogenous phospholipase activities (Waite, 1969; Victoria et al., 1971) which could be part of the regulatory process for controlling membrane function. For example, Keith et al. (1975) have found that hibernating ground squirrels have an increased content of lysophosphatides in mitochondrial membranes from their heart muscle. These lysophosphatides (presumably owing to their detergent properties) alter the properties not only of the membrane lipid phase but also of certain membrane-bound respiratory enzymes (see Section II.E). The factors modulating the activities of membrane phospholipases are unstudied, but in the absence of some sort of control there would be a continuous, rapid destruction of membrane as it is formed. In vivo changes in the rate of phospholipase-A-catalyzed hydrolysis of membrane phospholipids could potentially alter the properties of other membrane-bound enzymes. Another mechanism by which rapid regulation of membrane-bound enzymes could be effected is via the action of hormones. It has been shown already, in fact, that vasopressin alters the water permeability of artificial lipid membranes (Graziani and Livne, 1971). Further, aldosterone alters the lipid composition of membranes (Goodman et al., 1971), and it has been proposed that this accounts for the increased sensitivity to ouabain of ATPase from aldosterone-treated tissue (Goodman et al., 1969).

IX. Conclusions

Recent advances in our understanding of the function of phospholipids in membranes establish that these components have a more important role than acting simply as a static barrier to the passage of compounds into and out of various cell compartments. It is clear that the phospholipids themselves can influence the properties of the membrane from the viewpoint of its permeability and also that the behavior of a variety of membrane-bound proteins is modified on perturbation of the structure of the lipid phase. Evidence has accumulated, too, which indicates that dynamic changes in membrane lipids have important physiologic actions in intact systems. Further definition of the nature of lipid–protein interactions in biological membranes requires clearer understanding of the morphological basis for these interactions. Perhaps most exciting is that we have arrived at the time when the properties of membranes and membrane-bound enzymes will be subject to manipulation through the use of pharmacological agents which have well-known effects on the physical parameters of synthetic membrane systems.

References

Abou-Issa, H. M., and Cleland, W. W., 1969, Studies on the microsomal acylation of 1-glycerol-3-phosphate. II. The specificity and properties of the rat liver enzyme, *Biochim. Biophys. Acta* **176**:692.
Abramson, M. B., and Pisetsky, D., 1972, Thermal-turbidometric studies of membranes from *Acholeplasma laidlawii*, *Biochim. Biophys. Acta* **282**:80.
Arion, W. J., and Nordlie, R. C., 1967, Biological regulation of inorganic pyrophosphate-glucose phosphotransferase and glucose-6-phosphatase, *J. Biol. Chem.* **242**:2207.
Arvidsson, E. O., Green, F. A., and Laurell, S., 1971, Branching and hydrophobic bonding. Partition equilibria and serum albumin binding of palmitic and phytanic acids, *J. Biol. Chem.* **246**:5373.
Atkinson, D. E., 1966, Regulation of enzyme activity, *Annu. Rev. Biochem.* **35**:8.
Attwood, D., Graham, A. B., and Wood, G. C., 1971, The phospholipid dependence of uridine diphosphate glucuronyltransferase. Reactivation of phospholipase A-inactivated enzyme by phospholipids and detergents, *Biochem. J.* **123**:875.
Azzi, A., 1974, The use of fluorescent probes for the study of membranes, in: *Methods of Enzymology*, Vol. XXXII (S. Fleischer and L. Packer, eds.), pp. 234–246, Academic Press, New York.
Azzi, A., Chance, B., Radda, G. K., and Lee, C. P., 1969, A fluorescence probe of energy-dependent structure changes in fragmented membranes, *Proc. Natl. Acad. Sci. USA* **62**:612.
Beaufay, H., and de Duve, C., 1954, Le système hexose-phosphatasique. VI.Essais de démembrement des microsomes porteurs de glucose-6-phosphatase, *Bull. Ste. Chim. Biol.* **36**:1551.
Blazyk, J. F., and Steim, J. M., 1972, Phase transitions in mammalian membranes, *Biochim. Biophys. Acta* **266**:737.
Callahan, J. W., and Kosicki, G. W., 1967, The effect of lipid micelles on mitochondrial malate dehydrogenase, *Can. J. Biochem.* **45**:839.
Capaldi, R. A., and Vanderkooi, G., 1972, The polarity of many membrane proteins, *Proc. Natl. Acad. Sci. USA* **69**:930.
Chance, B., and Lee, C. P., 1969, Comparison of fluorescence probe and light scattering readout of structural states of mitochondrial membrane fragments, *FEBS Lett.* **4**:181.
Chapman, D., 1968, Recent physical studies of phospholipids and natural membranes, in: *Biological Membranes Physical Fact and Function* (D. Chapman, ed.), pp. 125–202, Academic Press, New York.
Chapman, D., and Collin, D. T., 1965, Differential thermal analysis of phospholipids, *Nature* **206**:189.

Chapman, D., and Dodd, G. H., 1971, Physiochemical probes of membrane structure, in: *Structure and Function of Biological Membranes* (L. I. Rothfield, ed.), pp. 13–83, Academic Press, New York.

Chapman, D., and Penkett, S. A., 1966, Nuclear magnetic resonance spectroscopic studies of the interaction of phospholipids with cholesterol, *Nature* **211**:1304.

Chen, L. F., Lund, D. B., and Richardson, T., 1971, Essential fatty acids and glucose permeability of lecithin membranes, *Biochim. Biophys. Acta* **225**:89.

Criddle, R. S., Bock, R. M., Green, D. E., Tisdale, H. D., 1961, Specific interaction of mitochondrial structural protein with cytochrome and lipids, *Biochem. Biophys. Res. Commun.* **5**:75.

Cunningham, C. C., and Hager, L. P., 1971, Crystalline pyruvate oxidase from *Escherichia coli*. II. Activation by phospholipids, *J. Biol. Chem.* **246**:1575.

Dark, A., Finer, E. G., Flook, A. G., and Phillips, M. C., 1972, Nuclear magnetic resonance studies of lecithin-cholesterol interactions, *J. Mol. Biol.* **63**:265.

DeGier, J., Haest, C. W. M., Mandersloot, J. G., and Van Deenen, L. L. M., 1970, Valinomycin-induced permeation of $^{86}Rb^+$ of liposomes with varying composition throught the bilayer, *Biochim. Biophys. Acta* **24**:373.

Dehlinger, P. J., Jost, P. C., and Griffith, O. H., 1974, Lipid binding to the amphipathic membrane protein cytochrome b_5, *Proc. Natl. Acad. Sci. USA* **71**:2280.

DeKruyff, B., Demel, R. A., and Van Deenen, L. L. M., 1972, The effect of cholesterol and epicholesterol incorporation on the permeability and on the phase transition of intact *Acholeplasma laidlawii* cell membranes and derived liposomes, *Biochim. Biophys. Acta* **255**:331.

Demel, R. A., Bruchdorfer, K. R., and Van Deenen, L. L. M., 1972a, Structural requirements of sterols for the intersection with lecithin at the air–water interface, *Biochim. Biophys. Acta* **255**:311.

Demel, R. A., Bruchdorfer, K. R., and Van Deenen, L. L. M., 1972b, The effect of sterol structure on the permeability of liposomes to glucose, glycerol, and Rb^+, *Biochim. Biophys. Acta* **255**:321.

Demel, R. A., Van Kessel, W. S. M. G., and Van Deenen, L. L. M., 1972c, The properties of polyunsaturated lecithins in monolayers and liposomes and the interactions of these lecithins with cholesterol, *Biochim. Biophys. Acta* **266**:26.

Duttera, S. M., Bryne, W. L., and Ganoza, M. C., 1968, Studies on the phospholipid requirement of glucose-6-phosphatase, *J. Biol. Chem.* **243**:2216.

Eletr, S., and Inesi, G., 1972, Phase changes in the lipid moieties of sarcoplasmic reticulum membranes induced by temperature and protein conformational changes, *Biochim. Biophys. Acta* **290**:178.

Eletr, S., and Keith, A. D., 1972, Spin label studies of lipid alkyl chain dynamics in biological membranes. Role of unsaturated sites, *Proc. Natl. Acad. Sci. USA* **69**:1353.

Eletr, S., Zakim, D., and Vessey, D. A., 1973, A spin-label study of the role of phospholipids in the regulation of membrane-bound microsomal enzyme, *J. Mol. Biol.* **78**:351.

Eley, D. D., and Hedge, D. G., 1956, Protein interactions with lecithin and cephalin monolayers, *J. Colloid Sci.* **11**:445.

Eley, D. D., and Hedge, D. G., 1957, Properties of biological membranes. The structure of films of proteins adsorbed on lipids, *Discuss. Faraday Soc.* **21**:221.

Emmelot, E., and Bos, C. J., 1967, Studies of plasma membranes. V. On the lipid dependence of some phosphohydrolases of isolated rat-liver plasma membranes, *Biochim. Biophys. Acta* **150**:341.

Esfahani, M., Limbrick, A. R., Knutton, S., Oka, T., and Wakil, S. J., 1971, The molecular organization of lipids in the membrane of *Escherichia coli*: Phase transitions, *Proc. Natl. Acad. Sci. USA* **68**:3180.

Featherstone, R. M., Muchlbaecher, C. A., DeBon, F. L., and Forsaith, J. A., 1961, Interaction of inert anesthetic gases with proteins, *Anesthesiology* **22**:977.
Fisher, D. B., and Kaufman, S., 1972, The stimulation of rat liver phenylalanine hydroxylase by phospholipids, *J. Biol. Chem.* **247**:2250.
Foster, D. W., and McWhorter, W. P., 1969, Microsomes, microsomal phospholipids and fatty acid synthesis, *J. Biol. Chem.* **244**:260.
Frenkel, E. P., 1973, Abnormal fatty acid metabolism in peripheral nerves of patients with pernicious anemia, *J. Clin. Invest.* **52**:1237.
Gaffney, B. J., 1974, Spin label measurements in membranes, *Methods Enzymol.* **32**:161.
Glaser, M., Ferguson, K. A., and Bayer, W. H., 1974, Manipulation of the lipid composition of mammalian cells, *Fed. Proc.* **33**:1296.
Goodman, D. B. P., Allen, J. E., and Rasmussen, H., 1969, On the mechanism of action of aldosterone, *Proc. Natl. Acad. Sci. USA* **64**:330.
Goodman, D. B. P., Allen, J. E., and Rasmussen, H., 1971, Studies on the mechanism of action of aldosterone: Hormone-induced changes in lipid metabolism, *Biochemistry* **10**:3825.
Gotterer, G. S., 1967a, Rat liver D-β-hydroxybutyrate dehydrogenase. I. Partial purification and general properties, *Biochemistry* **6**:2139.
Gotterer, G. S., 1967b, Rat liver D-β-hydroxybutyrate dehydrogenase. II. Lipid requirement, *Biochemistry* **6**:2147.
Gram, T. E., Guaviro, A. M., Schroder, D. H., and Gillette, J. R., 1969, Changes in certain kinetic properties of hepatic microsomal aniline hydroxylase and ethylmorphine demethylase associated with postnatal development and maturation in male rats, *Biochem. J.* **113**:681.
Graziani, Y., and Livne, A., 1971, Vasopressin and water permeability of artificial lipid membranes, *Biochem. Biophys. Res. Commun.* **45**:321.
Grover, A. K., Slotboom, A. J., De Haas, G. H., and Hammes, G. G., 1975, Lipid specificity of β-hydroxybutyrate dehydrogenase activation, *J. Biol. Chem.* **250**:31.
Haslam, J. M., 1971, The effects of depletion of unsaturated fatty acids on the energy dependent reactions of yeast mitochondria, *Biochem. J.* **123**:6P.
Hsia, J. C., and Boggs, J. M., 1972, Influence of pH and cholesterol on the structure of phosphatidylethanolamine multi bilayers, *Biochim. Biophys. Acta* **226**:18.
Inoue, K., 1974, Permeability properties of liposomes prepared from dipalmitoyl lecithin, dimyristoyl lecithin, egg lecithin, rat liver lecithin and beef brain sphingomyelin, *Biochim. Biophys. Acta* **339**:390.
Jones, P. D., Holloway, P. W., Peluffo, R. O., and Wakil, S. J., 1969, A requirement for lipids by the microsomal stearyl co-enzyme A desaturase, *J. Biol. Chem.* **244**:744.
Jost, P. C., Griffith, O. H., Capaldi, R. A., and Vanderkooi, G., 1973, Evidence for boundary lipid in membranes, *Proc. Natl. Acad. Sci USA* **70**:480.
Jurtshuk, P., Jr., Sekuzu, I., and Green, D. E., 1963, Studies on the electron transfer system LIV. On the formation of an active complex between the apo-D(−)-β-hydroxybutyric dehydrogenase and micellar lecithin, *J. Biol. Chem.* **238**:3595.
Kahlke, E., 1964, Refsum-Syndrome. Lipoidchemische Untersuchurgen bei 9 Fällen, *Klin. Wochenschr.* **42**:1011.
Kaufman, S., 1971, The phenylalanine hydroxylating system from mammalian liver, *Adv. Enzymol.* **35**:245.
Kauzman, W., 1959, Some factors in the interpretation of protein denaturation. *Adv. Protein Chem.* **14**:1.
Keith, A. D., Waggoner, A. S., and Griffith, O. H., 1968, Spin labelled mitochondrial lipids in *Neurospora crassa, Proc. Natl. Acad. Sci. USA* **61**:819.
Keith, A. D., Aloia, R. C., Lyons, J., Snipes, W., and Pengelley, E. T., 1975, Spin label

evidence for the role of lysoglycerophosphatides in cellular membranes of hibernating mammals, *Biochim. Biophys. Acta* **394**:204.

Keough, K. M., Oldfield, E., Chapman, D., and Beynon, P., 1973, Carbon-13 and proton nuclear magnetic resonance of unsonicated model and mitochondrial membranes, *Chem. Phys. Lipids* **10**:37.

Kishimoto, Y., Williams, M., Moser, H. W., Hignite, C., and Biemann, K., 1973, Branched-chain and odd-numbered fatty acids and aldehydes in the nervous system of a patient with deranged vitamin B_{12} metabolism, *J. Lipid Res.* **14**:69.

Koshland, D. E., 1970, The molecular basis for enzyme regulation, in: *The Enzymes* (P. D. Boyer, ed.), Vol. 1, pp. 342–396, Academic Press, New York.

Ladbrooke, B. D., and Chapman, D., 1969, Thermal analysis of lipids, proteins and biological membranes. A review and summary of some recent studies, *Chem. Phys. Lipids* **3**:304.

Leathes, J. B., 1925, Role of fats in vital phenomena, *Lancet* **208**:853.

Lee, A. G., Birdsall, N. J. M., and Metcalfe, J. C., 1974a, Nuclear magnetic relaxation and the biological membrane, in: *Methods in Membrane Biology*, Vol. 2 (E. Korn, ed.), pp. 1–156, Plenum, New York.

Lee, A. G., Birdsall, N. J. M., Metcalfe, J. C., Toon, P. A., and Warren, G. B., 1974b, Clusters in lipid bilayers and the interpretation of thermal effects in biological membranes, *Biochemistry* **13**:3699.

Long, R. A., Hruska, R., Geiser, H. D., Hsia, J. C., and Williams, R., 1970, Membrane condensing effect of cholesterol and the role of its hydroxyl group, *Biochem. Biophys. Res. Commun.* **41**:321.

Luzzati, V., 1968, X-ray diffraction studies of lipid–water systems, in: *Biological Membranes Physical Fact and Function* (D. Chapman, ed.), pp. 71–124, Academic Press, New York.

Luzzati, V., Gulik-Krzywicki, T., Tardieu, A., Rivas, E., and Reiss-Husson, F., 1969, Lipids and membranes, in: *The Molecular Basis of Membrane Function* (D. Tosteson, ed.), pp. 47–78, Prentice-Hall, Englewood Cliffs, N.J.

Martonosi, A., Donley, J., and Halpin, R. A., 1968, Sarcoplasmic reticulum. The role of phospholipids in the adenosine triphosphatase activity and Ca^{++} transport, *J. Biol. Chem.* **243**:61.

Machtiger, M. A., and Fox, C. F., 1973, Biochemistry of bacterial membranes, *Annu. Rev. Biochem.* **42**:575.

Mavis, R. D., and Vagelos, P. R., 1972, The effect of phospholipid fatty acid composition on membranous enzymes in *Escherichia coli*, *J. Biol. Chem.* **247**:652.

McElhaney, R. N., 1974, The effect of alterations in the physical state of the membrane lipids on the ability of *A. laidlawii* B to grow at various temperatures, *J. Mol. Biol.* **84**:145.

McElhaney, R. N., Degier, J., and Van Der Neut-Kok, E. C. M., 1973, The effect of alterations in fatty acid composition of the non-electrolyte permeability of *Achoplasma laidlawii* B cells and derived liposomes, *Biochim. Biophys. Acta* **248**:500.

Melchior, D. L., Morowitz, H. J., Sturtevant, J. M., and Tsong, T. Y., 1970, Characterization of the plasma membrane of *Mycoplasma laidlawii*. VII. Phase transitions of membrane lipids, *Biochim. Biophys. Acta* **219**:114.

Mookerjea, S., 1971, Action of choline on lipoprotein metabolism, *Fed. Proc.* **30**:143.

Moss, J., and Lane, M. D., 1971, The biotin dependent enzymes, *Adv. Enzymol.* **35**:321.

Nordlie, R. C., and Arion, W. J., 1964, Evidence for the common identity of glucose-6-phosphatase, inorganic pyrophosphatase, and pyrophosphate-glucose phosphotransferase, *J. Biol. Chem.* **239**:1680.

Nordlie, R. C., Arion, W. J., Hanson, T. L., Gilsdorf, J. R., and Horne, R. N., 1968, Biological regulation of liver microsomal inorganic pyrophosphate-glucose phosphotransferase, glucose-6-phosphatase, and inorganic pyrophosphatase. Differential effects of fasting on synthetic and hydrolytic activities, *J. Biol. Chem.* **243**:1140.

Oldfield, E., and Chapman, D., 1971, Effects of cholesterol and cholesterol derivatives on hydrocarbon chain mobility in lipids, *Biochem. Biophys. Res. Commun.* **43**:610.

Raison, J. K., and Lyons, J. M., 1971, Hibernation: Alteration of mitochondrial membranes as a requisite for metabolism at low temperatures, *Proc. Natl. Acad. Sci. USA* **68**:2092.

Raison, J. K., Lyons, J. M., Melborn, R. J., and Keith, A. D., 1971a, Temperature-induced phase changes in mitochondrial membranes detected by spin labeling, *J. Biol. Chem.* **246**:4036.

Raison, J. K., Lyons, J. M., and Thomson, W. W., 1971b, The influence of membranes on the temperature-induced changes in the kinetics of some respiratory enzymes of mitochondria, *Arch. Biochem. Biophys.* **142**:83.

Rand, R. P., and Luzzati, V., 1968, X-ray diffraction study in water of lipids extracted from human erythrocytes. The position of cholesterol in the lipid lamellae, *Biophys. J.* **8**:125.

Scarpa, A., and DeGier, J., 1971, Cation permeability of liposomes as a function of the chemical composition of the lipid bilayers, *Biochim. Biophys. Acta* **241**:789.

Schmid, R., 1972, Hyperbilirubinemia, in: *The Metabolic Basis of Inherited Diseases*, 3rd ed. (J. B. Stanbury, J. B. Wyngaarden, and D. S. Fredrickson, eds.), p. 1141, McGraw-Hill, New York.

Segal, H. L., and Washko, M. E., 1959, Studies of liver glucose-6-phosphatase. III. Solubilization and properties of the enzyme from normal and diabetic rats, *J. Biol. Chem.* **234**:1937.

Shah, D. O., and Schulman, J. H., 1967a, Influence of calcium, cholesterol, and unsaturation on lecithin monolayers, *J. Lipid Res.* **8**:215.

Shah, D. O., and Schulman, J. H., 1967b, The ionic structure of lecithin monolayers, *J. Lipid Res.* **8**:227.

Shechter, E., Letellier, L., and Gulik-Krzywicki, 1974, Relations between structure and function in cytoplasmic membrane vesicles isolated from an *E. coli* fatty acid auxotroph, *Eur. J. Biochem.* **49**:61.

Shimshick, E. J., and McConnell, H. M., 1973, Lateral phase separation in phospholipid membranes, *Biochemistry* **12**:2351.

Simpkins, H., Panko, E., and Tay, S., 1971, Structural changes in the phospholipid regions of axonal membrane produced by phospholipase C action, *Biochemistry* **10**:3851.

Singer, S. J., and Nicolson, G. L., 1972, The fluid mosaic model of the structure of cell membranes, *Science*, **175**:720.

Snart, R. S., and Sanyal, N. N., 1968, Interaction of polypeptide hormones with lipid monolayers, *Biochem. J.* **108**:369.

Spatz, L., and Strittmatter, P., 1971, A form of cytochrome b_5 that contains an additional hydrophobic sequence of 40 amino acid residues, *Proc. Natl. Acad. Sci. USA* **68**:1042.

Stadtman, E. R., 1966, Allosteric regulation of enzyme activity, *Adv. Enzymol.* **28**:41.

Steim, J. M., Tourtellotte, M. E., Reinert, J. C., McElhaney, R. N., and Radar, R. L., 1969, Calorimetric evidence for the liquid-crystalline state of lipids in a biomembrane, *Proc. Natl. Acad. Sci. USA* **63**:104.

Stetten, M. R., 1964, Metabolism of inorganic pyrophosphate. I. Microsomal inorganic pyrophosphate phosphotransferase of rat liver, *J. Biol. Chem.* **239**:3576.

Stetten, M. R., and Burnett, F. R., 1966, Activation of rat liver microsomal glucose-6-phosphatase, inorganic pyrophosphatase and inorganic pyrophosphate-glucose phosphotransferase by hydroxyl ion, *Biochim. Biophys. Acta* **128**:344.

Stevenson, I., Greenwood, D., and McEwen, J., 1968, Hepatic UDP-glucuronyltransferase

in Wistar and Gunn rats—in vitro activation by diethylnitrosamine, *Biochem. Biophys. Res. Commun.* **32**:866.
Strittmatter, P., 1967, Cytochrome b_5, in: *Methods in Enzymology*, Vol. 10 (R. W. Estabrook and M. E. Pullman, eds.), pp. 553–556, Academic Press, New York.
Van Deenen, L. L. M., 1966, Some structural investigations on phospholipids from membranes, *J. Am. Oil. Chem. Soc.* **43**:296.
Van Deenen, L. L. M., Houtsmuller, V. M. T., DeHaas, G. H., and Muller, E., 1962, Monomolecular layers of synthetic phosphatides, *J. Pharm. Pharmacol.* **14**:429.
Vessey, D. A., and Zakim, D., 1971, Regulation of microsomal enzymes by phospholipids. II. Activation of hepatic uridine diphosphate-glucuronyltransferase, *J. Biol. Chem.* **246**:4649.
Vessey, D. A., and Zakim, D., 1972a, Regulation of microsomal enzymes by phospholipids. V. Kinetic studies of hepatic uridine diphosphate-glucuronyltransferase, *J. Biol. Chem.* **247**:3023.
Vessey, D. A., and Zakim, D., 1972b, Regulation of microsomal enzymes by phospholipids. IV. Species differences in the properties of microsomal UDP-glucuronyltransferase, *Biochim. Biophys. Acta* **268**:61.
Vessey, D. A., and Zakim, D., 1974, Membrane fluidity and the regulation of membrane-bound enzymes, in: *Horizons in Biochemistry and Biophysics*, Vol. 2 (E. Quaglieriello, F. Palmieri, and T. P. Singer, eds.), pp. 137–174, Addison Wesley, Boston.
Vessey, D. A., Goldenberg, J., and Zakim, D., 1973, Kinetic properties of microsomal UDP-glucuronyltransferase. Evidence for cooperative kinetics and activation by UDP-N-acetylglucosamine, *Biochim. Biophys. Acta* **309**:58.
Victoria, E. J., Van Golde, L. M. G., Hostetler, K. Y., Scherphof, G. L., and Van Deenen, L. L. M., 1971, Some studies on the metabolism of phospholipids in plasma membranes from rat liver, *Biochim. Biophys. Acta* **239**:443.
Waite, M., 1969, Isolation of rat liver mitochondrial membrane fractions and localization of the phospholipase A, *Biochemistry* **8**:2536.
Wetlaufer, D. B., and Lovrien, R., 1963, Induction of reversible structural changes in proteins by nonpolar substances, *J. Biol. Chem.* **239**:596.
Williams, R. M., and Chapman, D., 1970, Phospholipids, liquid crystals and cell membranes, *Progr. Chem. Fats Other Lipids* **11**:3.
Wilson, G., and Fox, C. F., 1971, Biogenesis of microbial transport systems: Evidence for coupled incorporation of newly synthesized lipids and proteins into membrane, *J. Mol. Biol.* **55**:49.
Wishnia, A., 1962, The solubility of hydrocarbon gases in protein solutions, *Proc. Natl. Acad. Sci. USA* **48**:2200.
Zakim, D., 1970, Regulation of microsomal enzymes by phospholipids. I. The effect of phospholipases and phospholipids on glucose-6-phosphatase, *J. Biol. Chem.* **245**:4953.
Zakim, D., 1973, Influence of fructose on hepatic synthesis of lipids. *Prog. Biochem. Pharmacol.* **8**:161.
Zakim, D., and Vessey, D. A., 1971, Selective regulation of microsomal enzymes by alteration of membrane phospholipids, *Fed. Proc.* **30**:1060.
Zakim, D., and Vessey, D. A., 1972, Regulation of microsomal enzymes by phospholipids. III. The role of -SH groups in the regulation of microsomal UDP-glucuronyltransferase, *Arch. Biochem. Biophys.* **148**:97.
Zakim, D., and Vessey, D. A., 1973a, Multiple forms of UDP-glucuronyltransferase in liver microsomes, *Fed. Proc.* **31**:925.
Zakim, D., and Vessey, D. A., 1973b, Techniques for the characterization of UDP-glucuronyltransferase, glucose-6-phosphatase and other tightly-bound microsomal enzymes, *Methods Biochem. Anal.* **21**:1.

Zakim, D., and Vessey, D. A., 1974, Membrane dependence of UDP-glucuronyltransferase: Effect of the membrane on kinetic properties, *Biochem. Soc. Trans.* **2**:1165.

Zakim, D., Goldenberg, J., and Vessey, D. A., 1973a, Differentiation of homologous forms of hepatic microsomal UDP-glucuronyltransferase, *Biochim. Biophys. Acta* **309**:67.

Zakim, D., Goldenberg, J., and Vessey, D. A., 1973b, Regulation of microsomal enzymes by phospholipids. VII. Abnormal enzyme–lipid interactions in liver microsomes from Gunn rats, *Biochim. Biophys. Acta* **297**:497.

Zull, J. E., Grenoff, S., and Adam, H. K., 1968, Interaction of egg lecithin with cholesterol in the solid state, *Biochemistry* **7**:4172.

11

Membrane Structure and Transport Systems

Pamela D. McNamara and Božena Ožegović

> No human being is constituted to know the truth, the whole truth, and nothing but the truth; and even the best of men must be content with fragments, with partial glimpses, never the full fruition.
>
> William Osler, *The Student Life*

I. Introduction

The concept of a plasma or cell membrane has evolved in little over a century from the simple nineteenth century idea of a passive bag containing the cell "sap" to the complex and dynamic subcellular fraction we think of today. The work of Moldenhauer in 1812 [cited in Bresnick and Schwartz (1969)] made possible the separation of tissues into component parts and the recognition of a limiting unit in the cell, i.e., a boundary. It was not until the work of Nägeli in 1855 (Nägeli and Cramer, 1855), however, that the presence of a separate structure formed from the proteinaceous material of the cell was conceptualized. He first coined the term "Plasmamembran" and recognized that the membrane was responsible for conferring osmotic properties on the cell. Little

---

Pamela D. McNamara • Department of Metabolic Research, Children's Hospital of Philadelphia, University of Pennsylvania, Philadelphia, Pennsylvania 19104.   **Božena Ožegović** • Laboratory for Experimental Medicine, University of Zagreb, Yugoslavia.

Table 1. Multiple Functions of Membranes[a]

Function	Membrane
1. General and selective diffusion of small molecules and ions	
2. Active transport	All
3. Control influx and efflux of ions, substrates, and products between cell compartments	
4. Phagocytosis, pinocytosis	
5. Cell adhesion and mobility	Plasma membrane
6. Carry surface antigens	
7. Limit organ growth	
8. Electrical insulator	Myelin
9. Generate nervous impulses	Nerve plasma membrane
10. Conduct nervous impulses	Sarcoplasmic reticulum
11. Convert light to electrical impulses	Retinal rod disc membrane
12. Convert light to phosphate bond energy in ATP	Chloroplast membrane
13. Convert oxidation chemical energy to phosphate bond energy in ATP	Inner membrane of mitochondria
14. Move secretory products to outside of cell	Endoplasmic reticulum and Golgi membranes
15. Wall off autolytic enzymes	Lysosomal membrane

[a]Taken from D. F. Parsons (1967), Proceedings of the 7th Canadian Cancer Research Conference, p. 193, with kind permission.

attention was paid to the membrane or its properties until the work of Chambers [cited in Bresnick and Schwartz (1969)] in the 1920s when the semipermeable nature of the membrane was recognized. This realization that the membrane controls the availability of substrates to the cell led to further investigations into the nature of the plasma membrane. Today we think of it as a dynamic regulatory mechanism.

The primary roles of the surface membrane (Table 1) include control of cell growth and intracellular communication by surface contact phenomena, as well as control of metabolite transport between the cell and its environment. A third possible role of the membrane in free-living (nonaggregated) cells would be that of movement. Codes of recognition must exist within the cell membrane for the orderly development of cells within multicellular organisms. Proteins and nucleic acids are well accepted as informational macromolecules, and the addition of carbohydrate moieties to them enhances the informational capacity of these macromolecules (Whaley et al., 1972), making the presence of carbohydrates on the cell surface an important factor in cell aggregation and adhesion.

II. Contact Inhibition and Intercellular Communication

One form of cell regulation manifested in the membrane is density-dependent inhibition (DDI) of growth or contact inhibition. Normal cells in culture exhibit contact inhibition—i.e., limited growth (DDI) and limited cell movement as a result of cell–cell contact. These cells are not agglutinated by carbohydrate-binding plant agglutinins (lectins) such as concanavalin A (Con A), whereas cancer cells not exhibiting DDI are. When such inhibitors of protein synthesis as cycloheximide, pactomycin, and emetine are added to confluent cultures of chick embryo fibroblasts, the cells assume some properties of cancer cells, since they can be agglutinated by plant lectins (Baker and Humphreys, 1972). The exposure of agglutinin-receptor sites by protease is accompanied by the corresponding loss of DDI, indicating a functional relationship between cell surface proteins, their synthesis, agglutination, and growth control properties of the cell.

Working with human erythrocytes, Kornfeld and Kornfeld (1969) have solubilized and partially characterized a receptor glycopeptide which binds lectins. The sugar which determines the specificity of binding in the oligosaccharide portion of the glycopeptide is a galactose residue which is penultimate to an N-acetylneuraminic acid (sialic acid) residue in some cases and uncovered in others. Similarly, factors which cause cellular aggregation can be isolated from normal embryonic chick and mammalian cells grown in culture. Pessac and Defendi (1972) have implicated acid mucopolysaccharides with hyaluronic acid, chondrosulfates A or C, or a mixture of hyaluronic acid and one of the chondrosulfates as the active components of these aggregation factors, which they propose are cell ligands that bind to receptors on the cell surface. Protease destroys the receptors for these aggregation factors.

The action of aggregation factors (mucopolysaccharides) may help regulate in vitro phenomena such as DDI or contact inhibition. The role of intercellular adhesion in embryogenesis and other biological processes has been discussed by Marchase et al. (1976), along with methods used to measure adhesion in morphogenetic and pathological contexts. The mechanism of intercellular adhesion is not yet known, but many theories have been proposed. Involvement of Ca^{2+}, sialic acid, and hexosamine residues has been suggested (Roseman, 1970), and the importance of cell surface glycosyltransferases has been implicated (Shur and Roth, 1975). Curtis (1973) has reviewed the evidence supporting the theory that intercellular adhesion depends on the balance of two opposing electrical interactions between cells. On the other hand, mediation by macromolecular, multivalent ligands has been suggested and reviewed by Moscona (1976) and Glaser (1976).

Communication in the form of establishment of tight junctions allows passage of molecules up to 10,000 molecular weight between cells which have the capacity to form areas called "junctional seals" (Loewenstein, 1972). The formation of a junctional seal by contact between two compatible cells essentially encloses those membrane sections involved in the seal in an intracellular environment with low cytoplasmic Ca^{2+} concentration (10^{-6} M). This favors the detachment of Ca^{2+} from the enclosed membrane portions, which converts them from areas of low to high permeability. The establishment of tight junctions permits the maintenance of order in a cell system by the exchange of information between the units. Azarnia and Loewenstein (1971) have shown that some cancer cells which do not exhibit contact inhibition cannot form junctions either with other cancer cells or with normal cells, as evidenced by the failure of fluorescent dyes or electrical impulses—which are readily passed between junctions in normal cells—to pass from normal cells to cancer cells or between cancer cells. This basic lack of communication may distinguish the mechanism by which some cancer cells lose growth control mechanisms. For a fine overview of this area see the discussions by Burger (1971), Ashwell and Morell (1977), and Hughes (1977). Changes in cell surfaces associated with malignant transformation have been investigated and are summarized by Altman and Katz (1976).

Since cell recognition, contact inhibition, and intercellular communication seem to depend on carbohydrate-containing proteins on the cell surface, the roles of the endoplasmic reticulum and Golgi apparatus become important—the former in the elaboration of proteins or glycoproteins and the latter in the attachment of terminal sugars to the glycoprotein or glycolipoprotein. The final glycoproteins are then transported from the Golgi apparatus to the cell surface and determine certain characteristics of the plasma membrane (Whaley et al., 1972; Whaley, 1975).

It can be seen then that the metabolic state of the cell is an important factor influencing surface membrane functions. Where viral transformation causes cancer-like properties, metabolic control at the nucleic acid level is likely, although viral–host interactions seem more complex than first theorized (Altman and Katz, 1976). Receptors for enteroviruses have been reported and shown to be specific for various viral strains. Susceptibility to viral infection is correlated with the presence of receptor sites on intracellular membranes as well as on the cell surface. Chemically, virus receptors solubilized from plasma membranes have been determined to be lipoproteins, with the protein moiety being most important in determining receptor activity (McLaren et al., 1968). A review of cell membrane receptors for viruses, antigens and antibodies, hormones and small molecules presented at the Miles International Symposium has recently been published (Beers and Bassett, 1976).

III. Antigenic and Receptor Sites

Highly specialized immunological functions of mammalian cells are determined by specific cell membrane antigens and antigen receptors which initiate the immune response. Each cell taking part in any supracellular assembly that has a characteristic form is thought to bear a "surface display" which determines its position within the structure (Moscona, 1957; Sperry, 1965). Indeed, the cell surface appears to be an organized assembly of selected gene products (Boys et al., 1968; Boys and Old, 1970), and the antigenic specificities of proteins are determined by their special tertiary structure, which assumes specific spatial relationships not only among the antigenic molecules themselves, but also between the antigenic molecules and other molecules of the cell surface (Goodman, 1969).

There is great interest in determining the chemical nature of specific antigenic sites on the cell membrane because of the role these antigens play in a host of immunological interactions. However, this area of inquiry is still in its infancy, and blood group antigens have been more completely characterized than other mammalian surface antigens. Watkins (1966) has shown that blood group A is characterized by the presence of an N-acetylgalactosamine group, blood group B by D-galactose, and blood group O by fucose, in an oligosaccharide on the erythrocyte membrane. Using freeze-etching techniques, which involve fracture of the cell membrane to expose its inner matrix, DaSilva et al. (1971) have demonstrated the localization of A-antigen sites on the erythrocyte membrane.

The role of lipids in determining the structure and properties of cell membrane antigens is not fully understood, although the importance of lipids in the membrane for enzyme activities has been demonstrated (Chapters 9 and 10, this volume), and the role of inositol phospholipids in surface receptor function has been reviewed (Michell, 1975). A large part of the ABH determinants are generally agreed to be localized in the membrane glycolipids (Gardas and Koscielak, 1974; Slomiany and Slomiany, 1977). The dependence of Rh-antigen activity on the presence of phospholipid was shown by removal and subsequent replacement of phospholipids in erythrocyte membranes (Green, 1972). The relationship of glycolipids to antigenic functions has been reviewed by Lloyd (1975).

Specificity of the M and N blood groups apparently lies in the protein moieties of the cell surface. Wasniowska et al. (1977) have indicated that the major structural difference between the antigens of the M and N blood groups is in the N-terminal amino acids, which are serine and leucine, respectively. In addition, the amino acid presumably at position 5 of each peptide is glycine for the M group and glutamic acid for the N group.

Another important group of receptors on the membrane or in the cytoplasm is for hormones (Lübke et al., 1976). The active component of the hormonal receptor site is believed to be proteinaceous, although phospholipids apparently play a subtle role in regulating the coupling of hormone–receptor interactions to enzymatic catalysis (Birnbaumer, 1973). It was first proposed that receptor protein consists of two distinct, nonoverlapping sites—the catalytic and the regulatory site. Monod et al. (1963) introduced the term "allosteric" to describe the protein in which a conformational change results from the binding of a ligand to the regulatory site, thus altering the properties of the catalytic site of the protein. This change is termed an "allosteric transition." On the basis of this hypothesis, Jardetzky (1966) and Hill (1969) proposed that transport sites could be of an allosteric nature (see transport carrier theories, below).

The first demonstration of intracellular steroid-hormone receptors was by Jensen and Jacobson (1962) for estradiol. Levinson et al. (1972) showed that hormone binding to the glucocorticoid receptor in cultured rat hepatoma cells takes place inside the cell membrane rather than at the cell surface, but did not exclude the existence of other biological actions of steroids mediated by surface membrane receptors. Indeed, Ožegović et al. (1977) have shown that aldosterone binds to isolated kidney plasma membranes and have proposed that this steroid–membrane interaction may reflect an early event in the transmembrance movement of aldosterone.

Whether the steroid hormones enter the cell freely, as assumed by some investigators (O'Malley, 1971), or by carrier-mediated transport, it is generally accepted that they bind to receptor proteins in the cytoplasm. The steroid–receptor complex apparently undergoes a thermal activation process which allows the complex to enter the nucleus of the cell, where an incompletely understood interaction with chromatin takes place. As a result, specific DNA sequences are transcribed and the new mRNA is released to the cytoplasm, where corresponding translation results in the formation of protein products typical of the target-tissue response to the hormone (Lippman, 1976). This mechanism of hormone action relies almost exclusively for specificity on the presence and number of cytoplasmic hormone receptors. Sheridan (1975) has reviewed experimental evidence which suggests the presence of specific, saturable hormone receptors in or on the cell nucleus, which bind steroid hormones in the absence of cytoplasmic intervention, indicating another possible mechanism for steroid-hormone interaction with the cell.

Membrane-bound hormone receptors were detected in the late 1960s. The binding of insulin, glucagon, and epinephrine to isolated plasma membranes of the rat liver or to isolated fat cells and fat cell membranes has been reported (Tomasi et al., 1970; Rodbell et al., 1971; Cuatrecasas, 1971a,b; Freychet et al., 1971; Dunnick and Marinetti, 1971). Species-specific interaction between growth hormones and erythrocyte membranes has been shown by Cambiaso et al. (1971). Lefkowitz et al. (1971) have published a detailed description of the interaction of adrenocorticotropic hormone with its receptors in the adrenal cortex, which appears to be a membrane-associated interaction (Finn et al., 1972). The modes of action for polypeptide hormones and their receptors have been the subject of intense investigation, and a number of reviews on this subject have been published (Cuatrecasas, 1974; Kahn, 1975; Catt and Dufau, 1977).

Of great biological importance was the recognition of the action of adenyl cyclase in a cell-free system (Rall et al., 1957). It has been postulated that cAMP acts as a second messenger, mediating the effects of a number of hormones by interaction with membrane-bound hormone receptors. Rasmussen (1970) has proposed a model for hormonal interaction with the plasma membrane, suggesting that hormones stimulate adenyl cyclase, which in turn results in the production of cAMP. Cyclic AMP activates protein kinases in the cytoplasm, which phosphorylate several enzymes or contractile proteins within the cell, thereby regulating the activities of these macromolecules. Phospholipids may also play a significant role in hormonal stimulation of adenyl cyclase (Birnbaumer, 1973). Another controlling influence is postulated to be the calcium ion concentration. Rasmussen has suggested that some hormones interact with their receptors in the cell membrane, increasing calcium permeability and adenyl cyclase activity. Indeed, Ca^{2+} may even be considered as a second messenger itself (Rasmussen and Goodman, 1977).

It should be remembered, however, that, while adenyl cyclase is found in almost every mammalian tissue examined (excluding erythrocytes), this enzyme can be influenced by a limited number of hormones in any given tissue and often only by one. Thus, although glucagon, epinephrine, and ACTH act on the same adenyl cyclase system of the fat-cell membrane, there appear to be separate receptor sites for each hormone (Birnbaumer and Rodbell, 1969; Bär and Hechter, 1969).

Schramm et al. (1977) have shown, using cell fusion, that the hormone receptor for one animal species will stimulate the adenyl cyclase system of another animal species. They suggest that the coupling of β-androgen receptors to adenyl cyclase in eukaryotic cells may occur by a universal mechanism.

Although the mechanism of hormone receptor coupling to adenyl cyclase activity has not yet been elucidated, the involvement of gangliosides is strongly implicated in the determination of hormone receptor specificity. The thyrotropin (TSH) receptor on the plasma membrane of thyroid cells is postulated to be a complex containing both glycoproteins and gangliosides. The structural relationship among glycoprotein hormone receptors and the oligosaccharide determinants (i.e., G_{M1}, G_T, G_{M3}) in the receptor appears to be the key element for target-organ specificity, such that the ganglioside in a receptor binds to the hormone, producing a conformational change. This change puts a hormonal subunit in a position that favors membrane interaction and adenyl cyclase stimulation (Lee et al., 1977). For more detailed information on hormone interactions with cell membranes, the reader is referred to Chapter 14 of this book.

During recent years, significant attention has been given to drug receptors on the cell membrane surface. Efforts are being made to isolate some specific drug receptors, and this would permit the direct study of interactions between drugs and tissues. Principal theories on drug–receptor interaction are described by Rang (1971) and Seeman (1972).

In its role as a regulatory mechanism, the plasma membrane also controls the exchange of such metabolites as sugars, amino acids, and ions between the cell interior and extracellular fluids. Transport of such metabolites is thought to be mediated in large part by some carrier mechanism. Specific sites for carriers and pores for the exchange of metabolites have been postulated but have not been identified. Transport systems for sugars, amino acids, and ions will be discussed later in this chapter.

The importance of surface membrane functions differs in various cell and tissue types. Intestinal mucosal cells and kidney tubule cells represent cell systems highly specialized for moving substances across cell membranes. The transport processes in the intestine and kidney are essential for obtaining and reclaiming from the environment substrates needed for cellular metabolism. The function of these cells in absorption is reflected in the specialized structure of the cell surface, which is enormously increased by the presence of numerous microvilli forming a "brush border." Because of the specialized nature of the intestinal mucosal and kidney tubule cells, much data have come from the study of these tissues in an attempt to elucidate the regulatory mechanism of transport systems. "It goes without saying that the more we know about membrane structure and biosynthesis, the better we will be able to interpret the membrane's regulatory roles, for the units of explanation will eventually be the units of membrane structure" (Pardee, 1971).

IV. Membrane Structure

There are three definitions of the cell membrane which, it is generally agreed (Ponder, 1961; Dervichian, 1955), refer to one entity, presumably the biologically functional plasma membrane. These definitions are used by cell biologists in three different senses: anatomically, biochemically, and physiologically. In the anatomic sense, the membrane is the darkly-staining, electron-dense, limiting region of the cell seen in the light and electron microscopes. Biochemically, the membrane exists as a subcellular "fraction" that can be isolated by various techniques of cellular disruption and differential centrifugation. In the physiologic sense, the membrane is a hypothetical structure used to explain data on permeability, transport kinetics, and distribution of metabolites.

Although the structural hypothesis of the molecular conformation of biological membranes was introduced over five decades ago (Gorter and Grendel, 1925), we still do not know the real structure of the cell membrane. Early research into membrane structure dates back to 1895, when Overton proposed the existence of lipid components in biological membranes. On the basis of experiments with human red cells, Gorter and Grendel (1925) concluded that the lipid was spread over the red cell surface in a bimolecular layer, with hydrophobic tails directed toward the center of the lipid leaflet and polar, hydrophilic heads on the surface.

The existence of phospholipid bilayers in biological membranes has since been well established by numerous experimental data using newly improved methods. Melchior and Steim (1976) have observed thermotropic phase transitions in membrane lipids and have postulated that these transitions come from a melting of hydrocarbon chains associated with one another. While lipids might exist in one of several liquid-crystalline phases, the physical data indicate that a bilayer is the most probable configuration. Other physical data, obtained by differential thermal analysis, NMR spectroscopy, X-ray diffraction, and light microscopy, support the view that the reversible thermotropic-gel–liquid-crystal phase transition arises from the melting of the hydrocarbon interiors of lipid bilayers (Chapman, 1970; Oseroff et al., 1973).

Studying the relative permeability of certain cells to nonpolar and polar molecules, Danielli and Davson (1935) postulated a model which included layers of proteins over the bimolecular lipid leaflet (the paucimolecular model). The main idea of the Danielli–Davson model was the existence of one or more bimolecular lipid layers consisting mostly of phospholipids, separated by neutral lipids, and sandwiched between two layers of globular proteins (Fig. 1). In a later proposal, Danielli (1954) rejected the neutral lipid layer between the phospholipid bilayers. Polar groups of the phospholipids were presumed to be bound to proteins through electrostatic interactions, while the neutral lipids and hydrophobic tails of the phospholipids were held together by Van der Waals forces. This model assumed a type of symmetry on the part of the membrane, with both outer surfaces being virtually identical. Compelling evidence, however, for the asymmetric nature of biological membranes has been presented and is reviewed by Rothman and Lenard (1977) and Bergelson and Barsukov (1977).

On the basis of electron microscopic studies, Robertson (1967) has modified the Danielli–Davson model, presenting a new model—the unit membrane hypothesis—for all cell membranes. This model was similar to that of Danielli–Davson but with a few significant changes. The number of bimolecular phospholipid layers in Robertson's model was restricted to one, and the membrane was considered to be asymmetrical, with mucopolysaccharide or mucoprotein on the outside and unconjugated protein on the inner part of the membrane (Fig. 2). The

EXTERIOR

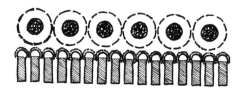

LIPOID

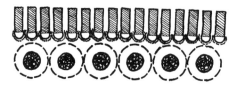

INTERIOR

Figure 1. The Danielli–Davson paucimolecular model of membrane structure. Globular proteins (circular) cover phospholipid layers (club-shaped figures) which are separated by neutral lipoid region. Taken, with kind permission, from Danielli and Davson (1935), p.498.

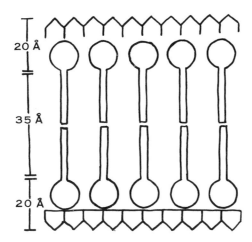

Figure 2. The unit membrane model (after Robinson). Extended protein layers (⌠,⌡) cover phospholipid bilayers.

membrane protein layer was assumed to be in the extended β form rather than in the globular (α) configuration incorporated into the Danielli–Davson model. Each of the two protein layers was thought to have a thickness of about 20 Å, while the bimolecular lipid leaflet was set at about 35–40 Å. Nonpolar portions were to be oriented toward the center of the membrane at right angles to the surface, while polar moieties comprised the external surface of the bilayer leaflet.

The main supporting evidence for Robertson's unit-membrane hypothesis came from electron micrographs of myelin showing three layers in the membrane after fixation of the membrane with $KMnO_4$ (Finean and Robertson, 1958). In addition, X-ray diffraction studies based on biophysical properties of myelin gave evidence for a lipid bilayer configuration in the unfixed myelin sheath, with spacing between the layers of 180 Å. During the preparation of nerve tissue for electron microscopy, Fernández-Morán and Finean (1957) demonstrated shrinkage of the myelin unit from 180 Å in the fresh tissue to 130–140 Å in the embedded specimen—the membrane thickness Robertson had proposed on the basis of electron micrographic data.

Criticism of the unit-membrane hypothesis began with the work of Sjöstrand (1963a,b), whose electron micrographs indicated differences in the dimensions of various cellular membranes, as well as the existence of globular subunits in the planes of mitochondrial membranes and smooth endoplasmic reticulum. A few years later, Korn (1966) presented a serious discussion of the validity of the unit-membrane theory. His main criticism was that much of the supporting evidence for the unit-membrane theory depends upon the interpretation of electron microscopic images. The value of the observation of triple-layered membranes comes under question in the work of Fleischer et al. (1965), who have shown the same electron microscopic images before and after lipid extraction in mitochondrial membranes. There is also the problem of molecular alteration of the protein with $KMnO_4$, OsO_4, and the glutaraldehyde used in fixing specimens for electron microscopy. In a recent reevaluation of the unit-membrane image, Luftig et al. (1977) point out that the "trilammelar" image upon which the unit-membrane hypothesis was based reflects only the basic, chemically "fixed" structure of a phospholipid bilayer devoid of protein. The biological membrane upon which this model was most often based—the myelin sheath—is functionally very different from most mammalian plasma membranes (Table 1), since it is essentially metabolically and structurally inert. Most animal plasma membranes contain more than 20 enzymes and a great deal more protein than myelin (Table 2).

In Korn's view (1966), membrane protein rather than membrane lipid is the important functional group, contrary to the implications in the paucimolecular models. Korn visualized the first step in membrane biosynthesis as the synthesis of protein, and, according to the primary structure of the protein, lipid would be bound sequentially, forming lipoprotein. The globular subunits seen in electron micrographs could then be lipoprotein molecules.

At the same time, Green and Perdue (1966) proposed a new interpretation of the ultrastructure of biological membranes. Their model was based on studies showing repeating globular units, with various sizes and shapes, and having molecular weights from 50,000 to over 1 million, in different biological membranes. All membranes were seen

Table 2. Components of Isolated Membranes

	% Dry weight					
				Carbohydrates		
Tissue	Protein	Lipids	Neutral sugars	Sialic acid	Hexosamines	References
Rat liver PM	51	44	4.3	0.7		Simon et al. (1970)
Rat liver PM	60	40				Benedetti and Emmelot (1968)
Rat liver PM	50,67,48	50,33,48	1.0	0.1		Takeuchi and Teryama (1965)
Human RBC	45	40		12% unidentified		Dodge et al. (1963)
Human RBC	60	40				Maddy (1966)
Human RBC	49.2	43.6	4.0	1.2	2.0	Rosenberg and Guidotti (1968)
Human myelin	22	78				O'Brien and Sampson (1965)
Human myelin	21.8	78	0.2% (combined)			O'Brien (1965)
Ox myelin	22.3					Norton and Autilio (1966)

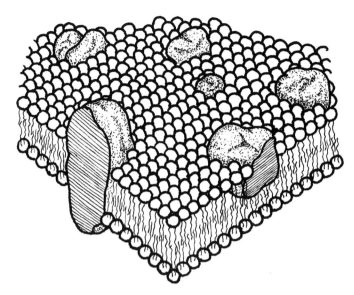

Figure 3. The fluid mosaic model of the cell membrane. Integral proteins (solid bodies with stippled surfaces) are seen penetrating the fluid lipid matrix. Taken, with kind permission, from Singer and Nicolson (1972), *Science* **175**:723. Copyright 1972 by the American Association for the Advancement of Science.

as two-dimensional continua consisting of lipoprotein repeating units, which, in any given membrane, are all complimentary in form but not in composition. Because the repeating unit model did not attempt to explain the relationship between protein and lipid molecules, a new model—the protein crystal model—was proposed as a supplement for the earlier subunit model (Vanderkooi and Green, 1970). The main idea of this model is the notion of bimodal proteins packed together with bimodal lipids in a membrane continuum, forming a highly stable hybrid system. The stability and biological consequences of this model are presented in a review by Green and Brucker (1972). Other models for membrane structure have been suggested and are the subject of a fine symposium in the *Annals of the New York Academy of Sciences* (Green, 1972).

In 1966, Lenard and Singer suggested a hydrophobic model—known as the fluid mosaic model—for membrane structure. This concept was further developed (Singer, 1971, 1974; Singer and Nicholson, 1972) and is now generally accepted as a working model for the molecular orientation of the membrane (Singer, 1977). According to the fluid mosaic model, molecules of globular proteins are dispersed in a fluid lipid matrix arranged as a discontinuous bilayer (Figs. 3 and 4). The protein molecules, which are integrally involved in the maintenance of

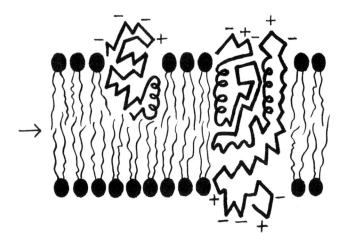

Figure 4. The fluid mosaic model showing charges on ionic residues of proteins in the lipid matrix. The integral proteins with heavy lines representing folded polypeptide chains are shown as globular proteins partially embedded in and protruding from the phospholipid matrix. Arrow shows the plane of cleavage expected in freeze-etching. Taken, with kind permission, from Singer and Nicolson (1972), *Science* **175**:722. Copyright 1972 by the American Association for the Advancement of Science.

membrane structure, are partially embedded in the interior of the membrane and partially protruding from the membrane. The sizes and structural properties of globular proteins determine the degree of their penetration into the interior of the membrane. The hydrophobic parts of lipids and nonpolar amino acid residues of the proteins are oriented into the hydrophobic interior of the membrane, while the ionic groups of lipids and charged residues of the proteins are largely exposed to the aqueous phase. The asymmetric nature of the membrane, then, results from the distribution of all its components, and not just from any "outer layers." The hydrophobic and hydrophilic interactions are considered most important to membrane stability, whereas electrostatic interactions are probably only of local importance to specific structural arrangements. Membrane proteins have been thought to diffuse freely in the lipid matrix, and phospholipids within the two monolayers of the bilayer have been shown to undergo rapid lateral diffusion (Wirtz, 1974).

A refinement of the fluid mosaic model recently presented by Israelachvili (1977) incorporates thermodynamic as well as molecular packing considerations to indicate a highly coupled organization of lipids and proteins, in which the lipids are considered to be in a fluid-like state with rigid, globular proteins interspersed. Important corollaries from this model indicate that: (1) proteins may not be thought of as "free-floating" in a sea of lipid, but structurally coupled to neighboring lipids; (2) lipids near the protein—boundary lipids—must be of a different hydrocarbon chain configuration than the rest of the lipid bilayer; (3) packing constraints affect the upper and lower lipid layers of the bilayer differently; (4) the lateral movement of a protein is affected by the lipid region which is structurally coupled to the protein; and (5) in multicomponent biological membranes, there may exist a phase separation of boundary lipids around a protein. Therefore, the fluid mosaic model presents a two-dimensionally oriented, liquid-like solution of integral, amphipathic proteins dissolved in a lipid bilayer, providing an ordered yet dynamic structure.

V. Membrane Composition

Lipids and proteins are the primary components dealt with in all membrane models. However, while chemical analysis of cell membranes has shown the bulk of the membrane material to be lipids and proteins, most biological membranes also contain a certain amount of carbohydrate. It has already been established that carbohydrates play an important role in cell–cell interactions and in the immunological properties of the cell. Most membrane models are not designed to include all of the components of cell membranes or their interactions. Indeed, the current state of the art of membranology puts severe limitations on the ability to detect all of the membrane components.

The membrane is a delicate yet sturdy structure which must be studied intact but cannot be isolated without disruption. In essence, membrane isolation involves the removal of all other parts of the cell. The purity of a membrane preparation is usually measured by the relative increase in specific activity of a membrane-bound marker enzyme in the membrane preparation over the specific activity of the marker enzyme in the homogenate. In cases where specific membrane markers are lacking, enzymatic activities associated exclusively or primarily with other subcellular components must be demonstrated to be absent in order to show the preparation free from cellular contaminants. Such data, combined with phase-contrast and electron microscopic images of the preparation, are usually accepted as adequate criteria for the measurement of purity. One additional technique, at least for membranes with antigenic properties, is preparation of fluorescent antibodies to the isolated membranes and observation of the antibody localization in whole cells (Kabat and Meyer, 1971). The use of fluorescent probes in determining membrane structure and function has been reviewed (Radda, 1975; Azzi, 1975).

Table 3. References to Some Common Membrane Isolation Methods

Tissue and membrane	Method	Reference
Red blood cell ghosts; human and rat	Hemolysis in hypotonic phosphate buffer; pH and ionic strength effects	Dodge et al. (1963)
Red blood cell ghosts; inside-out vesicles	Hemolysis in low ionic strength alkaline buffer; density gradient centrifugation	Steck et al. (1970)
Rabbit leucocytes	Homogenization; gradient centrifugation	Woodin and Wieneke (1966)
Pig lymphocytes and thymocytes	Stirring in 0.15 M KCl, 0.01 M Tris; heavy microsome fraction; isopycnic centrifugation	Allan and Crumpton (1970)
Rat liver plasma membranes	Homogenization and flotation on sucrose solution	Neville (1960)
Rat liver plasma membranes	Modified Neville technique	Emmelot and Bos (1962); Emmelot et al. (1964)
Rat liver plasma membranes	Homogenization; low speed zonal centrifugation	El-Aaser et al. (1966)
Rat liver plasma membranes	High resolution rate-zonal and isopycnic-zonal separation of purified membranes	Anderson et al. (1968)
Rat liver plasma membranes	Modified Neville technique; addition of $CaCl_2$ (0.5 mM)	Ray (1970)
Mouse liver plasma membranes	Homogenization and flotation on KBr solution	Hertzenberg and Hertzenberg (1961)
Rat liver, adipose, and striated muscle membranes	Cell contents separated from membranes by passage of cells between rollers	McCollester (1960)
Rat heart muscle plasma membrane	Homogenization, filtration, and gradient centrifugation	Kidwai et al. (1971)
Rat fat pad epithelial membranes	Hypotonic disruption of free cells; centrifugation (low speed)	Rodbell (1967)
Rat fat pad epithelial membranes	Homogenization of free cells; isopycnic sedimentation in Ficoll gradient	McKeel and Jarrett (1970)
Rat bladder epithelial membranes	Pretreatment with FMA[a]; homogenization and gradient centrifugation	Hicks and Ketterer (1970)
Calf adrenal and thyroid membranes	Technique of Emmelot et al. (1964) for liver	Turkington (1962)
Sheep thyroid membranes	DNAse treatment of blended tissue; separation at interface of 2M sucrose layer after centrifugation	Benabdelzlil et al. (1967)
Rat brain myelin	Homogenization and isopycnic centrifugation	Laatsch et al. (1962)
Ox myelin, peripheral nerve	Homogenization, separation at sucrose layer during centrifugation, ultracentrifugation, osmotic shock, gradient centrifugation	Autilio et al. (1964)
Ox myelin	Modified Autilio et al. technique	O'Brien et al. (1967)
Ox myelin	Modified Autilio et al. technique	Wolfgram and Kotorii (1968a,b)
Ox intradural spinal root myelin	Isopycnic zonal centrifugation	London (1972)
Rat brain synaptic plasma membranes	Lysis of synaptosomes in distilled water; isopycnic centrifugation	Rodríguez de Lores Arnaiz et al. (1967)

(conti

Table 3 (continued)

Tissue and membrane	Method	Reference
Rat brain synaptic plasma membranes	Synaptosome lysis; differential and zonal centrifugation	Cotman et al. (1968)
Rat kidney cell membranes	Technique of Emmelot et al. for liver	Coleman and Finean (1966)
Rat kidney plasma membranes	Homogenization, isopycnic flotation, and differential centrifugation	Fitzpatrick et al. (1969)
Rat kidney basolateral membranes	Homogenization and centrifugation on a discontinuous sucrose gradient	Ebel et al. (1976)
Rat kidney brush border and basolateral membranes	Homogenization, centrifugation, and free-flow electrophoresis	Heidrich et al. (1972)
Rat kidney brush border membranes	Homogenization and differential centrifugation	Kinne and Kinne-Saffran (1969)
Rat kidney brush border membranes	Homogenization, isopycnic flotation, and rate sedimentation	Wilfong and Neville (1970)
Rabbit kidney brush border membranes	Homogenization, Mg^{2+} precipitation, and differential centrifugation	Booth and Kenny (1974)
Rabbit kidney brush border membranes	Homogenization and differential centrifugation	Berger and Sacktor (1970)
Rabbit kidney basolateral membranes	Homogenization, centrifugation, and continuous sucrose density gradient centrifugations	Liang and Sacktor (1977)
Beef kidney glomerular basement membranes	Isolated glomeruli sieved and sonicated	Spiro (1967)
Rat kidney glomerular basement membranes	Sonication of isolated glomeruli in 1 M NaCl	Blau and Michael (1971)
Rat intestinal microvillar membranes	Hypotonic disruption of mucosal cell homogenates in presence of EDTA; low speed centrifugation	Miller and Crane (1961)
Ehrlich ascites cell membranes	Sonication and centrifugation in sucrose solution	Rajam and Jackson (1958)
Ehrlich ascites cell membranes	Homogenization and density gradient centrifugation	Wallach and Ullrey (1962); Wallach and Kamat (1964)
Mouse fibroblast plasma membranes (L cells)	Pretreatment of cells with various agents, homogenization, and gradient centrifugation	Warren et al. (1966)
HeLa cell membranes	Homogenization and isopycnic centrifugation	Bosmann et al. (1968)
HeLa cell membranes	Homogenization, isopycnic flotation, sonication and gradient centrifugation	Boone et al. (1969)
Krebs ascites cell membranes	Homogenization and separation on interface of layered sucrose	Stonehill and Huppert (1968)
Tissue culture cell plasma membranes	Hardened by Zn^{2+}, homogenized and separated by centrifugation on dextran–polyethylene glycol aqueous two-phase system	Brunette and Till (1971)
Cow milk gland membranes	Homogenization by whirling blade and low speed centrifugation	Keenan et al. (1970)
Nuclear membranes from Chinese hamster cells	Treatment of nuclear suspension with polyanion heparin–sucrose buffer	Hildebrand and Okinaka (1976)

[a]FMA, fluorescein mercuric acetate.

The most serious criticism or limitation in studying isolated membranes is the isolation procedure itself. Once the cell is destroyed, the biological integrity of the cell membrane comes into question, and, although isolated membranes are most easily analyzed, there is no doubt that different isolation procedures yield membranes which vary slightly in their composition (Dodge et al., 1963), as well as in their physiological characteristics (Kamat et al., 1972; Bramley et al., 1971; Hanahan and Ekholm, 1971). Homogenization must be gentle enough to yield large, uniform membrane segments so that repeated centrifugation will not select for higher density components which do not represent the whole membrane. Perhaps the most questionable component is the protein, since the loss of soluble protein from membranes does occur during the isolation procedure (Reynolds and Trayer, 1971); therefore, one should always take into account the isolation procedure when comparing membranes. Critical evaluations of membrane isolation and fractionation techniques may be found (Wallach and Lin, 1973; Maddy and Dunn, 1976).

Most easily obtained in "pure" form are membranes from the nonnucleate red blood cells of mammals. For this reason, many of the earlier studies on membrane composition were done with erythrocytes. In recent years, an increasing number of membrane types have been isolated from many sources using a variety of techniques. Some references to isolation procedures commonly employed may be found in Table 3.

Membranes from different sources have different concentrations of proteins, carbohydrates, and lipids (Table 2). Major differences between bacterial and mammalian membranes can be seen (Korn, 1968); however, it is the purpose of this review whenever possible to deal exclusively with mammalian systems. Among mammalian membranes, myelin represents the least complex system with respect to enzymatic or biosynthetic activity. Other mammalian membranes exhibit many functional and enzymatic properties, and it is therefore not surprising that these membranes are in a state of flux and that turnover of the components may be seen (Arias et al., 1969; Baker and Humphreys, 1972; Ferber, 1971; Warren and Glick, 1968; Siekevitz et al., 1967; Haddad et al., 1977; Doyle and Baumann, 1979). Glick and Warren (1969) have even demonstrated protein synthesis in isolated surface membrane vesicles from mouse fibroblasts, indicating sites of protein synthesis within the membrane structure.

With the exception of the carbohydrates, minor components of the cell membrane have not been studied in detail. Small quantities of RNA (10–25 μg/mg protein) could be found associated with isolated plasma membranes (Weiss, 1968; Benedetti and Emmelot, 1968). While absorption or adhesion of RNA to membranes can occur, the possibility that the RNA detected is a structural or functional constituent of the membrane—or both—should not be excluded, since no evidence of membrane contamination by single ribosomes or smooth ER has been found in rat liver membranes (Emmelot and Benedetti, 1967). If indeed the RNA is a constituent, there exists the exciting possibility of some form of informational capacity for in situ protein synthesis.

A. Carbohydrates

Carbohydrates, comprising up to 7% of the cell surface, are found almost exclusively in a bound form with proteins or lipids, as glycoproteins or glycolipids (Malhotra, 1970; Guidotti, 1972), although some investigators have favored the concept of an extraneous carbohydrate layer (Rambourg and Leblond, 1967). Those carbohydrates that are present may take the form of mucopolysaccharide moieties occurring on the exterior surface of the cell. Electrophoretic studies and enzymatic treatments of cells have indicated that the net negative surface charge is due primarily to the presence of sialic acids ($pK_a = 2.6$) in the sugar residues of the polysaccharides. The involvement of carbohydrates or mucopolysaccharides in membrane function has been mentioned previously in connection with antigenic properties, receptor sites, and intercellular contact and communication.

We have already indicated the role of the Golgi apparatus in the glycosylation of peptides or proteins before their release to the surface membrane. Schacter et al. (1970) have established the existence of three specific glycosyltransferases in the Golgi apparatus, which transfer N-acetylglucosamine, galactose, and sialic acid (N-acetylneuraminic acid) in sequence to an appropriate acceptor (oligosaccharides–protein), forming the most commonly occurring, terminal nonreducing trisaccharide moiety of serum glycoproteins. Roseman (1970) has demonstrated the synthesis of complex carbohydrates by glycosyltransferases in the Golgi apparatus and has postulated a mechanism for the involvement of such carbohydrates in the membrane function of intercellular adhesion.

Recent studies by Katz et al. (1977) have served to demonstrate a likely method for membrane glycoprotein synthesis—specifically, the insertion of the G protein of vesicular stomatitis virus (VSV) into pancreatic endoplasmic reticulum (PER). Their results are consistent with a scheme of membrane glycoprotein synthesis which postulates that the growing protein is extruded across the ER membrane with the amino terminus extruded first and is then translocated across the membrane. Asymmetric insertion, as well as glycosylation, requires the presence of membranes during protein synthesis, and glycosylation is restricted to the luminal surface of the ER.

Frequently occurring membrane sugar residues include glucose, galactose, glucosamine, galactosamine, mannose, fucose, and sialic acid. These sugars may or may not be present in their acetylated forms. All sialic acids are acylated derivatives of neuraminic acid ($C_9H_{17}O_8N$). Neuraminidase cleaves the O-glycosidic bond between the keto group of neuraminic acid and D-galactose or D-galactosamine. The structure of the most commonly occurring sialic acids and some discussion of their importance may be found in an article by Cook (1968). Per milligram of protein, most mammalian membranes studied contain more sialic acid and hexosamine than any other cell fraction (Benedetti and Emmelot, 1968), although the actual amount of sialic acid is variable. Part of this variation is related to growth rate and the cell cycle of cells in culture (Glick et al., 1971). The availability of the sialic acid residues also varies between organisms and treatments, but, for the most part, sialic acid is readily released by neuraminidase treatment (Benedetti and Emmelot, 1968; Heath, 1971). That portion of the membrane sialic acid not released by neuraminidase might be incorporated into glycolipids, and thus be buried in the lipid matrix or otherwise sterically inaccessible (Weinstein et al., 1970; Glick et al., 1970).

The functional significance of membrane sialic acids, the majority of which are recovered in membrane glycoprotein complexes, has been studied by a number of investigators. The asymmetric distribution of sialic acid on the outer portion of the membrane leaflet most likely has functional significance, since plasma membranes rather than inner membranes have higher contents of sialic acid per milligram of protein. A number of membrane processes which require the presence of sialic acid in some terminal sugar sequence include the functioning of viral (Cohen, 1963) and serotonin (Woolley and Gommi, 1964) receptors, transport of K^+ (Glick and Githens, 1965), secretion of protein from cells (Glick et al., 1966), and establishment of a negative surface charge (Cook et al., 1961, 1963; Wallach and DePerez Ensandi, 1964; Ambrose, 1965). Most of these functions are inhibited or destroyed by the action of neuraminidase.

Only in liver plasma membranes has it been demonstrated that sialic acid does not contribute to the net negative charge of the liver cell (Doljanski and Eisenberg, 1965). The manner in which sialic acid residues dictate the specific binding properties of liver cells has been revealed in studies of circulating glycoproteins. Cell surface receptors called hepatic binding proteins (HBP) have been isolated from liver plasma membranes. These HBPs recognize and bind circulating asialoglycoproteins, which are then transported to hepatocytes and catabolized in the lysosome. Stockert et al. (1977) suggest that sialic acid plays a protective role in glycoprotein metabolism by the liver. The sialic acid residues mask galactosyl residues in both circulating glycoproteins and in membrane-bound HBP. When exposed, the galactosyl residues of circulating glycoproteins are recognized and bound by membrane-bound HBP, thus removing the glycoprotein from circulation. Exposed galactosyl residues on HBP reduce the ability of HBP to bind circulating asialoglycoproteins by competing for nearby binding sites. Thus, the sialic acid "mask" prevents autoinhibition.

In a junctional complex, three types of areas have been identified: the tight junction (zonula occludens), the intermediate junction (zonula adherens), and the desmosome (macula adherens). Only in the tight junction has no sialic acid been found. The other two areas are suspected of participating in cell-to-cell adhesion, and the involvement of sialic acid and hexosamine residues in that process has been suggested (Roseman, 1970). The sialic acid may act as a binding site for the Ca^{2+} necessary for adhesion (Jaques et al., 1977). In binding to the sialic acid, Ca^{2+} is thought to "mask" the carboxyl groups of the sialic acid, rendering them insensitive to neuraminidase (Benedetti and Emmelot, 1968). The effect of EDTA in reversing this process by "opening" intermediate junctions and desmosomes and making them neuraminidase sensitive can then be understood. On tight junctions, the effect of EDTA is more subtle, since no sialic acid seems to be present. However, it still disrupts the intercellular passage of ions and molecules irreversibly and can be seen by electron microscopy to cause transverse densities in the zonula occludens. Ca^{2+} as well as sialic acid, then, plays a critical role in intermembrane cohesion and intercellular communication.

One of the most difficult aspects of studying regulatory mechanisms such as those involved with the cell membranes is the inability to examine all of the necessary components of the system involved. There is no doubt, however, that the minor carbohydrate components of cell membranes play major roles in immunologic behavior, electrical potentials, and transmembrane movement.

B. Proteins

A major component of biological membranes is protein. In characterizing membrane proteins, the first line of investigation has been to identify their enzymatic activities. However, with the use of solubilizing agents such as SDS and the development of amino acid analysis, gel electrophoresis, ORD, circular dichroism, and other methods, more information about structural proteins has evolved.

There are primarily two classes of membrane proteins: the extrinsic, peripheral, or soluble proteins; and the intrinsic, integral, or structural proteins. The extrinsic or peripheral proteins are those easily dissociated from the membrane by mild treatment such as washing or solubilization with dilute chaotropic agents. Singer (1971) has postulated that the amino acid sequence of these proteins allows for the maximum hydrophobic interactions internally while simultaneously satisfying hydrophilic interactions externally.

Proteins in the second class—the intrinsic or integral proteins—have a higher content of nonpolar amino acid residues, which favor hydrophobic interactions of the protein with the hydrophobic lipid interior of the membrane. Consequently, these proteins are less readily soluble (Singer, 1977). The degree of protein–lipid interaction can be measured by the ease of solubilization of the membrane proteins, which varies considerably. This involvement with the lipid component of the membrane is a major problem in protein characterization, since removal of the lipid usually denatures the protein (see Chapters 9 and 10 for further discussion). Kaplan and Criddle (1971) have written an excellent review of the problems involved and techniques used in handling and characterizing structural proteins. It is their suggestion that structural proteins, which may or may not have enzymatic activities, can have a role in defining the binding sites which determine the spatial relations in enzymes of a reaction sequence, can undergo conformational changes as a function of active transport, or can determine architectural sites for absorption of nonprotein components into the protein matrix.

In a review on membrane proteins, Guidotti (1972) has classified membranes into three types on the basis of their protein content. The first class is the simple, inert membrane represented by myelin. It consists primarily of lipid with little protein, acts as a permeability barrier and insulator, and has only three known enzymatic activities (Beck et al., 1968; Olafson et al., 1969; Kurihara and Tsukada, 1967; Cammer et al., 1976; Yandrasitz et al., 1976). The large second class of membranes which have a protein-to-lipid ratio of about 1:1 (w:w) are typified by most mammalian plasma membranes. They have many enzymatic activities and sophisticated transport systems associated with them, in addition to the permeability factor. The third class of membranes has bacterial and inner mitochondrial membranes as its models. These membranes have proportionately larger amounts of protein than lipid and have added functions such as oxidative phosphorylation and nucleic acid synthesis. In general, the specialization and enzyme function of the membrane increases in proportion to its protein content. Table 4 gives the amino acid composition of some isolated membrane proteins. Total membrane protein (intrinsic + extrinsic) often has an amino acid composition which falls into the range of other nonmembrane, "soluble" proteins (Vanderkooi and Capaldi, 1972).

It has been demonstrated that many membrane-bound enzymes are lipid-requiring and that the protein–lipid interaction within the membrane plays an important part in the biological activity of the

Table 4. Amino Acid Composition of Some Proteins from Isolated Membranes

	Residues/1000 residues				
				Bovine myelin (peripheral nerve)[d]	
Amino acid	Rat liver PM protein[a]	Rat glomerular basement membrane[b]	Human erythrocyte ghost[c]	First basic protein	Second basic protein
Hydroxylysine		18			
Lys	68	35	52	144	140
His	22	20	24	8	18
Arg	50	50	45	64	58
Asp	78	89	85	110	97
Thr	62	45	59	109	87
Ser	82	56	63	73	73
Glu	113	112	122	80	101
Pro	75	60	43	18	33
Gly	88	191	67	92	105
Ala	86	65	82	45	49
Val	62	44	71	86	68
Isoleu	48	37	53	48	39
Leu	98	73	113	77	87
Tyr	25	16	24	12	18
Phenylala	43	33	42	40	39
Hydroxyproline (3 and 4)		58			
Try			25	14	11
Met			20	20	16
Half-cystine			11		

[a]Simon et al. (1970)
[b]Blau and Michael (1971)
[c]Rosenberg and Guidotti (1968)
[d]London (1971)

membrane (Coleman, 1973; Gulik-Krzywicki, 1975). Thus, the interaction of proteins and lipids in the membrane is a source of controversy as far as a structural model for membranes is concerned. Indeed, there is no reason to be sure that the distribution of certain proteins or lipids is uniform in a membrane or that any one model will be accurate for an entire membrane.

C. Lipids

The recognition of lipid as one of the major components of biological membranes dates back to the late nineteenth and early twentieth centuries (Overton, 1895; Gorter and Grendel, 1925). The immiscibility of lipids and water has been a stimulus to the investigation of lipid components of the membrane since its nature as a permeability barrier was first recognized. Membranes also contain large amounts of phospholipids whereas other cell organelles do not, making membrane lipids an interesting subject for study. In addition, in the 1950s and 1960s, simple definitive methods were developed for lipid analysis, including silicic acid thin-layer chromatography and gas–liquid chromatography. The availability of synthetic and natural lipids at that time also prompted the investigation of lipid bilayers.

The natural predisposition of a lipid–water system to form stable bilayers was first demonstrated by Mueller et al. (1962a,b). The lipid molecules condensed, with their nonpolar ends on the inside and polar (hydrophilic) groups on the outside, in contact with the aqueous medium. The orientation of these bilayers supported the Danielli–Davson model of membrane structure. Further investigation using artificial membranes yielded much information on the ordered formation of multimolecular structures of lipids in aqueous environments, as well as on the fluidity of hydrocarbon chains in these structures. From such studies it was postulated that the most stable, tightly packed membranes resembling lipid bilayers would contain large amounts of cholesterol and long-chain saturated fatty acids. This is precisely what has been found in myelin, with its high lipid and cholesterol content (Tables 2 and 5). Inner mitochondrial membranes, however, contain virtually no cholesterol and seem more likely to be composed of lipoprotein subunits (Green and Perdue, 1966). Thus, membrane structure can be correlated with the type of lipid components present. Lee (1975) has compiled much of the data on the importance of lipids to membrane structure and function.

Membrane lipids may be classified in a variety of ways. In this review, membrane lipids have been divided into three groups: (1) neutral lipids; (2) phospholipids; and (3) glycolipids. The neutral lipids contain the simple lipids, such as the fatty-acid esters of glycerol (mono-, di-, and triglycerides), cholesterol and cholesterol esters, and free fatty acids. The mono-, di-, and triglycerides are glycerol molecules with one, two, or three fatty acids esterified through the hydroxyl groups of glycerol. Cholesterol is the major steroid found in membranes and may be free or occasionally in an esterified form. All of these neutral lipids are found in a chloroform–methanol (1:1) extraction of membranes. The polar lipids appear in the methanol phase and include phospholipids and glycolipids. A scheme of such an isolation procedure is presented by Weinstein et al. (1969).

Glycerophosphatides are the most abundant phospholipids. They are derivatives of glycerol phosphate, containing one glycerol molecule, two fatty acids, a phosphate group, and a nitrogen or organic base (either choline, serine, ethanolamine, or inositol). These glycerophosphatides are named for their corresponding base, e.g., phosphatidylcholine, phosphatidylserine, etc. Lyso derivatives contain only one fatty acid. Phosphatidic acids do not contain the nitrogen or organic base but otherwise resemble the glycerophosphatides mentioned above. Phosphoglycerides have one glycerol residue, two fatty acids, and one phosphate group, while cardiolipins have two glycerol, four fatty acid, and two phosphate residues. Plasmalogens are a small class of glycerophosphatides which resemble phosphatidylcholine and phosphatidylethanolamine except for the presence of an α,β-unsaturated ether linkage. The other major phospholipid is sphingomyelin, which contains the nitrogen base sphingosine, one fatty acid, a phosphate group, and choline. The molecular structures of these lipid classes may be found in Chapman (1970).

The glycolipids are those lipids containing sugar moieties and are classed mainly as gangliosides, cerebrosides, or ceramides, in order of decreasing complexity. The ceramides are simply sphingosine moieties with one esterified fatty acid. The cerebrosides are formed by the addition of a galactose molecule to a ceramide, and the gangliosides contain glucose, neuraminic acid, and/or a hexosamine, in addition to the components of cerebrosides. Sulfatides are sulfur-containing glycolipids. The lipid composition of some isolated mammalian membranes is presented in Table 5 to give the reader some idea of the variability of these components according to membrane type, species specificity, environmental factors, preparation methods, and analytical methods.

The distribution of fatty acids varies considerably between membranes, tissues, species, and phospholipid classes, but those which appear most commonly have ratios of carbon atoms to double bonds of 16:0, 18:0, 18:2, 20:4, 22:5, and 22:6 in the glycerophosphatides, as well as 24:0 and 24:1 in sphingomyelin (Dewey and Barr, 1970). Fatty acids are usually attached by ester linkages to the hydroxyl groups of the phosphate and glycerol molecules of phospholipids. In the case of plasmalogens, α,β-unsaturated ether links exist. Reviews and papers which cover the fatty acid composition in various lipid classes include those by Dewey and Barr (1970) and by Chapman (1970). It now seems evident that the fatty acid composition of a particular membrane provides the correct fluidity under the particular environmental temperature and conditions to fulfill the required rate for diffusion or metabolism in each tissue (Chapman et al., 1966). Changes in lipid class composition have been demonstrated with age and in certain pathological states (Rouser et al., 1971).

The interaction of lipid, protein, and carbohydrate in the total membrane is complex and cannot be clearly defined for all cells. Some aspects, however, are clear. Lipids, although they are nonantigenic, are

Table 5. Lipid Components of Some Mammalian Membranes. Results Presented as % Total Lipid.[a]

Compound[b]	Plasma membranes								Erythrocyte ghosts			
	Rat liver[d]	Rat liver[e]	Rat liver[f]	Rat liver[g]	L-cells[h] (mouse) TRIS	L-cells FMA	Human[i]	Human[j]	Human[k]	Human[l]	Ox[j]	Rat[j]
NL	34.6	32.5	34	27	42.3	41.3	23	29	28		33	28
C+CE	15.7	32.5	17	22.5	21.9	20.0	23	23	23	21.5	31.5	28
TG	4.7	trace	17	2.2	18.4	17.1	trace		5			
FFA	6.8			2.3	2.0	3.2						
PL	39.0	65	59	47	57.7	58.7	60	58	59		56	61
PA					8.2	7.7	6.0					
PG	2.9			1.7	0	0						
CL			—		0	0				15.9		
PI	2.8		3.5	6.9	2.9	3.2	trace					
PS	3.2		3.5		2.3	2.6	6.0	1.7	16	5.2	3.9	0.6
PE	6.0		6.5	10.0	7.5	8.1	8.5	14–18.6 (13.9–20.3)	15	14.3	18–19 (17.9–23)	11–18 (11–18.3)
PM												
EPM												
CPM												
PC	14.3	20	24.2	19.2	18.6	18.0	16.8	22.6	16	16.3	3.9	34.2
LPC	1.4			1.2	2.9	3.3		21.5 (38–45)			34.2 (33–38)	15.9 (43–50)
SM	7.2	10	19.5	8.0	14.1	13.6	9.0		12	9.3		
GL	26.4		7	12	0.7	0.8	trace		13	12 trace		
G					0	0						
CB												
CA										—		
S					0	0						
U	7.5	2.5			0	0				5.5	2.2	2.4
P								2.3				

Membrane Structure and Transport Systems

Compound[b]	Myelin						Microsomes				Mitochondria Rat liver[p]	
	Human[c,m]	Human[n] A	B	Ox[o]	Ox[n]	Rat[n]	Human[n]	Rat brain[n]	Rat liver[p]	Rat liver[f]	Outer	Inner
NL												
C+CE	24.9	27.0	32.8	28.1	24	25.9	22.8	21.7	8	19	3	2
TG									1	4	—	—
FFA												
PL										15		
PA												
PG										80		
CL		0.7	—		—	5.1	5.7	7.5		8	2	14
PI		0.8	9.8	1.0	1.6	2.5	5.4	2.2	0	7.2		
PS	6.7	8.9		6.4	8.9	8.0	8.7	6.7	10	8.0	14	4
PE	14.1	23.2	19.4	18.7	15.9	22.0	13.0	20.0	15	18.4	25	27
PM												
EPM		14	18	15.2	13.4	17.0	10.3	8.3				
CPM						4.1						
PC	10.5	9.9	9.1	11.1	9.2	13.5	29.3	21.8	60	39.2	52	45
LPC									4	—	3	3
SM	5.6	8.3	6.3	6.3	6.6	4.0	8.7	4.2				
GL												
G												
CB	20.2	20.5	16.1	23.5	23.4	18.3	8.3	6.4				
CA	1.5											
S	5.0	4.6	—	3.7	—	4.1						
U	11.5											
P								0.9	2		1	5

[a] The reader is advised to consider the results tabulated above as estimates only, given to emphasize differences and similarities among different membrane preparations, since calculations used in the preparation of the table have often required assumptions which could not be verified from the papers containing original data. Where molar percent was converted to percent total lipid, the molecular weights used, unless otherwise stated were CL = 2251; PA = 732; PG = 508; PC = 819; CPM = 817; EPM = 716; PE = 777; PS = 820; PI = 895; CA = 580—all of the above assuming all fatty acids were palmitic. Molecular weights for SM = 742 and C + CE = 386 + 668 were calculated assuming all fatty acids were oleic; while CB = 826, S = 905, and G = 1928 were calculated assuming all fatty acids to be arachidonic. Unidentified lipid components were assigned a molecular weight of 800.

[b] NL, total neutral lipids; C + CE, cholesterol and cholesterol esters; TG, mono-, di-, and triglycerides; FFA, free fatty acids; PL, total phospholipids; PA, phosphatidic acids; PG, phosphoglycerides; CL, cardiolipin; PI, phosphatidylinositol; PS, phosphatidylserine; PE, phosphatidylethanolamines, including lyso-PE; PM, total plasmalogens; EPM, plasmalogens containing ethanolamines; CPM, plasmalogens containing choline; PC, phosphatidylcholines (lecithins); LPC, lysophosphatidylcholine (lysolecithin); SM, sphingomyelin; GL, total glycolipids; G, gangliosides; CB, cerebrosides; CA, ceramides; S, sulfatides; U, unidentified; P, phosphorus.

[c,d] Molecular weights to convert molar percent to percent total lipid were those found in O'Brien and Sampson (1965).

References: [e]Skipski et al. (1965); [f]Benedetti and Emmelot (1968); [g]Dod and Gray (1968); [h]Pfleger et al. (1968); [i]Weinstein et al. (1969, 1970); [j]Maddy (1966); [k]De Gier and Van Deenen (1961); [l]Van Deenen (1968); [m]As found in O'Brien (1967); [n]O'Brien (1965); [o]Cuzner et al. (1965); [p]Norton and Autilio (1966); [q]As found in Siekevitz (1970).

apparently necessary for stabilizing membrane structure and for the normal functioning of antigenic protein (Popp, 1971). The asymmetric distribution of phospholipids in the erythrocyte membrane apparently serves to regulate hemostasis and to avoid thrombosis through the localization of procoagulant phospholipids only at the cytoplasmic surface of the membrane (Zwaal et al., 1977). Removal of most of the lipid component of membranes has been found to inactivate membrane-bound enzymes (Fleischer et al., 1962) and antigens (Green, 1972). Subsequent addition of certain lipids (usually phospholipids) has partially restored these activities (Fleischer et al., 1962; Kimelberg and Papahadjopoulos, 1972; Dean and Tanford, 1977). The involvement of inositol phospholipids in the essential active functions of membranes is strongly suggested (Michell, 1975). The stability of the membrane matrix may be due primarily to the hydrophobic interactions between proteins and lipids, and the destruction of enzymatic activity by delipidation may be the consequence of altered tertiary protein structure because of loss of the lipid backbone. The fatty acid composition of the membrane and changes in this composition produced by controlled fat diets have been shown to change the allosteric behavior of membrane-bound enzymes (Farías et al., 1975). It has also been hypothesized that membrane proteins partially influence the lipid composition of a given membrane because of the preferential binding of certain lipids to isolated membrane proteins (Kramer et al., 1972). The importance of mono- and divalent cations in the recombination of lipid–protein subunits indicates one vital role of inorganic ions in membrane function (Razin et al., 1965). The association of changes in carbohydrate components (glycolipid and glycoprotein) with altered antigenic properties is well documented, especially in virally transformed cells (Altman and Katz, 1976; Weber et al., 1976); the biochemical basis for this has been demonstrated to be the impairment, in at least one cell strain, of N-acetylgalactosamine transferase activity (Cumar et al., 1970). An excellent volume devoted to methods of studying membrane lipids has been compiled (Korn, 1977). Although unresolved questions remain when correlating membrane structure with composition, continuing investigations in this area should improve our understanding not only of membrane structure but of membrane function as well.

VI. Transport Systems

As mentioned previously, among many other functions important in the biological activity of the plasma membrane is its control over the movement of various cellular components into and out of the cell. Permitting the passage of some substances and restricting the transport of others, the plasma membrane represents one of the main regulatory units in the functioning of the cell. An overview of transport regulation is given by Kalckar (1976), Guidotti et al. (1978), and Ussing (1978).

Although there is a natural tendency toward equilibrium of the solute concentration on both sides of the membrane, such an equilibrium is rare in a living system, and selective permeability of the plasma membrane therefore assures the required distribution of metabolically important material inside and outside the cell. Kinetic studies of solute transport often permit characterization of the type of transmembrane movement involved (Neame and Richards, 1972). As outlined by Csáky (1965), a given substance can cross the cell membrane in several different ways: free diffusion, diffusion through pores, pinocytosis, and carrier-mediated transport.

A. Free Diffusion

Uninfluenced by other factors, all solutes will diffuse down a concentration gradient. In the same way, solutes will move across a membrane from an area of high concentration to one of lower concentration in an attempt to equilibrate. The rate of diffusion should then behave according to Fick's first law of diffusion, $dn/dt = -DA(dc/dx)$, where dn/dt is the number of molecules per unit time crossing area A in the interface at a concentration difference of dc over the distance dx. The cell membrane, however, acts as a selectively permeable mechanism owing to its lipid layers. In general, the rate of permeation of a substance is related to its lipid solubility, and the membrane thus acts as a solvent for nonpolar molecules.

Penetration of a substance is measured by the permeability coefficient, P, which could be converted to a measurable diffusional coefficient, D, if Fick's law applied strictly. In the more complex situation of a membrane barrier, Kedem and Katchalsky (1958, 1961) have shown that under rigidly controlled conditions there exist at least three parameters which must be considered when characterizing the behavior of a membrane toward a particular solute: (1) the interaction between membrane and solvent; (2) the interaction between solute and membrane; and (3) the interaction between solute and solvent. The reflection coefficient, σ, measures relative rates of solute and solvent permeabilities in the system (Staverman, 1952) and is therefore a measure of semipermeability. L_p is the mechanical coefficient of filtration or pressure filtration coefficient, and ω is the solute mobility or solute diffusional coefficient. In the case of living membranes, conditions such as volume flow, osmotic gradients, and cell volume can be manipulated in order to measure the phenomenological coefficients σ, ω, and L_p. Detailed discussions of the theories, methods, and problems involved in such measurements can be found elsewhere (Stein, 1967; Kotyk and Janacek, 1970; Dowben, 1969).

Charged materials influenced by electrical fields may tend to diffuse against their concentration gradient if exposed to a high potential gradient. The charge on an ion adds complexity to the description of its movement across a membrane, and it becomes a necessary condition that the sum total of positive and negative charges be equal on the same side of the membrane. For a simple salt such as NaCl, the cation movement then becomes dependent on Fick's law (cation concentration) and on the local concentration of the charged anion. More complex systems consisting of a number of ions can be expressed in terms of the activities of the components in equilibria (Daniels and Alberty, 1961). We shall omit further discussion of this area since a comprehensive review on electrodiffusion and membrane transport phenomena has previously been published (Lakshminarayanaiah, 1969).

B. Diffusion through Pores

Water and small water-soluble particles may penetrate the proteolipid barrier of the cell membrane by passing through hypothetical pores in the membrane structure. Most cells other than erythrocytes are more permeable to cations than to anions. The passive diffusion of cations through negatively charged pores has been offered as an explanation for these ion transport phenomena. Smythies et al. (1971) have constructed a model pore for the diffusion of Na^+ to explain the existing permeability data. Hille (1971) proposes a model of an oxygen-lined cavity 3×5 Å, in order to explain cationic selectivity. A review (Spurway, 1972) concerning anionic movement and the "fixed pore concept" refers to water-filled channels (pores) in the membrane whose permselectivity is due partly to pore radius and partly to the charged groups fixed in their walls. The high anionic permselectivity of the red blood cell has been attributed to the positively charged NH_4^+ groups in the superficial layers of the membrane (Schnel, 1972). More recently, Dutton et al. (1976) have provided evidence supporting the hypothesis that Ca^{2+} transport in sarcoplasmic reticulum occurs via a "pore-type" mechanism.

An alternative method for the movement of water-soluble material across a membrane would include pinocytosis or some carrier-mediated process.

C. Pinocytosis

Although pinocytosis has not been observed in most mammalian cells, it has been seen in kidney tubular cells, intestinal epithelial cells, leucocytes, reticuloendothelial cells, the Kupffer (phagocytic) cells of the liver, and some malignant cells. A variety of substances can induce vesicular formation, among them proteins, viruses, amino acids, and ions. It is interesting to note, however, that carbohydrates and nucleic acids cannot induce pinocytosis.

Pinocytosis itself is an active process (Gross, 1967), requiring the expenditure of metabolic energy for vesicle formation. It is temperature dependent and inhibited by anaerobiosis, 2,4-dinitrophenol, fluoroacetate, cyanide, and carbon monoxide (Dowben, 1969). The only selective phase of pinocytosis appears to be the binding of inducer molecules to the cell surface, which may be independent of the energy requirement. The membrane of the pinocytotic vesicle in rat kidney has been shown to differ from the brush border, basolateral plasma membrane, and lysosomal membranes of the tubule cell (Bode et al., 1976), indicating that it acts as a separate entity from the rest of the cell membrane. Indeed, the formation of pinocytotic vesicles may be a means of taking into the cell those portions of the membrane which have adsorbed specific substances. Acidic mucopolysaccharides of the cell surface may be the specific binding sites involved (Dowben, 1969). Upon interaction and adhesion of the inducer substances, it has been noted that the permeability of cell membranes—at least to glucose—increases (Holter, 1959). Pinocytosis may play a role in the transport of proteins into embryonic cells (Giese, 1962), the transport of B_{12}-intrinsic factor in the intestine (Wilson, 1963), and the transport of ions (Ussing, 1965), but it remains a complicated process defying functional classification as a transport mechanism.

D. Carrier-Mediated Transport and Ion Pumps

The penetration of water-soluble material into the cell is thought to occur via a hypothetical membrane-transport "carrier," since a significant part of the cell membrane is composed of lipids and is therefore somewhat restrictive to the passive diffusion of metabolically important substances in aqueous solution. The lattice-pore model has been proposed by Naftalin (1972) as an alternative to the carrier model for sugar transport in erythrocyte membranes. However, the concept of carrier-mediated transport has been postulated by many investigators, with the general scheme consisting of three steps: (1) binding of the penetrant to the "carrier" site; (2) translocation of the penetrant across the membrane; and (3) release of the penetrant on the other side of the membrane.

The nature of the carrier represents an open question, and only its isolation and identification can provide an appropriate answer. Considering substrate specificity, which is characteristic for carrier-mediated transport, it is very likely that the substrate binding site is proteinaceous, because protein possesses selective behavior. This presumption is supported by evidence that puromycin, an inhibitor of protein synthesis, inhibits the transport of amino acids in rat kidney slices (Elsas and Rosenberg, 1967). In further support of this concept are results obtained in studies of bacterial membranes, which have shown that the kinetic constants for transport of amino acids parallel the binding activities of isolated "binding" proteins. Carafoli and Crompton (1976) have reviewed the work in the area of binding proteins and membrane transport in both bacterial and mammalian systems. In this chapter, we will focus on transport systems in the plasma membrane of mammalian cells rather than intracellular membrane systems.

Studies in animal membranes have not reached as advanced a stage as have those in bacteria, but certain progress has been made. The binding of glucose or phlorizin, a competitive inhibitor of glucose, to human erythrocyte membranes or renal brush border membranes has been reported (Bobinski and Stein, 1966; Bode et al., 1970). At present, the distinction between binding to the membrane and uptake by the membrane vesicle for a given substance depends on reaction of the membrane preparation to osmotic perturbations. Uptake of sugars and amino acids has been studied in isolated membrane preparations (Table 8), but isolation and characterization of a mammalian carrier protein is difficult. The ADP–ATP carrier in mitochondria has been studied extensively (Klingenberg, 1976); however, Semenza (1976) and coworkers were the first to isolate a soluble protein—the intestinal sucrase–isomaltase complex—and convincingly demonstrated its transport capabilities in reconstituted artificial membranes.

In regard to the mode of action of the carrier in the membrane, various proposals have been made, such as the theory of sliding membranes (Booij, 1962), the invagination hypothesis (Crane, 1966), allosteric transition between two states (Vidaver, 1966), and a rotating model (Kotyk and Janáček, 1970), to mention only a few. Carrier mediation occurs in several types of transport: (1) facilitated diffusion; (2) exchange diffusion; and (3) active transport, which requires an energy source. The carrier does not, however, provide the energy for transport, and, where energy is required, it must come from the coupling of transport to metabolic reactions, to the flow of matter in the cell, or both (Wilbrandt, 1975). This immediately brings to mind the role of proton flux in the chemiosmotic theory (see Chapter 9).

1. Facilitated Diffusion

Certain physiologically important water-soluble molecules penetrate cell membranes at a faster rate than would be expected from the knowledge of membrane structure. To explain the rapid penetration of such substances, a type of carrier-mediated transport—"facilitated diffusion"—has been proposed. Several criteria for the identification of facilitated diffusion systems have been outlined:

1. The system has no requirement for free energy other than that needed to maintain the cell membrane structure. Facilitated diffusion is driven by an existing electrochemical gradient of the permeant and eventually leads to the disappearance of this gradient.
2. The rate of penetration of permeant is highly stereospecific. Isomers and optical enantiomorphs of the penetrant will most likely diffuse at markedly different rates.
3. The rate of penetration is not directly proportional to concentration (as in Fick's law) but reaches a saturation value as concentration increases, thus indicating a limited number of carrier sites.
4. Structural analogs of the permeant can compete for carrier sites, thus reducing the rate of permeant penetration.
5. Inhibitor substances which often act as enzyme poisons and are chemically different from the permeant may reduce the rate of penetration. Different inhibitors will "poison" various facilitated diffusion systems specifically. This criterion also applies to active transport systems.
6. The rate of permeant penetration measured as net transfer of permeant may be considerably less than the unidirectional flux measured by use of isotopically-labeled permeant.
7. The high temperature coefficients characteristic of chemical reactions ($Q_{10} \sim 3$) are frequently found in facilitated diffusion systems.

While the characteristics of this type of transport have been studied and identified in a number of cell systems and for a number of substances (Table 6), little is known of the molecular mechanism involved, although kinetic analysis of facilitated diffusion may be found (Stein, 1967; Neame and Richards, 1972). However, because the proposed mechanism involves a reversible reaction of the substrate with a membrane carrier to form a complex which traverses the membrane and releases the substrate at the other side, it may be that facilitated diffusion mechanisms do not differ from those involved in active transport of the same molecule (Csáky, 1965; Wilbrandt, 1972). Specifically, the active transport of D-glucose may simply require the presence and cotransport of sodium ions (Crane, 1962; Stein, 1967). This theory has recently received support from studies of Na^+-gradient-dependent uptake of D-glucose by isolated intestinal and renal brush border membranes (Murer and Hopfer, 1974; Kinne et al., 1975). However, the electrogenic nature of D-glucose transport is probably more accurately class-

Table 6. Facilitated Diffusion Systems for Amino Acids and Sugars

Tissue	Substrate	References[a]
Human erythrocytes	Glc	1,2,3,4
	Gal	1,4,5,6
	Man	2,4,5,6
	Xyl	5,6
	Rib	5,6
	D and L-Ara	5
	Sorbose	8
	Fru	8
	Leu	9
	Phe	9
	Met	9
	Val	9
	Ala	9
	Gly	9
Guinea pig erythrocytes, fetal	Glc	10
Rabbit erythrocyte	Glc	11
	3-O-MG[b]	11
	Rib	12
	Leu	13
	D and L-Val	13
Rat heart muscle	Glc	14,15
	Ara	16
	Xyl	16
Guinea pig brain slices	Xyl	17
	Glc	17
	2-dGlc	17
Mouse fibroblasts (L-cells)	Glc	18
	Gal	18
	2-dGlc	18
Ehrlich ascites tumor cells	3-O-MG	19
	Gal	19,20
	Rib	19,20
	Sorbose	19
	Glc	20
	Xyl	20
	D and L-Ara	20

[a](1) LeFevre (1954); (2) Sen and Widdas (1962a,b); (3) Britton (1964); (4) Miller (1965); (5) LeFevre (1962); (6) Widdas (1957); (7) LeFevre (1963); (8) Miller (1966); (9) Winter and Christensen (1964); (10) Dawson and Widdas (1964); (11) Regen and Morgan (1964); (12) Steinbrecht and Hofmann (1964); (13) Winter and Christensen (1965); (14) Post et al. (1961); (15) Morgan et al. (1964); (16) Fisher and Zachariah (1961); (17) Gilbert (1965); (18) Rickenberg and Maio (1961); (19) Crane et al. (1957); (20) Kolber and LeFevre (1967).
[b]3-O-Methyl-D-glucose.

ified as secondary active transport or flow-coupled active transport (Wilbrandt, 1975). The kinetic treatment of carrier-mediated transport is, in general, analogous to that of the Michaelis–Menten kinetics of enzyme–substrate interactions.

2. Exchange Diffusion

In some cases, the facilitated movement of a permeant down its electrochemical gradient may be linked with the movement of its structural analog in the opposite direction. This produces an apparent "uphill" transport of the analog against its electrochemical gradient and is often called "counterflow," "counter-transport," or "exchange diffusion." Exchange diffusion was first proposed by Ussing (1952) to explain the situation where no net transfer is observed but where there is a one-to-one exchange of substrate across the membrane. The phenomenon of exchange diffusion will occur when two substrates share a common facilitated diffusion system, when a concentration of the driving substrate must be made or maintained across the membrane, and when the concentration of the substrate on one side of the membrane is high in relation to its K_m. The influx and efflux pathways for the molecules must not interfere with one another in exchange diffusion (counterflow) or competitive exchange diffusion (Dowben, 1969; Stein, 1967). According to Rosenberg and Wilbrandt (1957), the membrane component must be a movable "carrier" for the specific process of "countertransport." Different experimental conditions are used to demonstrate counterflow, countertransport, and exchange diffusion as opposed to competitive exchange diffusion. However, it has been shown that these variations of a basic exchange phenomenon are kinetically equivalent (Neame and Richards, 1972) and that a mobile carrier mechanism or a system of unidirectional pores specifically for influx or efflux could serve equally as well for all. Exchange diffusion has been observed for a variety of amino acids in Ehrlich ascites cells (Heinz and Walsh, 1958; Jacquez and Sherman, 1965), rat kidney cortex (Schwartzman et al., 1967), and mouse pancreas (Clayman and Scholefield, 1969).

3. Active Transport

The process which maintains a difference in the concentration of solute inside and outside the cell is energy requiring and is known as active transport. According to Rosenberg (1954), active transport is the term used when movement of substances occurs in an "uphill" manner, or against a combined difference in concentration and electrical potential of the substrate. If the compound crossing the membrane is uncharged, only the concentration gradient is taken into account. On the basis of the thermodynamics of irreversible processes, active transport is described as an interaction between a transmembrane passage and a metabolic reaction (Kedem, 1961). Substances which are actively transported in some tissue systems may be found in Table 7.

Primary active transport occurs when the transport of a substrate is coupled to an energy-yielding metabolic reaction. The energy required may come from several different sources: (a) the high-energy compound ATP used by a specific ATPase (ATPase pump); (b) energy from the electron transport system released as electrons that flow down the cytochrome chain (redox-pump); and (c) the electric field produced by free radicals. Implicit in these three theories is the participation of ions and ion transport. Secondary active transport is a term often used to denote the transport of one substrate linked to the flow of a second substrate. Wilbrandt (1975) refers to this as flow-coupled active transport; it may be this form of transport that is most often involved in the active uptake of sugars and amino acids. A review of some models of carrier-mediated active transport transport has recently been presented by Crane (1977).

Table 7. Some Active Transport Systems for Certain Compounds

Substrate	Tissue	Reference[a]
Ions		
Sodium	Rat ileum	1
	Rabbit ileum	2,3
	Human erythrocytes (efflux)	4,5
Potassium	Erythrocytes (sheep, beef)	6
	Human erythrocytes	4,5
Ammonium	Human erythrocytes	4
Calcium	Erythrocytes	7
	Sarcoplasmic reticulum of striated muscle cells	8
Sugars		
Glc	Hamster ileum	9
	Rabbit ileum	10
	Rat ileum	11
	Rabbit kidney cortex	12
Gal	Hamster ileum	9
	Rabbit ileum	10
	Rat ileum	13
	Rabbit kidney cortex	12,14,15
	Rat kidney cortex	16
α-Methyl-D-glucoside	Hamster intestine	17
	Rabbit kidney cortex	18

(continued)

Table 7 (*continued*)

Substrate	Tissue	Reference[a]
	Rat kidney cortex	19,20
3-O-Methyl glucose	Hamster ileum	17
	Rabbit ileum	10
	Rat skeletal muscle	21
2-dGlc	Rabbit kidney cortex	18
2-dGal	Rabbit kidney cortex	18
Myoinositol	Rat kidney cortex	22
Amino acids		
α-Aminoisobutyric acid	Whole rat kidney	23
	Rat kidney cortex	24
	Rat glomeruli	25
	Rat renal papilla	26
	Rat diaphragm	27
	Rat liver	28
Cycloleucine	Whole rat kidney	23
	Rat diaphragm	27
	Rat renal papilla	26
Arg	Rat kidney cortex	29
Lys	Rat kidney cortex	29,30,31
	Rat renal papilla	26
CyS	Rat kidney cortex	29,32,33
Cys	Rat kidney cortex	33
Val	Rat kidney cortex	34
Gly	Rat kidney cortex	24
	Rabbit kidney tubules	35
His	Rat kidney cortex	24
	Dog kidney	36
Phe	Rat kidney cortex	24
Pro	Rat kidney glomeruli	25
	Rabbit kidney tubules	37
Tyr	Dog kidney	36
Ala	Ehrlich ascites cells	38
	Mouse fibroblasts	39
Ser	Ehrlich ascites cells	38
Leu	Ehrlich ascites cells	40
	Mouse fibroblasts	39

[a] (1) Curran and Solomon (1957); (2) Schultz and Zalusky (1964a); (3) Love et al. (1965); (4) Post and Jolly (1957); (5) Glynn (1956); (6) Shaw (1955); (7) Schatzmann (1966); (8) Hasselbach and Makinose (1961, 1963); (9) Alvarado and Crane (1964); (10) Schultz and Zalusky (1964b); (11) Fisher and Parsons (1953a); (12) Kleinzeller et al. (1967a); (13) Fisher and Parsons (1953b); (14) Krane and Crane (1959); (15) Kolinska (1970); (16) McNamara and Segal (1972); (17) Crane (1960); (18) Kleinzeller et al. (1967b); (19) Segal et al. (1968); (20) McNamara et al. (1971); (21) Kohn and Clausen (1972); (22) Hauser (1968); (23) Wedeen and Thier (1971); (24) Rosenberg et al. (1961); (25) Mackensie and Scriver (1971); (26) Lowenstein et al. (1968); (27) London and Segal (1967); (28) Crawhall and Segal (1968); (29) Rosenberg et al. (1962); (30) Segal et al. (1967); (31) Ausiello et al. (1972); (32) Greth et al. (1971); (33) Segal and Crawhall (1968); (34) Smith and Segal (1968); (35) Hillman et al. (1968b); (36) Doty (1941); (37) Hillman and Rosenberg (1969); (38) Christensen et al. (1967); (39) Oxender et al. (1977); (40) Oxender and Christensen (1963).

a. ATPase Pumps

i. $(Na^+ + K^+)$-ATPase. Energy from ATP is utilized by ATPase, an enzyme whose activity depends on the integrity of the lipoprotein portion of the cell membrane (Schatzmann, 1962). This enzyme has been found in many sodium-transporting tissues from various species (Bonting et al., 1962). In 1957, Skou isolated an $(Na^+ + K^+)$-stimulated ATP hydrolyzing enzyme from crab nerve membranes and provided evidence for the enzymatic nature of the active transport of ions. Subsequently, the stimulation of ATPase activity with Na^+ and K^+ ions was described in erythrocyte membranes (Post et al., 1960) and in the crude nuclear fraction of kidney homogenate containing the bulk of the cell membranes (Whittam and Wheeler, 1961). The activity of ATPase is greatly reduced by digoxin, ouabain, and other drugs known to inhibit active ion transport, as well as by compounds such as p-chloromercuribenzoate (PCMB) and N-ethyl maleimide (NEM), which bind to -SH groups (Skou, 1962, 1963).

Whittam (1962a) and Whittam and Ager (1962) have shown that ATP hydrolysis within the erythrocyte ghost is stimulated by K^+ ions in the medium and by Na^+ ions inside the ghost. The effect of Na^+ on ATPase depends on the sodium localization in relation to the inner and outer surface of the membrane, since, at low external potassium concentrations, external sodium decreases the activity of the enzyme. Thus, the competitive inhibition of K^+ by Na^+ externally opposes the synergistic effect of internal Na^+ and external K^+ upon the activity of ATPase. Also, some alkali metal ions, such as lithium, rubidium, and cesium, show synergistic and directional stimulation of ATP breakdown; i.e., internal lithium and sodium, as well as external lithium, potassium, rubidium, and cesium, stimulate ATP hydrolysis, while internal potassium, rubidium, and cesium or external sodium do not (Whittam, 1962b).

The following scheme for the reaction between ATP, its hydrolyzing enzyme, and ions (Mg^{2+}, Na^+, and K^+) has evolved (Albers, 1976):

$$E_1 + ATP + Na_i^+ \rightleftharpoons E_1\text{-}P(Na^+) + ADP \quad (1)$$
$$E_1\text{-}P(Na^+) + Mg^{2+} \rightleftharpoons E_2\text{-}P + Na_o^+ \quad (2)$$
$$E_2\text{-}P + K_o^+ \rightleftharpoons E_2(K^+) + P \quad (3)$$
$$E_2(K^+) \rightleftharpoons E_1 + Mg^{2+} + K_i^+ \quad (4)$$

According to this proposal, formation of an enzyme–phosphate–sodium complex is the first step in the reaction. Enzyme phosphorylation requires the presence of Na^+ and Mg^{2+}, although the role of Mg^{2+} is not well defined [step (2)]. Phosphorylation of the enzyme [step (1)] is strictly dependent on Na^+. The release of Na^+ from the phosphorylated enzyme complex [step (2)] results in the movement of Na^+ outside the cell. The demonstration of the existence of a K^+-activated phosphatase associated with $(Na^+ + K^+)$-ATPase (Rega and Garrahan, 1976) provides additional evidence for the dephosphorylation step [step (3)] of this scheme, involving a phosphorylated intermediate. The final step results in the movement of K^+ into the cell. From the above scheme for ion translocation, the 2:3 ratio for K^+/Na^+ movement is not apparent. The reader is referred to the detailed discussion of stoichiometry presented by Albers (1976). Recently, Goldin (1977) has reconstituted highly purified $(Na^+ + K^+)$-ATPase from dog kidney in phospholipid vesicles and has demonstrated an in vitro Na^+/K^+-active countercurrent transport system having a stoichiometry of 3:2:1 for $Na^+:K^+:ATP$ hydrolyzed, and which is mediated by no protein component other than the two polypeptides of $(Na^+ + K^+)$-ATPase. Thus, the ATPase pump may, theoretically, provide a method for the linked active transport of sodium and potassium across the cell membrane.

In 1961, Eisenman derived a theory for cationic selectivity owing to differences in electrostatic field strength (EFS). On the basis of this theory, Skou (1964) suggested a sodium–potassium-linked transport system driven by an ATP-induced electron transfer. By altering the distribution of electrons on a macromolecule in a membrane pore, ATP could induce the change from Na^+ affinity to K^+ affinity of one charged group while inducing a corresponding change from K^+ affinity to Na^+ affinity of another group. The net result is transport of Na^+ in one direction and K^+ in the other.

The aforementioned theories on the ATPase pump assume the presence of a fixed-site carrier as distinguished from a mobile carrier, i.e., one which moves through the membrane. A mobile carrier mechanism was described by Hokin and Hokin (1963). According to their theory, cyclic phosphorylation and dephosphorylation of phosphatidic acid to diglyceride would result in a conformational change in a lipoprotein carrier, leading to the formation and destruction of specific binding sites that would bring about active transport of ions across the membrane. A spatial rotation of sites of 180° could be involved. Therefore, this cyclic conversion would function as a transducing mechanism using the chemical energy of ATP for active transport. Yager (1977) has proposed a mechanism involving only a single hydrophobic polypeptide helix spanning the membrane, translocation occurring as a result of interconversion between two helical forms. For further clarification and detail on the action of the $(Na^+ + K^+)$-ATPase system and its relation to transport, see reviews by Skou (1975, 1977), Albers (1976), Whittam and Chipperfield (1975), and Schultz (1978).

An entirely different view of the involvement of ATP in Na^+ and K^+ transport is presented by Ling (1977). This theory involves the interaction of water, ions, and ATP with a key protein in such a way that the association of ATP with the protein induces a change in the state of ion and water adsorption resulting in the K^+/Na^+ distribution found in biological systems. According to this theory, the energetics of ion transport do not reside in the high-energy-bond of ATP but in its specific electronic interaction with protein molecules.

ii. Ca^{2+}-ATPase. Calcium transport via a Ca^{2+}-ATPase found in the plasma membrane of many cell types and in the sarcoplasmic reticulum has led to the proposal of a Ca^{2+} pump which may well be found to be universal. Details of the Ca^{2+}-ATPase of sarcoplasmic reticulum can be found in Chapter 9 of this volume and in a review by MacLennan and Holland (1976). Plasma membrane transport of calcium has been reviewed by Vincenzi and Hinds (1976).

b. The Redox Pump. This theory proposes that respiratory enzymes act as ion carriers during the oxidation–reduction cycle (Conway, 1963, 1964). Using energy derived from electron transfer, one ion would move for each transmitted electron. Therefore, the mediation through an ATPase system would be unnecessary. This proposed system for ion transport in mitochondria has been reviewed by Meijer and Van Dam (1974).

c. The Free-Radical Theory. Although ATPase pump and redox pump theories have been generally accepted, other theories have been postulated to explain the transport of ions. The free-radical theory assumes that the transport of ions could be influenced by an electronic field caused by free radicals which are produced during metabolism (Kometiani, 1963).

Regardless of the energy mechanism involved, active transport of select ions, including Na^+ and K^+, remains one of the most striking features of the cell membrane and may provide the energy for active transport of other substances.

E. Transport of Amino Acids and Sugars

Although the transport of amino acids and sugars has been studied for a number of years, the mechanisms by which this movement occurs remain unknown. In mammalian cells such as those of the kidney, intestine, and muscle, and in erythrocytes, transport of most amino acids is an active process involving a proposed mobile "carrier" that binds its substrate before translocating it across the membrane. Sugars are usually actively transported only in the kidney and intestine, while in erythrocytes and muscle, sugar translocation takes place predominantly by facilitated diffusion. The basic characteristic of the active transport of both amino acids and sugars is the energy-dependent movement of solute against a concentration gradient. Some amino acids, such as the dibasic amino acids, carry a positive charge at a neutral (physiological) pH, and their accumulation against a concentration gradient could be the result of an electrical driving force since most cell interiors have a net negative charge at the same pH (Munch and Schultz, 1969). The active accumulation (energy-dependent) of these amino acids against a combined electrochemical gradient must then be demonstrated in order to identify active transport systems. Since sugars do not carry any charge, their transport against a concentration gradient is equivalent to their accumulation in the face of a combined electrochemical potential difference and may be considered a demonstration of active transport.

As we have mentioned previously, the existence and nature of a specific carrier involved in active transport in mammalian cells is not known but has been inferred. Scriver and Mohyuddin (1968) have reported heterogeneous uptake of α-aminoisobutyric acid (AIB) by kidney cortex slices over a wide range of substrate concentrations. This led them to suggest multiple systems which function at the same time but possess different binding affinities for amino acids. The alternative explanation could be a single system whose binding site can be altered conformationally with increasing substrate concentration, causing a change in affinity (Segal and Thier, 1973).

Meister (1973) has suggested that transport of many amino acids in the kidney and probably in other tissues could be mediated by a cycle of enzymatic reactions involving the membrane-bound enzyme γ-glutamyl transpeptidase as a binding site for amino acids. Meister does not, however, exclude the possibility of other transport pathways for the active transport of amino acids, nor is he able to define the extent to which the cycle could function. At the present time, the identification of the γ-glutamyl transpeptidase cycle as a mediator of amino acid transport represents only an interesting hypothesis, and attempts to correlate γ-glutamyl transpeptidase activity with amino acid transport activity have been fruitless.

On the basis of evidence obtained both in vivo and in vitro, Young and Freedman (1971) have proposed the existence of at least four different systems for amino acid transport in mammalian kidney, according to the amino acid structure, i.e., for neutral, basic, and acidic amino acids, and for imino acids and glycine. Using Ehrlich ascites tumor cells, Oxender and Christensen (1963) have characterized several transport "agencies" mediating the transport of neutral amino acids: (1) a sodium-dependent system with high affinity for alanine and glycine; (2) a sodium-dependent system that prefers alanine, serine, and cysteine; (3) a sodium-dependent system with high affinity for leucine and phenylalanine (i.e., amino acids with large nonpolar chains); and (4) a sodium-independent system which is not saturable but does not seem to involve simple diffusion. It appears that many of the active processes of amino acid movement in animal cells require sodium, which supports the proposal of Riggs et al. (1958) that active transport of amino acids is coupled to the movement of ions. Non-energy-requiring carrier-mediated transport of amino acids (facilitated diffusion or exchange diffusion) shows only a slight requirement for sodium ion (Wheeler and Christensen, 1967). Various kinetic models for amino acid movement coupled to sodium ion transport have been proposed; the general concept of having two substrates interacting at the same transport site in the membrane is referred to as cotransport. As the result of such interaction, both substrates are transported through the membrane in the same direction. Reviews on the interaction between sodium and organic solutes have been written by Schultz and Curran (1970) and by Ullrich (1979), and a comprehensive volume dealing with flow-coupled systems in the intestine and kidney has been published (Robinson, 1976). Some cotransport systems which have been identified can be found in Table 8.

Metabolic poisons and uncouplers such as dinitrophenol, cyanide, and cardiac glycosides, as well as anaerobiosis, can inhibit active transport (Segal et al., 1968; Hillman et al., 1968a; Chez et al., 1967), although it has been shown that aerobic metabolism is not always necessary for the active transport of certain amino acids (Freedman and Young, 1969; Silbernagel and Deetjen, 1969). The nature of the above inhibitions of active transport is not known, but they could result from an effect of the inhibitor directly on the transport system, on the supply of ATP, or on the sodium pump, which would result in an increase in the sodium concentration inside the cell, causing a disappearance of the sodium gradient (Curran, 1972).

A number of sugars have been demonstrated to interact with amino acid transport. Using rat kidney cortex slices, Thier et al. (1964) have shown inhibition of accumulation of some neutral amino acids by glucose and fructose. Since sugars were not effective inhibitors in the presence of 2,4-dinitrophenol and during anaerobiosis, the authors suggested that the basis for the interaction between amino acids and sugars could be at the level of the sodium-dependent ATPase system which provided the energy for active transport. Mutual inhibition of transport between certain amino acids, AIB, and α-methyl-D-glucoside

(αMG) in rat kidney was explained as competition of given substances for a common energy supply, or as allosteric inhibition by related but separate "carrier" systems (Genel et al., 1971). Alvarado (1966) has observed that active transport of neutral amino acids in the hamster intestine could be partially inhibited by D-galactose, L-arginine, and their actively transported analogs. It was proposed that amino acids and sugars could share the same polyfunctional carrier, consisting of a series of separate binding sites for sugars, neutral amino acids, basic amino acids, and sodium. In the carrier, allosteric interaction between associated binding sites would occur.

It is interesting that the active transport of sugars in intestine and kidney is largely sodium dependent, although not to the same extent as for amino acids. As demonstrated by Crane (1962, 1966), the active transport of glucose in the intestine is coupled to the movement of sodium ion down its concentration gradient. It is postulated that the energy for the active transport of sugars in the intestine is provided by the difference in the sodium concentration between the cytoplasm and external medium. The electrogenic nature of sugar transport has received support from work in isolated membrane preparations (Murer and Hopfer, 1974; Kinne et al., 1975). Kleinzeller et al. (1967a,b) have found that the transport of some, but not all, monosaccharides tested in rabbit renal cortex was sodium dependent. In rat kidney cortex, two components of the galactose transport system have been observed, one saturable and the other nonsaturable (McNamara and Segal, 1972). Only about 25% of total galactose transport was sodium dependent.

Although active transport systems for sugars have been observed in the kidney and intestine, it is important to mention that in almost all other mammalian cells, most sugars cross the cell membrane by facilitated diffusion. As postulated by Lieb and Stein (1970, 1971), the specific carrier involved in the facilitated diffusion of glucose in erythrocytes is considered to be a tetrameric protein, embedded in the cell membrane, which has two binding sites, one with high affinity for substrate and the other with low affinity for substrate. Combination of the substrate with either of these sites would cause the tetramer to undergo conformational changes, so that the substrate molecule enters the intramembranous space. Inside the membrane, interconversion between the two energetic forms can occur. The substrate in the intramembranous space distributes itself between the two forms according to energetic considerations. Upon a second conformational change, substrate bound to the binding site other than that on which it entered the membrane is then released, thus effecting its transmembrane migration. LeFevre (1973) has suggested a model for sugar transport across the erythrocyte membrane based on substrate-conditioned "introversion" of binding

Table 8. Cotransport Systems for Amino Acids and Sugars[a]

Tissue	Substrate	References[b]
Rabbit reticulocytes	Gly	1
	Ala	1
Rat intestine	Ser	2,3
	Asn	2
	Gln	2
	Asp	2
	Glu	2
	Gly	2,3
	His	2,3
	Leu	2,3
	Phe	2,3
	Ala	2,3
	Ile	2,3
	Met	2,3
	D and L-Pro	2,3
	Val	2,3
	Arg	2,3
	D and L-Orn	2,3
	Lys	2,3
	AIB[c]	3,4
	D and L-Trp	3
	Thr	3
	Tyr	3
	Glc	5
	Gal	6
Hamster intestine	D and L-Met	7,8
	Pro	8
	Val	8,9
	Ala	9
	AIB	9
	Gly	9
	Leu	9
	1-dGlc	10
	6-dGlc	10
	Arbutin	10
	Glc	10
	Gal	10
Rabbit intestine	Ala	11
	Glc	12
	Gal	12
	3-O-MG[d]	12
Rabbit kidney cortex (slices)	Glc	13,14
	Gal	13,14,15
	αMG[e]	15,16
	Fru	15
	Xyl	15
	βMG[e]	17
	Glucosamine	17
	Galactosamine	17
Rat kidney cortex (slices)	AIB	18
	His	18

(*continued*)

Table 8 (*continued*)

Tissue	Substrate	References[b]
Rat kidney cortex (*continued*)	Gly	18
	Lys	18,19
	Arg	19
	CyS	19
	Gal	20
	αMG	21
Rat diaphragm	AIB	22
Ehrlich ascites tumor cells	AIB	23
	Gly	23
	Met	23
	Pro	23
	Ser	23
	D and L-Ala	23,24
	D and L-Val	23,24
	Leu	24
Sheep thyroid gland	Iodide	25
	Bromide	25
	Cyanide	25
	Nitrate	25
	Nitrite	25
	Thiocyanate	25
Isolated rabbit kidney brush border vesicles	Ala	26
	Pro	27
Isolated rat kidney brush border membrane vesicles	Gly	28
	Pro	28
	Phe	29
	Glu	30
	Gln	30
	Glc	31
	αMG	31
Isolated rat intestinal brush border vesicles	Glc	32
Isolated mouse fibroblast membrane vesicles	AIB	33,34
	Leu	34
	Gly	35
	Ala	35
	Gln	35
	Met	35

[a] A system in which transport (active or by facilitated diffusion) takes place only when a complex between substrate (amino acid, sugar, or anion), sodium ion, and carrier is formed. Within the cell, dissociation of substrate and sodium ion from the carrier occurs freely.

[b] (1) Winter and Christensen (1965); (2) Finch and Hird (1960); (3) Larsen et al. (1964); (4) Akedo and Christensen (1962); (5) Fisher and Parsons (1953a); (6) Fisher and Parsons (1953b); (7) Lin et al. (1962); (8) Hagihira et al. (1962); (9) Matthews and Laster (1965); (10) Alvarado and Crane (1964); (11) Schultz and Zalusky (1965); (12) Schultz and Zalusky (1964b); (13) Krane and Crane (1959); (14) Kleinzeller et al. (1967a); (15) Kleinzeller et al. (1967b); (16) Kleinzeller (1970b); (17) Kleinzeller (1970a); (18) Rosenberg et al. (1963); (19) Segal et al. (1967); (20) McNamara and Segal (1972); (21) Segal et al. (1973); (22) Parrish and Kipnis (1964); (23) Inui and Christensen (1966); (24) Oxender and Christensen (1963); (25) Wolff (1964); (26) Fass et al. (1977); (27) Hammerman and Sacktor (1977); (28) McNamara et al. (1976); (29) Evers et al. (1976); (30) Weiss et al. (1978); (31) Kinne et al. (1975); (32) Murer and Hopfer (1974); (33) Lever (1976); (34) Quinlan et al. (1976); (35) Lever (1977).

[c] AIB, α-aminoisobutyric acid.

[d] 3-O-Methyl-D-glucose.

[e] αMG, α-methylglucoside; βMG, β-methylglucoside.

sites. According to this theory, fixed binding sites are in equilibrium between an "introverted" and an "extroverted" conformation, neither form being in contact with either the cell interior or the outside medium. The binding sites indiscriminately absorb all of the suitable substrate molecules, but the sugar can move only when both sites are "introverted." The state of occupancy of the sites and the identity of the occupant determine the equilibrium distinction between the introverted and the extroverted orientation. Since introversiveness alone expresses the equilibrium characteristic for each state, no affinity differences between substrates need be postulated.

A general carrier mechanism involving subunit aggregates for all active or facilitated transport of hydrophilic ligands has been proposed by Singer (1977). This "aggregate rearrangement mechanism" is similar to that previously proposed by Jardetzky (1966) and involves water-filled pores lined with transport proteins which translocate ligands via a quaternary rearrangement of the binding protein–ligand complex. As yet, there is scant experimental evidence regarding this theory.

F. Water Transport

Transmembrane movement of water in the absence of or against a limited osmotic gradient may theoretically occur by a number of mechanisms, since classical osmosis cannot be involved. Active transport of water has not been observed in vertebrates (Pappius, 1964). During pinocytosis (see above) a small quantity of the aqueous medium surrounding the membrane may be engulfed in the pinocytotic vesicle. The subsequent secretion of the water from the vesicle results in a net movement of the fluid from one side of the membrane to the other.

Most of the data on water permeability indicate that the rate of water penetration is about fifty times faster than expected from the number of hydrogen bonds presumably holding it in an aqueous lattice. A very plausible explanation for the high rate of water penetration is that of pores in the membrane. In the charged linings of the pore channels, ions or solute particles moving through the pores sweep with them some of the fluid in the pores. Whether the driving force for solute movement is electrical or active transport of the solute by a metabolic process, the net result of water movement through the pores is the same. Recent evidence for the presence of unstirred water layers adjacent to the membrane surface poses a problem to the more classical interpretation of water transport through pores. A review of two metabolically dependent processes which participate in the cellular transport of water has been published (Kleinzeller, 1972). Schafer and Andreoli (1972) and Dick (1971) cover the subject of water movement in detail.

VII. Summary

We have seen how the cell membrane participates in the regulation of cellular processes such as contact inhibition of growth, intracellular communication, and transport of metabolic substrates and intermediates. In its structure and composition, the plasma membrane provides a barrier to external forces while acting as a receptor for informational and regulatory molecules such as hormones. Its antigenic properties characterize the cell, as do various membrane-bound enzymes. Malfunctions of the transport systems located in the membrane can cause a variety of clinical conditions such as cystinuria, generalized aminoacidurias, Hartnup disease, renal acidurias, and Fanconi syndrome. Most recently shown are the changes and relationships in membrane characteristics associated with cancer. We can appreciate, then, that the state of the cell membrane plays an important role in the general regulation of the mammalian cell.

References

Akedo, H., and Christensen, H. N., 1962, Transfer of amino acids across the intestine: A new model amino acid, *J. Biol. Chem.* **237**:113.

Albers, R. W., 1976, The (sodium plus potassium)-transport ATPase, in: *The Enzymes of Biological Membranes*, Vol. 3, *Membrane Transport* (A. Martonosi, ed.), pp. 283–301, Plenum Press, New York.

Allan, D., and Crumpton, M. J., 1970, Preparation and characterization of the plasma membrane of pig lymphocytes, *Biochem. J.* **120**:133.

Altman, P. L., and Katz, D. D., eds., 1976, Changes in cell surface membranes associated with malignant transformation, in: *Cell Biology*, pp. 134–138, Federation of American Societies for Experimental Biology, Bethesda, Md.

Alvarado, F., 1966, Transport of sugars and amino acids in the intestine: Evidence for a common carrier, *Science* **151**:1010.

Alvarado, F., and Crane, R. K., 1964, Studies on the mechanism of intestinal absorption of sugars. VII. Phenylglycoside transport and its possible relationship to phlorizin inhibition of the active transport of sugars by the small intestine, *Biochim. Biophys. Acta* **93**:116.

Ambrose, E. J., 1965, *Cell Electrophoresis*, Churchill, London.

Anderson, N. G., Lansing, A. I., Lieberman, I., Rankin, C. T., Jr., and Elrod, H., 1968, Isolation of rat liver cell membranes by combined rate and isopycnic zonal centrifugation, in: *Biological Properties of Mammalian Surface Membranes*, Wistar Monograph No. 8, pp. 23–30, Wistar Press, Phila.

Arias, I. M., Doyle, D., and Schimke, R. T., 1969, Studies on the synthesis and degradation of proteins of the endoplasmic reticulum of rat liver, *J. Biol. Chem.* **244**:3303.

Ashwell, G., and Morell, A. G., 1977, Membrane glycoproteins and recognition phenomena, *TIBS* **2**:76.

Ausiello, D. A., Segal, S., and Thier, S. O., 1972, Cellular accumulation of L-lysine in rat kidney cortex in vivo, *Am. J. Physiol.* **222**:1473.

Autilio, L. A., Norton, W. T., and Terry, R. D., 1964, The preparation and some properties of purified myelin from the central nervous system, *J. Neurochem.* **11**:17.
Azarnia, R., and Loewenstein, W. R., 1971, Intercellular communication and tissue growth. V. A cancer cell strain that fails to make permeable membrane junctions with normal cells, *J. Membrane Biol.* **6**:368.
Azzi, A., 1975, The application of fluorescent probes in membrane studies, *Q. Rev. Biophys.* **8**:237.
Baker, J. B., and Humphreys, T., 1972, Turnover of molecules which maintain the normal surfaces of contact-inhibited cells, *Science* **175**:905.
Bär, H. P., and Hechter, O., 1969, Adenyl cyclase and hormone action. I. Effects of adrenocorticotropic hormone, glucagon and epinephrine on the plasma membrane of rat fat cells, *Proc. Natl. Acad. Sci. USA* **63**:350.
Beck, C. S., Hasinoff, C. W., and Smith, M. E., 1968, L-Alanyl-β-naphthaylamidase in rat spinal cord myelin, *J. Neurochem.* **15**:1297.
Beers, R. J., Jr., and Bassett, E. G., 1976, *Cell Membrane Receptors for Viruses, Antigens and Antibodies, Peptide Hormones, and Small Molecules*, Raven Press, New York.
Benabdelzlil, C., Michel-Bechel, M., and Lissitzky, S., 1967, Isolation and iodinating ability of apical poles of sheep thyroid epithelial cells, *Biochem. Biophys. Res. Commun.* **27**:74.
Benedetti, E. L., and Emmelot, P., 1968, Structure and function of plasma membranes isolated from liver, in: *Ultrastructure in Biological Systems*, Vol. 4, *The Membranes* (A. J. Dalton and F. Haguenau, eds.), pp. 33–121, Academic Press, New York.
Bergelson, L. D., and Barsukov, L. I., 1977, Topological asymmetry of phospholipids in membranes, *Science* **197**:224.
Berger, S. J., and Sacktor, B., 1970, Isolation and biochemical characterization of brush borders from rabbit kidney, *J. Cell. Biol.* **47**:637.
Birnbaumer, L., 1973, Hormone-sensitive adenylyl cyclases: Useful models for studying hormone receptor functions in cell-free systems. *Biochim. Biophys. Acta* **300**:129.
Birnbaumer, L., and Rodbell, M., 1969, Adenyl cyclase in fat cells. II. Hormone receptors, *J. Biol. Chem.* **244**:3477.
Blau, E., and Michael, A. F., 1971, Rat glomerular basement membrane composition and metabolism in aminonucleoside nephrosis, *J. Lab. Clin. Med.* **77**:97.
Bobinski, H., and Stein, W. D., 1966, Isolation of a glucose-binding component from human erythrocyte membranes, *Nature* **211**:1366.
Bode, F., Baumann, K., Frasch, W., and Kinne, R., 1970, Die Bindung von Phlorrhizin an die Bürstensaumfraktion der Rattenniere, *Pflügers Arch.* **315**:53.
Bode, F., Baumann, K., and Kinne, R., 1976, Analysis of the pinocytic process in rat kidney. II. Biochemical composition of pinocytic vesicles compared to brush border microvilli, lysosomes and basolateral plasma membranes, *Biochim. Biophys. Acta* **433**:294.
Bonting, S. L., Caravaggio, L. L., and Hawkins, N. M., 1962, Studies on sodium-potassium-activated adenosinetriphosphatase. IV. Correlation with cation transport sensitive to cardiac glycosides, *Arch. Biochem. Biophys.* **98**:413.
Booij, H. L., 1962, as quoted in *Cell Membrane Transport—Principles and Techniques* (A. Kotyk and K. Janáček, eds.), 1970, p. 185, Plenum Press, New York.
Boone, C. W., Ford, L. E., Bond, H. E., Stuart, D. C., and Lorenz, D., 1969, Isolation of plasma membrane fragments from HeLa cells, *J. Cell. Biol.* **41**:378.
Booth, A. G., and Kenny, A. J., 1974, A rapid method for the preparation of microvilli from rabbit kidney, *Biochem. J.* **142**:575.
Bosmann, H. B., Hagopian, A., and Eylar, E. H., 1968, Cellular membranes: The isolation and characterization of the plasma and smooth membranes of HeLa cells, *Arch. Biochem. Biophys.* **128**:51.

Boys, E. A., and Old, L. J., 1970, Organization and modulation of cell membrane receptors, in: *Immune Surveillance* (R. T. Smith and M. Landy, eds.), pp. 1–82, Academic Press, New York.

Boys, E. A., Old, L. J., and Stockert, E., 1968, An approach to the mapping of antigens on the cell surface, *Proc. Natl. Acad. Sci. USA* **60**:886.

Bramley, T. A., Coleman, R., and Finean, J. B., 1971, Chemical, enzymological, and permeability properties of human erythrocyte ghosts prepared by hypotonic lysis in media of different osmolarities, *Biochim. Biophys. Acta* **241**:752.

Bresnick, E., and Schwartz, A., 1969, *Functional Dynamics of the Cell*, Academic Press, New York.

Britton, H. G., 1964, Permeability of the human red cell to labelled glucose, *J. Physiol. (London)* **170**:1.

Brunette, D. M., and Till, J. E., 1971, A rapid method for the isolation of L-cell surface membranes using an aqueous two-phase polymer system, *J. Membrane Biol.* **5**:215.

Burger, M. M., 1971, Cell surfaces in neoplastic transformation, in: *Current Topics in Cellular Regulation*, Vol. 3 (B. L. Horecker and E. R. Stadtman, eds.), pp. 135–193, Academic Press, New York.

Cambiaso, C. L., Dellacha, J. M., Santomé, J. A., and Paladini, A. C., 1971, Species-specific interaction of growth hormones with erythrocyte membranes, *FEBS Lett.* **12**:236.

Cammer, W., Fredman, T., Rose, A. L., and Norton, W. T., 1976, Brain carbonic anhydrase: Activity in isolated myelin and the effect of hexachlorophene, *J. Neurochem.* **27**:165.

Carafoli, E., and Crompton, M., 1976, Binding proteins and membrane transport, in: *The Enzymes of Biological Membranes*, Vol. 3, *Membrane Transport* (A. Martonosi, ed.), pp. 193–220, Plenum Press, New York.

Catt, K. J., and Dufau, M. L., 1977, Peptide hormone receptors, *Annu. Rev. Physiol.* **39**:529.

Chapman, D., 1970, The chemical and physical characteristics of biological membranes, in: *Membranes and Ion Transport*, Vol. 1 (E. E. Bittar, ed.), pp. 23–63, John Wiley and Sons, New York.

Chapman, D., Byrne, P., and Shipley, G. G., 1966, The physical properties of phospholipids I. Solid state and mesomorphic properties of some 2,3-diacyl-DL-phosphatidylethanolamines, *Proc. Roy. Soc. (London) Ser. A* **290**:115.

Chez, R. A., Palmer, R. R., Schultz, S. G., and Curran, P. F., 1967, Effect of inhibitors on alanine transport in isolated rabbit ileum, *J. Gen. Physiol.* **50**:2357.

Christensen, H. N., Liang, M. and Archer, E. G., 1967, A distinct Na^+-requiring transport system for alanine, serine, cysteine, and similar amino acids, *J. Biol. Chem.* **242**:5237.

Clayman, S., and Scholefield, P. G., 1969, The uptake of amino acids by mouse pancreas in vitro. IV. The role of exchange diffusion, *Biochim. Biophys. Acta* **173**:277.

Cohen, A., 1963, Mechanisms of virus infection, in: *Mechanisms of Cell Infection. Part I. Virus Attachment and Penetration* (W. Smith, ed.), pp. 153–190, Academic Press, New York.

Coleman, R., 1973, Membrane-bound enzymes and membrane ultrastructure, *Biochim. Biophys. Acta* **300**:1.

Coleman, R., and Finean, J. B., 1966, Preparation and properties of isolated plasma membranes from guinea-pig tissues, *Biochim. Biophys. Acta* **125**:197.

Conway, E. J., 1963, Significance of various factors including lactic dehydrogenase on the active transport of sodium ions in skeletal muscle, *Nature* **198**:760.

Conway, E. J., 1964, New light on the active transport of sodium ions from skeletal muscle, *Fed. Proc.* **23**:680.

Cook, G. M. W., 1968, Chemistry of membranes, *Br. Med. Bull.* **24**:118.
Cook, G. M. W., Heard, D. H. and Seaman, G. V. F., 1961, Sialic acids and the electrokinetic charge of the human erythrocyte, *Nature* **191**:44.
Cook, G. M. W., Seaman, G. V. F., and Weiss, L., 1963, Physicochemical differences between ascitic and solid forms of sarcoma 37 cells, *Cancer Res.* **23**:1813.
Cotman, C., Mahler, H. R., and Anderson, N. G., 1968, Isolation of a membrane fraction enriched in nerve-end membranes from rat brain by zonal centrifugation, *Biochim. Biophys. Acta* **163**:272.
Crane, R. K., 1960, Intestinal absorption of sugars, *Physiol. Rev.* **40**:789.
Crane, R. K., 1962, Hypothesis for mechanism of intestinal active transport of sugars, *Fed. Proc.* **21**:891.
Crane, R. K., 1966, Structural and functional organization of an epithelial cell brush border, in: *Intracellular Transport* (K. B. Warren, ed.), pp. 71–102, Academic Press, New York.
Crane, R. K., 1977, The gradient hypothesis and other models of carrier-mediated active transport, *Rev. Physiol. Biochem. Pharmacol.* **78**:99.
Crane, R. K., Field, R. A., and Cori, C. F., 1957, Studies of tissue permeability, I. The penetration of sugars into the Ehrlich ascites tumor cell, *J. Biol. Chem.* **224**:649.
Crawhall, J. C., and Segal, S., 1968, Transport of some amino acids and sugars in rat-liver slices, *Biochim. Biophys. Acta* **163**:163.
Csáky, T. Z., 1965, Transport through biological membranes, *Annu. Rev. Physiol.* **27**:415.
Cuatrecasas, P., 1971a, Insulin-receptor interactions in adipose tissue cells: Direct measurement and properties, *Proc. Natl. Acad. Sci. USA* **68**:1264.
Cuatrecasas, P., 1971b, Properties of the insulin receptor of isolated fat cell membranes, *J. Biol. Chem.* **246**:7265.
Cuatrecasas, P., 1974, Membrane receptors, *Annu. Rev. Biochem.* **43**:169.
Cumar, F. A., Brady, R. O., Kolodny, E. H., McFarland, V. W., and Mora, P. T., 1970, Enzymatic block in the synthesis of gangliosides in DNA virus-transformed tumorigenic mouse cell lines, *Proc. Natl. Acad. Sci. USA* **67**:757.
Curran, P. F., 1972, Active transport of amino acids and sugars, *Arch. Intern. Med.* **129**:258.
Curran, P. F., and Solomon, A. K., 1957, Ions and water fluxes in the ileum of rats, *J. Gen. Physiol.* **41**:143.
Curtis, A. S. G., 1973, Cell adhesion, *Prog. Biophys. Mol. Biol.* **27**:315.
Cuzner, M. L., Davison, A. N., and Gregson, N. A., 1965, The chemical composition of vertebrate myelin and microsomes, *J. Neurochem.* **12**:469.
Danielli, J. F., 1954, The present position in the field of facilitated diffusion and selective active transport, *Proc. Symp. Colston Res. Soc.* **7**:1.
Danielli, J. F., and Davson, H., 1935, A contribution to the theory of permeability of thin films, *J. Cell. Comp. Physiol.* **5**:495.
Daniels, F., and Alberty, R. A., 1961, *Physical Chemistry*, 2nd ed., John Wiley and Sons, New York.
Da Silva, P. P., Douglas, S. D., and Branton, D., 1971, Localization of A antigen sites on human erythrocyte ghosts, *Nature* **232**:194.
Dawson, A. C., and Widdas, W. F., 1964, Variations with temperature and pH of the parameters of glucose transfer across the erythrocyte membrane in the foetal guinea pig, *J. Physiol. (London)* **172**:107.
Dean, W. L., and Tanford, C., 1977, Reactivation of lipid-depleted Ca^{++}-ATPase by a nonionic detergent, *J. Biol. Chem.* **252**:3551.
De Gier, J., and Van Deenen, L. L. M., 1961, Some lipid characteristics of red cell membranes of various animal species, *Biochim. Biophys. Acta* **49**:286.

Dervichian, D. G., 1955, in: *Exposes Actuels: Problems de Structures, d'Ultrastructures et de Functions Cellulaires,* Chapter 4, Masson, Paris.

Dewey, M. M., and Barr, L., 1970, Some consideration about the structure of cellular membranes, in: *Current Topics in Membranes and Transport,* Vol. 1 (F. Bronner and A. Kleinzeller, eds.), pp. 1–33, Academic Press, New York.

Dick, D. A. T., 1971, Water movements in cells, in: *Membranes and Ion Transport,* Vol. 3 (E. E. Bittar, ed.), pp. 211–250, John Wiley and Sons, New York.

Dod, B. J., and Gray, G. M., 1968, The lipid composition of rat-liver plasma membranes, *Biochim. Biophys. Acta* **150**:397.

Dodge, J. T., Mitchell, C. D., and Hanahan, D. J., 1963, The preparation and chemical characteristics of hemoglobin-free ghosts of human erythrocytes, *Arch. Biochem. Biophys.* **100**:119.

Doljanski, F., and Eisenberg, S., 1965, The action of neuraminidase on the electrophoretic mobility of liver cells, in: *Cell Electrophoresis* (E. J. Ambrose, ed.), pp. 78–84, Churchill, London.

Doty, J. R., 1941, Reabsorption of certain amino acids and derivatives by the kidney tubule, *Proc. Soc. Exp. Biol. Med.* **46**:129.

Dowben, R. M., 1969, *General Physiology—A Molecular Approach,* Harper and Row, New York.

Doyle, D., and Baumann, H., 1979, Turnover of the plasma membrane of mammalian cells, *Life Sci.* **24**:951.

Dunnick, J. K., and Marinetti, G. V., 1971, Hormone action at the membrane level: III. Epinephrine interaction with the rat liver plasma membrane, *Biochim. Biophys. Acta* **249**:122.

Dutton, A., Rees, E. D., and Singer, S. J., 1976, An experiment eliminating the rotating carrier mechanism for the active transport of Ca ion in sarcoplasmic reticulum membranes, *Proc. Natl. Acad. Sci. USA* **73**:1532.

Ebel, H., Aulber, E., and Merker, H. J., 1976, Isolation of the basal and lateral plasma membranes of rat kidney tubule cells, *Biochim. Biophys. Acta* **433**:531.

Eisenman, G., 1961, On the elementary atomic origin of equilibrium ionic specificity, in: *Membrane Transport and Metabolism* (A. Kleinzeller and A. Kotyk, eds.), pp. 163–179, Academic Press, New York.

El-Aaser, A. A., Fitzsimons, J. T. R., Hinton, R. H., Reid, E., Klucis, E., and Alexander, P., 1966, Zonal centrifugation of crude nuclear fractions from rat liver, *Biochim. Biophys. Acta* **127**:553.

Elsas, L. J., and Rosenberg, L. E., 1967, Inhibition of amino acid transport in rat kidney cortex by puromycin, *Proc. Natl. Acad. Sci. USA* **57**:371.

Emmelot, P., and Benedetti, E. L., 1967, On the possible involvement of the plasma membrane in the carcinogenic process, in: *Carcinogenesis, A Broad Critique,* M. D. Anderson Hosp. Tumor Inst. Symp., Univ. of Texas, pp. 471–533, Williams and Wilkins, Baltimore, Maryland.

Emmelot, P., and Bos, C. J., 1962, A stimulatory effect of potassium cyanide on the glycerophosphate dehydrogenase of normal and neoplastic tissues, *Biochim. Biophys. Acta* **59**:495.

Emmelot, P., Bos, C. J., Benedetti, E. L., and Rümke, P. H., 1964, Studies on plasma membranes: I. Chemical composition and enzyme content of plasma membranes isolated from rat liver, *Biochim. Biophys. Acta* **90**:126.

Evers, J., Murer, H., and Kinne, R., 1976, Phenylalanine uptake in isolated renal brush border vesicles, *Biochim. Biophys. Acta.* **426**:598.

Fariás, R. N., Bloj, B., Morero, R. D., Siñeriz, F., and Trucco, R. E., 1975, Regulation of allosteric membrane-bound enzymes through changes in membrane lipid composition, *Biochim. Biophys. Acta* **415**:231.

Fass, S. J., Hammerman, M. R., and Sacktor, B., 1977, Transport of amino acids in renal brush border membrane vesicles: Uptake of the neutral amino acid L-alanine, *J. Biol. Chem.* **252**:583.
Ferber, E., 1971, Membrane phospholipid metabolism during cell activation and differentiation, from 22. Colloquium der Gesellschaft für Biologische Chemie, 15–17, April 1971, in Mosbach/Baden, in: *The Dynamic Structure of Cell Membranes,* Springer-Verlag, Berlin.
Fernándes-Morán, H., and Finean, J. B., 1957, Electron microscope and low-angle X-ray diffraction studies of the nerve myelin sheath, *J. Biophys. Biochem. Cytol.* **3**:725.
Finch, L. R., and Hird, F. J. R., 1960, The uptake of amino acids by isolated segments of rat intestine. II. A survey of affinity for uptake from rates of uptake and competition for uptake, *Biochim. Biophys. Acta* **43**:278.
Finean, J. B., and Robertson, J. D., 1958, Lipids and the structure of myelin, *Br. Med. Bull.* **14**:267.
Finn, F. M., Widnell, C. C., and Hofmann, K., 1972, Localization of an adrenocorticotropic hormone receptor on bovine adrenal cortical membranes, *J. Biol. Chem.* **247**:5695.
Fisher, R. B., and Parsons, D. S., 1953a, Glucose movements across the wall of the rat small intestine, *J. Physiol. (London)* **119**:210.
Fisher, R. B., and Parsons, D. S., 1953b, Galactose absorption from the surviving small intestine of the rat, *J. Physiol. (London)* **119**:224.
Fisher, R. B., and Zachariah, P., 1961, The mechanism of the uptake of sugars by the rat heart and the action of insulin on this mechanism, *J. Physiol. (London)* **158**:73.
Fitzpatrick, D. F., Davenport, G. R., Forte, L., and Landon, E. L., 1969, Characterization of plasma membrane proteins in mammalian kidney. 1. Preparation of a membrane fraction and separation of the protein, *J. Biol. Chem.* **244**:3561.
Fleischer, S., Brierley, G., Klouwen, H., and Slautterback, D. B., 1962, Studies of the electron transfer system, *J. Biol. Chem.* **237**:3264.
Fleischer, S., Fleischer, B., and Stoeckenius, W., 1965, Fine structure of whole and fragmented mitochondria after lipid depletion, *Fed. Proc.* **24**:296 (Abstr. #922).
Freedman, B., and Young, J. A., 1969, Microperfusion study of L-histidine transport by rat nephron, *Aust. J. Exp. Biol. Med. Sci.* **47**:10.
Freychet, P., Roth, J., and Neville, D. N., Jr., 1971, Monoiodoinsulin: Demonstration of its biological activity and binding to fat cells and liver membranes, *Biochem. Biophys. Res. Commun.* **43**:400.
Gardas, A., and Koscielak, J., 1974, Megaloglycolipids—unusually complex glycosphingolipids of human erythrocyte membrane with A, B, H, and I blood group specificity, *FEBS Lett.* **42**:101.
Genel, M., Rea, C. F., and Segal, S., 1971, Transport interactions of sugars and amino acids in mammalian kidney, *Biochim. Biophys. Acta* **241**:779.
Giese, A. C., 1962, *Cell Physiology,* 2nd ed., W. B. Saunders, Philadelphia.
Gilbert, J. C., 1965, Mechanism of sugar transport in brain slices, *Nature* **205**:87.
Glaser, L., 1976, Cell–cell recognition, *TIBS* **1**:84.
Glick, J. L. and Githens, S., III, 1965, Role of sialic acid in potassium transport of L1210 leukaemia cells, *Nature* **208**:88.
Glick, J. L., Goldberg, A. R., and Pardee, A. B., 1966, The role of sialic acid in the release of proteins from L1210 leukemia cells, *Cancer Res.* **26**:1774.
Glick, M. C., and Warren, L., 1969, Membranes of animal cells. III. Amino acid incorporation by isolated surface membranes, *Proc. Natl. Acad. Sci. USA* **63**:563.
Glick, M. C., Comstock, C., and Warren, L., 1970, Membranes of animal cell: VII. Carbohydrates of surface membranes and whole cells, *Biochim. Biophys. Acta* **219**:290.

Glick, M. C., Gerner, E. W., and Warren, L., 1971, Changes in the carbohydrate content of the KB cell during the growth cycle, *J. Cell Physiol.* **77**:1.
Glynn, I. M., 1956, Sodium and potassium movements in human red cells, *J. Physiol. (London)* **134**:278.
Goldin, S. M., 1977, Active transport of sodium and potassium ions by the sodium and potassium ion-activated adenosine triphosphatase from renal medulla, *J. Biol. Chem.* **252**:5630.
Goodman, J. W., 1969, Immunochemical specificity: Recent conceptual advances, *Immunochemistry* **6**:139.
Gorter, E., and Grendel, F., 1925, On bimolecular layers of lipoids on the chromocytes of the blood, *J. Exp. Med.* **41**:439.
Green, D. E., ed., 1972, Membrane structure and its biological applications, *Ann. N.Y. Acad. Sci.* **195**.
Green, D. E., and Brucker, R. F., 1972, The molecular principles of biological membranes, *BioScience* **22**:13.
Green, D. E., and Perdue, J. F., 1966, Membranes as expressions of repeating units, *Proc. Natl. Acad. Sci. USA* **55**:2195.
Green, F. A., 1972, Erythrocyte membranes, lipids and Rh antigen activity, *J. Biol. Chem.* **247**:881.
Greth, W. E., Thier, S. O., and Segal, S., 1971, Transport of cystine in rat kidney cortex: Independent luminal and contraluminal mechanisms, *Clin. Res.* **19**:742.
Gross, L., 1967, Active membranes for active transport, *J. Theor. Biol.* **15**:298.
Guidotti, G., 1972, Membrane proteins, *Annu. Rev. Biochem.* **41**:731.
Guidotti, G. G., Borghetti, A. F. and Gazzola, G. C., 1978, The regulation of amino acid transport in animal cells, *Biochim. Biophys. Acta* **515**:329.
Gulik-Krzywicki, T., 1975, Structural studies of the associations between biological membrane components, *Biochim. Biophys. Acta* **415**:1.
Haddad, A., Bennett, G., and LeBlond, C. P., 1977, Formation and turnover of plasma membrane glycoproteins in kidney tubules of young rats and adult mice, as shown by radioautography after an injection of ^{3}H-fucose, *Am. J. Anat.* **148**:241.
Hagihira, H., Wilson, T. H., and Lin, E. C. C., 1962, Intestinal transport of certain N-substituted amino acids, *Am. J. Physiol.* **203**:637.
Hammerman, M. R., and Sacktor, B., 1977, Transport of amino acids in renal brush border membrane vesicles—Uptake of L-proline, *J. Biol. Chem.* **252**:591.
Hanahan, D. J., and Ekholm, J., 1972, Changes in erythrocyte membranes during preparation, as expressed by ATPase activity, *Biochim. Biophys. Acta* **255**:413.
Hasselbach, W., and Makinose, M., 1961, Die Calciumpumpe der "Erschlaffungsgrana"des Muskels und ihre Abhängigkeit von der ATP-Spaltung, *Biochem. Z.* **333**:518.
Hasselbach, W., and Makinose, M., 1963, Über den Mechanismus des Calciumtransportes durch die Membranen des sarkoplasmatischen Reticulums, *Biochem. Z.* **339**:94.
Hauser, G., 1968, Myo-inositol transport in slices of rat kidney cortex. I. Effect of incubation conditions and inhibitors, *Biochim. Biophys. Acta* **173**:257.
Heath, E. C., 1971, Complex polysaccharides, *Annu. Rev. Biochem.* **40**:29.
Heidrich, H. G., Kinne, R., Kinne-Saffran, E., and Hannig, K., 1972, The polarity of the proximal tubule cell in rat kidney. Different surface charges for the brush-border microvilli and plasma membranes from the basal infoldings, *J. Cell Biol.* **54**:232.
Heinz, E., and Walsh, P. M., 1958, Exchange diffusion, transport and intracellular level of amino acids in Ehrlich carcinoma cells, *J. Biol. Chem.* **233**:1488.
Hertzenberg, L. A., and Hertzenberg, L. A., 1961, Association of H-2 antigens with the cell membrane fraction of mouse liver, *Proc. Natl. Acad. Sci. USA* **47**:762.

Hicks, R. M., and Ketterer, B., 1970, Isolation of the plasma membrane of the luminal surface of rat bladder epithelium, and the occurrence of a hexagonal lattice of subunits both in negatively stained whole mounts and in sectioned membranes, *J. Cell Biol.* **45**:542.

Hildebrand, C. E., and Okinaka, R. T., 1976, A rapid method for preparation of nuclear membranes from mammalian cells, *Anal. Biochem.* **75**:290.

Hill, T. L., 1969, A proposed common allosteric mechanism for active transport, muscle contraction, and ribosomal translocation, *Proc. Natl. Acad. Sci. USA* **64**:267.

Hille, B., 1971, The permeability of the sodium channel to organic cations in myelinated nerve, *J. Gen. Physiol.* **58**:599.

Hillman, R. E., and Rosenberg, L. E., 1969, Amino acid transport by isolated mammalian renal tubules. II. Transport systems for L-proline, *J. Biol. Chem.* **244**:4494.

Hillman, R. E., Albrecht, I., and Rosenberg, L. E., 1968a, Transport of amino acids by isolated rabbit renal tubules, *Biochim. Biophys. Acta* **150**:528.

Hillman, R. E., Albrecht, I., and Rosenberg, L. E., 1968b, Identification and analysis of multiple glycine transport systems in isolated mammalian renal tubules. *J. Biol. Chem.* **243**:5566.

Hokin, L. E., and Hokin, M. R., 1963, Phosphatidic acid metabolism and active transport of sodium, *Fed. Proc.* **22**:8.

Holter, H., 1959, Pinocytosis, *Intern. Rev. Cytol.* **8**:481.

Hughes, R. C., 1977, *Membrane Glycoproteins, A Review of Structure and Function*, Butterworths, London.

Inui, Y., and Christensen, H. N., 1966, Discrimination of single transport systems: The Na^+ sensitive transport of neutral amino acids in the Ehrlich cell, *J. Gen. Physiol.* **50**:203.

Israelachvili, J. H., 1977, Refinement of the fluid-mosaic model of membrane structure, *Biochim. Biophys. Acta* **469**:221.

Jacquez, J. A., and Sherman, J. H., 1965, The effect of metabolic inhibitors on transport and exchange of amino acids in Ehrlich ascites cells, *Biochim. Biophys. Acta* **109**:128.

Jaques, L. W., Brown, E. B., Barrett, J. M., Brey, W. S., Jr., and Weltner, W., Jr., 1977, Sialic acid, a calcium-binding carbohydrate, *J. Biol. Chem.* **252**:4533.

Jardetzky, O., 1966, Simple allosteric model for membrane pumps, *Nature* **211**:969.

Jensen, E. V., and Jacobson, H. I., 1962, Basic guides to the mechanism of estrogen action, *Recent Prog. Horm. Res.* **18**:387.

Kabat, E. A., and Meyer, M. M., 1971, *Experimental Immunochemistry*, Charles C Thomas, Springfield, Ill.

Kahn, C. R., 1975, Membrane receptors for polypeptide hormones, *Meth. Membrane Biol.* **3**:81.

Kalckar, H. M., 1976, Cellular regulation of transport and uptake of nutrients: An overview, *J. Cell. Physiol.* **89**:503.

Kamat, V. B., Wyatt, A. J., and Davis, M. A. F., 1972, Erythrocyte membranes—some effects of sonication, *Chem. Phys. Lipids* **8**:341.

Kaplan, D. M., and Criddle, R. S., 1971, Membranes structural proteins, *Physiol. Rev.* **51**:249.

Katz, F. N., Rothman, J. E., Lingappa, V. R., Blobel, G., and Lodish, H. F., 1977, Membrane assembly in vitro: Synthesis, glycosylation, and asymmetric insertion of a transmembrane protein, *Proc. Natl. Acad. Sci. USA* **74**:3278.

Kedem, O., 1961, Criteria of active transport, in: *Membrane Transport and Metabolism* (A. Kleinzeller and A. Kotyk, eds.), pp. 87–93, Academic Press, New York.

Kedem, O., and Katchalsky, A., 1958, Thermodynamic analysis of the permeability of biological membranes to non-electrolytes, *Biochim. Biophys. Acta* **27**:229.

Kedem, O., and Katchalsky, A., 1961, A physical interpretation of the phenomenological coefficients of membrane permeability, *J. Gen. Physiol.* **45**:143.

Keenan, T. W., Morre, D. J., Olson, D. E., Yunghans, W. N., and Patton, S., 1970, Biochemical and morphological comparison of plasma membrane and milk fat globule membrane from bovine mammary gland, *J. Cell Biol.* **44**:80.

Kidwai, A. M., Radcliffe, M. A., Duchon, G., and Daniel, E. E., 1971, Isolation of plasma membrane from cardiac muscle, *Biochem. Biophys. Res. Commun.* **45**:901.

Kimelberg, H. K., and Papahadjopoulos, D., 1972, Phospholipid requirements for ($Na^+ + K^+$)-ATPase activity: Head group specificity and fatty acid fluidity, *Biochim. Biophys. Acta* **282**:277.

Kinne, R., and Kinne-Saffran, E., 1969, Isolierung und enzymatische Charakterisierung einer Bürstensaum-fraktion des Rattenniere, *Pflügers Arch.* **308**:1.

Kinne, R., Murer, H., Kinne-Saffran, E., Thees, M., and Sachs, G., 1975, Sugar transport by renal plasma membrane vesicles: Characterization of the systems in the brush-border microvilli and basal-lateral plasma membranes, *J. Membrane Biol.* **21**:375.

Kleinzeller, A., 1970a, The specificity of the active sugar transport in kidney cortex cells, *Biochim. Biophys. Acta* **211**:264.

Kleinzeller, A., 1970b, Active sugar transport in renal cortex cells: The electrolyte requirement, *Biochim. Biophys. Acta* **211**:277.

Kleinzeller, A., 1972, Cellular transport of water, in: *Metabolic Transport*, Vol. VI (L. E. Hokin, ed.), pp. 92–131, Academic Press, New York.

Kleinzeller, A., Kolínská, J., and Beneš, I., 1967a, Transport of glucose and galactose in kidney-cortex cells, *Biochem. J.* **104**:843.

Kleinzeller, A., Kolínská, J., and Beneš, I., 1967b, Transport of monosaccharides in kidney-cortex cells, *Biochem. J.* **104**:852.

Klingenberg, M., 1976, The ADP-ATP carrier in mitochondrial membranes, in: *The Enzymes of Biological Membranes*, Vol. 3, *Membrane Transport* (A. Martonosi, ed.), pp. 383–438, Plenum Press, New York.

Kohn, P. G., and Clausen, T., 1972, The relationship between the transport of glucose and cations across cell membranes in isolated tissues. VII. The effects of extracellular Na^+ and K^+ on the transport of 3-O-methylglucose and glucose in rat soleus muscle, *Biochim. Biophys. Acta* **255**:798.

Kolber, A. R., and LeFevre, P. G., 1967, Evidence for carrier-mediated transport of monosaccharides in the Ehrlich ascites tumor cell, *J. Gen. Physiol.* **50**:1907.

Kolínská, J., 1970, Kinetics of sugar transport in rabbit kidney cortex, in vitro: Movement of D-galactose, 2-deoxy-D-galactose and α-methyl-D-glucoside, *Biochim. Biophys. Acta* **219**:200.

Kometiani, Z. P., 1963, Interrelation of free radicals with active transfer of sodium into the skin of frogs, *Biofizika* **8**:40.

Korn, E. D., 1966, Structure of biological membranes, *Science* **153**:1491.

Korn, E. D., 1968, Structure and function of the plasma membrane. *J. Gen. Physiol.* **52**:257.

Korn, E. D., 1977, *Methods in Membrane Biology*, Vol. 8, Plenum Press, New York.

Kornfeld, S., and Kornfeld, R., 1969, Solubilization and partial characterization of a phytohemagglutinin receptor site from human erythrocytes, *Proc. Natl. Acad. Sci. USA*, **63**:1439.

Kotyk, A., and Janáček, K., 1970, *Cell Membrane Transport—Principles and Techniques*, Plenum Press, New York.

Kramer, R., Schlatter, C., and Zahler, P., 1972, Preferential binding of sphingomyelin by membrane proteins of the sheep red cell, *Biochim. Biophys. Acta* **282**:146.

Krane, S. M., and Crane, R. K., 1959, The accumulation of D-galactose against a concentration gradient by slices of rabbit kidney cortex, *J. Biol. Chem.* **234**:211.

Kurihara, T., and Tsukada, Y., 1967, The regional and subcellular distribution of 2',3'-cyclic nucleotide 3'-phosphohydrolase in the central nervous system, *J. Neurochem.* **14**:1167.

Laatsch, R. H., Kies, M. W., Gordon, S., and Alvord, E. C., Jr., 1962, The encephalomyelitic activity of myelin isolated by ultracentrifugation, *J. Exp. Med.* **115**:777.

Lakshminarayanaiah, N., 1969, *Transport Phenomena in Membranes*, Academic Press, New York.

Larsen, P. R., Ross, J. E., and Tapley, D. F., 1964, Transport of neutral, dibasic and N-methyl-substituted amino acids by rat intestine. *Biochim. Biophys. Acta* **88**:570.

Lee, A. G., 1975, Functional properties of biological membranes: A physical–chemical approach, *Prog. Biophys. Molec. Biol.* **29**:3.

Lee, G., Aloj, S. M., and Kohn, L. D., 1977, The structure and function of glycoprotein hormone receptors: Ganglioside interactions with lutenizing hormone, *Biochem. Biophys. Res. Commun.* **77**:434.

LeFevre, P. G., 1954, The evidence for active transport of monosaccharides across the red cell membrane, *Symp. Soc. Exp. Biol.* **8**:118.

LeFevre, P. G., 1962, Rate and affinity in human red blood cell sugar transport, *Am. J. Physiol.* **203**:286.

LeFevre, P. G., 1963, Absence of rapid exchange component in a low-affinity carrier transport, *J. Gen. Physiol.* **46**:721.

LeFevre, P. G., 1973, A model for erythrocyte sugar transport based on substrate-conditioned "introversion" of binding sites, *J. Membrane Biol.* **11**:1.

Lefkowitz, R. J., Roth, J., and Pastan, I., 1971, ACTH-receptor interaction in the adrenal: A model for the initial step in the action of hormones that stimulate adenyl cyclase, *Ann. N. Y. Acad. Sci.* **185**:195.

Lenard, D. J., and Singer, S. J., 1966, Protein conformation in cell membrane preparations as studied by optical rotatory dispersion and circular dichroism, *Proc. Natl. Acad. Sci. USA* **56**:1828.

Lever, J. E., 1976, Regulation of active α-aminoisobutyric acid transport expressed in membrane vesicles from mouse fibroblasts, *Proc. Natl. Acad. Sci. USA* **73**:2614.

Lever, J. E., 1977, Active amino acid transport in plasma membrane vesicles from Simian virus 40-transformed mouse fibroblasts: Characteristics of electrochemical Na^+ gradient-stimulated uptake, *J. Biol. Chem.* **252**:1990.

Levinson, B. B., Baxter, F. D., Rousseau, G. G., and Tomkins, G. M., 1972, Cellular site of glucocorticoid-receptor complex formation, *Science* **175**:189.

Liang, C. T., and Sacktor, B., 1977, Preparation of renal cortex basal-lateral and brush border membranes. Localization of adenylate cyclase and guanylate cyclase activities, *Biochim. Biophys. Acta* **466**:474.

Lieb, W. R., and Stein, W. D., 1970, Quantitative predictions of a non-carrier model for glucose transport across the red cell membrane, *Biophys. J.* **10**:585.

Lieb, W. R., and Stein, W. D., 1971, New theory for glucose transport across membranes, *Nature (New Biol.)* **230**:108.

Lin, E. C. C., Hagihira, H., and Wilson, T. H., 1962, Specificity of the transport system for neutral amino acids in the hamster intestine, *Am. J. Physiol.* **202**:919.

Ling, G. N., 1977, The physical state of water and ions in living cells and a new theory of the energization of biological work performance of ATP, *Molec. Cell. Biochem.* **15**:159.

Lippman, M., 1976, Minireview: Steroid hormone receptors in human malignancy, *Life Sci.* **18**:143.

Lloyd, C. W., 1975, Sialic acid and the social behavior of cells, *Biol. Rev.* **50**:325.

Loewenstein, W. R., 1972, Cellular communication through membrane junctions, *Arch Intern. Med.* **129**:299.

London, D. R., and Segal, S., 1967, Differences in the uptake and efflux of two nonutilizable amino acids, α-aminoisobutyric and 1-aminocyclopentane carboxylic acid in the "cut" rat diaphragm, *Biochim. Biophys. Acta* **135**:179.

London, Y., 1971, Ox peripheral nerve myelin membrane. Purification and partial characterization of two basic proteins, *Biochim. Biophys. Acta* **249**:188.

London, Y., 1972, Preparation of purified myelin from ox intradural spinal roots by rate-isopycnic zonal centrifugation, *Biochim. Biophys. Acta* **282**:195.

Love, A. H. G., Mitchell, T. G., and Neptune, E. M., 1965, Transport of sodium and water by rabbit ileum, in vitro and in vivo, *Nature* **206**:1158.

Lowenstein, L. M., Smith, I., and Segal, S., 1968, Amino acid transport in the rat renal papilla, *Biochim. Biophys. Acta* **150**:73.

Lübke, K., Schillinger, E., and Töpert, M., 1976, Hormone receptors, *Angew. Chem. (Engl.)* **15**:741.

Luftig, R. B., Wehrli, E., and McMillan, P. N., 1977, Minireview: The unit membrane image: A re-evaluation, *Life Sci.* **21**:285.

Mackensie, S., and Scriver, C. R., 1971, Transport of L-proline and α-aminoisobutyric acid in isolated rat kidney glomerulus, *Biochim. Biophys. Acta* **241**:725.

MacLennan, D. H., and Holland, P. C., 1976, The calcium transport ATPase of sarcoplasmic reticulum, in: *The Enzymes of Biological Membranes*, Vol. 3, *Membrane Transport* (A. Martonosi, ed.), pp. 221–259, Plenum Press, New York.

Maddy, A. H., 1966, as quoted in, *Current Topics in Membranes and Transport*, Vol. 1 (F. Bronner and A. Kleinzeller, eds.), 1970, pp. 6–7, Academic Press, New York.

Maddy, A. H., and Dunn, M. J., 1976, The solubilization of membranes, in: *Biochemical Analysis of Membranes* (A. H. Maddy, ed.) p. 177, John Wiley and Sons, New York.

Malhotra, S. K., 1970, Organization and biogenesis of cellular membranes, in: *Membranes and Ion Transport*, Vol. 1 (E. E. Bittar, ed.), pp. 1–22, John Wiley and Sons, New York.

Marchase, R. B., Vosbeck, K., and Roth, S., 1976, Intercellular adhesive specificity, *Biochim. Biophys. Acta* **457**:385.

Matthews, D. M., and Laster, L., 1965, Kinetics of intestinal active transport of five neutral amino acids, *Am. J. Physiol.* **208**:593.

McCollester, D. L., 1960, A device for obtaining cell membranes, *Biochim. Biophys. Acta* **41**:160.

McKeel, D. W., and Jarrett, L., 1970, Preparation and characterization of a plasma membrane fraction from isolated fat cells, *J. Cell Biol.* **44**:417.

McLaren, L. C., Scaletti, J. V., and James, C. G., 1968, Isolation and properties of enterovirus receptors, in: *Biological Properties of the Mammalian Surface Membrane* (L. A. Manson, ed.), pp. 123–136, The Wistar Institute Press, Philadelphia.

McNamara, P. D., and Segal, S., 1972, Transport and metabolism of galactose in rat kidney cortex, *Biochem. J.* **129**:1109.

McNamara, P. D., Rea, C., and Segal, S., 1971, Sugar transport: Effect of temperature on concentrative uptake of α-methylglucoside by kidney cortex slices, *Science* **172**:1033.

McNamara, P. D., Ožegović, B., Pepe, L. M., and Segal, S., 1976, Proline and glycine uptake by renal brushborder membrane vesicles, *Proc. Natl. Acad. Sci. USA* **73**:4521.

Meijer, A. J., and VanDam, K., 1974, The metabolic significance of anion transport in mitochondria, *Biochim. Biophys. Acta* **346**:213.

Meister, A., 1973, On the enzymology of amino acid transport, *Science* **180**:33.

Melchior, D. L., and Steim, J. M., 1976, Thermotropic transitions in biomembranes, *Ann. Rev. Biophys. Bioeng.* **5**:205.

Michell, R. H., 1975, Inositol phospholipids and cell surface receptor function, *Biochim. Biophys. Acta* **415**:81.

Miller, D. M., 1965, The kinetics of selective biological transport. I. Determination of transport constants for sugar movements in human erythrocytes, *Biophys. J.* **5**:407.

Miller, D. M., 1966, A re-examination of Stein's dimer theory of sugar transport in human erythrocytes, *Biochim. Biophys. Acta* **120**:156.

Miller, D., and Crane, R. K., 1961, The digestive function of the epithelium of the small intestine, II. Localization of disaccharide hydrolysis in the isolated brush border portion of intestinal epithelial cells, *Biochim. Biophys. Acta* **52**:293.

Monod, J., Changeux, J. P., and Jacob, F., 1963, Allosteric proteins and cellular control system, *J. Mol. Biol.* **6**:306.

Morgan, H. E., Regen, D. M., and Park, C. R., 1964, Identification of a mobile carrier-mediated sugar transport system in muscle, *J. Biol. Chem.* **239**:369.

Moscona, A., 1957, The development in vitro of chimeric aggregates of dissociated embryonic chick and mouse cells, *Proc. Natl. Acad. Sci. USA* **43**:184.

Moscona, A. A., 1976, Cell recognition in embryonic morphogenesis and the problem of neuronal specificities, in: *Neuronal Recognition* (S. H. Barondes, ed.), pp. 205–226, Plenum Press, New York.

Mueller, P., Rudin, D. O., Tien, H. T., and Wescott, W. C., 1962a, Reconstitution of excitable cell membrane structure in vitro, *Circulation* **26**:1167.

Mueller, P., Rudin, D. O., Tien, H. T., and Wescott, W. C., 1962b, Reconstitution of cell membrane structure in vitro and its transformation into an excitable system, *Nature* **194**:979.

Munck, B. G., and Schultz, S. G., 1969, Lysine transport across rabbit ileum, *J. Gen. Physiol.* **53**:157.

Murer, H., and Hopfer, U., 1974, Demonstration of electrogenic Na^+-dependent D-glucose transport in intestinal brush border membranes, *Proc. Natl. Acad. Sci. USA* **71**:484.

Naftalin, R. J., 1972, An alternative to the carrier model for sugar transport across red cell membranes, in: *Biomembranes*, Vol. 3 (F. Kreuzer and J. F. G. Slegers, eds.), pp. 117–126, Plenum Press, New York.

Nägeli, C., and Cramer, C., 1855, *Pflanzenphysiologische Untersuchungen*, 1 Heft, F. Schultess, Zurich.

Neame, K. D., and Richards, T. G., 1972, *Elementary Kinetics of Membrane Carrier Transport*, John Wiley and Sons, New York.

Neville, D. M., 1960, The isolation of a cell membrane fraction from rat liver, *J. Biophys. Biochem. Cytol.* **8**:413.

Norton, W. T., and Autilio, L. A., 1966, The lipid composition of purified bovine brain myelin, *J. Neurochem.* **13**:213.

O'Brien, J. S., 1965, Stability of the myelin membrane, *Science* **147**:1099.

O'Brien, J. S., 1967, Cell membranes—composition:structure:function, *J. Theoret. Biol.* **15**:307.

O'Brien, J. S., and Sampson, E. L., 1965, Lipid composition of the normal human brain: Gray matter, white matter, and myelin, *J. Lipid Res.* **6**:537.

O'Brien, J. S., Sampson, E. L., and Stern, M. B., 1967, Lipid composition of myelin from the peripheral nervous system; intradural spinal roots, *J. Neurochem.* **14**:357.

Olafson, R. W., Drummond, G. I., and Lee, J. F., 1969, Studies on 2',3'-cyclic nucleotide-3'-phosphohydrolase from brain, *Can. J. Biochem.* **47**:961.

O'Malley, B. W., 1971, Unified hypothesis for early biochemical sequence of events in steroid hormone action, *Metabolism* **20**:981.

Oseroff, A. R., Robbins, P. W., and Burger, M. M., 1973, The cell surface membrane: Biochemical aspects and biophysical probes, *Annu. Rev. Biochem.* **42**:647.

Overton, E., 1895, Über die osmotischem Eigenschaften der lebenden Pflanzen and Tierzelle, *Vierteljahresschr. Naturforsch Ges. Zürich* **40**:159.

Oxender, D. L., and Christensen, H. N., 1963, Distinct mediating systems for the transport of neutral amino acids by the Ehrlich cell, *J. Biol. Chem.* **238**:3686.

Oxender, D. L., Lee, M., Moore, P. A., and Cecchini, G., 1977, Neutral amino acid transport systems of tissue culture cells, *J. Biol. Chem.* **252**:2675.
Ožegović, B., Schön, E., and Milković, S., 1977, Interaction of [^{3}H]-aldosterone with rat kidney plasma membranes, *J. Steroid Biochem.* **8**:815.
Pappius, H. M., 1964, Water transport at cell membranes, *Can. J. Biochem.* **42**:945.
Pardee, A. B., 1971, Membranes and the coordination of cellular activities, in: *Biomembranes*, Vol. 2 (L. A. Manson, ed.), pp. 1–2, Plenum Press, New York.
Parrish, J. E., and Kipnis, D. M., 1964, Effects of Na$^+$ on sugar and amino acid transport in striated muscle, *J. Clin. Invest.* **43**:1994.
Parsons, D. F., 1967, Ultrastructural and molecular aspects of cell membranes, in: *Proceedings of the 7th Canadian Cancer Research Conference*, Pergamon Press, New York.
Pessac, B., and Defendi, V., 1972, Cell aggregation: Role of acid mucopolysaccharides, *Science* **175**:898.
Pfleger, R. C., Anderson, N. G., and Snyder, F., 1968, Lipid class and fatty acid composition of rat liver plasma membranes isolated by zonal centrifugation, *Biochemistry* **7**:2826.
Ponder, E., 1961, The cell membrane and its properties, in: *The Cell*, Vol. 2 (J. Brachet and A. E. Mirsky, eds.), pp. 1–84, Academic Press, New York.
Popp, R. A., 1971, Chemistry of specific antigenic sites on cell surfaces, in: *Biomembranes*, Vol. 2 (L. A. Manson, ed.), pp. 223–245, Plenum Press, New York.
Post, R. L., and Jolly, P. C., 1957, The linkage of sodium, potassium, and ammonium active transport across the human erythrocyte membrane, *Biochim. Biophys. Acta* **25**:118.
Post, R. L., Meritt, C. R., Kinsolving, C. R., and Albright, C. D., 1960, Membrane adenosine triphosphatase as a participant in the active transport of sodium and potassium in the human erythrocyte, *J. Biol. Chem.* **235**:1796.
Post, R. L., Morgan, H. E., and Park, C. R., 1961, Regulation of glucose uptake in muscle. III. The interaction of membrane transport and phosphorylation in the control of glucose uptake, *J. Biol. Chem.* **236**:269.
Quinlan, D. C., Parnes, J. R., Shalom, R., Garvey, T. Q., III., Isselbacher, K. J., and Hochstadt, J., 1976, Sodium-stimulated amino acid uptake into isolated membrane vesicles from Balb/c 3T3 cells transformed by simian virus 40, *Proc. Natl. Acad. Sci. USA* **73**:1631.
Radda, G. K., 1975, Fluorescent probes in membrane studies, *Meth. Membr. Biol.* **8**:97.
Rajam, P. C., and Jackson, A. L., 1958, A cytoplasmic membrane-like fraction from cells of the Ehrlich mouse ascites carcinoma, *Nature* **181**:1670.
Rall, T. W., Sutherland, E. W., and Berthet, J., 1957, The relationship of epinephrine and glucagon to liver phosphorylase. IV. The effect of epinephrine and glucagon on the reactivation of phosphorylase in liver homogenates, *J. Biol. Chem.* **224**:463.
Rambourg, A., and Leblond, C. P., 1967, Electron microscope observations on the carbohydrate-rich cell coat present at the surface of cells in the rat, *J. Cell. Biol.* **32**:27.
Rang, H. P., 1971, Drug receptors and their function, *Nature* **231**:91.
Rasmussen, H., 1970, Cell communication, calcium ion and cyclic adenosine monophosphate, *Science* **170**:404.
Rasmussen, H., and Goodman, B. P., 1977, Relationships between calcium and cyclic nucleotides in cell activation, *Physiol. Rev.* **57**:421.
Ray, T. K., 1970, A modified method for the isolation of the plasma membrane from rat livers, *Biochim. Biophys. Acta* **196**:1.
Razin, S., Morowitz, H. J., and Terry, T. H., 1965, Membrane subunits of *Mycoplasma laidlawii* and their assembly to membrane-like structures, *Proc. Natl. Acad. Sci. USA* **54**:219.
Rega, A. F., and Garrhan, P. J., 1976, Potassium-activated phosphatase, in: *The Enzymes*

of Biological Membranes, Vol. 3, *Membrane Transport* (A. Martonosi, ed.), pp. 303–314, Plenum Press, New York.

Regen, D. M., and Morgan, H. E., 1964, Studies of the glucose transport system in the rabbit erythrocyte, *Biochim. Biophys. Acta* **79:**151.

Reynolds, J. A., and Trayer, H., 1971, Solubility of membrane proteins in aqueous media, *J. Biol. Chem.* **246:**7337.

Rickenberg, H. V., and Maio, J. J., 1961, The accumulation of galactose by mammalian tissue culture cells, in: *Membrane Transport and Metabolism* (A. Kleinzeller and A. Kotyk, eds.), pp. 409–422, Academic Press, New York.

Riggs, T. R., Walker, L. M., and Christensen, H. N., 1958, Potassium migration and amino acid transport, *J. Biol. Chem.* **233:**1479.

Robertson, J. D., 1967, Origin of the unit membrane concept, *Protoplasma* **63:**218.

Robinson, J. W. L., ed., 1976, *Intestinal Ion Transport,* University Park Press, Baltimore.

Rodbell, M., 1967, Metabolism of isolated fat cells. V. Preparation of "ghosts" and their properties; adenyl cyclase and other enzymes, *J. Biol. Chem.* **242:**5744.

Rodbell, M., Birnbaumer, L., Pohl, S. L., and Sundby, F., 1971, The reaction of glucagon with its receptor; evidence for discrete regions of activity and binding in the glucagon molecule, *Proc. Natl. Acad. Sci. USA* **68:**909.

Rodríguez de Lores Arnaiz, G., Alberici, M., and DeRobertis, E., 1967, Ultrastructural and enzymic studies of cholinergic and non-cholinergic synaptic membranes isolated from brain cortex, *J. Neurochem.* **14:**215.

Roseman, S., 1970, The synthesis of complex carbohydrates by multiglycosyltransferase systems and their potential function in intercellular adhesion, *Chem. Phys. Lipids* **5:**270.

Rosenberg, L., Blair, A., and Segal, S., 1961, The transport of amino acids by rat kidney cortex slices, *Biochim. Biophys. Acta* **54:**479.

Rosenberg, L., Downing, S., and Segal, S., 1962, Competitive inhibition of dibasic amino acid transport in rat kidney, *J. Biol. Chem.* **237:**2265.

Rosenberg, L. E., Fox, M., Thier, S., and Segal, S., 1963, In vitro and in vivo studies of amino acid transport in rat and human kidney, *Excerpta Medica International Congress Series* **78:**563.

Rosenberg, S. A., and Guidotti, G., 1968, The protein of human erythrocyte membranes: I. Preparation, solubilization, and partial characterization, *J. Biol. Chem.* **243:**1985.

Rosenberg, T., 1954, The concept and definition of active transport, *Symp. Soc. Exp. Biol.* **8:**27.

Rosenberg, T., and Wilbrandt, W., 1957, Uphill transport induced by counter-flow, *J. Gen. Physiol.* **41:**289.

Rothman, J. E., and Lenard, J., 1977, Membrane asymmetry, *Science* **195:**743.

Rouser, G., Yamamoto, A., and Kritchevsky, G., 1971, Cellular membranes: Structure and regulation of lipid class composition, species differences, changes with age, and variations in some pathological states, *Arch. Intern. Med.* **127:**1105.

Schacter, H., Jabbal, I., Hudgin, R. L., Pinteric, L., McGuire, E. J., and Roseman, S., 1970, Intracellular localization of liver sugar nucleotide glycoprotein glycosyltransferases in a Golgi-rich fraction, *J. Biol. Chem.* **245:**1090.

Schafer, J. A., and Andreoli, T. E., 1972, Water transport in biological and artificial membranes, *Arch. Intern. Med.* **129:**279.

Schatzmann, H. J., 1962, Lipoprotein nature of red cell adenosine triphosphatase, *Nature* **196:**677.

Schatzmann, H. J., 1966, ATP-dependent Ca^{++}-extrusion from human red cells, *Experientia* **22:**364.

Schnell, K. F., 1972, On the mechanism of inhibition of the sulfate transfer across the human erythrocyte membrane, *Biochim. Biophys. Acta* **282:**265.

Schramm, M., Orly, J., Eimerl, S., and Korner, M., 1977, Coupling of hormone receptors to adenylate cyclase of different cells by cell fusion, *Nature* **268**:310.

Schultz, S. G., 1978, Is a coupled Na–K exchange "pump" involved in active transepithelial Na$^+$ transport? A status report, in: *Membrane Transport Processes*, Vol. 1 (J. F. Hoffman, ed.), pp. 213–227, Raven Press, New York.

Schultz, S. G., and Curran, P. F., 1970, Coupled transport of sodium and organic solutes, *Physiol. Rev.* **50**:637.

Schultz, S. G., and Zalusky, R., 1964a, Ion transport in isolated rabbit ileum. I. Short-circuit current and Na$^+$ fluxes, *J. Gen. Physiol.* **47**:567.

Schultz, S. G., and Zalusky, R., 1964b, Ion transport in isolated rabbit ileum. II. The interaction between active sodium and active sugar transport, *J. Gen. Physiol.* **47**:1043.

Schultz, S. G., and Zalusky, R., 1965, Interactions between active sodium transport and active amino acid transport in isolated rabbit ileum, *Nature* **205**:292.

Schwartzman, L., Blair, A., and Segal, S., 1967, Exchange diffusion of dibasic amino acids in rat kidney cortex slices, *Biochim. Biophys. Acta* **135**:120.

Scriver, C. R., and Mohyuddin, F., 1968, Amino acid transport in kidney: Heterogeneity of α-aminoisobutyric acid uptake, *J. Biol. Chem.* **243**:3207.

Seeman, P., 1972, The membrane actions of anesthetics and tranquilizers, *Pharmacol. Rev.* **24**:583.

Segal, S., and Crawhall, J. C., 1968, Characteristics of cystine and cysteine transport in rat kidney cortex slices, *Proc. Natl. Acad. Sci. USA* **59**:231.

Segal, S., and Thier, S. O., 1973, The renal handling of amino acids, in: *Handbook of Physiology* (R. W. Berliner and J. Orloff, eds.), pp. 653–676, American Physiological Society, Washington, D.C.

Segal, S., Schwartzman, L., Blair, A., and Bertoli, D., 1967, Dibasic amino acid transport in rat kidney cortex slices, *Biochim. Biophys. Acta* **135**:127.

Segal, S., Genel, M., and Smith, I., 1968, Effect of storage at 4°C on alpha-methylglucoside transport by rat kidney cortex slices, *J. Lab. Clin. Med.* **72**:778.

Segal, S., Rosenhagen, M., and Rea, C., 1973, Developmental and other characteristics of α-methyl-D-glucoside transport by rat kidney cortex slices, *Biochim. Biophys. Acta* **291**:519.

Semenza, G., 1976, Small intestinal disaccharidases: Their properties and role as sugar translocators across natural and artificial membranes, in: *The Enzymes of Biological Membranes*, Vol. 3, *Membrane Transport* (A. Martonosi, ed.), pp. 349–382, Plenum Press, New York.

Sen, A. K., and Widdas, W. F., 1962a, Determination of the temperature and pH dependence of glucose transfer across the human erythrocyte membrane measured by glucose exit, *J. Physiol. (London)* **160**:392.

Sen, A. K., and Widdas, W. F., 1962b, Variations of the parameters of glucose transfer across the human erythrocyte membrane in the presence of inhibitors of transfer, *J. Physiol. (London)* **160**:404.

Shaw, T. I., 1955, Potassium movement in washed erythrocytes, *J. Physiol. (London)* **129**:464.

Sheridan, P. J., 1975, Minireview: Is there an alternative to the cytoplasmic receptor model for the mechanism of action of steroids?, *Life Sci.* **17**:497.

Shur, B. D., and Roth, S., 1975, Cell surface glycosyltransferases, *Biochim. Biophys. Acta* **415**:473.

Siekevitz, P., 1970, The organization of biologic membranes, *N. Engl. J. Med.* **283**:1035.

Siekevitz, P., Palade, G. E., Dallner, G., Ohad, I., and Omura, T., 1967, The biogenesis of intracellular membranes, in: *Organizational Biosynthesis* (H. J. Vogel, J. O. Lampen, and V. Bryson, eds.), pp. 331–362, Academic Press, New York.

Silbernagel, S., and Deetjen, P., 1969, Microperfusionuntersuchungen zur Resorption von Glycin im proximalen Tubules, *Pflügers Arch. Europ. J. Physiol.* **312**:R82.

Simon, F. R., Blumenfeld, O. O., and Arias, I. M., 1970, Two protein fractions obtained from hepatic plasma membranes; studies of their composition and differential turnover, *Biochim. Biophys. Acta* **219**:349.

Singer, S. J., 1971, The molecular organization of biological membranes, in: *Structure and Function of Biological Membranes* (L. I. Rothfield, ed.), pp. 145–222, Academic Press, New York.

Singer, S. J., 1974, The molecular organization of membranes, *Annu. Rev. Biochem.* **43**:805.

Singer, S. J., 1977, The fluid mosaic model of membrane structure, in: *The Structure of Biological Membranes* (S. Abrahamsson and I. Pascher, eds.), pp. 443–461, Plenum Press, New York.

Singer, S. J., and Nicolson, G. L., 1972, The fluid mosaic model of the structure of cell membranes, *Science* **175**:720.

Sjöstrand, F. S., 1963a, A new ultrastructural element of the membranes in mitochondria and of some cytoplasmic membranes, *J. Ultrastruct. Res.* **9**:340.

Sjöstrand, F. S., 1963b, A comparison of plasma membrane, cytomembranes and mitochondrial membrane elements with respect to ultrastructural features, *J. Ultrastruct. Res.* **9**:561.

Skipski, V. P., Barclay, M., Archibald, F. M., Terebus-Kekish, O., Reichman, E. S., and Good, J. J., 1965, Lipid composition of rat liver cell membranes, *Life Sci.* **4**:1673.

Skou, J. C., 1957, The influence of some cations on an adenosine triphosphatase from peripheral nerves, *Biochim. Biophys. Acta* **23**:394.

Skou, J. C., 1962, Preparation from mammalian brain and kidney of the enzyme system involved in active transport of Na^+ and K^+, *Biochim. Biophys. Acta* **58**:314.

Skou, J. C., 1963, Studies on the $Na^+ + K^+$ activated ATP hydrolyzing enzyme system. The role of SH groups, *Biochem. Biophys. Res. Commun.* **10**:79.

Skou, J. C., 1964, Enzymatic aspects of active linked transport of Na^+ and K^+ through the cell membrane, *Progr. Biophys. Mol. Biol.* **14**: 133.

Skou, J. C., 1975, The $(Na^+ + K^+)$ activated enzyme system and its relationship to transport of sodium and potassium, *Q. Rev. Biophys.* **7**:401.

Skou, J. C., 1977, Coupling of chemical reaction to transport of sodium and potassium, in: *The Structure of Biological Membranes* (S. Abrahamsson and I. Pascher, eds.), pp. 463–478, Plenum Press, New York.

Slomiany, B. L., and Slomiany, A., 1977, Complex glycosphingolipids with blood-group A specificity, *FEBS Lett.* **73**:175.

Smith, I., and Segal, S., 1968, The influence of size of rat kidney cortex slices on the accumulation of amino acids, *Biochim. Biophys. Acta* **163**:281.

Smythies, J. R., Benington, F., and Morin, R. D., 1971, Model for the action of tetrodotoxin and batrachotoxin, *Nature* **231**:188.

Sperry, R., 1965, Embryogenesis of behavioral nerve nets, in: *Organogenesis* (R. L. DeHaan and H. Ursprung, eds.), pp. 161–186, Holt, Rinehart and Winston, New York.

Spiro, R. G., 1967, Studies on the renal glomerular basement membrane: Preparation and chemical composition, *J. Biol. Chem.* **242**:1915.

Spurway, N. C., 1972, Mechanisms of anion permeation, in: *Biomembranes*, Vol. 3 (F. Kreuzer and J. F. G. Slegers, eds.), pp. 363–380, Plenum Press, New York.

Staverman, A. J., 1952, Non-equilibrium thermodynamics of membrane processes, *Trans. Faraday Soc.* **48**:176.

Steck, T. L., Weinstein, R. S., Straus, J. H., and Wallach, D. F. H., 1970, Inside-out red cell membrane vesicles: Preparation and purification, *Science* **168**:255.

Stein, W. D., 1967, *The Movement of Molecules across Cell Membranes*, Academic Press, New York.
Steinbrecht, I., and Hofmann, E., 1964, Die Permeabilität von kaninchen Erythrocyten für 2-Desoxy-D-Ribose, *Z. Physiol. Chem.* **339**:194.
Stockert, R. J., Morell, A. G., and Scheinberg, I. H., 1977, Hepatic binding protein: The protective role of its sialic acid residues, *Science* **197**:667.
Stonehill, E. H., and Huppert, J., 1968, Ribonuclease activity associated with mammalian cell walls, *Biochim. Biophys. Acta* **155**:353.
Takeuchi, M., and Terayama, H., 1965, Preparation and chemical composition of rat liver cell membranes, *Exp. Cell. Res.* **40**:32.
Thier, S., Fox, M., Rosenberg, L., and Segal, S., 1964, Hexose inhibition of amino acid uptake in the rat kidney cortex slice, *Biochim. Biophys. Acta* **93**:109.
Tomasi, V., Koretz, S., Ray, T. K., Dunnick, J., and Marinetti, G. V., 1970, Hormone action at the membrane level. II. The binding of epinephrine and glucagon to the rat liver plasma membrane, *Biochim. Biophys. Acta* **211**:31.
Turkington, R. W., 1962, Thyrotropin-stimulated adenosine triphosphatase in isolated thyroid cell membranes, *Biochim. Biophys. Acta* **65**:386.
Ullrich, K. J., 1979, Sugar, amino acid, and Na^+ cotransport in the proximal tubule, *Annu. Rev. Physiol.* **41**:181.
Ussing, H. H., 1952, Some aspects of the application of tracers in permeability studies, *Adv. Enzymol.* **13**:21.
Ussing, H. H., 1965, as quoted in Csáky, T. Z., 1965, Transport through biological membranes, *Annu. Rev. Physiol.* **27**:415.
Ussing, H. H., 1978, Physiology of transport regulation, *J. Membrane Biol.* **40**:5.
Van Deenen, L. L. M., 1968, Membrane lipids, in: *Regulatory Functions of Biological Membranes* (J. Järnefelt, ed.), pp. 72–86, Elsevier, Amsterdam.
Vanderkooi, G., and Capaldi, R. A., 1972, A comparative study of the amino acid composition of membrane and other proteins, *Ann. N.Y. Acad. Sci.* **195**:135.
Vanderkooi, G., and Green, D. E., 1970, Biological membrane structure. I. the protein crystal model for membranes, *Proc. Natl. Acad. Sci. USA* **66**:615.
Vidaver, G. A., 1966, Inhibition of parallel flux and augmentation of counter flux shown by transport models not involving a mobile carrier, *J. Theor. Biol.* **10**:301.
Vincenzi, F. F., and Hinds, T. R., 1976, Plasma membrane calcium transport and membrane-bound enzymes, in: *The Enzymes of Biological Membranes*, Vol. 3, *Membrane Transport* (A. Martonosi, ed.), pp. 261–281, Plenum Press, New York.
Wallach, D. F. H., and DePerez Ensandi, M. V., 1964, Sialic acid and the electrophoretic mobility of three tumor cell types, *Biochim. Biophys. Acta* **83**:363.
Wallach, D. F. H., and Kamat, V. B., 1964, Plasma and cytoplasmic membrane fragments from Ehrlich ascites carcinoma, *Proc. Natl. Acad. Sci. USA* **52**:721.
Wallach, D. F. H., and Lin, P. S., 1973, A critical evaluation of plasma membrane fractionation, *Biochim. Biophys. Acta* **300**:211.
Wallach, D. F. H., and Ullrey, D., 1962, Studies on the surface and cytoplasmic membranes of Ehrlich ascites-carcinoma cells. I. The hydrolysis of ATP and related nucleotides by microsomal membranes, *Biochim. Biophys. Acta* **64**:526.
Warren, L., and Glick, M. C., 1968, Membranes of animal cells. II. The metabolism and turnover of the surface membrane, *J. Cell Biol.* **37**:729.
Warren, L., Glick, M. C., and Nass, M. K., 1966, Membranes of animal cells. I. Methods of isolation of the surface membrane, *J. Cell Physiol.* **68**:269.
Wasniowska, K., Drzeniek, Z., and Lisowska, E., 1977, The amino acids of M and N blood group glycopeptides are different, *Biochem. Biophys. Res. Commun.* **76**:385.
Watkins, W. M., 1966, Blood-group substances, *Science* **152**:172.
Weber, M. J., Buckman, T., Hale, A. H., Yau, T. M., Brady, T. M., and La Rossa, D. D.,

1976, Cell-surface structure and function in Rous sarcoma virus-transformed cells, in: *Biogenesis and Turnover of Membrane Macromolecules* (J. S. Cook, ed.), pp. 251–276, Raven Press, New York.
Wedeen, R. P., and Thier, S. O., 1971, Intrarenal distribution of non-metabolized amino acids in vivo, *Am. J. Physiol.* **220**:507.
Weinstein, D. B., Marsh, J. B., Glick, M. C., and Warren, L., 1969, Membranes of animal cells. IV. Lipids of the L cell and its surface membrane, *J. Biol. Chem.* **244**:4103.
Weinstein, D. B., Marsh, J. B., Glick, M. C., and Warren, L., 1970, Membranes of animal cells. VI. The glycolipids of the L cell and its surface membrane, *J. Biol. Chem.* **245**:3928.
Weiss, L., 1968, Some comments on RNA as a component of the cell periphery, in: *Biological Properties of the Mammalian Surface Membrane* (L.A. Manson, ed.), pp. 73–76, The Wistar Institute Press, Philadelphia.
Weiss, S. D., McNamara, P. D., Pepe, L. M., and Segal, S., 1978, Glutamine and glutamic acid uptake by rat renal brush border membrane vesicles, *J. Membrane Biol.* **43**:91.
Whaley, W. G., 1975, *The Golgi Apparatus,* Springer-Verlag, Vienna.
Whaley, W. G., Dauwalder, M., and Kephart, J. E., 1972, Golgi apparatus: Influence on cell surfaces, *Science* **175**:596.
Wheeler, K. P., and Christensen, H. N., 1967, Role of Na^+ in the transport of amino acids in rabbit red cells, *J. Biol. Chem.* **242**:3782.
Whittam, R., 1962a, The asymmetrical stimulation of a membrane adenosine triphosphatase in relation to active cation transport, *Biochem. J.* **84**:110.
Whittam, R., 1962b, Directional effects of alkali metal ions on adenosine triphosphate hydrolysis in erythrocyte ghosts, *Nature* **196**:134.
Whittam, R., and Ager, M. E., 1962, Dual effects of sodium ions on an erythrocyte-membrane adenosine triphosphatase, *Biochim. Biophys. Acta* **65**:383.
Whittam, R., and Chipperfield, A. R., 1975, The reaction mechanism of the sodium pump, *Biochim. Biophys. Acta* **415**:149.
Whittam, R., and Wheeler, K. P., 1961, Sensitivity of a kidney ATPase to ouabin and to sodium and potassium, *Biochim. Biophys. Acta* **51**:622.
Widdas, W. F., 1957, as quoted in Bowyer, F., 1957, The kinetics of the penetration of non-electrolytes into the mammalian erythrocyte, *Int. Rev. Cytol.* **6**:469.
Wilbrandt, W., 1972, Carrier diffusion, in: *Biomembranes,* Vol. 3 (F. Kreuzer and J. F. G. Slegers, eds.), pp. 79–99, Plenum Press, New York.
Wilbrandt, W., 1975, Minireview: Recent trends in membrane transport research, *Life Sci.* **16**:201.
Wilfong, R. F., and Neville, D. M., Jr., 1970, The isolation of a brush border membrane fraction from rat kidney, *J. Biol. Chem.* **245**:6106.
Wilson, T. H., 1963, Seventh Bowditch Lecture: Intestinal absorption of vitamin B_{12}, *Physiologist* **6**:11.
Winter, C. G., and Christensen, H. N., 1964, Migration of amino acids across the membrane of the human erythrocyte, *J. Biol. Chem.* **239**:872.
Winter, C. G., and Christensen, H. N., 1965, Contrasts in neutral amino acid transport by rabbit erythrocytes and reticulocytes, *J. Biol. Chem.* **240**:3594.
Wirtz, K. W. A., 1974, Transfer of phospholipids between membranes, *Biochim. Biophys. Acta* **344**:95.
Wolff, J., 1964, Transport of iodide and other anions in the thyroid gland, *Physiol. Rev.* **44**:45.
Wolfgram, F., and Kotorii, K., 1968a, The composition of the myelin proteins of the central nervous system, *J. Neurochem.* **15**:1281.
Wolfgram, F., and Kotorii, K., 1968b, The composition of the myelin proteins of the peripheral nervous system, *J. Neurochem.* **15**:1291.

Woodin, A. M., and Wieneke, A. A., 1966, The interaction of leucocidin with the cell membrane of the polymorphonuclear leucocyte, *Biochem. J.* **99**:479.

Woolley, D. W., and Gommi, B. W., 1964, Serotonin receptors:V. Selective destruction by neuraminidase plus EDTA and reactivation with tissue lipids, *Nature* **202**:1074.

Yager, P., 1977, A novel mechanism for the Na^+-K^+ ATPase, *J. Theor. Biol.* **66**:1.

Yandrasitz, J. R., Ernst, S. A., and Salganicoff, L., 1976, The subcellular distribution of carbonic anhydrase in homogenates of perfused rat brain, *J. Neurochem.* **27**:707.

Young, J. A., and Freedman, B. S., 1971, Renal tubular transport of amino acids, *Clinical Chem.* **17**:245.

Zwaal, R. F. A., Comfurius, P., and Van Deenen, L. L. M., 1977, Membrane asymmetry and blood coagulation, *Nature* **268**:358.

12

Cellular Mechanisms of Secretion

Harold I. Friedman

I. Introduction

Prior to the advent of electron microscopy, studies of secretory processes were limited primarily to identifying secretory cells and the materials they produce, determining the gross mechanisms of secretion, and delineating the stimuli which cause the release of material to the extracellular environment. The development and refinement of ultrastructural techniques provided an entirely new dimension to the understanding of secretory processes as investigations were made possible into the assembly, intracellular transport, and release of secretory material. Subsequently, from the examination of a wide variety of secretory cells, certain common principles evolved regarding the mechanisms of production and release of secretory products. The importance of cellular membranes in the synthesis, packaging, and intracellular transport of secretory material was recognized, and defects in secretory function were traced to abnormalities in cellular mechanisms. In the ensuing discussion, an attempt will be made to outline the common pathways shared by many different types of secretory cells for producing and releasing their products. Several representative examples of secretory cells and the cellular organelles involved in the manufacture of precursor material will be analyzed in depth. Finally, an effort will be made to describe some of the current controversies, theories, and

---

Harold I. Friedman • Department of Surgery, Letterman Army Medical Center, Presidio of San Francisco, California 94129. The opinions or assertions contained herein are the private views of the author and are not to be construed as official or as reflecting the views of the Department of the Army or the Department of Defense.

supporting experimental evidence regarding certain aspects of the intracellular mechanisms of secretion.

II. Representative Secretory Cells

A. The Pancreatic Exocrine Cell

Although it is possible to list all of the mammalian secretory cells and to note in some detail the biochemical nature of both precursor material and final secretory product, as well as to describe the intracellular pathway of secretory product formation, this would defeat the purpose of the current discourse. Rather, it will be more instructive to examine one secretory cell in detail and to compare it to several other representative cell types. From this analysis, one may derive an overview of the general mechanisms of secretion. For this purpose, I have chosen as a starting point the pancreatic exocrine cell because this cell has been one of the most thoroughly investigated, and its secretory pattern provides an ideal model on which to build broad concepts.

1. Acinar Cell Morpohology

The pancreatic acinar cells are serous cells forming tubuloacinar or tubuloalveolar glands in the body of the pancreas. The acinus is composed of a single layer of pyramidal cells, with the broad bases of the

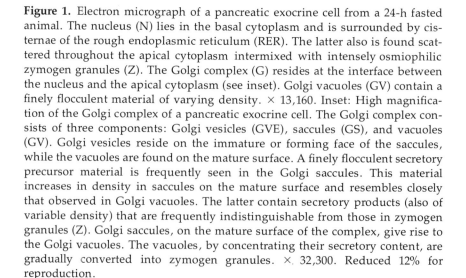

Figure 1. Electron micrograph of a pancreatic exocrine cell from a 24-h fasted animal. The nucleus (N) lies in the basal cytoplasm and is surrounded by cisternae of the rough endoplasmic reticulum (RER). The latter also is found scattered throughout the apical cytoplasm intermixed with intensely osmiophilic zymogen granules (Z). The Golgi complex (G) resides at the interface between the nucleus and the apical cytoplasm (see inset). Golgi vacuoles (GV) contain a finely flocculent material of varying density. × 13,160. Inset: High magnification of the Golgi complex of a pancreatic exocrine cell. The Golgi complex consists of three components: Golgi vesicles (GVE), saccules (GS), and vacuoles (GV). Golgi vesicles reside on the immature or forming face of the saccules, while the vacuoles are found on the mature surface. A finely flocculent secretory precursor material is frequently seen in the Golgi saccules. This material increases in density in saccules on the mature surface and resembles closely that observed in Golgi vacuoles. The latter contain secretory products (also of variable density) that are frequently indistinguishable from those in zymogen granules (Z). Golgi saccules, on the mature surface of the complex, give rise to the Golgi vacuoles. The vacuoles, by concentrating their secretory content, are gradually converted into zymogen granules. × 32,300. Reduced 12% for reproduction.

Cellular Mechanisms of Secretion

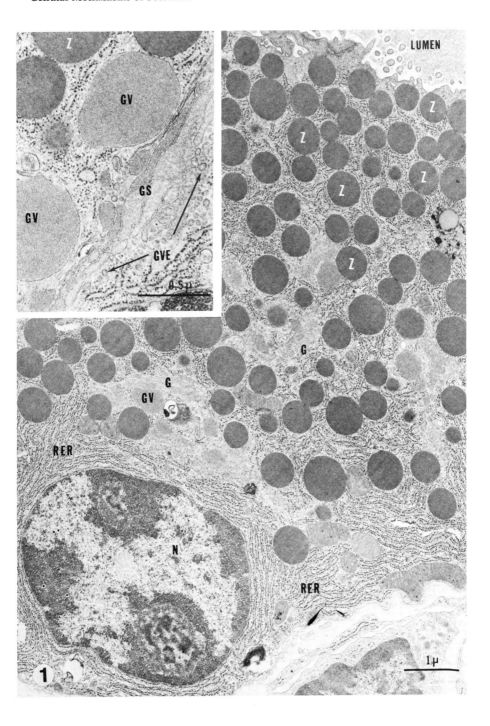

cells resting on a thin basement membrane which has underlying reticular connective tissue. The narrow, apical ends of the cells border on a lumen continuous with the terminal portions of the pancreatic duct system. In the fasted individual, the apical cytoplasm is laden with many acidophilic zymogen granules. These granules contain at least nine specific enzymes for digesting nutrient material, including pepsinogen, chymotrypsinogen, carboxypeptidase, amylase, lipase, lecithinase, ribonuclease, and deoxyribonuclease. Pancreatic juice also contains varying quantities of water, bicarbonate, and salts, added to the secretory product by intralobular and interlobular duct cells.

At the ultrastructural level (Fig. 1), the exocrine cell is characterized by a prominent nucleus located in the basal portion of the cytoplasm, and a relatively large quantity of rough endoplasmic reticulum (RER) scattered throughout the cytoplasm, but most readily observed in the region of the nucleus. The RER appears as paired membranous lines studded on their outer surfaces by numerous dark-staining ribonucleic acid (RNA)-containing granules—the ribosomes. A well-developed Golgi complex is usually found at the interface of the nucleus with the apical cytoplasm (Fig. 1 and inset). The Golgi complex consists of three major components: Golgi saccules, vesicles, and vacuoles (Fig. 1 inset). The saccules appear as stacks of flattened membranes with varying lengths and numbers of saccules per stack. The vacuoles reside on the concave side or mature surface of the Golgi saccules, vary in size, and contain material of differing density. Often the material within the vacuoles is dense enough to render it indistinguishable from zymogen granules. Golgi vesicles are considerably smaller than the vacuoles and are found predominantly on the convex or forming surface of the Golgi complex. The apical cytoplasm of the cell in the fasted animal contains numerous osmiophilic or dark-staining zymogen granules (Fig. 1). These granules are separated from the surrounding cytoplasm by smooth-surfaced membranes similar to those of the Golgi vacuoles. At the luminal border of the cell, the plasma membrane frequently exhibits short, blunt microvilli, and tight junctions are found between adjacent exocrine cells just below the luminal surface.

2. The Secretory Pathway

Our current knowledge of the intracellular sequence of events regarding the formation of pancreatic zymogen granules is largely a result of the investigative efforts of Palade and his co-workers. In a classic electron microscopic radioautographic study, Caro and Palade (1964) followed the uptake and intracellular distribution of a pulse label of tritiated leucine. The label was administered intravenously to rats subjected to a 24-h fast and then fed 1 h prior to injection so as to initiate the secretory cycle. At the light-microscopic level, the label was observed first over the basilar aspects of the exocrine cells. With time, the bulk of the label was seen sequentially in the central portion of the cell and then eventually over the apical zymogen granules. Upon observing these cells with the electron microscope 4 min after injection, the investigators initially noted the appearance of radioactive leucine in the RER of the basal cytoplasm, particularly within the cisternae of this organelle. Six minutes after injection, labeled leucine was localized in small vesicles of the Golgi complex, and by 20 min most of the label had left the RER to enter the larger condensing vacuoles of the Golgi complex. By 45 min, the radioactive grains were distributed equally between Golgi vacuoles and apical zymogen granules. Finally, at 4 h, the labeled leucine was predominantly found in the zymogen granules, while the RER and Golgi complex were relatively free of labeled material.

Combining previous cell fractionation studies (Siekevitz and Palade, 1958a,b,c: Siekevitz and Palade, 1960) with the results of the above investigation, the authors hypothesized a sequence in the formation of exocrine secretory product (Fig. 2). In response to the appropriate stimulation, protein enzymes are synthesized by ribosomes attached to the membranes of the endoplasmic reticulum (ER). The newly formed enzymes are sequestered into the cisternae of the RER and are then transferred, possibly by way of smooth-surfaced vesicles, to condensing vacuoles and saccules on the mature surface of the Golgi complex. Within the vacuoles the precursor material is concentrated, probably by removal of water, and the zymogen granule is formed. As can be seen in the inset of Fig. 1, Golgi vacuoles contain material of varying density, representing the zymogen precursors. The final step in the release of zymogen granules involves fusion of the surrounding Golgi-derived membrane with the apical plasma membrane and extrusion of the zymogen granule content to the lumen of the secretory duct.

In subsequent studies, using incubated pancreatic slices, [^{14}C]leucine, and improved differential centrifugation techniques, Jamieson and Palade (1967a,b) demonstrated that a pulse label was incorporated first into the rough microsomal fraction of the cell, corresponding to the RER. At later time intervals, the label appeared in the smooth microsomal fraction and eventually reached and was concentrated in the zymogen granules. The authors concluded that the smooth microsomal fraction corresponded to the small, smooth-surfaced vesicles associated with the peripheral elements of the Golgi complex, and that the vesicles were responsible for the transportation of secretory proteins from the RER to the Golgi vacuoles. The fact that the labeled

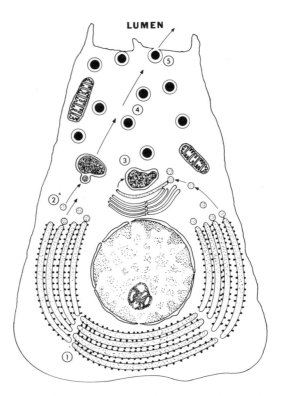

Figure 2. Diagrammatic representation of a pancreatic exocrine cell. Secretory precursor proteins are synthesized by the RER and sequestered in the cisternae of that organelle (1). The material is then transferred by way of small vesicles to mature Golgi saccules and vacuoles (2). Within the vacuoles the secretory precursors are concentrated with the removal of water (3) and the Golgi vacuole is transformed into a mature zymogen granule (4). In response to the appropriate stimulus, the membrane of the zymogen granule fuses with the apical plasma membrane (5), and the final secretory product is released to the lumen of the gland.

leucine actually reflected the course of secretory enzymes in the various cell compartments was supported by a previous observation that the rate of labeling of secretory proteins was five to seven times faster than that of proteins which were retained by the exocrine cell (Siekevitz and Palade, 1960). Furthermore, most of the label could be extracted by saline-bicarbonate from the different cellular fractions. Saline-bicarbonate was known to solubilize digestive enzymes and their precursors from zymogen granules (Greene et al., 1963). The authors, therefore, speculated that any label incorporated into nonexportable proteins (e.g., membrane proteins) could be considered negligible within the time limits of their experiments. Their studies also clarified the question of the site of enzyme protein synthesis. It had previously been postulated that the Golgi complex might be involved in the de novo synthesis of secretory proteins (Sjöstrand, 1962). The results of Jamieson and Palade's work (1967a,b) failed to support this contention, and instead proved that the RER was the only site of secretory protein synthesis. More specifically, based on other investigations (Siekevitz and Palade, 1960; Redman et al., 1966; Redman and Sabatini, 1966), it was possible to conclude that the enzymes actually were synthesized by the ribosomes attached to the RER and that the resultant macromolecules were sequestered within the RER cisternae.

The mechanism by which the smooth-surfaced vesicles acquired secretory product and transferred it to the Golgi vacuoles was unclear. Electron microscopic images were obtained which suggested fusion of vesicles with Golgi vacuoles, as well as a budding process of vesicles from the RER. It was proposed, therefore, that the vesicles might form from the RER, pinch off from it, carry their protein content to the vacuoles, and release the secretory material into the vacuole lumen by fusing with the vacuole membrane. In any event, it was certain that the secretory proteins traversed several organelles, always exclusively separated from the cellular cytoplasm by the limiting membranous walls of these organelles. There was never any increase in the specific radioactivity of proteins in the postmicrosomal supernatant, which corresponded to the cytoplasmic matrix (Jamieson and Palade, 1967b).

3. Metabolic Requirements

The intracellular transport of secretory material from the RER via smooth vesicles to Golgi condensing vacuoles is an energy-requiring process that can be blocked by inhibitors of mitochondrial respiration, including sodium cyanide, antimycin A, and anaerobic conditions, as well as by uncouplers of oxidative phosphorylation such as oligomycin and 2,4-dinitrophenol (Jamieson and Palade, 1968b). Inhibitors of protein synthesis and glycolysis have no effect (Jamieson and Palade, 1968a). The enzymatic nature of the transport process is indicated by its inhibition at low temperature (Jamieson and Palade, 1968b). Not only is the transport of secretory material to Golgi vacuoles dependent on mitochondrial ATP, but the discharge of completed zymogen granules to the duct lumen is also dependent on mitochondrial ATP (Jamieson and Palade, 1968b). Both processes involve the fusion of one intracellular membrane with another. In the latter case, this involves the fusion of zymogen granule membranes derived from Golgi vacuoles with the apical plasma membrane. The actual formation of zymogen granules from Golgi condensing vacuoles does not appear to require energy and is not associated with an ion pump (Jamieson and Palade, 1968b). Rather, it has been postulated that during the continuing accumulation of pancreatic enzymes in Golgi vacuoles, the enzymes form molecular aggregates which are osmotically "inactive" (Jamieson and Palade, 1971a). This phenomenon, in turn, may result in a net movement of water from the vacuole lumen to the more osmotically active cellular cytoplasm, thereby concentrating the zymogen material.

We have thus briefly traced the pathway of secretory material in the pancreatic acinar cell from its synthesis in the RER to its transference to Golgi vacuoles and eventual release from zymogen granules into the duct lumen. Almost identical pathways of secretion have been demonstrated for salivary gland cells and the gastrin-producing cells of the stomach. From these comparative studies, it is apparent that the Golgi complex plays a pivotal role between the synthesis of precursor material and the release of secretory product to the extracellular space. In the pancreatic acinar cell, the Golgi complex is responsible for packaging the final secretory product and for transporting it, in Golgi vacuoles, to the luminal plasma membrane. The Golgi complex does not seem to be involved in the synthesis of any of the exportable material. In this fashion it differs somewhat from that of the other secretory cells. Therefore, in the following section we will examine another type of secretory cell, the intestinal goblet cell, and analyze the contribution of the Golgi complex to the final secretory product.

B. The Intestinal Goblet Cell

1. Carbohydrates in Secretory Products

Ramon y Cajal (1914) was the first to identify the Golgi complex as the site of formation of mucigen granules in goblet cells. This occurred at a time when the actual identity of the Golgi apparatus as a discrete organelle was in some doubt. Subsequently, with the advent of electron microscopic radioautography, the Golgi complex clearly was demonstrated to be the site of segregation or packaging of a variety of secretory proteins (Jamieson and Palade, 1971a; Nadler et al., 1964; Revel and Hay, 1963; Ross and Benditt, 1965), as exemplified by the pancreatic exocrine cell.

Most secretory cells produce a product which is a complex of both proteins and carbohydrates. These cells include parotid acinar cells, chondrocytes, epididymis epithelial cells, and mucous cells of the alimentary and respiratory tracts. The complex carbohydrate secretions fall into two main categories, the mucopolysaccharides, consisting of long carbohydrate chains loosely bound to protein moieties and the predominant product of connective tissue secretions, and the glycoproteins, which are shorter-chained carbohydrates firmly bound to protein and mainly derived from epithelial cell secretions. The site of addition or "complexing" of carbohydrates to proteins during the formation of these secretory products was first elucidated by Neutra and Leblond (1966a,b). Using electron microscopic radioautography, they studied the fate of intraperitoneally injected tritiated glucose in colonic goblet cells at various times after its administration to rats.

2. Goblet Cell Morphology

The goblet cell is a columnar epithelial cell characterized by a relatively dark-staining cytoplasm. The nucleus is elongated and resides at the basal portion of the cytoplasm, surrounded by cisternae of the RER (Fig. 3). At the luminal border of the nucleus is a relatively large Golgi complex (Fig. 3 and inset). The Golgi complex here forms a cuplike shape in three dimensions, with its base situated just above the nucleus and its walls extending upward along the lateral aspects of the cytoplasm toward the lumen of the intestine or colon. The Golgi saccules form the outline of the cup, while the Golgi vacuoles are located in the center. The vacuoles usually vary in size and contain a mucigen material of varying density (Fig. 3 inset). The apical cytoplasm of the cell is distended with mucigen granules and is separated at the cellular apex from the intestinal lumen by only a thin layer of cytoplasm.

3. The Golgi Complex and Carbohydrate Secretion

Five minutes after the intraperitoneal injection of tritium-labeled glucose, Neutra and Leblond (1966a), using radioautography, observed silver grains over the stacks of Golgi saccules, but none were noted over the mucigen granules. At later time periods, the labeled glucose was seen sequentially over mucigen-containing Golgi vacuoles, and finally within the mucigen granules of the apical cytoplasm. Enzymatic digestion of the tissue with either α-amylase or hyaluronidase failed to remove any label; this suggested that the material was not glycogen, hyaluronic acid, or chondroitin sulfate. In contrast, treatment with peracetic acid–β-glucuronidase removed a considerable amount of label; thus, as this indicates, the secretory product consisted in some part of glycoproteins. Therefore, glucose incorporation into the secretory product while in the Golgi complex suggested an additional role for this

448   Chapter 12

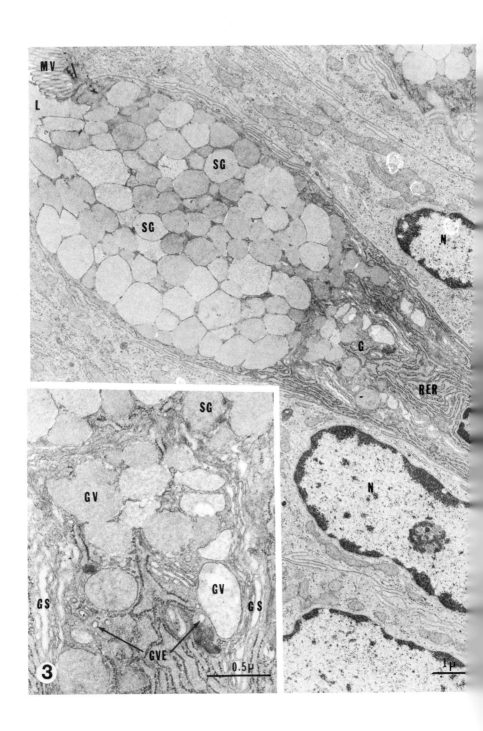

organelle beyond that of packaging secretory material. In further studies, Neutra and Leblond (1966b) demonstrated tritiated galactose incorporation into secretory products by Golgi complexes of certain goblet cells, pancreatic acinar cells, hepatocytes, intestinal absorptive cells, proximal tubular cells of the kidney, epididymal epithelial cells, and chondrocytes. In the case of the pancreatic exocrine cell, ribonuclease (RNase) (one of the zymogen granule enzymes) was shown to contain 2% carbohydrate (Plummer and Hirs, 1963), and it was suggested that this carbohydrate moiety was added to the zymogen granule as the secretory precursor material traversed the Golgi complex. Hepatocytes secrete plasma glycoproteins and glycolipoproteins containing galactosyl residues. Again, the Golgi complex was implicated in the addition of these moieties to the final secretory product. Duodenal columnar absorptive cells and proximal tubular cells of the kidney both have apical surface coats of mucopolysaccharides covering their microvillous borders. Tritiated galactose localized in the Golgi complexes of these cells and then, at later intervals, it appeared in the surface coat. Finally, it is well established that the matrix of cartilage contains both collagen and the sulfated mucopolysaccharide, chondroitin sulfate. Electron microscopic radioautography demonstrated that [^{35}S]sulfate and both tritiated glucose and galactose were incorporated into the Golgi complex prior to their appearance in the cartilage matrix (Fewer et al., 1964; Godman and Lane, 1964; Neutra and Leblond, 1966b). The Golgi complex in some cells, therefore, is also responsible for the addition of sulfated moieties, as well as carbohydrate components, to secretory material.

Figure 3. Electron micrograph of the apical portions of a goblet cell and several intestinal absorptive cells. The goblet cell is characterized by a relatively dark-staining cytoplasm. The nucleus resides in the basal aspect of the cell, surrounded by the RER. The Golgi complex (G) forms a cup-shaped arrangement between the nucleus and the goblet, or secretory granule (SG)-containing portion of the cytoplasm. Nucleus (N); microvillus border (MV); lumen of the intestine (L). × 9880. Inset: High magnification of the Golgi complex of an intestinal goblet cell. The Golgi complex contains the usual three components: Golgi vesicles (GVE), saccules (GS), and vacuoles (GV). Secretory precursors are synthesized in the RER and transferred by way of the Golgi vesicles to the Golgi saccules and vacuoles. Here the secretory material is concentrated and a carbohydrate moiety is added. With the further concentration of secretory material, the Golgi vacuoles are converted into secretory granules (SG). Note the variable density of the secretory material in the Golgi vacuoles and its frequent similarity to that in secretory granules. × 29,700. Reduced 12% for reproduction.

On the basis of the above morphological observations, it is not surprising that in more recent studies, when Golgi-rich fractions have been isolated from intact cells, it has been found that they contained relatively high activities of several glycosyltransferase enzymes (Keenan et al., 1974; Morré et al., 1969; Schachter et al., 1970). These enzymes not only catalyze the coupling of carbohydrate to protein, but also are responsible for catalyzing the final stage of ganglioside synthesis (i.e., lipid glycosylation). Enzymes of phospholipid biosynthesis have also been localized to the Golgi complex (Morré et al., 1970), as have several enzymes commonly found in lysosomal particles (e.g., acid phosphatases).

4. Turnover of Golgi Membranes

From the above considerations, it appears that the Golgi complex serves three main functions in the elaboration of secretory products: (1) it provides a cellular site for the complexing of certain moieties to secretory precursors which are derived from the cisternae of the RER; (2) it is involved in the final concentration of secretory material; and (3) it provides a vehicle for the net transport of intracellular secretory material to the extracellular space. A fourth function of the Golgi complex involves the formation of primary lysosomes. Since the lysosome is a product retained intracellularly rather than secreted to the extracellular environment, it will not be included in the ensuing discussion of cellular secretory mechanisms. The reader, therefore, is referred to an excellent and thorough review of lysosomes edited by Dingle and Fell (1969a,b).

The above Golgi functions, particularly the first and third, raise two important questions. Since the Golgi complex is not known to have the machinery for protein synthesis, what is the origin of the enzymes required for the covalent linkage of carbohydrates and other moieties to the secretory precursors, and how do they enter the Golgi complex? The second question involves the use of intracellular membranes, particularly those of Golgi vacuoles, in the transport of secretory products.

In addition to their study of the intracellular pathway of mucigen granule formation and release, Neutra and Leblond (1966a) made several interesting observations regarding the turnover of Golgi membranes. They noted that Golgi saccules were labelled within 20 min after injection of radioactive glucose. The label subsequently disappeared from the Golgi saccules and entered mucigen-containing vacuoles 40 min after its administration. It was apparent that the Golgi vacuoles were formed from the innermost or central Golgi saccules on the concave surface of the stack of saccules, yet the total numbers of Golgi saccules remained constant. These findings implied that with the use of Golgi saccule membranes to form Golgi vaccuoles, there was a replacement or addition of new Golgi saccules on the convex surface of the stacks. The turnover time of the entire stack of saccules was in the range of 20 min.

Since the Golgi complex contained approximately 10 saccules per stack, Neutra and Leblond (1966a) conjectured that a saccule was released in the form of a mucigen-containing vacuole as often as every 2–4 min. In contrast to the release of zymogen granules from the pancreatic exocrine cell, where the membrane surrounding the granule fuses with the apical plasma membrane, in the goblet cell the membrane of the mucigen granule is actually extruded with the granule through small gaps in the apical cytoplasm and into the lumen of the gut. This phenomenon would naturally result in a net loss of membrane from the goblet cell, whereas one could postulate that in the exocrine pancreatic cell the membrane is retained. In any event, the loss of Golgi-derived membranes from the goblet cell necessitates some mechanism for the de novo synthesis of new Golgi membranes in order to maintain that organelle. Again, there is no evidence that the Golgi complex contains the synthetic machinery for de novo membrane production. Thus, the second main question concerning the Golgi complex is how this organelle acquires new saccular membranes to replace those depleted by the secretory process.

In an effort to answer both questions, we will examine in some detail the intestinal absorptive cell, which ordinarily is not considered a secretory cell, but which does produce a secretory product. During intestinal fat absorption, chylomicra are produced and released from the absorptive cells in a manner similar to secretory products in other cell types.

C. The Intestinal Absorptive Cell

1. Absorptive Cell Morphology

Intestinal absorptive cells are simple columnar, highly polarized cells (Fig. 4). The luminal plasma membrane is modified to form numerous fingerlike projections, the microvilli, which markedly increase the total absorptive surface of the cell. Directly beneath the microvillous border is an area relatively devoid of cellular organelles, but containing numerous filaments extending from the microvilli into the cytoplasm. This region is termed the terminal web zone of the cell. Below the terminal web area is the apical cytoplasm, which extends to the level of the nucleus and Golgi complex. The apical cytoplasm, in the fasted

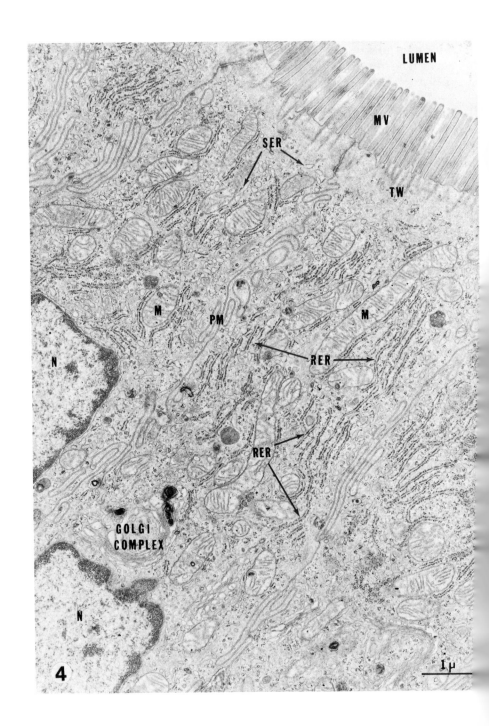

animal, contains a large quantity of RER whose cisternae characteristically are directed parallel to the long axis of the cell (Fig. 4). Similarly aligned filamentous mitochondria form a close association with the RER. The smooth endoplasmic reticulum (SER) occurs interspersed among the RER and mitochondria, but is most clearly observed in the distal cytoplasm directly beneath the terminal web (Fig. 4). It consists of membranous, smooth-surfaced, convoluted tubules and vesicles which vary both in diameter (10–100 mμ) and direction. The SER often appears as isolated segments; however, continuity of SER with RER cisternae frequently is found.

The Golgi complex resides in a supranuclear position (Figs. 4 and 5), and exhibits the characteristic structure of stacks of parallel flattened sacs (saccules), small vesicles, and large vacuoles. These structures usually appear empty in the fasted rat, but occasionally the vacuoles contain droplets 500–1000Å in diameter, which resemble the very-low-density lipoprotein particles (VLDL) previously described by others in both intestinal absorptive (Tygat et al., 1971) and liver parenchymal cells (Jones et al., 1968). A maximum of six to eight parallel saccules per stack occurs in the Golgi complex. The saccules are approximately 200–300 Å apart, and their lumens vary in width. Golgi vacuoles reside on one side of the saccules, while small spherical vesicles lie on the other (Fig. 5). In order to facilitate discussions of the Golgi complex, and to be consistent with the literature, the vesicular side of the Golgi complex is referred to as the immature or forming face, and the side containing vacuoles, as the mature face. The reason for this nomenclature will become apparent in the forthcoming discussion of fat absorption and membrane flow.

A relatively close association exists between elements of the SER and RER with the Golgi complex. Frequently, RER profiles are seen adjacent to the outermost Golgi saccules of the forming face. These profiles display a peculiar configuration, with ribosomes absent on the

⬅

Figure 4. Electron micrograph of the apical portions of several intestinal absorptive cells in the fasted state. Absorptive cells are highly polar. The plasma membrane bordering on the lumen of the intestine is modified to form numerous microvilli (MV). Beneath the microvillus border is the terminal web (TW) zone of the cytoplasm which contains relatively few cytoplasmic organelles. The smooth endoplasmic reticulum (SER) is most prominent just below the terminal web. The RER extends from the region of the terminal web all the way to the supranuclear Golgi complex. Mitochondria (M); nucleus (N); plasma membrane (PM). × 15,940. Reduced 12% for reproduction. From Friedman and Cardell (1977) with permission of the publisher.

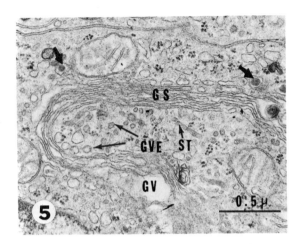

Figure 5. High magnification of the Golgi complex from an intestinal absorptive cell in the fasted state. The Golgi complex is similar in composition to that observed in other cells, consisting of Golgi saccules (GS), vacuoles (GV) and vesicles (GVE). There are usually six to eight saccules per stack and the vesicles reside on the forming face of the stack of saccules, while the vacuoles are found on the mature surface. Smooth-surfaced tubules (ST) and elements of the SER are observed frequently in association with the forming Golgi saccules and are thought to be important in providing membranes for their formation. The unmarked arrows indicate lipoprotein particles in small Golgi vacuoles. × 27,300. From Friedman and Cardell (1977) with permission of the publisher.

membrane surface opposite the forming face of the Golgi saccules (Fig. 6). Furthermore, in fortuitous sections, the RER membranes show continuity with membranes that lack ribosomes on both surfaces and are closely opposed to Golgi saccules. Randomly arranged tubular elements of the SER, independent of the RER, also are observed adjacent to the forming face of the Golgi saccules (Fig. 5). This association of ER and Golgi complex is similar to that described by Claude (1970) in the liver, and it is interpreted to be important in providing membranes for the formation and maintenance of the Golgi complex. The remainder of the absorptive cell does not participate morphologically in the absorption or transport of lipids, and it consists of an elongated nucleus and an area of basal cytoplasm containing many free ribosomes and spherical mitochondria.

2. Intestinal Fat Absorption

The biochemical and physiological events in the transport of fat are well known (Borgstrom, 1962; Senior, 1964). Free fatty acids and monoglycerides, which result from the intraluminal digestion of triglycerides, diffuse passively into the apical cytoplasm of the absorptive cells. Within the cells, these products are reesterified to form triglycerides and are then complexed with several moieties to produce chylomicra. The chylomicron is the final secretory product released to the extracellular space. It consists of 85–90% triglyceride, 6–9% phospholipid, 3% cholesterol, 0.5–1.0% protein, and a small carbohydrate component. The enzymes for triglyceride resynthesis are located in the microsomal fraction of the cell, and shortly after the administration of a lipid meal, droplets of reesterified triglyceride can be seen in bulbous expansion of the SER in the apical cytoplasm below the terminal web (Fig. 7). These SER expansions frequently are continuous with the RER (Fig. 7). With time, the SER continues to dilate and eventually loses its attachment to the RER, and it thereby forms discrete lipid-containing vesicles (Fig. 7). As more and more SER vesicles accumulate, the quantity of RER in the cell decreases. The loss of RER begins in the most distal cytoplasm near the terminal web and proceeds toward the nucleus (Fig. 8). Using morphometric techniques, Friedman and Cardell (1977) have quantitated the conversion of RER cisternae into SER vesicles (Fig. 9). As in other cells (Dallner et al., 1966; Cardell, 1977), the SER in the absorptive cell proliferates by both loss of ribosomes from the RER and by de novo synthesis and incorporation of new membranes into the RER itself.

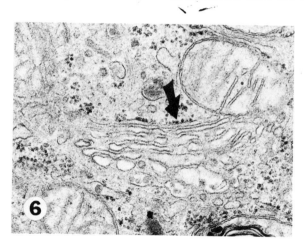

Figure 6. High magnification of the Golgi complex from an intestinal absorptive cell in the fasted state. Note the close association between the profile of the RER (arrow) and the immature or forming Golgi saccules. The RER lacks ribosomes on the surface adjacent to the Golgi saccules. Images such as this suggest the direct conversion of RER into Golgi saccules through the loss of ribosomes. Alternatively, these RER profiles may synthesize new protein membrane components for incorporation into forming Golgi saccules. × 38,470. From Friedman and Cardell (1977) with permission of the publisher.

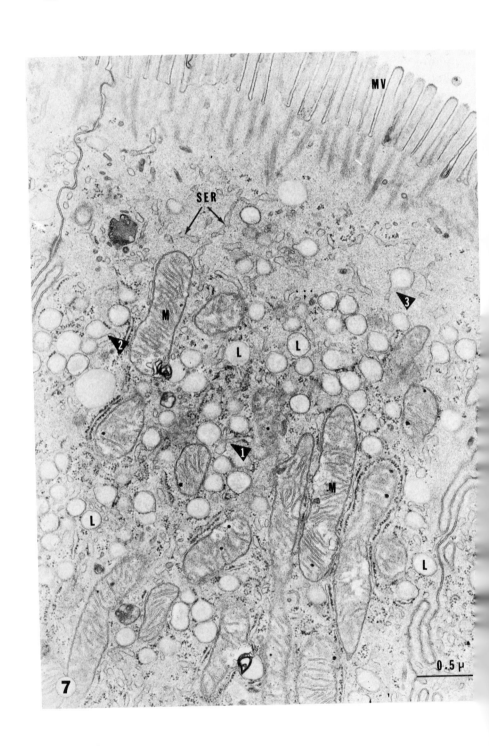

Obviously, the synthesis of new membrane is not sufficient to keep pace with its conversion to SER vesicles; thus, a gradual reduction in quantity of RER occurs as fat absorption continues.

In the Golgi complex, lipid accumulates in dilated Golgi vacuoles on the mature surface of the Golgi saccules. The vacuoles acquire this lipid through the fusion of SER lipid-containing vesicles with the vacuole membrane. In time, the Golgi vacuoles and their lipid content increase in both size and number (Fig. 10) at the expense of Golgi saccules (compare Figs. 11 and 5). One hour after administration of a corn-oil meal to rats, three or four Golgi saccules in a stack of membranes are observed, a reduction from the six to eight seen in the fasted condition (Figs. 11 and 12). Similarly, the lengths of Golgi saccules decrease from the fasted (2.2 μm) to the fed condition (1.0 μm) (Fig. 13). Unlike the goblet cell, therefore, the absorptive cell appears unable to maintain a full complement of Golgi saccules during the packaging and secretion of chylomicra. The fate of completed chylomicra in Golgi vacuoles is similar to that of zymogen granules. The Golgi vacuole membrane fuses with the lateral plasma membrane via a process of reverse pinocytosis (Fig. 14), and the completed chylomicron is thereby exteriorized from the cell to the intercellular space (Friedman and Cardell, 1972b). It is uncertain where the additional components of the chylomicra are added, but it may be speculated that cholesterol, protein, and phospholipid are complexed to the triglyceride while it is in the endoplasmic reticulum. The carbohydrate moiety most likely is added by the Golgi complex.

3. Membrane Utilization in Secretory Transport

As in the other secretory cells, precursors and products are sequestered from the cellular cytoplasm in their passage through the absorp-

◀──────────────────────────────────────

Figure 7. Apical cytoplasm of an intestinal absorptive cell 30 min after administration of a corn oil meal. This micrograph illustrates the sequence of events in the reesterification of triglycerides and the formation of SER lipid-containing vesicles. Free fatty acids and monoglycerides passively diffuse across the microvillus border (MV) and enter the SER of the apical cytoplasm. The latter frequently is in continuity with the RER (arrow 1). As the free fatty acids and monoglycerides are converted into triglycerides, the SER dilates and lipid droplets are formed within the SER cisternae (arrow 2). Eventually the SER loses its continuity with the RER (arrow 3), and discrete, lipid-containing vesicles (L) bud off from the SER to lie free in the cytoplasm. Mitochondria (M). × 26,060. Reduced 12% for reproduction.

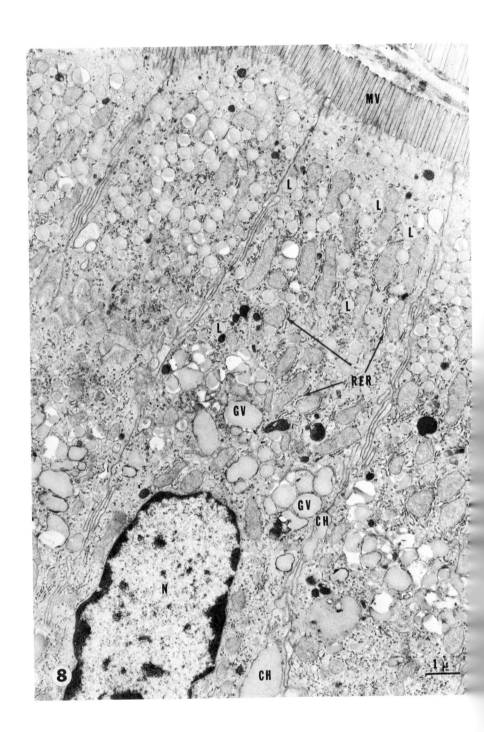

tive cell, by a series of membrane compartments. The main difference is that the interconversion and subsequent depletion of membranes (RER and Golgi saccules) is not compensated entirely by new membrane formation. The reason for this phenomenon is not well understood. However, under the experimental conditions in which fat absorption and chylomicron formation have been studied (Friedman and Cardell, 1972a), the quantities of lipid instilled into the gastrointestinal tract have been relatively large. It is conceivable, therefore, that the stimulus for lipid transport was equally large and spread over a relatively long period of time (several hours), in contrast to the situation observed with smaller quantities of lipid normally ingested in the diet. The end result may be that the absorptive cells are taxed beyond their maximal protein- or membrane-synthetic capability. It has been postulated that, as free fatty acids and monoglycerides passively diffuse into the apical cytoplasm, the SER serves as a sort of "sink" for the collection of these precursors (Cardell et al., 1967). With the conversion of these molecules into triglycerides, more and more precursors enter the SER. Eventually, the bulbous dilatation of SER pinches off from the RER and forms a discrete vesicle, and a new portion of SER is made available from the RER to accept triglyceride precursors and reesterify them. Assuming that there is no feedback inhibition to this process, it would continue as long as appropriate ER membranes are available. Continuous stimulation of secretory product formation and transport normally is not observed in either goblet or pancreatic exocrine cells and has not been induced experimentally. Possibly for this reason one does not see depletion of RER and Golgi membranes during the secretory cycles of these cells.

An alternative explanation for the decrease of RER and Golgi membranes of absorptive cells may lie with the simple assumption that de novo membrane synthesis does not keep pace with membrane utilization or turnover in this cell type. The importance of membrane containers and de novo membrane synthesis in the intestinal transport of lipid was emphasized by a study in which ribosomal protein synthesis was inhibited with puromycin (Friedman and Cardell, 1972a). In this

---

Figure 8. Electron micrograph of an intestinal absorptive cell 1 h after administration of a corn oil meal. The apical cytoplasm contains numerous SER-bound droplets of resynthesized lipid (L). The RER has receded from the apical cytoplasm to predominate in the region of the cell above the Golgi complex. Golgi vacuoles (GV) are now numerous and filled with lipid (compare with Fig. 4). Completed chylomicra (CH) are observed in the intercellular spaces between adjacent absorptive cells. The chylomicra closely parallel the size of lipid droplets observed in Golgi vacuoles. Microvillus border (MV); nucleus (N). $\times$ 9355.

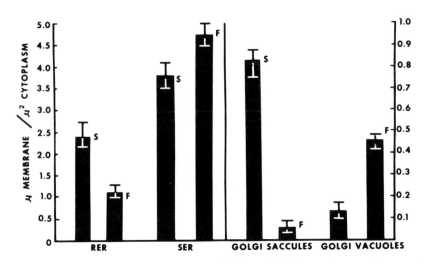

Figure 9. A morphometric comparison of ER and Golgi-complex membranes in the starved (S) and fat-fed (F) states (±SD, $n = 5$). One hour after administering corn oil to the fasted animal, the quantity of SER membranes increases at the expense of RER. Similarly, the Golgi saccule membranes appear to be converted into Golgi vacuoles. From Friedman and Cardell (1977) with permission of the publisher.

case, fat absorption proceeded normally during early time intervals after lipid administration. However, as fat absorption progressed, both the RER and Golgi saccule membranes eventually disappeared from the cytoplasm. Concomitantly, lipid droplets, which lacked surrounding SER or Golgi membranes, were found in the cytoplasm. From these observations, several conclusions were drawn: (1) the inhibition of protein synthesis did not affect the enzymes necessary for triglyceride resynthesis, presumably because these enzymes are long-lived; (2) protein synthesis inhibition resulted in a termination of de novo membrane synthesis; and (3) maintenance of the Golgi complex and ER was dependent on continued protein and, therefore, on membrane synthesis by the RER.

The latter conclusion was supported by the work of a number of investigators. Reid et al. (1970) noted a decrease in size and number of both Golgi vacuoles and saccules in hepatocytes after puromycin treatment, as well as a loss of ribosomes from the membranous profiles of RER. In a similar manner, Weinstock (1970) demonstrated marked alterations in the Golgi complexes of ameloblasts, hepatocytes, and pancreatic acinar cells at a puromycin dosage sufficient to inhibit protein synthesis by approximately 98%. Finally, in an elegant study, Flickinger (1968) recorded that after enucleation of *Amoeba proteus*, Golgi bodies declined in size and number, and eventually disappeared from the

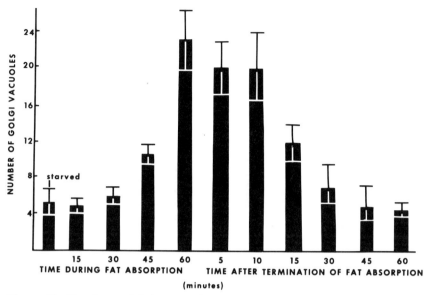

Figure 10. Numbers of Golgi vacuoles per cell during fat absorption and after abrupt termination of fat absorption by flushing the lumen of the bowel with isotonic saline (±SD, $n = 5$). During absorption, Golgi vacuoles containing lipid progressively increase in both size and number; a maximum is reached at 1 h. With the abrupt termination of lipid transport, there is a gradual decline in the quantity of Golgi vacuoles. From Friedman and Cardell (1977) with permission of the publisher.

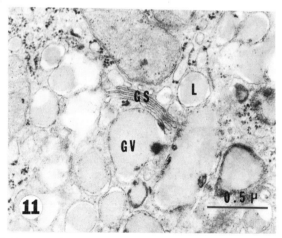

Figure 11. High magnification of the Golgi complex 1 h after corn oil administration. Both the lengths and numbers of Golgi saccules (GS) have decreased from the fasted condition (compare with Fig. 5). At the same time, Golgi vacuoles (GV) have increased in both size and number, presumably through the conversion of Golgi saccules into vacuoles. Lipid in SER vesicles (L). × 27,300. From Friedman and Cardell (1977) with permission of the publisher.

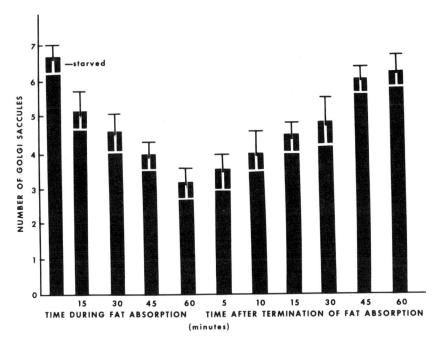

Figure 12. Number of Golgi saccules per cell during fat absorption and after abrupt termination of lipid transport by flushing the lumen of the bowel with isotonic saline (±S D, n = 5). The number of Golgi saccules in a stack of membranes progressively decreases in response to a lipid meal, from approximately six or seven in the fasted state to three or four 1 h after lipid administration. With the termination of lipid transport, the Golgi saccules are replenished and almost reach the number observed in the fasted animal 1 h after termination of fat absorption. From Friedman and Cardell (1977) with permission of the publisher.

cytoplasm. After renucleation, the Golgi bodies reappeared in the cytoplasm. If renucleated cells were treated with emetine, an inhibitor of protein synthesis, the formation of Golgi bodies was blocked almost completely (Flickinger, 1971a). In addition, Flickinger (1971b) demonstrated that maintenance of preexisting Golgi bodies also depended on continued RER protein synthesis.

From these studies of intestinal absorptive and other cells, it is apparent that the membranes of both the ER and Golgi complex are synthesized by the RER and, in the latter case, transferred to the Golgi complex for its maintenance. Interestingly, when fat absorption is terminated abruptly by flushing the lumen of the intestine with saline, there is a prompt return of Golgi saccules, vacuoles, and RER to the levels seen in the fasted animal (Friedman and Cardell, 1977) (Figs. 10,

12, and 13). This observation is interpreted as evidence that, once the stimulus for the interconversion and depletion of membranes is removed, de novo membrane production by the RER exceeds membrane utilization, and the full complement of ER and Golgi membranes is returned to the cell (Friedman and Cardell, 1977). The fact that membranes are required for the intracellular transport and release of secretory material is supported by the marked accumulation of triglyceride droplets in the cytoplasm after puromycin treatment. These lipid droplets lack membranous containers and presumably are therefore unable to leave the intestinal absorptive cell (Friedman and Cardell, 1972a).

The actual mechanism by which the RER transfers membrane to the Golgi complex is unclear. During fat absorption, SER vesicles containing lipid are seen in what appears to be a process of fusion of the SER membrane with vacuoles on the mature surface of the Golgi complex. Possibly, the SER membrane is subsequently incorporated into the Golgi vacuole membrane. This suggestion would seem unlikely in

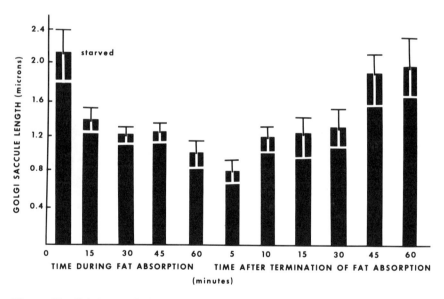

Figure 13. Golgi saccule length during fat absorption and after abrupt termination of lipid transport by flushing the lumen of the bowel with isotonic saline ($\pm$S D, n = 5). The lengths of Golgi saccules progressively decrease from the fasted to the fat-fed states and, in this fashion, parallel the decrease seen in numbers of Golgi saccules. This diminution of Golgi saccule membranes is a direct result of their conversion into Golgi vacuoles during fat absorption. After termination of lipid transport, the lengths of Golgi saccule membranes progressively return to the level observed in the fasted condition. From Friedman and Cardell (1977) with permission of the publisher.

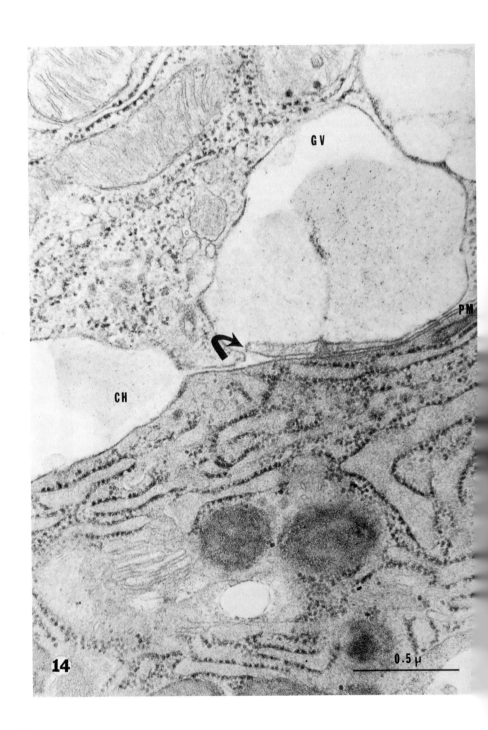

light of the fact that the total quantity of SER membrane available for the process would be far greater than is necessary to form new Golgi vacuoles, and it would still not explain the loss of Golgi saccule components. It would be more reasonable to hypothesize that, after SER vesicles deposit their lipid content in Golgi vacuoles, the SER membrane is degraded in some fashion. Therefore, another explanation, which more readily fits the available observations, would be that the RER synthesizes Golgi membranes at a location other than the site of lipid resynthesis (i.e., at that portion of the RER closest to the Golgi complex), and delivers that membrane to the Golgi complex either by intermediate smooth-surfaced tubules or vesicles. The new membrane then would be incorporated by Golgi saccules on the forming face of the complex, rather than by vacuoles on the mature surface. During fat absorption, these saccules would eventually form the new Golgi vacuoles. Indeed, small, smooth-surfaced vesicles often are observed in various stages of fusion with the saccules on the forming face of the Golgi complex, while smooth-surfaced tubules form a very close association with these saccules. A similar, but still alternative explanation may lie with the direct conversion of RER membranes into Golgi vacuoles. This possibility is raised by the finding of cisternae of the RER which are in opposition to the first saccule of the forming face of the Golgi complex and which lack ribosomes on one of their surfaces. Images such as these suggest an intermediate stage in the conversion of RER membranes into Golgi saccules. Figure 15 outlines the intracellular pathway of fat transport, membrane utilization, and the effects of protein synthesis inhibition on membrane formation.

The RER may not only provide the Golgi complex with new membranes but, in all likelihood it also produces the enzymes used by the Golgi complex for the addition of various moieties to the final secretory product. The method for the transference of these enzymes to the Golgi complex would be the same as for the movement of membrane material. The enzymes may be incorporated or attached to the membranes which are transferred from the RER to the Golgi. Thus, both membrane and enzyme reach the Golgi complex simultaneously.

Figure 14. High magnification electron micrograph of the lateral plasma membrane regions of an intestinal absorptive cell (above) and an adjacent goblet cell (below). This micrograph demonstrates the process of exocytosis or reverse pinocytosis between a Golgi vacuole (GV) and the plasma membrane. Note the extrusion of lipid into the intercellular space at the point of fusion between the vacuole and plasma membranes (arrow). A completed chylomicron (CH) is observed in the intercellular space between the adjacent cells. × 52,800. Reduced 12% for reproduction. From Friedman and Cardell (1972b) with permission of the publisher.

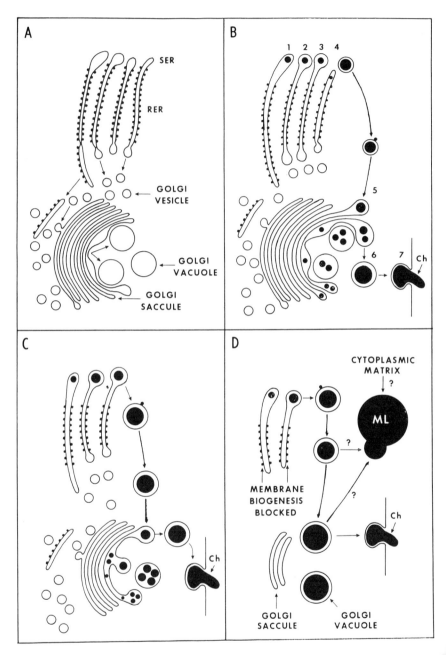

Figure 15. Summary of the observations on intestinal absorptive cells. (A) After 24 h fasting; (B) 15 min after fat administration; (C) 60 min after fat administration; and (D) 60 min after fat administration to puromycin-treated rats. In the fasted condition, (A) the RER occupies a considerable portion of the apical cytoplasm; it extends from the SER to the Golgi complex. The apical portions of the

The fate of the Golgi vacuole membrane is equally uncertain once it fuses with the plasma membrane. There are four possibilities. The membrane may be released with the secretory product to the extracellular space, as in the case of the goblet cell. Alternatively, the membrane may be incorporated into the plasma membrane. During an active secretory phase, this addition of Golgi vacuole membrane would even-

---

cisternae of the RER are continuous with the tubules of the SER, while the opposite ends of the RER near the Golgi complex show modified cisternae devoid of ribosomes on one surface, and many forming vesicles. The vesicles and modified cisternae associate with the immature or forming face of the Golgi complex, which suggests that they represent vehicles for the transfer of membranes from the RER to the Golgi complex. Thus, membranes added to the forming face of the Golgi complex are incorporated into saccules which are progressively converted to Golgi vacuoles at the mature face. During fat absorption (B and C), the SER increases in amount and contains lipid droplets within its lumen; the RER decreases in amount. These observations are interpreted to mean that the RER is converted to SER during fat absorption. The SER vesicles containing fat droplets migrate to the Golgi complex where they fuse with Golgi vacuoles or saccules adjacent to the vacuoles. This observation suggests that alterations occur in the membranes of the Golgi saccules as they move toward the mature face of the Golgi complex. These maturation changes apparently are required for the SER membranes to fuse with the Golgi membranes. Moreover, the alterations in the Golgi membranes may be important for the fusion of Golgi vacuoles with plasma membranes, thus allowing discharge of chylomicra (Ch) from the cell. A possible sequence of these events (1–7) is shown in (B). Puromycin treatment causes the intestinal absorptive cell to accumulate lipid and to show a striking decrease in intracellular membranes. After 60 min of exposure to fat (D), the RER is almost entirely absent and the Golgi complex shows only isolated saccules. It is argued that the cell becomes deficient in membranes because their replacement through synthesis is blocked by puromycin, whereas their utilization during fat absorption continues. Accumulation of lipid may be related also to a deficiency of Golgi membranes. As pointed out above, appropriate Golgi membranes are required for fusion of SER vesicles containing fat. If such membranes are lacking in the cell, the SER vesicles are unable to fuse with the Golgi complex and may remain in the cytoplasm or coalesce to form matrix droplets (D). It is further possible that membranes of the Golgi vacuoles that form in the puromycin-treated animals are qualitatively different from those of untreated cells and are unable to fuse with the lateral plasma membranes; they either accumulate in the cell or fuse to form matrix droplets (D). Three possibilities have been proposed for the origin of matrix lipid droplets (ML): (1) from fusion of SER vesicles containing fat, with subsequent dissolution of the SER membrane; (2) from a similar fusion and dissolution of Golgi vacuoles containing fat, or (3) from synthesis via enzymes located in the cytoplasmic matrix. From Friedman and Cardell (1972a) with permission of the publisher.

tually result in a markedly enlarged plasma membrane unless some mechanism existed for the degradation of plasma membrane at a rate similar to the addition of Golgi membrane. Jamieson and Palade (1971b), in studies of the pancreatic acinar cell, have suggested a third possibility; that there is a recycling of Golgi membrane components from the plasma membrane back to the Golgi complex. This mechanism, although reasonable for the exocrine cell, would be inconsistent with the observation of diminished Golgi saccule membranes in the intestinal absorptive cell. The fourth explanation invokes a degradation of Golgi membranes into constituent amino acids and lipids once the Golgi vacuole membrane has fused with the plasma membrane and released its secretory content. The amino acids and lipids then would be reutilized by the RER to produce new membranes. This hypothesis appears to be the most reasonable explanation of events in the absorptive cell (Friedman and Cardell, 1972a, 1977) and is compatible with the findings of membrane depletion when puromycin is administered. If membrane recycling were dominant in the absorptive cell, then protein synthesis inhibition should not affect the intracellular transport of lipid in the way that it does.

The overall issues of membrane formation, utilization, interaction, and degradation are complex. The fact that there is a unidirectional flow of secretory material from one cellular compartment to the next, often along a discontinuous pathway made continuous only by the fusions of the different participating membranes, has led to a number of concepts regarding the flux of membrane components in the secretory cell. The experimental evidence supporting each hypothesis varies considerably in content and is not always mutually exclusive. Although the data derived from the study of intestinal fat transport favor one hypothesis, those gained from investigations of pancreatic acinar cells suggest another. In the following sections, an attempt will be made to outline two of the major hypotheses and the supporting experimental evidence for each concerning membrane differentiation, interaction, and utilization.

III. Membrane Flow and Differentiation

A. The Hypothesis

The concepts of membrane flow and differentiation were first emphasized by Morré and his associates (Grove et al., 1968; Morré et al., 1974). By the term "membrane flow," these investigators mean the actual process of physical transference of membranes from one cellular organelle to another. "Membrane differentiation" refers to a change in either the organization or composition of membranes as they progress through the various cellular compartments. In essence, the hypothesis proposed by this group of investigators is as follows: cellular membrane proteins are synthesized de novo by ribosomes on the RER. The various membrane components are then assembled and incorporated into the membranes of the RER itself. The SER serves as a "transitional element" between the RER and Golgi complex in two ways: it delivers small packages of membrane from the RER to the forming face of the Golgi complex, and it transports secretory precursors from the RER to Golgi vacuoles and dilated saccules on the mature surface of the Golgi complex. The SER is in physical continuity with the RER and most likely is formed through the loss of ribosomes from the surface of the RER. SER vesicles which carry membrane or secretory material to the Golgi complex are created, in turn, by a process of pinching off of membranes from the ends of SER or RER cisternae.

The SER membrane, once it fuses with the saccule membrane on the forming face of the Golgi complex, is actually incorporated into that membrane. A process of membrane maturation then occurs across the stack of Golgi saccules. During this maturation, Golgi saccule membranes progressively move toward the mature surface of the complex and acquire both biochemical and physical attributes of Golgi vacuoles. The Golgi vacuoles form from the outermost saccule of the stack of membranes by a process of either direct conversion of the saccule into a vacuole, or by a pinching or budding off of the vacuole from the saccule in a fashion similar to the derivation of SER vesicles. The vacuoles, as well as the saccules on the mature face of the Golgi complex, receive secretory material from the RER via intermediate SER vesicles. It is unclear whether these SER-derived vesicle membranes actually are incorporated into the vacuole wall—as they are in the forming Golgi saccules—or whether they just deposit their secretory content into the vacuole lumen and are degraded. The mature Golgi vacuole then fuses with the plasma membrane and is incorporated into it. Thus, membrane flows from the RER through the Golgi complex and is eventually added to the plasma membrane. The physical and biochemical characteristics of the membrane vary as it sequentially acquires properties of each of the organelles in which it is located.

B. Supportive Experimental Data

1. Membrane Characteristics

The observations recorded for intestinal epithelial cells during fat absorption (Friedman and Cardell, 1972a, 1977) would tend to support a hypothesis of membrane flow and differentiation similar to that outlined above. However, an additional body of evidence lends further strength to this argument. In many cell types there is a differential in staining characteristics and membrane thickness amongst the saccules of the Golgi complex (Berlin, 1967; Flickinger and Wise, 1970; Grove et al., 1968; Meek and Bradbury, 1963; Mollenhauer and Whaley, 1963), such that saccules on the forming face of the complex resemble membranes of the ER, while vacuoles and saccules on the mature surface have characteristics of plasma membranes. The variation in membrane thickness is confirmed by low-angle X-ray diffraction analysis (Morré et al., 1974). In hepatocytes, isolated ER fragments have a center-to-center spacing between phospholipid polar groups of 40 Å. In Golgi membranes, the average value is 45 Å, while that for plasma membranes is approximately 50 Å. These findings suggest that the Golgi complex occupies a pivotal or transitional position between membranes of the ER and plasma membrane. Similarly, the lipid composition of Golgi membranes appears intermediate in quantity between that of ER and plasma membranes. With regard to the five major phospholipid classes, the Golgi fraction contains relative percentages of phosphatidylcholine, phosphatidylserine, and sphingomyelin that are intermediate between those for the ER and those for plasma membrane (Yunghams et al., 1970). Only phosphatidylethanolamine and phosphatidylinositol are higher in Golgi fractions than in either ER or plasma membranes. Sterol content is highest in plasma membranes, with progressively lower quantities observed in Golgi and ER membranes, respectively. Disc-gel electrophoretic patterns of total protein content in each cell fraction reveals a decreasing complexity of major bands, from the ER to the Golgi complex to the plasma membrane. The Golgi complex and ER share all but three bands, while the plasma membrane has only one major band in common with the ER (Yunghans et al., 1970).

2. Enzyme Markers

Further evidence for the transitional nature of Golgi membranes between ER and plasma membranes rests with various membrane markers (Table 1). For example, several enzymes, including 5'-nucleotidase, magnesium ATPase, and uridine diphosphoryl galactose hydrolase are found in highest concentration in plasma membrane fractions, while the Golgi fraction contains less and the ER the least relative activity. In contrast, enzyme markers highest in activity in the ER (i.e., glucose-6-phosphatase, arylsulfatase C, gulonolactone oxidase, electron-transport enzymes, and cytochromes) are lower in the Golgi apparatus and either have even lower activities in the plasma membrane fraction or are absent (Morré et al., 1974). Only thiamine pyrophosphatase and

Table 1. Membrane Enzyme Markers

Endoplasmic reticulum	Golgi complex	Plasma membrane
Glucose-6-phosphatase	N-Acetylglucosamine transferase	5'-Nucleotidases
NADH-cytochrome c reductase	Other specific glycosyl transferases	Alkaline phosphatases
NADPH-cytochrome c reductase	Thiamine pyrophosphatase	ATPases
Cytochrome b_5		NADase
Cytochromes P_{450}, P_{420}		NAD-pyrophosphatase
Arylsulfatase C		Nucleoside triphosphate-pyrophosphohydrolase
L-Gulonolactone oxidase		Phosphodiesterase

certain glycosyltransferases are found more concentrated in the Golgi complex than the other two cellular fractions. These data suggest that, as membranes traverse the various cellular compartments, they differentiate or alter their physical and chemical characteristics in a progressive manner so that they assume the properties of the organelle in which they are located. The mechanisms responsible for this differentiation remain unclear.

3. Isotope Experiments

Morré et al. (1974) studied the incorporation of radioisotopically-labeled leucine and arginine into both membranes and secretory products (VLDL) of the rat hepatocyte, using cell fractionation, gel electrophoresis, and spectrophotometry. They found that labeled material initially was incorporated into ER and Golgi membranes almost simultaneously, while there was a significant lag between the migration of secretory proteins from the ER to the Golgi complex. Complete transfer of labeled membranes from the RER to the Golgi complex took approximately 10–15 min after injection of the isotope. Labeled secretory vacuole membranes were released 20–30 min after injection, and an increase in plasma membrane label occurred 30–60 min after injection. Thus, newly synthesized membranes reached the Golgi complex prior to the secretory product. However, the maturation of Golgi saccule membranes across the stack of membranes slowed the exit of new membranes from the Golgi complex, so that both membrane and secretory product were released from the Golgi simultaneously. These investigators also studied the appearance of a membrane-linked enzyme, hepatic-O-demethylase, in response to an injection of phenobarbital (Morré et al., 1974). Again, activity was first found in cell fractions corresponding to the RER, and then sequentially in SER and Golgi membranes. Both labeled amino acid incorporation into membranes and hepatic-O-demethylase induction support the hypothesis of a flow of membrane components from the RER, through the Golgi complex, to the plasma membrane.

The concept of membrane flow and differentiation suggests that both secretory precursors and the intracellular membrane which contains them are produced simultaneously in the RER by de novo protein synthesis. If membranes, instead of traversing the pathway described above, were recycled or reutilized, then one would expect the incorporation of labeled material into membranes to lag far behind that of the secretory product. Amsterdam et al. (1971) investigated this problem in the exocrine cell of the parotid gland. They found that, at all times after radioisotope administration, the specific radioactivity of the secretory granule membrane was almost equal to that of the granule's exportable content. Thus, membrane and secretory material probably were produced simultaneously. When the data from the studies of Morré et al. (1974) and others are added to the work previously mentioned concerning the importance of protein synthesis by the RER for maintenance of membranes of both the ER and Golgi complex, it would appear that the experimental evidence is overwhelmingly in favor of an hypothesis of membrane flow and differentiation in the production and release of secretory material. However, there is an alternate explanation for the events occurring in the intracellular transport of secretory products, and this deserves careful consideration. As mentioned above, the hypothesis of Morré et al. assumes that membrane turnover is relatively rapid, and that there is a constant degradation and de novo synthesis of membranes in secretory cells. However, on the basis of experimental evidence derived from a study of pancreatic exocrine secretions, a number of investigators have suggested that membrane de novo synthesis and degradation are not important aspects of secretory cell function (Jamieson and Palade, 1971a; Meldolesi and Cova, 1971; Meldolesi, 1974). Rather, there is a recycling of membranes which occurs during secretion and is independent of membrane synthesis and degradation. Thus, membranes are reutilized constantly by the cell for the intracellular packaging and transport of secretory material.

IV. Membrane Reutilization

A. Supportive Experimental Data

In order to develop a theory of membrane recycling, rather than flow and differentiation, it is worthwhile to review briefly the experimental evidence favoring this alternate explanation of membrane utilization.

1. Isotope Experiments

When pancreatic tissue slices were pulse-labeled in vitro with tritiated leucine and then incubated in chase medium, the specific radioactivity of all membrane fractions was relatively low compared to the radioactivities of secretory products (Meldolesi, 1974). The majority of radioactivity associated with the membrane fractions was the result of a small quantity of contaminating secretory material. Throughout the period of pulse-label study, the specific activities of the various membrane proteins did not change (Meldolesi and Cova, 1971). These data suggested that biogenetic relationships among the various membranes were not demonstrable during the secretory cycle. Furthermore, since the concentration of leucine was approximately the same in secretory and membrane proteins, the data also indicated that the rate of synthesis of membrane proteins was slow relative to the rate for secretory proteins. To confirm the latter finding, Meldolesi (1974) and coworkers employed a double-label technique in which [^{3}H] leucine was injected into animals approximately 4–6 days prior to sacrifice. Thirteen hours prior to sacrifice, the same animals were injected with [^{14}C] leucine, and the incorporation rates of both labels into membrane and secretory proteins were observed. The investigators found that label incorporation into membrane proteins proceeded much more slowly than into secretory material. The average half-life of rough microsomal membrane proteins varied from approximately 5 to 28 days, depending on the chain length of the specific protein, while the half-life of secretory proteins was found to be on the order of 10–20 h. These findings obviously are in direct conflict with the results of Amsterdam et al. (1971) and Morré et al. (1974), which indicate that membrane and secretory proteins are synthesized concomitantly. However, to reconcile this discrepancy, Meldolesi (1974) noted that in both liver and parotid gland there is a sizeable contamination of microsomal membrane fractions with secretory products, the latter being tightly bound to the membrane fraction and difficult to remove. Therefore, the results of Amsterdam et al. (1971) might reflect this contamination, although Amsterdam and his associates went to great lengths to insure the purity of their membrane fractions.

2. Inhibition of Protein and Membrane Synthesis

If membranes are recycled during the process of secretion, then inhibitors of protein synthesis, such as puromycin or cycloheximide, should not interfere with intracellular transport. Indeed, Jamieson and Palade (1968a) demonstrated that with a cycloheximide concentration sufficient to inhibit protein synthesis by 98%, the intracellular transport and release of zymogen material proceeded at 80% of control levels. The 20% disparity was believed to result from an effect of the drug on mitochondrial oxidation. Similar results were obtained by Meldolesi and Cova (1971). These data also would seem to be in conflict with the experimental observations of Friedman and Cardell (1972a), which revealed an inhibition of chylomicron secretion in intestinal absorptive cells after treatment with puromycin. However, one might argue that in the pancreas both the membrane container and secretory products were already present prior to the administration of cycloheximide. Thus, new membrane production was unnecessary during the time course of these experiments.

According to the hypothesis of Meldolesi (1974), once SER vesicles fuse with the Golgi vacuoles, the vesicle membrane would recycle back to the ER for reincorporation, rather than becoming part of the vacuole wall or being degraded into constituent amino acids and lipids. Similarly, after Golgi vacuoles fuse with the plasma membrane, the Golgi membranes would be recycled back to the Golgi complex. In this fashion, the membrane of each organelle retains its identity, instead of being incorporated and altered to conform to a new organelle location. The precise mechanism by which the membrane would be recycled is unclear. It could be degraded into macromolecules for reassembly at the parent organelle site. Alternatively, it could be recycled as discrete membranes, such as in the form of small vesicles. This hypothesis is diagrammatically illustrated in Fig. 16.

B. Reutilization versus Flow and Differentiation of Membranes

It is extremely difficult to reconcile the discrepancies in data which support the two different hypotheses of membrane utilization. The concept of membrane recycling leaves no room for the observation that, in intestinal goblet cells, Golgi vacuole membranes are actually lost to the lumen of the gut and must be replaced by new membrane synthesis. Nor does it effectively explain the loss of RER and Golgi saccules during fat absorption and puromycin treatment. Similarly, the concept of membrane flow and differentiation appears inconsistent with the findings of a very slow membrane turnover in pancreatic exocrine cells. It would seem possible that there is more than one mechanism for the production and maintenance of membranes, both during active secretion and while the cell is at rest. One or another of these mechanisms may be dominant in a given secretory cell during any phase of a secretory cycle. Thus, although there is an active release of zymogen granules in response to appropriate hormones, the level of enzyme synthesis remains constant in pancreatic acinar cells (Jamieson and Palade,

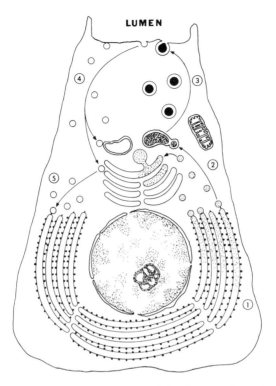

Figure 16. Diagrammatic representation of the theory of membrane recycling. The right-hand portion of the cell illustrates the pathway of transport of secretory material, while the left hand portion shows the pathway of membrane recovery. Thus, secretory material is synthesized in the RER (1) and transported via small, smooth-surfaced vesicles to the Golgi complex (2). After conversion of Golgi vacuoles into zymogen granules, the latter fuse with the plasma membrane to release the completed secretory material (3). According to the recycling hypothesis, Golgi vacuole membranes are not degraded or extensively incorporated into the plasma membrane; rather, they return to the Golgi complex, possibly in the form of small, smooth-surfaced vesicles (4). A similar concept of membrane recycling may occur for the endoplasmic reticulum (ER), in which membranes return to the ER after depositing their secretory material in Golgi saccules and vacuoles (5).

1971b). A pattern of membrane recycling therefore may best reflect the requirement of this cell for the continual manufacture of secretory material. In contrast, the absorptive cell is not constantly active in the intracellular transport and synthesis of chylomicra. In addition, the absorptive cell has a relatively short lifespan. These factors may make membrane conservation by reutilization unnecessary. The goblet cell resembles the pancreatic exocrine cell in that the requirement for the synthesis of secretory material is continual. However, it differs from the exocrine cell in the fact that release of secretory material, although constant, is intimately tied to a loss of Golgi membranes. Hence a need exists for the continual synthesis of new membranes along with secretory precursors.

C. Unidirectional Secretory Product Transport

Although the mechanisms for the production and utilization of the membranous vehicles of intracellular transport may vary among different secretory cells, the pathways of synthesis and movement of secretory precursors and products are remarkably similar. Despite differences in membrane utilization, the intracellular transport of secretory precursors occurs along a unidirectional, but discontinuous pathway of membranous organelles. The pathway is made continuous only by the fusion of membranes. A natural question is: what insures the proper fusion of one membrane with the next appropriate membrane in the secretory pathway? For example, in the intestinal absorptive cell, Golgi vacuoles appear to fuse only with the plasma membrane at the lateral border of the cell. In the pancreatic exocrine cell, fusion of zymogen granule membranes and plasma membrane occurs only at the apex of the cell. Friedman and Cardell (1972a,b) speculated that a specificity or recognition factor was built into each membrane and that therefore only certain membranes are allowed to fuse with one another in a precise and controlled manner. Support for this hypothesis comes from the examination of freeze-fracture replicas of pancreatic acinar cells, which reveals that the luminal portion of the plasma membrane (where exocytosis of secretory material takes place) has a structure which differs from the rest of the plasma membrane but which is similar to the structure of zymogen granule membranes (DeCamilli et al., 1974). Similarly, the lateral plasma membrane of absorptive cells clearly has different structural properties than the membrane which forms the luminal border. A more precise modeling of the fusion interaction between biological membrane systems must await further resolution of the biochemical and molecular configurations of different types of intracellular membranes.

V. The Clinical Importance of Intracellular Membranes for Secretion

Considerable discussion has been devoted to the function of intracellular membranes in the production of secretory products. The importance of these membranes in the transport of secretory material is underscored by the results of a defect in membrane formation. For example, as mentioned previously, if protein synthesis in the intestinal absorptive cell is inhibited, there is a consequent loss of intracellular membranes and an inability to transport triglycerides out of the cell. This research model may well have a clinical counterpart in the disease known as abetalipoproteinemia (Bassen–Kornzweig syndrome), which is characterized by steatorrhea, retinitis pigmentosa, spiny-shaped red cells (acanthocytes) in the peripheral blood, neurological alterations similar to Freidreich ataxia, and a deficiency or absence of serum low-density (β) lipoproteins. β-Lipoproteins are synthesized by both hepatocytes and intestinal absorptive cells in much the same way that chylomicra are produced. A primary defect in abetalipoproteinemia patients is the inability to export both lipoproteins and chylomicra from the intestine. Dobbins (1966) studied the ultrastructural pattern of intestinal fat absorption in several patients with the disease and observed abnormalities similar to those seen after protein synthesis inhibition. Large lipid droplets, which appeared to lack definitive surrounding membranes, accumulated in the apical cytoplasm of the cells, while the Golgi vacuoles remained empty. Dobbins (1966) attributed the defect in lipid transport to an abnormality in either the membranes of the ER or the Golgi complex and, hence, to an interruption in the normal pathway of chylomicron formation and release. The membrane defect was hypothetically caused by the administration of puromycin. Interestingly, the ultrastructural appearance of fat absorption in adrenal insufficiency also resembles that observed after puromycin treatment (Friedman and Cardell, 1969). The steatorrhea observed in Addison disease, therefore, may be related to similar defects in membrane function.

Another area where membrane deformities may give rise to a pathological process is in fatty infiltration of the liver. There is a broad spectrum of entities which frequently result in hepatic steatosis, incuding chronic alcoholism, essential fatty acid deficiency, jejunoileal bypass procedures, Wilson disease, marasmus, and kwashiorkor. The liver is normally involved in the processing of both endogenous and exogenous sources of lipid. Bruni and Hegsted (1970) have shown that when rats are made progressively choline deficient, there is an intracellular accumulation of large lipid droplets in the cytoplasm of the hepatocytes. This cytoplasmic lipid is devoid of the normal surrounding membranes of the ER and Golgi complex. The hypothesis has been advanced that, since choline is an integral component of membrane structures, its deficiency results in a generalized defect in hepatocyte membrane production. Consequently, resynthesized hepatic triglycerides can no longer move along their normal pathway of formation and release, and instead accumulate in the cytoplasm.

Enzymes for the resynthesis of triglycerides exist within both the microsomal fraction and the soluble cytoplasm of intestinal absorptive and hepatic cells (Johnston, 1968; Kern and Borgstrom, 1965). The K_m for the synthesis of triglycerides over each of these routes favors the use of enzymes in the microsomal fraction (Johnston, 1968). It is suggested, therefore, that when membrane synthesis is inhibited or defective, the buildup of precursor molecules shunts more of the lipid through the cytoplasmic, or L-α-glycerophosphate, pathway. It would not be unrealistic to postulate a similar defect as underlying many of the disease processes which lead to hepatic steatosis. Patients suffering from alcoholism frequently consume diets lacking essential nutrient materials, one of which may be required for appropriate membrane synthesis. Similarly, in essential fatty acid deficiency, important membrane building blocks are lacking.

As more is learned about the normal mechanisms of cellular secretion, it may be possible, in future studies of certain diseases, to link the pathologic process to some abnormality in the intracellular transport of secretory material.

VI. Microtubules

No discussion of the mechanisms of cellular secretion would be complete without a consideration of another ubiquitous cytoplasmic organelle, the microtubule. At the outset, it should be mentioned that the precise role played by microtubules in secretory processes is uncertain. However, current evidence suggests that microtubules exert a profound regulatory influence on both intracellular transport of secretory precursors and on the final release of secretory products in response to extracellular stimuli. Investigations into the functions of microtubules have expanded greatly in the last decade, and only now are we beginning to gain true insight into their importance in a wide variety of cellular activities.

A. Structure

Microtubules appear as hollow, nonbranching cylinders approximately 240 Å in diameter and of a variable length, often approaching several microns. The wall of the cylinder is usually straight or slightly curved and is composed of 40-Å globular protein subunits, precisely arranged, with longitudinal rows of subunits appearing as protofilaments. The walls also possess a certain degree of inherent rigidity as they tend to break, rather than bend, when isolated from the cell cytoplasm. The core of the microtubule is apparently patent, which permits the ingress of negative stains.

B. Function

Microtubules are found in all prokaryotic and eukaryotic cells, and a considerable number of functions have been ascribed to them. One of the first of these was the maintenance of cell shape, i.e., the microtubules form the cytoskeleton of cells as well as operating in cell shape changes, such as elongation. Microtubules also function in ciliary and flagellar motility, and in the intracellular movement of almost all cytoplasmic constituents including chromosomes, nuclei, mitochondria, synaptic vesicles, cytoplasmic granules, lysozomes, melanin pigment, and ribosomes. It is uncertain whether the microtubules actually provide the motive force for this intracellular transport or merely function as structural guides along which cytoplasmic components are propelled. Alignment of microtubules is necessary for axon development, and they form the basis for complex organelles such as cilia, centrioles, and the achromatic mitotic apparatus. Microtubules appear to form a framework for the proper alignment of developing myofibrils of skeletal muscle, and even form the core of certain mechanoreceptors in arthropods. Finally, a considerable body of evidence has linked them to the normal and appropriate functioning of secretory cells, and it is to this function that further discussion is addressed. An excellent and more complete review of our current knowledge concerning the structure, synthesis, and functions of microtubules may be found in a recent symposium (Soifer, 1975).

C. Microtubules and Secretion

The importance of microtubules in cellular secretory processes was first brought to light when it was noted that, in a wide variety of cells, disrupters of tubule structure inhibit cellular secretion, including insulin release from pancreatic endocrine cells, secretion of plasma protein by hepatocytes, mobilization of thyroid hormone, and release of catecholamines from the adrenal medulla. In fact, a great deal of our knowledge concerning the functioning of microtubules is derived from experiments in which microtubular structure has been altered by a number of specifically-acting chemical agents.

1. Microtubular Function in Pancreatic Endocrine Cells

An excellent example of the role of microtubules in secretory product release is provided by the work of Malaisse et al. (1975) in the pancreatic endocrine cell. In this cell type, the microtubules are found scattered throughout the cytoplasm; they are particularly conspicuous in the ectoplasm and near both the nucleus and Golgi complex. At the apex of the cell, just below the plasma membrane, is a region termed the cell web. It is relatively devoid of cytoplasmic organelles but contains numerous bundles of microfilaments 4–7 nm in diameter, which frequently appear to insert on several organelles, including the plasma membrane, microtubules, small vesicles, and the surface of secretory granules. The latter granules are membrane-bound vesicles containing insulin, which release their secretory product to the extracellular space by a process of reverse pinocytosis or exocytosis with the plasma membrane. In the general scheme proposed by Malaisse et al. (1975), the microtubules form a guiding pathway which directs the secretory granules toward the appropriate plasma membrane, while the microfibrils of the cytoplasmic web control the access of these granules to the plasma membrane. To test this hypothesis, the investigators studied the secretory response of the endocrine cell under the conditions of microtubule and microfibril dysfunction induced by the drugs cytocholasin B, vincristine, colchicine, and D_2O.

Treatment of the cells with cytocholasin B, which interferes with the contractile function of microfilaments, resulted in marked alterations in the cellular web region. The microfilaments appeared as dense clumps of material beneath the plasma membrane rather than as a continuous band. At the same time, the stimulation threshold of the pancreas with insulinotopic agents (i.e, glucose, leucine, sulfonylurea, etc.) was found to be enhanced. These observations supported the concept that microfilaments serve as both "a barrier to and an effector of exocytotic insulin release" (Malaisse et al., 1975, p. 639). Both vincristine and colchicine produced a loss of microtubules from the cytoplasm, and, although they did not affect the early component of the secretory response (i.e., the characteristic, immediate peak-shaped release of insulin), they drasticaly reduced the late response (progressive and sustained high secretion rate) normally seen during prolonged glucose stimulation.

Deuterium oxide was known to stabilize microtubules and, when applied to pancreatic endocrine cells, resulted in a dose-dependent inhibition of secretion which eventually reached complete cessation of insulin release. Pretreatment of cells with D_2O protected them from the effects of colchicine or vinblastine when D_2O was removed from the incubation media. Similarly, pretreatment of cells with cytocholasin B protected the cells from the effects of D_2O.

In an effort to explain these observations, Malaisse et al. (1975) suggested a sequence of events in the endocrine cells. During the early phase of glucose-induced insulin release, secretory vesicles already near the cellular web and plasma membrane are discharged. This secretion phase is independent of microtubule function and hence is affected only by cytocholasin B. The late phase, with its progressive buildup of secretory rate, is dependent on microtubules, and, therefore, is inhibited by colchicine and vinblastine. It corresponds to the movement of secretory vesicles along microtubular pathways to the cellular periphery. The final movement of secretory granules to the plasma membrane, and their subsequent release in both phases, is under the control of the microfilament system of the cell web. This microfilament system is believed to provide a motive force for the transference of the secretory granules to the plasma membrane. Several pieces of evidence support this assumption; actin-like material has been localized to this cell area, the process is ATP-dependent, it is associated with a cellular depolarization, and also with an accumulation of calcium in the cell. All of these properties are similar to those observed with excitation–contraction coupling in muscle.

The force which propels secretory granules along the microtubules is less clear. It is known that the microtubular system exists in at least two states: the fully polymerized form represented by intact microtubules, and the disintegrated form represented by a pool of depolymerized globular proteins (tubulin) in the cytoplasm. In order for microtubules to function properly, a dynamic state of equilibrium must exist between the fully-formed tubules and the tubule constituent pool. Thus, colchicine and other antimitotic agents bind to specific sites on the microtubular subunits. It has been proposed that they exert their effect by inactivating the free subunits and thereby shift the equilibrium between the associated and dissociated states of the microtubules so that eventually no intact microtubules remain and secretion is inhibited. Similarly, stabilization of microtubules in the polymerized form with D_2O also inhibits cellular secretion of insulin. From this, one can hypothesize that if the secretory vesicles were somehow attached to the microtubules, possibly by way of microfilaments, a constant cycle of depolymerization near the cell periphery, with a repolymerization at the central area of the cell, would advance the secretory vesicle from the cell center to the cell web. In addition, if tubulin actually contains an actin-like contractile protein, then this contractile property may well contribute to the intracellular movement of secretory materials.

2. Relationship between Tubulin and Actin in Adrenal Medullary Cells

Poisner and Cooke (1975) studied the role of microtubules in the release of chromaffin granules from the adrenal medulla. As in the endocrine pancreas, both colchicine and vinblastine affected microtubules and thus inhibited the release of secretory granules. Of further interest was the fact that a cytoplasmic precipitate, formed from the action of vinblastine, possessed an antigenicity similar to that of both the tubulin protein (which comprises microtubules) and that of actomyosin-like proteins. These observations suggest both a structural and functional relatedness between tubulin and actin. Furthermore, there was evidence that chromaffin granules themselves either contained or were closely associated with tubulin-like proteins. Thus, the motive force for movement of secretory granules from the central portion of the cell to the plasma membrane may well be related to an interaction between the microtubules and the tubulin or actin-like material of the granule vesicle wall.

3. Microtubules and Endocytosis of Thyroglobin

In the thyroid, colchicine and vinblastine inhibit the usual secretory response to both thyroid stimulating hormone (TSH) and cyclic AMP (cAMP). Wolff and Bhattacharyya (1975) specifically demonstrated that endocytosis of thyroglobulin from the colloid lumen of the gland was blocked by these antimitotic agents; thereby endocytosis was linked to microtubular function. In the normal thyroid gland, both TSH and cAMP stimulation led to an increase in the number of observable microtubules, while D_2O, although stabilizing microtubules, inhibited secretory release in a manner similar to that observed in pancreatic endocrine cells. These observations provide further support for the concept that the constant remodeling of microtubules—i.e., from the polymerized to the depolymerized forms—is essential for the proper functioning of microtubules in secretion. To complicate the situation further, however, colchicine binding to the plasma membrane fraction of thyroid cells also was noted. This implies that there is a tubulin-like protein component to the plasma membrane similar to that in secretory vesicles of the adrenal medullary cells. The functional significance of this membrane tubulin activity is unclear, but one might speculate that it provides an effector component of TSH-induced secretory stimulation, either alone or in concert with microtubules.

4. Microtubules and Intracellular Transport in the Hepatocyte

In an elegant study of liver secretion, Redman et al. (1975) found that colchicine inhibited the secretion of albumin, other plasma proteins, and VLDLs from rat hepatocytes. Specifically, colchicine blocked the release of secretory material from Golgi vacuoles to the extracellular space without interfering with synthesis, transference from the RER to the SER, or incorporation into Golgi vacuoles; nor did colchicine inhibit oxidative phosphorylation. Thus, in the liver, functioning microtubules appear necessary for the final intracellular transport of secretory products to the plasma membrane. A similar mechanism is postulated for the release of chylomicra from Golgi vacuoles in the intestinal absorptive cell (Reaven and Reaven, 1977).

Obviously, many questions remain unanswered regarding microtubule function, particularly how these organelles physically participate in the translocation of secretory materials within the cell. Do they actually move secretory vesicles and vacuoles intracellularly, or do they just proivde sets of tracks over which secretory containers passively glide? What is the true significance of the localization of colchicine binding sites and tubulin proteins in the cellular membranes? Finally, how do the microtubules interact with various secretory stimuli to effect the release of secretory products? In the ensuing discussion, a brief consideration of this last question will be made, with a presentation of some of the hypotheses and evidence concerning the influence of secretory stimulators on microtubular function.

VII. Mechanisms of Secretory Activation

The release of secretory material from many cells usually is coupled to one or more fairly specific stimuli. During this time, the cell is activated and secretory contents are discharged to the extracellular space. Thus, some mechanism must exist for translating the arrival of the secretory stimulus at the cell surface into a chain of events, the sequel to which is the net movement of secretory products from the cytoplasm. The possible role of microtubules in this process has already been mentioned. However, it is apparent that without a means for controlling the activity of microtubules, secretion would continue unabated. Recent evidence has implicated two interrelated molecules which serve as second messengers and exert a profound influence on the level of activity of many cellular functions, including secretion (Rasmussen and Goodman, 1975). These molecules are Ca^{2+} and cyclic AMP (cAMP).

A. Calcium and cAMP

The intracellular cytoplasmic concentration of calcium is under the influence of several ion pumps. One exists at the level of the mitochondrial membrane, which actively transports Ca^{2+} into the mitochondrial matrix. Release of calcium from the mitochondria occurs by passive diffusion. A second pump exists in many cells in the endoplasmic reticulum. Here, at the expense of ATP, calcium is brought into the ER cisternae. Leakage of calcium back into the cytoplasm also occurs by passive diffusion. The third ion pump is located in the plasma membrane, where the asymmetric distribution of Ca^{2+} across the membrane is maintained by the Na^+ gradient and an Na^+-Ca^{2+} exchange mechanism. The net result of the action of these pumps is to maintain a relatively low concentration of cytoplasmic Ca^{2+}, on the order of $10^{-7}-10^{-6}$ M.

As pointed out by Rasmussen and Goodman (1975), one could envision only small perturbations in the concentration of cytoplasmic calcium resulting in an equivalent change in several other ions, thereby altering the level of ion-dependent activity of a variety of cellular enzymes. Thus, control of certain metabolic functions in many cells, including those involved in secretion, parallels that observed in muscle during excitation–contraction coupling, when calcium is released from the sarcoplasmic reticulum in response to excitation. The participation of each of the membrane compartments and ion pumps in the control of Ca^{2+} flux varies among the different types of secretory cells.

Like Ca^{2+}, cAMP has also been found to be a universal agent in excitation/activation of cells, and, in most cells, the two molecules have interrelated functions in activating the response to extracellular messengers. Rasmussen and Goodman (1975) have summarized many of the known relationships between Ca^{2+} and cAMP. In general, the concentrations of Ca^{2+} and cAMP change in the cytosol as a result of activation by specific extracellular messengers. This alteration usually takes one of two forms, depending on the particular cell under consideration: Ca^{2+} and cAMP both increase in concentration or, alternatively, the cAMP level increases while the Ca^{2+} concentration decreases. The cellular response triggered by concentration changes of Ca^{2+} and cAMP may be a consequence of the effect of one or both of these molecules. The result is either inhibition or stimulation of an enzyme reaction, again depending on the specific cell type. Of importance to the functioning of secretory cells is that some of the effects of Ca^{2+} and cAMP are mediated through specific Ca^{2+} or cAMP-activated protein kinases. It should also be mentioned that Ca^{2+} and cAMP levels are closely regulated by one another through a series of complex interactions. Thus, increases in cAMP effect the efflux of Ca^{2+} from mitochondria and increase the Ca^{2+} uptake by microsomes. On the other hand, an increase in Ca^{2+} will often inhibit adenyl cyclase activity, as well as either inhibit or stimulate phosphodiesterase activity, depending on cell type.

The functional interreaction of Ca^{2+}, cAMP, and microtubules is equally complex. Gillespie (1975) has proposed and tested a model

Figure 17. Diagram of the hypothetical relationships between calcium, cAMP, and the various forms of microtubules as proposed by Gillespie (1975).

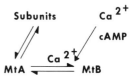

which may explain some of the experimental findings (Fig. 17). She postulates that microtubules exist in two forms (A and B), and that the repeated interconversion of the two forms results in a sustained activity of the microtubular system. The actual nature or differences between the two microtubular forms (A and B) is uncertain, but it is postulated they may represent phosphorylated and dephosphorylated material. Alternatively, the forms may be microtubules with or without crossbridge attachments. Calcium and cAMP control the intercoversion of the A and B forms. With high concentrations of either Ca^{2+} or cAMP, one of the two forms would predominate; the result would be microtubular inactivity. An important assumption in this scheme is that microtubules serve as a point of control of secretory processes. If this is true, then there are at least four ways to initiate a secretory event either by raising or lowering the concentrations of cAMP and Ca^{2+}. Which of these processes actually induces microtubular activity would depend on the relative concentrations of each in the specific tissue to which a stimulus is applied. Thus, raising cAMP levels in thyroid, parotid, and pituitary tissue causes the respective discharge of ^{131}I, amylase, and growth hormone, while a decrease in cAMP stimulates antigen-induced release of histamine from human leukocytes. Similarly, calcium appears to be the major agent in adrenal medullary catecholamine and posterior pituitary vasopression release. As predicted by the model, calcium opposed and cAMP facilitated the effects of colchicine in various experimental secretory systems (Gillespie, 1975). Furthermore, not only do Ca^{2+} and cAMP concentrations effect one another, but they also have opposing influences on the state of microtubules.

The question arises as to exactly how Ca^{2+}, cAMP, or both influence the functional form of microtubules. In partial answer to this question, Soifer et al. (1975) have demonstrated that a protein kinase resides as either a component part or in close association to the microtubule proteins. The protein kinase has as its most favorable substrate the protein tubulin itself, and therefore plays an important role in self-phosphorylation of microtubules. Furthermore, cAMP increases the affinity of the enzyme for ATP and thereby increases self-phosphorylation two- to threefold, while 5'AMP inhibits the reaction.

B. A Hypothesis of Secretory Activation

In an attempt to form an eclectic hypothesis regarding the activation of the secretory response, one might postulate that a hormone or some other first-order messenger interacts with a specific binding site on the plasma membrane of the secretory cell. The nature of these binding sites has not been characterized, but is has been demonstrated that a given homogeneous cell population may have several different sites, each specific for one or more different hormones (Klaeveman et al., 1975). As a result of this binding, several events may occur. Adenyl cyclase is activated, and the intercellular concentration of cAMP is increased. Alternatively, some change in the Ca^{2+} pump occurs, which facilitates entry of Ca^{2+} into the soluble cytoplasm. Cyclic AMP may affect the influx of calcium, or the calcium itself may influence the production of cAMP. In any event, either cAMP or Ca^{2+} facilitates the phosphorylation of tubulin by a protein kinase in the microtubules; the result is microtubular polymerization and also simultaneous depolymerization. Microtubular activation, in turn, induces the transport of secretory vesicles and vacuoles to the plasma membrane where they fuse with that membrane and release their secretory content to the extracellular space.

This scenario of events in the secretory cell obviously is oversimplified and contains many ambiguities. Unfortunately, it reflects our current lack of specific knowledge concerning many of the events in the stimulation and control of secretion. It is to be hoped that investigators in the near future will be able to describe precisely, and with confidence, events in the activation and release of secretory products.

VIII. Types of Secretory Discharge

In many types of secretory cells, the final discharge of secretory products occurs via fusion of a Golgi vacuole membrane with the

---

Figure 18. Electron micrograph of a prostatic secretory epithelial cell. The nucleus (N) is elongated and resides in the basal cytoplasm. The Golgi complex (G) usually occupies the cellular region between the nucleus and apical cytoplasm. The latter contains numerous secretory granules (SG) of varying osmiophilic density. This cell type is capable of merocrine secretion, in which portions of the apical cytoplasm, above the level of the tight junctions between adjacent cells (arrows), are extruded into the lumen of the gland. Lysosome (L). × 11,930. Reduced 12% for reproduction.

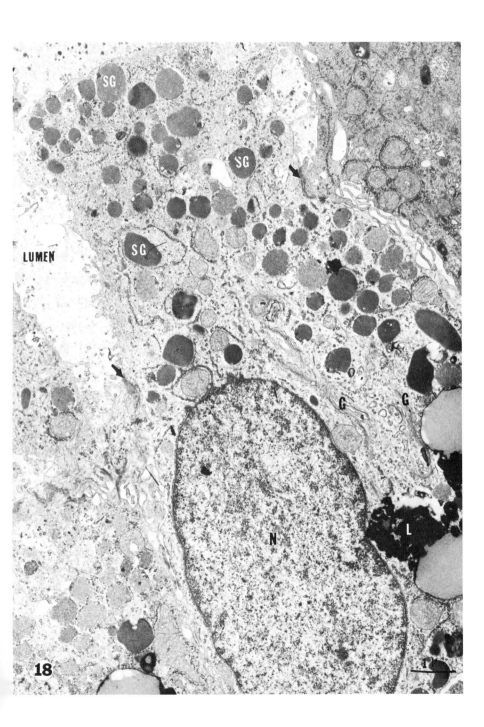

plasma membrane, resulting in extrusion of the contents of the vacuole to the extracellular space. This process has been variously termed reverse pinocytosis or exocytosis; it corresponds to the apocrine-type secretion observed by light microscopists. A second type of secretion was illustrated by the intestinal goblet cell, in which Golgi vacuoles pass intact through small breaks in the plasma membrane. Attention will now be drawn to a third major form of secretion exemplified by the prostatic secretory cell.

A double layer of epithelial cells line the acini of the human prostate gland. The basal cells are cuboidal, rest on a prominent basement membrane, and are generally devoid of secretory products. They provide a source of new secretory cells to replace those desquamated into the lumen. The cells adjacent to the lumen are columnar, and like intestinal absorptive cells, are highly polarized (Fig. 18). The nuclei are elongated and reside at the base of the cell, surrounded by cisternae of the RER. The Golgi complex occupies the supranuclear region of the cell. The area of the cell from the Golgi complex to the lumen constitutes the apical cytoplasm and is usually filled with osmiophilic secretory granules of varying density (Fig. 18). Adjacent cells are joined by tight junctions. However, in contrast to other columnar cells, these junctions are not found near the cell apex. Rather, a considerable amount of the apical cell cytoplasm extends into the lumen above the tight junctions (Fig. 18).

The pathway of secretory product formation is identical to that in other secretory cells, with synthesis of enzyme precursors in the ER and condensation of final products in Golgi vacuoles. What differentiates this cell type from other secretory cells is that release of the final products to the lumen of the gland can occur in both of two ways (Brandes, 1966). The Golgi-derived vacuoles may undergo exocytosis with the plasma membrane, as in other cells. Alternatively, entire portions of the apical cytoplasm may bud off from the cell and be released to the lumen. This latter phenomenon was termed merocrine secretion by light microscopists. Thus, within the lumen of the gland may be found portions of cellular cytoplasm containing intact mitochondria, ER, secretory granules, and lysosomes. Merocrine secretion by prostatic epithelial cells may explain why many of the enzyme activities normally associated with cytoplasmic organelles—but not with the secretory granules—are found in the secretory fluid from this gland. The mechanism and control of merocrine secretion is unknown. However, there is an abundance of microtubules in the cytoplasm, and it may be speculated that these tubules play an integral role in both types of prostatic epithelial cell secretion. Clearly, this is another area that is open for further investigative efforts.

IX. Summary

It should be apparent from this discussion of secretory cell function that as investigations have answered some of the questions regarding how secretory material is made and transported, more complicated but still basic mysteries have been brought to the surface for further clarification. The major pathways in the synthesis and intracellular transport of secretory materials appear roughly similar in many cell types. Yet we still do not have a clear understanding of the interaction which occurs between the membranes of the various cell organelles. Is membrane flow and differentiation an important means of replenishing and distributing membranes in all cells, or do many cells recycle major membrane components? Microtubules, tubular protein, cAMP, and Ca^{2+} are obviously key components for the regulation or control of release of secretory materials, yet we still lack a clear understanding of how each component exerts it influence. Are there other molecules which also participate in the regulation of secretory product release? Hormones and other primary messengers often exert their influence by binding in some fashion to specific membrane receptors. However, little and only indirect evidence about the physical and functional properties of these receptors is known.

In the last decade-and-a-half, the field of cell biology has witnessed a growth and development of research tools and knowledge of relatively massive proportions. It would seem only likely, therefore, that the answers to many of the questions raised in this chapter will soon be forthcoming.

ACKNOWLEDGEMENT. The author wishes to gratefully acknowledge Sue Davis and Thomas J. Nemeth for their kind assistance in the preparation of this manuscript.

References

Amsterdam, A., Schramm, M., Ohad, I., Salomon, T., and Sefinger, Z., 1971, Concomitant synthesis of membrane protein and exportable protein of the secretory granule in rat parotid gland, *J. Cell Biol.* **50**:187.

Berlin, T. O., 1967, The localization of acid mucopolysaccharides in the Golgi complex of intestinal goblet cells, *J. Cell Biol.* **32**:760.

Borgstrom, B., 1962, Digestion and absorption of fat, *Gastroenterology* **43**:216.

Brandes, D., 1966, The fine structure and histochemistry of prostatic glands in relation to sex hormones, *Int. Rev. Cytology* **20**:207.

Bruni, C., and Hegsted, D. M., 1970, Effects of choline-deficient diets on the rat hepatocyte, *Am. J. Pathol.* **61**:413.

Cardell, R. R., 1977, The smooth endoplasmic reticulum in rat hepatocytes during glycogen deposition and depletion, *Int. Rev. Cytol.* **48**:221.

Cardell, R. R., Badenhausen, S., and Porter, K. R., 1967, Intestinal triglyceride absorption in the rat. An electron microscopical study, *J. Cell Biol.* **34**:123.

Caro, L. G., and Palade, G. E., 1964, Protein synthesis, storage, and discharge in the pancreatic exocrine cell. An autoradiographic study, *J. Cell Biol.* **20**:473.

Claude, A., 1970, Growth and differentiation of cytoplasmic membranes in the course of lipoprotein granule synthesis in the hepatic cell. I. Elaboration of elements of the Golgi complex, *J. Cell Biol.* **47**:745.

Dallner, G., Siekevitz, P., and Palade, G. E., 1966, Biogenesis of endoplasmic reticulum membranes. I. Structural and chemical differentiation in developing rat hepatocyte, *J. Cell Biol.* **30**:73.

DeCamilli, P., Pelnchetti, D., and Meldolesi, J., 1974, Structural difference between luminal and lateral plasmalemma in pancreatic acinar cells, *Nature* **248**:245.

Dingle, J. T., and Fell, H. B., 1969a, *Lysosomes in Biology and Pathology*, Vol. I, in: *Frontiers of Biology* (A. Newberger and E. L. Tatum, eds.), Vol. 14A, pp. 3–543, John Wiley and Sons, New York.

Dingle, J. T., and Fell, H. B., 1969b, *Lysosomes in Biology and Pathology*, Vol. II, in: *Frontiers of Biology* (A. Newberger and E. L. Tatum, eds.), Vol. 14B, pp. 3–668, John Wiley and Sons, New York.

Dobbins, W. O., 1966, An ultrastructural study of the intestinal mucosa in congenital β-lipoprotein deficiency with particular emphasis upon the intestinal absorptive cell, *Gastroenterology* **50**:195.

Fewer, P., Threadgold, J., and Sheldon, H., 1964, Studies on cartilage: Electron microscopic observations on the autoradiographic localization of S^{35} in cells and matrix, *J. Ultrastruct. Res.* **11**:166.

Flickinger, C. J., 1968, The effects of enucleation on the cytoplasmic membranes of *Amoeba proteus*, *J. Cell Biol.* **37**:300.

Flickinger, C. J., 1971a, Decreased formation of Golgi bodies in amoebae in the presence of RNA and protein synthesis inhibitors, *J. Cell Biol.* **49**:221.

Flickinger, C. J., 1971b, Alteration in the Golgi apparatus of amoebae in the presence of an inhibitor of protein synthesis, *Exp. Cell Res.* **68**:381.

Flickinger, C. J., and Wise, G. E., 1970, Cytochemical staining of the Golgi apparatus in *Amoeba proteus*, *J. Cell Biol.* **46**:620.

Friedman, H. I., and Cardell, R. R., 1969, Effects of adrenalectomy and puromycin on rat intestinal epithelial cells during fat absorption, *Anat. Rec.* **163**:336.

Friedman, H. I., and Cardell, R. R., 1972a, Effects of puromycin on the structure of rat intestinal epithelial cells during fat absorption, *J. Cell Biol.* **52**:15.

Friedman, H. I., and Cardell, R. R., 1972b, Morphological evidence for the release of chylomicra from intestinal absorptive cells, *Exp. Cell Res.* **75**:57.

Friedman, H. I., and Cardell, R. R., 1977, Alterations in the endoplasmic reticulum and Golgi complex of intestinal epithelial cells during fat absorption and after termination of this process: A morphological and morphometric study, *Anat. Rec.* **188**:77.

Gillespie, E., 1975, Microtubules, cyclic AMP, calcium, and secretion, *Ann. N.Y. Acad. Sci.* **253**:711.

Godman, G. C., and Lane, N., 1964, On the site of sulfation in the chondrocyte, *J. Cell Biol.* **21**:353.

Greene, L. J., Hirs, C. H. W., and Palade, G. E., 1963, On the protein composition of bovine pancreatic zymogen granules, *J. Biol. Chem.* **238**:2054.

Grove, S. N., Bracker, C. E., and Morré, D. J., 1968, Cytomembrane differentiation in the endoplasmic reticulum–Golgi apparatus–vesicle complex, *Science* **161**:171.

Jamieson, J. D., and Palade, G. E., 1967a, Intracellular transport of secretory proteins in the pancreatic exocrine cell. I. Role of the peripheral elements of the Golgi complex, *J. Cell Biol.* **34**:577.

Jamieson, J. D., and Palade, G. E., 1967b, Intracellular transport of secretory proteins in the pancreatic exocrine cell. II. Transport to condensing vacuoles and zymogen granules, *J. Cell Biol.* **34**:597.

Jamieson, J. D., and Palade, G. E., 1968a, Intracellular transport of secretory proteins in the pancreatic exocrine cell. III. Dissociation of intracellular transport from protein synthesis, *J. Cell Biol.* **39**:580.

Jamieson, J. D., and Palade, G. E., 1968b, Intracellular transport of secretory proteins in the pancreatic exocrine cell. IV. Metabolic requirements, *J. Cell Biol.* **39**:589.

Jamieson, J. D., and Palade, G. E., 1971a, Condensing vacuole conversion and zymogen granule discharge in pancreatic exocrine cells: Metabolic studies, *J. Cell Biol.* **48**:503.

Jamieson, J. D., and Palade, G. E., 1971b, Synthesis, intracellular transport, and discharge of secretory proteins in stimulated pancreatic exocrine cells, *J. Cell Biol.* **50**:135.

Johnston, J. M., 1968, Mechanism of fat absorption, in: *Handbook of Physiology* (J. Field, ed.), pp. 1353–1375, American Physiological Society, Washington.

Jones, A. L., Ruderman, N. B., and Herrera, M. G., 1968, Electron microscopic and biochemical study of lipoprotein synthesis in the isolated perfused rat liver, *J. Lipid Res.* **8**:429.

Keenan, T. W., Morré, D. J., and Basu, S., 1974, Ganglioside biosynthesis. Concentration of glycosphingolipid glycosyltransferases in Golgi apparatus from rat liver, *J. Biol. Chem.* **249**:310.

Kern, F., and Borgstrom, B., 1965, Quantitative study of the pathway of triglyceride synthesis by hamster intestinal mucosa, *Biochim. Biophys. Acta* **98**:520.

Klaeveman, H. L., Conlon, T. P., and Gardner, J. D., 1975, Effects of gastrointestinal hormones on adenylate cyclase activity in pancreatic exocrine cells, in: *Gastrointestinal Hormones. A Symposium* (J. C. Thompson, ed.), pp. 321–344, University of Texas Press, Austin.

Malaisse, W. J., Malaisse-Lagae, F., Van Obbenghen, E., Somers, G., Davis, G., Ravazzola, M., and Orci, L., 1975, Role of microtubules in the phasic pattern of insulin release, *Ann. N.Y. Acad. Sci.* **253**:630.

Meek, G. A., and Bradbury, S., 1963, Localization of thiamine pyrophosphatase activity in the Golgi apparatus of the mollusk, *Helix aspersa, J. Cell Biol.* **18**:73.

Meldolesi, J., 1974, Secretory mechanisms in pancreatic acinar cells. Role of the cytoplasmic membranes, in: *Advances in Cytopharmacology*, Vol. 2 (B. Ceccarelli, F. Clementi, and J. Meldolesi, eds.), Raven Press, New York.

Meldolesi, J., and Cova, D., 1971, in vitro stimulation of enzyme secretion and the synthesis of microsomal membranes in the pancreas of the guinea pig, *J. Cell Biol.* **51**:396.

Mollenhauer, H. H., and Whaley, W. G., 1963, An observation on the functioning of the Golgi complex, *J. Cell Biol.* **17**:222.

Morré, D. J., Merlin, M. L., and Keenan, T. W., 1969, Localization of glycosyl transferase activities in a Golgi apparatus-rich fraction isolated from rat liver, *Biochem. Biophys. Res. Commun.* **37**:813.

Morré, D. J., Nyquist, S. E., and Rivera, E., 1970, Lecithin biosynthetic enzymes of onion stem and the distribution of phosphorylcholine-cytidylic transferase among cell fractions, *Plant Physiol.* **45**:800.

Morré, D. J., Keenan, T. W., and Huang, C. M., 1974, Membrane flow and differentiation:

Origin of Golgi apparatus membranes from endoplasmic reticulum, in: *Advances in Cytopharmacology*, Vol. 2 (B. Ceccarelli, F. Clementi, and J. Meldolesi, eds.), pp. 107–125, Raven Press, New York.

Nadler, N. J., Young, B. A., Leblond, C. P., and Mitmaker, B., 1964, Elaboration of thyroglobulin in the thyroid follicle, *Endocrinology* **74**:333.

Neutra, M., and Leblond, C. P., 1966a, Synthesis of the carbohydrate of mucus in the Golgi complex as shown by electron microscope radioautography of goblet cells from rats injected with glucose-H^3, *J. Cell Biol.* **30**:119.

Neutra, M., and Leblond, C. P., 1966b, Radioautographic comparison of the uptake of galactose-H^3 and glucose-H^3 in the Golgi region of various cells secreting glycoproteins or mucopolysaccharides, *J. Cell Biol.* **30**:137.

Plumer, T. H., and Hirs, C. H. W., 1963, Isolation of ribonuclease β, a glycoprotein, from bovine pancreatic juice, *J. Biol. Chem.* **238**:1396.

Poisner, A. M., and Cooke, P., 1975, Microtubules and the adrenal medulla, *Ann. N.Y. Acad. Sci.* **253**:653.

Ramon y Cajal, S., 1914, Algunas variaciones fisiologicas y patologicas del aparto reticular de Golgi, *Trabajos Inst. Cajal Invest. Biol. (Madrid)* **12**:127.

Rasmussen, H., and Goodman, D. B. P., 1975, Calcium and cAMP as inter-related intracellular messengers, *Ann. N.Y. Acad. Sci.* **253**:789.

Reaven, E. P., and Reaven, G. M., 1977, Distribution and content of microtubules in relation to the transport of lipid. An ultrastructural quantitative study of the absorptive cell of the small intestine, *J. Cell Biol.* **75**:559.

Redman, C. M., and Sabatini, D. D., 1966, Vectorial discharge of peptides released by puromycin from attached ribosomes, *Proc. Natl. Acad. Sci. USA* **56**:608.

Redman, C. M., Siekevitz, P., and Palade, G. E., 1966, Synthesis and transfer of aminase in pigeon pancreatic microsomes, *J. Biol. Chem.* **241**:1150.

Redman, C. M., Banerjee, D., Howell, K., and Palade, G. E., 1975, The step at which colchicine blocks the secretion of plasma protein by rat liver, *Ann. N.Y. Acad. Sci.* **253**:1975.

Reid, I. M., Shinozuka, H., and Didransky, H., 1970, Polyribosomal disaggregation induced by puromycin and its reversal with time, *Lab. Invest.* **23**:119.

Revel, J. P., and Hay, E. D., 1963, An autoradiographic and electron microscopic study of collagen synthesis in differentiating cartilage, *Z. Zellforsch. Mikrosk. Anat.* **61**:110.

Ross, R., and Benditt, E. P., 1965, Wound healing and collagen synthesis. V. Quantitative electron microscope radioautographic observations of proline-H^3 utilization by fibroblasts, *J. Cell Biol.* **27**:83.

Schachter, H., Jabbal, I., Hudgin, R. L., Pinkric, L., McGuire, E. J., and Roseman, S., 1970, Intracellular localization of liver sugar nucleotide glycoprotein glycosyltransferases in a Golgi-rich fraction, *J. Biol. Chem.* **245**:1090.

Senior, J. R., 1964, Intestinal absorption of fats, *J. Lipid Res.* **5**:495.

Siekevitz, P., and Palade, G. E., 1958a, A cytochemical study on the pancreas of the guinea pig. I. Isolation and enzymatic activities of cell fractionations, *J. Biophys. Biochem. Cytol.* **4**:203.

Siekevitz, P., and Palade, G. E., 1958b, A cytochemical study on the pancreas of the guinea pig. II. Functional variations in the enzymatic activity of microsomes, *J. Biophys. Biochem. Cytol.* **4**:309.

Siekevitz, P., and Palade, G. E., 1958c, A cytochemical study on the pancreas of the guinea pig. III. In vivo incorporation of leucine-1-C^{14} into the proteins of cell fractions, *J. Biophys. Biochem. Cytol.* **4**:557.

Siekevitz, P., and Palade, G. E., 1960, A cytochemical study on the pancreas of the guinea pig. V. In vivo incorporation of leucine-C^{14} into chymotrypsinogen of various cell fractions, *J. Biophys. Biochem. Cytol.* **7**:619.

Sjöstrand, F. S., 1962, The fine structure of the exocrine pancreas, in: *Ciba Foundation Symposium, The Exocrine Pancreas, Normal and Abnormal Function* (A. V. S. deReuck, and M. P. Cameron, eds.), pp. 1–22, Little, Brown and Company, Boston.

Soifer, D., ed., 1975, The biology of cytoplasmic microtubules, *Ann. N.Y. Acad. Sci.* **253**.

Soifer, D., Laszlo, A., Mack, K., Scotto, J., and Siconolfi, L., 1975, The association of a cyclic AMP-dependent protein kinase activity with microtubule protein, *Ann. N.Y. Acad. Sci.* **253**:598.

Tytgat, G. N., Rubin, C. E., and Saunders, D. R., 1971, Synthesis and transport of lipoprotein particles by intestinal absorptive cells in man, *J. Clin. Invest.* **50**:2065.

Weinstock, A., 1970, Cytotoxic effects of puromycin on the Golgi apparatus of pancreatic acinar cells, hepatocytes, and ameloblasts, *J. Histochem. Cytochem.* **18**:875.

Wolff, J., and Bhattacharyya, B., 1975, Microtubules and thyroid hormone mobilization, *Ann. N. Y. Acad. Sci.* **253**:763.

Yunghans, W. N., Keenan, T. W., and Morré, D. J., 1970, Isolation of Golgi apparatus from rat liver. III. Lipid and protein composition, *Exp. Mol. Pathol.* **12**:36.

13

Compartmentation and Its Role in Metabolic Regulation

Ifeanyi J. Arinze and Richard W. Hanson

I. Introduction

In writing on a subject as broad and varied as the role of intracellular compartmentation in metabolic regulation, one is forced to come to grips with the question of which of the many aspects of this subject is central to a clear understanding of the field. Our answer to this question reflects, to a great extent, our personal view of metabolic regulation. However, since the general aim of this volume is to present principles rather than simply a catalog of facts, we have chosen to limit this chapter to a discussion of intracellular compartmentation in mammalian cells. This is not meant to negate the very important role of intercellular compartmentation in the regulation of mammalian metabolism. Space and the scope of coverage do not permit the inclusion of this material here. When considering the content of this chapter, it also became clear that most of the better understood examples of the important role of compartmentation within the cell involve the mitochondria. A number of recent reviews have considered various aspects of this subject (Greenbaum et al., 1971; Williamson, 1976; Denton and Pogson, 1976;

Ifeanyi J. Arinze • Department of Biochemistry and Nutrition, Meharry Medical College, Nashville, Tennessee 37208.   **Richard W. Hanson** • Department of Biochemistry, Case Western Reserve University School of Medicine, Cleveland, Ohio 44106. This work was supported in part by grants AM-16008, AM-18034, CA-12227, and HD-08792 from the National Institutes of Health, and by a Basil O'Connor Starter Research Grant, No. 5-89, from the National Foundation-March of Dimes.

Söling and Kleineke, 1976; Zuurendonk et al., 1976). Of particular interest is the excellent and still timely review on compartmentation by the late Guy Greville (1969a), which lucidly brings together many of the specific aspects of the subject that will be covered here in a more general manner. Compartmentation involving other subcellular organelles such as lysosomes, endoplasmic reticulum, the nucleus, and the Golgi apparatus has been studied, but the involvement of these organelles in metabolic control is, in most cases, less well characterized.

II. Nature of Intracellular Compartments

The segregation of materials within the cell can be broadly defined as compartmentation. While such a defintion is technically correct, it is of little value in considering the way in which compartmentation can affect the complex interactions involved in regulating metabolic processes. To render the definition useful, specific examples are needed as a framework on which to build a more general discussion of this topic. We know, for example, that the citric acid cycle provides intermediates for biosynthetic processes such as lipogenesis and gluconeogenesis. Since this cycle occurs in the mitochondria, the movement of various anions across the mitochondrial membrane is of primary importance. Then, in seeking whether the mitochondrial membrane acts to compartment these compounds, and what general mechanisms govern their flux across this membrane, we can be led from more specific to more general considerations about the nature of compartmentation.

There are, within the mammalian organism, examples of tissues with the *specific function* of compartmenting compounds. Adipose tissue contains largely triglyceride, and the segregation of this bulk lipid within the fat cell is of primary importance, since the segregation separates the insoluble lipid component from the cytosol of the adipocyte. To the extent to which removal of fatty acids, with their well-characterized detergent properties, can alter the flux of intermediates through metabolic pathways, this type of compartmentation might also be considered important in metabolic regulation. Various secretory products such as zymogen granules, formed in the acinar cells of the pancreas, and milk protein, in the epithelial cells of the mammary gland, are compartmented within the Golgi apparatus for subsequent release into the lumen of the secretory ducts. Another type of compartmentation important in regulating cellular function may be loosely termed "chemical compartmentation." An example of this occurs in the case of zymogens which possess the complete amino acid sequence of the active enzyme plus additional amino acid residues. These are converted to active enzyme by the cleavage of a portion of the protein by specific proteolytic enzymes. We may then consider the zymogen as chemically compartmented from the other proteins of the pancreatic acinar cell in which it is synthesized. More will be said on several specific aspects of this type of compartmentation in the next section of this chapter.

In a broad sense, interconvertible forms of the same enzyme are compartmented in being rendered temporarily inactive by mechanisms such as phosphorylation by a specific kinase, as is the case with pyruvate dehydrogenase (Reed, 1976). Other examples, such as keto–enol tautomerization, can be mentioned as chemical compartmentation. Pogson and Wolfe (1972) have demonstrated that, at pH 7.4, 74% of oxaloacetate is in the keto form, and that a number of enzymes for which oxaloacetate is a substrate bind this form. However, the active species of oxaloacetate responsible for inhibiting succinate dehydrogenase and citrate lyase is the enol form. The balance between the two forms is maintained by a specific tautomerase, and this type of chemical compartmentation must be considered a potentially important factor in regulating metabolic pathways. Since there is little information relating to this point, however, we will not discuss it in detail in this review.

III. Zymogen Activation and Compartmentation

An increasing number of specific regulatory systems involving the activation of zymogens by selective enzymatic cleavage have been described. The principle of this type of compartmented regulatory system is the existence of precursor molecules (zymogens) which are activated by proteolytic cleavage. The cleavage of the zymogen by a protease is an example of a system that is rapid and irreversible and which has the capacity of responding to a wide variety of physiological stimuli. The number of regulatory processes known to involve zymogen activation is growing rapidly. Examples such as hormone synthesis, blood coagulation, the formation of proteolytic enzymes, and, perhaps of unique importance, the basic process by which secretory proteins are selected for ultimate movement through the Golgi apparatus are now thought to involve a form of zymogen activation. While it is impossible to describe each of these examples in detail, we shall select several features of zymogen activation as examples of the unique way in which compartmentation can regulate metabolism. More details of the overall subject of zymogen activation are contained in an excellent review by Neurath and Walsh (1976).

A. Enzymes as Zymogens in Their Own Degradation

The half-lives of intracellular proteins vary remarkably, from 11 min for ornithine decarboxylase (Russell and Snyder, 1968) to 16 days for isozyme 5 of lactate dehydrogenase (Fritz et al., 1969), and within this broad range is the approximate 5-day half-life for general cytosol proteins (Arias et al. 1969). What regulates this diversity in the rate of enzyme turnover is still poorly understood, but several ideas concerning the nature of the selection processes have been presented. It is possible that intracellular proteins are ultimately degraded within the lysosomes, although direct evidence in support of this contention is still lacking (there is no doubt that extracellular proteins, once taken in by a cell, are degraded in the lysosomes). Lysosomes contain the bulk of the cellular proteases, the cathepsins (A through E), all of which have a pH optimum of from 3 to 5. These protein degradation enzymes, therefore, are compartmented within the lysosomes, which use energy-dependent processes to maintain the acid pH necessary for the activity of these proteases (Barrett and Dingle, 1971). The link between the turnover of intracellular proteins and the compartmented proteases of the lysosomes may be the action of group-specific proteases which act on specific classes of proteins via a limited proteolysis. By this process, the protein itself would act as a zymogen and would be "activated" for subsequent degradation by lysosomal proteases.

Katunuma et al. (1975) and Kominami and Katunuma (1976) have described several group-specific proteases which act on specific enzymes. The best studied of these is the protease which acts on pyridoxal enzymes, including the well-studied tyrosine aminotransferase and ornithine transaminase. Limited proteolysis may proceed by selection of the enzymes to be degraded, thus rendering the "nicked" enzyme susceptible to subsequent degradation within the lysosome. This is a unique form of compartmentation involving both a zymogen activation process and the physical separation of the ultimate protein degradation system safely within the lysosome. A further discussion of the overall process of protein degradation is presented in the extensive reviews by Goldberg and Dice (1974) and Goldberg and St. John (1976).

B. Compartmentation of Secretory Proteins—The Signal Peptide Theory

One recurring theme in any discussion of compartmentation is the importance of physical separation of compounds by membranes. In most instances, mechanisms exist to ensure that secretory products remain segregated from the cytosol of the cell that synthesizes these products. A classic example of this type of compartmentation is the segregation of secretory products in the lumen of the endoplasmic reticulum of the pancreatic acinar cell. Trypsinogen and chymotrypsinogen are normally formed as zymogens and are activated by proteolytic cleavage in the intestine. It now seems probable that secretory products such as these proteolytic enzymes may themselves pass through a prezymogen stage subsequent to their secretion from the cell.

Recently, Blobel and Dobberstein (1975a,b) presented evidence which indicated that the immunoglobulin light chain, synthesized in a cell-free, protein-synthesizing system from partially purified mRNA from the murine myeloma MOPC 41, was larger than the authentic light chain. The occurrence of a unique sequence of codons, located immediately to the right of the initiation codon, is probably present only in those mRNA species which when translated will have products destined to be transported across membranes. Since no other mRNA species contain the codon for this specific protein, the segment to the right of the initiation codon is unique. The resulting amino acid sequence— the so-called "signal peptide"—is therefore a unique peptide sequence which can confer some ordered recognition upon the preprocessed zymogen molecule. In an interesting series of experiments, Blobel and Dobberstein (1975a,b) proposed that "the emergence of the signal sequence of the nascent chain from within a space in the large ribosomal subunit triggers attachment of the ribosome to the membrane, thus providing the topological conditions for the transfer of the nascent chain across the membrane." Subsequent work by Thibodeau, Gagnon, and Palmiter (personal communication), using highly purified mRNA for ovomucoid and lysozyme, has shown that these two proteins contain a signal peptide. The significance of the signal peptide theory and its applicability to all secretory proteins remains to be established. It is, however, an exciting concept that adds a new dimension to our understanding of the functional basis of compartmentation.

IV. Membrane Permeability and the Movement of Molecules in the Cell

The separation of compounds within the cell is one of the most basic methods for controlling cellular function. The presence of membrane-bound compartments results in the creation of gradients across which molecules move either passively (by simple diffusion) or by transport on specific carriers. Since the subject of this chapter is compartmentation with regard to metabolic regulation, we shall consider examples most clearly related to this aspect of compartmentation and neglect other possibly equally important points.

Of all the intracellular organelles, the mitochondrion has been the most extensively studied with respect to the compartmentation of compounds within its boundaries. In part, this results from the ease of separation of mitochondria from mammalian tissues (most notably the liver), as well as from the key role mitochondria play in a number of metabolic processes. The mitochondrial membrane is capable of transporting metabolites on specific transporters and of segregating metabolites from the cytosol. It is important to note that some metabolites apparently move across the mitochondrial membrane in an unspecific or non-carrier-linked manner. For example, ketone bodies, water, CO_2, and oxygen appear to freely diffuse into and out of mitochondria. In the following sections we will discuss specific aspects of the transport mechanisms, followed by a more general discussion of their role in regulating major metabolic pathways. We will start with the most important result of intracellular compartmentation—oxidative phosphorylation—as viewed by the chemiosmotic theory.

A. Compartmentation and Oxidative Phosphorylation

Since ATP meets the energy requirement for biological synthesis, an understanding of what controls ATP synthesis is of central importance to any discussion of metabolic regulation. Oxidative phosphorylation is associated with enzymes which are an integral part of the inner mitochondrial membrane. This membrane is also known to selectively segregate H^+ and OH^- in a manner related to energy conservation. During periods of oxidative phosphorylation, isolated mitochondria extrude H^+ ions, acidifying the incubation medium. This acidification can be measured using a specially designed pH electrode and is a convenient indicator of the kinetics of ATP formation. The separation of H^+ and OH^- ions by the mitochondrial membrane is an important and fundamental biological phenomenon, since, as long as these ions remain apart, they are a potential source of energy. Mitchell (1961, 1966, 1968) proposed that the separation of charge by mitochondria causes the synthesis of ATP from ADP and inorganic phosphate. The mechanism for this reaction was thought to result from the reversal of an ATPase, an enzyme which is part of the inner mitochondrial membrane. This ATPase (Fig. 1) is a spherical protein, termed F_1 by Racker,

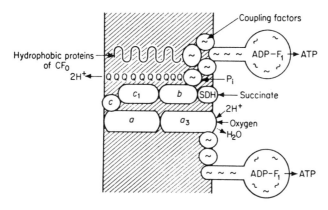

Figure 1. Topography of the inner mitochondrial membrane. This figure shows the folding of the respiratory chain in the inner mitochondrial membrane and its association with the ATPase of the F1. The figure is from a review by Racker (1970) and presents the author's reconstruction of the membrane in a way that has appeal in terms of the chemiosmotic hypothesis of Mitchell.

and is located on the inside of the inner mitochondrial membrane [the entire subject of mitochondrial "sidedness" has been reviewed by Racker (1970)].

The inner mitochondrial membrane also contains the respiratory chain arranged in such a way as to move H^+ out and electrons in, and to maintain OH^- within the mitochondria (Fig. 2). Since H^+ is being moved out of the mitochondria, the inner membrane is negatively charged, with a membrane potential of about -250 mV. The membrane is impermeable to H^+ and OH^-, so that this potential gradient is maintained. The separation of this charge is maintained by oxidation–reduction reactions which utilize the NADH formed in the citric acid cycle. The reentrance of H^+ into the mitochondria can occur only at the points at which the spherical F_1 proteins containing the ATPase are attached to the inner mitochondrial membrane. According to the chemiosmotic theory, it is this influx of H^+ which, in the presence of ADP and inorganic phosphate, drives the synthesis of ATP.

The placement of the components of the respiratory chain in the inner mitochondrial membrane is of considerable importance to the mechanism of the chemiosmotic theory. Mitchell (1967) calls this the "coupling membrane"; H^+, citric-acid-cycle anions, amino acids, and cations can be transported across this membrane by specific carriers imbedded in it. The respiratory chain is folded into three "loops"

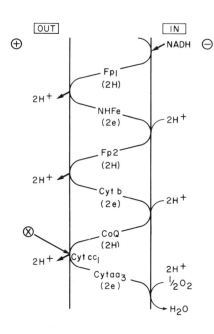

Figure 2. Folding of the respiratory chain in the inner mitochondrial membrane with resultant proton translocations for NAD-linked substrates. In this scheme six protons are translocated for each pair of reducing equivalents oxidized in the chain. The abbreviations are F_p, flavoprotein; NHFe, nonheme iron; Cyt, cytochrome. X is the site of interaction with ferricyanide and TMPD is N,N,N',N'-tetramethyl-p-phenylenediamine. This figure is redrawn from Greville (1969b).

which correspond to the coupling sites. An individual loop is composed of a respiratory-chain component capable of being reduced by the addition of two H atoms, and of a component whose oxidized and reduced forms are made up of metal ions with differing valences which act as electron carriers. The nature of the folding of the oxidation–reduction loops is also important, since it aligns the specific components of the chain at fixed positions within the inner membrane (see Fig. 2). The reaction of NADH at loop 1 of the respiratory chain occurs within the mitochondrion, and its subsequent oxidation to NAD results in the ejection of two H^+ ions on the outside of the inner membrane. Since loop 2 contains flavoprotein oriented toward the inside of the inner mitochondrial membrane, the FAD-linked dehydrogenations would result in the ejection of 4 H^+ ions rather than the 6 for NAD-linked oxidation. The mechanism of oxidative phosphorylation as proposed by the chemiosmotic theory would therefore involve a cyclical process of moving H^+ ions to the outside of the mitochondrial inner-membrane, followed by a translocation of protons back into the mitochondrion at the specific ATPase site. The ATPase reaction has an equilibrium far in the direction of ATP hydrolysis,

$$ATP + H_2O \rightarrow ADP + P_i \qquad (1)$$

and the translocation of protons back across the mitochondrial membrane provides the energy for ATP formation. According to the chemiosmotic theory of oxidative phosphorylation, the mitochondrial inner membrane may be regarded "as an oxido–reduction cell generating a proton current which drives ATP synthesis in the ATPase system" (Mitchell, 1968).

It is something of an understatement to point out that the chemiosmotic theory was not enthusiastically embraced by all of the workers in the field of oxidative phosphorylation (Slater, 1967). Greville (1969b) has scrutinized the theory and compared the evidence in its favor with that for the other major mechanism proposed for oxidative phosphorylation, the chemical intermediate hypothesis. The weight of evidence to date seems to support, at least in part, the chemiosmotic theory (Boyer et al., 1977), and there is in progress active research which should either substantiate or modify the major tenets of the theory (see Chapter 9). What Mitchell has succeeded in doing, however, is to present a clear and testable model to explain oxidative phosphorylation. His theory also provides a unifying mechanism by which the movement of ions within the cell can be coupled to energetic processes. It is this aspect of the chemiosmotic theory that allows its integration with the movement of anions across the mitochondrial membrane, a process vital to a clear understanding of metabolic control. We might visualize, as does Mitchell, that "the uptake of an anionic substrate such as succinate occurs as a proton–anion symport, and the work required to maintain the effective flow of protons into the mitochondria with the anions is done by the respiratory system, by the ATPase system or by the exit of another anionic substrate (e.g., malate as malic acid or bicarbonate as CO_2) in the protonated state." As a basis for our discussion of the role of compartmentation in metabolic regulation, we may consider the membrane-bound compartments, with their highly specific transport mechanisms, as an important factor in control processes.

B. ATP Translocation across the Inner Mitochondrial Membrane

The mitochondrial inner membrane acts as a barrier to the free movement of ATP and ADP [see Klingenberg (1972) for a review]. Since ATP is formed on the inside of the inner mitochondrial membrane, whereas most of the ATP-utilizing reactions are in the cytosol, the mechanism for ATP translocation is of considerable importance. Klingenberg and his associates (1966, 1972) were the first to clearly describe the transport process and to point out the nature of the specific carrier involved. The outer compartment of the mitochondrion contains adenylate kinase and a low activity of ATPase, while activities in the matrix include a dinitrophenol-stimulated ATPase, the oxidative phosphorylation of ADP to ATP, and a transphosphorylating ATP–ADP exchange reaction (see Klingenberg, 1972, for a review). It is these processes which govern the exchange of ATP across the inner mitochondrial membrane.

Freshly isolated, intact mitochondria contain considerable amounts of adenine nucleotides which are resistant to removal by repeated washings with isotonic sucrose. This indicates that these compounds are in a compartment—presumably within the inner mitochondrial membrane—which is inaccessible to the sucrose solution. When exogenous adenine nucleotide is added to the mitochondria, there is a rapid exchange with endogenous adenine nucleotides with no net increase in the concentration of adenine nucleotides in the mitochondria. ADP exchanges most rapidly, followed by ATP and then by AMP, which is relatively impermeable. It is the inner mitochondrial membrane through which the adenine nucleotides do not permeate and which contains the specific adenine-nucleotide transporting system. The movement of ATP across the inner mitochondrial membrane (and hence out of the mitochondria) depends directly on the translocation of ADP in the presence of adenylate kinase in the outer compartment of the mitochondria.

$$ATP + AMP \longleftrightarrow 2\ ADP \qquad (2)$$

This affords a mechanism for balancing the relative concentration of either of the permeant adenine nucleotides.

The adenine nucleotide carrier or translocase is specifically inhibited by atractyloside (competitive with respect to adenine nucleotide) and bongkrekic acid (noncompetitive). Atractyloside has been known for some time to be highly poisonous and its mechanism of action attests to the importance of the adenine nucleotide translocase. Other nucleotides, such as GTP, must first be converted to ATP by nucleoside diphosphokinase prior to transport out of the mitochondria. The outer mitochondrial compartment also contains nucleoside diphosphokinase for the conversion of ATP to GTP.

The exchange of ADP and ATP across the mitochondrial inner membrane, mediated by the adenine nucleotide translocase, is thought to be rate-limiting for oxidative phosphorylation (Heldt and Pfaff, 1969). This translocase functions asymmetrically, i.e., it favors a rapid entry of ADP dependent on the availablity of intramitochondrial ATP and results in an asymmetric distribution of ATP and ADP on either side of the membrane (Heldt et al., 1972, Slater et al., 1973). Thus the

ATP/ADP ratio is greater outside than inside the membrane. Furthermore, Heldt et al. (1972) have shown that the maintenance of this gradient—the so-called phosphorylation potential (ATP/ADP + P_i)—will consume metabolic energy. Since the inward movement of ADP in exchange for ATP will markedly alter the charge balance within the mitochondria, there is a concomitant efflux of H^+ or entry of P_i. One can predict, therefore, that changes in membrane potential can greatly influence the net exchange of ADP and ATP via the translocase.

An excellent example of an experimental approach to the problem of adenine nucleotide compartmentation and its role in regulating energy transduction is a paper by Davis and Lumeng (1975). They have developed a system in which the ATP/ADP concentration ratio in the mitochondrial incubation medium is held constant by the addition of a partially purified ATPase. With this system it is possible to maintain approximate physiological steady states of respiration, and to hold the concentration of adenine nucleotides and inorganic phosphate in the incubation medium within the intracellular range. This allows the direct measurement of changes in the ratio of ATP/ADP in the mitochondria as a function of the respiratory state of the adenine nucleotides, and the ratio can be varied experimentally by adjusting the ATPase added to the incubation medium (see Fig. 3). Thus, if the respiratory state of mitochondria, metabolizing glutamate plus malate, decreases from state 4 to about 80% of full state 3 respiration, the ATP/ADP ratio outside the mitochondria decreases from 80 to about 20, whereas the ATP/ADP in the matrix remains constant. This technique requires the rapid separation of mitochondria from the incubation

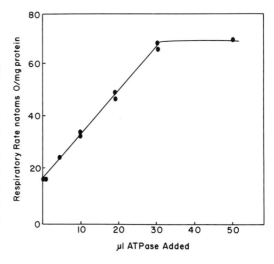

Figure 3. The effect of purified ATPase on the rate of respiration of isolated rat liver mitochondria incubated with ATP. In this experiment, the respiration rate was monitored in the presence of ATP for 2 min, followed by measured amounts of added mitochondrial ATPase. This work is taken from Davis and Lumeng (1975) and redrawn with permission of the authors.

medium, in this case by centrifuging the mitochondria through silicone fluid layered over 2 M perchlorate [see Davis and Lumeng (1975) for a description of this method]. This procedure, if properly performed, allows the rapid separation of mitochondria from incubation medium with only a marginal degree of anaerobiosis.

The independence of the ATP/ADP ratio in the mitochondria from the adenine nucleotide concentration ratio in the incubation medium dramatically illustrates the importance of mitochondrial compartmentation in the control of respiration. The rate of phosphorylation-coupled respiration is limited by the activity of the adenine nucleotide translocase and is markedly influenced by the asymmetric kinetic characteristics of the exchange system. By maintaining two distinct pools of adenine nucleotides in the mitochondria and cytosol, a higher phosphate potential can be established in the cytosol. This is due to the high specificity of the adenine nucleotide exchange for ADP. Since the stoichiometry of ATP–ADP exchange is 1:1, there occurs a rate-limiting system which is related directly to the membrane potential. Compartmentation of adenine nucleotides, therefore, offers one of the best examples of regulation of vital cellular processes by highly specialized membranes.

C. Compartmentation of Citric-Acid-Cycle Intermediates and Other Anions

It is now well established that the compartmentation of whole segments of the citric acid cycle, between the mitochondria and the cytosol, is of major regulatory significance. In a later section of this chapter we will discuss the role of such systems in the control of major biosynthetic and biodegradative pathways.

In a number of biosynthetic pathways, such as fatty acid synthesis and gluconeogenesis, the citric acid cycle provides key intermediates. Citric-acid-cycle anions such as malate and citrate move across the inner mitochondrial membrane into the cytosol by poorly understood translocation processes (Chappell and Haarhoff, 1967; Chappell, 1968). Figure 4 is a list of these anion translocators taken from an excellent review of anion transport by Williamson (1976) to which the reader is directed for a more detailed treatment of this subject. Most anion translocations from cytosol to mitochondria involve an exchange for anions moving in the opposite direction. It is presumed that this translocation process is carrier mediated, although specific carrier molecules have not been purified and characterized. The exchange–translocation processes can be grouped into three types.

1. Electrogenic. The translocation of anions results in an alteration of charge distribution across the inner mitochondrial membrane. An

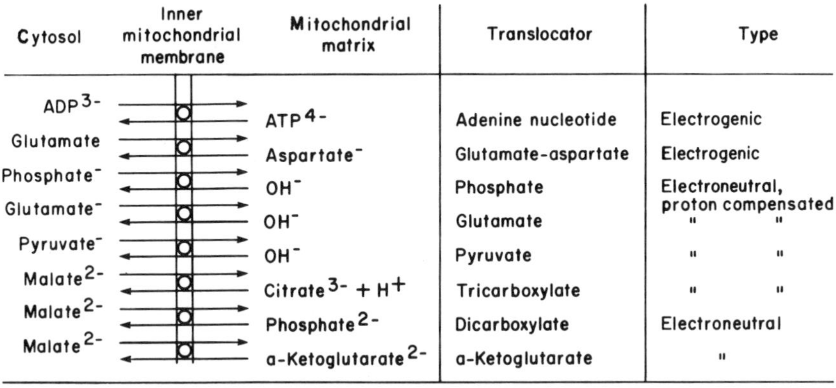

Compartmentation and Its Role in Metabolic Regulation

Cytosol	Inner mitochondrial membrane	Mitochondrial matrix	Translocator	Type
ADP^{3-}	⇌	ATP^{4-}	Adenine nucleotide	Electrogenic
Glutamate	⇌	Aspartate$^-$	Glutamate-aspartate	Electrogenic
Phosphate$^-$	⇌	OH^-	Phosphate	Electroneutral, proton compensated
Glutamate$^-$	⇌	OH^-	Glutamate	" "
Pyruvate$^-$	⇌	OH^-	Pyruvate	" "
Malate^{2-}	⇌	Citrate^{3-} + H^+	Tricarboxylate	" "
Malate^{2-}	⇌	Phosphate^{2-}	Dicarboxylate	Electroneutral
Malate^{2-}	⇌	α-Ketoglutarate^{2-}	α-Ketoglutarate	"

Figure 4. Mitochondrial anion translocators. This figure is redrawn from Williamson (1976).

example of this type of translocation is the aspartate–glutamate exchange, which is energy consuming (LaNoue et al., 1974). Aspartate efflux from the mitochondria is stimulated by an electrical potential gradient across the mitochondrial membrane. There is a 1:1 stoichiometry between glutamate entry and aspartate efflux from the mitochondria, accompanied by a stoichiometric uptake of protons from the medium.

2. Electroneutral, proton compensated. The transport of pyruvate is accompanied by an exchange with hydroxyl ion but is equivalent to cotransport with protons, in which citrate plus a proton will translocate in exchange for malate, maintaining electroneutrality within the mitochondria.

3. Electroneutral. This type of exchange is typified by the dicarboxylate shuttle. The basic requirement for anion shuttles of the type discussed above is the presence of an enzyme in the mitochondrial matrix and one in the cytosol, both of which catalyze the same reaction.

Two key enzymes in the movement of reducing equivalents from the inside to the outside of mitochondria are NAD^+ malate dehydrogenase and aspartate aminotransferase. These enzymes in both the cytosol and the mitochondria allow the interconversion of intermediates to form a cycle (Fig. 5). In this sequence, NADH generated inside the mitochondria is used to reduce oxaloacetate to malate. Malate moves across the inner mitochondrial membrane on the dicarboxylate carrier, where it is oxidized to oxaloacetate by cytosolic NAD^+ malate dehydrogenase. Oxaloacetate has a limited rate of flux across the inner mitochondrial membrane (Haslam and Krebs, 1968) and is converted to aspartate by aspartate aminotransferase, with the latter compound

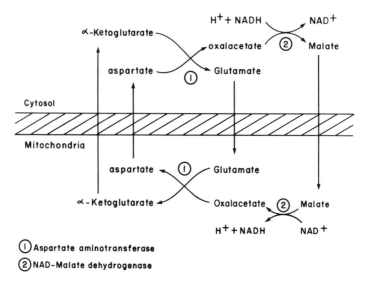

Figure 5. The aspartate–malate shuttle.

being transported back into the mitochondria to be reconverted to oxaloacetate. Glutamate is required in the cytosol for transamination with oxaloacetate, and this glutamate moves out of the mitochondria on the electrogenic glutamate–aspartate transporter in exchange for the aspartate moving in. The α-ketoglutarate in the cytosol must also return to the mitochondria to complete the cycle, presumably in exchange for malate.

The malate–aspartate shuttle is a balanced mechanism for the transport of reducing equivalents across a membrane through which NADH itself does not move. The shuttle can be used to provide both carbon and NADH for gluconeogenesis, as presented in detail in Section V.A of this chapter. The major experimental verification of the functioning of the malate–aspartate shuttle comes from experiments with inhibitors. One of the most important of these is the work of Rognstad and Katz (1970), which introduced the use of aminooxyacetic acid as an inhibitor of aspartate aminotransferase in rat kidney cortex slices. Aminooxyacetate blocks gluconeogenesis from lactate, which must use aspartate as a vehicle for anion movement from the mitochondria to the cytosol, but it does not inhibit pyruvate conversion to glucose since this involves the efflux of malate to the cytosol. Several other inhibitors have been used, most notably n-butylmalonate, to block the malate–phosphate exchange (Robinson and Chappell, 1967).

There are numerous other shuttle mechanisms that function to move small molecules across the inner mitochondrial membrane. The malate–aspartate shuttle is perhaps the best studied and provides an excellent example of the overall strategy of these shuttles. It is important to stress that shuttle systems have very different functions in different tissues. In muscle, for example, the citric acid cycle functions predominantly as a cycle and does not support the biosynthesis of a variety of compounds, as it does in the liver. In cardiac muscle, the di- and tricarboxylate carriers are virtually absent, perhaps because this tissue does not actively move citric acid cycle anions for lipogenesis or gluconeogenesis. As with all physiological processes, when considering the role of shuttle mechanisms in metabolic control, one must avoid generalizing from tissue to tissue.

D. Species Differences in Compartmentation

Most of our concepts on the nature and functional regulation of compartmented sequences derive from studies using the rat. It is natural that one species be used as a "standard" since this allows a unified approach to the already complex task of understanding metabolic regulation. Concepts derived from studies with the rat have been widely applied to a variety of species, and often with excellent results. There are important exceptions, however, which indicate that caution must be exercised in extending our use of a single species too uncritically.

A number of recent studies have focused on the compartmentation of key enzymes of gluconeogenesis and lipogenesis and on the differences which occur among mammalian species (for a general review of this work see Hanson, 1974). Some of the specific aspects of the consequences of compartmentation in various species will be discussed later in this chapter. At this point, it is important to draw several general principles which have emerged from studies with animals other than the rat.

1. The Loss of Enzyme Sequences in Certain Species

Part of the adaptation of animals to their environment has involved emphasizing certain reactions and deleting others. An excellent example of such an adaptation is the pathway of fatty acid synthesis in ruminant animals. Ruminants live on cellulose, which is degraded by the microflora of their rumen to propionate, butyrate, acetate, and amino acids, but not to glucose. These animals are thus dependent on gluconeogenesis to maintain their blood glucose concentration within a critical range. It is not surprising, then, that fatty acid synthesis in the liver of these animals uses as precursors the short-chain fatty acids—notably

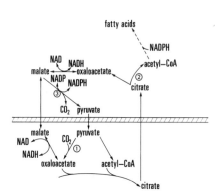

Figure 6. Fatty acid synthesis in mammalian tissues showing the transhydrogenation cycle. Pyruvate is generated from glucose in the cytosol (upper portion of figure) and converted to fatty acids by a reaction sequence involving enzymes in the mitochondrial matrix (lower portion of figure). 1, pyruvate carboxylase; 2, ATP citrate lyase; 3, NADP-malate dehydrogenase. These reactions were absent in adipose tissue from ruminant animals (Hanson and Ballard, 1967).

acetate—with glucose being carefully conserved. The normal pathway of fatty acid synthesis in nonruminant animals such as the rat involves the flow of *glucose* carbon to fatty acids by a process shown in Fig. 6. In ruminant liver and adipose tissue (Fig. 6), the entire sequence for the generation of cytosolic acetyl CoA is absent. The key enzymes ATP-citrate lyase and NADP-malate dehydrogenase are also absent; this was shown for cow and sheep liver (Hanson and Ballard, 1967, 1968). In the cow, for example, citrate does not move across the mitochondrial membrane to support fatty acid synthesis from glucose. Models based on our knowledge of compartmentation of metabolites from rat studies are useful as a starting point but must be modified for different species. A more detailed treatment of this subject can be found in reviews by Ballard et al. (1969) and Bauman (1976).

2. Differences in the Distribution of Phosphoenolpyruvate Carboxykinase

One of the major metabolic differences between the enzyme sequence of gluconeogenesis in animal species is the variation in intracellular location of hepatic and renal phosphoenolpyruvate (P-enolpyruvate) carboxykinase. Table 1 (Hanson and Garber, 1972) shows how different are the various species studied to date with regard to the location of this enzyme in the liver. The rat, mouse, and hamster contain 90% of the enzyme in the cytosol and 10% in the mitochondria. On the other extreme, the rabbit, pigeon, and chicken contain 95% of the enzyme in the mitochondria and 5% in the cytosol. The most widely occurring distribution is in species such as the guinea pig, cow, sheep, dog, cat, and human, in which P-enolpyruvate is evenly distributed between the cytosol and mitochondria. There have been numerous reviews discussing the significance of this distribution relative to the regulation of gluconeogenesis (Hanson, 1974; Söling and Kleineke 1976; Ray, 1976). In this section we will emphasize the role that compartmentation of a key enzyme can have on the normally accepted patterns of metabolite flux in tissues.

Marco et al. (1969) have described oxaloacetate as being at "metabolic cross-roads," an apt description of an important metabolite which is a central intermediate in a variety of reactions, all of which compete for the low concentration of this compound present in the mitochondria. Mitochondria from guinea pig liver have been intensively studied in an effort to determine the effect of the added competition for oxaloacetate by P-enolpyruvate carboxykinase on the net flux of carbon for biosynthetic reactions, such as gluconeogenesis, and for the citric acid cycle. Four major mitochondrial enzymes, NAD^+ malate dehydrogenase, aspartate aminotransferase, citrate synthetase, and P-enolpyruvate carboxykinase can convert oxaloacetate to malate, aspartate, citrate or P-enolpyruvate, respectively (Fig. 7). The major factor regulating the flux of oxaloacetate via these four reactions is the energy state of the mitochondria as reflected in the $NAD^+/NADH$ ratio in the matrix space (Garber and Salganicoff, 1973). Normally, isolated guinea pig liver mitochondria will convert added substrates, such as malate, α-ketoglutarate, or pyruvate to P-enolpyruvate when the NAD^+ to NADH ratio is relatively oxidized [about 40 (Garber and Hanson, 1971a)]. When fatty acids are added, there is a reduction of the mitochondrial redox

Table 1. The Intracellular Location of Hepatic P-Enolpyruvate Carboxykinase in Various Species[a]

Species	% of total activity	
	Cytosol	Mitochondria
Rat	90	10
Mouse	95	5
Hamster	90	10
Chicken	<5	95
Pigeon	5	95
Rabbit	10	90
Guinea pig	40	60
Dog	40	60
Cat	50	50
Sheep	40	60
Cow	40	60
Human	40	60

[a]From Hanson and Garber (1972).

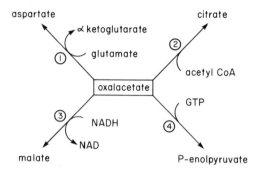

Figure 7. Metabolic fate of oxaloacetate in mammalian tissues. 1, Aspartate aminotransferase; 2, citrate synthetase; 3, malate dehydrogenase; 4, P-enolpyruvate carboxykinase.

with a resultant decrease in the net synthesis of P-enolpyruvate and an increased conversion of oxaloacetate carbon to malate (Jomain-Baum and Hanson, 1975). This alteration in flux of oxaloacetate toward malate and away from P-enolpyruvate in guinea pig liver mitochondria illustrates the importance of the compartmentation of NADH within the mitochondrial matrix and its effect on the synthesis and subsequent release of citric-acid-cycle anions used in biosynthetic sequences. The significance of this in the regulation of hepatic gluconeogenesis will be discussed in a later section.

E. Measurement of the Compartmentation of Intermediates—Limitations of Available Methods

It is implicit in our discussion of compartmentation that we be able to accurately determine the concentration of intermediates within the various intracellular organelles, especially the mitochondria. Without this ability it is impossible to determine the flux rate of key compounds across membranes. Is it possible, for example, to accurately determine the intramitochondrial concentration of oxaloacetate during periods of enhanced gluconeogenic flux in order to relate its levels to the known kinetic properties of the enzymes which compete for oxaloacetate? Clearly, the formulation of any model for the control of a complex metabolic sequence must consider the actual concentration of all intermediates in the mitochondria if it is to achieve a meaningful interpretation of the kinetic properties of key enzymes. A simple measurement of the whole-cell concentration of an intermediate can be totally misleading, since intracellular gradients occur in various compartments.

For the measurement of large molecules, such as enzymes and coenzymes, which do not readily move across the mitochondrial membrane, it is possible to use marker enzymes of known location to correct for breakage of mitochondria during isolation. An excellent example of this approach occurs in a paper of Bottger et al. (1969), in which the intracellular location of pyruvate carboxylase in adipose tissue was determined. Even this procedure is not completely free of complication, however, considering the longstanding question of the intracellular location of pyruvate carboxylase in rat liver. For many years the proponents of a cytosolic—as well as a mitochondrial—localization for this enzyme presented experimental data derived from intracellular localization studies using exactly the same techniques as were employed by others to demonstrate the exclusively mitochondrial localization of the enzyme. The recent review of this area by Walter (1976) makes interesting reading and graphically demonstrates the problems associated with attempting to determine the intracellular localization of even large molecules, whose intracellular permeability would normally be expected to be restricted.

In view of the above discussion, it should be apparent that the localization within the cell of small molecules such as citric-acid-cycle anions is a major problem. There have been two approaches used to determine levels of key intermediates in the cytosol and mitochondria of mammalian cells; one is indirect, involving the measurement of whole-cell levels of intermediates, followed by partitioning as judged by a set of calculation; the second involves the rapid lysis of isolated cells, followed by their separation via centrifugation through silicone oil. The following is a brief critique of both of these procedures.

1. Partitioning by Calculation

This method is based on the use of equilibrium enzymes to determine the concentration of free NAD^+ and NADH in both the cytosol and mitochondria. Since NADH is normally bound to intracellular protein, the use of enzymes which are normally in equilibrium with NAD^+ and NADH, and with a substrate couplet that can be measured directly, was first suggested by Holzer et al. (1956). The procedure was refined and broadly applied to a variety of biological systems by Williamson et al. (1967) and Krebs (1968). It is assumed that equilibrium enzymes such as lactate dehydrogenase (LDH) reflect the NAD^+/NADH ratio in the cytosol, and others such as β-hydroxybutyrate dehydrogenase or glutamate dehydrogenase reflect the mitochondrial redox state. If the equilibrium constant is known for the enzyme at intracellular pH (about 7.0), and the concentration of lactate and pyruvate or β-hydroxybutyrate and acetoacetate are measured directly, a calculation of the NAD^+/NADH ratio can be made using the following equations:

$$\text{lactate} + \text{NAD}^+ \longleftrightarrow \text{pyruvate} + \text{NADH} \quad (3)$$

$$\frac{[\text{NADH}][\text{pyruvate}]}{[\text{NAD}^+][\text{lactate}]} = K_{eq} \quad (4)$$

$$\frac{[\text{NAD}^+]}{[\text{NADH}]} = \frac{[\text{pyruvate}]}{[\text{lactate}]} \cdot \frac{1}{K_{eq}} \quad K_{eq}\text{ LDH} = 1.11 \times 10^{-4} \quad (5)$$

The values for lactate and pyruvate given in Table 2 can be used to calculate the NAD$^+$/NADH ratio:

$$\frac{\text{NAD}^+}{\text{NADH}} = \frac{0.13}{1.62} \times \frac{1}{1.1 \times 10^{-4}} \quad (6)$$

$$\frac{\text{NAD}^+}{\text{NADH}} = 725 \quad (7)$$

The assumption is also made that this calculated ratio represents the value for the cytosolic redox state, since lactate dehydrogenase is cytosolic. Similar calculations can be made for the mitochondrial NAD$^+$/NADH ratio using the glutamate dehydrogenase system. See Table 2 for an example of such calculations made on livers from fed and starved rats and guinea pigs.

In practice, the tissue being studied is freeze-clamped, using metal tongs which have been precooled in liquid N$_2$. It is important that the tissue be rapidly frozen without an interruption in blood supply; periods of anoxia of as little as 0.1 s can dramatically alter the NAD$^+$/NADH ratio. The use of specially designed tongs allows rapid freezing of the tissue in situ without disturbing the blood flow, and, with practice, the liver can be freeze-clamped in less than 10 s after cervical dislocation of a rat. The tissue is pressed between the frozen blocks of the

Table 2. The Effect of Dietary Changes on the Cytosolic and Mitochondrial Redox State of Rat, Guinea Pig, and Rabbit Livers[a,b]

Species	Dietary state	Mitochondria				Cytosol		
		Glutamate	α-Ketoglutarate	NH$_3$	NAD$^+$/NADH	Lactate	Pyruvate	NAD$^+$/NADH
Rat	Fed	2.41	0.145	0.47	7.3	1.62	0.130	725
	Starved (48 h)	2.64	0.086	0.56	4.7	0.78	0.047	528
Guinea Pig	Fed	3.57	0.31	0.37	8.3	1.38	0.045	293
	Starved (48 h)	2.76	0.19	0.63	28.3	0.34	0.013	344
Rabbit	Fed	2.85	0.170	0.29	4.0	1.68	0.09	482
	Starved (96 h)	2.11	0.200	0.70	17.4	0.79	0.03	342

[a] Values for the rat are from Williamson et al. (1967), for the guinea pig from Garber and Hanson (1971a) and for the rabbit Garber and Hanson (1971b). The equilibrium constants used to calculate the NAD$^+$/NADH ratios were 1.11×10^{-4} M for lactate dehydrogenase and 3.38×10^{-3} M for glutamate dehydrogenase (Williamson et al., 1967).
[b] All values are given in μmol/g liver.

tongs and placed immediately in liquid N_2. The frozen liver is then pulverized in a mortar to a fine powder, with the frequent addition of liquid nitrogen. From this powder the various intermediates can be extracted into $HClO_4$, which is then neutralized and the intermediates determined by established methods (Bergmeyer 1973). The paper by Williamson et al. (1967) carefully outlines the procedures to be used and is generally considered the fundamental reference for this approach.

A test for the usefulness of this procedure has been presented by Krebs (1968), employing as an example the inhibitory effect of ethanol on gluconeogenesis in rat liver. The infusion of ethanol into isolated rat liver results in a profound alteration in the $NAD^+/NADH$ ratio of the cytosol (and a marked decrease in hepatic gluconeogenesis). This results from the oxidation of ethanol in the liver to acetaldehyde by alcohol dehydrogenase and of acetaldehyde to acetate by aldehyde dehydrogenase. The result is the release of 70% of the acetate from the liver and a marked increase in hepatic NADH levels. The calculated $NAD^+/NADH$ ratios accurately reflect the marked reduction of the cytosol NAD^+ which accompanies alcohol oxidation in the liver. Krebs (1968) goes on to relate these changes to the regulation of gluconeogenesis in a paper well worth careful consideration for its skillful mixture of basic applications of biochemical techniques with relevant physiological questions.

This method can be extended to determine the distribution of certain metabolites within the cell. The major requirements for this approach are that the metabolite be one of an equilibrium reaction which involves adenine nucleotides and that it be present in the organelle in which the concentration of the metabolite is to be measured. As an example we will use the method of Williamson (1969), which requires that the total cellular concentration of malate and oxaloacetate be determined and that the redox state of both the mitochondria and cytosol be calculated. Three equations are then solved for each metabolite:

$$[OAA]_c = [malate]_c \times \frac{[pyruvate]}{[lactate]} \times \frac{K_{MDH}}{K_{LDH}} \quad (8)$$

$$[OAA]_m = [malate]_m \times \frac{[AcAc]}{[BOHB]} \times \frac{K_{MDH}}{K_{BOHD}} \quad (9)$$

$$[OAA]_{total} = [OAA]_c + [OAA]_m \quad (10)$$

This procedure depends on an accurate measurement of oxaloacetate, a compound notoriously labile and thus difficult to determine. Also, the final figures for the compartmentation of either malate or oxaloacetate will be only as valid as the underlying assumptions concerning the equilibrium status of the specific dehydrogenases, the accu-

Table 3. A Comparison of Two Methods for Determining the Concentration of Malate and Oxaloacetate in Rat Liver Cytosol and Mitochondria[a]

Method	Conditions	Oxaloacetate		Malate	
		Cytosol (μM)	Mitochondria (μM)	Cytosol (μM)	Mitochondria (μM)
Williamson et al. (1969)	Perfused rat liver metabolizing lactate and oleate	8.8	0.36	866	1952
Zuurendonk and Tager (1974)	Isolated rat liver cells metabolizing lactate, pyruvate and oleate	5	70	730	4600

[a]The concentration of malate and oxaloacetate were determined, either by calculation from equilibrium enzymes (Williamson, 1969) or by the direct separation procedure (Zuurendonk and Tager, 1974). For these calculations cytosolic water content was taken as 2.0 mg/g dry weight and the mitochondrial matrix water content as 0.21 ml/g dry weight of liver (Williamson, 1969).

racy of the calculated equilibrium constant, and the true partitioning of the reference couplets for the mitochondrial and cytosolic redox states. We do not know, for example, what portion of the β-hydroxybutyrate and acetoacetate, which are used to calculate the mitochondrial redox state, are in the cytosol and are therefore not in equilibrium with β-hydroxybutyrate dehydrogenase. There are other procedures which employ a variety of reactions to partition substrates using the above method. Of particular interest is the article by Greenbaum et al. (1971), which compares the available methods and draws many timely conclusions about the methods employed. Table 3 gives the calculated distribution of malate and oxaloacetate in the mitochondria and cytosol of livers from animals in various nutritional and hormonal states and deserves careful study. The low concentration of oxaloacetate in the mitochondria as compared with its concentration in the cytosol will have major significance in later sections of this chapter which discuss the regulation of gluconeogenesis.

2. Direct Measurement of Intermediates Following Rapid Cell Breakage and Separation of Subcellular Organelles

This method was developed by Zuurendonk and Tager (1974) and involves the treatment of isolated rat liver cells with low concentrations of digitonin to lyse the plasma membrane selectively. Since the liver cell membrane contains a relatively high concentration of cholesterol, it is selectively lysed, whereas the low cholesterol concentration of the mitochondrial membrane renders it relatively impervious to digitonin attack. The cells are lysed in a digitonin-containing buffer which has been placed on top of silicone oil layered on $HClO_4$. The total exposure of the isolated liver cells to digitonin is as short as 15 s, after which time the resultant lysate is centrifuged through silicone oil at $5000g$ for 50 s. After centrifugation of the digitonin-treated cells, two fractions are obtained. The top fraction, above the oil, contains cytosol and extracellular metabolites, while the pellet fraction contains cell particulate elements. It is assumed that the metabolites found in the latter fraction were derived from mitochondria and not from other cellular components which happen to sediment in the pellet.

This method is an ingenious example of an attempt to disrupt the cell in a gentle manner and to rapidly separate intracellular components and is finding increasing application for the determination of intracellular metabolite concentrations. Table 3 presents a comparison of the distribution of oxaloacetate and malate in the cytosol and mitochondria of rat liver as determined by the method of Zuurendonk and Tager (1974) and by the partitioning (by calculation) technique of Williamson (1969). There are marked differences in the concentration of oxaloacetate in the mitochondria calculated by the two methods, with the technique of Zuurendonk and Tager indicating 14 times more oxaloacetate in the mitochondria than in the cytosol as compared with the partitioning calculation procedure, which gives values of 25 times more oxaloacetate in the cytosol. Both methods agree fairly well on the cytosolic oxaloacetate concentration, however, as well as on both the concentration and intracellular distribution of malate.

It was our aim in this section of the chapter to present two of the methods presently used to determine the distribution of metabolites as examples of current approaches used in this area. It is important to be aware of the limitations of these approaches and to realize that a definitive method of localizing the concentration of small molecules within cells has yet to be devised. The data presently available, however, provide some insight into what is a recurring problem in metabolic regulation i.e., how to determine the concentration of a compound within a compartment of a functioning cell without so severely disrupting the cell that the information obtained is useless.

V. Examples of the Role of Compartmentation in the Regulation of Energy Metabolism

In the following sections we will deal with three specific metabolic processes—gluconeogenesis, fatty acid oxidation, and lipogenesis—which are regulated in part by the compartmentation of key enzyme systems.

A. Gluconeogenesis

The pathways for the oxidation of glucose to lactate and for gluconeogenesis are illustrated in Fig. 8. For the purpose of this discussion the metabolic region of interest is that segment of the gluconeogenic pathway which immediately precedes the formation of P-enolpyruvate.

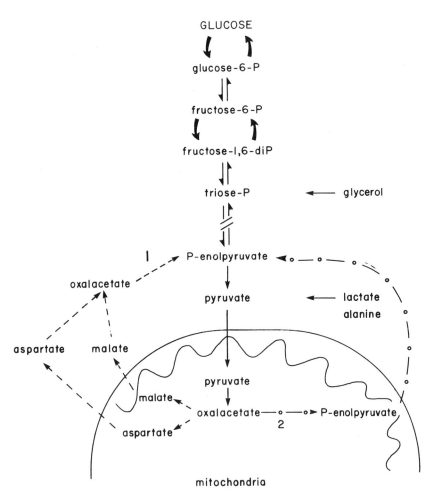

Figure 8. The pathway of gluconeogenesis in mammalian liver. This scheme shows the major reactions for gluconeogenesis and indicates the cytosol (1) and mitochondrial (2) forms of P-enolpyruvate carboxykinase. (From Tilghman, 1975).

Unlike glycolysis, which occurs strictly in the cell cytosol, gluconeogenesis involves a complex interaction between the mitochondrion and the cytosol. This interaction is necessitated by the irreversibility of the pyruvate kinase reaction, by the relative impermeability of the inner mitochondrial membrane to oxaloacetate, and by the specific mitochondrial location of pyruvate carboxylase. Compartmentation within the cell has led to the distribution of a number of enzymes (aspartate and alanine aminotransferases, and NAD^+-malate dehydrogenase) in *both* the mitochondria and the cytosol. In the classical situation represented by the rat, mouse, or hamster hepatocyte, the indirect "translocation" of oxaloacetate—the product of the pyruvate carboxylase reaction—into the cytosol is effected by the concerted action of these enzymes. Within the mitochondria oxaloacetate is converted either to malate or aspartate, or both. Following the exit of these metabolites from the mitochondria, oxaloacetate is regenerated by essentially similar reactions in the cytosol and is subsequently decarboxylated to P-enolpyruvate by P-enolpyruvate carboxykinase. Thus the presence of a membrane barrier to oxaloacetate leads to the functioning of the malate–aspartate shuttle as an important element in gluconeogenesis.

A major piece of evidence for the scheme was provided by Rognstad and Katz (1970), who showed that a specific inhibitor of aspartate aminotransferase—aminooxyacetic acid—diminished gluconeogenesis from lactate by more than 80% in rat kidney cortex slices but had no effect on pyruvate conversion to glucose. They also showed that D-malate, a competitive inhibitor of L-malate in the NAD^+-malate dehydrogenase reaction, inhibited gluconeogenesis from pyruvate but not from lactate, indicating that both malate and aspartate are transported out of the mitochondria during gluconeogenesis, with the metabolite transferred depending on the starting gluconeogenic substrate. These findings are also true for the perfused liver (Arinze et al., 1973). Scrutiny of the shuttle scheme shows that for maximum effectiveness, the exit of aspartate should be accompanied by a stoichiometric amount of α-ketoglutarate, which in turn should be accompanied by entry of glutamate into the mitochondria.

An important function of the scheme is that it provides a mechanism for supplying not only carbon, but also reducing equivalents in the cytosol during gluconeogenesis. Since synthesis of glucose from P-enolpyruvate requires NADH which is formed intramitochondrially, and since NADH does not readily pass through the inner mitochondrial membrane, the transfer of malate out of the mitochondria and its subsequent oxidation to oxaloacetate in the cytosol provides the necessary reducing equivalents as well as the carbon skeleton for gluconeogenesis.

On the basis of the intracellular distribution of P-enolpyruvate carboxykinase among species, a variation of the malate–aspartate cycle must be expected in species which have the potential to form P-enolpyruvate within the mitochondrial compartment. Stated another way, one may ask whether the capability for the mitochondrial generation of P-enolpyruvate, an obligatory intermediate in gluconeogenesis, renders the operation of the malate–aspartate cycle superfluous. This problem has been studied with the aid of specific inhibitors which allow a useful decision to be made of the functionality of the malate–aspartate cycle in systems having an inherent capacity for mitochondrial formation of P-enolpyruvate. In the perfused cat (Arinze and Hanson, 1973) or guinea pig liver (Arinze et al., 1973), for example, the intramitochondrial formation of P-enolpyruvate and its subsequent transfer to the cytosol occur in parallel with the operation of the malate–aspartate cycle during active gluconeogenesis from such substrates as lactate and pyruvate. The relative contribution to glucose carbon by either route, as demonstrated by these inhibitor studies, depends on the gluconeogenic precursor. With lactate as substrate, 40–50% of the new glucose is synthesized via direct P-enolpyruvate formation in the mitochondria (Arinze et al., 1973).

Figure 9 is a schematic representation of the mitochondrial–cytosolic interactions occurring during hepatic gluconeogenesis in such species as the guinea pig, cat, dog, rabbit, various birds, and man, all of which have both mitochondrial P-enolpyruvate carboxykinase and a cytosolic form of the enzyme. In these species, intramitochondrial P-enolpyruvate synthesis appears to be an integral part of gluconeogenesis, and the pattern of regulation of hepatic gluconeogenesis reflects to a large extent the compartmentation of P-enolpyruvate carboxykinase (Hanson, 1974; Söling and Kleineke, 1976). This concept is important, since the two forms of P-enolpyruvate carboxykinase are neither identical nor are they usually influenced by the same set of experimental or physiological factors.

The small molecules involved in these interactions—malate, aspartate, P-enolpyruvate, etc.—move across the mitochondrial membrane by way of the specific carrier systems described in Section IV. These carrier systems could become important sites for the regulation of gluconeogenesis, but at present, little is known of their control.

1. Relationship of Fatty Acid Oxidation to Gluconeogenesis

The oxidation of fatty acids occurs in the mitochondrial matrix and, in rat liver at least, fatty acid oxidation is thought to enhance gluconeogenesis by two mechanisms. It provides acetyl CoA—a known allo-

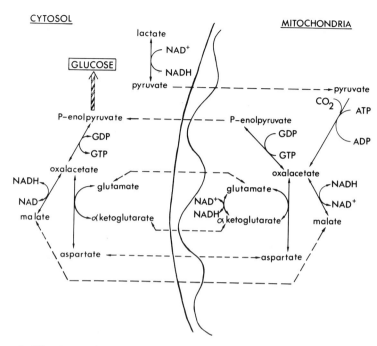

Figure 9. The interrelationship between the cytosol and mitochondria in the generation of intermediates for gluconeogenesis.

steric activator of pyruvate carboxylase in the mitochondria—and NADH, which if not oxidized through the electron transport chain is translocated to the cytosol via malate, where it is used in the subsequent reactions of gluconeogenesis. The energy generated in the form of ATP from β-oxidation may also be used to drive gluconeogenesis, a considerable portion of the ATP being transferred across the mitochondrial membrane to the cell cytosol where it is needed. Thus fatty acid oxidation, a strictly mitochondrial event, is related to gluconeogenesis, a compartmented process, in a stimulatory fashion.

In a variety of species, however, this relationship is not always parallel. Fatty acid oxidation results in an increased ratio of NADH/NAD$^+$ within the mitochondrial compartment, a phenomenon which has been shown to be inhibitory to glucose production in in vitro preparations of cat, guinea pig, and rabbit liver. Suppression of P-enolpyruvate synthesis occurs in tightly coupled guinea pig or human liver mitochondria oxidizing fatty acid (Jomain-Baum and Hanson, 1972). These experiments are interpreted to mean that the inhibition of gluconeogenesis by fatty acid oxidation in these species is related to the compartmentation of P-enolpyruvate carboxykinase and to the redox control of mitochondrial P-enolpyruvate synthesis. Thus, in these species, the overall pattern of regulation of the gluconeogenic process seems to be influenced by the compartmentation of a key enzyme.

Another potential relationship of fatty acid oxidation with gluconeogenesis is the role that long-chain fatty-acyl CoAs may play in the translocation of ATP across the mitochondrial membrane during gluconeogenesis. CoA derivatives of long- (but not short-) chain fatty acids inhibit adenine nucleotide translocation in a manner roughly similar to that demonstrated for actractyloside or bongkrekic acid (Shrago et al., 1976). Since P-enolpyruvate also appears to be transported by a carrier system(s) similar to that for ADP–ATP exchange, it has been argued that the fatty acid inhibition of gluconeogenesis may be localized at the site of P-enolpyruvate transport (Shrago et al., 1974). This argument implies that the species-selective effect of fatty acids may be unrelated to any changes in the oxidation–reduction state within the mitochondrial compartment. It fails, however, to account for the lack of effect of short-chain acyl CoA. Moreover, the experiments have not ruled out the detergent properties of long-chain fatty acids. Morel et al. (1974) have reviewed the evidence for the possible regulatory role of long-chain acyl CoA, and, although the physiologic importance of the effect in relation to gluconeogenesis is not established, it is an interesting phenomenon, implicating the separation of molecules across membranes, and needs further delineation.

2. Compartmentation-Related Interactions of Enzyme Modifiers

The widespread occurrence of mitochondria in eukaryotic cells indicates that segregation of some of the most important aspects of energy metabolism has been evolutionarily useful. Such segregation has been accompanied by modifications of regulatory interrelationships which depend upon the movement of metabolites into and out of mitochondria and upon the specific modification of regulatory enzymes by specific allosteric modifiers. For example, the activity of pyruvate carboxylase, an enzyme which fulfills an anaplerotic function in a wide variety of cells, depends upon allosteric activation by acetyl CoA. Since the inner mitochondrial membrane is virtually impermeable to coenzyme A and its esters, the activation of pyruvate carboxylase may be looked upon as a compartmentation phenomenon in which this acetyl CoA is separated from the cytosol acetyl CoA pool and must be formed intramitochondrially. In this case movement across the membrane is not involved, since both the activation and the formation of the activator occur within the same compartment. Although a distinct possibility, there is little evidence to suggest a major role for microcompartmentation within the matrix, and the efficiency of the activation process would therefore appear to depend, among other things, on the localized migration of the allosteric effector to the enzyme site(s), a parameter on which there is very little information. Other known modifiers of pyruvate carboxylase in vitro include acetoacetyl CoA and β-hydroxybutyryl CoA (Utter and Fung, 1971), both of which are intermediates in β-oxidation. The former is inhibitory whereas the latter is stimulatory. This subject has recently been reviewed by Barritt et al. (1976).

A different kind of enzyme–modifier interaction is that represented by the modulation of phosphofructokinase by citrate and ATP and by the regulation of pyruvate kinase by ATP. These interactions are different from the acetyl CoA modulation of pyruvate carboxylase in the sense that the modifiers are formed in a cell compartment totally different from that containing the enzyme to be regulated. Consequently their effect depends upon a preliminary transfer of an intermediate to the appropriate compartment. Citric acid is a well known negative modifier of phosphofructokinase, a cytosolic enzyme considered to catalyze a rate-limiting reaction in glycolysis. The tricarboxylate carrier and the adenine nucleotide translocase catalyze the transfer, from mitochondria to cytosol, of citrate and ATP, respectively. High levels of ATP inhibit not only phosphofructokinase but also pyruvate kinase, although the inhibition of the former enzyme is also modulated by ADP and AMP (Mansour, 1972). These modulations are important in the overall regulation of energy metabolism and gluconeogenesis, since any tendency for a decreased translocation of ATP should reduce gluconeogenesis. A reduction in energy level could increase the activity of pyruvate kinase or phosphofructokinase, or both, by relaxing the inhibition by ATP and citrate.

3. A Possible Role of Compartmentation in "Futile Cycles"

Three cycles frequently described as "futile" are presumed to occur during gluconeogenesis, at the pyruvate $\longleftrightarrow$ phosphoenolpyruvate, fructose-1,6-diphosphate $\longleftrightarrow$ fructose-6-phosphate, and glucose-6-phosphate $\longleftrightarrow$ glucose interconversions. Although the stoichiometry of the reactions at these steps suggests a wastage of energy, there have been recent suggestions that these "futile cycles" constitute an important regulatory mechanism for attenuating the rate of gluconeogenesis. In two of the three cycles, the role of compartmentation is minimal. However, the pyruvate–P-enolpyruvate interconversion involves the malate–aspartate cycle, and the rate of carrier-mediated transport of metabolites across the mitochondrial membrane could therefore be a crucial determinant of the overall rate of recycling. The possible control of gluconeogenesis by this recycling step must be closely linked to the regulation of the malate–aspartate cycle and to the translocation of phosphoenolpyruvate. The tracer experiments of Friedmann et al. (1971) and Williamson et al. (1971) suggest that recycling of lactate carbon through the P-enolpyruvate–pyruvate interconversion occurs at up to 30% of the rate of lactate conversion to glucose, although it has not been possible to undertake these studies under conditions where recycling is totally prevented.

4. Movement of Molecules which Provide Energy for Gluconeogenesis

The synthesis of 1 mol of glucose from 2 mol of alanine requires the expenditure of about 6 mol of ATP. One-third of this energy is spent within the mitochondria for the carboxylation of pyruvate, the other two-thirds are spent in the segment of the gluconeogenic pathway outside the mitochondria. Since the P-enolpyruvate carboxykinase reaction utilizes GTP rather than ATP, about 50% of the energy expended in the cytosol must be in the form of GTP. The energy for most biosynthetic processes is provided by citric-acid-cycle oxidations in the mitochondria, and the movement of molecules which provide energy for gluconeogenesis is therefore a particularly important function of the mitochondrial membrane.

As discussed in Section IV, Klingenberg et al. (1972, 1976) have elucidated the characteristics of adenine nucleotide translocase, which is associated with the inner mitochondrial membrane. The specific role of this enzyme is to transport ATP out of the mitochondria in exchange for ADP, while AMP is not transported. Two interesting questions arise from this work: the first concerns how adenine nucleotide translocation is regulated; the second relates to how GTP is moved out of the mitochondria. As demonstrated with the aid of uncouplers (Pfaff and Klingenberg, 1968), adenine nucleotide translocation is influenced by the intramitochondrial ratio of ATP/ADP. Undoubtedly there are other regulatory factors, but—other than the phosphorylation state—we know relatively little about the intracellular regulation of adenine nucleotide translocation, although the possible regulatory influence of long-chain acyl CoA has been suggested (see Section IV.B of this chapter). No carrier system has yet been discovered for the guanine nucleotides. However, within the mitochondrial matrix, GTP is formed by either substrate-level phosphorylation or by the transphosphorylation between di-and trinucleotides which is catalyzed by nucleoside diphosphate kinase. The data of Bücher and Scholz (1965) and of Garber and Ballard (1970) indicate that the intramitochondrial GTP pool rapidly equilibrates with the adenine nucleotide pool by virtue of the high activity of nucleoside diphosphate kinase. This [enzyme is located both inside and] outside the inner mitochondrial [membrane, and this trans-] phosphorylation, in conjunction with [the nucleotide translo-] case, regenerates in the cytosol the GTP [needed] for the P-enolpyru- vate carboxykinase reaction.

B. Lipid Biosynthesis

De novo fatty acid synthesis involves the carb[oxyla]tion of acetyl CoA to malonyl CoA and the subsequent addition of [acetyl] units by the fatty acid synthetase complex. All of these reactions [occur in the] cytosol. Although the initial enzymatic step is the carboxy[lation of ace-] tyl CoA, the substrate for lipogenesis in mammals is gluco[se. In the] tissues of normal animals fed carbohydrate, pyruvate generate[d from] glucose is converted in the mitochondria, preferentially to acetyl CoA. The diffusion of acetyl CoA out of the mitochondria is too slow to meet the demands of rapid lipogenesis. This fact led several years ago to a great deal of experimentation and debate as to how the acetyl CoA is supplied in the cytosol for lipogenesis. It is now clear that citrate efflux out of the mitochondria and the cleavage of citrate in the cytosol supply the necessary acetyl CoA (Greville, 1969a).

In the citrate cleavage pathway (Fig. 6), oxaloacetate formed as a result of pyruvate carboxylation is condensed with intramitochondrial acetyl CoA to yield citrate. The citrate is then transferred, via the tricarboxylic acid carrier, to the cytosol, where it is cleaved by the citrate cleavage enzyme to oxaloacetate and acetyl CoA. The latter then becomes accessible to the fatty acid synthetase. Oxaloacetate is subsequently reduced to malate by NAD-malate dehydrogenase, and pyruvate is regenerated from malate via malic enzyme. The reentry of pyruvate into the mitochondria is important for the regeneration of oxaloacetate in that compartment. It should be noted that fatty acid biosynthesis is intimately related to the citric acid cycle not only because the cycle provides citrate—which acts as a carrier for the acetyl moiety required for lipogenesis—but also because the cycle provides the energy needed to drive the overall process. Citrate also serves to regulate glycolysis by modulating the activity of phosphofructokinase and can further function as an activator of acetyl CoA carboxylase. As demonstrated by Gregolin et al. (1968), citrate and other tricarboxylic acids (notably isocitrate) promote aggregation of protomers of acetyl CoA carboxylase in vitro, the aggregation leading to increased activity.

The experimental evidence for the citrate cleavage pathway is based largely on studies of the intracellular compartmentation of various enzymes in the pathway, the reversibility of the reactions they catalyze, and the incorporation of certain specifically labeled interme-

as fatty acids and CO_2 in the liver and adipose tissue (Flatt et al., 1971). These studies emphasize the anaplerotic function of pyruvate carboxylase, i.e., to regenerate four-carbon intermediates used for biosynthesis. For example, lipogenesis from pyruvate in adipose tissue incubated in vitro is directly related to the concentration of pyruvate in the incubation medium (Schmidt and Katz, 1969; Patel et al., 1971), and the normally observed block in lipogenesis in adipose tissue from starved animals can be largely overcome in vitro by increasing the concentration of pyruvate.

1. Movement of Reducing Equivalents

The movement of reducing equivalents during fatty acid biosynthesis is as important as the pathway of carbon flow and also has been studied extensively. The NADPH required in the reductive steps in lipogenesis is generated mainly in the cytosol, up to 50–60% coming from the pentose pathway (Flatt and Ball, 1963; Katz et al., 1966) and the remainder from the malic enzyme reaction. This enzyme completes a transhydrogenation cycle in which the hydrogen of NADH formed by the malate dehydrogenase reaction is transferred to NADP (Fig. 6).

In contrast to the corresponding organs in the rat, bovine mammary gland and adipose tissue have a negligible activity of malic enzyme, while the activity of cytosolic NADP-linked isocitrate dehydrogenase is substantial (Hanson and Ballard, 1967, 1968). It is conceivable that in these tissues either isocitrate or citrate or both leave the mitochondria, where conversion to α-ketoglutarate results in the generation of NADPH. It is of interest that the ruminant adipose tissue lacks ATP citrate lyase (Hanson and Ballard, 1967), so that operation of the classical citrate cleavage pathway is either nonexistent or minimal. Although not studied in any detail, estimates of carbon flow during lipogenesis in ruminant mammary or adipose tissue are consistent with a role for cytosol isocitrate dehydrogenase as a major supplier of NADPH for this process (Ballard et al., 1969). This is an appealing suggestion because, in ruminants, glucose is not absorbed to any significant extent, so that the major lipogenic substrates for fatty acid synthesis are acetate and other short-chain fatty acids which are rapidly activated in the cytosol.

2. Compartmentation and Cholesterol Metabolism

In animal cells, 3-hydroxymethylglutaryl CoA (HMG-CoA) occurs at a metabolic branch-point from which either ketone bodies or cholesterol may be synthesized. This intermediate is formed from the condensation of acetoacetyl CoA and acetyl CoA in a reaction catalyzed by

HMG-CoA synthetase but can also be formed intramitochondrially as a degradation product of leucine. The synthesis of ketone bodies occurs in the mitochondria, whereas that of cholesterol occurs in the endoplasmic reticulum. For these processes at least two pools of HMG-CoA are necessary, because HMG-CoA, like other CoA derivatives, does not readily pass into or out of the mitochondria. The existence of two forms of HMG-CoA synthetase, one mitochondrial and the other cytosolic, ensures the formation of HMG-CoA in each compartment. Further metabolism of HMG-CoA depends upon the presence of functionally unrelated enzymes in each compartment. In the cytosol, a reductase converts HMG-CoA to mevalonate, a precursor of cholesterol, whereas in the mitochondria, HMG-CoA lyase cleaves this metabolite to acetoacetate and acetyl CoA. The pathways of ketogenesis and cholesterogenesis are therefore not only physically separated but are also independently regulated. For example, dietary cholesterol depresses cholesterogenesis, presumably by depressing the activity of HMG-CoA reductase (Bortz, 1973), but has no effect on HMG-CoA lyase, the mitochondrial enzyme which metabolizes HMG-CoA. On the other hand, conditions which promote ketogenesis—such as starvation—do not have a parallel effect on cholesterogenesis.

Another aspect of compartmentation in cholesterol metabolism is the intracellular localization of steroid hydroxylation reactions in the adrenal gland. This organ, whose major function is the elaboration of steroid hormones, contains a battery of mixed-function oxidases which attack the cholesterol side-chain with subsequent addition of oxygen to the ring. The enzymes which hydroxylate the 20 and 22 positions of cholesterol, leading to the eventual formation of pregnenolone, are mitochondrial, whereas the 21-hydroxylase is associated with the endoplasmic reticulum. Similarly, steroid 11-hydroxylase is mitochondrial, but steroid 17-hydroxylase occurs in the endoplasmic reticulum. Obviously, such differential localization of enzymes of apparently similar functions involves intricate interactions between the mitochondrion and the endoplasmic reticulum, and could account for the formation of a diversity of steroid hormones by this gland. As with most mixed function oxidations, NADPH, O_2 and cytochrome P-450 are involved. The latter is known to occur in the mitochondria, as well as the endoplasmic reticulum, of the adrenal cortex. Critical interactions between these compartments are bound to be essential elements in the regulation of steroidogenesis in this tissue.

C. Fatty Acid Oxidation

The oxidation of fatty acids in animal cells occurs in the mitochondria primarily by the process of β-oxidation, in which two-carbon units are consecutively removed from the carboxyl end of the molecule. The enzymes of β-oxidation are associated with the mitochondrial matrix. However, the fatty acids to be oxidized are first activated in the cytosol and are then transferred into the mitochondria by an elaborate transfer mechanism involving at least two acyltransferases (depending on the chain length of the fatty acid) and carnitine. Short-chain fatty acids such as hexanoic or octanoic acid readily pass through the inner mitochondrial membrane as free acids. Long-chain fatty acids, e.g., oleate and palmitate, do not. Within the matrix and in the presence of CoA, a GTP-dependnt fatty-acid thiokinase (Rossi and Gibson, 1964), which is specific for short-chain fatty acids, catalyzes the formation of short-chain fatty acyl CoA which then undergoes β-oxidation. On the other hand, the activation of long-chain fatty acids is ATP-dependent and occurs in the cytosol. The activated fatty acid subsequently passes through the outer membrane to the intermembrane space where it encounters a barrier (the inner mitochondrial membrane) to direct entry into the matrix.

Two carnitine acyltransferases, designated I and II, which are bound to the inner membrane have been described. Their functional orientation on the membrane is such that carnitine acyltransferase I catalyzes the formation of long-chain fatty acyl carnitine from (−)-carnitine and long-chain fatty acyl CoA on the outer membrane side of the inner membrane, and carnitine acyltransferase II catalyzes the same reaction but in the reverse direction on the matrix side of the membrane. For a variety of reasons the transferase step has been regarded as a control site in the oxidation of long-chain fatty acids. It is the first step specific to long-chain fatty acid oxidation, and is therefore subject to metabolic regulation. Second, the entry into the mitochondria of medium-chain fatty acids bypasses the acyltransferase, and medium-chain fatty acids are known to be more ketogenic than long-chain fatty acids.

McGarry et al. (1971) showed that, unlike oleate, octanoate was oxidized to ketone bodies at similar rates in livers from fasted and diabetic rats. Furthermore, studies with isolated mitochondria are consistent with a limitation of fatty acid oxidation at the transferase step (Fritz and Marquis, 1965). Carnitine acyltransferase I is probably as important as transferase II. The physiolgocial significance of the transferase step becomes more apparent when one considers that the majority of the free fatty acids circulating in the plasma during starvation are of the long-chain variety. Their oxidation to CO_2 or to ketone bodies within the mitochondria ultimately depends on passage of the acylcarnitine across the inner mitochondrial membrane.

In another respect the physical separation of the initial activation step from the subsequent oxidative steps in fatty acid oxidation enables the partitioning of acyl CoAs between the esterification pathway (cytosolic, endoplasmic reticulum) and the β-oxidation pathway. This partitioning is important in the overall regulation of disposal of free fatty acids.

VI. Conclusions

In this chapter we have presented a brief view of compartmentation using examples aimed at emphasizing the unique role of this process in controlling metabolite flux over various pathways. The examples we have chosen have seemed to us to most clearly illustrate a particular point; but there are other, equally good illustrations which might have been used. The goal of this book is to stress *principles* of metabolic regulation, not simply to present facts. This is an admirable goal but one which probably will never be completely attainable.

Compartmentation is not really a principle in itself, but rather reflects an integration of functional necessities with the structural elements of the cell. The spatial separation of reactants probably evolved as a functional necessity and through evolution was elevated to a real virtue. The compartmentation of mitochondrial processes is an excellent example of this point. The citric acid cycle, functioning as a cycle, is totally mitochondrial and is closely integrated with the major energy-generating process of oxidative phosphorylation. Some of the enzymes of the citric acid cycle are totally integrated, as are components of the inner mitochondrial membrane. This membrane is capable of maintaining a proton gradient, and can use this gradient to drive the phosphorylation of ADP. The positioning of citric-acid-cycle enzymes and the respiratory chain enzymes within the inner mitochondrial membrane is probably a functional necessity.

However, many of the enzymes of the citric acid cycle are found in the mitochondrial matrix and also have forms present in the cytosol. This is a departure of the strict functional coupling of two processes, the citric acid cycle and oxidative phosphorylation, which are involved in energy transduction. We are beginning to understand the importance of the separation of functionally identical enzymes in both the mitochondria and the cytosol as a key factor in the movement of small molecules across the mitochondrial membrane. In order for a tissue such as the liver to make glucose (from lactate or alanine) or fatty acids (from glucose), the citric acid cycle must be involved in the formation of key intermediates. Compartmentation thus allows regulation at an exceedingly sophisticated level. Exchange across the inner mitochondrial membrane occurs with a delicate balance which insures that the net charge on either side of the mitochondrial membrane is not altered. Thus, the very fact of compartmentation allows this control.

We have not discussed the mechanism of metabolite *movement* across membranes. How does a molecule actually move from one location to another in a manner that is sufficiently directed to insure rapid response of reactions on either side of the membrane? We know little about the real nature of carrier molecules, probably because their association with the membrane, of which they are an integral component, makes it difficult to purify and study them. We talk about *transporters* only because molecules are transported by intact mitochondria, not because we have isolated and characterized the transporting entities. As these transporters are purified and studied, we will know more about their specific regulatory properties.

Finally, we can safely predict that the number of compartmented processes which will be recognized and studied in the future will increase greatly. Molecules do not simply move freely in the cell; their movement is directed and tightly controlled. It is probable that no major metabolic pathways remain to be described. The challenge is to understand the regulation of the pathways that already have been elucidated. We have attempted herein to help focus attention on compartmentation as a key factor in this regulation.

ACKNOWLEDGMENTS. The authors wish to thank Dr. Mireille Jomain-Baum for critically reading this review and suggesting numerous changes which have greatly strengthened it. We also are endebted to Ms. Rhona Smith for her excellent assistance in all phases of the preparation of the manuscript; without her help this chapter would never have reached completion. Thanks are also owed to Drs. Efraim Racker and E. Jack Davis for granting us their permission to use figures from their previous work, and to Dr. Richard Palmiter for sharing unpublished studies.

References

Arias, I. M., Doyle, D., and Schimke, R. T., 1969, Studies on the synthesis and degradation of proteins of the endoplasmic reticulum of rat liver, *J. Biol. Chem.* **244**:534.

Arinze, I. J., and Hanson, R. W., 1973, Mitochondrial redox state and the regulation of gluconeogenesis in the isolated, perfused cat liver, *FEBS Lett.* **31**:280.

Arinze, I. J., Garber, A. J., and Hanson, R. W., 1973, The regulation of gluconeogenesis in mammalian liver. The role of mitochondrial P-enolpyruvate carboxykinase, *J. Biol. Chem.* **248**:2266.

Ballard, F. J., Hanson, R. W. and Kronfeld, D. S., 1969, Pathways of gluconeogenesis and lipogenesis in ruminant tissues, *Fed. Proc.* **28**:218.

Barrett, A. J., and Dingle, J. T., 1971, *Tissue Proteinases,* American Elsevier, New York.

Barritt, G. J., Zander, G. L. and Utter, M. F., 1976, The regulation of pyruvate carboxylase activity in gluconeogenic tissues, in: *Gluconeogenesis: Its Regulation in Mammalian Species* (R. W. Hanson and M. A. Mehlman, eds.), pp. 3–41, John Wiley and Sons, New York.

Bauman, D. E., 1976, Intermediary metabolism of adipose tissue, *Fed. Proc.* **35**:2308.

Bergmeyer, H. A., 1973, *Methods in Enzymatic Analysis*, Academic Press, New York.

Blobel, G., and Dobberstein, B., 1975a, Transfer of proteins across membranes. I. Presence of proteolytically processed and unprocessed nascent immunoglobulin light chains on membrane-bound ribosomes of murine myeloma, *J. Cell Biol.* **67**:835.

Blobel, G., and Dobberstein, B., 1975b, Transfer of proteins across membranes II. Reconstruction of functional rough microsomes from heterologous components, *J. Cell Biol.* **67**:852.

Bortz, W. M., 1973, On the control of cholesterol synthesis, *Metabolism* **22**:1507.

Bottger, I., Wieland, O., Bridiczka, D. and Pette, D., 1969, Intracellular localization of pyruvate carboxylase and P-enolpyruvate carboxykinase in rat liver, *Eur. J. Biochem.* **8**:113.

Boyer, P. D., Chance, B., Ernster, L., Mitchell, P., Racker, E., and Slater, E. C., 1977, Oxidative phosphorylation and photophosphorylation, *Annu. Rev. Biochem.* **46**:955.

Bücher, T., and Scholz, R., 1965, *The Control of Energy Metabolism* (B. Chance, R. Estabrook, and J. R. Williamson, eds.), p. 339, Academic Press, New York.

Chappell, J. B., 1968, Systems used for the transport of substrates into mitochondria, *Br. Med. Bull.* **24**:150.

Chappell, J. B., and Haarhoff, K. N., 1967, The penetration of the mitochondrial membrane by anions and cations, in: *Biochemistry of Mitochondria* (E. C. Slater, Z. Kaniuga, and L. Wojtczak, eds.), Academic Press, New York.

Davis, E. J., and Lumeng, L., 1975, Relationships between the phosphorylation potential generated by liver mitochondria and respiratory state under conditions of adenosine diphosphate control, *J. Biol Chem.* **250**:2275.

Denton, R. M., and Pogson, C. I., 1976, *Metabolic Regulation*, John Wiley and Sons, New York.

Flatt, J. P., 1970, Conversion of carbohydrate to fat in adipose tissue: An energy-yielding and, therefore, self-limiting process, *J. Lipid Res.* **11**:131.

Flatt, J. P., and Ball, E. G., 1963, Studies on the metabolism of adipose tissue. XV. An evaluation of the major pathways of glucose catabolism as influenced by insulin and epinephrine, *J. Biol. Chem.* **239**:675.

Freidmann, B., Goodman, E. H., Jr., Saunders, H., Kostos, V., and Weinhouse, S., 1971, An estimation of pyruvate recycling during gluconeogenesis in the perfused rat liver, *Arch. Biochem. Biophys.* **143**:566.

Fritz, I. B., and Marquis, N. R., 1965, The role of acylcarnitine esters and carnitine palmityltransferase in the transport of fatty acyl groups across the mitochondrial membrane, *Proc. Natl. Acad. Sci. USA* **54**:1226.

Fritz, P. J., Vessell, E. S., White, E. L., and Pruitt, K. M., 1969, The roles of synthesis and degradation in determining tissue concentrations of lactate dehydrogenase-5, *Proc. Natl. Acad. Sci. USA* **62**:558.

Garber, A. J., and Ballard, F. J., 1970, Regulation of P-enolpyruvate metabolism in mitochondria from guinea pig liver, *J. Biol. Chem.* **245**:2229.

Garber, A. J., and Hanson, R. W., 1971a, The interrelationship of the various pathways forming gluconeogenic precursors in guinea pig liver mitochondria, *J. Biol. Chem.* **246**:589.

Garber, A. J., and Hanson, R. W., 1971b, The control of P-enolpyruvate formation by rabbit liver mitochondria, *J. Biol. Chem.* **246**:5555.

Garber, A. J., and Salganicoff, L., 1973, Regulation of oxalacetate metabolism in liver mitochondria, *J. Biol. Chem.* **248**:1520.

Goldberg, A. L., and Dice, J. F., 1974, Intracellular protein degradation in mammalian and bacterial cells. Part 1, *Annu. Rev. Biochem.* **43**:835.

Goldberg, A. L., and St. John, A. C., 1976, Intracellular protein degradation in mammalian and bacterial cells. Part 2, *Annu. Rev. Biochem.* **45**:747.

Greenbaum, A. L., Gumaa, K. A., and McLean, P., 1971, The distribution of hepatic metabolites and the control of the pathways of carbohydrate metabolism in animals of different dietary and hormonal status, *Arch. Biochem. Biophys.* **143**:617.

Gregolin, C., Ryder, E., Warner, R. C., Kleinschmidt, A. K., Chang, H. C., and Lane, M. D., 1968, Liver acetyl CoA carboxylase. II. Further molecular characterization, *J. Biol. Chem.* **243**:4326.

Greville, G. D., 1969a, Intracellular compartmentation and the citric acid cycle, in: *Citric Acid Cycle, Control and Compartmentation* (J. M. Lowenstein, ed.), pp. 1–136, Marcel Dekker, New York.

Greville, G. D., 1969b, A scrutiny of Mitchell's chemiosmotic hypothesis of respiratory chain and photosynthetic phosphorylation, in: *Current Topics in Bioenergetics*, Vol. 3 (D. Sanadi, ed.), Academic Press, New York.

Hanson, R. W., 1974, The choice of animal species for studies of metabolic regulation, *Nutr. Rev.* **32**:1.

Hanson, R. W., and Ballard, F. J., 1967, The relative significance of acetate and glucose as precursors for lipid synthesis in liver and adipose tissue from ruminants, *Biochem. J.* **105**:529.

Hanson, R. W., and Ballard, F. J., 1968, The metabolic fate of the products of citrate cleavage, *Biochem. J.* **108**:705.

Hanson, R. W., and Garber, A. J., 1972, P-enolpyruvate carboxykinase I. Its role in gluconeogenesis, *Am. J. Clin. Nutr.* **25**:1010.

Haslam, J. M., and Krebs, H. A., 1968, Permeability of mitochondria to oxalacetate and malate, *Biochem. J.* **107**:659.

Heldt, H. W., and Pfaff, E., 1969, Adenine nucleotide translocation in mitochondria. Quantitative evaluation of the correlation between the phosphorylation of endogenous and exogenous ADP in mitochondria, *Eur. J. Biochem.* **10**:494.

Heldt, H. W., Klingenberg, M., and Milovancev, M., 1972, Differences between the ATP/ADP ratios in the mitochondrial matrix and in the extra-mitochondrial space, *Eur. J. Biochem.* **30**:434.

Holzer, H., Scholtz, G., and Lynen, F., 1956, Bestimmung des Quotienten DPNH/DPN in lebenden Hefezellen durch Analyse stationärer Alkohol und Acetaldehyd Konzentration, *Biochem. Z.* **328**:252.

Jomain-Baum, M., and Hanson, R. W., 1972, The effect of fatty acids on the synthesis of P-enolpyruvate by human liver mitochondria, *FEBS Lett.* **29**:145.

Jomain-Baum, M., and Hanson, R. W., 1975, The regulation of hepatic gluconeogenesis in the guinea pig by fatty acids and ammonia, *J. Biol. Chem.* **250**:8978.

Katunuma, N., Kominami, E., Kobayashi, K., Hamagushi, Y., Banno, Y., Chichibu, K., Katsunuma, T. and Shiotani, T., 1975, Initiating mechanisms of intracellular enzyme degradation and new special proteases in various organs, in: *Intracellular Protein Turnover* (R. T. Schimke and N. Katunuma, eds.), pp. 187–204, Academic Press, New York.

Katz, J. Landau, B. R., and Barsch, G. E., 1966, Pentose cycle, triose phosphate isomerization and lipogenesis in rat adipose tissue, *J. Biol. Chem.* **241**:727.

Klingenberg, M., 1972, Metabolite transport in mitochondria: An example of intracellular membrane function, *Essays Biochem.* **6**:119.

Klingenberg, M., 1976, The ADP–ATP carrier in mitochondrial membranes, in: *The Enzymes of Biological Membranes*, Vol. 3, *Membrane Transport* (A. Martonosi, ed.), pp. 383–438, Plenum Press, New York.

Klingenberg, M., and Pfaff, E., 1966, Structural and functional compartmentation in mitochondria, in: *Regulation of Metabolic Process in Mitochondria* (J. M. Tager, S. Papa, E. Quaghariello, and E. C. Slater, eds.), B. B. A. Library Vol 7, p. 180, Elsevier, Amsterdam.

Kominami, E., and Katunuma, N., 1976, Studies on new intracellular proteases in various organs of rats, *Eur. J. Biochem.* **62**:425.

Krebs, H. A., 1968, The effects of ethanol on the metabolic activities of the liver, in: *Advances in Enzyme Regulation* Vol. 6 (G. Weber, ed.), pp.467–480, Pergamon Press, New York.

La Noue, K., Bryla, J. and Bassett, D. J. P., 1974, Energy driven aspartate efflux from heart and liver mitochondria, *J. Biol. Chem.* **249**:7514.

McGarry, J. D. and Foster, D. W., 1971, The regulation of ketogenesis from octanoic acid. The role of the tricarboxylic acid cycle and fatty acid synthesis, *J. Biol. Chem.* **246**:1146.

Mansour, T. E., 1972, Phosphofructokinase, in: *Current Topics in Cellular Regulation* Vol. 5 (B. L. Horecker and E. R. Stadtman, eds.), pp. 1–46, Academic Press, New York.

Marco, R., Sebasitian, J., and Sols, A., 1969, Localization of the enzymes of the oxalacetate metabolic cross-roads in rat liver mitochondria, *Biochem. Biophys. Res. Commun.* **34**:725.

Mitchell, P., 1961, Coupling of phosphorylation to electron and hydrogen transfer by a chemi-osmotic type of mechanism, *Nature* **191**:144.

Mitchell, P., 1966, *Chemiosmotic Coupling in Oxidative and Photosynthetic Phosphorylation*, Glynn Research, Bodmin, Cornwell, England.

Mitchell, P., 1967, Proton-translocation phosphorylation in mitochondria, chloroplasts and bacteria: Natural fuel cells and solar cells, *Fed. Proc. Fed. Am. Soc. Exp. Biol.* **26**:1370.

Mitchell, P., 1968, *Chemiosmotic Coupling and Energy Transduction*, Glynn Research, Bodmin, Cornwell, England.

Mitchell, P., and Moyle, J., 1967, Proton-transport phosphorylation: Some experimental tests, in: *Biochemistry of Mitochondria* (E. C. Slater, Z. Kaniuga, and L. Wojtczak, eds.), pp. 53–74, Academic Press, New York.

Morel, F., Lauguin, G., Lunardi, J., Duszynski, J., and Vignais, P. V., 1974, An appraisal of the functional significance of the inhibitory effect of long-chain acyl CoAs on mitochondrial transport, *FEBS Lett.* **39**:133.

Neurath, H., and Walsh, K., 1976, Role of proteolytic enzymes in biological regulation (a review), *Proc. Natl. Acad. Sci. U.S.A.* **73**:3825.

Patel, M. S., Jomain-Baum, M., Ballard, F. J., and Hanson, R. W., 1971, Pathways of carbon flow during fatty acid synthesis from lactate and pyruvate in rat adipose tissue, *J. Lipid Res.* **12**:179.

Pfaff, E., and Klingenberg, M., 1968, Adenine nucleotide translocation of mitochondria I. Specificity and control, *Eur. J. Biochem.* **6**:66.

Pogson, C. I., and Wolfe, R. G., 1972, Oxaloacetic acid tautomeric and hydrated forms in solution, *Biochem. Biophys. Res. Commun.* **46**:1048.

Racker, E., 1970, The two faces of the inner mitochondrial membrane, *Essays Biochem.* **6**:1.

Ray, P., 1976, Hepatic gluconeogenesis in the rabbit, in: *Gluconeogenesis: Its Regulation in Mammalian Species* (R. W. Hanson and M. A. Mehlman, eds.), pp. 293–335, John Wiley and Sons, New York.

Reed, L., 1976, Regulation of mammalian pyruvate dehydrogenase complex by phosphorylation and dephosphorylation, in: *Thiamine* (C. J. Gubler, M. Fujiwara and P. M. Dreyfus, eds.), pp. 19–27, John Wiley and Sons, New York.

Robinson, B. H. and Chappell, J. B., 1967, The inhibition of malate, tricarboxylate and oxoglutarate entry into mitochondria by α-n-butylmalonate, *Biochem. Biophys. Res. Commun.* **28**:249.

Rognstad, R., and Katz, J., 1970, Gluconeogenesis in the kidney cortex, Effects of D-malate and aminoxyacetic acid, *Biochem. J.* **116:**403.

Rossi, C. R., and Gibson, D. M., 1964, Activation of fatty acids by a guanosine triphosphate thiokinase from liver mitochondria, *J. Biol. Chem.* **239:**1694.

Russell, D., and Snyder, S. H., 1968, Amine synthesis in rapidly growing tissues: Ornithine decarboxylase activity in regenerating rat liver, chick embryo and various tissues, *Proc. Natl. Acad. Sci. U.S.A.* **62:**558.

Schmidt, K., and Katz, J., 1969, Metabolism of pyruvate and L-lactate in rat adipose tissue, *J. Biol. Chem.* **244:**2125.

Shrago, E., Shug, A., Elson, C., Spennetta, T., and Crosby, C., 1974, Regulation of metabolite transport in rat and guinea pig liver mitochondria by long-chain fatty acyl coenzyme A esters, *J. Biol. Chem.* **249:**5269.

Shrago, E., Shug, A., and Elson, C., 1976, Regulation of cell metabolism by mitochondrial transport systems, in: *Gluconeogenesis, Its Regulation in Mammalian Species* (R. W. Hanson and M. A. Mehlman, eds.), pp. 221-239, John Wiley and Sons, New York.

Slater, E. C., 1967, An evaluation of the Mitchell hypothesis of chemiosmotic coupling in oxidative and photosynthetic coupling, *Eur. J. Biochem.* **1:**317.

Slater, E. C., Rosing, J., and Mol, A., 1973, The phosphorylation potential generated by respiring mitochondria. *Biochim. Biophys. Acta* **292:**534.

Söling, H. D., and Kleineke, J., 1976, Species dependent regulation of hepatic gluconeogenesis in higher animals, in *Gluconeogenesis: Its Regulation in Mammalian Species* (R. W. Hanson and M. A. Mehlman, eds.), pp. 369-463, John Wiley and Sons, New York.

Tilghman, S. M., 1975, Hormonal regulation of P-enolpyruvate carboxykinase synthesis, Ph.D. Thesis, Temple University School of Medicine, Philadelphia.

Utter, M. F., and Fung, C. H., 1971, Possible control mechanisms of liver pyruvate carboxylase, in: *Regulation of Gluconeogenesis* (H. D. Söling and B. Willms, eds.), pp. 1–10, Academic Press, New York.

Walter, P., 1976, Pyruvate carboxylase: Intracellular localization and regulation, in: *Gluconeogenesis: Its Regulation in Mammalian Species* (R. W. Hanson and M. A. Mehlman, eds.), pp. 239-269, John Wiley and Sons, New York.

Williamson, D. H., Lund, P. and Krebs, H. A., 1967, The redox state of free nicotinamide adenine dinucleotide in the cytoplasm and mitochondria of rat liver, *Biochem J.* **103:**514.

Williamson, J. R., 1969, Transport of reducing equivalents across the mitochondrial membrane in rat liver, in: *The Energy Level and Metabolic Control in Mitochondria* (S. Papa, J. M. Tager, E. Quagliariello, and E. C. Slater, eds.), pp. 385–397, Adriatica Editrice, Bari.

Williamson, J. R., 1976, Role of anion transport in the regulation of metabolism, in: *Gluconeogenesis, Its Regulation in Mammalian Species* (R. W. Hanson and M. A. Mehlman, eds.), pp. 165–220, John Wiley and Sons, New York.

Williamson, J. R., Jakob, A., and Scholz, R., 1971, Energy cost of gluconeogenesis in rat liver, *Metabolism* **20:**13.

Zuurendonk, P. F., and Tager, J. M., 1974, A rapid separation of particulate components and soluble cytoplasm of isolated rat-liver cells, *Biochim. Biophys. Acta* **333:**393.

Zuurendonk, P. F., Akerboom, T. P. M., and Tager, J. M., 1976, Metabolite distribution in isolated rat-liver cells and equilibrium relationships of mitochondrial and cytosolic dehydrogenases, in: *Use of Isolated Liver Cells and Kidney Tubules in Metabolic Studies* (J. M. Tager, H. D. Söling, and J.R. Williamson, eds.), pp.17–27, North Holland Publishing Co., Amsterdam.

14

The Mechanism of Action of Hormones

Robert H. Herman and O. David Taunton

I. Introduction

The ability to receive and process information for initiating metabolic processes and coordinating the activities of the specialized cells comprising an organ is essential to the orchestrated activities of multicellular organisms. Informational molecules can bind at stereospecific receptor sites either on the membrane or within the cell and can start a chain of events eliciting a biochemical response. The endocrine system, with its myriad hormones, provides a level of control of metabolic processes that permits the coordinate responses of many cells, enabling the resultant biochemical response to be expressed at a physiologic level. In this chapter, we will focus on the biochemical activities of hormones.

The term "hormone" was coined by W. B. Hardy and was first used by Starling in 1905 (Sawin, 1969), although secretin is recognized as being the first hormone described—in 1902—by Bayliss and Starling. The existence of gastrin, discovered in 1905 by Edkins, required more than 50 years to be proved definitively (Gregory and Tracy, 1964; Gregory et al., 1964). Since then a great number of hormones in almost all body tissues have been discovered. This chapter will not treat all the actions of all the mammalian hormones but will rather attempt to deduce general principles of hormone action from what is currently known about the different classes of hormones.

Robert H. Herman • Endocrine Metabolic Service, Letterman Army Medical Center, Presidio of San Francisco, California 94129.   **O. David Taunton** • Department of Medicine, University of Alabama School of Medicine, Birmingham, Alabama 35205.

Since hormones are substances elaborated by the cells of one tissue (endocrine tissue) to act on and control the function of cells of another tissue (target tissue) which are distant from the endocrine tissue, it seems necessary to distinguish them from those substances now classified as paracrines, chalones, and neurotransmitters. These latter compounds are hormone-like substances which are also elaborated by one tissue and which also act on different adjacent tissues. The principal difference between hormones and paracrines, chalones, and neurotransmitters is that the former have a circulating form while the latter do not. However, in many cases, neurotransmitter substances are found in nonnervous tissues where they appear to act as hormones, and thus the distinction between hormones and the other classes of signal molecules is not always sharp.

The vast literature on hormones can only be touched upon to a limited degree. In this chapter, we have attempted to extract, from the vast knowledge of endocrine action, general principles that appear to govern endocrine function. While certain specific examples will be used to illustrate these principles, it is not possible to cover the endocrine literature in any degree of completeness. The mechanism of action of hormones and other extracellular molecules involves (at least) considerations of protein synthesis (Chapter 5), secretory mechanisms (Chapter 12), membrane structure (Chapter 11), metabolic networks (Chapter 1), servomechanisms (Chapter 8), and compartmentation (Chapter 13). We will cover only certain aspects of these subjects, or omit their consideration entirely.

A. Definitions

Extracellular signal molecules are those substances produced by the cells of one tissue which act on the cells of a different tissue. In order to act, the signal molecule must pass out from the cell of origin into an extracellular phase and must then bind to the surface of the target tissue or pass into the interior of the target cell.

1. Hormones

Hormones are extracellular signal molecules that are produced and secreted by the cells (endocrine cells) of one tissue (endocrine tissue), that circulate in the vascular system either freely, bound to a plasma protein, or both, and that act on the cells (target cells) of a distant tissue (target tissue) by binding to the cell surface or by entering into the interior of the cell.

2. Paracrines

Paracrines are extracellular signal molecules that are produced and secreted by the cells (paracrine cells) of one tissue (paracrine tissue) and act on adjacent cells (target cells) of adjacent tissue (target tissue) by binding to the cell surface or by entering into the interior of the cell. Paracrines do not enter into the circulation but only traverse the extracellular space. In some cases an extracellular signal molecule may act as a hormone in one site and as a paracrine in another site. Although it is possible that paracrines may pass directly from one cell to another by traversing membranes of adjacent cells, there is little evidence to support this possibility. Some examples of paracrine hormones include somatostatin, histamine, and certain of the prostaglandins. The problems related to paracrines will not be included in this chapter.

3. Chalones

Chalones are extracellular signal molecules that are produced and secreted by differentiated tissue and that act on adjacent cells (target cells) of adjacent tissue (target tissue) which are mitotically active. The chalones act to inhibit mitosis (see Chapter 7). Chalones are a special category of paracrine hormones and are thus not found in a circulating form. There is controversy concerning the nature and existence of chalones.

4. Neurotransmitters

Neurotransmitters are extracellular signal molecules that are produced and secreted from the synaptosome portion of neuronal axons, traverse a synaptic gap, and act on another neuron, muscle cell, or glandular cell by binding to the surface or by entering the interior of the cell. In some cases, neurotransmitters are found in nonnervous tissue, where they appear to act as hormones. Notwithstanding their similarities to hormones, we shall not discuss neurotransmitters except to draw certain analogies to hormones.

B. Classification of Hormones

It is convenient to classify hormones into circulating and noncirculating hormones. Paracrines and chalones belong in the latter category. The circulating hormones can be divided into two categories: (1) polypeptide hormones; and (2) nonpolypeptide hormones. Each of

Table 1. Classification of Hormones

A. Polypeptide hormones
  1. Membrane receptor binding
     a. Adenylate cyclase–cAMP mechanism
     b. Non-adenylate-cyclase–cAMP mechanism
B. Nonpolypeptide hormones
  1. Membrane receptor binding
     a. Adenylate cyclase–cAMP mechanism
     b. Non-adenylate-cyclase–cAMP mechanism
  2. Intracellular receptor binding protein
     a. Hormone–receptor action in cell nucleus
     b. Extranuclear action of hormone–receptor complex

these hormone categories can be further divided into (1) those hormones that act by binding to the cell surface; and (2) those hormones that act by entering into the interior of the cell. A large number of hormones bind to the surfaces of target cells, thereby activating adenyl cyclase, which generates cyclic-3',5'-adenosine monophosphate (cAMP), which in turn acts as the intracellular signal molecule (the so-called "second messenger"; the hormone being the "first messenger"). However, not all hormones which bind to the cell surface act in this way. Many of the nonpeptide hormones enter into the cell interior, bind to a receptor protein, and finally enter into the cell nucleus. However, the mechanism of action of many hormones is still unknown.

Many substances which appear on first investigation to be hormones have not been definitely proved to have hormonal function. Nevertheless, such substances may tentatively be considered to be hormones, pending further information. Table 1 presents a convenient classification of hormones.

C. The Criteria for Establishing a Substance as a Hormone

In order to determine if a substance is a hormone, it must be shown to be contained in the presumed endocrine tissue, to be present in a circulating form, and to have a biological action; and administration of the purified substance must demonstrate that it exerts its action in doses that produce physiological, circulating levels of the substance. Furthermore, removal or loss of the presumed endocrine tissue should result in the absence of the circulating form of the substance, with loss of the biological action. Administration of the material purified from the tissue, or of synthetic material, should restore circulating levels and biological action. Although doses greater than those that produce physiological circulating levels of the given substance may produce certain biological actions, these may not be physiological actions. Additional evidence can be gained by showing that suitable antibodies to the presumed hormone block biological action. The occurrence of disease related to deficiency of a hormone, and its treatment by administration of the hormone, is final proof of the hormone's biological importance. The occurrence of disease related to excess production of the hormone, and treatment of the disease by removal of the hormone-secreting tissue, provides further evidence of the function of the hormone.

The criteria used to determine the hormonal nature of substances isolated from the gastrointestinal tract have been as follows: (1) A stimulus applied to a part of the gastrointestinal tract produces a response in a distant target tissue. (2) The effect persists after cutting all nerves connecting the site of stimulation and the target. (3) The effect is produced by an extract of the part to which the stimulus was applied, but not by extracts from other tissues. (4) The effect is produced by infusing exogenous hormone in amounts and molecular forms that reflect the increase in blood concentration seen after the stimulus for endogenous release (Grossman, 1977). These criteria apply to the physiological action which led to the identification of gastrin; in order to consider other biological effects of the hormone as physiologic, however, other criteria must be satisfied. (5) The other biological effects must occur during endogenous release of the hormone. (6) The other biological effects must occur during infusion of the exogenous hormone. (7) The D_{50} for the additional action must not be greater than the D_{50} (dose required to produce half-maximal response) of the original biological action. In the case of gastrointestinal hormones, the criteria are satisfied for certain actions of gastrin, secretin, cholecystokinin, and gastric inhibitory peptide. However, the physiological roles for vasoactive intestinal polypeptide, motilin, and chymodenin have not been established (Grossman, 1977).

D. The Need for Hormones and Other Signal Molecules

Hormonal systems function as regulatory mechanisms. One tissue is able to regulate the function of a distant tissue by elaborating a circulating hormonal substance. In response to extracellular environmental changes, the endocrine tissue can elaborate hormones. The target tissue may in turn elaborate another hormone which regulates the first hormone, as well as carrying out its own biological action on a second target tissue. Hormonal systems provide mechanisms for time-ordered responses to environmental changes that cannot be mediated by neural mechanisms. Nerve function is quite rapid (milliseconds), whereas hormones function in time ranges of minutes to hours, or even over periods of days to months. These different time frames of biological action probably have been important for the development of each of these systems.

II. Hormones

A. The Basic Principles of Hormone Function

1. The Universality of Hormones

a. Every Tissue Produces at Least One Hormone. At one time it was thought that endocrine glands were a special set of tissues that elaborated hormones while the remainder of the body tissues served only as target tissues. But, as an ever increasing number of hormonal substances have been discovered, it has become evident that hormones are elaborated by many tissues not previously considered to be endocrine in nature. Thus, it now appears that almost all tissues produce at least one hormone. Although there appear to be some exceptions, production of at least one hormone by a tissue is the rule rather than the exception. Table 2A illustrates the role of various organs and tissues in hormone production.

b. Every Tissue is the Target of at Least One Hormone. Just as most tissues and organs produce at least one hormone, it has become apparent that almost all organs and tissues serve as the targets for at least one hormone. This is shown in Table 2B.

Table 2. Some Examples of Production of Hormones by Organs and Tissues and the Target Tissues of Certain Hormones

A. Production of hormones by tissues and organs	
Tissue or organ	Hormone produced
1. Hypothalamus	Oxytocin, vasopressin, thyroid releasing hormone, LH-RH, somatostatin
2. Anterior pituitary	Growth hormone, melanocyte stimulating hormone, ACTH, FSH, LH, prolactin, TSH
3. Pineal	Melatonin
4. Lung	Prostacyclin
5. Thyroid	Thyroxine, triiodothyronine
6. Parafollicular cells	Calcitonin
7. Parathyroid glands	Parathormone
8. Adrenal cortex	Aldosterone, cortisol, corticosterone
9. Adrenal medulla	Epinephrine

(continued)

Table 2 *(continued)*

A. Production of hormones by tissues and organs

Tissue or organ	Hormone produced
10. Testes	Testosterone
11. Ovaries	Estrogen, progesterone, relaxin
12. Stomach (antrum)	Gastrin
13. Small intestine	Secretin, cholecystokinin, serotonin, enteroglucagon, gastric inhibitory peptide, motilin
14. Pancreas	Insulin, glucagon, somatostatin, pancreatic polypeptide
15. Liver	Somatomedin, 25-hydroxycholecalciferol, renin substrate
16. Salivary glands	Nerve growth factor, epidermal growth factor
17. Kidney	Renin, erythropoietin, 1,25-dihydroxycholecalciferol
18. Skin	Vitamin D_3
19. Placenta	Chorionic gonadotropin, placental lactogen

B. Target organs and tissues of some hormones

Hormone	Target tissue or organ
1. TRH, LH-RH, somatostatin	Anterior pituitary
2. ACTH	Adrenal cortex
3. TSH	Thyroid
4. Melanocyte stimulating hormone	Melanocytes in skin
5. FSH, LH	Gonads
6. Prolactin	Mammary gland
7. Vasopressin	Kidney
8. Oxytocin	Uterus, mammary gland
9. Thyroid hormone	Liver, kidney, skeletal muscle, heart, skin
10. Prostacyclin	Platelets
11. Cortisol	Liver, bone, skeletal muscle, brain
12. Estrogen, progesterone	Uterus, liver, skeletal muscle, adipose tissue, genitalia, mammary glands
13. Testosterone	Genitalia, prostate, skeletal muscle, bone, hair follicles
14. Relaxin	Pelvic ligaments
15. Nerve growth factor	Neurons in ganglia
16. Erythropoietin	Bone marrow
17. Angiotensin II	Kidney, vascular smooth muscle
18. Aldosterone	Kidney
19. Growth hormone	Skeletal muscle, bone, liver, kidneys
20. 1,25-Dihydroxycholecalciferol	Kidney, small intestine
21. Insulin	Liver, skeletal muscle, adipose tissue, renal medulla

2. The Opposing Action of Hormones

a. The Inverse or Reciprocal Action of Hormones. Many hormones appear to have antagonistic biological actions. Not all hormones function in opposition to some other hormone. Table 3A demonstrates the opposing or reciprocal action of various hormone *pairs*. Table 3B demonstrates that different hormone *groups* may have opposing actions. Contrary to what is seen with hormone pairs, the hormone groups have a variety of physiologic actions, so that not all functions of each hormone in the group are affected by the other members of the group. Thus, although insulin and glucagon have reciprocal actions with regard to each other, insulin and cortisol, epinephrine, or growth hormone antagonize one another only with regard to certain functions. Therefore, there are hormones which share certain but not all functions, and which antagonize one another only with regard to the shared functions. Opposing hormone pairs appear to function via inverse actions on the same general hormone mechanism. Each hormone of the opposing hormone groups may act via different mechanisms, although the biological function that is mediated may be the same.

b. Hormone Imbalance and Endocrine Disease. Because of the opposing action of the various hormones, a well-regulated physiological state is maintained. If one hormone is increased or decreased beyond the physiological range, its functions are exaggerated or diminished, respectively. In these circumstances the action of a particular hormone is insufficiently opposed or excessively opposed. In either case, the result is an endocrine disorder where the clinical manifestations are the result of hypo- or hyperfunction. For example, the absence of insulin, as seen in juvenile diabetes mellitus, results in inadequate control of glycogenesis, glycogenolysis, glycolysis, gluconeogenesis, and poor control of blood glucose levels outside of the normal physiological range. The opposing action of the other hormones which regulate blood glucose becomes overwhelming. Similarly, excess production of insulin by a tumor of the pancreatic islet cells leads to inadequate control of glycogenesis, glycogenolysis, glycolysis, gluconeogenesis, and, hence, blood glucose levels. In this case, however, blood glucose levels are excessively low, with resulting impaired function of the central nervous system. The hormonal mechanisms which act to increase blood glucose levels are unable to compensate for the increased insulin production. Hormone imbalance results in endocrine disease. In the case of hormone deficiency, hormonal replacement is the treatment of choice. In the case of excess hormone production, ablation of the hormone-producing tissue is necessary.

Table 3. The Opposing of Reciprocal Action of Certain Hormone Pairs and Groups

A. Hormone pair	Reciprocal function
1. Insulin	Decreases blood glucose, enhances glycogenesis, enhances glycolysis, decreases glycogenolysis and gluconeogenesis
Glucagon	Increases blood glucose, decreases glycogenesis and glycolysis, increases glycogenolysis and gluconeogenesis
2. Parathormone	Increases blood calcium
Calcitonin	Decreases blood calcium
3. Prostacyclin	Decreases platelet aggregation, vasodilatation
Thromboxane A_2	Increases platelet aggregation, vasoconstriction

B. Hormone group	Reciprocal function
1. Insulin	Decreases blood glucose
Glucagon, Epinephrine, Growth Hormone	Increase blood glucose
2. Prostacyclin, Histamine	Vasodilatation
Thromboxane A_2, Angiotensin II, Norepinephrine	Vasoconstriction
3. Insulin, Cortisol	Increase glycogenesis, decrease glycogenolysis
Glucagon, Epinephrine	Increase glycogenolysis, decrease glycogenesis
4. Cortisol, Glucagon	Increase gluconeogenesis, decrease glycolysis
Insulin	Increases glycolysis, decreases gluconeogenesis

3. Hormone Biosynthesis

a. Polypeptide Hormones. Polypeptide hormones are synthesized through the general mechanisms of protein biosynthesis (see Chapter 5). As far as is known, the biosynthetic mechanism involves synthesis of messenger ribonucleic acid (mRNA), movement of the mRNA from the nucleus into the cytoplasm, translation of the mRNA on the ribosome, which involves the assembly of the aminoacyl transfer ribonucleic acids (AA-tRNA) in the proper sequence according to the arrangement of codons on the mRNA, and resulting synthesis of the polypeptide hormone. The synthesis of the mRNA is preceded by the synthesis of heterogeneous RNA (hnRNA), from which mRNA appears to be derived.

In the case of insulin, the gene serves as a template for the synthesis of mRNA for preproinsulin. The mRNA consists of a preregion (leader sequence), the segment that codes for the B chain, the segment that codes for the C peptide, and the segment that codes for the A chain. The fully formed insulin (actually preproinsulin) mRNA also contains poly(A) at its 3'-terminus (Chan et al., 1976; Lomedico and Saunders, 1976; Permutt et al., 1976).Three RNA fractions, with molecular weights of 200,000, 280,000, and 360,000, respectively, have been identified as the major polyadenylated transcription products derived from a transplantable rat pancreatiic islet-cell tumor that synthesizes insulin. The lowest-molecular-weight component appears to be the mRNA that directs the synthesis of one of the rat insulins. The nature of the higher molecular weight components is not certain, but they share common structural features with the lowest-molecular-weight component (Duguid and Steiner, 1978). Translation of the mRNA results in the biosynthesis of preproinsulin. The preregion is thought to aid in the binding of the polyribosomes to the membranes of the endoplasmic reticulum, and to enable proinsulin to be packaged into its secretory granules (Steiner, 1977).

In some cases the resulting polypeptide contains more than one hormone. The polypeptide must be cleaved at specific points to form the circulating form of each hormone. For example, in the pituitary gland, there is formed a polypeptide, now called "pro-opiocortin," containing adrenocorticotropin (ACTH) and lipotropin (LTP). This polypeptide must be hydrolyzed to produce each hormone in the free state (Mains and Eipper, 1978).

b. Nonpolypeptide Hormones. Nonpolypeptide hormones are synthesized via steady-state, enzymatically-catalyzed chemical reactions (see Chapter 1). The enzymatic reactions occur in sequence, starting with an initial substance and ending with the nonpolypeptide hormone. Thus, tyrosine is the precursor for epinephrine and norepinephrine in the adrenal medulla. The sequence of events is tyrosine → 3,4-dihydroxyphenylalanine (dopa) → 3,4-dihydroxyphenethylamine (dopamine) → norepinephrine → epinephrine (Moskowitz and Wurtman, 1975). Similarly, enzymatic pathways exist for the various nonpolypeptide hormones (thyroxine, triiodothyronine, 1,25-dihydroxyvitamin D_3, serotonin, histamine, estrogen, progesterone, testosterone, aldosterone, cortisol, prostaglandins, melatonin). In a sense, the nonpolypeptide hormones are defined by the genes which encode the enzymes controlling the production of each hormone.

4. Prepro- and Prohormones

a. Posttranslational Modifications of Polypeptide Hormones. In many instances hormones are not synthesized directly; instead, larger-molecular-weight peptides with little or no hormone activity are first synthesized. These hormonal precursor forms are termed prohormones. If there is a precursor to the prohormone, this is termed a preprohormone. This terminology, unfortunately, may be confusing, since there are also *pre*hormones. Prehormones are those circulating forms of hormones which are transformed in the target tissue or in intermediate tissues into the final, active hormonal form. Prehormones generally are nonpolypeptide hormones. They are discussed in Section II.A.8 of this chapter. Preprohormones must be distinguished from prehormones.

Some of the prepro- and prohormones are given in Table 4. Proinsulin is the prototype of the prohormone. This was originally discovered in insulin-producing islet cell tumors of the pancreas (Steiner et al., 1967). Proinsulin is synthesized, and portions of the polypeptide chain are then excised, leaving the final structure that forms insulin: two peptide chains (A and B) connected by two disulfide bridges. A precursor form of proinsulin has been found. This consists of 109 amino acids arranged in such a way that there is a preregion at the N-terminus, followed by the B-chain sequence, the C-peptide sequence, and the A-chain sequence. Cleavage of the preregion results in the forma-

Table 4. Posttranslational Modification of Polypeptide Hormones[a]

Hormone	Prepro- form or its equivalent	Pro- form	Reference
Insulin	+	+	Steiner et al. (1967); Steiner (1977)
Renin		+	Morris (1978)
Angiotensin II	+	+	Morris (1978); Oparil and Haber (1974)
Relaxin		+	Kwok et al. (1978)
Nerve growth factor		+	Berger and Shooter (1977)
Parathormone	+	+	MacGregor et al. (1978); Habener et al. (1975); Habener and Potts (1978a)
Gastrin, G-17		+	Walsh and Grossman (1975)
Cholecystokinin		+	Mutt and Jorpes (1968)
Calcitonin		+	Moya et al. (1975)
Placental lactogen	+	0	Birken et al. (1977)
Growth hormone	+	0	Seeburg et al. (1977); Lingappa et al. (1977)
Prolactin		+	Lingappa et al. (1977)
Chorionic gonadotropin	+		Birken et al. (1978)

[a] The symbols are: +, form is present; 0, form is absent. No symbol indicates that no information is available.

tion of proinsulin. Cleavage of the C-peptide converts proinsulin into insulin (Steiner, 1977). More recently a family has been described in whom proinsulin or a proinsulin-like substance comprises the major fraction of circulating insulin immunoreactivity. It has been hypothesized that this represents a deficiency of the proinsulin-cleaving enzyme within the beta cell, or an abnormal species of proinsulin (Gabbay et al., 1976).

The kidney produces the enzyme renin, which functions in the renin–angiotensin–aldosterone hormonal system. Renin has a prorenin form which is inactive in producing angiotensin from its precursor. The prorenin (mol. wt. 60,000) can be converted to active renin by cathepsin D (Morris, 1978). Renin can form angiotensin I by proteolytic action on an α_2-globulin (angiotensinogen) secreted by the liver. Angiotensin I is converted to angiotensin II by an angiotensin cleavage enzyme. Since angiotensin I (a decapeptide) is a precursor of angiotensin II it has been suggested that angiotensin I be termed proangiotensin (IUPAC–IUB Commission, 1975). Among its other functions, angiotensin II is the trophic hormone that controls the elaboration of aldosterone by the zona glomerulosa of the adrenal cortex. In this system, angiotensinogen is the preprohormone and angiotensin I is the prohormone (Oparil and Haber, 1974). A prohormone has also been reported for relaxin (Kwok et al., 1978). And in mouse submaxillary glands, a pro-β-nerve growth factor (mol. wt. 22,000) is synthesized which is a precursor to β-nerve growth factor (mol. wt. 13,260) (Berger and Shooter, 1977). The parathyroid hormone, parathormone, is formed in the parathyroid glands from proparathormone by cleavage of the first six amino acids as an intact hexapeptide (MacGregor et al., 1978).

Preproparathormone—which serves as the precursor to proparathormone (Habener et al., 1975; Habener and Potts, 1978a)—has been described. Preproparathormone contains 115 amino acids. Cleavage of the first 25 amino acids at the N-terminus converts preproparathormone into proparathormone. Cleavage of the next six amino acids converts proparathormone into parathormone.

Other types of posttranslational modifications of hormones occur. Gastrin I has 17 amino acids and a molecular weight of 2098. Gastrin II is identical to gastrin I, except that gastrin II has a sulfate group covalently linked to the tyrosine at position 12. Both gastrin I and gastrin II are termed G-17. Gastrin G-17 appears to be derived from a 34-aminoacid precursor, big gastrin, or G-34. The 17 amino acids of G-17 are identical in sequence to amino acids 18 through 34 of big gastrin. Big gastrin exists in a sulfated form (Walsh and Grossman, 1975). Cholecystokinin is composed of 33 amino acids (CCK-33) and is sulfated on amino acid 27. Loss of the sulfate group results in loss of hormonal activity (Rayford et al., 1976). There also exists a larger form of cholecystokinin, composed of 39 amino acids. It has been suggested that this larger form of cholecystokinin is a precursor form (prohormone) of the smaller form (Mutt and Jorpes, 1968). It has been demonstrated that an enzyme isolated from canine and porcine cerebral cortical extracts can convert CCK-33 to smaller immunoreactive forms (Straus et al., 1978).

A procalcitonin has been described in chicken ultimobranchial glands incubated in vitro (Moya et al., 1975). Procalcitonin has a molecular weight of about 13,000 as compared to a molecular weight of about 3,500 of calcitonin. In the case of some hormones, the equivalent of the preprohormone form is cleaved to its final hormone form without going to an intermediate, prohormone form. Such seems to be the case with placental lactogen (Birken et al., 1977) and growth hormone (Spielman and Bancroft, 1977). In such cases, these primary polypeptide forms are called "pre"-hormone forms (Habener and Potts, 1978a,b), but these must not be confused with the prehormone form that is further processed in target tissue.

Some hormones have been found to exist in larger-molecular-weight forms that are structurally related. It is not clear if these larger hormonal forms represent the pro- or prepro- precursors. For example, a large form of glucagon has been described (Rigopoulou et al., 1970; O'Connor and Lazarus, 1976; Danforth et al., 1976). In a patient with a pancreatic alpha-cell tumor, a circulating glucagon was found that had a molecular weight greater than that of pancreatic glucagon (Danforth et al., 1976). Corticotropin-releasing factor exists in at least two different forms in the bovine hypophyseal stalk. It can be converted to a lower-molecular-weight form by heating in vitro at 100°C at pH 1–2 for 15 min. The lower-molecular weight corticotropin-releasing factor retains its biological activity (Yasuda and Greer, 1978). Prolactin-releasing factor also exists in bovine hypophyseal-stalk extracts in a larger-molecular-weight form. It too can be converted to a biologically active lower-molecular-weight form by heating in dilute acid in vitro (Yasuda and Greer, 1978).

b. Nonpolypeptide Hormones. Of the nonpolypeptide hormones (thryroxine, triiodothyronine, estrogen, progesterone, aldosterone, cortisol, melatonin, epinephrine, norepinephrine, vitamin D_3, testosterone, histamine, serotonin, prostacyclin, thromboxane), only vitamin D_3 (which is not the active form of the hormone) is considered to exist in a prohormone form. However, in a sense, it could be argued that all of the precursor forms for each hormone constitute prohormones. Thus, thyroglobulin, which is the immediate precursor to thyroxine and triiodothyronine, could be considered to be the prohormone of the thyroid hormone. Vitamin D_3 is synthesized in the skin from 7-dehydrocholesterol, under the influence of ultraviolet irradiation. The biosynthetic sequence is 7-dehydrocholesterol → previtamin D_3 → vitamin D_3. In this case, 7-dehydrocholesterol represents the preprohormone, previtamin D_3 is equivalent to the prohormone, and vitamin D_3 is the hormone—although it must undergo further transformations in other tissues (Holick et al., 1977).

5. Secretion of Hormones

a. Secretion Mechanisms. In order for hormones to enter into the circulation, they must undergo secretion; this is discussed in detail in Chapter 12. Here we will discuss the secretory mechanism as it involves hormones only.

i. Polypeptide hormones. The details of secretion are not known for all polypeptide hormones, but data thus far accumulated point to certain general features of the process shared by all polypeptide hormones.

In the case of parathormone, the primary protein formed during synthesis is believed to be preproparathormone. As the preproparathormone is formed on the polyribosomes attached to the endoplasmic reticulum (ER), the initial portion of the growing polypeptide chain enters the membrane of the ER. This peptide sequence, which contains methionine at its N-terminus, enters the lumen of the ER (the intracisternal space). The "leader sequence" (Habener and Potts, 1978b) contains 20–23 amino acids, with a large proportion of the amino acids being hydrophobic in nature, thus facilitating penetration into the intracisternal space. This is characteristic of proteins associated with membranes (Segrest and Feldman, 1974). The leader sequence has also been termed the NH_2-terminal extension, extra piece, presequence, and signal sequence (Habener and Potts, 1978b). Specific membrane receptor sites on the ER may exist, but no information is available concerning such receptor sites. The leader sequence permits the entrance of the covalently bonded prohormone. Once the preproparathormone has entered the intracisternal space, the leader sequence is removed and proparathormone is formed. The proparathormone moves through the ER to the Golgi complex (Habener et al., 1977a), where the hexapeptide at the N-terminus is removed by proteolytic enzymes (Habener et al., 1977b). The hormone polypeptide is incorporated into secretory granules, which discharge their contents by the process of emiocytosis (exocytosis) (see Chapter 12).

A similar sequence of events takes place with regard to insulin secretion (Steiner, 1977; Ostlund, 1977; Renold, 1972). The leader sequence of the preproinsulin enables the polypeptide to enter the intracisternal space of the rough ER where cleavage of the leader sequence transforms the preproinsulin into proinsulin. By means of microvesicles, the proinsulin is transferred from the rough ER to the Golgi complex, where the proinsulin is converted to insulin and C-peptide through protease action, although some proinsulin persists. The insulin is packaged into secretory granules, together with zinc, forming a crystaloid. The secretory granules are discharged by the process of emiocytosis. The secreted products consist of 94% insulin and C-peptide and 6% proinsulin and other insulin intermediates. Zinc ion is also included as one of the secretory products.

A similar mechanism of secretion is thought to occur in the anterior and posterior pituitary gland and in the parafollicular cells of the thyroid (Trifaro, 1977). The process of exocytosis occurs in the neurohypophysis, with the release of oxytocin, vasopressin, and neurophysins I and II (Uttenthal et al., 1971; Nordmann et al., 1971); in the anterior pituitary, with the release of thyrotropin (Farquhar, 1969), adrenocorticotropin (Rennels and Shiino, 1968), and growth hormone (Farquhar, 1961); and in the alpha (Gomez-Acebo et al., 1968), beta (Lacy, 1961), and delta (Gomez-Acebo et al., 1968) cells of the pancreas.

ii. Nonpolypeptide hormones. Catecholamines—mainly epinephrine—are packaged in secretory granules in the Golgi complex of the adrenal medulla (Trifaro, 1977). The secretory granules (chromaffin granules) serve as storage depots until the appropriate stimulatory signal results in secretion. Extrusion of the secretory granule occurs by exocytosis (Trifaro, 1977). Also stored within the secretory granule are dopamine-β-hydrochromogranin A, and ATP. These substances are released along with epinephrine (Douglas, 1966).

Exocytosis requires the presence of Ca^{2+} and is an energy-dependent process (Trifaro, 1977). A protein called synexin has been isolated from the bovine adrenal medulla, and this protein causes the aggregation of isolated chromaffin granules when incubated in the presence of free calcium at concentrations greater than 6 μM (Creutz et al., 1978). It seems likely that the membrane of the secretory granule is retained and reutilized. The mast cell undergoes a similar secretory mechanism, releasing histamine, heparin, and protein from the secretory granule (Fillion et al., 1970). The release of hormones from discrete secretory granules provides a quantal release mechanism for the hormones, rather than a continuous secretory outflow. This results in an episodic or pulselike secretion of hormones.

b. Stimulus–Secretion Coupling. There are a large variety of stimuli which trigger the quantal release of the different hormones. In a sense, the stimulus–secretion coupling represents the target tissue response to a hormone, neurotransmitter, endogenous metabolite, or exogenous nutrient. In many cases, the stimulation of secretion of a hormone from endocrine tissue is mediated by a hormone which forms part of that particular endocrine system. Thus, stimulus–secretion coupling could logically be treated as part of the general mechanism of hormonal effects on target tissues. Not all stimulation–secretion coupling is mediated by hormonal substances or neurotransmitters, however. In certain instances, exogenous nutrients or endogenous metabolites act as secretory stimulants.

The secretion of insulin from the β cell of the pancreas is a well-studied example of stimulus–secretion coupling. It is well-known that glucose ingested orally will result in insulin secretion into the blood (Gerich et al., 1976). But oral glucose is more effective than intravenous glucose, and the process is stereospecific since D-glucose is effective while L-glucose is not. Furthermore, the α-anomer of D-glucose is more active than the β-anomer (Grodsky et al., 1974; Malaisse et al., 1976). Other substances will also cause secretion of insulin. These include orally ingested amino acids, particularly leucine (Fajans et al., 1967), arginine, and lysine. And insulin secretion will be caused by such other sugars than glucose as mannose (Zawalich et al., 1977), pentoses (ribose), and pentitols (xylitol) (Asano et al., 1977; Deery and Taylor, 1973).

There are certain species differences in the release of insulin. Acetoacetate stimulates insulin secretion in the dog (Madison et al., 1963), and short-chain fatty acids are stimulatory in ruminants (Manns and Boda, 1967; Hertelendy et al., 1969). But acetoacetate and short-chain fatty acids are ineffective in releasing insulin in man. Various hormones may also stimulate insulin secretion. These include glucagon (Samols et al., 1965; Montague and Cook, 1971), secretin (Chisholm et al., 1969), cholecystokinin (Dupre et al., 1969), growth hormone in cats and dogs (Randle and Young, 1956; Cambell and Rastogi, 1966; Altszuler et al., 1968), and placental lactogen (Beck and Daughaday, 1967). Epinephrine and norepinephrine inhibit insulin release (Coore and Randle, 1964; Porte, 1969), as does the β-adrenergic blocking agent, propranolol. Glucagon is thought to stimulate insulin release by the generation of cAMP (Montague and Cook, 1971). Epinephrine decreases cAMP in the islets of Langerhans (Montague and Cook, 1971). Inhibitors of cAMP degradation, such as caffeine and theophylline, may stimulate insulin release in animals and in isolated islets. Somatostatin inhibits the release of insulin, as well as of many other hormones (Gerich et al., 1975). More recently it has been shown that the opiate peptide, β-endorphin, will inhibit the secretion of somatostatin-like immunoreactivity from isolated perfused pancreas and that this peptide stimulates the release of both insulin and glucagon (Ipp et al., 1978). Opioid peptides are present in the gastrointestinal tract as well as the brain (Polak et al., 1977).

Although a great deal of information has been collected about substances which stimulate and inhibit insulin secretion, the kinetics of the process, the effects of the autonomic nervous system, and other details, the molecular events whereby glucose and the different hormones stimulate insulin secretion (Gerich et al., 1976) remain unknown. Glucose causes an acute elevation of islet cAMP (Gerich et al., 1976), and it is believed that a glucose metabolite is responsible for this. Because of the finding that α-D-glucose is more potent than β-D-glucose in stimulating insulin release, it has been postulated that a specific glucose receptor exists in the β-cell of the pancreas (Grodsky et al., 1974; Grodsky et al., 1975). This is supported by the finding that xylitol is a potent stimulator of insulin release in dogs and shares a structure in common with α-D-glucose (Asano et al., 1977). An alternate explanation suggests that α-D-glucose has a greater glycolytic flux than β-D-glucose and has a greater stimulation of net calcium uptake (Malaisse et al., 1976). Currently, it is hypothesized that cAMP and calcium interact to regulate insulin secretion. Glucose promotes calcium influx into the cell and inhibits calcium efflux. cAMP is believed to facilitate the mobilization of intracellular calcium from storage depots. A similar mechanism may be operating in the case of glucagon secretion (Gerich et al., 1976). The interaction between cAMP and calcium in the stimulus–secretion coupling of insulin thus appears to be a special case of the general system by which cAMP-dependent hormone action is mediated.

The secretion of the posterior pituitary hormones takes place from the axon terminals of the neurons of the supraoptic and paraventricular nuclei of the hypothalamus. Vasopressin and oxytocin and their carrier proteins—neurophysins I and II—are synthesized by the usual protein-synthesizing mechanisms in the neuronal cell bodies. There are data to suggest that a larger-molecular-weight precursor of neurophysin (20,000) is synthesized in the supraoptic nucleus in the rat and is then converted to neurophysin (mol. wt. 12,000) during axonal transport (Gainer et al., 1977). There is other evidence demonstrating that rat neurophysin I is converted to neurophysin I', and that rat neurophysin II can be converted to neurophysin III (North et al., 1977) within the neurosecretory granules. The hormones and the neurophysins bind together and move through the axonal cytoplasm within the confines of neurosecretory granules. At the axonal terminal within the posterior pituitary, the neurosecretory granules release their contents into the perivascular space via the process of exocytosis (McKelvy, 1975; Dreifuss, 1975). Oxytocin may be stored with neurophysin I and vasopressin with neurophysin II in separate neurosecretory granules in bovine posterior pituitaries (Dean et al., 1968).

Various stimuli control the release of vasopressin. These include changes in blood osmolality, changes in blood volume, and psychological stimuli (Edwards, 1977). The osmoreceptors that mediate the osmolar response are located in the hypothalamus. The volume receptors are located in the left atrium and in baroreceptors in the aortic arch and carotid sinus. Afferent impulses from these receptors travel via the vagus, glossopharyngeal, and aortic nerves to the diencephalon and lead to a decrease in vasopressin secretion. The neuronal pathways release neurotransmitters (norepinephrine, acetylcholine) which cause the propagation of a nerve impulse in the axon. The axon potential at the axonal terminal causes an influx of calcium into the secretory endings. Newly formed granules located at the periphery of the endings are preferentially released (Dreifuss, 1975). Although protein kinase activity exists within the neurosecretory granules and the neurosecretory-granule membrane proteins are capable of being phosphorylated (McKelvy, 1975), other data suggest that cAMP is not involved in the secretory process of the posterior pituitary hormones (Dreifuss, 1975). On the other hand, the secretion of parathormone from dispersed bovine parathyroid cells is closely correlated with the rise in cAMP content of the cells (Brown et al., 1976a). Cholera toxin was found to cause a dose-and time-dependent increase in cellular cAMP and PTH release (Brown et al., 1979). PTH secretion and cAMP accumulation owing to a variety of agonists is inhibited by prostaglandin $F_{2\alpha}$ and by α-adrenergic agonists (Brown et al., 1978). These studies demonstrate that the release of PTH is related to cAMP generation. However, the exact mechanism whereby cAMP activates the PTH secreting system remains to be determined.

Neurophysin release is stimulated by nicotine, and this nicotine-stimulated neurophysin is secreted under conditions in which one would expect vasopressin to be secreted. Similarly, an estrogen-stimulated neurophysin is secreted under conditions in which one would expect oxytocin to be secreted (during estrogen administration, in pregnancy, in newborns) (Robinson et al., 1977). Oxytocin release involves an afferent arc in which stimulation of the nipple triggers nerve impulses which travel via peripheral nerves and the spinal cord to the hypothalamus, resulting in the release of oxytocin from the posterior pituitary (Edwards, 1977). Since vasopressin and oxytocin are released in different physiological situations, they are generally released independent of one another. Present evidence indicates that oxytocin and vasopressin are produced in different neurons (Dean et al., 1968; Sokol et al., 1976; Pickup et al., 1973).

The adrenal medulla contains chromaffin granules which contain epinephrine. Epinephrine is released from the chromaffin granules by the process of exocytosis. Adrenal secretion of catecholamines is prompted by various stimuli. These include emotional stress, hypotension, hypoglycemia, asphyxia, anoxia, cold temperature, and physical exercise. These stimuli operate through neural pathways in the central nervous system, via the sympathetic innervation of the adrenal medulla (Perlman and Chalfie, 1977). The adrenal medulla is innervated exclusively by cholinergic nerves. Acetylcholine is released at the splanchnic nerve–chromaffin cell junction in the adrenal medulla. The released acetylcholine crosses the space between the nerve terminus and the chromaffin cell, and binds to both nicotinic and muscarinic cholinergic receptors, with the number of nicotinic receptors predominating significantly. The acetylcholine increases the permeability of the chromaffin cell to calcium (Douglas, 1966). Calcium uptake by the chromaffin cells of the adrenal medulla results in the release of epinephrine from the chromaffin granules by the process of exocytosis (Perlman and Chalfie, 1977).

Calcium is obligatory for the secretory process and also for the release of acetylcholine from the sympathetic nerve terminal in the adrenal medulla; there is a low basal rate of secretion of catecholamines from the denervated adrenal glands. The mechanism for this is unknown. Various substances reaching the adrenal medulla via the circulation can elicit adrenal medullary secretion, although this action does not appear to have a physiological role. These secretion-stimulating substances include glucagon (Scian et al., 1960), histamine (Burn and Dale, 1926), angiotensin II, and bradykinin (Feldberg and Lewis, 1964). It has also been demonstrated that cAMP increases in chromaffin cells as a consequence of splanchnic nerve and acetylcholine stimulation. The cAMP activates a cAMP-dependent protein kinase (Guidotti et al., 1975a), but what role this plays in the stimulus–secretion coupling mechanism is unknown. The release of epinephrine is primarily under the control of acetylcholine acting via the nicotinic cholinergic receptor. The generation of cAMP seems to be related to the regulation of tyrosine hydroxylase synthesis (Guidotti et al., 1975b), but the relationship of cAMP to secretion is unclear. The activation of the muscarinic receptors by acetylcholine stimulates guanylate cyclase, with the generation of cGMP, but the role of cGMP in enzyme synthesis or secretion remains to be established (Costa et al., 1974).

6. Circulating Forms of Hormones

a. Hormonal Plasma-Binding Proteins. Most of the polypeptide hormones are sufficiently soluble in plasma to circulate in the blood in a free form. This is true for the pituitary hormones (ACTH, MSH, TSH, FSH, LH, prolactin, growth hormone, vasopressin), the pancreatic hormones (insulin, glucagon), parathormone, calcitonin, angiotensin II, and the gastrointestinal hormones (gastrin, cholecystokinin). Certain of the nonpolypeptide hormones (viz., catecholamines) also are quite soluble and circulate in the free form. These soluble hormones turn over rapidly and have half-lives of about 20 min or less. Their concentrations may change within seconds in response to various stimuli (Sterling, 1979). A notable exception to the polypeptides circulating in a free form are the somatomedins—insulin-like growth factors which are polypeptides but which circulate bound to specific plasma carrier proteins (Schalch et al., 1978). Other hormones are poorly soluble in plasma and generally circulate bound to one or more specific—and, in some cases nonspecific—plasma carrier proteins.

The steroid hormones and other nonpolypeptide hormones utilize carrier proteins for their circulation. The hormones which require carrier proteins are cortisol, estradiol, progesterone, testosterone, vitamin D_3 and its metabolites, thyroxine and triiodothyronine, and prostacyclin. Testosterone and estradiol are carried by a sex hormone-binding globulin (TeBG) (Rosenbaum et al., 1966; Anderson, 1974), cortisol by transcortin [corticosteroid-binding globulin (CBG)] (Mills, 1962; Rosner, 1969), thyroxine and triiodothyronine by thyroxine-binding α-globulin (TBG), prealbumin and albumin (Goodman, 1974; Ingbar, 1963; Oppenheimer, 1968), vitamin D_3 and its metabolites (25-hydroxycholecalciferol and 1,25-dihydroxycholecalciferol) by group-specific component (Gc) [vitamin-D-binding globulin (VDBG)] (Brissenden and Cox, 1978; Svasti and Bowman, 1978), and progesterone by a progesterone-binding protein (blastokinin or uteroglobin) (Popp et al., 1978). Estrogen increases the serum concentration of thyroxine-binding globulin (Barbosa et al., 1973) by stimulating the hepatic synthesis of TBG (Glinoer et al., 1977a). Estrogen also increases the serum concentration of CBG and TeBG (Glinoer et al., 1977b). CBG has relatively low affinity for aldosterone (Rosner, 1969). There is little information concerning the nature of any prostaglandin-binding proteins in the circulation.

(b) Intermediate Transformations of Hormones. After they are secreted into the circulation, some hormones undergo transformations either in other organs or in the plasma before the final active hormonal form is generated. Vitamin D_3 is manufactured in the skin under the influence of ultraviolet irradiation. It circulates bound to a specific plasma protein and is transformed into 25-hydroxycholecalciferol in the liver. The 25-hydroxycholecalciferol enters the circulation and enters the kidney, where it is transformed into 1,25-dihydroxycholecalciferol. This latter substance circulates in the blood attached to a specific plasma protein and represents the final active hormonal form. The transformation of

25-hydroxycholecalciferol to the 1,25-form is regulated, in part, by parathormone (Haussler and McCain, 1977; DeLuca, 1978).

In the renin–angiotensin–aldosterone system, the juxtaglomerular cells in the kidney secrete the acid proteolytic enzyme, renin, from cytoplasmic granules (Davis and Freeman, 1976; Oparil and Haber, 1974; Peart, 1975; Erdos, 1977). Renin acts on its substrate, angiotensinogen, which is a 56,800-molecular-weight α-globulin secreted by the liver, producing the decapeptide, angiotensin I. Renin may accumulate in the cells in peripheral arterioles, and angiotensin I may thus be formed intracellularly (Thurston, 1976). In its passage through the lung, angiotensin I is converted to angiotensin II by angiotensin-I-converting enzyme (peptidyl dipeptide hydrolase), through cleavage of the dipeptide histidyl-leucine (Ng and Vane, 1967; Ryan et al., 1972; Erdos, 1977). Angiotensin-I-converting enzyme is contained in the luminal surface of vascular endothelial cells, where it has direct access to the circulation (Hayes et al., 1978). The conversion of angiotensin I to angiotensin II in extrapulmonary vascular beds occurs in situ and does not contribute to circulating angiotensin II levels (Erdos, 1977). Thus, renin, which is a nonhormonal enzyme, transforms angiotensinogen (a preprohormone) into angiotensin I (a prohormone), which is in turn converted into angiotensin II (the hormonal substance) both in the plasma and in vascular tissue. Angiotensin II is also converted by aminopeptidase A into *des*-Asp angiotensin II (angiotensin III), which has many of the same functions as angiotensin II (Goodfriend and Peach, 1975; Tsai et al., 1975a).

7. Plasma Membrane-Mediated Systems

a. Requirements of the System. Most of the polypeptide hormones and the β-adrenergic agonists seem to produce their metabolic effects through similar mechanisms. The sequential steps involved in the mechanism of action of these hormones include:

1. Binding to a specific membrane cellular receptor.
2. Activation of an effector (usually adeylate cyclase), with the subsequent increase in intracellular concentration of a signal molecule (second messenger) (usually cyclic adenosine-3',5'-monophosphate—cAMP).
3. Second-messenger mediated intracellular effects resulting in specific tissue responses.

Studies relating to these three topics are voluminous, and space does not permit a detailed presentation of the data. Instead, we shall attempt a summary of the general concepts relating to the mechanism of action of the polypeptide hormones and the β-adrenergic agonists.

b. Binding of Hormones to Specific Membrane Receptors. The initial step in the action of all known polypeptide hormones and of the β-adrenergic agonists, is binding to a specific membrane receptor (Roth, 1973; Lefkowitz, 1973; Desbuquois and Cuatrecasas, 1973) located in the outer layer of the plasma membrane. This interaction has been studied extensively in several laboratories utilizing whole cell preparations, isolated membrane fractions, and "solubilized" receptor preparations. In most studies, isotopically labeled hormones (^{125}I or ^{3}H) or hormone antagonists (in the case of the β-adrenergic receptor) have been used to characterize this interaction. A few studies have utilized photoaffinity probes (Galardy et al., 1974) and ferritin-labeled hormones (Jarett and Smith, 1974). Recent reviews concerning polypeptide hormone receptors and their interaction with specific hormones should be consulted for a more detailed treatment (Kahn, 1975, 1976).

c. Membrane Hormone Receptors. A hormone receptor may be dinfined as a cellular component which specifically binds a particular hormone, with the resulting complex having the potential to initiate a series of reactions leading to a biological response. Some investigators would limit the use of the term only to those conditions in which a biological response is actually produced by the hormone-receptor complex (Birnbaumer et al., 1974). This discussion will be restricted to membrane receptors localized primarily in the plasma membrane of mammalian cells, although, where necessary, nonmammalian systems will be considered.

The general relationship between the membrane-bound hormone receptor and intracellular events produced by hormone binding to this receptor is depicted in Fig. 1. This model was suggested by Robison et al. (1967) and is composed of two subunits. The subunit in contact with the external enviroment is the hormone receptor and is the regulatory portion of the complex. Located in the inner portion of the plasma membrane and in contact with the intracellular environment is the effector or catalytic component which, in most cases, is adenylate cyclase. There is a third portion of the model which recent studies have elucidated and which is discussed in section B (p. 564).

Following the interaction of a hormone (first messenger or extracellular signal molecule) with its specific membrane receptor, the effector is activated and generates a second messenger (an intracellular signal molecule which is usually cAMP); this second messenger then mediates the action of the hormone. Although not completely established for all polypeptide hormones, this model implies that it is not necessary for the hormone to gain entrance into the cell to produce its metabolic

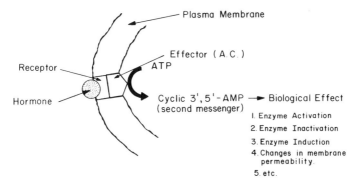

Figure 1. General mechanism of action of polypeptide hormones and β-adrenergic agonists. The hormone receptor is in the external portion of the plasma membrane and the effector [usually adenylate cyclase (A.C.)] is in the inner portion in contact with the cell cytoplasm.

effects. (In the case of insulin there is evidence that insulin can enter cells and bind to intracellular structures. The significance of this concept is presently unknown. See Section II.C.1 of this chapter.) Evidence in support of this concept includes: (1) the ability of hormones conjugated to cellulose or sepharose particles to produce their metabolic effects; (2) the reversal of hormone effects on tissue samples in vitro after rapid exposure to specific antibodies; and (3) the loss of hormone responsiveness after trypsin treatment of some tissues under conditions which did not disrupt cells or interfere with other metabolic functions.

The specific receptors for hormones seem to be located primarily in the outer portion of the plasma membrane, in contact with the external environment (Orci et al., 1975; Bennett and Cuatrecasas, 1973). In broken-cell preparations, the plasma-membrane fraction generally contains the highest concentration of receptors for the peptide hormones and the catecholamines. However, several investigators have found binding sites for some of these hormones on intracellular organelles such as the smooth and rough ER, the Golgi apparatus, and nuclei (Kahn, 1976). It is unclear whether the binding sites on these cellular components represent contamination with plasma-membrane fragments, true receptors undergoing degradation or synthesis, or some other type of binding site.

The distribution of hormone receptors within the plasma membrane may not be homogeneous. Clustering of insulin receptors in fat- and liver-cell membranes has been observed using ferritin-labeled insulin (Jarett and Smith, 1974; Orci et al., 1975). However, the clustering may have been produced following insulin binding, and the distribution may be different when the receptors are unoccupied. The number of specific receptor sites per cell for peptide hormones and the catecholamines has ranged from 500 to 250,000 (Kahn, 1975).

Although no intact hormone receptor has yet been isolated and completely characterized chemically, some progress has been made in this area. All of the receptors studied thus far are degraded by proteases, indicating that protein is an essential part of the receptor complex (Kahn, 1975). Specific membrane phospholipids also seem to play an important role in the activity of ACTH, glucagon, TRH, LH-hCG, and β-adrenergic receptors (Kahn, 1975, 1976), since phospholipase treatment of the membrane fractions studied resulted in decreased binding of these hormones. The role of phospholipids in the binding of a hormone to its receptor is not clear, but they may be important in maintaining the receptor protein in the proper conformation so as to render it capable of interacting with the hormone (e.g., α-helical conformation). Carbohydrate also seems to be an important constituent of TSH, insulin, and LH-hCG receptors (Kahn, 1976). Partial purification of some polypeptide hormone receptors suggests that they may exist as oligomeric structures having monomeric molecular weights of approximately 20,000 to 90,000 (Kahn, 1976). However, see Section B.1.C.

d. The Interaction of Hormones with Receptors. The ability of a specific tissue or cell type to respond to a given hormone is probably determined by whether or not the cell membrane contains the specific hormone receptor. Specific hormone receptors bind only one type of hormone, exhibit characteristics of saturability, and of time and temperature dependence, and are of high affinity in nature (Kahn, 1976). The ability of polypeptide hormone analogs to bind to these receptors generally correlates with the ability of the hormones to produce a biological response. In contrast, cells also have nonspecific hormone binding sites which are not saturable and are of low affinity. These nonspecific, low-affinity receptor sites may have little or no physiological role, since at usual in vivo hormone concentrations, little, if any, hormone is bound to these receptors.

Although not completely understood, hormone binding to the specific receptors is rapid, reversible and may be depicted as a single, bimolecular reaction as follows:

$$[H] + [R] \underset{k_d}{\overset{k_a}{\rightleftharpoons}} [H-R] \qquad (1)$$

where [H] is the concentration of unbound hormone, [R] is the concentration of unbound receptors, [H–R] is the concentration of hormone–receptor complexes, and k_a and k_d are the association and dissociation constants, respectively. From this equation, K (the equilibrium or affinity constant) may be derived as follows:

$$K = \frac{k_a}{k_d} = \frac{[\text{H--R}]}{[\text{H}][\text{R}]} \qquad (2)$$

For these equations to be valid, a number of conditions must be met, as outlined by Kahn (1975). Actually, the interaction between hormone and receptor is usually much more complex than can be explained by this simple bimolecular equation. However, this simple bimolecular relationship seems to exist with regard to the receptors for growth hormone (Lesniak et al., 1974), prolactin (Shiu and Friesen, 1974), LH-hCG (Rao and Saxena, 1973), calcitonin (Marx et al., 1974), and the β-adrenergic receptor in some tissues (Brown et al., 1976b; Aurbach et al., 1974).

Thus, in order for the interaction between a hormone and a cellular binding site to be considered a receptor interaction, it is necessary to fulfull the criteria of a hormonal response, exhibiting specificity, high affinity, reversibility, and saturability. Furthermore, the kinetic characteristics of the binding should correlate with those of the hormonal effect (Cuatrecasas and Hollenberg, 1975). With these criteria, many hormone and neurotransmitter receptors have been characterized (Cuatrecasas, 1974). Receptors have been demonstrated for insulin, glucagon, catecholamines, acetylcholine, TSH, angiotensin II, ACTH, calcitonin, FSH, luteinizing-hormone releasing factor, prolactin, vasopressin, TRH, human chorionic gonadotropin, luteinizing hormone, vasoactive intestinal polypeptide, human growth hormone, oxytocin, epidermal growth factor, serotonin, etc. (Cuatrecasas, 1974).

Hormones with similar biologic activities may or may not interact with the same receptors. LH and hCG interact with the same receptors, whereas closely related hormones such as glucagon, vasoactive intestinal polypeptide, and secretin appear to interact with different receptors (Kahn, 1976). Enteroglucagon can bind to some but not all of the pancreatic glucagon receptors in liver. Where two hormones have overlapping biological activities, the situation appears even more complex. For example, growth hormone and human placental lactogen seem to interact with at least two different receptor sites, one specific for the growth aspects of the peptides and another specific for the lactogenic effect of the hormones. Similarly, the overlapping metabolic activities of insulin, serum nonsuppressible insulin-like activity (NSILA, somatomedin), and multiplication stimulating factor (MSF) seem to involve at least two separate membrane receptors. One has a high affinity for insulin and a low affinity for the other substances. The other receptor, which seems to mediate the growth effects of the hormones, has a low affinity for insulin but a high affinity for MSF and the somatomedins (Kahn, 1976). Finally, the hormone-specific membrane receptor does not appear to be involved in the degradation of the hormone, in most cases.

e. Cooperativity Effects. The equilibrium binding data of several hormones to their receptors suggest two or more classes of binding sites which differ in affinity for the hormones. These include binding sites for ACTH (Lefkowitz et al., 1971), TSH (Moore and Wolff, 1974), TRH (Grant et al., 1973), insulin (Kahn et al., 1974), glucagon (Shlatz and Marinetti, 1972), oxytocin (Jard et al., 1975), and catecholamines in some tissues (Lefkowitz et al., 1973). Site–site interaction—or cooperativity—has been suggested to explain some of the kinetic data on hormone binding to receptors in cases in which the reaction is not one of simple equilibrium (Kahn, 1976). This cooperativity is called positive if binding of hormones to some receptors increases the affinity of the remaining receptors for the hormone, and negative if receptor affinity is decreased after some hormone is bound. Homotropic hormone-induced negative cooperativity has been suggested to explain the kinetic binding data of the interaction of the following hormones with their receptors (Kahn, 1976): insulin, TSH, TRH, β-adrenergic agonists in some tissues, and nerve growth factor. In most instances of negative cooperativity, the decrease in affinity observed at high receptor occupancy appears to result from an increase in the dissociation rate of the hormone–receptor complex (Kahn, 1976). Positive cooperativity has been suggested in some cases of vasopressin–receptor interactions.

Divalent cations and pH have been shown to have profound effects on the binding of some hormones to their receptors, and it has been suggested that these factors may act as heterotropic regulators of receptor affinity. Bashford et al. (1975) have suggested that the cation effects may result from their ability to decrease the fluidity of the membrane.

f. Hormone Degradation. Hormones attached to plasma membrane receptors undergo degradation. Insulin is degraded by intact rat epididymal fat cells, by plasma membranes from fat cells, and by a soluble extract from fat-cell plasma membranes (Crofford et al., 1972). Treatment of fat cells with trypsin not only abolishes insulin action but also inactivates the insulin degradative system (Crofford et al., 1972). It is suggested that the degradative system is one mechanism by which the cellular response to insulin is terminated. Human granulocytes have an active insulin-degrading mechanism which is due to exocytotically released, extracellular, and cell-membrane-bound enzymes. Degradation was enhanced by calcium and thiols and was inhibited by protease inhibitors and by sulfhydryl-blocking agents (Fussganger et al., 1976).

Different membrane-bound enzymes of the liver and kidney have been described which degrade insulin (Burghen et al., 1972; Katz and Rubenstein, 1973). Liver and kidney also actively degrade insulin by means of a glutathione-insulin transhydrogenase present in the microsomal fraction (Thomas, 1973; Ansorge et al., 1973) and by an "insulin specific protease" found in the cytosol (Brush, 1971). It has been suggested that insulin is cleaved at the disulfide bonds by the glutathione-insulin transhydrogenase, forming A and B chains, with proteolysis of the polypeptide chains occurring as a second step (Varandani, 1974). Glutathione-insulin transhydrogenase and "insulin specific protease" occur in lysosomes (Grisolia and Wallace, 1976). Thus, it has been suggested that degradation of the insulin bound to its receptor on the cell surface occurs as a consequence of a release of lysosomal insulin-degrading enzymes at the cell surface (Grisolia and Wallace, 1976). Insulin degradation is facilitated by receptor binding (Dial et al., 1977). Liver and kidney cell membranes degrade glucagon (Duckworth, 1978), kidney cell membranes being more than 20 times as active as liver cell membranes. Glucagon degrading activity in the kidney is inhibited by glutathione and EDTA but is unaffected by N-ethylmaleimide, ACTH, or insulin (Duckworth, 1978). In the liver, glucagon degradation is inhibited by N-ethylmaleimide, ACTH, and insulin. Insulin degradation by kidney membranes is inhibited by N-ethylmaleimide but not by EDTA (Duckworth, 1978).

The enzyme dipeptidyl aminopeptidase I of rat liver or bovine spleen degrades several hormones in vitro. These include β-corticotropin, the B chain of oxidized insulin, angiotensin II amide, the carboxy-terminal tetrapeptide of gastrin (McDonald et al., 1969a), glucagon, and secretin (McDonald, 1969b). Dipeptidyl aminopeptidase I requires chloride ion and sulfhydryl activators (McDonald, 1969b). Purified plasma membranes from rat liver rapidly inactivate glucagon, but this does not appear to be due to the presence of dipeptidyl aminopeptidase I (Pohl et al., 1972). The degradative process seems to be distinct from the process of adenyl cyclase activation (Pohl et al., 1972). An enzyme that degrades both insulin and glucagon has been isolated from skeletal muscle and appears to be highly specific for these two enzymes (Baskin et al., 1975).

8. Autoregulation of Receptors by Hormones

The concentration of hormone receptors in the plasma membrane is in a state of flux and is dependent on the rates of receptor synthesis and degradation. Although little is known about these processes, recent studies have yielded some information concerning the degradation of

Table 5. Hormones Showing Evidence of Autoregulation of Their Receptors

Hormone	Tissue
Insulin	Many
Calcitonin	Rat kidney
β-Adrenergic agonists	Frog erythrocytes and rat pineal cells
Human GH	Cultured human lymphocytes
TRH	Rat pituitary cells
FSH	Mouse ovary
LH	Rat ovary and testis
hCG	Rat testis

receptors. In cultured human lymphocytes, the $t_{1/2}$ for growth hormone and insulin receptors has been shown to be about 10 h, and 30–40 h, respectively (Kahn, 1976).

With several hormones, one of the major determinants of the number of specific cellular receptors seems to be the concentration of the hormone to which the cell is exposed. This has been referred to as homologous autoregulation of the receptor. Insulin receptors have been the most extensively studied in this regard. Human and animal studies in vitro and in vivo, have shown an inverse relationship between the concentration of insulin to which the cell is exposed and the number of specific insulin receptors present [see Kahn (1976) for an excellent summary of these data].

Several other hormones have also shown similar "self-regulation" of their receptors; these are listed in Table 5. All have shown negative feedback control except for prolactin, for which there is positive control, with an increasing number of receptors present as the prolactin concentration increases. The mechanism of self-regulation of hormone receptors is unknown but seems to involve a loss of receptors. This loss of receptors could result from endocytosis, degradation by membrane proteases, or by shedding of the hormone–receptor complex (Kahn, 1976). The negative feedback, in the case of insulin, seems to depend on new protein synthesis, since cycloheximide inhibits this effect (Gavin et al., 1974). Finally, the self-regulation of membrane receptors is not limited to hormone receptors but seems to be a common biologic phenomenon (Raff, 1976).

Heterologous hormonal regulation of membrane receptors has also been observed. FSH, given in vivo, increases the number of LH receptors in the ovary (Zeleznik et al., 1974). Prolactin receptors in rat liver membranes were increased following estrogen injection in vivo and were decreased by testosterone, progesterone, and hydrocortisone

(Sherman et al., 1975). Adrenal steroids affect the number of vasopressin receptors in rat kidney (Rajerison et al., 1974), and the number of β-adrenergic receptors in rat liver (Posmer et al., 1974). LH and human chorionic gonadotropin receptors in the rat testicular Leydig cells are diminished in experimentally induced diabetes in the rat and restored by the administration of insulin (Charreua et al., 1978). Both thyroxine and triiodothyronine increase the number of β-adrenergic receptors in rat heart cell membranes (Williams et al., 1977). In tissue culture of rat pituitary cells, hydrocortisone increases the number of thyrotropin-releasing hormone receptors (Tashjian et al., 1977). It requires 6–8 h for the effect to become significant and 18–24 h for the maximal effect (Tashjian et al., 1977). TRH decreases the number but not the affinity of its own receptors in the same rat pituitary cells (Hinkle and Tashjian, 1975).

9. Hormonal Action of Receptor Antibodies

In certain pathological situations, antibodies bind to cell-membrane receptors and act as hormones. Although antibodies to receptors generally block the action of the hormone that specifically binds to the receptor, the antibody itself usually does not have hormonal action. A number of disorders appear to be related to the presence of receptor antibodies which prevent hormones or neurotransmitters from binding effectively to the receptor. But, in some cases the antibody has hormonal action, which then becomes part of the pathological process.

In thyrotoxicosis due to diffuse hyperplasia (Graves disease) an antibody to the TSH receptor stimulates the receptor in a manner similar or identical to that of TSH (Mukhtar et al., 1975; Burke, 1969). The immunoglobulins found in Graves disease were originally termed long-acting thyroid stimulators (LATS) but are now referred to as thyroid-stimulating immunoglobulins (TSI). Recently, direct evidence has been obtained proving that TSI binds to the TSH receptor on the thyroid and stimulates adenylate cyclase (Mehdi and Kriss, 1978). Serum from patients with Graves disease may contain—in addition to LATS, which stimulates mouse thyroid cells—other immunoglobulins (LATS protector) which stimulate human but not mouse thyroid cells (Adams and Kennedy, 1967; Shishiba et al., 1973).

In a few patients with insulin resistance and acanthosis nigricans, the insulin resistance has been attributed to circulating IgG antibodies to insulin receptors (Kahn et al., 1977). The insulin antibodies have insulin-like activity and block the action of insulin (Kahn et al., 1976). Rabbit antibodies to purified rat-liver-membrane insulin receptors have been prepared and have been shown to have insulin-like activity

(Jacobs et al., 1978). The rabbit antibodies bind to the receptor at a site distinct from that of insulin but are nevertheless hormonally active.

10. The Enzyme Action of Hormones

Renin is an acidic protease which acts on its substrate, angiotensinogen, to form angiotensin I (Oparil and Haber, 1974). In a sense, renin is not a hormone since it does not act solely on a target organ. But renin does enter into vascular endothelial cells, where it can transform angiotensinogen into angiotensin I. In this latter case, renin does fulfill the criteria of a hormone. Thus, we have an instance of an enzyme functioning as a hormone. Nerve growth factor purified from the mouse submandibular salivary gland is capable of converting plasminogen to plasmin (Orenstein, et al., 1978). Here we have the case of a hormone functioning as an enzyme. Nerve growth factor enhances the growth of sympathetic ganglia in vitro and in vivo and elicits neurite outgrowth from explanted embryonic sympathetic and sensory ganglia in vitro (Levi-Montalcini and Angeletti, 1963; Mobley, et al., 1977). It is also capable of increasing the specific activities of tyrosine hydroxylase and dopamine β-hydroxylase in the superior cervical ganglia of young rats (Thoenen et al., 1971). Recently, it has been shown that in a clonal cell line of sympathetic ganglia, nerve growth factor increases the intracellular level of cAMP, which causes an increase in calcium mobilization, resulting in a structural change in the limiting cell membrane. The membrane alteration enhances cell–substratum adhesion and neurite extension (Schubert et al., 1978). Thus we have an instance of a protein which has both hormonal function and enzymatic activity.

B. The Membrane Receptor–Adenylate Cyclase–cAMP–Cyclic Nucleotide Systems

1. The General cAMP System

a. Sutherland's Criteria. A large number of polypeptide and some nonpeptide hormones and neurotransmitters are capable of stimulating adenylate cyclase and generating cAMP (Robison et al., 1971). The following criteria were established by Sutherland and his colleagues to demonstrate that a given hormone produces its physiological action through the stimulation of adenylate cyclase and the generation of cAMP:

1. The hormone should be capable of stimulating adenyl cyclase in broken-cell preparations from the appropriate cells, while hor-

mones which do not produce the physiological response should not stimulate adenylate cyclase.
2. The hormone should be capable of increasing the intracellular level of cAMP in intact cells, while inactive hormones should not increase cAMP levels. It should be demonstrated that the effect on the level of cAMP occurs at dose levels of the hormone which are at least as small as the smallest levels capable of producing a physiologic response. The increase in the level of cAMP should precede or at least not follow the physiological response.
3. It should be possible to potentiate the hormone (i.e., increase the magnitude of the physiological response) by administering the hormone together with theophylline or other phosphodiesterase inhibitors. The hormone and the phosphodiesterase inhibitor should act synergistically.
4. It should be possible to mimic the physiological effect of the hormone by the addition of exogenous cAMP.

These criteria have been fulfilled for many polypeptide hormones, catecholamines, and certain neurotransmitters. Thus, it is clear that the mechanism of action of a large number of extracellular signal molecules occurs via the adenylate cyclase–cAMP system, by first binding to a membrane receptor. As is pointed out by Sutherland (Robison et al., 1971), the major problems remaining are to determine the molecular events whereby a specific hormone activates adenylate cyclase to generate cAMP, and to determine how cAMP mediates the various physiological functions attributed to the specific hormones.

b. Model of the Receptor–Adenylate Cyclase System. The mechanism by which the hormone–receptor complex increases the activity of adenylate cyclase has not been defined. One of the first models suggested that adenylate cyclase and the hormone receptor were part of a single peptide chain which spanned the entire plasma membrane, with the receptor portion located in the outer portion of the plasma membrane and adenylate cyclase restricted to the inner portion. It was posulated that, on binding of hormone to its receptor, a conformational change occured in adenylate cyclase which increased its catalytic activity. As more data were obtained, this model became untenable. This model implied one adenylate cyclase per hormone receptor and postulated the existence of a direct relationship between the amount of hormone bound to the receptor and adenylate cyclase activity. This has been shown not to be the case with several hormones, where maximal activation of adenylate cyclase has been achieved when only a small percent of hormone receptors were occupied. Furthermore, in tissues such

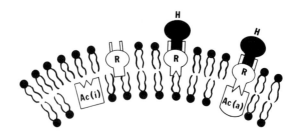

Figure 2. General model of activation of adenylate cyclase by hormones. The hormone receptor (R) is located in the outer portion of the phospholipid bilayer while the adenylate cyclase is in the inner part of the bilayer. Ac (i) represents inactive adenylate cyclase; Ac(a) represents hormone-activated adenylate cyclase; H represents the hormone.

as adipose tissue, which contains specific receptors for several hormones, when more than one hormone is present in vitro at concentrations giving maximal stimulation of adenylate cyclase for that hormone, the combined effect of the hormones was not additive. Under these conditions adenylate cyclase activity was usually no greater than that of the hormone giving the greatest activity when present alone. These data suggest that the adenylate cyclase was rate limiting. Finally, the β-adrenergic receptor and adenylate cyclase from frog erythrocytes have recently been shown to partition separately using gel exclusion chromatography after solubilization with digitonin (Linbard and Lefkowitz, 1977). This suggests that the hormone receptor and adenylate cyclase are probably two distinct membrane proteins.

The model shown in Fig. 2 seems to offer the best explanation for activation of adenylate cyclase by hormone–receptor complexes (McGuire and Barber, 1976). In this model the hormone receptor and adenylate cyclase are separate membrane proteins which are capable of interacting or being coupled after binding of the hormone to its receptor. With the current concept of the plasma membrane as a dynamic fluid structure in which the phospholipid and membrane proteins diffuse laterally, the hormone receptor is thought to be restricted to the outer portion of the plasma membrane and the adenylate cyclase to the inner portion. After the hormone–receptor interaction, the complex formed is capable of interaction (coupling) with the adenylate cyclase molecule owing to a change in receptor conformation. Following coupling of the hormone–receptor complex with inactive adenylate cyclase, the enzyme is activated, presumably through a conformational change induced by the coupling. The rate limiting factor is, presumably, the number of adenylate cyclase molecules in the membrane. The ability of some hormone analogs (such as β-receptor antagonists) to bind to the receptor, but not activate adenylate cyclase, probably results from the inability of these substances to properly alter the conformation of the receptor so that it can couple with the adenylate cyclase and activate it; alternatively, the substance may permit the receptor to couple with but not to activate the adenylate cyclase. This model is also consistent with the recent findings that the fluidity of the membrane is important in hormone binding to its receptor and that the phospholipids must be in a "fluid" or liquid crystalline state before maximum binding occurs (Bashford et al., 1975).

Most of the polypeptide hormones and catecholamines seem to exert most, if not all, of their physiological effects by increasing the activity of adenylate cyclase, with a subsequent increase in intracellular cAMP concentration in target tissues (Table 6) (Robison et al., 1971). The two most notable exceptions to this are growth hormone and insulin, although there is still some debate regarding the effect of insulin on adenylate cyclase activity.

c. The Nature of the Receptor: Gangliosides. The hormone receptors of the plasma membrane are generally ill-defined. In a few cases, the receptor has been identified as a ganglioside. A series of studies has shown that cholera toxin acts on small-intestinal adenylate cyclase to generate cAMP, resulting in the diarrheal state in cholera (Field, 1971), by binding to gangliosidic small-intestinal surface receptors (Holmgren et al., 1973, 1975; Cuatrecasas, 1973; Mullin et al., 1976a). Subsequently, similar studies have been applied to the receptors of other hormones,

Table 6. Polypeptide Hormone and β-Adrenergic Agonists that Increase Adenylate Cyclase Activity and/or cAMP Concentrations

Adrenocorticotropin (ACTH)	β-Melanocyte stimulating hormone
Angiotensin II	Norepinephrine
Antidiuretic hormone	Oxytocin
Calcitonin	Pancreozymin
Dopamine	Parathormone (PTH)
Epinephrine	Prostaglandins
Follicle stimulating hormone (FSH)	Secretin
Glucagon	Serotonin
Growth hormone releasing hormone	Somatostatin
Gut glucagon (enteroglucagon)	Thyrotropin releasing hormone (TRH)
Histamine	Thyrotropin (thyroid stimulating hormone) (TSH)
Human chorionic gonadotropin (hCG)	Vasoactive intestinal polypeptide
Luteinizing hormone (LH)	
Luteinizing hormone releasing hormone	Vasopressin

toxins, and other substances (Fishman and Brady, 1976). Even before these findings, several investigators considered the possibility that the binding of serotonin to its receptor involved binding to a ganglioside (van Heyningen, 1963; Woolley and Gommi, 1965; Gielen, 1968). More recent studies implicate a ganglioside as the receptor for thyrotropin (thyroid stimulating hormone, TSH) (Mullin et al., 1976b), as well as serotonin (van Heyningen, W. E., 1974).

Mixed gangliosides in solution will bind serotonin, tryptamine, lysergic acid diethylamide, and ergometrine, but not epinephrine, norepinephrine, dopamine, histamine, or γ-aminobutyric acid (van Heyningen, 1963). Serotonin is not bound by ganglioside–cerebroside complexes which bind tetanus toxin (van Heyningen and Mellanby, 1968). Rat stomach strips contract when treated with serotonin, but the contraction is decreased when the strips are pretreated with sialidase. Responsiveness is restored if ganglioside is added to the sialidase-treated strips. Similar results were obtained with strips from ganglioside-deficient stomachs from rats reared on a galactose-rich diet (Woolley and Gommi, 1965). The serotonin-reactive ganglioside was shown to be sialidase-labile (Gielen, 1968). The structure of certain gangliosides is shown in Fig. 3.

Gangliosides inhibit the binding of TSH to thyroid membrane; the inhibition is specific and greatest with G_{D1b}. The gangliosides bind TSH and alter its conformation, as shown by changes in fluorescence. The change in fluorescence is proportioal to the degree of inhibition of binding (Mullin et al., 1976b). TSH is a glycoprotein composed of two subunits, α and β. The α subunit is common to many glycoprotein hormones, while the β subunit confers target-organ specificity (Liao and Pierce, 1970) and appears to be necessary for the binding of the hormone to membranes (Wolff et al., 1974). There is amino acid sequence homology between the β chains of TSH, LH, hCG, and FSH (Mullin et al., 1976b; Ledley et al., 1976). Cholera toxin can partially inhibit the binding of TSH to thyroid membranes; suggesting that there are several types of TSH receptors (Mullin et al., 1976a). At low concentrations of cholera toxin, the binding of TSH is increased because of cooperativity effects (Mullin et al., 1976a). Since cholera toxin binds to ganglioside G_{M1}, this ganglioside may also be a TSH receptor on thyroid membranes.

Gangliosides inhibit the stimulation of thyroid membrane adenylate cyclase by cholera toxin or TSH. But although cholera toxin requires NAD to stimulate adenylate cyclase, there is no such requirement of TSH (Mullin et al., 1976a). Rat thyroid tumor membranes which bind TSH poorly contain only G_{M3}. On the other hand, rat thyroid membranes which effectively bind TSH contain more complex gangliosides

The Mechanism of Action of Hormones

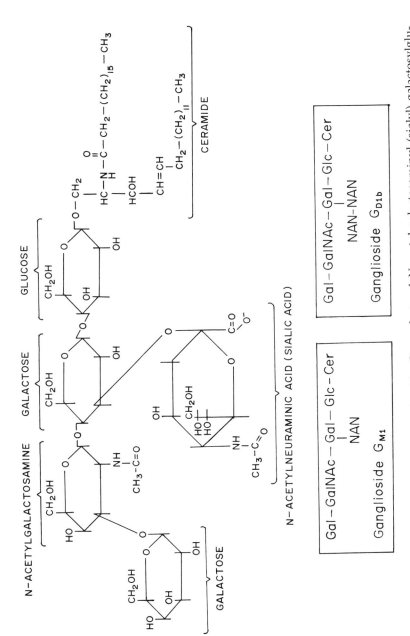

Figure 3. Structure of the monosialoganglioside, G_{M1}, galactosyl-N-acetyl-galactosaminyl-(sialyl)-galactosylglucosylceramide. The abbreviations used are: Cer, ceramide; Glc, glucose; Gal, galactose; GalNAc, N-acetylgalactosamine; NAN, N-acetylneuraminic acid (sialic acid).

(Meldolesi et al., 1976). More recently, it has been found that a ganglioside present in trace amounts in cell membranes from bovine thyroid gland is the most potent inhibitor of TSH binding to thyroid gland cell membranes (Mullin et al., 1978). This not yet structurally characterized ganglioside contains at least two sialic acid residues. It constitutes 0.015% of the total thyroid gangliosides, or about 10,000 molecules per thyroid cell.

Gangliosides inhibit the binding of human chorionic gonadotropin to rat testicular membranes (Lee et al., 1976). However, rat ovarian cell protein kinases were activated by human chorionic gonadotropin and LH, whether gangliosides were added to the preparation or not. Under the same experimental conditions, the same ganglioside mixture (obtained from bovine brain) inhibited the action of cholera toxin (Azhar and Menon, 1978). It is possible that rat ovarian cell membranes contain a hormone-specific ganglioside not present in the bovine brain preparation, as seems to be the case for bovine thyroid gland cell membranes (Mullin et al., 1978).

Although cholera toxin will stimulate lipolysis in fat cells, and this activity is enhanced by crude brain gangliosides, no G_{M1} could be detected in adipocyte cell membranes (Kanfer et al., 1976). The action of purified G_{M1} was different from that of crude brain gangliosides. These studies suggest that cholera toxin acts on adipocytes through some receptor other than G_{M1} and that treatment of adipocytes with G_{M1} creates additional receptors (Kanfer et al., 1976). It was suggested by Kanfer that the crude brain gangliosides contained some ganglioside other than G_{M1} which accounted for the action of cholera toxin. The studies imply that hormonal receptors in adipocytes may be other than the G_{M1} ganglioside.

Despite studies such as those described above, however, it is not entirely clear that gangliosides serve as the only component of the hormone receptor. There is evidence that the membrane receptor of TSH is a glycoprotein which can be solubilized by detergents and trypsin treatment of the membrane (Tate et al., 1975). The complex formed between human chorionic gonadotropin and its receptor has been solubilized and found to have a molecular weight of the order of 200,000 or more (Dufau et al., 1973; Bellisario and Bahl, 1975). It has also been suggested that glycoprotein hormones such as TSH bind to a glycoprotein receptor and then interact with a specific ganglioside, causing a conformational change in the hormone such that the α-subunit stimulates adenylate cyclase (Fishman and Brady, 1976). It appears, for instance, that the TSH receptor in thyroid plasma membranes is composed of a glycoprotein and a ganglioside (Tate et al., 1975; Winand and Kohn, 1975; Meldolesi et al., 1977). Tetanus toxin, which causes thyroid hyperfunction in mice (Habig et al., 1978), binds to thyroid plasma membranes (Ledley et al., 1977) and is inhibited by TSH (Meldolesi et al., 1977). If a TSH-receptor defect is present in the membranes, tetanus toxin will not bind (Habig et al., 1978). When tetanus toxin is incorporated into liposomes, it binds to the glycoprotein component of the bovine thyroid TSH receptor (Lee et al., 1978). This interaction can be modified by changes in the fatty acid side chains of phosphatidylcholine, which is the major phospholipid component of the plasma membrane.

Treatment of thyroid cell membranes with phospholipids alters TSH binding. Phosphatidylglycerol and phosphatidylinositol, for example, inhibit TSH binding to bovine thyroid membranes. The effect seems to be related to the interaction of the phospholipids with the membranes rather than with TSH, and this contrasts with gangliosides, which interact with TSH. Furthermore, treatment of thyroid cell membranes with phospholipase A results in increased binding of TSH by the membranes (Omodeo-Sale et al., 1978). When a glucagon-sensitive adenylate cyclase was solubilized with Lubrol® PX, it lost its sensitivity to glucagon. The activity was restored by addition of phospholipids; however, phosphatidyl serine was effective, while phosphatidyl inositol was not (Levey et al., 1975a). Phospholipase A treatment of liver membranes abolished the effect of the glucagon on the adenylate cyclase; addition of membrane phospholipids restored hormone activity (Pohl et al., 1971a,b).

d. Guanyl Nucleotide Regulatory Subunit. Purine nucleosides (e.g., adenosine) and nucleotides interact with adenylate cyclase in vitro and produce variable degrees of stimulation or inhibition, depending on tissue origin and the type of nucleoside or nucleotide. The interaction of GTP and an analog, 5'-guanylylimidodiphosphate (Gpp(NH)p), with adenylate cyclase has been studied and shown to be complex, involving not only basal but also hormonal stimulated activity. GTP stimulates the activity of several hormone-dependent adenylate cyclases (Rodbell et al., 1971a,b; Londos et al., 1974; Schramm and Rodbell, 1975). In this stimulation, it is believed that GTP interacts with a specific component in the cell membrane, distinct from the hormone receptor and adenylate cyclase. This GTP component has been called the guanyl nucleotide regulatory subunit (Levitzki, 1977). The GTP analogs Gpp(NH)p (Londos et al., 1974) and GTPγS activate hormone-dependent adenylate cyclase in a quasi-irreversible fashion and, in the presence of hormone, induce the formation of a highly active and extremely stable adenylate cyclase. The turkey erythrocyte possesses a specific β-receptor-dependent GTPase incapable of hydrolyzing Gpp(NH)p and which is blocked irreversibly by GTPγS (Cassel and Selinger, 1976). It has been proposed that, when hormone binds to the receptor site, adenylate cyclase is activated. Hydrolysis of the GTP bound to the guanyl nucleotide regulatory site leads to inactivation of adenylate cyclase and termination of hormone action (Cassel and Selinger, 1976). Since the GTP analogs Gpp(NH)p and GTPγS are poorly hydrolyzed, the adenylate cyclase has greatly increased, continued activity (Levitzki, 1977).

Propranolol cannot reduce the activity of the highly active adenylate cyclase formed with Gpp(NH)p but can inhibit further activation of a partially active enzyme (Levitzki et al., 1976). In the turkey red blood cell, Gpp(NH)p and Ca^{2+} have no direct effect on the adenylate cyclase, but the Gpp(NH)p will activate the adenylate cyclase slowly in the absence of hormone. In the absence of a β-agonist and at low Mg^{2+} concentration, the activation of adenylate cyclase by Gpp(NH)p is insignificant. With high Mg^{2+} concentration, the conversion of adenylate cyclase to its active form by Gpp(NH)p is slow. To produce high persistent adenylate cyclase activity, both the hormone and Gpp(NH)p must be present (Sevilla et al., 1976). In the turkey erythrocyte, adenosine is an adenylate cyclase stimulator and acts synergistically with Gpp(NH)p, as does epinephrine (Sevilla et al., 1977).

Adenosine also stimulates the adenylate cyclase of other tissues, including guinea pig cerebral cortex slices (Sattin and Rall, 1970), isolated bone cells (Peck et al., 1974), platelets (Mills and Smith, 1971), fetal rat brain cell cultures (Gilman and Schrier, 1972), and lymphocytes (Wolberg et al., 1975). It is of interest that adenosine increases the concentration of cAMP in lymphocytes, which inhibits their in vitro cytolysis of tumor cells. Inhibition of adenosine deaminase potentiates the effect of adenosine. It is supposed that the impairment of lymphocyte function in adenosine deaminase deficiency is possibly related to the stimulatory effect of adenosine on lymphocyte adenylate cyclase (Wolberg et al., 1975). In brain slices, adenosine is released during depolarization (Shimizu et al., 1970; Schultz and Daly, 1973a,b).

The ability of adenosine to stimulate adenylate cyclase suggests that, in certain areas of the brain, adenosine may function as a neurotransmitter. However, although adenosine may stimulate adenylate cyclase in several tissues, there are no data to demonstrate that it functions as a hormone under physiological conditions, although it is possible that it may have local tissue effects, i.e., function as a paracrine substance.

It is well known that fluoride increases the cAMP concentration in particulate fractions from many different tissues (Robison et al., 1971; Manganiello and Vaughan, 1976). Fluoride may act on the guanyl nucleotide regulatory subunit, although the possibility that it functions through a different and as yet unknown component cannot be excluded (Johnson et al., 1978).

There appear to be at least two modes of interaction between a hormone receptor on the outer portion of the cell membrane and adenylate cyclase on the inner portion of the membrane. Either the receptor and the enzyme are separately mobile and float in the membrane, or they are permanently coupled together (Rimon et al., 1978). In the first case ("collision" coupling model), activation of the adenylate cyclase occurs when the hormone–receptor complex encounters the adenylate cyclase. The role of the guanyl nucleotide regulatory subunit is uncertain, but it could conceivably act to link the two separate components. It is also possible that the guanyl nucleotide regulatory subunit is permanently bound to the adenylate cyclase. In the second case (precoupling model), the receptor and adenylate cyclase are permanently coupled, and activation of the adenylate cyclase occurs when the hormone binds to the receptor.

The coupling of turkey erythrocyte adenylate cyclase to the β-adrenergic receptor is of the first type, where the receptor and adenylate cyclase are separate (Tolkovsky and Levitzki, 1978a). All of the adenylate cyclase is activated by the β-adrenergic receptor, whereas only 60–70% of the adenylate cyclase can be activated by the adenosine receptor (Tolkovsky and Levitzki, 1978b). In the turkey erythrocyte (Tolkovsky and Levitzki, 1978b) the receptor for adenosine is permanently coupled to adenylate cyclase. Epinephrine-dependent adenylate cyclase activation in this system is modified by changes in membrane fluidity, while the adenosine-dependent adenylate cyclase activation is independent of membrane fluidity (Rimon et al., 1978). Adenosine and l-epinephrine seem to activate adenylate cyclase through a common guanyl nucleotide subunit (Tolkovsky and Levitzki, 1978b).

This mobile receptor theory of hormone receptor–adenylate cyclase interactions explains a number of hitherto unexpained findings (Pohl, 1977; Bennett et al., 1975). It accounts for the apparent large excess of glucagon binding sites compared to the number of sites that must be occupied in order to activate adenylate cyclase (Birnbaumer and Pohl, 1973). It explains how multiple hormones including glucagon can activate adenylate cyclase in isolated fat cells (Pohl, 1977). In the rat reticulocyte, stimulation of β-adrenergic receptors by β-adrenergic agonists increases phospholipid methylation in the plasma membrane and decreases membrane microviscosity, resulting in increased lateral movement of the β-adrenergic receptors and enhancing the coupling of the receptor with adenylate cyclase (Hirata et al., 1979). It is not known how the binding of agonists with the β-adrenergic receptor stimulates phospholipid methylation; it is speculated that the membrane methyltransferases bind to the unoccupied β-adrenergic receptors, suppressing their activity. When the agonist binds to the receptor, the suppression of enzyme activity is reversed.

  e. Adenylate Cyclase. Adenylate cyclase is a ubiquitous membrane-

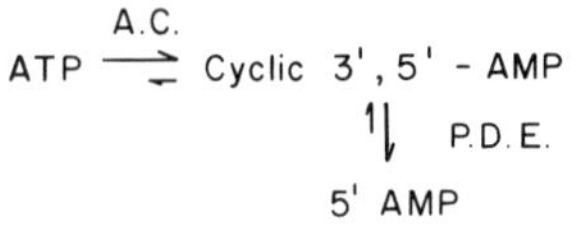

$$\text{ATP} \underset{}{\overset{A.C.}{\rightleftarrows}} \text{Cyclic 3',5' - AMP}$$
$$\downarrow \text{P.D.E.}$$
$$\text{5' AMP}$$

Figure 4. Cyclic AMP synthesis and degradation. Abbreviations used are: ATP, adenosine triphosphate; A.C., adenylate cyclase, P.D.E., phosphodiesterase; 5'-AMP, 5'-adenosine monophosphate.

bound enzyme whose substrate is MgATP. Evidence indicates that binding of Mg^{2+} to a site separate from the catalytic site is required for activity of the enzyme (Birnbaumer, 1973). Although Mn^{2+} may substitute for Mg^{2+} with variable effectiveness in different tissues, other divalent cations, such as Zn^{2+}, Cu^{2+}, and Ca^{2+} (at concentrations from 0.1 to 5.0 nM) are inhibitory. On the other hand, trace amounts of Ca^{2+} (submicromolar) seem to be essential for hormonal stimulation of adenylate cyclase activity in some tissues. These include stimulation by ACTH in fat and adrenal tissues, by oxytocin in frog bladder epithelium, and by (Arg^8) vasopressin in pig renal medullary membranes (Birnbaumer, 1973). The mechanism of this Ca^{2+} effect remains to be established.

Recently, an inhibitor of adenylate cyclase activity which seems to be a small peptide has been isolated from rat liver by Levey and co-workers (Levey et al., 1975b). The mechanism of inhibition and the physiological importance of this inhibition remain to be determined.

f. Regulation of Intracellular cAMP Concentration. The intracellular concentration of cAMP depends upon its rates of synthesis and degradation, as shown in Fig. 4. The major control point seems to be in the regulation of adenylate cyclase activity by the various hormones. The by-product of the reaction is inorganic pyrophosphate.

Nucleotide phosphodiesterase (PDE) catalyzes the breakdown of cAMP to 5'-AMP, and, although a potential control point in the regulation of cellular cAMP concentration, the activity of the PDE is rarely affected by hormones. (However, see Section II.B.3 concerning the role of calcium.) Nucleotide PDE is found in multiple forms in different tissues and is found in the soluble and membrane fractions from broken-cell preparations. Some forms hydrolyze only cGMP or cAMP, while others are capable of hydrolyzing both. Further work is required to understand the significance of the different PDE forms. A protein activator of PDE has been isolated from several tissues and partially purified (Cheung, 1971; Teo et al., 1973). The activator requires the presence of Ca^{2+} for activity. The effect of hormones on this activator has not been evaluated. A number of in vitro inhibitors of PDE have been described, including adenosine and a number of drugs such as the methylxanthines (especially theophylline), papaverine, tolbutamide, and chlorpropamide. Insulin has been shown to increase the activity of membrane-bound PDE in rat adipose tissue, and this may account for some of the actions of insulin in this tissue (Manganiello and Vaughan 1973; Zinnam and Hollenberg, 1974).

g. *Protein Kinase and Protein Phosphorylation and Dephosphorylation.* cAMP is the classic second messenger—or intracellular signal molecule—described by Sutherland and coworkers over 20 years ago. However, whether or not all of the metabolic effects of a given polypeptide hormone or β-adrenergic agonist result from changes in the intracellular cAMP concentration remains to be determined.

At present, the best understood action of cAMP is the activation of specific intracellular protein kinases, as depicted in Fig. 5. Cyclic AMP binds to the regulatory portion of a specific protein kinase complex, causing the regulatory subunit to dissociate from the catalytic portion, and thereby activating it. The active protein kinase is an Mg^{2+}-dependent enzyme which phosphorylates other proteins (some of which are also protein kinases), with the γ-phosphate of ATP being transferred to serine or threonine hydroxyl groups within the protein (Williams, 1976). Phosphorylation of various proteins, both membrane-bound and soluble, results in enzyme activation, enzyme inactivation, alteration in mRNA, alterations in enzyme synthesis, and changes in membrane permeability, depending on the specific protein that is phosphorylated.

It is outside the scope of this chapter to discuss the phosphorylation–dephosphorylation mechanisms of individual proteins. It is only possible here to discuss the system in general terms.

It has been proposed that hormones and neurotransmitters as well as other types of regulatory agents control physiological processes by means of protein phosphorylation (Greengard, 1975, 1976). Some of the protein phosphorylation occurs as a consequence of a hormone receptor–adenylate cyclase–cAMP–protein kinase sequence of action. In other cases, different cyclic nucleotides may replace cAMP, while with cyclic nucleotide-independent protein kinases, other signal molecules,

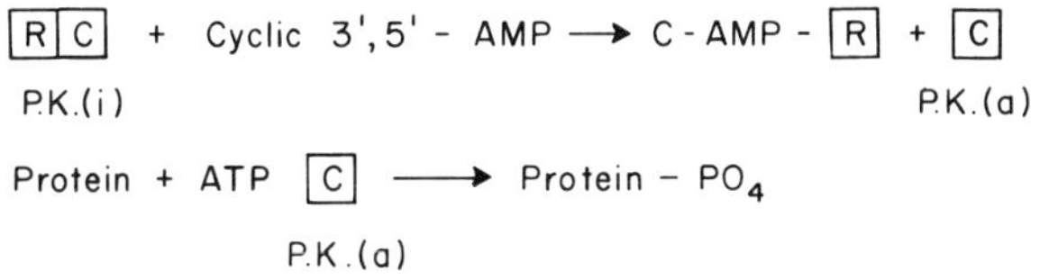

$$\boxed{R\,C} + \text{Cyclic } 3',5' - \text{AMP} \longrightarrow \text{C-AMP-}\boxed{R} + \boxed{C}$$
P.K.(i) \hspace{6cm} P.K.(a)

$$\text{Protein} + \text{ATP} \boxed{C} \longrightarrow \text{Protein} - PO_4$$
P.K.(a)

Figure 5. General mechanism of cAMP metabolic effects. R, regulatory subunit of protein kinase; C, catalytic subunit of protein kinase; P.H. (i), inactive protein kinase; P.K. (a), active protein kinase.

such as Ca^{2+}, may control protein kinase activity. As a general proposition, protein kinases translate information received at the cell membrane into specific responses of enzymes on other proteins, which themselves subsequently control an enzymatic function.

Protein kinase is actually composed of two regulatory subunits and two catalytic subunits. It combines with two moles of cAMP to produce two free and enzymatically active catalytic moieties (Hofman et al., 1975):

$$R_2C_2 + 2cAMP \rightleftharpoons R_2(cAMP)_2 + 2C \qquad (3)$$

The type I and type II protein kinase isozymes have the same catalytic subunit but different regulatory subunits, which accounts for the differences in their properties (Corbin and Keely, 1977). Protein kinase I is more characteristic of skeletal muscle and easily dissociates into its subunits by interacting with cAMP (Hofman et al., 1975). The type II kinase is more characteristic of bovine heart muscle, and readily dissociates only after the regulatory subunits have been phosphorylated. Type I protein kinse is not phosphorylated by MgATP, but the regulatory subunits of type II protein kinase are rapidly phosphorylated by MgATP (Hofman et al., 1975).

Several examples of hormonal control of enzymes by phosphorylation are considered in Chapter 4 under covalent modification of enzymes. Although protein phosphorylation–dephosphorylation may be a general regulatory mechanism for physiological processes, not all protein phosphorylation is cAMP mediated, or even hormonally mediated.

h. Cholera Toxin: The General Adenylate Cyclase Stimulator. Cholera toxin is capable of stimulating the adenylate cyclase of a great variety of tissues (Holmgren and Lönnroth, 1976). Cholera toxin (choleragen, mol. wt. 84,000) is a protein produced by *Vibrio cholerae* (van Heyningen, 1976) which causes diarrhea by activating adenylate cyclase in the intestinal mucosal epithelial cells (Field, 1971) after binding to receptors on the cell surface (Holmgren et al., 1975). The membrane receptor for choleragen in the small intestine is the ganglioside G_{M1} (Fig. 3), and there is a direct correlation between the cellular G_{M1} content and the number of choleragen receptors (Holmgren et al., 1975). G_{M1} binds and inactivates cholera toxin in equimolar proportions (Holmgren et al., 1973). On the other hand, chemical modifications of choleragen affect its binding to cells and to isolated G_{M1} equally (Hansson et al., 1977). The incorporation of G_{M1} into cell membranes increases the number of binding receptors and the biological responsiveness to the toxin (Holmgren et al., 1975; Cuatrecasas, 1973).

The binding of choleragen to cell membranes prevents the tritiation of membrane G_{M1} by galactose oxidase treatment followed by reduction with sodium [^{3}H] borohydride (Mullin et al., 1976a). It has been demonstrated by an immunoelectron microscopic method that membrane G_{M1} ganglioside is located on the external side of the membrane exclusively (Hansson et al., 1977). Transformed mouse fibroblasts (NCTC 2071 cells) are deficient in gangliosides and do not respond to choleragen, but, after ganglioside G_{M1} is taken up by the cells from an artificial medium (Fishman et al., 1976) or from serum (Fishman et al., 1978), the cells respond to choleragen.

As a consequence of its stimulation of adenylate cyclase in various cells, cholera toxin causes the same type of physiological action as does adenylate cyclase and the various hormones that stimulate adenylate cyclase. The toxin is composed of two subunits, A (mol. wt. 29,000) and B, in proportions ranging from 1:6 to 1:4, respectively (Cuatrecasas et al., 1973; Holmgren and Lönnroth, 1975; Mendez et al., 1975). The B subunits bind to the ganglioside receptor (Cuatrecasas, 1973; Holmgren et al., 1973; van Heyningen, S., 1974), which allows the A subunit to penetrate the cell membrane and activate the adenylate cyclase (Gill and King, 1975; Bennett et al., 1975). Following binding of the B subunits to G_{M1}, the choleragen–ganglioside complex undergoes a conformational change that facilitates the dissociation of the A subunit. This may involve the binding of one toxin molecule to more than one ganglioside receptor, with redistribution of the surface components by lateral movement of the toxin–ganglioside complex in the fluid phase of the membrane (Bennett et al., 1975). Such redistribution (patching and capping) has been demonstrated in lymphocytes using fluorescein-labeled cholera toxin (Craig and Cuatrecasas, 1975), or with immunofluorescence (Revesz and Greaves, 1975).

There is usually a lag between the binding of choleragen to the cell surface of an intact cell and the activation of adenylate cyclase. The lag is eliminated if only the A subunit is used (Bennett et al., 1975) or if cell membranes are treated with either ganglioside G_{M1} (Moss et al., 1976a) or the A subunit (Gill and King, 1975). The lag period apparently represents the time necessary for the A subunit to dissociate from the cholera toxin and to penetrate the cell membrane. The A subunit consists of two peptides linked by disulfide bridges (Gill and King, 1975; van Heyningen, S., 1974). The A subunit undergoes cleavage of these disulfide bonds by interacting with glutathione to form an A_1 fragment (mol. wt. 24,000) which activates adenylate cyclase (Gill and King, 1975; Moss et al., 1976b), and an A_2 fragment (mol. wt. 5,000). The A_1 fragment binds to the guanyl nucleotide regulatory subunit (mol. wt. 42,000), which is located on the inner surface of the plasma membrane (Johnson and Bourne, 1977), and activates adenylate cyclase by ADP-ribosylation (Moss and Vaughan, 1977a):

$$\text{NAD}^+ + \text{adenylate cyclase} \xrightarrow{\text{A fragment}} \text{ADP-ribose–adenylate cyclase} + \text{nicotinamide} + \text{H}^+ \quad (4)$$

Cholera toxin is an ADP-ribosyl transferase in which the A_1 component is the active moiety (Moss et al., 1976b). For a general review of the ADP-ribosylation of proteins see Hayaishi and Ueda (1977). The A protomer of choleragen can catalyze the ADP-ribosylation of L-arginine. The guanido group is the structural determinant of acceptor specificity (Moss and Vaughan, 1977a). In the pigeon erythrocyte, cholera toxin fragment A_1 catalyzes the transfer of ADP-ribose from NAD to the adenylate-cyclase-associated GTP-binding protein of molecular weight 42,000 (presumably the guanyl nucleotide regulatory subunit) (Gill and Meren, 1978). Adenylate cyclase activity increases with the addition of ADP-ribose to this protein. The protein is only accessible to toxin A subunits if the pigeon erythrocytes are lysed. In addition to NAD^+ and GTP, a cytoplasmic protein is required for activity of the cholera toxin A_1 subunit (Gill and King, 1975; Gill and Meren, 1978).

Since the B subunit of choleragen contains an animo acid sequence similar to the one in the β subunits of glycopeptide hormones (e.g., hCG and TSH), and both the β subunit of choleragen and the β subunits of the hormones are involved in the binding of the substances to the cell membrane receptor (Ledley et al., 1976; Liao and Pierce, 1970; Kurosky et al., 1977), it might be thought that the glycopeptide hormones activate adenylate cyclase in the same manner as does cholera toxin. Although NAD glycohydrolase activity was detected in highly purified preparations of hCG and bovine TSH, this apparently was due to a contaminant and was not an intrinsic property of the α or β subunits of either hormone. Thus, neither hormone activates adenylate cylase through an NAD-dependent mechanism analogous to that of cholera toxin (Moss et al., 1978).

Choleragen binds to liposomes containing G_{M1} but not other gangliosides (Moss et al., 1976c), and releases glucose only from G_{M1}-liposomes. Choleragen can induce permeability changes in liposomes containing G_{M1} ganglioside. However, although the B subunit (choleragenoid) is as effective as choleragen in releasing glucose from liposomes, the A subunit is ineffective (Moss et al., 1977). In artificial membranes, cholera toxin reacts with the sugar residues in G_{M1} to form ion-conducting channels (Tosteson and Tosteson, 1978).

Escherichia coli produces a heat labile toxin (LT) and a heat stable toxin (ST). The LT has a mode of action similar to that of cholera toxin (Evans et al., 1972; Field, 1979), but the ST does not (Field, 1979).

2. Other Cyclic Nucleotides

Because of the success in explaining the mechanism of action of many polypeptide and certain nonpolypeptide hormones by means of the hormone–cAMP-mediated protein phosphorylation, it has been tempting to postulate the action of other cyclic nucleotides in a parallel or analogous system. In particular, the role of guanylate cyclase and cGMP (guanosine 3′,5′-monophosphate) has been investigated. Since the role of adenylate cyclase and cAMP is well understood in terms of protein phosphorylation and dephosphorylation, the possible role of cGMP in the inverse function has been examined in certain systems. In the liver glycogen metabolic system, the sequence of events is hormone (glucagon, eprinephrine) → receptor → adenylate cyclase → cyclic AMP → protein kinase → phosphorylase b kinase → phosphorylase b → phosphorylase a → hydrolysis of glycogen, to produce glucose-1-phosphate. At the same time that glycogenolysis is activated, glycogenesis is deactivated by the same phosphorylation sequence (Schlender et al., 1969; Soderling et al., 1970). Insulin stimulates the inverse operation, i.e., enhances glycogenesis and decreases or inhibits glycogenolysis. During these processes, phosphorylase a is converted to phosphorylase b, and glycogen synthetase is converted from the D form to the I form (see Fig. 4, Chapter 4). Thus, phosphorylase is deactivated and glycogen synthetase is activated. Involved in these reactions are phosphorylase phosphatase, phosphorylase b kinase phosphatase, and glycogen synthetase phosphatase (Walsh et al., 1971). There are no data to support the hypothesis that insulin activates a guanylate cyclase and generates a cGMP so that it functions by the process of dephosphorylation. The exact mechanism of activation of the various phosphatases is presently unknown.

Although a variety of hormones increase cGMP in isolated rat liver cells, it seems unlikely that cGMP affects glycogenolysis (Pointer et al., 1976). The increases in cGMP owing to epinephrine, isoproterenol, or phenylephrine were inhibited by phenoxybenzamine but not by propranolol. Although carbachol and insulin also increase cGMP, carbachol has little effect on glycogenolysis, and insulin inhibits glycogenolysis. The action of carbachol was inhibited by atropine. All of these hormones required extracellular Ca^{2+} in order to be effective. The glycogenolytic effect of glucagon and catecholamines was only slightly inhibited by the omission of calcium.

Nonetheless, it is tempting to speculate that some symmetrical but opposing system is responsible for the opposite physiological responses of the hormones. This symmetrical and opposing system of cAMP and cGMP is termed the yin-yang theory (Goldberg et al., 1975) after the Far Eastern philosophical concept of equal but opposite forces of nature. The yin-yang hypothesis postulates that there is a reciprocal biological dualism between cAMP and cGMP, so that the ratio of the two nucleotides within cells regulates inverse functions. The yin-yang theory is supported by the functional antagonism between cAMP and cGMP in muscular tissue and neurons (George et al., 1970; Lee et al., 1972; Schultz et al., 1973a; Greengard and Kebabian, 1974). And in the adrenal medulla and superior cervical ganglion, the ratio of cAMP to cGMP seems to control the synthesis of tyrosine hydroxylase (Guidotti et al., 1975b). In addition, despite the generally unsuccessful attempts to prove the yin-yang theory, there are other studies which suggest that cGMP and perhaps other cyclic nucleotides may be involved in the regulation of certain hormonal and neurotransmitter systems.

It has been suggested that the metabolic effects of some of the polypeptide hormones which alter cAMP levels may also relate to their effects on cGMP levels. Rubin et al. (1977) have suggested that this is the case with corticotropin, since they showed reciprocal changes in cAMP and cGMP levels in isolated bovine adrenal cells prior to measurable changes in steroid production.

Guanylate cyclase is both soluble and membrane bound (Kimura and Murad, 1974; Chrisman et al., 1975) and converts guanosine triphosphate (GTP) to cyclic GMP (Goldberg et al., 1973). Adenylate cyclase requires either Mg^{2+} or Mn^{2+}, while Ca^{2+} is usually inhibitory (Drummond and Duncan, 1970). On the other hand, guanylate cyclase requires Mn^{2+} (Hardman and Sutherland, 1969), with Mg^{2+} or Ca^{2+} being poor substitutes. Guanylate cyclase is not stimulated by fluoride in broken-cell preparations (Hardman and Sutherland, 1969), in contrast to most adenylate cyclase enzymes, which are stimulated by fluoride. However the adenylate cyclases of human sperm (Hardman et al., 1971) and newborn rat brain are unresponsive to fluoride (Perkins and Moore, 1971).

It appears that β-adrenergic responses are mediated by adenylate cyclase, while cholinergic agents cause an increase in cGMP (George et al., 1970; Stoner et al., 1974). Some hormones and neurotransmitters can increase the concentration of cGMP. These include acetylcholine (George et al., 1970; Stoner et al., 1974; Clyman et al., 1975), fibroblast growth factor (Rudland et al., 1974), secretin (Thompson et al., 1972), histamine, serotonin, bradykinin (Clyman et al., 1975), norepinephrine (Butcher et al., 1976; O'Dea and Zatz, 1976), epinephrine (O'Dea and Zatz, 1976), and glutamic acid (Beam et al., 1977).

The acetylcholine-generated increase in cGMP is mediated by muscarinic receptors (Kuo et al., 1972; George et al., 1973) and requires Ca^{2+} in the extracellular medium (Schultz et al., 1973b). Most hormones that stimulate adenylate cyclase, however, have no effect on cGMP (Hardman et al., 1971) or guanylate cyclase. The guanylate cyclase activity in cell-free systems from liver and heart are unaffected by glucagon, insulin, or epinephrine (Hardman and Sutherland, 1969).

cGMP can be hydrolyzed by cyclic nucleotide phosphodiesterases. The phosphodiesterases are not entirely specific for cGMP, since cAMP can also be hydrolyzed. The different isozymes of phosphodiesterase have different relative affinities for the various cyclic nucleotides (Thompson and Appleman, 1971; Beavo et al., 1971; Miller et al., 1973; Russell et al., 1973; Brostrom and Wolff, 1976). cGMP has been reported to increase the activity of cAMP phosphodiesterase by binding to an allosteric site (Beavo et al., 1971; Franks and MacManus, 1971). cGMP is capable of producing a number of biological responses similar to those produced by cAMP, but the doses required far exceed physiological levels (Posternak, 1974). Since guanylate cyclase is not stimulated and increases in cGMP are not produced by most hormones, the physiological significance of the guanylate cyclase–cGMP system is uncertain at best.

A characteristic membrane-bound protein of mammalian smooth muscle cells is selectively phosphorylated by a cGMP-dependent protein kinase (Casnellie and Greengard, 1974). Acetylcholine, acting at the muscarinic receptors of smooth muscle, causes a marked increase in cGMP (Lee et al., 1972; Schultz et al., 1973c), elevating the physiological level of cGMP to the dose level necessary for the production of biological responses. Although cGMP-dependent protein kinases exist in many tissues, smooth muscle (Greengard, 1976), cerebellum (Schlichter et al., 1978), and intestinal brush border epithelium (De Jonge, 1976) are the only tissues in which any endogenous substrate proteins have so far been found. cGMP-dependent protein kinase is mainly found in lung, heart, small intestine, and cerebellum (Lincoln et al., 1976).

It has been reported that partially purified cGMP-dependent protein kinase does not activate phosphorylase kinase or inactivate glycogen synthase (Takai et al., 1975; Kuo et al., 1976; Inoue et al., 1976). More recent studies, however, have demonstrated that purified cGMP-dependent protein kinase catalyzes phosphorylation, activation, or both, of hormone-sensitive lipase, rabbit skeletal muscle phosphorylase kinase, rabbit skeletal muscle glycogen synthase, rat liver pyruvate kinase, rat liver fructose-1,6-bisphosphatase, and histone (Lincoln and Corbin, 1977; Khoo et al., 1977; Hashimoto et al., 1976). However, the cAMP-dependent protein kinase was a more effective catalyst than the cGMP-dependent protein kinase (Lincoln and Corbin, 1977). As has been pointed out, cAMP-dependent protein kinase is ubiquitous, but cGMP-dependent protein kinase is only present to an appreciable degree in lung, heart, small intestine, and cerebellum (Lincoln et al., 1976). Furthermore, in cAMP-dependent protein kinase, separate regulatory and catalytic subunits exist, while the binding and kinase activities in cGMP-dependent protein kinase are located on the same polypeptide chain. However, the sites of phosphorylation of protein seem to be the same for each protein kinase (Lincoln and Corbin, 1977). It has been suggested that each protein kinase is evolutionarily related to the other, with the cGMP-dependent protein kinase being the more primitive (Lincoln and Corbin, 1977).

cGMP and cAMP are both able to phosphorylate the inhibitory subunit (TN-I) of purified cardiac troponin (Blumenthal et al., 1978; Lincoln and Corbin, 1978), which is the Ca^{2+} regulatory protein of cardiac myofibrils. Its inhibitory subunit inhibits actomyosin ATPase activity. β-Adrenergic stimulation elevates cAMP concentrations (England, 1976), while cholinergic stimulation causes elevations in cGMP content (England, 1976) and depression of the contractile state of the heart (George et al., 1973). The V_{max} value for the cAMP-dependent protein kinase-catalyzed reaction exceeded that for the cGMP-dependent protein kinase-catalyzed reaction by 12-fold. This, plus the apparent 11-fold greater amounts of cAMP-dependent protein kinase present in the heart, indicates that phosphorylation of TN-I of troponin by the cAMP-dependent protein kinase in the intact heart would occur at a rate 130 times greater than by the cGMP-dependent protein kinase (Blumenthal et al., 1978). This explains why elevations in the cGMP of the intact heart are not associated with TN-I phosphorylation (England, 1976). Increases in cGMP concentration in the intact heart are related instead to the dephosphorylation of TN-I (England, 1976). The physiological significance of TN-I phosphorylation remains to be clarified.

There has been isolated from mammalian tissue a protein kinase modulator which seems to inhibit cAMP protein kinases and to activate cGMP kinases (Kuo, 1975). Although not completely established, the modulator may be a single protein which alters the activity of both types of protein kinases in concert. This modulator seems to be, at least in part, under hormonal control, since alloxan-induced diabetes caused changes in its activity in rat pancreas and adipose tissue.

There have been isolated from mammalian tissues two separate protein kinases which modulate the actions of the cAMP- and cGMP-dependent protein kinases (Shoji et al., 1978a). One of these two proteins is termed the inhibitory modulator of cAMP-dependent protein kinase, and the other is called the stimulatory modulator of cGMP-dependent protein kinase. Neither protein affects the opposite cyclic-nucleotide-dependent protein kinase. The stimulatory modulator is necessary for the maximum activity of the cGMP-dependent protein kinase found in many tissues. A homogeneous preparation of stimulatory modulator, obtained from dog heart, augmented the cGMP-dependent phosphorylation of histones, but not of protamine, myelin basic protein, troponin, or glycogen synthetase.

Nuclei of rat liver contain a guanylate cyclase (Earp et al., 1977), whose activity is increased by 35% in nuclei isolated from rats 30 min after glucagon administration. This coincides with maximal nuclear-membrane cGMP immunofluorescence. In vitro incubation of isolated rat liver nuclei with cAMP produced a 25% increase in nuclear guanylate cyclase activity. Apparently, cAMP activates the nuclear guanylate cyclase; but the mechanism and significance of these findings are both uncertain.

cGMP (as the dibutyryl derivative) injected into the lateral ventricles or other areas of the rat brain in pharmacological amounts abolishes the response of the animal to various noxious stimuli, including pain. The activity was not blocked by naloxone. Dose-related analgesia without sedation was produced (Cohn et al., 1978). Enkephalin and morphine increase cGMP levels in rat striatal slices (Minneman and Iversen, 1976), and morphine increases cGMP concentrations in vivo in the rat neostriatum (Racagni et al., 1976).

In human platelets, aggregating agents such as thrombin or collagen act on the platelet plasma membrane and stimulate guanylate cyclase, most of which is soluble. Arachidonic acid stimulates the soluble guanylate cyclase (Glass et al., 1977), as do unsaturated fatty acid peroxides (Hidaka and Asano, 1977). It is suggested that the aggregating agents release arachidonic acid from the platelet plasma membrane and that this arachidonic acid then stimulates guanylate cyclase and can be converted to prostaglandin endoperoxides (Glass et al., 1977). Various prostaglandins (E_1, E_2, $F_{2\alpha}$) had no effect on guanylate cyclase.

In Irish Setter dogs with inherited rod-cone dysplasia, there is a deficiency of cGMP phosphodiesterase in the affected visual cells

(Aguirre et al., 1978). This results in elevated levels of cGMP but no change in the levels of cAMP. A similar finding has been observed in mice (Aguirre et al., 1978) with an inherited retinal degeneration. The data support a causal relationship between the elevated cGMP levels and retinal degeneration.

Lymphocytes treated with phytohemagglutinin demonstrate an increase in cGMP (Hadden et al., 1972). In rabbit small-intestinal epithelial cells, treatment with cholera toxin not only activates adenylate cyclase but inactivates guanylate cyclase (Kiefer et al., 1975). There is also depression of cGMP phosphodiesterase. The heat stable toxin of *E. coli* increases mucosal cGMP levels in rabbit intestinal tissue (Hughes et al., 1978; Field, 1979). Endotoxin, contaminating erythropoietin preparations, is responsible for the increase in cGMP levels in rat fetal liver cells (Graber et al., 1979).

Much less is known about other cyclic nucleotides. cCMP (cytidine 3′,5′-monophosphate) can be formed by mouse liver homogenate from CTP (cytidine triphosphate) (Cech and Ignarro, 1977, 1978). It is argued, however, that the products formed were 5′-CMP and CDP, rather than cCMP (Gaion and Krishna, 1979a). This problem remains controversial (Gaion and Krishna, 1979b; Ignarro, 1979). A cCMP phosphodiesterase has been demonstrated in rat tissues, particularly liver, kidney, and intestine. The enzyme was best activated by Fe^{2+} and less well by Mn^{2+} and Mg^{2+}, and exhibited kinetics typical of a high K_m phosphodiesterase (Kuo et al., 1978). The enzyme had decreased activity in fetal liver and regenerating liver (Shoji et al., 1978b).

3. The Role of Calcium

It has been suggested that calcium acts as an intracellular signal molecule or messenger (Rasmussen, 1970; Rasmussen et al., 1972). Calcium is involved in a number of important biological functions, including blood clotting, glycogenolysis, gluconeogenesis, muscle contraction, stimulation and inhibition of enzyme activity, exocytosis, nerve conduction, calcification, fertilization, platelet function, and mitosis. Many of these functions are explicable in terms of the role of calcium in the phosphorylation–dephosphorylation of proteins. In particular, calcium ions regulate the level of phosphorylation of a number of endogenous proteins in intact nerve terminals through activation of a calcium-sensitive protein kinase (Krueger et al., 1977). Calcium may operate through the activation of various calcium-dependent protein kinases. In the glycogen metabolic system, calcium ion activates phosphorylase kinase (Walsh et al., 1971). Calcium ion also activates phosphorylase, transforming phosphorylase *b* into phosphorylase *a*, independent of phosphorylase kinase (Walsh et al., 1971). Whether the activation of phosphorylase *b* kinase by calcium is of physiologic significance is not known. It is known that electrical stimulation of muscle does not activate phosphorylase kinase (Drummond et al., 1969) and does not increase cAMP levels (Posner et al., 1965). Yet, electrical stimulation causes rapid formation of phosphorylase *a*. This is believed to result from the release of Ca^{2+} as a consequence of the contractile process (Walsh et al., 1971). The release of calcium during muscle contraction is thought to activate phosphorylase *b* kinase (Brostrom et al., 1971). The activation of phosphorylase *b* kinase increases its sensitivity to low concentrations of calcium (Brostrom et al., 1971).

There has been described a cyclic-nucleotide-independent protein kinase of rat liver which is produced from its inactive proenzyme by limited proteolysis with a Ca^{2+}-dependent protease (Takai et al., 1977). The proenzyme had a molecular weight of 7.7×10^4. The protease required a Ca^{2+} concentration of 2–3 mM, and had a molecular weight of about 9.3×10^4. A similar protein kinase has been isolated from rat brain (Inoue et al., 1977). The latter enzyme is termed protein kinase M in order to distinguish it from cAMP-dependent protein kinase (protein kinase A) and cGMP-dependent protein kinase (protein kinase G). Protein kinase M from rat brain was capable of phosphorylating rabbit skeletal muscle glycogen phosphorylase kinase, which markedly increased the enzymatic activity of the latter (Kishimoto et al., 1977). It has been found also that the proenzyme and the calcium-dependent protease also occur in other rat tissues including kidney, lung, heart, skeletal muscle, and adipose tissue (Kishimoto et al., 1977).

The endogenous activator of phosphodiesterase (Cheung, 1971; Weiss and Hait, 1977) is a heat-stable, calcium-dependent protein with a molecular weight of 18,920 (Liu and Cheung, 1976), which increases the activity of both cAMP and cGMP phosphodiesterases. This endogenous activator is termed calcium-dependent regulator (CDR) or modulator protein (MP). In addition to its stimulating effect on phosphodiesterases, CDR activates adenylate cyclase of the brain (Brostrom et al., 1975; Cheung et al., 1975), activates Ca^{2+}, Mg^{2+}-ATPase in plasma membranes of human red blood cells (Gopinath and Vincenzi, 1977; Jarrett and Penniston, 1977), is a component of myosin light-chain kinase in chicken gizzard (Dabrowska et al., 1978) and rabbit skeletal muscle (Yagi et al., 1978; Waisman et al., 1978), and activates Ca^{2+}-dependent protein kinase (Schulman and Greengard, 1978). A CDR from rat testis containing 151 amino acids was found to have the unusual amino acid ϵ-*N*-trimethyllysine (Jackson et al., 1977). Although in micromolar concentration, Ca^{2+} inhibits adenylate cyclase, it can also be stimulatory by forming a complex with CDR (Brostrom et al., 1975). The stimulation is due to the formation of a CDR–Ca^{2+} complex which activates CDR by inducing a conformational change (Liu and Cheung, 1976). The calcium-dependent regulator mediates Ca^{2+}-dependent phosphorylation in the membranes of many tissues (Schulman and Greengard, 1978). CDR is a subunit of two protein kinases (Cohen et al., 1978), specifically muscle phosphorylase kinase and myosin light-chain kinase (Barylko et al., 1978).

CDR is stored in the membranes and synaptic structures of the brain and can be released into the cytosol following phosphorylation of the membranes by cAMP-dependent phosphorylation (Gnegy et al., 1976) or when the concentration of cAMP increases after M adenylate cyclase is activated (Uzunov et al., 1975). It has been proposed that the release of CDR and its consequent activation of cyclic nucleotide phosphodiesterase serves to limit the duration of the cyclic nucleotide action (Uzunov et al., 1976).

A heat-labile inhibitor of CDR has been partially purified from bovine brain. This inhibitor, called modulator binding protein, combines with CDR, thereby causing an inhibition of phosphodiesterase activation (Wang and Desai, 1977). A different inhibitor of CDR has also been found in bovine brain. This inhibitor is heat-stable, has a molecular weight of about 70,000, and counteracts the stimulation of cyclic nucleotide phosphodiesterase by CDR (Sharma et al., 1978). This heat-stable inhibitor has no effect on calcium or CDR-independent cyclic nucleotide phosphodiesterase. The exact physiological significance of this inhibitor is unknown.

A cyclic-nucleotide-independent protein kinase has been found to transform glycogen synthase I (also termed a_4 synthase) into glycogen synthase D (b_4 synthase). This cyclic-nucleotide-independent protein kinase is stimulated by CDR (Srivastava et al., 1979). Conceivably, this enzyme mediates the inactivation of glycogen synthase by α-adrenergic agents (Assimacopoulos-Jeannet et al., 1977).

The adenylate cyclase of brain particulate fraction, solubilized with Triton X-100, required CDR, Ca^{2+}, and GTP in order to be activated by choleragen (Moss and Vaughan, 1977b). Use of Gpp(NH)p, CDR, and Ca^{2+} increased the adenylate cyclase activity, but choleragen had no effect. It appears that multiple protein factors may be involved in the activation of adenylate cyclase and that there may be multiple subunits of the adenylate cyclase system.

In the turkey erythrocyte, Ca^{2+} is a potent inhibitor of epinephrine-stimulated adenylate cyclase (Steer et al., 1975; Hanski et al., 1977). In the intact erythrocyte, catecholamines increased calcium efflux and decreased calcium influx, which resulted in a net decrease in intracellular calcium concentration.

In isolated rat hepatocytes, α-adrenergic stimulation of glycogenolysis depends on the activation of phosphorylase by calcium (Assimacopoulos-Jeannet et al., 1977). The primary action of calcium is on phosphorylase b kinase.

C. The Membrane Receptor–Non-Adenylate-Cyclase System

Certain polypeptide hormones, namely insulin and growth hormone, bind to cell membrane receptors (Czech, 1977; Kahn, 1976; Fain, 1974); yet, despite intensive study, their mode of action is still not known. A variety of mechanisms have been proposed to explain their actions, but as yet no satisfactory unifying theory accounts for their biological functions. It is generally agreed that each hormone binds to its specific receptor. The intervening metabolic sequence of events, however, remains obscure. It has been proposed that an intracellular signal molecule is generated, which then, by some as yet unknown pathway, leads to the well-known physiological function of each hormone. It has been tempting to postulate that cGMP, in accord with the yin-yang hypothesis, is the insulin intracellular signal molecule. However, attempts to demonstrate this link have not been successful. And, although much is known about the insulin receptor, little is known about how the insulin–receptor complex leads to the final physiological response attributed to insulin.

1. Insulin and Its Mechanism of Action

Insulin binds to a specific receptor in the cell membrane (Czech, 1977). One possible mode of action of insulin that has been proposed is that insulin causes a decrease in cAMP levels. However, this does not seem to be an obligatory step in the mechanism of action of insulin (Steinberg, 1976). None of the actions of insulin are mediated by cAMP or cGMP (Czech, 1977). Although insulin inhibits the increased levels of cAMP due to submaximal concentrations of glucagon, and blocks the action of exogenous cAMP in the perfused liver and isolated hepatocytes, the action of insulin can be dissociated from the effects of intracellular cAMP. The effects of insulin on cAMP appear to be secondary. And, although insulin may cause an increase in cGMP levels in isolated fat cells and liver slices, other agents which cause an increase in cGMP levels in the same tissues display none of the responses of insulin. Clearly, an increase in cGMP levels cannot be related directly to the physiological actions of insulin.

So far, it has proven impossible to link any of the physiological actions of insulin to changes in membrane electrical potential (Czech, 1977). It also has not been possible to link changes in monovalent cation concentration to insulin action. The effects of insulin on cellular Na^+ and K^+ are opposite to those necessary to mimic insulin action: insulin causes hyperpolarization and increases intracellular K^+ concentration, whereas depolarization and decreased K^+ concentration are required to mimic insulin action.

It has been suggested that insulin releases Ca^{2+} from the cell membrane and that this Ca^{2+} is then the intracellular signal molecule (the "third messenger") which mediates insulin action (Clausen, 1975; Clausen et al., 1974; Kissebah et al., 1975; Fraser, 1975). There is a body of information consistent with this view, but, there also are discrepancies between the action of intracellular Ca^{2+} and the physiological action of insulin. Thus, if some of the actions are mediated by the release of calcium from the cell membrane, this does not explain all of the actions of insulin. In the fat cell, calcium appears to promote lipolysis, which is contrary to insulin action (Park et al., 1972; Exton et al., 1972; Schimmel, 1976).

Several investigators have found that insulin promoted the phosphorylation of specific cytoplasmic proteins in fat cells (Benjamin and Singer, 1974, 1975; Avruch et al., 1976a,b). This effect of insulin was antagonized by epinephrine, which caused phosphorylation of another specific protein (Benjamin and Singer, 1974, 1975). Dibutyryl cAMP mimicked the effect of epinephrine, and insulin was antagonistic (Benjamin and Singer, 1975). No effect of insulin on dephosphorylation was found. Similar results were found by other investigators in fat cells (Avruch et al., 1976a,b) and hepatocytes (Avruch et al., 1978). In the hepatocytes, glucagon induced a rapid, selective, and transient increase in phosphorylation of from 4 to 12 major cytoplasmic phosphopeptides of varying molecular weights. cAMP content, histone kinase activity, and glycogen phosphorylase activity were all transiently increased. Insulin, on the other hand, increased the phosphorylation of one major phosphopeptide of molecular weight 46,000. At the same time, glycogen phosphorylase activity was inhibited and cAMP and histone kinase activity remained unchanged. In the presence of submaximal amounts of glucagon, insulin inhibited the glucagon-stimulated phosphorylation of two of the peptides, but still caused phosphorylation of its 46,000-molecular-weight protein. Glucagon did not inhibit the insulin-stimulated phosphorylation. The physiological significance of the protein phosphorylation in the fat cells and hepatocytes is presently unknown.

Insulin treatment of adipocytes results in activation of pyruvate dehydrogenase as a consequence of dephosphorylation (Weiss et al., 1971; Taylor, et al., 1973). No consistent change has been reported in the activity of the pyruvate dehydrogenase phosphate phosphatase (Czech, 1977). It is of considerable interest that calcium activates pyruvate dehydrogenase phosphate phosphatase (Severson et al., 1976). No increase in intramitochondrial calcium incorporation could be found despite the activation of pyruvate dehydrogenase. It is possible, however, that insulin may increase a labile calcium pool within the mitochondria. So, although the total calcium content remains constant, there may be a shift in intramitochondrial calcium from a stable to a more labile pool (McDonald et al., 1976). Insulin promotes protein dephosphorylation (Nimmo and Cohen, 1977; Larner and Villar-Palasi, 1970; Corbin et al., 1975a,b), as well as the dephosphorylation and consequent activation of glycogen synthase and pyruvate dehydrogenase. However, insulin injected intravenously into rats produced no change in liver glycogen synthase phosphatase, although glucose could reverse the action of glucagon on the enzyme (Gilboe and Nuttall, 1978).

It has been demonstrated (Larner, 1975) that insulin specifically activates glycogen synthase by converting the D form into the I form by dephosphorylation. Insulin acts within 5 min, and the activity continues for at least 30 min. The enzyme phosphorylase kinase kinase (cAMP-dependent protein kinase of glycogen synthase I kinase) converts glycogen synthase I to glycogen synthase D, but insulin has no effect on this mechanism. Insulin does not change cAMP levels. However, insulin does alter the sensitivity of the kinase to cAMP. Insulin converts the active protein kinase into an inactive form, allowing the glycogen synthase phosphatase to dephosphorylate glycogen synthase, to form the I form (active form) of the enzyme. How insulin is able to transform the glycogen synthase I kinase into its inactive D form is yet unknown.

Other possible mechanisms for insulin action have been proposed. It has been proposed that membrane phosphorylation regulates transport activity across membranes. Thus, if membrane phosphorylation decreases transport, it would be expected that dephosphorylation would enhance transport (Randle and Smith, 1958). However, insulin does not appear to phosphorylate plasma membrane proteins (Avruch et al., 1976b,c). The insulin activation of glucose transport in fat cells has been postulated to involve the oxidation of certain membrane sulfhydryl groups to the disulfide form (Czech, 1977). A rat liver plasma membrane subfraction has been isolated containing cAMP phosphodiesterase activity that could be stimulated by nanomolar amounts of insulin in vitro (House et al., 1972). About 40% stimulation was obtained. Similar results were obtained with a fat cell plasma membrane cAMP phosphodiesterase (Kono et al., 1975). This insulin-stimulated phosphodiesterase has certain sulfhydryl groups which can be oxidized by certain oxidants including molecular oxygen, which stabilizes the activity of the enzyme. The presence of this enzyme in the adipocyte plasma membrane may account for the observation that insulin lowers the cAMP level in fat cells (Butcher et al., 1966; Kono and Barham, 1973).

A number of recent observations have shown that insulin also penetrates cells and localizes to various subcellular structures. Thus, insulin has been demonstrated to localize in the endoplasmic reticulum and nuclear membrane of human lymphocytes (Goldfine et al., 1978), and the nuclear membrane of rat hepatocytes (Vigneri et al., 1978). Insulin will bind to other subcellular structures such as the Golgi apparatus (Bergeron, 1978). The significance of these findings is unknown. Conceivably this may represent an intracellular degradative pathway which results in the termination of insulin action, as has been suggested (Terris and Steiner, 1975). Nevertheless, it has been proposed that insulin has two related mechanisms of action. One action of insulin is mediated by its binding to the plasma membrane where it stimulates membrane transport. The other action of insulin is mediated by the entrance of insulin into the cell, where it binds to subcellular organelles as the first step in its intracellular actions (Goldfine, 1977). The latter mechanism would account for the proposal that insulin enhances the binding of hexokinase to mitochondria (Bessman and Gots, 1975).

2. Growth Hormone

As incomplete as the picture is for insulin, still less is known about the mechanism of action of growth hormone. Growth hormone binds to the cell membrane (Kahn, 1976) and physiologically antagonizes the action of insulin. In the absence of growth hormone, there is increased insulin sensitivity, lowered glucose levels, and decreased glucose production. The administration of growth hormone increases the levels of blood glucose and insulin, increases hepatic glucose production, inhibits the disposal of glucose loads, and can cause diabetes. Yet, despite the purification and characterization of growth hormone, identification of its receptor sites, and knowledge of its physiological actions, its mechanism of action is unknown.

Growth hormone stimulates growth, protein synthesis, lipolysis, and ketogenesis. The main effect of growth hormone is primarily through the stimulation of amino acid transport. Although theophylline inhibits the effect of growth hormone on amino acid uptake and protein synthesis, dibutyryl cAMP does not duplicate growth hormone action. The lipolytic action of growth hormone is blocked by the inhibition of protein synthesis. It has not been possible to implicate adenylate cyclase and cAMP in the action of growth hormone (Schreiber, 1974).

Growth hormone promotes the appearance in tissues of a factor originally called sulfation factor but later called somatomedin (Hall and Luft, 1974). It is now believed that somatomedin is in part responsible for the nonsuppressible insulin-like activity found in serum. Even more recently, the term insulin-like growth factor has been employed (Ginsberg et al., 1979). It is not possible here to discuss the many problems involved in characterizing and investigating these substances. It is clear, however, that growth hormone does function via the formation of somatomedin. But the mechanism by which somatomedin forms remains unknown. It is known that somatomedin has insulin-like actions and that insulin and somatomedin compete for the same receptor. A purified preparation of somatomedin A, for example, stimulates the transport and uptake of amino acids into proteins, and the glucose uptake by diaphragms from hypophysectomized rats (Hall and Luft, 1974). The mechanism of action of somatomedin is unknown.

D. Intracellular Hormone-Binding Proteins

1. The General System

The steroid hormones and thyroid hormones have a mechanism of action distinct from that of the polypeptide hormones and catecholamines. The steroid and thyroid hormones do not act at the cell membrane by binding to a specific receptor which subsequently generates an intracellular signal molecule, although they may interact with the cell membrane in other capacities (Chapter 11). Rather, steroid and thyroid hormones enter the cell by traversing the cell membrane and bind to an intracellular carrier or binding protein. This hormone-binding protein complex enters the nucleus from the cytoplasm. Once in the nucleus, the hormones affect DNA in such a way that protein synthesis is stimulated. In some cases, specific protein products which are synthesized as a consequence of hormone stimulation of DNA can be identified. Thus, the specific hormones themselves act to stimulate protein synthesis by acting directly on the protein synthetic mechanism. How these hormones influence DNA and the protein synthetic mechanism is not entirely clear. A great deal of information has been obtained about the presence and characteristics of the binding proteins in various tissues for each of the hormones. Although details differ from tissue to tissue and between the particular hormones concerned in the system, in principle the general mechanism is quite similar in the different tissues and animals.

a. Prehormones. In at least two cases, hormones that reach their target cells must be transformed into their final active forms. Thus, the circulating hormones that are transformed are prehormones. These prehormones must not be confused with the preprohormones or their equivalents formed in endocrine-secreting cells (see Section II.A.4).

The thyroid hormones thyroxine (T_4) and 3,5,3'-triiodothyronine (T_3) are released from the thyroid gland, circulate in the blood bound to thyroid binding globulin from which they are released, and enter into their target cells. Once in the target cells, T_4 is deiodinated to form T_3. T_4 is the major source of T_3 in man (Surks et al., 1973). And, since T_3 is considered to be the active form of the hormone, T_4 is thus a prehormone (Ingbar and Braverman, 1975). T_4 can also be deiodinated to form 3,3',5'-triiodothyronine (reverse T_3, RT_3), which has little hormonal activity (Chopra, 1974; Gavin, et al., 1977). The peripheral conversion of thyroid hormones has been demonstrated in liver, kidney, fibroblast cultures, and other tissues (Schimmel and Utiger, 1977). Reverse T_3 is deiodinated to form 3,3'-diiodothyronine (T_2). A 5'-deiodinase converts T_4 into T_3 and RT_3 into T_2. The 5'-deiodinase appears to be located in the plasma membrane (Leonard and Rosenberg, 1978). A number of conditions appear to inhibit the 5'-deiodinase activity and thus to diminish the conversion of T_4 into T_3 and RT_3 into T_2. Inhibition occurs during fasting, various illnesses, the administration of drugs (glucocorticoids, propylthiouracil), and in the fetus (Schimmel and Utiger, 1977). When 5'-deiodinase is inhibited, a separate enzyme, 5-deiodinase, converts T_4 to RT_3 (Cavalieri et al., 1977), which then accumulates because its further metabolism is inhibited. How the various factors inhibit the 5'-deiodinase is not clear. During fasting, reduced glutathione decreases in the liver. This hepatic decrease in reduced glutathione accounts for only part of the loss of 5'-deiodinase activity (Kaplan, 1979). RT_3 can also inhibit the conversion of T_4 to T_3 (Chopra, 1977).

Testosterone is secreted from the testes, circulates in the blood bound to its specific carrier protein (sex-hormone-binding globulin), and enters into its target tissues. Testosterone is converted into 5α-dihydrotestosterone (DHT), which is considered to be the active hormone in some target tissues such as the accessory sex organs (Baker et al., 1977). Thus, testosterone is a prehormone. Testosterone appears to be the active hormone in certain other tissues, however, such as the mouse kidney (Bullock and Bardin, 1974), male rat perineal muscles (Krieg et al., 1974), rat pituitary (Naess et al., 1975), and fetal Wolffian duct derivatives (Goldstein and Wilson, 1975). The transformation of testosterone to DHT is carried out by an NADPH:Δ^4-3-ketosteroid 5α-oxido-reductase (5α-reductase)—which is located in microsomal (Ofner, 1968) and nuclear membranes (Moore and Wilson, 1972).

b. *Intracellular Binding Proteins.* Steroid and thyroid hormones bind to intracellular binding proteins or receptors. The hormone–receptor complex is translocated to the cell nucleus where it influences the DNA to synthesize specific mRNAs. The characteristics of this system have been studied for the various hormones including cortisol, testosterone, dihydrotestosterone, 17β-estradiol, aldosterone, corticosterone, progesterone, 1,25-dihydroxycholecalciferol, and triiodothyronine (Baxter and Forsham, 1972; Feldman et al., 1972; O'Malley et al., 1975; Buller and O'Malley, 1976; Chan and O'Malley, 1978; Oppenheimer and Surks, 1975; Sterling, 1979; Lubke et al., 1976; Mainwaring, 1975; Gorski and Gannon, 1976; Williams, 1974; Baulieu et al., 1975). All of the systems follow the same general principles. An idealized system is shown in Fig. 6. Intracellular binding proteins have been described for estrogens (Gorski and Gannon, 1976), progesterone (Baulieu et al., 1975), dihydrotestosterone (Liao, 1975), cortisol (Yamamoto and Alberts, 1976), aldosterone (Anderson and Fanestil, 1978), 1,25-dihydroxycholecalciferol (Haussler and McCain, 1977; DeLuca, 1978), and triiodothyronine (Oppenheimer and Surks, 1975; Sterling, 1979).

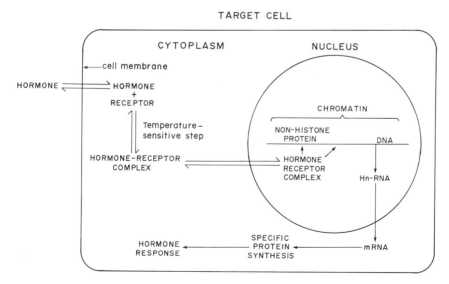

Figure 6. Idealized scheme of the mechanism of action of steroid and thyroid hormones using an intracellular binding protein or receptor. DNA, deoxyribonucleic acid; RNA, ribonucleic acid; Hn-RNA, heterogeneous RNA; mRNA, messenger RNA.

The intracellular binding proteins or receptors have the characteristics well known for hormone receptors. The receptors have high specific affinity for their respective hormones, are saturable, and the binding of the receptor to the hormone corresponds to the appearance of physiological responses of the tissue to the hormone. The hormone receptors are proteins which have a high affinity for their specific hormones, with a K_d ranging from 10^{-10} to 10^{-8} M.

The hormone receptors have various structures. The progesterone receptor in the chick oviduct is composed of two subunits, A and B (Chan and O'Malley, 1978). The A subunit has a molecular weight of 79,000 daltons and the B unit 117,000 daltons. In low ionic strength, the progesterone receptor sediments at 6 S and 8 S. The 8-S form is believed to be a nonspecific aggregate of 4-S components. Both the A and B subunits sediment at 4 S despite their difference in molecular weight.

The estrogen receptor of the rat uterus has a sedimentation coefficient of 8 S. Depending on conditions, the sedimentation coefficient can be altered (Chan and O'Malley, 1978). The estrogen receptor was the first intracellular receptor to be discovered (Baulieu et al., 1975). Receptors for estrogen have been found in the uterus, vagina, and anterior pituitary (Baulieu et al., 1975). In the various species studied (rat, calf, mouse, pig, rabbit, human), the sedimentation coefficient has ranged from 8 to 10 S. A high ionic strength (0.3 M KCl) causes a reversible dissociation of the receptor to a 4- to 5-S component (Baulieu et al., 1975).

The hormone–receptor complex, once formed, moves to the nucleus, where the complex interacts with the chromatin of the nucleus. The translocation of the complex is the main process whereby the hormone reaches the nucleus. Neither the cytoplasmic receptor nor the hormone alone enters the nucleus (Baulieu et al., 1975). The binding of the estrogen and the cytoplasmic hormone receptor is temperature dependent. Binding occurs at 37°C but is inhibited at 0°C.

Within the nucleus, the hormone–receptor complex binds to the chromatin. The complex binds poorly to purified DNA, however (Chan and O'Malley, 1978). The complex appears to be specific for its chromatin, since little binding is observed when non-target-cell chromatin is used. The free hormone also binds poorly to chromatin. Using nonhistone chromosomal proteins from target-tissue chromatin, it is possible to produce hormone–receptor complex binding in non-target-tissue chromatin (Spelsberg et al., 1972).

The B subunit of the progesterone receptor binds to nonhistone protein–DNA complexes of oviduct chromatin but not to purified DNA, while the A subunit binds to purified DNA but poorly to chromatin (Schrader and O'Malley, 1972). It has been postulated, therefore, that the B subunit acts as a site-specific binder to localize the hormone–receptor complex to certain regions of the chromatin and that the A subunit alters the conformation of a localized region of the chromatin DNA so that the synthesis of mRNA (or hnRNA) can be initiated (O'Malley and Schrader, 1976). Protein synthesis is stimulated as a consequence of hormone–receptor complex activation of specific genes (Chapter 5).

Treatment of the immature chick with estrogen causes growth and differentiation of the oviduct. The tubular gland cells proliferate and a variety of proteins are synthesized, including ovalbumin, conalbumin, ovomucoid, and lysozyme (Chan et al., 1978). The mRNAs for these proteins have been characterized (Chan et al., 1978). Progesterone is able to induce the synthesis of avidin, ovalbumin, ovomucoid, conalbumin, and lysozyme in the chick oviduct. Estrogen also induces vitellogenin and transferrin in the chick liver. Other steroid hormones can induce protein synthesis in their target tissues. Androgen induces α_{2u}-globulin in rat liver and aldolase in rat prostate. Glucocorticoids induce tyrosine aminotransferase and tryptophan oxygenase in rat liver, phosphoenolpyruvate carboxykinase in rat kidney, and growth hormone in rat pituitary cell tissue culture, and decrease the synthesis of adrenocorticotropin in rat pituitary (Chan et al., 1978).

Withdrawal of estrogen administration from chicks leads to a decrease in the size of the oviduct and in the synthesis of the various proteins. Readministration (secondary stimulation) induces specific protein synthesis which is independent of cell proliferation. Ovalbumin accounts for about 60% of the total oviduct protein in the fully stimulated oviduct.

It has been shown that, during secondary stimulation with estrogen, there is an increase in the concentration of nuclear receptor molecules followed by an increase in the available initiation sites for RNA synthesis on the chromatin (Tsai et al., 1975b; Kalimi et al., 1976; Anderson et al., 1972). Using hybridization techniques, it has been shown that control of the expression of the ovalbumin gene by estrogen occurs at the transcriptional level (Harris et al., 1976).

Purification of the ovalbumin gene has demonstrated that it is discontinuous (Breathnach et al., 1977; Doel et al., 1977; Weinstock et al., 1978; Lai et al., 1978). The culmination of these recent studies has shown that there are seven intragenic spacers or interrupted sequences (inserts, introns) in the ovalbumin gene (Dugaiczyk et al., 1978). Because of the additional material in the ovalbumin gene (the intragenic spacers), the ovalbumin gene is about four times longer than its corresponding mRNA. The primary transcript of the ovalbumin gene is a large RNA molecule (>40 S) which may constitute part of the hnRNA.

Further processing, such as capping, the addition of poly(A), and the excision of the intragenic spacers (naturally occurring genetic recombination) is necessary in order to transform the primary transcript RNA into the final mRNA (Chapter 5).

The exact mechanism whereby the steroid hormones and thyroid hormones cause specific genes to initiate RNA synthesis, however, remains unknown.

2. Extranuclear Action of Steroids and Thyroid Hormones

Although it is generally agreed that steroid and thyroid hormones act in the nucleus of the target cell to induce specific protein synthesis, there are data which suggest that thyroid hormone (Sterling, 1979) and 1,25-dihydroxycholecalciferol (Bikle et al., 1979) may also act via extranuclear mechanisms.

Thyroid hormone binds to its appropriate receptor in the cytoplasm and is translocated to the nucleus, where it induces the synthesis of specific proteins (Sterling, 1979; Oppenheimer and Surks, 1975). The thyroid hormone receptor does not enter into the nucleus; instead, the thyroid hormone (T_3) enters and binds to a separate receptor in the chromatin (Sterling, 1979). In addition, T_3 binds to a specific receptor in mitochondria which is separate from the cytoplasmic thyroid hormone receptor (Sterling, 1979). T_3 stimulates the appearance of the protein α_{2u}-globulin, which also requires testosterone for its synthesis (Kurtz et al., 1976; Chan et al., 1978). Furthermore, cells cultured from a rat pituitary tumor are stimulated by T_3 to produce the mRNA for growth hormone (Shapiro et al., 1978). There are a number of problems in ascribing all of the actions of T_3 to induction of protein synthesis by the action of T_3 at the nuclear level. Similarly, it is not possible to account for all of the actions of T_3 at the mitochondrial level.

Other possible primary loci of action have been suggested by a number of investigators (Sterling, 1979). These include action of T_3 on the membrane ($Na^+ + K^+$)-ATPase (Philipson and Edelman, 1977), effects at the plasma membrane (Pliam and Goldfine, 1977), interaction with the β-adrenergic receptor (Malbon et al., 1978), function as a tyrosine analog (Dratman, 1974), or a combination of some or all of these proposed mechanisms (Sterling, 1979). There are data to support all of the contentions, with the concept that T_3 binds in the nucleus and stimulates protein synthesis seeming to have the preponderance of data so far. But since receptor sites have been located in mitochondria (Sterling, 1979) and in the plasma membrane (Pliam and Goldfine, 1977), it is possible that T_3 operates to some degree at various intracellular sites. It is not yet possible to determine the precise mechanism of action of the thyroid hormone.

There is general agreement that 1,25-dihydroxycholecalciferol—the final active hormone of the vitamin D series—affects its target organs through the same general mechanism as the other steroids, namely, by binding to a cytoplasmic receptor, translocation to the cell nucleus, activation of DNA-dependent RNA synthesis, formation of specific mRNAs, and, finally, synthesis of specific proteins (Haussler and McCain, 1977; DeLuca, 1978). Calcium binding protein is one such protein which originally was thought to mediate the membrane transport of calcium. However, calcium binding protein is now known not to be involved in the membrane transport of calcium but rather is involved in the intracellular regulation of calcium to prevent accumulation of toxic levels of calcium or perhaps to facilitate the movement of calcium in and out of mitochondria (Bikle et al., 1979). It is clear that 1,25-dihydroxycholecalciferol enhances the absorption of calcium in the small intestine; however, the mechanism is not yet clear.

In addition to calcium binding protein, alkaline phosphatase activity is stimulated by 1,25-dihydroxycholecalciferol. This stimulation is inhibited by cycloheximide but not by actinomycin D or cordycepin. In fact, the latter two substances stimulate alkaline phosphatase activity (Bikle et al., 1979). Other enzymes are stimulated by 1,25-dihydroxycholecalciferol, including RNA polymerase and calcium-activated ATPase.

It has been found that 1,25-dihydroxycholecalciferol stimulates calcium and phosphate transport in the small intestine and calcium accumulation within the cells by means that do not require RNA or protein synthesis. Yet, many of the other changes in protein and enzyme activity require the induction of protein synthesis, which is compatible with the activation of DNA-dependent RNA synthesis in the nucleus by 1,25-dihydroxycholecalciferol. The stimulation of the brush border enzyme alkaline phosphatase by inhibitors of DNA-dependent RNA synthesis is puzzling. These various results suggest, but do not prove, that 1,25-dihydroxycholecalciferol has extranuclear actions (Bikle et al., 1979).

References

Adams, D. D., and Kennedy, T. H., 1967, Occurrence in thyrotoxicosis of a gamma globulin which protects LATS from neutralization by an extract of thyroid glands, *J. Clin. Endocrinol. Metab.* **27**:173.

Aguirre, G., Farber, D., Lolley, R., Fletcher, R. T., and Chader, G. J., 1978, Rod-cone dysplasia in Irish Setters: A defect in cyclic GMP metabolism in visual cells, *Science* **201**:1133.

Altszuler, N., Rathgeb, I., Winkler, B., DeBodo, R. C., and Steele, R., 1968, The effects of growth hormone on carbohydrate and lipid metabolism in the dog, *Ann. N.Y. Acad. Sci.* **148**: 441.

Anderson, D. C., 1974, Sex-hormone-binding globulin, *Clin. Endocrinol.* **3**:69.
Anderson, J., Clark, J. H., and Peck, E. J., Jr., 1972, Oestrogen and nuclear binding sites: Determination of specific sites by [^{3}H]oestradiol exchange, *Biochem. J.* **126**:561.
Anderson, N. S., III, and Fanestil, D. D., 1978, Biology of mineralocorticoid receptors, in: *Receptors and Hormone Action*, Vol. II (B. W. O'Malley and L. Birnbaumer, eds.), pp. 323–351, Academic Press, New York.
Ansorge, S., Bohley, P., Kirschke, H., Langner, J., Wiederanders, B., and Hanson, H., 1973, Metabolism of insulin and glucagon. Glutathione-insulin transhydrogenase from microsomes of rat liver, *Eur. J. Biochem.* **32**:27.
Asano, T., Greenberg, B. Z., Wittmers, R. V., and Goetz, F. C., 1977, Xylitol, a partial homologue of α-D-glucopyranose: Potent stimulator of insulin release in dogs, *Endocrinology* **100**:339.
Assimacopoulos-Jeannet, F. D., Blackmore, P. F., and Exton, J. H., 1977, Studies on α-adrenergic activation of hepatic glucose output. Studies on role of calcium in α-adrenergic activation of phosphorylase, *J. Biol. Chem.* **252**:2662.
Aurbach, G. D., Fedak, S. A., Woodward, C. J., Palmer, J. S., Hauser, D., and Troxler, G., 1974, β-Adrenergic receptor: Stereospecific interaction of iodinated β-blocking agent with high affinity site, *Science* **186**:1223.
Avruch, J., Leone, G. R., and Martin, D. B., 1976a, Identification and subcellular distribution of adipocyte peptides and phosphopeptides, *J. Biol. Chem.* **251**:1505.
Avruch, J., Leone, G. R., and Martin, D. B., 1976b, Effects of epinephrine and insulin on phosphopeptide metabolism in adipocytes, *J. Biol. Chem.* **251**:1511.
Avruch, J., Fairbanks, G., and Crapo, L. M., 1976c, Regulation of plasma membrane protein phosphorylation in two mammalian cell types, *J. Cell. Physiol.* **89**:815.
Avruch, J., Witters, L. A., Alexander, M. C., and Bush, M. A., 1978, Effects of glucagon and insulin on cytoplasmic protein phosphorylation in hepatocytes, *J. Biol. Chem.* **253**:4754.
Azhar, S., and Menon, K. M. J., 1978, Differential actions of gangliosides on gonadotropin and cholera enterotoxin stimulated adenosine 3':5' cyclic monophosphate dependent protein kinase in isolated rat ovarian cells, *Biochem. Biophys. Res. Commun.* **81**:205.
Baker, H. W. G., Bailey, D. J., Feil, P. D., Jefferson, L. S., Santen, R. J., and Bardin, C. W., 1977, Nuclear accumulation of androgens in perfused rat accessory sex organs and testes, *Endocrinology* **100**:709.
Barbosa, J., Seal, U. S., and Doe, R.P., 1973, Anti-estrogens and plasma proteins. II. Contraceptive drugs and gestagens, *J. Clin. Endocrinol. Metab.* **36**:706.
Barylko, B., Kuźnicki, J., and Drabikowski, W., 1978, Identification of Ca^{2+}-binding subunit of myosin light chain kinase from skeletal muscle with modulator protein, *FEBS Lett.* **90**:301.
Bashford, C. L., Harrison, S. J., and Radda, G. K., 1975, The relation between lipid mobility and the specific hormone binding of thyroid membranes, *Biochem. J.* **146**:473.
Baskin, F. K., Duckworth, W. C., and Kitabchi, A. E., 1975, Sites of cleavage of glucagon by insulin-glucagon protease, *Biochem. Biophys. Res. Commun.* **67**:163.
Baulieu, E.-E., Atger, M., Best-Belpomme, M., Corvol, P., Courvalin, J.-C., Mester, J., Milgrom, E., Robel, P., Rochefort, H., and De Catalogne, D., 1975, Steroid hormone receptors, *Vitam. Horm.* **33**:649.
Baxter, J. D., and Forsham, P. H., 1972, Tissue effects of glucocorticoids, *Am. J. Med.* **53**:573.
Bayliss, W. M., and Starling, E. H., 1902, The mechanism of pancreatic secretion, *J. Physiol.* **28**:325.
Beam, K. G., Nestler, E. J., and Greengard, P., 1977, Increased cyclic GMP levels associated with contraction in muscle fibres of the giant barnacle, *Nature* **267**:534.

Beavo, J. A., Hardman, J. G., and Sutherland, E. W., 1971, Stimulation of adenosine 3',5'-monophosphate hydrolysis by guanosine 3',5'-monophosphate, *J. Biol. Chem.* **246**:3841.

Beck, P., and Daughaday, W. H., 1967, Human placental lactogen: Studies of its acute metabolic effects and disposition in normal man, *J. Clin. Invest.* **46**:103.

Bellisario, R., and Bahl, O. P., 1975, Human chorionic gonadotropin. V. Tissue specificity of binding and partial characterization of soluble human chorionic gonadotropin-receptor complexes, *J. Biol. Chem.* **250**:3837.

Benjamin, W. B., and Singer, I., 1974, Effect of insulin on the phosphorylation of adipose tissue protein, *Biochim. Biophys. Acta* **351**:28.

Benjamin, W. B., and Singer, I., 1975, Actions of insulin, epinephrine, and dibutyryl cyclic adenosine 5'-monophosphate on fat cell protein phosphorylations. Cyclic adenosine 5'-monophosphate dependent and independent mechanisms, *Biochemistry* **14**:3301.

Bennett, V., and Cuatrecasas, P., 1973, Preparation of inverted plasma membrane vesicles from isolated adipocytes, *Biochim. Biophys. Acta* **311**:362.

Bennett, V., O'Keefe, E., and Cuatrecases, P., 1975, Mechanism of action of cholera toxin and the mobile receptor theory of hormone receptor-adenylate cyclase interactions, *Proc. Natl. Acad. Sci. USA* **72**:33.

Berger, E. A., and Shooter, E. M., 1977, Evidence for pro-β-nerve growth factor, a biosynthetic precursor to β-nerve growth factor, *Proc. Natl. Acad. Sci. USA* **74**:3647.

Bergeron, J. J. M., Posner, B. I., Josefsberg, Z., and Sikstrom, R., 1978, Intracellular polypeptide hormone receptors. The demonstration of specific binding sites for insulin and human growth hormone in Golgi fractions isolated from the liver of female rats, *J. Biol. Chem.* **253**:4058.

Bessman, S. P., and Gots, R. E., 1975, The hexokinase acceptor theory of insulin action—hormonal control of functional compartmentation, *Life Sci.* **16**:1215.

Bikle, D. D., Morrissey, R. L., and Zolock, D. T., 1979, The mechanism of action of vitamin D in the intestine, *Am. J. Clin. Nutr.* **32**:2322.

Birken, S., Smith, D. L., Canfield, R. E., and Boime, I., 1977, Partial amino acid sequence of human placental lactogen precursor and its mature hormone form produced by membrane-associated enzyme activity, *Biochem. Biophys. Res. Commun.* **74**:106.

Birken, S., Fetherston, J., Desmond, J., Canfield, R., and Boime, I., 1978, Partial amino acid sequence of the pre-protein form of the alpha subunit of human choriogonadotropin and identification of the site of subsequent proteolytic cleavage, *Biochem. Biophys. Res. Commun.* **85**:1247.

Birnbaumer, L., 1973, Hormone-sensitive adenylate cyclase, useful models for studying hormone receptor functions in cell free systems, *Biochim. Biophys. Acta* **300**:129.

Birnbaumer, L., and Pohl, S. L., 1973, Relation of glucagon-specific binding sites to glucagon-dependent stimulation of adenylyl cyclase activity in plasma membranes of rat liver, *J. Biol. Chem.* **248**:2056.

Birnbaumer, L. S., Pohl, S. L., and Kaumann, A. J., 1974, Receptors and acceptors, a necessary distinction in hormone binding studies, *Adv. Cyclic Nucleotide Res* **4**:239.

Blumenthal, D. K., Stull, J. T., and Gill, G. N., 1978, Phosphorylation of cardiac troponin by guanosine 3':5'-monophosphate-dependent protein kinase, *J. Biol. Chem.* **253**:334.

Breathnach, R., Mandel, J. L., and Chambon, P., 1977, Ovalbumin gene is split in chicken DNA, *Nature* **270**:314.

Brissenden, J. E., and Cox, D. W., 1978, Electrophoretic and quantitative assessment of vitamin D-binding protein (group-specific component) in inherited rickets, *J. Lab. Clin. Med.* **91**:455.

Brostrom, C. O., and Wolff, D. J., 1976, Calcium-dependent cyclic nucleotide phosphodiesterase from brain: Comparison of adenosine 3',5'-monophosphate and guanosine 3',5'-monophosphate as substrates, *Arch. Biochem. Biophys.* **172**:301.

Brostrom, C. O., Hunkeler, F. L., and Krebs, E. G., 1971, The regulation of skeletal muscle phosphorylase kinase by Ca^{2+}, *J. Biol. Chem.* **246**:1961.

Brostrom, C. O., Huang, Y.-C., Breckenridge, B. McL., and Wolff, D. J., 1975, Identification of a calcium-binding protein as a calcium-dependent regulator of brain adenylate cyclase, *Proc. Natl. Acad. Sci. USA* **72**:64.

Brown, E. M., Hurwitz, S., and Aurbach, G. D., 1976a, Preparation of viable isolated bovine parathyroid cells, *Endocrinology* **99**:1582.

Brown, E. M., Hauser, D., Troxler, F., and Aurbach, G. D., 1976b, β-Adrenergic receptor interactions: characterization of iodohydroxybenzylpindolol as a specific ligand, *J. Biol. Chem.* **251**:1232.

Brown, E. M., Hurwitz, S., and Aurbach, G. D., 1978, α-Adrenergic inhibition of cAMP accumulation and PTH release from dispersed bovine parathyroid cells, *Endocrinology* **103**:893

Brown, E. M., Gardner, D. G., Windeck, R. A., and Aurbach, G. D., 1979, Cholera toxin stimulates 3′,5′-adenosine monophosphate accumulation and parathyroid hormone release from dispersed bovine parathyroid glands, *Endocrinology* **104**:218.

Brush, J. S., 1971, Purification and characterization of a protease with specificity for insulin from rat muscle, *Diabetes* **20**:140.

Buller, R. E., and O'Malley, B. W., 1976, The biology and mechanism of steroid hormone receptor interaction with the eukaryotic nucleus, *Biochem. Pharmacol.* **25**:1.

Bullock, L. P., and Bardin, C. W., 1974, Androgen receptors in mouse kidney: A study of male, female and androgen-insensitive (tfm/y) mice, *Endocrinology* **94**:746.

Burghen, G. A., Kitabchi, A. E., and Brush, J. S., 1972, Characterization of a rat liver protease with specificity for insulin, *Endocrinology* **91**:633.

Burke, G., 1969, The cell membrane: A common site of action of thyrotrophin (TSH) and long-acting thyroid stimulator (LATS), *Metabolism* **18**:720.

Burn, J. H., and Dale, H. H., 1926, The vaso-dilator action of histamine, and its physiological significance, *J. Physiol.* **61**:185.

Butcher, R. W., Sneyd, J. G. T., Park, C. R., and Sutherland, E. W., Jr., 1966, Effect of insulin on adenosine 3′,5′-monophosphate in the rat epididymal fat pad, *J. Biol. Chem.* **241**:1651.

Butcher, F. R., Rudich, L., Emler, C., and Nemerovski, M., 1976, Adrenergic regulation of cyclic nucleotide levels, amylase release, and potassium efflux in rat parotid gland, *Mol. Pharmacol.* **12**:862.

Campbell, J., and Rastogi, K. S., 1966, Augmented insulin secretion due to growth hormone. Stimulating effects of glucose and food in dogs, *Diabetes* **15**:749.

Casnellie, J. E., and Greengard, P., 1974, Guanosine 3′,5′-cyclic monophosphate dependent phosphorylation of endogenous substrate proteins in membranes of mammalian smooth muscle, *Proc. Natl. Acad. Sci. USA* **71**:1891.

Cassel, D., and Selinger, Z., 1976, Catecholamine-stimulated GTPase activity in turkey erythrocyte membranes, *Biochim. Biophys. Acta* **452**:538.

Cavalieri, R. R., Gavin, L. A., Bui, F., McMahon, F., and Hammond, M., 1977, Conversion of thyroxine to 3,3′,5′-triiodothyronine (reverse-T_3) by a soluble enzyme system of rat liver, *Biochem. Biophys. Res. Commun.* **79**:897.

Cech, S. Y., and Ignarro, L. J., 1977, Cytidine 3′,5′-monophosphate (cyclic CMP) formation in mammalian tissues, *Science* **198**:1063.

Cech, S. Y., and Ignarro, L. J., 1978, Cytidine 3′,5′-monophosphate (cyclic CMP) formation by homogenates of mouse liver, *Biochem. Biophys. Res. Commun.* **80**:119.

Chan, L., and O'Malley, B. W., 1978, Steroid hormone action: Recent advances, *Ann. Intern. Med.* **89**(Part 1):694.

Chan, L., Means, A. R., and O'Malley, B. W., 1978, Steroid hormone regulation of specific gene expression, *Vitam. Horm.* **36**:259.

Chan, S. J., Keim, P., and Steiner, D. F., 1976, Cell-free synthesis of rat preproinsulins:

Characterization and partial amino acid sequence determination, *Proc. Natl. Acad. Sci. USA* **73**:1964.

Charreau, E. H., Calvo, J. C., Tesone, M., De Souza Valle, L. B., and Baranao, J. L., 1978, Insulin regulation of Leydig cell luteinizing hormone receptors, *J. Biol. Chem.* **253**:2504.

Cheung, W. Y., 1971, Cyclic 3',5'-nucleotide phosphodiesterase. Evidence for and properties of a protein activator, *J. Biol. Chem.* **246**:2859.

Cheung, W. Y., Bradham, L. S., Lynch, T. J., Lin, Y. M., and Tallant, E. A., 1975, Protein activator of cyclic 3':5'-nucleotide phosphodiesterase of bovine or rat brain also activates its adenylate cyclase, *Biochem. Biophys. Res. Commun.* **66**:1055.

Chisholm D. J., Young, J. D., and Lazarus, L., 1969, The gastrointestinal stimulus to insulin release. I. Secretin, *J. Clin. Invest.* **48**:1453.

Chopra, I. J., 1974, A radioimmunoassay for measurement of 3,3',5' triiodothyronine (reverse T_3), *J. Clin. Invest.* **54**:583.

Chopra, I. J., 1977, A study of extrathyroidal conversion of thyroxine (T_4) to 3,3',5-triiodothyronine (T_3) in vitro, *Endocrinology* **101**:453.

Chrisman, T. D., Garbers, D. L., Parks, M. A., and Hardman, J. G., 1975, Characterization of particulate and soluble guanylate cyclases from rat lung, *J. Biol. Chem.* **250**: 374.

Clausen, T., 1975, Effect of insulin on glucose transport in muscle cells, *Curr. Top. Membr. Transp.* **6**:169.

Clausen, T., Elbrink, J., and Martin, B. R., 1974, Insulin controlling calcium distribution in muscle and fat cells, *Acta Endocrinol.* **77**(Suppl.191):137.

Clyman, R. I., Sandler, J. A., Manganiello, V. C., and Vaughan, M., 1975, Guanosine 3',5'-monophosphate and adenosine 3',5'-monophosphate content of human umbilical artery. Possible role in perinatal arterial patency and closure, *J. Clin. Invest.* **55**:1020.

Cohen, P., Burchell, A., Foulkes, J. G., Cohen, P. T. W., Vanaman, T. C., and Nairn, A. C., 1978, Identification of the Ca^{2+}-dependent modulator protein as the fourth subunit of rabbit skeletal muscle phosphorylase kinase, *FEBS Lett.* **92**:287.

Cohn, M. L., Cohn, M., and Taylor, F. H., 1978, Guanosine 3',5'-monophosphate: A central nervous system regulator of analgesia, *Science* **199**:319.

Coore, H. G., and Randle, P. J., 1964, Regulation of insulin secretion studied with pieces of rabbit pancreas incubated in vitro, *Biochem. J.* **93**:66.

Corbin, J. D., and Keely, S. L., 1977, Characterization and regulation of heart adenosine 3':5'-monophosphate-dependent protein kinase isozymes, *J. Biol. Chem.* **252**:910.

Corbin, J. D., Keely, S. L., and Park, C. R., 1975a, The distribution and dissociation of cyclic adenosine 3':5'-monophosphate-dependent protein kinases in adipose, cardiac, and other tissues, *J. Biol. Chem.* **250**:218.

Corbin, J. D., Keely, S. L., Soderling, T. R., and Park, C. R., 1975b, Hormonal regulation of adenosine 3',5'-monophosphate-dependent protein kinase, *Adv. Cyclic Nucleotide Res.* **5**:265.

Costa, E., Guidotti, A., and Hanbauer, I., 1974, Cyclic nucleotides and trophism of secretory cells: Study of adrenal medulla, in: *Cyclic Nucleotides and Disease* (B. Weiss, ed.), pp. 167–186, University Park Press, Philadelphia.

Craig, S. W., and Cuatrecasas, P., 1975, Mobility of cholera toxin receptors on rat lymphocyte membranes, *Proc. Natl. Acad. Sci. USA* **72**:3844.

Creutz, C. E., Pazoles, C. J., and Pollard, H. B., 1978, Identification and purification of an adrenal medullary protein (synexin) that causes calcium-dependent aggregation of isolated chromaffin granules, *J. Biol. Chem.* **253**:2858 .

Crofford, O. B., Rogers, N. L., and Russell, W. G., 1972, The effect of insulin on fat cells. An insulin degrading system extracted from plasma membranes of insulin responsive cells, *Diabetes* **21**(Suppl. 2):403.

Cuatrecasas, P., 1973, Gangliosides and membrane receptors for cholera toxin, *Biochemistry* **12**:3558.
Cuatrecasas, P., 1974, Membrane receptors, *Annu. Rev. Biochem.* **43**:169.
Cuatrecasas, P., and Hollenberg, M. D., 1975, Binding of insulin and other hormones to non-receptor material. Saturability, specificity and apparent negative cooperativity, *Biochem. Biophys. Res. Commun.* **62**:31.
Cuatrecasas, P., Parikh, I., and Hollenberg, M. D., 1973, Affinity chromatography and structural analysis of *Vibrio cholerae* enterotoxin-ganglioside agarose and the biological effects of ganglioside-containing soluble polymers, *Biochemistry* **12**:4253.
Czech, M. P., 1977, Molecular basis of insulin action, *Annu. Rev. Biochem.* **46**:359.
Dabrowska, R., Sherry, J. M. F., Aromatorio, D. K., and Hartshorne, D. J., 1978, Modulator protein as a component of the myosin light chain from chicken gizzard, *Biochemistry* **17**:253.
Danforth, D. N., Jr., Triche, T., Doppman, J. L., Beazley, R. M., Perrino, P. V., and Recant, L., 1976, Elevated plasma proglucagon-like component with a glucagon-secreting tumor. Effect of streptozotocin, *N. Engl. J. Med.* **295**:242.
Davis, J. O., and Freeman, R. H., 1976, Mechanisms regulating renin release, *Physiol. Rev.* **56**:1.
Dean, C. R., Hope, D. B., and Kažić, T., 1968, Evidence for the storage of oxytocin with neurophysin-I and of vasopressin with neurophysin-II in separate neurosecretory granules, *Br. J. Pharmacol.* **34**:192P.
Deery, D. J., and Taylor, K. W., 1973, Effects of azaserine and nicotinamide on insulin release and nicotinamide adenine dinucleotide metabolism in isolated rat islets of Langerhans, *Biochem. J.* **134**:557.
De Jonge, H. R., 1976, Cyclic nucleotide-dependent phosphorylation of intestinal epithelium proteins, *Nature* **262**:590.
DeLuca, H. F., 1978, Vitamin D metabolism and function, *Arch. Intern. Med.* **138**:836.
Desbuquois, B., and Cuatrecasas, P., 1973, Insulin receptors, *Annu. Rev. Med.* **24**:233.
Dial, L. K., Miyamoto, S., and Arquilla, E. R., 1977, Modulation of ^{125}I-insulin degradation by receptors in liver plasma membranes, *Biochem. Biophys. Res. Commun.* **74**:545.
Doel, M. T., Houghton, M., Cook, E. A., and Carey, N. H., 1977, The presence of ovalbumin mRNA coding sequences in multiple restriction fragments of chicken DNA, *Nucleic Acid Res.* **4**:3701.
Douglas, W. W., 1966, The mechanism of release of catecholamines from the adrenal medulla, *Pharmacol. Rev.* **18**:471.
Dratman, M. B., 1974, On the mechanism of action of thyroxin, an amino acid analog of tyrosine, *J. Theor. Biol.* **46**:255.
Dreifuss, J. J., 1975, A review on neurosecretory granules: Their contents and mechanisms of release, *Ann. N.Y. Acad. Sci.* **248**:184.
Drummond, G. I., and Duncan, L., 1970, Adenyl cyclase in cardiac tissue, *J. Biol. Chem.* **245**:976.
Drummond, G. I., Harwood, J. P., and Powell, C. A., 1969, Studies on the activation of phosphorylase in skeletal muscle by contraction and by epinephrine, *J. Biol. Chem.* **244**:4235.
Duckworth, W. C., 1978, Insulin and glucagon binding and degradation by kidney cell membranes, *Endocrinology* **102**:1766.
Dufau, M. L., Charreau, E. H., and Catt, K. J., 1973, Characteristics of a soluble gonadotropin receptor from the rat testis, *J. Biol. Chem.* **248**:6973.
Dugaiczyk, A., Woo, S. L. C., Lai, E. C., Mace, M. L., Jr., McReynolds, L., and O'Malley, B. W., 1978, The natural ovalbumin gene contains seven introns, *Nature* **274**:328.
Duguid, J. R., and Steiner, D. F., 1978, Identification of the major polyadenylated transcription products and the genes active in their synthesis in a rat insulinoma, *Proc. Natl. Acad. Sci. USA* **75**:3249.

Dupre, J., Curtis, J. D., Unger, R. H., Waddell, R. W., and Beck, J. C., 1969, Effects of secretin, pancreozymin, or gastrin on the response of the endocrine pancreas to administration of glucose or arginine in man, *J. Clin. Invest.* **48**:745.

Earp, H. S., Smith, P., Ong, S.-H. H., and Steiner, A. L., 1977, Regulation of hepatic nuclear guanylate cyclase, *Proc. Natl. Acad. Sci. USA* **74**:946.

Edkins, J. S., 1905, On the chemical mechanism of gastric secretion, *Proc. R. Soc. London Ser. B* **76**:376.

Edwards, C. R. W., 1977, Vasopressin and oxytocin in health and disease, *Clin. Endocrinol. Metab.* **6**:223.

England, P. J., 1976, Studies on the phosphorylation of the inhibitory subunit of troponin during modification of contraction in perfused rat heart, *Biochem. J.* **160**:295.

Erdos, E. G., 1977, The angiotensin I converting enzyme, *Fed. Proc.* **36**:1760.

Evans, D. J., Chen, L. C., Curlin, G. T., and Evans, D. G., 1972, Stimulation of adenyl cyclase by *Escherichia coli* enterotoxin, *Nature (New Biol.)* **236**:137.

Exton, J. H., Friedmann, N., Wong, E. H.-A., Brineaux, J. P., Corbin, J. D., and Park, C. R., 1972, Interaction of glucocorticoids with glucagon and epinephrine in the control of gluconeogenesis and glycogenolysis in liver and of lipolysis in adipose tissue, *J. Biol. Chem.* **247**:3579.

Fain, J. N., 1974, Mode of action of insulin, *MTP Internat. Rev. Sci., Biochem. of Horm., Series One*, **8**:1.

Fajans, S. S., Floyd, J. C., Jr., Knopf, R. F., and Conn, J. W., 1967, Effect of amino acids and proteins on insulin secretion in man, *Recent Prog. Horm. Res.* **23**:617.

Farquhar, M. G., 1961, Origin and fate of secretory granules in cells of the anterior pituitary gland, *Trans. N.Y. Acad. Sci.* **23**:346.

Farquhar, M. G., 1969, Lysosome function in regulating secretion: Disposal of secretory granules in cells of the anterior pituitary gland, in: *Lysosomes in Biology and Pathology*, Vol. 2 (J. T. Dingle and H. B. Fell, eds.), pp. 462–482, North-Holland, Amsterdam.

Feldberg, W., and Lewis, G. P., 1964, The action of peptides on the adrenal medulla. Release of adrenaline by bradykinin and angiotensin, *J. Physiol.* **171**:98.

Feldman, D., Funder, J. W., and Edelman, I. S., 1972, Subcellular mechanisms in the action of adrenal steroids, *Am. J. Med.* **53**:545.

Field, M., 1971, Intestinal secretion: Effect of cyclic AMP and its role in cholera, *N. Engl. J. Med.* **284**:1137.

Field, M., 1979, Mechanisms of action of cholera and *Escherichia coli* enterotoxins, *Am. J. Clin. Nutr.* **32**:189.

Fillion, G. M., Slorach, S. A., and Junäs, B., 1970, The release of histamine, heparin, and granule protein from rat mast cells treated with compound 48-80 in vitro, *Acta Physiol. Scand.* **78**:547.

Fishman, P. H., and Brady, R. O., 1976, Biosynthesis and function of gangliosides, *Science* **194**:906.

Fishman, P. H., Moss, J., and Vaughan, M., 1976, Uptake and metabolism of gangliosides in transformed mouse fibroblasts. Relationship of ganglioside structure to choleragen response, *J. Biol. Chem.* **251**:4490.

Fishman, P. H., Bradley, R. M., Moss, J., and Manganiello, V. C., 1978, Effect of serum on ganglioside uptake and choleragen responsiveness of transformed mouse fibroblasts, *J. Lipid Res.* **19**:77.

Fleischer, N., Rosen, O. M., and Reichlin, M., 1976, Radioimmunoassay of bovine heart protein kinase, *Proc. Natl. Acad. Sci. USA* **73**:54.

Franks, D. J., and MacManus, J. P., 1971, Cyclic GMP stimulation and inhibition of cyclic AMP phosphodiesterase from thymic lymphocytes, *Biochem. Biophys. Res. Commun.* **42**:844.

Fraser, T. R., 1975, Is insulin's second messenger calcium? *Proc. R. Soc. Med.* **68**:785.

Fussganger, R. D., Kahn, C. R., Roth, J., and De Meyts, P., 1976, Binding and degradation of insulin by human peripheral granulocytes, *J. Biol. Chem.* **251**:2761.

Gabbay, K. H., DeLuca, K., Fisher, J. N., Jr., Mako, M. E., and Rubenstein, A. H., 1976, Familial hyperproinsulinemia. An autosomal dominant defect, *N. Engl. J. Med.* **294**:911.

Gainer, H., Sarne, Y., and Brownstein, M. J., 1977, Neurophysin biosynthesis: Conversion of a putative precursor during axonal transport, *Science* **195**:1354.

Gaion, R. M., and Krishna, G., 1979a, Cytidylate cyclase: The product isolated by the method of Cech and Ignarro is not cytidine 3',5'-monophosphate, *Biochem. Biophys. Res. Commun.* **86**:105.

Gaion, R. M., and Krishna, G., 1979b, Cytidylate cyclase: Possible artifacts in the methodology, *Science* **203**:672.

Galardy, R. E., Craig, L. C., Jamieson, J. D., and Printz, M. P., 1974, Photo-affinity labeling of peptide hormone binding sites, *J. Biol. Chem.* **249**:3510.

Gavin, J. R., Roth, J., Neville, D. M., Jr., De Meyts, P., and Buell, D. N., 1974, Insulin-dependent regulation of insulin receptor concentrations: A direct demonstration in cell culture, *Proc. Natl. Acad. Sci. USA* **71**:84.

Gavin, L., Castle, J., McMahon, F., Martin, P., Hammond, M., and Cavalieri, R. R., 1977, Extrathyroidal conversion of thyroxine to 3,3',5'-triiodothyronine (reverse-T_3) and 3,5,3'-triiodothyronine (T_3) in humans, *J. Clin. Endocrinol. Metab.* **44**:733.

George, W. J., Polson, J. B., O'Toole, A. G., and Goldberg, N. D., 1970, Elevation of guanosine 3',5'-cyclic phosphate in rat heart after perfusion with acetylcholine, *Proc. Natl. Acad. Sci. USA* **66**:398.

George, W. J., Wilkerson, R. D., and Kadowitz, P. J., 1973, Influence of acetylcholine on contractile force and cyclic nucleotide levels in the isolated perfused rat heart, *J. Pharmacol. Exp. Ther.* **184**:228.

Gerich, J. E., Lovinger, R., and Grodsky, G. M., 1975, Inhibition by somatostatin of glucagon and insulin release from the perfused rat pancreas in response to arginine, isoproterenol, and theophylline: Evidence for a preferential effect on glucagon secretion, *Endocrinology* **96**:749.

Gerich, J. E., Charles, M. A., and Grodsky, G. M., 1976, Regulation of pancreatic insulin and glucagon secretion, *Annu. Rev. Physiol.* **38**:353.

Gielen, W., 1968, On the function of gangliosides: The distribution of serotonin receptors, *Z. Naturf. (B)* **23**:117.

Gilboe, D. P., and Nuttall, F. Q., 1978, In vivo glucose-, and glucagon-, and cAMP-induced changes in liver glycogen synthase phosphatase activity, *J. Biol. Chem.* **253**:4078.

Gill, D. M., and King, C. A., 1975, The mechanism of action of cholera toxin in pigeon erythrocyte lysates, *J. Biol. Chem.* **250**:6424.

Gill, D. M., and Meren, R., 1978, ADP-ribosylation of membrane proteins catalyzed by cholera toxin: Basis of the activation of adenylate cyclase, *Proc. Natl. Acad. Sci. USA* **75**:3050.

Gilman, A. G., and Schrier, B. K., 1972, Adenosine cyclic 3',5'-monophosphate in fetal rat brain cell cultures. I. Effect of catecholamines, *Mol. Pharmacol.* **8**:410.

Ginsberg, B. H., Kahn, C. R., Roth, J., Megyesi, K., and Baumann, G., 1979, Identification and high yield purification of insulin-like growth factors (nonsuppressible insulin-like activities and somatomedins) from human plasma by use of endogenous binding proteins, *J. Clin. Endocrinol. Metab.* **48**:43.

Glass, D. B., Frey, W., II, Carr, D.W., and Goldberg, N. D., 1977, Stimulation of human platelet guanylate cyclase by fatty acids, *J. Biol. Chem.* **252**:1279.

Glinoer, D., Gershengorn, M. C., Dubois, A., and Robbins, J., 1977a, Stimulation of thyroxine-binding globulin synthesis by isolated rhesus monkey hepatocytes after in vivo β-estradiol administration, *Endocrinology* **100**:807.

Glinoer, D., McGuire, R. A., Gershengorn, M. C., Robbins, J., and Berman, M., 1977b, Effects of estrogen on thyroxine-binding globulin metabolism in rhesus monkeys, *Endocrinology* **100**:9.

Gnegy, M., Costa, E., and Uzunov, P., 1976, Regulation of transsynaptically elicited increase of 3':5'-cyclic AMP by endogenous phosphodiesterase activation, *Proc. Natl. Acad. Sci. USA* **73**:352.

Goldberg, N. D., O'Dea, R. F., and Haddox, M. K., 1973, Cyclic GMP, *Adv. Cyclic Nucleotide Res.* **3**:155.

Goldberg, N. D., Haddox, M. K., Nicol, S. E., Glass, D. B., Sanford, C. H., Kuehl, F. A., Jr., and Estensen, R., 1975, Biologic regulation through opposing influences of cyclic GMP and cyclic AMP: The yin yang hypothesis, *Adv. Cyclic Nucleotide Res.* **5**:307.

Goldfine, I. D., 1977, Does insulin need a second messenger? *Diabetes* **26**:148.

Goldfine, I. D., Jones, A. L., Hradek, G. T., Wong, K. Y., and Mooney, J. S., 1978, Entry of insulin into human cultured lymphocytes: Electron microscope autoradiographic analysis, *Science* **202**:760.

Goldstein, J. L., and Wilson, J. D., 1975, Genetic and hormonal control of male sexual differentiation, *J. Cell. Physiol.* **85**:365.

Gomez-Acebo, J., Parrilla, R., and Candela, L. R., 1968, Fine structure of the A and D cells of the rabbit endocrine pancreas in vivo and incubated *in vitro*. I. Mechanism of secretion of the A cells, *J. Cell. Biol.* **36**:33.

Goodfriend, T. L., and Peach, M. J., 1975, Angiotensin III (des-aspartic acid-1)-angiotensin II. Evidence and speculation for its role as an important agonist in the renin-angiotensin system, *Circ. Res. (Suppl.)* **36**:38.

Goodman, DeW. S., 1974, Vitamin A transport and retinol-binding protein metabolism, *Vitam. Horm.* **32**:167.

Gopinath, R. M., and Vincenzi, F. F., 1977, Phosphodiesterase protein activator mimics red blood cell cytoplasmic activator of (Ca^{2+}-Mg^{2+}) ATPase, *Biochem. Biophys. Res. Commun.* **77**:1203.

Gorski, J., and Gannon, F., 1976, Current models of steroid hormone action: A critique, *Annu. Rev. Physiol.* **38**:425.

Graber, S. E., Bomboy, J. D., Jr., Salmon, W. D., Jr., and Krantz, S. B., 1979, Evidence that endotoxin is the cyclic 3':5'-GMP-promoting factor in erythropoietin preparations, *J. Lab. Clin. Med.* **93**:25.

Grant, G., Vale, W., and Guillemin, R., 1973, Characteristics of the pituitary binding sites for thyrotropin-releasing factor, *Endocrinology* **92**:1629.

Greengard, P., 1975, Cyclic nucleotides, protein phosphorylation, and neuronal function, *Adv. Cyclic Nucleotide Res.* **5**:585.

Greengard, P., 1976, Possible role for cyclic nucleotides and phosphorylated membrane proteins in postsynaptic actions of neurotransmitters, *Nature* **260**:101.

Greengard, P., and Kebabian, J. W., 1974, Role of cyclic AMP in synaptic transmission in the mammalian peripheral nervous system, *Fed. Proc.* **33**:1059.

Gregory, H., Hardy, P. M., Jones, D. S., Kenner, G. W., and Sheppard, R. C., 1964, The antral hormone gastrin. Structure of gastrin, *Nature* **204**:931.

Gregory, R. A., and Tracy, H. J., 1964, The constitution and properties of two gastrins extracted from hog antral mucosa, *Gut* **5**:103.

Grisolia, S., and Wallace, R., 1976, Insulin degradation by lysosomal extracts from rat liver; model for a role of lysosomes in hormone degradation, *Biochem. Biophys. Res. Commun.* **70**:22.

Grodsky, G. M., Fanska, R., West, L., and Manning, M., 1974, Anomeric specificity of glucose-stimulated insulin release: Evidence for a glucoreceptor? *Science* **186**:536.

Grodsky, G. M., Fanska, R., and Lundquist, I., 1975, Interrelationships between α and β anomers of glucose affecting both insulin and glucagon secretion in the perfused rat pancreas. II, *Endocrinology* **97**:573.

Grossman, M. I., 1977, Physiological effects of gastrointestinal hormones, *Fed. Proc.* **36**:1930.
Guidotti, A., Kurosawa, A., Chuang, D. M., and Costa, E., 1975a, Protein kinase activation as an early event in the trans-synaptic induction of tyrosine 3-monooxygenase in adrenal medulla, *Proc. Natl. Acad. Sci. USA* **72**:1152.
Guidotti, A., Hanbauer, I., and Costa, E., 1975b, Role of cyclic nucleotides in the induction of tyrosine hydroxylase, *Adv. Cyclic Nucleotide Res.* **5**:619.
Habener, J. F., and Potts, J. T., Jr., 1978a, Biosynthesis of parathyroid hormone, *N. Engl. J. Med.* **299**: 580.
Habener, J. F., and Potts, J. T., Jr., 1978b, Biosynthesis of parathyroid hormone, *N. Engl. J. Med.* **299**:635.
Habener, J. F., Kemper, B., Potts, J. T., Jr., and Rich, A., 1975, Pre-proparathyroid hormone identified by cell-free translation of messenger RNA from hyperplastic human parathyroid tissue, *J. Clin. Invest.* **56**:1328.
Habener, J. F., Amherdt, M., and Orci, L., 1977a, Subcellular organelles involved in the conversion of biosynthetic precursors of parathyroid hormone, *Trans. Assoc. Am. Physicians* **90**:366.
Habener, J. F., Chang, H. T., and Potts, J. T., Jr., 1977b, Enzymic processing of proparathyroid hormone by cell-free extracts of parathyroid glands, *Biochemistry* **16**:3910.
Habig, W. H., Grollman, E. F., Ledley, F. D., Meldolesi, M. F., Aloj, S. M., Hardegree, M. C., and Kohn, L. D., 1978, Tetanus toxin interactions with the thyroid: Decreased toxin binding to membranes from a thyroid tumor with a thyrotropin receptor defect and *in vivo* stimulation of thyroid function, *Endocrinology* **102**:844.
Hadden, J. W., Hadden, E. M., Haddox, M. K., and Goldberg, N. D., 1972, Guanosine 3′:5′-cyclic monophosphate: A possible intracellular mediator of mitogenic influences in lymphocytes, *Proc. Natl. Acad. Sci. USA* **69**:3024.
Hall, K., and Luft, R., 1974, Growth hormone and somatomedin, *Adv. Metab. Dis.* **7**:1.
Hanski, E., Sevilla, N., and Levitzki, A., 1977, The allosteric inhibition by calcium of soluble and partially purified adenylate cyclase from turkey erythrocytes, *Eur. J. Biochem.* **76**:513.
Hansson, H.-A., Holmgren, J., and Svennerholm, L., 1977, Ultrastructural localization of cell membrane G_{M1} ganglioside by cholera toxin, *Proc. Natl. Acad. Sci. USA* **74**:3782.
Hardman, J. G., and Sutherland, E. W., 1969, Guanyl cyclase, an enzyme catalyzing the formation of guanosine 3′,5′-monophosphate from guanosine triphosphate, *J. Biol. Chem.* **244**:6363.
Hardman, J. G., Robison, G. A., and Sutherland, E. W., 1971, Cyclic nucleotides, *Ann. Rev. Physiol.* **33**:311.
Harris, S. E., Schwartz, R. J., Tsai, M.-J., and O'Malley, B. W., 1976, Effect of estrogen on gene expression in the chick oviduct. In vitro transcription of the ovalbumin gene in chromatin, *J. Biol. Chem.* **251**:524.
Hashimoto, E., Takeda, M., Nishizuka, Y., Hamana, K., and Iwai, K., 1976, Studies on the sites in histones phosphorylated by adenosine 3′:5′-monophosphate-dependent and guanosine 3′:5′-monophosphate-dependent protein kinases, *J. Biol. Chem.* **251**:6287.
Haussler, M. R., and McCain, T. A., 1977, Basic and clinical concepts related to vitamin D metabolism and action, *N. Engl. J. Med.* **297**:974 and 1041.
Hayaishi, O., and Ueda, K., 1977, Poly (ADP-ribose) and ADP-ribosylation of proteins, *Annu. Rev. Biochem.* **46**:95.
Hayes, L. W., Goguen, C. A., Ching, S.-F., and Slakey, L. L., 1978, Angiotensin-converting enzyme: Accumulation in medium from cultured endothelial cells, *Biochem. Biophys. Res. Commun.* **82**:1147.
Hertelendy, F., Machlin, L., and Kipnis, D. M., 1969, Further studies on the regulation of insulin and growth hormone secretion in the sheep, *Endocrinology* **84**:192.

Hidaka, H., and Asano, T., 1977, Stimulation of human platelet guanylate cyclase by unsaturated fatty acid peroxides, *Proc. Natl. Acad. Sci. USA* **74**:3657.

Hinkle, P. M., and Tashjian, A. H., Jr., 1975, Thyrotropin-releasing hormone regulates the number of its own receptors in the GH_3 strain of pituitary cells in culture, *Biochemistry* **14**:3845.

Hirata, F., Strittmatter, W. J., and Axelrod, J., 1979, β-Adrenergic receptor agonists increase phospholipid methylation, membrane fluidity, and β-adrenergic receptor-adenylate cyclase coupling, *Proc. Natl. Acad. Sci. USA* **76**:368.

Hofman, F., Beavo, J. A., Bechtel, P. J., and Krebs, E. G., 1975, Comparison of adenosine 3':5'-monophosphate-dependent protein kinases from rabbit skeletal and bovine heart muscle, *J. Biol. Chem.* **250**:7795.

Holick, M. F., Frommer, J. E., McNeill, S. C., Richtand, N. M., Henley, J. W., and Potts, J. T., Jr., 1977, Photo-metabolism of 7-dehydrocholesterol to previtamin D_3 in skin, *Biochem. Biophys. Res. Commun.* **76**:107.

Holmgren, J., and Lönnroth, I., 1975, Oligomeric structure of cholera toxin: Characteristics of the H and L subunits, *J. Gen. Microbiol.* **86**:49.

Holmgren, J., and Lönnroth, I., 1976, Cholera toxin and the adenylate cyclase-activating signal, *J. Infect. Dis.* **133**(Suppl.):S64.

Holmgren, J., Lönnroth, I., and Svennerholm, L., 1973, Tissue receptor for cholera exotoxin: Postulated structure from studies with G_{M1} ganglioside and related glycolipids, *Infect. Immun.* **8**:208.

Holmgren, J., Lönnroth, I., Månsson, J. E., and Svennerholm, L., 1975, Interaction of cholera toxin and membrane G_{M1} ganglioside of small intestine, *Proc. Natl. Acad. Sci. USA* **72**:2520.

House, P. D. R., Poulis, P., and Weidemann, M. J., 1972, Isolation of a plasma-membrane subfraction from rat liver containing an insulin-sensitive cyclic-AMP phosphodiesterase, *Eur. J. Biochem.* **24**:429.

Hughes, J. M., Murad, F., Chang, B., and Guerrant, R. L., 1978, Role of cyclic GMP in the action of heat-stable enterotoxin of *Escherichia coli*, *Nature* **271**:755.

Ignarro, L. J., 1979, Cytidylate cyclase: Possible artifacts in the methodology (reply), *Science* **203**:673.

Ingbar, S. H., 1963, Observations concerning the binding of thyroid hormones by human serum prealbumin, *J. Clin. Invest.* **42**:143.

Ingbar, S. H., and Braverman, L. E., 1975, Active form of the thyroid hormone, *Annu. Rev. Med.* **26**:443.

Inoue, M., Kishimoto, A., Takai, Y., and Nishizuka, Y., 1976, Guanosine 3':5'-monophosphate-dependent protein kinase from silkworm. Properties of a catalytic fragment obtained by limited proteolysis, *J. Biol. Chem.* **251**:4476.

Inoue, M., Kishimoto, A., Takai, Y., and Nishizuka, Y., 1977, Studies on a cyclic nucleotide-independent protein kinase and its proenzyme in mammalian tissues. II. Proenzyme and its activation by calcium-dependent protease from rat brain, *J. Biol. Chem.* **252**:7610.

Ipp, E., Dobbs, R., and Unger, R. H., 1978, Morphine and β-endorphin influence the secretion of the endocrine pancreas, *Nature* **276**:190.

IUPAC-IUB Commission on biochemical nomenclature, 1975, The nomenclature of peptide hormones. Recommendations (1974), *J. Biol. Chem.* **250**:3215.

Jackson, R. L., Dedman, J. R., Schreiber, W. E., Bhatnager, P. K., Knapp, R. D. and Means, A. R., 1977, Identification of ε-N-trimethyllysine in a rat testis calcium-dependent regulatory protein of cyclic nucleotide phosphodiesterase, *Biochem. Biophys. Res. Commun.* **77**:723.

Jacobs, S., Chang, K.-J., and Cuatrecasas, P., 1978, Antibodies to purified insulin receptor have insulin-like activity, *Science* **200**:1283.

Jard, S., Roy, C., Barth, T., Rajerison, R., and Bochaert, J., 1975, Antidiuretic hormone-sensitive kidney adenylate cyclase, *Adv. Cyclic Nucleotide Res.* **5**:31.

Jarett, L., and Smith, R. M., 1974, Electron microscopic demonstration of insulin receptors on adipocyte plasma membranes utilizing a ferritin insulin conjugate, *J. Biol. Chem.* **249**:7024.

Jarrett, H. W., and Penniston, J. T., 1977, Partial purification of the Ca^{2+}-Mg^{2+} ATPase activator from human erythrocytes: Its similarity to the activator of 3':5'-cyclic nucleotide phosphodiesterase, *Biochem. Biophys. Res. Commun.* **77**:1210.

Johnson, G. L., and Bourne, H. R., 1977, Influence of cholera toxin on the regulation of adenylate cyclase by GTP, *Biochem. Biophys. Res. Commun.* **78**:792.

Johnson, G. L., Kaslow, H. R., and Bourne, H. R., 1978, Reconstitution of cholera toxin-activated adenylate cyclase, *Proc. Natl. Acad. Sci. USA* **75**:3113.

Kahn, C. R., 1975, Membrane receptors for polypeptide hormones, in: *Methods in Membrane Biology*, Vol. 3 (E. D. Korn, ed.), pp. 81–146, Plenum, New York.

Kahn, C. R., 1976, Membrane receptors for hormones and neurotransmitters, *J. Cell Biol.* **70**:261.

Kahn, C. R., Freychet, P., Neville, D. M., Jr., and Roth, J., 1974, Quantitative aspects of the insulin-receptor interaction in liver plasma membranes, *J. Biol. Chem.* **249**:2249.

Kahn, C. R., Flier, J. S., Bar, R. S., Archer, J. A., Gorden, P., Martin, M. M., and Roth, J., 1976, The syndromes of insulin resistance and acanthosis nigricans. Insulin-receptor disorders in man, *N. Engl. J. Med.* **294**:739.

Kahn, C. R., Baird, K., Flier, J. S., and Jarrett, D. B., 1977, Effects of autoantibodies to the insulin receptor on isolated adipocytes. Studies of insulin binding and insulin action, *J. Clin. Invest.* **60**:1094.

Kalimi, M., Tsai, S. Y., Tsai, M.-J., Clark, J. H., and O'Malley, B. W., 1976, Effect of estrogen on gene expression in the chick oviduct. Correlation between nuclear-bound estrogen receptor and chromatin initiation sites for transcription, *J. Biol. Chem.* **251**:516.

Kanfer, J. N., Carter, T. P., and Katzen, H. M., 1976, Lipolytic action of cholera toxin on fat cells. Re-examination of the concept implicating G_{M1} ganglioside as the native membrane receptor, *J. Biol. Chem.* **251**:7610.

Kaplan, M. M., 1979, Subcellular alterations causing reduced hepatic thyroxine-5'-monodeiodinase activity in fasted rats, *Endocrinology* **104**:58.

Katz, A. I., and Rubenstein, A. H., 1973, Metabolism of proinsulin, insulin, and C-peptide in the rat, *J. Clin. Invest.* **52**:1113.

Khoo, J. C., Sperry, P. J., Gill, G. N., and Steinberg, D., 1977, Activation of hormone-sensitive lipase and phosphorylase kinase by purifed cyclic GMP-dependent protein kinase, *Proc. Natl, Acad. Sci. USA* **74**:4843.

Kiefer, H. C., Atlas, R., Moldan, D., and Kantor, H. S., 1975, Inhibition of guanylate cyclase and cyclic GMP phosphodiesterase by cholera toxin, *Biochem. Biophys. Res. Commun.* **66**:1017.

Kimura, H., and Murad, F., 1974, Evidence for 2 different forms of guanylate cyclase in rat heart, *J. Biol. Chem.* **249**:6910.

Kishimoto, A., Takai, Y., and Nishizuka, Y., 1977, Activation of glycogen phosphorylase kinase by a calcium-activated, cyclic nucleotide-independent protein kinase system, *J. Biol. Chem.* **252**:7449.

Kissebah, A. H., Tulloch, B. R., Hope-Gill, H. F., Clarke, P. V., Vydelingum, N., and Fraser, T. R., 1975, The mode of action of insulin, *Lancet* **1**:144.

Kono, T., and Barham, F. W., 1973, Effects of insulin on the levels of adenosine 3':5'-monophosphate and lipolysis in isolated rat epididymal fat cells, *J. Biol. Chem.* **248**:7417.

Kono, T., Robinson, F. W., and Sarver, J. A., 1975, Insulin-sensitive phosphodiesterase.

Its localization, hormonal stimulation, and oxidative stabilization, *J. Biol. Chem.* **250**:7826.

Krieg, M., Szalay, R., and Voigt, K. D., 1974, Binding and metabolism of testosterone and of 5α-dihydrotestosterone in bulbocavernosus/levator ani (BCLA) of male rats. In vivo and in vitro studies, *J. Steroid Biochem.* **5**:453.

Krueger, B. K., Forn, J., and Greengard, P., 1977, Depolarization-induced phosphorylation of specific proteins, mediated by calcium ion influx, in rat brain synaptosomes, *J. Biol. Chem.* **252**:2764.

Kuo, J.-F., 1975, Changes in activities of modulators of cyclic AMP-dependent and cyclic GMP-dependent protein kinases in pancreas and adipose tissue from alloxan-induced diabetic rats, *Biochem. Biophys. Res. Commun.* **65**:1214.

Kuo, J.-F., Lee, T.-P., Reyes, P. L., Walton, K. G., Donnelly, T. E., Jr., and Greengard, P., 1972, Cyclic nucleotide-dependent protein kinases. X. An assay method for the measurement of guanosine 3′,5′-monophosphate in various biological materials and a study of agents regulating its levels in heart and brain, *J. Biol. Chem.* **247**:16.

Kuo, J.-F., Kuo, W.-N., Shoji, M., Davis, C. W., Serry, V. L., and Donnelly, T. E., Jr., 1976, Purification and general properties of guanosine 3′:5′-monophosphate-dependent protein kinase from guinea pig fetal lung, *J. Biol. Chem.* **251**:1759.

Kuo, J.-F., Brackett, N. L., Shoji, M., and Tse, J., 1978, Cytidine 3′:5′-monophosphate phosphodiesterase in mammalian tissues. Occurrence and biological involvement, *J. Biol. Chem.* **253**:2518.

Kurosky, A., Markel, D. E., Peterson, J. W., and Fitch, W. M., 1977, Primary structure of cholera toxin B-chain: A glycoprotein hormone analog? *Science* **195**:299.

Kurtz, D. T., Sippel, A., Ansah-Yiadom, R., and Feigelson, P., 1976, Effects of sex hormones on the level of the messenger RNA for the rat hepatic protein α 2u globulin, *J. Biol. Chem.* **251**:3594.

Kwok, S. C. M., Chamley, W. A., and Bryant-Greenwood, G. D., 1978, High molecular weight forms of relaxin in pregnant sow ovaries, *Biochem. Biophys. Res. Commun.* **82**:997.

Lacy, P. E., 1961, Electron microscopy of the beta cell of the pancreas, *Am. J. Med.* **31**:851.

Lai, E. C., Woo, S. L. C., Dugaiczyk, A., Caterall, J. F., and O'Malley, B. W., 1978, The ovalbumin gene: Structural sequences in native chick DNA are not contiguous, *Proc. Natl. Acad. Sci. USA* **75**:2205.

Larner, J., 1975, Four questions times two: A dialogue on the mechanism of insulin action dedicated to Earl W. Sutherland, *Metabolism* **24**:249.

Larner, J., and Villar-Palasi, C., 1970, Glycogen synthase and its control, *Curr. Top. Cell. Regul.* **3**:195.

Ledley, F. D., Mullin, B. R., Lee, G., Aloj, S. M., Fishman, P. H., Hunt, L. T., Dayhoff, M. O., and Kohn, L. D., 1976, Sequence similarity between cholera toxin and glycoprotein hormones. Implications for structure activity relationship and mechanism of action, *Biochem. Biophys. Res. Commun.* **69**:852.

Ledley, F. D., Lee, G., Kohn, L. D., Habig, W. H., and Hardegree, M. C., 1977, Tetanus toxin interactions with thyroid plasma membranes. Implications for structure and function of tetanus toxin receptors and potential patho-physiological significance, *J. Biol. Chem.* **252**:4049.

Lee, G., Aloj, S. M., Brady, R. O., and Kohn, L. D., 1976, The structure and function of glycoprotein hormone receptors: Ganglioside interactions with human chorionic gonadotropin, *Biochem. Biophys. Res. Commun.* **73**:370.

Lee, G., Consiglio, E., Habig, W., Dyer, S., Hardegree, C., and Kohn, L. D., 1978, Structure:function studies of receptors for thyrotropin and tetanus toxin: Lipid modulation of effector binding to the glycoprotein receptor component, *Biochem. Biophys. Res. Commun.* **83**:313.

Lee, T. P., Kuo, J. F., and Greengard, P., 1972, Role of muscarinic cholinergic receptors in regulation of guanosine 3',5'-cyclic monophosphate content in mammalian brain, heart, muscle, and intestinal smooth muscle, *Proc. Natl. Acad. Sci. USA* **69**:3287.

Lefkowitz, R. J., 1973, Isolated hormone receptors, physiologic and clinical implications, *N. Engl. J. Med.* **288**:1061.

Lefkowitz, R. J., Roth, J., and Pastan, I., 1971, ACTH-receptor interaction in the adrenal: A model for the initial step in the action of hormones that stimulate adenyl cyclase, *Ann. N.Y. Acad. Sci.* **185**:195.

Lefkowitz, R. J., Sharp, G. W. G., and Haber, E., 1973, Specific binding of beta-adrenergic catecholamines to a subcellur fraction from cardiac muscle, *J. Biol. Chem.* **248**:342.

Leonard, J. L., and Rosenberg, I. N., 1978, Subcellular distribution of thyroxine 5'-deiodinase in the rat kidney: A plasma membrane location, *Endocrinology* **103**:274.

Lesniak, M. A., Gorden, P., Roth, J., and Gavin, J. R., III, 1974, Binding of I-human growth hormone to specific receptors in human cultured lymphocytes, *J. Biol. Chem.* **249**:1661.

Levey, G. S., Fletcher, M. A., and Klein, I., 1975a, Glucagon and adenylate cyclase: Binding studies and requirements for activation, *Adv. Cyclic Nucleotide Res.* **5**:53.

Levey, G. S., Lehotay, D. C., Canterbury, J. M., Bricker, L. A., and Meltz, G. J., 1975b, Isolation of a unique peptide inhibitor of hormone-responsive adenylate cyclase, *J. Biol. Chem.* **250**:5730.

Levi-Montalcini, R., and Angeletti, P. U., 1963, Essential role of the nerve growth factor in the survival and maintenance of dissociated sensory and sympathetic embryonic nerve cells in vitro, *Dev. Biol.* **7**:653.

Levitzki, A., 1977, The role of GTP in the activation of adenylate cyclase, *Biochem. Biophys. Res. Commun.* **74**:1154.

Levitzki, A., Sevilla, N., and Steer, M. L., 1976, The regulatory control of β-receptor dependent adenylate cyclase, *J. Supramol. Struct.* **4**:405.

Liao, S., 1975, Cellular receptors and mechanisms of action of steroid hormones, *Int. Rev. Cytol.* **41**:87.

Liao, T.-H., and Pierce, J. G., 1970, The presence of a common type of subunit in bovine thyroid-stimulating and luteinizing hormones, *J. Biol. Chem.* **245**:3275.

Linbard, L. E., and Lefkowitz, R. J., 1977, Resolution of β-adrenergic receptor binding and adenylate cyclase activity by gel exclusion chromatography, *J. Biol. Chem.* **252**:799.

Lincoln, T. M., and Corbin, J. D., 1977, Adenosine 3':5'-cyclic monophosphate- and guanosine 3':5'-cyclic monophosphate-dependent protein kinases: Possible homologous proteins, *Proc. Natl. Acad. Sci. USA* **74**:3239.

Lincoln, T. M., and Corbin, J. D., 1978, Purified cyclic GMP-dependent protein kinase catalyzes the phosphorylation of cardiac troponin inhibitory subunit (TN-I), *J. Biol. Chem.* **253**:337.

Lincoln, T. M., Hall, C. L., Park, C. R., and Corbin, J. D., 1976, Guanosine 3':5'-cyclic monophosphate binding proteins in rat tissues, *Proc. Natl. Acad. Sci. USA* **73**:2559.

Lingappa, V. R., Devillers-Thiery, A., and Blobel, G., 1977, Nascent prehormones are intermediates in the biosynthesis of authentic bovine pituitary growth hormone and prolactin, *Proc. Natl. Acad. Sci. USA* **74**:2432.

Liu, A. Y.-C., and Greengard, P., 1976, Regulation by steroid hormones of phosphorylation of specific protein common to several target organs, *Proc. Natl. Acad. Sci. USA* **73**:568.

Liu, Y. P., and Cheung, W. Y., 1976, Cyclic 3':5'-nucleotide phosphodiesterase. Ca^{2+} confers more helical conformation to the protein activator, *J. Biol. Chem.* **251**:4193.

Ljungström, O., and Ekman, P., 1977, Glucagon-induced phosphorylation of pyruvate kinase (type L) in rat liver slices, *Biochem. Biophys. Res. Commun.* **78**:1147.

Lomedico, P. T., and Saunders, G. F., 1976, Preparation of pancreatic mRNA: Cell-free translation of an insulin-immunoreactive polypeptide, *Nucleic Acid Res.*, **3**:381.

Londos, C., Salomon, Y., Lin, M. C., Harwood, J. P., Schramm, M., Wolff, J., and Rodbell, M., 1974, 5'-Guanylylimidodiphospate, a potent activator of adenylate cyclase systems in eukaryotic cells, *Proc. Natl. Acad. Sci. USA*, **71**:3087.

Lübke, K., Schillinger, E., and Töpert, M., 1976, Hormone receptors, *Angew. Chem. (Engl.)* **15**:741.

MacGregor, R. R., Hamilton, J. W., and Cohn, D. V., 1978, The mode of conversion of proparathormone to parathormone by a particulate converting enzymic activity of the parathyroid gland, *J. Biol. Chem.* **253**:2012.

Madison, L. L., Mebane, D., and Lochner, A., 1963, Evidence for a stimulatory feedback of ketone acids on pancreatic beta cells, *J. Clin. Invest.* **42**:955.

Mains, R. E., and Eipper, B. A., 1978, Coordinate synthesis of corticotropins and endorphins by mouse pituitary tumor cells, *J. Biol. Chem.* **253**:651.

Mainwaring. W. I. P., 1975, Steroid hormone receptors: A survey, *Vitam. Horm.* **33**:223.

Malaisse, W. J., Sener, A., Koser, M., and Herchuelz, A., 1976, Stimulus–secretion coupling of glucose-induced insulin release, *J. Biol. Chem.* **251**:5936.

Malbon, C. C., Moreno, F. J., Cabelli, R. J., and Fain, J. N., 1978, Fat cell adenylate cyclase and β-adrenergic receptors in altered thyroid states, *J. Biol. Chem.* **253**:671.

Manganiello, V., and Vaughan, M., 1973, An effect of insulin on cyclic adenosine 3',5'-monophosphate phosphodiesterase activity in fat cells, *J. Biol. Chem.* **248**:7164.

Manganiello, V. C., and Vaughan, M., 1976, Activation and inhibition of fat cell adenylate cyclase by fluoride, *J. Biol. Chem.* **251**:6205.

Manns, J. G., and Boda, J.M., 1967, Insulin release by acetate, propionate, butyrate, and glucose in lambs and adult sheep, *Am. J. Physiol.* **212**:747.

Marx, S. J., Aurbach, G. D., Gavin, J. R., III, and Buell, D. W., 1974, Calcitonin receptors on cultured human lymphocytes, *J. Biol. Chem.* **249**:6812.

McDonald, J. K., Zeitman, B. B., Reilly, T. J., and Ellis, S., 1969a, New observations on the substrate specificity of cathepsin C (dipeptidyl aminopeptidase I), *J. Biol. Chem.* **244**:2693.

McDonald, J. K., Callahan, P. X., Zeitman, B. B., and Ellis, S., 1969b, Inactivation and degradation of glucagon by dipeptidyl aminopeptidase I (cathepsin C) of rat liver. Including a comparative study of secretin degradation, *J. Biol. Chem.* **244**:6199.

McDonald, J. M., Bruns, D. E., and Jarett, L., 1976, The ability of insulin to alter the stable calcium pools of isolated adipocyte subcellular fractions, *Biochem, Biophys. Res. Commun.* **71**:114.

McGuire, R. F., and Barber, R., 1976, Hormone receptor mobility and catecholamine binding in membranes, a theoretical model, *J. Supramol. Struc.* **4**:259.

McKelvy, J. F., 1975, Phosphorylation of neurosecretory granules by cAMP-stimulated protein kinase and its implication for transport and release of neurophysin proteins, *Ann. N.Y. Acad. Sci.* **248**:80.

Mehdi, S. Q., and Kriss, J. P., 1978, Preparation of radiolabeled thyroid-stimulating immunoglobulins (TSI) by recombining TSI heavy chains with ^{125}I-labeled light chains: Direct evidence that the product binds to the membrane thyrotropin receptor and stimulates adenylate cyclase, *Endocrinology* **103**:296.

Meldolesi, M. F., Fishman, P. H., Aloj, S. M., Kohn, L. D., and Brady, R. O., 1976, Relationship of gangliosides to the structure and function of thyrotropin receptors: Their absence on plasma membranes of a thyroid tumor defective in thyrotropin receptor activity, *Proc. Natl. Acad. Sci. USA* **73**:4060.

Meldolesi, M. F., Fishman, P. H., Aloj, S. M., Ledley, F. D., Lee, G., Bradley, R. M., Brady, R. O., and Kohn, L. D., 1977, Separation of the glycoprotein and ganglioside components of the thyrotropin receptor activity in plasma membranes, *Biochem. Biophys. Res. Commun.* **75**:581.

Mendez, E., Lai, C. Y., and Wodnar-Filipowicz, A., 1975, Location and the primary structure around the disulfide bonds in cholera toxin, *Biochem. Biophys. Res. Commun.* **67**:1435.

Miller, J. P., Boswell, K. H., Muneyama, K., Simon, L. N., Robins, R. K., and Shuman, D. A., 1973, Synthesis and biochemical studies of various 8-substituted derivatives of guanosine 3′,5′-cyclic phosphate, inosine 3′,5′-cyclic phosphate and xanthosine 3′,5′-cyclic phosphate, *Biochemistry* **12**:5310.

Mills, D. C. B., and Smith, J. B., 1971, The influence on platelet aggregation of drugs that affect the accumulation of adenosine 3′:5′-cyclic monophosphate in platelets, *Biochem. J.* **121**:185.

Mills, I. H., 1962, Transport and metabolism of steroids, *Br. Med. Bull.* **18**:127.

Minneman, K. P., and Iversen, L. L., 1976, Enkephalin and opiate narcotics increase cyclic GMP accumulation in slices of rat neostriatum, *Nature* **262**:313.

Mobley, W. C., Server, A. C., Ishii, D. N., Riopelle, R. J., and Shooter, E. M., 1977, Nerve growth factor, *N. Engl. J. Med.* **297**:1096 and 1149.

Montague, W., and Cook, J. R., 1971, The role of adenosine 3′:5′-cyclic monophosphate in the regulation of insulin release by isolated rat islets of Langerhans, *Biochem. J.* **122**:115.

Moore, R. J., and Wilson, J. D., 1972, Localization of the reduced nicotinamide adenine dinucleotide phosphate:Δ^4-3-ketosteroid 5α-oxidoreductase in the nuclear membrane of the rat ventral prostate, *J. Biol. Chem.* **247**:958.

Moore, W. V., and Wolff, J., 1974, TSH binding to beef thyroid membranes: Relation to adenylate cyclase activity, *J. Biol. Chem.* **249**:6255.

Morris, B. J., 1978, Activation of human inactive ("pro-") renin by cathepsin D and pepsin, *J. Clin. Endocrinol. Metab.* **46**:153.

Moskowitz, M. A., and Wurtman, R. J., 1975, Catecholamines and neurologic diseases, *N. Engl. J. Med.* **293**:274.

Moss, J., and Vaughan, M., 1977a, Mechanism of action of choleragen. Evidence for ADP-ribosyltransferase activity with arginine as an acceptor, *J. Biol. Chem.* **252**:2455.

Moss, J., and Vaughan, M., 1977b, Choleragen activation of solubilized adenylate cyclase: Requirement for GTP and protein activator for demonstration of enzymatic activity, *Proc. Natl. Acad. Sci. USA* **74**:4396.

Moss, J., Fishman, P. H., Manganiello, V. C., Vaughan, M., and Brady, R. O., 1976a, Functional incorporation of ganglioside into intact cells: Induction of choleragen responsiveness, *Proc. Natl. Acad. Sci. USA* **73**:1034.

Moss, J., Manganiello, V. C., and Vaughan, M., 1976b, Hydrolysis of nicotinamide adenine dinucleotide by choleragen and its A protomer: Possible role in the activation of adenylate cyclase, *Proc. Natl. Acad. Sci. USA* **73**:4424.

Moss, J., Fishman, P. H., Richards, R. L., Alving, C. R., Vaughan, M., and Brady, R. O., 1967c, Choleragen-mediated release of trapped glucose from liposomes containing ganglioside G_{M1}, *Proc. Natl. Acad. Sci. USA* **73**:3480.

Moss, J., Richards, R. L., Alving, C. R., and Fishman, P. H., 1977, Effect of the A and B protomers of choleragen on release of trapped glucose from liposomes containing or lacking ganglioside G_{M1}, *J. Biol. Chem.* **252**:797.

Moss, J., Ross, P. S., Agosto, G., Birken, S., Canfield, R. E., and Vaughan, M., 1978, Mechanism of action of choleragen and the glycopeptide hormones: Is the nicotinamide adenine dinucleotide glycohydrolase activity observed in purified hormone preparations intrinsic to the hormone? *Endocrinology* **102**:415.

Moya, F., Nieto, A., and R-Candela, J. L., 1975, Calcitonin biosynthesis: Evidence for a precursor, *Eur. J. Biochem.* **55**:407.

Mukhtar, E. D., Smith, B. R., Pyle, G. A., Hall, R., and Vice, P. , 1975, Relation of thyroid-stimulating immunoglobulins to thyroid function and effects of surgery, radioiodine and antithyroid drugs, *Lancet* **1**:713.

Mullin, B. R., Aloj, S. M., Fishman, P. H., Lee, G., Kohn, L. D., and Brady, R. O., 1976a,

Cholera toxin interactions with thyrotropin receptors on thyroid plasma membranes, *Proc. Natl. Acad. Sci. USA* **73**:1679.

Mullin, B. R., Fishman, P. H., Lee, G., Aloj S. M., Ledley, F. D., Winand, R. J., Kohn, L. D., and Brady, R. O., 1976b, Thyrotropin-ganglioside interactions and their relationship to the structure and function of thyrotropin receptors, *Proc. Natl. Acad. Sci. USA* **73**:842.

Mullin, B. R., Pacuszka, T., Lee, G., Kohn, L. D., Brady, R. O., and Fishman, P. H., 1978, Thyroid gangliosides with high affinity for thyrotropin: Potential role in thyroid regulation, *Science* **199**:77.

Mutt, V., and Jorpes, J. E., 1968, Structure of porcine cholecystokinin-pancreozymin. 1. Cleavage with thrombin and trypsin, *Eur. J. Biochem.* **6**:156.

Naess, O., Attramadal, A., and Aakvaag, A., 1975, Androgen binding proteins in the anterior pituitary, hypothalamus, preoptic area and brain cortex of the rat, *Endocrinology* **96**:1.

Ng, K. K. F., and Vane, J. R., 1967, Conversion of angiotensin I to angiotensin II, *Nature* **216**:762.

Nimmo, H. G., and Cohen, P., 1977, Hormonal control of protein phosphorylation, *Adv. Cyclic Nucleotide Res.* **8**:145.

Nordmann, J. J., Dreifuss, J. J., and Legros, J. J., 1971, A correlation of release of 'polypeptide hormones,' and of immunoreactive neurophysin from isolated rat neurohypophysis. *Experientia* **27**:1344.

North, W. G., Valtin, H., Morris, J. F., and La Rochelle, F. T., Jr., 1977, Evidence for metabolic conversions of rat neurophysins within neurosecretory granules of the hypothalamo-neurohypophysial system, *Endocrinology* **101**:110.

O'Connor, K. J., and Lazarus, N. R., 1976, The purification and biological properties of pancreatic big glucagon, *Biochem. J.* **156**:265.

O'Dea, R. F., and Zatz, M., 1976, Catecholamine-stimulated cyclic GMP accumulation in the rat pineal: Apparent presynaptic site of action, *Proc. Natl. Acad. Sci. USA* **73**:3398.

Ofner, P., 1968, Effects and metabolism of hormones in normal and neoplastic prostate tissue, *Vitam Horm.* **26**:237.

O'Malley, B. W., and Schrader, W. T., 1976, The receptors of steroid hormones, *Sci. Am.* **234**:32.

O'Malley, B. W., Woo, S. L.C., Harris, S. E., Rosen, J. M., and Means, A. R., 1975, Steroid hormone regulation of specific messenger RNA and protein synthesis in eucaryotic cells, *J. Cell. Physiol.* **85**:343.

Omodeo-Sale, F., Brady, R. O., and Fishman, P. H., 1978, Effect of thyroid phospholipids on the interaction of thyrotropin with thyroid membranes, *Proc. Natl. Acad. Sci. USA* **75**:5301.

Oparil, S., and Haber, E., 1974, The renin-angiotensin system, *N. Engl. J. Med.* **291**:389.

Oppenheimer, J. H., 1968, Role of plasma proteins in the binding, distribution and metabolism of the thyroid hormones, *N. Engl. J. Med.* **278**:1153.

Oppenheimer, J. H., and Surks, M. I., 1975, Biochemical basis of thyroid hormone action, *Biochem. Actions of Horm.* **3**:119.

Orci, L., Ruerner, C., Malaisse-Lagae, F., Blondel, B., Amherdt, M., Bataille, D., Freychet, P., and Perrelet, A., 1975, A morphological approach to surface receptors in islet and liver cells, *Isr. J. Med. Sci.* **11**:639.

Orenstein, N. S., Dvorak, H. F., Blanchard, M. H., and Young, M., 1978, Nerve growth factor: A protease that can activate plasminogen, *Proc. Natl. Acad. Sci. USA* **75**:5497.

Ostlund, R. E., Jr., 1977, Contractile proteins and pancreatic beta-cell secretion, *Diabetes* **26**:245.

Park, C. R., Lewis, S. B., and Exton, J. H., 1972, Relationship of some hepatic actions of insulin to the intracellular level of cyclic adenylate, *Diabetes* **21**(Suppl. 2):439.

Peart, W. S., 1975, Renin-angiotensin system, *N. Engl. J. Med.* **292**:302.

Peck, W. A., Carpenter, J., and Messinger, K., 1974, Cyclic 3':5'-adenosine monophos-

phate in isolated bone cells. II. Responses to adenosine and parathyroid hormone, *Endocrinology* **94**:148.
Perkins, J. P., and Moore, M. M., 1971, Adenyl cyclase of rat cerebral cortex. Activation by solium fluoride and detergents. *J. Biol. Chem.* **246**:62.
Perlman, R. L., and Chalfie, M., 1977, Catecholamine release from the adrenal medulla, *Clin. Endocrinol. Metab.* **6**:551.
Permutt, M. A., Biesbroeck, J., Chyn, R., Boime, I., Szczesna, E., and McWilliams, D., 1976, Isolation of a biologically active messenger RNA: Preparation from fish pancreatic islets by oligo (2'-deoxythymidylic acid) affinity chromatography, in: *Polypeptide Hormones: Molecular and Cellular Aspects*, pp. 97–116, Ciba Found. Symp. 41, Elsevier/Excerpta Medica/North-Holland, Amsterdam.
Philipson, K. D., and Edelman, I. S., 1977, Thyroid hormone control of Na^+-K^+-ATPase and K^+-dependent phosphatase in rat heart, *Am. J. Physiol.* **232**:C196.
Pickup, J. C., Johnston, C. I., Nakamura, S., Uttenthal, L. O., and Hope, D. B., 1973, Subcellular organization of neurophysins, oxytocin, [8-lysine]-vasopressin and adenosine triphosphatase in porcine posterior pituitary lobes, *Biochem. J.* **132**:361.
Pilkis, S. J., Pilkis, J., and Claus, T. H., 1978, The effect of fructose diphosphate and phosphoenolpyruvate on cyclic AMP-mediated inactivation of rat hepatic pyruvate kinase, *Biochem. Biophys. Res. Commun.* **81**:139.
Pliam, N. B., and Goldfine, I. D., 1977, High affinity thyroid hormone binding sites on purified rat liver plasma membranes, *Biochem. Biophys. Res. Commun.* **79**:166.
Pohl, S. L., 1977, The glucagon receptor and its relationship to adenylate cyclase, *Fed. Proc.* **36**:2115.
Pohl, S. L., Krans, H. M. J., Kozyreff, V., Birnbaumer, L., and Rodbell, M., 1971a, The glucagon-sensitive adenyl cyclase system in plasma membranes of rat liver. VI. Evidence for a role of membrane lipids, *J. Biol. Chem.* **246**:4447.
Pohl, S. L., Birnbaumer, L., and Rodbell, M., 1971b, The glucagon-sensitive adenyl cyclase system in plasma membranes of rat liver. I. Properties, *J. Biol. Chem.* **246**:1849.
Pohl, S. L., Krans, M. J., Birnbaumer, L., and Rodbell, M., 1972, Inactivation of glucagon by plasma membranes of rat liver, *J. Biol. Chem.* **247**:2295.
Pointer, R. H., Butcher, F. R., and Fain, J. N., 1976, Studies on the role of cyclic guanosine 3':5'-monophosphate and extracellular Ca^{2+} in the regulation of glycogenolysis in rat liver cells, *J. Biol. Chem.* **251**:2987.
Polak, J. M., Bloom, S. R., Sullivan, S. N., Facer, P., and Pearse, A. G. E., 1977, Enkephalin-like immunoreactivity in the human gastrointestinal tract, *Lancet* **1**:972.
Popp, R. A., Foresman, K. R., Wise, L. D., and Daniel, J. C., Jr., 1978, Amino acid sequence of a progesterone-binding protein, *Proc. Natl. Acad. Sci. USA* **75**:5516.
Porte, D., Jr., 1969, Sympathetic regulation of insulin secretion. Its relation to diabetes mellitus, *Arch. Intern. Med.* **123**:252.
Posner, B. I., Kelly, P. A., and Friesen, H. G., 1974, Induction of a lactogenic receptor in rat liver: Influence of estrogen and the pituitary, *Proc. Natl. Acad. Sci. USA* **71**:2407.
Posner, J. B., Stern, R., and Krebs, E. G., 1965, Effects of electrical stimulation and epinephrine on muscle phosphorylase, phosphorylase b kinase, and adenosine 3',5'-phosphate, *J. Biol. Chem.* **240**:982.
Posternak, T., 1974, Cyclic AMP and cyclic GMP, *Annu. Rev. Pharmacol.* **14**:23.
Racagni, G., Zsilla, G., Guidotti, A., and Costa, E., 1976, Accumulation of cGMP in striatum of rats injected with narcotic analgesics: Antagonism by naltrexone, *J. Pharm. Pharmacol.* **28**:258.
Raff, M., 1976, Self regulation of membrane receptors, *Nature* **259**:265.
Rajerison, R., Marchetti, J., Roy, C., Bockaert, J., and Jard, S., 1974, The vasopressin-sensitive adenylate cyclase of the rat kidney: Effect of adrenalectomy and corticosteroids on hormonal receptor-enzyme coupling, *J. Biol. Chem.* **249**:6390.

Randle, P. J., and Smith, G. H., 1958, Regulation of glucose uptake by muscle. 1. The effects of insulin, anaerobiosis and cell poisons on the uptake of glucose and release of potassium by isolated rat diaphragm, *Biochem. J.* **70**:490.

Randle, P. J., and Young, F. G., 1956, The influence of pituitary growth hormone on plasma insulin activity, *J. Endocrinol* **13**:335.

Rao, C. V., and Saxena, B. B., 1973, Gonadotropin receptors in the plasma membranes of rat luteal cells. *Biochim. Biophys. Acta* **313**:372.

Rasmussen, H., 1970, Cell communication, calcium ion, and cyclic adenosine monophosphate, *Science* **170**:404.

Rasmussen, H., Goodman, D. B. P., and Tenenhouse, A., 1972, The role of cyclic AMP and calcium in cell activation, *CRC Crit. Rev. Biochem.* **1**:95.

Rayford, P. L., Miller, T. A., and Thompson, J. C., 1976, Secretin, cholecystokinin and newer gastrointestinal hormones, *N. Engl. J. Med.* **294**:1093.

Rennels, E. G., and Shiino, M., 1968, Ultrastructural manifestations of pituitary release of ACTH in the rat, *Arch. Anat. Histol. Embryol.* **51**:575.

Renold, A. E., 1972, The beta cell and its responses, *Diabetes* **21**:619.

Revesz, T., and Greaves, M., 1975, Ligand-induced redistribution of lymphocyte membrane ganglioside G_{M1}, *Nature* **257**:103.

Rigopoulou, D., Valverde, I., Marco, J., Faloona, G., and Unger, R. H., 1970, Large glucagon immunoreactivity in extracts of pancreas, *J. Biol. Chem.* **245**:496.

Rimon, G., Hanski, E., Braun, S., and Levitzki, A., 1978, Mode of coupling between hormone receptors and adenylate cyclase elucidated by modulation of membrane fluidity, *Nature* **276**:394.

Riou, J.-P., Claus, T. H., Flockhart, D. A., Corbin, J. D., and Pilkis, S. J., 1977, In vivo and in vitro phosphorylation of rat liver fructose-1,6-bisphosphatase, *Proc. Natl. Acad. Sci. USA* **74**:4615.

Riou, J.-P., Claus, T. H., and Pilkis, S. J., 1978, Stimulation by glucagon of *in vivo* phosphorylation of rat hepatic pyruvate kinase, *J. Biol. Chem.* **253**:656.

Robinson, A. G., Haluszczak, C., Wilkins, J. A., Huellmantel, A. B., and Watson, C. G., 1977, Physiologic control of two neurophysins in human, *J. Clin. Endocrinol. Metab.* **44**:330.

Robison, G. A., Butcher, R. W., and Sutherland, E. W., 1967, Adenyl cyclase as an adrenergic receptor, *Ann. N.Y. Acad. Sci.* **139**:703.

Robison, G. A., Butcher, R. W., and Sutherland, E. W., 1971, *Cyclic AMP*, pp. 36–44, Academic Press, New York.

Rodbell, M., Krans, H. M. J., Pohl, S. L., and Birnbaumer, L., 1971a, The glucagon-sensitive adenyl cyclase system in plasma membranes of rat liver. IV. Effects of guanyl nucleotides on binding of ^{125}I-glucagon, *J. Biol. Chem.* **246**:1872.

Rodbell, M., Birnbaumer, L., Pohl, S. L., and Krans, H. M. J., 1971b, The glucagon-sensitive adenyl cyclase system in plasma membranes of rat liver. V. An obligatory role of guanyl nucleotides in glucagon action, *J. Biol. Chem.* **246**:1877.

Rosen, O. M., and Erlichman, J., 1975, Reversible autophosphorylation of a cyclic 3':5'-AMP-dependent protein kinase from bovine cardiac muscle, *J. Biol. Chem.* **250**:7788.

Rosenbaum, W., Christy, N. P., and Kelly, W. G., 1966, Electrophoretic evidence for the presence of an estrogen-binding-beta-globulin in human plasma, *J. Clin. Endocrinol. Metab.* **26**:1399.

Rosner, W., 1969, Interaction of adrenal and gonadal steroid with proteins in human plasma, *N. Engl. J. Med.* **281**:658.

Roth, J., 1973, Peptide hormone binding to receptors: A review of direct studies *in vitro*, *Metabolism* **22**:1059.

Rubin, R. P., Laychock, S. G., and End, D. W., 1977, On the role of cyclic AMP and cyclic GMP in steroid production by bovine cortical cells, *Biochim. Biophys. Acta* **496**:329.

Rudland, P. S., Seifert, W., and Gospodarowicz, D., 1974, Growth control in cultured mouse fibroblasts: Induction of the pleiotypic and mitogenic responses by a purified growth factor, *Proc. Natl. Acad. Sci. USA* **71**:2600.

Russell, T. R., Terasaki, W. L., and Appleman, M. M., 1973, Separate phosphodiesterases for the hydrolysis of cyclic adenosine 3',5'-monophosphate and cyclic guanosine 3',5'-monosphosphate in rat liver, *J. Biol. Chem.* **248**:1334.

Ryan, J. W., Smith, U., and Niemeyer, R. S., 1972, Angiotensin I: Metabolism by plasma membrane of lung, *Science* **176**:64.

Samols, E., Mari, G., and Marks, V., 1965, Promotion of insulin secretion by glucagon, *Lancet* **2**:415.

Sattin, A., and Rall, T. W., 1970, Effect of adenosine and adenine nucleotides on the cyclic adenosine 3',5'-phosphate content of guinea pig cerebral cortex slices, *Mol. Pharmacol.* **6**:13.

Sawin, C. T., 1969, Hormonology, *N. Engl. J. Med.* **280**:388.

Schalch, D., Draznin, B., and Miller, L. L., 1978, Effect of diabetes mellitus on the release of somatomedin IGF (insulin-like growth factor) and its CP (carrier protein) from the isolated perfused rat liver, *Clin. Res.* **26**:722A.

Schimmel, M., and Utiger, R. D., 1977, Thyroidal and peripheral production of thyroid hormones, *Ann. Intern. Med.* **87**:760.

Schimmel, R. J., 1976, The role of calcium ion in epinephrine activation of lipolysis, *Horm. Metabol. Res.* **8**:195.

Schlender, K., K., Wei, S. H., and Villar-Palasi, C., 1969, UDP-glucose:glycogen α-4-glucosyltransferase I kinase activity of purified muscle protein kinase. Cyclic nucleotide specificity, *Biochim. Biophys. Acta.* **191**:272.

Schlichter, D. J., Casnellie, J. E., and Greengard, P., 1978, An endogenous substrate for cGMP-dependent protein kinase in mammalian cerebellum, *Nature* **273**:61.

Schrader, W. T., and O'Malley, B. W., 1972, Progesterone-binding components of chick oviduct. IV. Characterization of purified subunits, *J. Biol. Chem.* **247**:51.

Schramm, M., and Rodbell, M., 1975, A persistent active state of the adenylate cyclase system produced by the combined actions of isoproterenol and guanylylimidodiphosphate in frog erythrocyte membranes, *J. Biol. Chem.* **250**:2232.

Schreiber, V., 1974, Adenohypophysial hormones: Regulation of their secretion and mechanisms of their action, *MTP Internat. Rev. Sci., Biochem. of Hormones, Series One* **8**:61.

Schubert, D., LaCorbiere, M., Whitlock, C., and Stallcup, W., 1978, Alteration in the surface properties of cells responsive to nerve growth factor, *Nature* **273**:718.

Schulman, H., and Greengard, P., 1978, Ca^{2+}-dependent protein phosphorylation system in membranes from various tissues, and its activation by "calcium-dependent regulator," *Proc. Natl. Acad. Sci. USA* **75**:5432.

Schultz, G., Hardman, J. G., Schultz, K., Baiard, C. E., and Sutherland, E. W., 1973a, The importance of calcium ions for the regulation of guanosine 3',5' monophosphate levels, *Proc. Natl. Acad. Sci. USA* **70**:3889.

Schultz, G., Hardman, J. G., Hurwitz, L., and Sutherland, E. W., 1973b, Importance of calcium for the control of cyclic GMP, *Fed. Proc.* **32**:773.

Schultz, G., Hardman, J. G., Schultz, K., Davis, J. W., and Sutherland, E. W., 1973c, A new enzymatic assay for guanosine 3':5'-cyclic monophosphate and its application to the ductus deferens of the rat, *Proc. Natl. Acad. Sci. USA* **70**:1721.

Schultz, J., and Daly, J. W., 1973a, Cyclic adenosine 3',5'-monophosphate in guinea pig cerebral cortical slices. I. Formation of cyclic adenosine 3',5'-monophosphate from endogenous adenosine triphosphate and from radioactive adenosine triphosphate formed during a prior incubation with radioactive adenine, *J. Biol. Chem.* **248**:843.

Schultz, J., and Daly, J. W., 1973b, Cyclic adenosine 3',5'-monophosphate in guinea pig cerebral cortical slices. II. The role of phosphodiesterase activity in the regulation of levels of cyclic adenosine 3',5'-monophosphate, *J. Biol. Chem.* **248**:853.

Scian, L. F., Westermann, C. D., Verdesea, A. S., and Hilton, J. G., 1960, Adrenocortical and medullary effects of glucagon, *Am. J. Physiol.* **199**:867.

Seeberg, P. H., Shine, J., Martial, J. A., Baxter, J. D., and Goodman, H. M., 1977, Nucleotide sequence and amplification in bacteria of structural gene for rat growth hormone, *Nature* **270**:486.

Segrest, J. P., and Feldman, R. J., 1974, Membrane proteins: Amino acid sequence and membrane penetration, *J. Mol. Biol.* **87**:853.

Severson. D. L., Denton, R. M., Bridges, B. J., and Randle, P. J., 1976, Exchangeable and total calcium pools in mitochondria of rat epididymal fat pads and isolated fat cells. Role in the regulation of pyruvate dehydrogenase activity. *Biochem. J.* **154**:209.

Sevilla, N., Steer, M. L., and Levitzki, A., 1976, Synergistic activation of adenylate cyclase by guanylylimidophosphate and epinephrine, *Biochemistry* **15**:3493.

Sevilla, N., Tolkovsky, A. M., and Levitzki, A., 1977, Activation of turkey erythrocyte adenylate cyclase by two receptors: Adenosine and catecholamines, *FEBS Lett.* **81**:339.

Shapiro, L. E., Samuels, H. H., and Yaffe, B. M., 1978, Thyroid and glucocorticoid hormones synergistically control growth hormone mRNA in cultured GH cells, *Proc. Natl. Acad. Sci. USA* **75**:45.

Sharma, R. K., Wirch, E., and Wang, J. H., 1978, Inhibition of Ca^{2+}-activated cyclic nucleotide phosphodiesterase reaction by a heat-stable inhibitor protein from bovine brain, *J. Biol. Chem.* **253**:3575.

Sherman, B. S., Stagner, J. I., and Zamudi, R., 1975, Hormonal regulation of lactogenic hormone binding in rat liver: Interaction of estrogen, progesterone, testosterone and hydroxycortisone, *Clin. Res.* **23**:479A.

Shimizu, H., Creveling, C. R., and Daly, J. W., 1970, Stimulated formation of adenosine 3',5'-cyclic phosphate in cerebral cortex: Synergism between electrical activity and biogenic amines, *Proc. Natl. Acad. Sci. USA* **65**:1033.

Shishiba, Y., Shimizu, T., Yoshimura, S., and Shizume, K., 1973, Direct evidence for human thyroidal stimulation by LATS-protector, *J. Clin. Endocrinol. Metab.* **36**:517.

Shiu, R. P. C., and Friesen. H. G., 1974, Solubilization and purification of a prolactin receptor from the rabbit mammary gland, *J. Biol. Chem.* **249**:7902.

Shla , L., and Marinetti, G. V., 1972, Hormone-calcium interactions with the plasma membrane of rat liver cells, *Science* **176**:175.

Shoji, M., Brackett, N. L., Tse, J., Shapira, R., and Kuo, J. F., 1978a, Molecular properties and mode of action of homogeneous preparation of stimulatory modulator of cyclic GMP-dependent protein kinase from the heart, *J. Biol. Chem.* **253**:3427.

Shoji, M., Brackett, N. L., and Kuo, J. F., 1978b, Cytidine 3',5'-monophosphate phosphodiesterase: Decreased activity in the regenerating and developing liver, *Science* **202**:826.

Soderling, T. R., Hickenbottom, J. P., Reimann, E., Hunkeler, F. L., Walsh, D. A., and Krebs, E. G., 1970, Inactivation of glycogen synthetase and activation of phosphorylase kinase by muscle adenosine 3',5'-monophosphate-dependent protein kinases, *J. Biol. Chem.* **245**:6317.

Soderling, T. R., Jett, M. F., Hutson, N. J., and Khatra, B. S., 1977, Regulation of glycogen synthase. Phosphorylation specificities of cAMP-dependent and cAMP-independent kinases for skeletal muscle synthase, *J. Biol. Chem.* **252**:7517.

Sokol, H. W., Zimmerman, E. A., Sawyer, W. H., and Robinson, A. G., 1976, The hypothalamic-neurohypophysial system of the rat: Localization and quantitation of neurophysin by light microscopic immunocytochemistry in normal rats and Brattleboro rats deficient in vasopressin and a neurophysin, *Endocrinology* **98**:1176.

Spelsberg, T. C., Steggles, A. W., Chytil, F., and O'Malley, B. W., 1972, Progesterone-binding components of chick oviduct. V. Exchange of progesterone-binding capacity from target to nontarget tissue chromatins, *J. Biol. Chem.* **247**:1368.

Spielman, L. L., and Bancroft, F. C., 1977, Pregrowth hormone: Evidence for conversion to growth hormone during synthesis on membrane-bound polysomes, *Endocrinology* **101**:651.

Srivastava, A. K., Waisman, D. M., Brostrom, C. O., and Soderling, T. R., 1979, Stimulation of glycogen synthase phosphorylation by calcium-dependent regulator protein, *J. Biol. Chem.* **254**:583.

Steer, M. L., Atlas, D., and Levitzki, A., 1975, Interrelations between β-adrenergic receptors, adenylate cyclase and calcium, *N. Engl. J. Med.* **292**:409.

Steinberg, D., 1976, Interconvertible enzymes in adipose tissue regulated by cyclic AMP-dependent protein kinase, *Adv. Cyclic Nucleotide Res.* **7**:157.

Steiner, D. F., 1977, Insulin today, *Diabetes* **26**:322.

Steiner, D. F., Cunningham, D. D., Spigelman, L., and Aten, B., 1967, Insulin biosynthesis: Evidence for a precursor, *Science* **157**:697.

Sterling, K., 1979, Thyroid hormone action at the cell level, *N. Engl. J. Med.* **300**:117.

Stoner, J., Manganiello, V. C., and Vaughan, M., 1974, Guanosine cyclic 3',5'-monophosphate and guanylate cyclase activity in guinea pig lung. Effects of acetylcholine and cholinesterase inhibitors, *Mol. Pharmacol.* **10**:155.

Straus, E., Malesci, A., and Yalow, R. S., 1978, Characterization of a nontrypsin cholecystokinin converting enzyme in mammalian brain, *Proc. Natl. Acad. Sci. USA* **75**:5711.

Surks, M. I., Schadlow, A. R., Stock, J. M., and Oppenheimer, J. H., 1973, Determination of iodothyronine absorption and conversion of L-thyroxine (T_4) to L-triiodothyronine (T_3) using turnover rate techniques, *J. Clin. Invest.* **52**:805.

Svasti, J., and Bowman, B. H., 1978, Human group-specific component. Changes in electrophoretic mobility resulting from vitamin D binding and from neuraminidase digestion. *J. Biol. Chem.* **253**:4188.

Takai, Y., Nishiyama, K., Yamamura, H., and Nishizuka, Y., 1975, Guanosine 3':5'-monophosphate-dependent protein kinase from bovine cerebellum, *J. Biol. Chem.* **250**:4690.

Takai, Y., Yamamoto, M., Inoue, M., Kishimoto, A., and Nishizuka, Y., 1977, A proenzyme of cyclic nucleotide-independent protein kinase and its activation by calcium-dependent neutral protease from rat liver, *Biochem. Biophys. Res. Commun.* **77**:542.

Tashjian, A. H., Jr., Osborne, R., Maina, D., and Knaian, A., 1977, Hydrocortisone increases the number of receptors for thyrotropin-releasing hormone on pituitary cells in culture, *Biochem. Biophys. Res. Commun.* **79**:333.

Tate, R. L., Holmes, J. M., Kohn, L. D., and Winand, R. J., 1975, Characteristics of a solubilized thyrotropin receptor from bovine thyroid plasma membranes, *J. Biol. Chem.* **250**:6527.

Taylor, S. I., Mukherjee, C., and Jungas, R. L., 1973, Studies on the mechanism of activation of adipose tissue pyruvate dehydrogenase by insulin, *J. Biol. Chem.* **248**:73.

Teo, T. S., Wang, T. H., and Wang, J. H., 1973, Purification and properties of the protein activator of bovine heart cyclic adenosine 3',5'-monophosphate phosphodiesterase, *J. Biol. Chem.* **248**:588.

Terris, S., and Steiner, D. F., 1975, Binding and degradation of ^{125}I-insulin by rat hepatocytes, *J. Biol. Chem.* **250**:8389.

Thoenen, H., Angeletti, P. U., Levi-Montalcini, R., and Kettler, R., 1971, Selective induction by nerve growth factor of tyrosine hydroxylase and dopamine-β-hydroxylase in the rat superior cervical ganglia, *Proc. Natl. Acad. Sci. USA* **68**:1598.

Thomas, J. H., 1973, The role of "insulinase" in the degradation of insulin, *Postgrad. Med. J.* **49**(Suppl. 7):940.

Thompson, J. C., Reeder, D. D., Bunchman, H. H., Becker, H. D., and Brandt, E. M., Jr., 1972, Effect of secretin on circulating gastrin, *Ann. Surg.* **176**:384.

Thompson, W. J., and Appleman, M. M., 1971, Characterization of cyclic nucleotide phosphodiesterase of rat tissue, *J. Biol. Chem.* **246**:3145.

Thurston, H., 1976, Vascular angiotensin receptors and their role in blood pressure control, *Am. J. Med.* **61**:768.

Tolkovsky, A. M., and Levitzki, A., 1978a, Mode of coupling between the β-adrenergic receptor and adenylate cyclase in turkey erythrocytes, *Biochemistry* **17**:3795.

Tolkovsky, A. M., and Levitzki, A., 1978b, Coupling of a single adenylate cyclase to two receptors: Adenosine and catecholamine, *Biochemistry* **17**:3811.

Tosteson, M. T., and Tosteson D. C., 1978, Bilayers containing gangliosides develop channels when exposed to cholera toxin, *Nature* **275**:142.

Trifaró, J. M., 1977, Common mechanisms of hormone secretion, *Annu. Rev. Pharmacol. Toxicol.* **17**:27.

Tsai, B.-S., Peach, M. J., Khosla, M. C., and Bumpus, F. M., 1975a, Synthesis and evaluation of [des-Asp^1]angiotensin I as precursor for [des-Asp^1]angiotensin II ("angiotensin III"), *J. Med. Chem.* **18**:1180.

Tsai, S. Y., Tsai, M. J., Schwartz, R., Kalimi, M., Clark, H. H., and O'Malley, B. W., 1975b, Effect of estrogen on gene expression in the chick oviduct: Nuclear receptor levels and initiation of transcription, *Proc. Natl. Acad. Sci. USA* **72**:4228.

Uttenthal, L. O., Livett, B. G., and Hope, D. B., 1971, Release of neurophysin together with vasopressin by a Ca^{2+}-dependent mechanism, *Philos. Trans. R. Soc. London Ser. B.* **261**:379.

Uzunov, P., Revuelta, A., and Costa, E., 1975, Role for the endogenous activator 3',5'-nucleotide phosphodiesterase in rat adrenal medulla, *Mol. Pharmacol.* **11**:506.

Uzunov, P., Gnegy, M. E., Revuelta, A., and Costa, E., 1976, Regulation of the high K_M cyclic nucleotide phosphodiesterase of adrenal medulla by the endogenous calcium-dependent-protein activator, *Biochem. Biophys. Res. Commun.* **70**:132.

van Heyningen, S., 1974, Cholera toxin: Interaction of subunits with ganglioside G_{M1}, *Science* **183**:656.

van Heyningen, S., 1976, The subunits of cholera toxin: Structure, stoichiometry and function, *J. Inf. Dis.* **133**(Suppl.):S5.

van Heyningen, W. E., 1963, The fixation of tetanus toxin, strychnine, serotonin and other substances by ganglioside, *J. Gen. Microbiol.* **31**:375.

van Heyningen, W. E., 1974, Gangliosides as membrane receptors for tetanus toxin, cholera toxin, and serotonin, *Nature* **249**:415.

van Heyningen, W. E., and Mellanby, J., 1968, The effect of cerebroside and other lipids on the fixation of tetanus toxin by gangliosides, *J. Gen. Microbiol.* **52**:447.

Varandani, P. T., 1974, Insulin degradation in insulinoma: Evidence for the occurrence of an active form of glutathione-insulin transhydrogenase and for the absence of insulin A and B chains degrading proteases, *Biochem. Biophys. Res. Commun.* **60**:1119.

Vigneri, R., Goldfine, I. D., Wong, K. Y., Smith, G. J., and Pezzino, V., 1978, The nuclear envelope. The major site of insulin binding in rat liver nuclei, *J. Biol. Chem.* **253**:2098.

Waisman, D. M., Singh, T. J., and Wang, J. H., 1978, The modulator-dependent protein kinase. A multifunctional protein kinase activatable by the Ca^{2+}-dependent modulator protein of the cyclic nucleotide system, *J. Biol. Chem.* **253**:3387.

Walsh, D. A., Perkins, J. P., Brostrom, C. O., Ho, E. S., and Krebs, E. G., 1971, Catalysis of the phosphorylase kinase activation reaction, *J. Biol. Chem.* **246**:1968.

Walsh, J. H., and Grossman, M. I., 1975, Gastrin, *N. Engl. J. Med.* **292**:1324.

Walter, U., Uno, I., Liu, A. Y.-C., and Greengard, P., 1977, Study of autophosphorylation of isoenzymes of cyclic AMP-dependent protein kinases, *J. Biol. Chem.* **252**:6588.

Wang, J. H., and Desai, R., 1977, Modulator binding protein. Bovine brain protein exhibiting the Ca^{2+}-dependent association with the protein modulator of cyclic nucleotide phosphodiesterase, *J. Biol. Chem.* **252**:4175.

Weinstock, R., Sweet, R., Weiss, M., Cedar, H., and Axel, R., 1978, Intragenic DNA spacers interrupt the ovalbumin gene, *Proc. Natl. Acad. Sci. USA* **75**:1299.

Weiss, B., and Hait, W. N., 1977, Selective cyclic nucleotide phosphodiesterase inhibitors as potential therapeutic agents, *Annu. Rev. Pharmacol. Toxicol.* **17**:441.

Weiss, L., Loeffler, G., Schirmann, A., and Wieland, O., 1971, Control of pyruvate dehydrogenase interconversion in adipose tissue by insulin, *FEBS Lett.* **15**:229.

Williams, D. L., 1974, The estrogen receptor: A minireview, *Life Sci,* **15**:583.

Williams, L. T., Lefkowitz, R. J., Watanabe, A. M., Hathaway, D. R., and Besch, H. R., Jr., 1977, Thyroid hormone regulation of β-adrenergic receptor number, *J. Biol. Chem.* **252**:2787.

Williams, R. E., 1976, Phosphorylated sites in substrates of intracellular protein kinases: A common feature in amino acid sequences, *Science* **192**:473.

Winand, R. J., and Kohn, L. D., 1975, Thyrotropin effects on thyroid cells in culture. Effects of trypsin on the thyrotropin receptor and on thyrotropin-mediated cyclic 3':5'-AMP changes, *J. Biol. Chem.* **250**:6534.

Wolberg, G., Zimmerman, T. P., Hiemstra, K., Winston, M., and Chu, L.-C., 1975, Adenosine inhibition of lymphocyte-mediated cytolysis: Possible role of cyclic adenosine monophosphate, *Science* **187**:957.

Wolff, J., Winand, R. J., and Kohn, L. D., 1974, The contribution of subunits of thyroid stimulating hormone to the binding and biological activity of thyrotropin, *Proc. Natl. Acad. Sci. USA* **71**:3460.

Woolley, D. W., and Gommi, B. W., 1965, Serotonin receptors. VII. Activities of various pure gangliosides as the receptors, *Proc. Natl. Acad. Sci. USA* **53**:959.

Yagi, K., Yazawa, M., Kakiuchi, S., Ohshima, M., and Uenishi, K., 1978, Identification of an activator protein for myosin light chain kinase as the Ca^{2+}-dependent modulator protein, *J. Biol. Chem.* **253**:1338.

Yamamoto, K. R., and Alberts, B. M., 1976, Steroid receptors: Elements for modulation of eukaryotic transcription, *Annu. Rev. Biochem.* **45**:721.

Yasuda, N., and Greer, S. E., 1978, Conversion of high molecular weight CRF (corticotropin-releasing factor) or PRF (prolactin-releasing factor) into lower molecular weight forms by boiling at low pH, *Biochem. Biophys. Res. Commun.* **85**:1291.

Zawalich, W. S., Rognstad, R., Pagliara, A. S., and Matschinsky, F. M., 1977, A comparison of the utilization rates and hormone-releasing actions of glucose, mannose, and fructose in isolated pancreatic islets, *J. Biol. Chem.* **252**:8519.

Zeleznik, A. J., Midgley, A. R., Jr., and Reichert, L. E., Jr., 1974, Granulosa cell maturation in the rat: Increased binding of human chorionic gonadotropin following treatment with follicle stimulating hormone in vivo, *Endocrinology* **95**:818.

Zinnam, B., and Hollenberg, C. H., 1974, Effect of insulin and lipolytic agents on rat adipocyte low K_m cyclic adenosine 3',5'-monophosphate phosphodiesterase, *J. Biol. Chem.* **249**:2182.

15

The Biochemical Basis of Disease

Robert H. Herman and Robert M. Cohn

I. Introduction

A. The Biochemical Nature of Disease

The physician sees patients who complain of symptoms, and, in order to provide rational therapy, he must arrive at a correct diagnosis. Accordingly, the physician takes a medical history, performs a physical examination, and obtains whatever laboratory studies are deemed necessary. The physician uses his findings to provide a diagnosis, therapy, and a prognosis. This classical approach to the practice of medicine is depicted in Fig. 1.

When we begin to inquire into the nature of disease, however, it is not sufficient to study the problem via the classical approach. Disease is more than a collection of symptoms and laboratory findings. To study the nature of disease we must understand the underlying basis of disease. The symptoms, signs, and abnormal laboratory findings of the patient are the clinical expressions of underlying abnormal physiology. We may define physiology as *inter*cellular events and biochemistry (intermediary metabolism or physiological chemistry) as *intra*cellular events. Hence, physiology is the study of functional relationships between the cells that constitute tissues, organs, and organ systems. Normal physiology is the expression of normal biochemistry. As a con-

---

Robert H. Herman • Endocrine-Metabolic Service, Letterman Army Medical Center, Presidio of San Francisco, California 94129. **Robert M. Cohn** • Department of Metabolic Research, Children's Hospital of Philadelphia, University of Pennsylvania, Philadelphia, Pennsylvania 19104.

Figure 1. The classical approach to the practice of medicine.

sequence, abnormal physiology is the result of abnormal biochemistry. Fig. 2 depicts these relationships.

The logical extension of the foregoing is that all disease is biochemical. We may consider that abnormal biochemistry leads to abnormal physiology, which is expressed as symptoms, signs, and abnormal laboratory findings. If so, then, in theory, every disease can be defined in terms of some biochemical aberration. We have not yet arrived at that state where every disease can be so defined. The theoretical approach which defines all disease in terms of abnormal biochemistry at least provides a methodological concept for studying disorders which are as yet of unknown etiology.

From these considerations we may derive the fourth fundamental theorem of theoretical biology. This fourth theorem states that all disease is biochemical in nature, i.e., all disease results from the dysfunctional biology of intracellular events. Disease may be operationally defined as one or more abnormalities of intermediary metabolism manifested by abnormal physiology and giving rise to symptoms, signs, and abnormal laboratory findings. Suffice it to say that the language of cell expression is biochemistry, and therefore that when cell biochemistry is deranged, disease supervenes.

Table 1 lists several broad categories of disease. Although this is a very simplified listing and obviously omits many disease categories, it serves to demonstrate the biochemical basis of disease. In the light of our present knowledge, it is easy to recognize that an abnormal gene can give rise to an abnormally functioning protein. Whatever function is subserved by that protein will be impaired. Thus, there are genetic diseases characterized by defects in enzymes, structural proteins, hormones, antibodies, cell membranes, transport mechanisms, blood clot-

Figure 2. The pathogenesis of the clinical manifestations of disease.

ting proteins, carrier proteins, etc. The so-called inborn errors of metabolism are clearly recognizable diseases that are biochemical. If a gene is absent and a protein is not synthesized, then its absence results in a biochemical defect and the resulting disease is biochemical in nature.

Multiple gene defects may arise in the various chromosomal disease syndromes. In many cases the relationship between the chromosomal abnormality and the clinical expression is quite clear. In other

Table 1. A Simplified Listing of Broad Categories of Disease[a]

A. Congenital defects owing to abnormal genes
  1. Abnormal protein: structural defects
  2. Abnormal enzyme: inborn error of metabolism
  3. Abnormal hormonal state: hyper- or hypoendocrine function
  4. Abnormal antibody state: immunoglobin defect
B. Nutritional deficiency
  1. Calories
  2. Proteins and specific amino acids
  3. Essential fatty acids
  4. Vitamins
  5. Minerals
C. Inflammation
  1. Trauma
    a. Impact
    b. Burns
  2. Infection
  3. Allergy
  4. Drugs and chemicals
D. Cardiovascular disease
  1. Atherosclerosis
  2. Hypertension
  3. Rheumatic heart disease
  4. Other
E. Neoplasia
  1. Genetic etiology
  2. Radiation
  3. Chemical carcinogen
  4. Viral etiology
F. Degenerative diseases and aging
G. Psychiatric disorders
  1. Neurosis
  2. Psychosis
  3. Mental retardation
  4. Behavior problems

[a]Disease is operationally defined as one or more abnormalities of intermediary metabolism that are manifested by abnormal physiology, giving rise to symptoms, signs, and abnormal laboratory findings.

cases, however, the chromosomal abnormality is not clearly connected with the clinical manifestations. For instance, although a chromosomal abnormality may be linked to a clinical syndrome, the exact biochemical abnormality may not be known. Thus, while trisomy 21 is related to the clinical manifestations of Down syndrome, the pathogenesis of this disease is unknown. Nevertheless, the clinical syndromes that are related to chromosomal abnormalities are biochemical in nature.

If a nutritional deficiency or excess occurs, the intermediary metabolic pathways may be affected sufficiently to result in an abnormal physiological state and ultimately in disease. Thus, a protein deficiency leads to kwashiorkor, while an excess of vitamin D leads to a toxic state. Without the proper substrates in the complex metabolic pathways which constitute the intermediary metabolic processes, the entire biochemical system will be unable to function despite the intrinsic normality of the pathways. Clearly, nutritional deficiencies and excesses can lead to disease which is biochemical in nature.

Inflammatory processes, be they traumatic or due to infection, allergy, or drugs and chemicals, are biochemical since they involve the complex chemistry of the inflammatory process. Not only is tissue destruction and injury part of the inflammatory process, but also certain substances which are released from the damaged tissue serve to orchestrate the inflammatory response and tissue repair mechanisms. Cardiovascular disease, although expressed in overt physiologic terms which are sometimes quite dramatic, is also biochemical in nature. The basic aspects of atherosclerosis, including the appearance of foam cells, deposition of cholesterol and atherosclerotic plaques, and intimal proliferation, involve biochemical processes. Essential hypertension involves the arteriolar smooth muscle. Contraction of muscle of any type is a complex biochemical event. Why arteriolar smooth muscle should have increased tension which results in hypertension and its attendant complications is unknown. Regardless of what theory one invokes (abnormal smooth muscle transport of sodium, potassium, etc.; renal defect in sodium and potassium reabsorption; prostaglandin abnormality; hypersensitivity to catecholamines; abnormality of sympathetic nerve function; etc.) the problem of essential hypertension is clearly related to the biochemistry of smooth muscle contraction. Rheumatic heart disease involves the generation of antibodies to the teichoic acid in streptococcal cell walls, and these antibodies also act against heart valves and synovial membranes. Such immunologic events are distinctly biochemical because of the informational role served by proteins.

Neoplastic disease involves the uncontrolled proliferation of cells. This is a disorder of the cell cycle (see Chapter 7) and hence of mitosis and DNA replication. Because of the biochemical nature of the cell cycle, neoplastic disease must be considered to be biochemical in nature regardless to which etiology one subscribes (carcinogenic viruses, irradiation, genetic predisposition, chemical carcinogens, etc.).

Aging is a process which involves the gradual atrophy of various tissues with a seeming diminution of protein biosynthesis. Although the exact biochemical mechanisms involved are unknown, the gradual alteration in structure and function of aging tissue is a result of biochemical change. The manifestations of the aging process are well known and need not be listed here. Not all individuals age at the same rate, nor do all changes occur in all tissues in the same individual. Associated degenerative diseases such as atherosclerosis, osteoarthritis, osteoporosis, senile dementia, etc., accompany the aging process, but may also occur alone. Whether the structural and biochemical changes seen in aging are reversible or can be prevented remains to be determined. Aging may be a consequence of the inherent instability of biological systems (see Chapter 1).

Psychiatric disorders must be considered to have a biochemical basis. Psychosis, neurosis, mental retardation, and behavior problems must, in the last analysis, be complex manifestations of some aberration of biochemical functioning of the central nervous system. Fig. 3 diagrams how one may conceptualize mental function in terms of intermediary metabolism. Mental function can be expressed in terms of an individual's intelligence, personality, and patterns of behavior. These outward manifestations of mental function can be considered to be complex expressions of consciousness (self-awareness), sensory perception (awareness of reality of the outside world), volition (decision making), memory, learning, creative ability (problem solving), and emotional reactivity (which is partially subconscious). All of these components are some function of the nerve network. How the nerve network gives rise to these components is unknown. It is apparent from

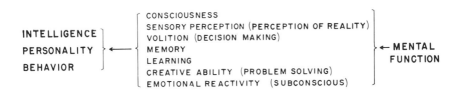

Figure 3. The relationship of neuronal intermediary metabolism to the expression of mental function.

considerations of organic lesions and the effects of drugs and chemicals that some function(s) (in the mathematical sense) of the nerve network is responsible. Each neuron in the nerve network functions as a consequence of its intermediary metabolism. If enough neurons are affected by abnormal intermediary metabolism, the nerve network is impaired and, consequently, derangement of brain function appears.

It is easy to comprehend changes in brain function when a diffuse viral process involves neurons, or when tumors, vascular impairment, or traumatic injury occur. It is less easy to understand the nature of the process when mental retardation occurs without any demonstrable lesion or biochemical abnormality. Even more difficult to envision is the nature of the defect when psychosis or severe neurosis occurs. Most puzzling is psychopathic or antisocial destructive behavior, criminal behavior, or addictive self-destructive behavior. Yet, because mental function is an emergent expression of the biochemistry of the nerve network, we must conclude that all abnormalities of mental function are ultimately explicable in biochemical terms. Although this may be difficult for some to accept, we must consider the enormous number of possible circuits in the central nervous system which may be affected. If a neuronal circuit is all possible combinations of the 10^{10}–10^{12} neurons in the CNS, the number of circuits becomes factorial 10^{12}. It may well be that various components of mental function (consciousness, sensory perception, volition, etc.) are only possible when the nerve network reaches a certain size in order to allow persistence of the flow of nerve impulses through the circuit. If outward behavior is an expression of the neuronal circuits, impairment of circuit flow or connections could lead to abnormal behavior. What the nature of such impairment could be is as yet unknown. Hence, we recognize the immensity of the problem confronting neurochemists in relating neuronal dysfunction to disorders of mentation.

From the foregoing, we can derive the fifth fundamental theorem of theoretical biology, which states that mental function is a function of the nerve network and is a consequence of the continued flow of nerve impulses through the neuronal circuits of the central nervous system.

Every disease entity known to exist can be analyzed in a similar manner. And a careful analysis of the basic biochemistry and resulting physiology will demonstrate that every disease is biochemical in nature. Such an approach should be helpful in those diseases in which the etiology is unknown. There still remain a number of puzzling and ill-defined syndromes about which we know very little. It is predicted that as these entities are solved, they will be found to have a biochemical basis.

In order to understand the nature of disease, then, it is necessary to understand intermediary metabolism. Since the previous chapters in this book deal with the nature of intermediary metabolism and its regulatory controls, it is unnecessary to define, other than in a cursory manner, what we mean by intermediary metabolism.

B. The Definition of Intermediary Metabolism

As pointed out in Chapter 1, intermediary metabolism is the totality of enzymatic and nonenzymatic chemical and physical reactions that occur within the cell. The enzymatic reactions consist of the transformation of substrates into products. The chemical reactions are linked so that the product of one reaction serves as the substrate of the next, until a final end product is reached. In this way, a metabolic pathway is formed, with substrates entering into the pathway at the first enzymatic reaction and being removed by the last enzymatic reaction. Since there is continual movement, in sequence, of intermediates through each enzymatic reaction, levels of intermediates never become sufficiently high for equilibrium to occur. Thus, the passage of substrates through each intermediate step to the final product occurs in a steady-state fashion. The various pathways are linked so that a metabolic network is established. The operation of the network is controlled by regulatory mechanisms which determine the direction and rate of flow of intermediates through the network. In addition to the enzymatic reactions, certain nonenzymatic chemical and physical reactions of physiological importance exist, including photochemical reactions, spontaneous chemical reactions, intra- and extracellular binding of ligands to soluble carrier proteins, and the binding of ligands to membrane receptors. These systems are described in previous chapters.

Intermediary metabolism, then, consists of the metabolic network, under regulatory control, operating in a steady-state fashion within the confines of the cell.

II. The Biochemical Basis of Disease

A. Implications of the Basic Thesis

In contrast to the classical view of medicine, we propose the following approach to understanding disease.
1. All disease is biochemical.
2. The normal biochemical systems permit the individual to adapt

to his environment. Fig. 4 illustrates the interaction between genetic and environmental factors and the ability to adapt.

3. The failure to adapt to environmental factors results in disease. A disease-producing environmental factor is termed "a stress factor" or "a stress."

4. The failure to adapt to an environmental factor implies failure of metabolic control, i.e., the presence of a metabolic defect. As has already been developed in the previous section, the metabolic defect should be traceable to some abnormality in a specific enzyme or protein.

5. A metabolic defect may be so profound that it is lethal regardless of environmental conditions. A metabolic defect may be latent so that it does not become manifest until there is exposure to an unusual or novel stress. A metabolic defect may be trivial so that it never causes clinical disease in any environment.

6. In any given tissue there are a finite number of cells. When a

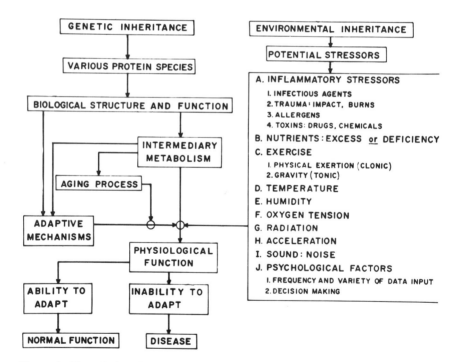

Figure 4. The relationship between the environment and the adaptive response of a biological system.

sufficient number of defective cells occurs, illness results. With age, the number of defective cells increases. Defective cells may occur because of viral infection, antibodies directed against the cell, mutation of cellular genes, accumulation of toxic substances, etc. It is possible that, although cells become defective or die and some if not all are removed, cellular regeneration could replace the defective or dead cells. It is also possible that the defective cells are not removed, and regeneration gradually fails. Ultimately, in either case, there would either be an insufficient number of normal cells to carry out physiological function or an excessive number of defective cells.

B. The Adaptive Response

As is shown in Fig. 4, there is an adaptive response to environmental factors. Failure to adapt to these environmental factors leads to disease. The metabolic system must be able to adapt to environmental factors, or it would be unable to respond to a sudden change in the environment. With a sudden change in the environment, two conditions can occur.

1. If there is a sudden input into the metabolic system, a toxic state can occur due to accumulation of metabolites. For example, starvation followed by refeeding can cause illness since the gastrointestinal-tract cells are unable to change enzyme levels rapidly enough to handle the sudden input of nutrients. Feeding galactose to a patient with galactosemia (uridyl transferase deficiency) causes illness because of the toxic metabolites that accumulate due to the deficient activity of the enzyme in the galactose metabolic pathway. Similar problems exist with fructose or sucrose feeding in patients with hereditary fructose intolerance (fructose-1-phosphate aldolase deficiency) or fructose diphosphatase deficiency. These are examples of input overload. In the case of enzyme deficiencies, substrates and their metabolites accumulate because of the slow rate of function of the deficient enzyme. This is a so-called damming effect. The accumulation of the substrate, its metabolites, or both may lead to toxic levels of substances that interfere with other critical reactions. Because of the slow transformation of the substrate into its product, there may also be the absence of a critical material necessary for the proper functioning of other portions of the cell. The accumulation of toxic metabolites or absence of a critical product, or both, is a common consequence of input overload in enzyme deficiency states. Some examples of input overload are given in Table 2.

2. The metabolic system may be operating at a greater than normal rate and not be able to de-adapt. The input of substrates may be insuf-

Table 2. Examples of the Clinical Manifestations of Disease Caused by Input Overload

Dietary substance	Disease	Accumulated intermediate	Toxic effect[a]
Galactose	Galactosemia	Galactose-1-phosphate	Hypoglycemia (1)
Fructose	Herediary fructose intolerance	Fructose-1-phosphate	Hypoglycemia (2)
Fructose	Fructosediphosphatase deficiency	α-Glycerophosphate, fructosediphosphate	Hypoglycemia (3)
Lactose	Lactase deficiency	Lactose and fermented products	Osmotic diarrhea (4)
Sucrose	Sucrase–isomaltase deficiency	Sucrose and fermented products	Osmotic diarrhea (5)
Protein	Urea cycle enzyme deficiencies	Ammonia	Hyperammonemia (6)
Phenylalanine	Phenylketonuria	Phenylalanine and its metabolites	Mental retardation (7)
Phytanic acid	Refsum disease	Phytanic acid	Progressive neurological syndrome (8)

[a]References: (1) Segal (1972); (2) Hue (1974); (3) Taunton et al. (1978); (4) Gudmand-Høyer (1971); (5) Greene et al. (1972); (6) Scriver and Rosenberg (1973); (7) Knox (1972); (8) Steinberg et al. (1967).

ficient to meet the requirements imposed by the increased metabolic rate. For example, in thyrotoxicosis it may not be possible to consume enough nutrients to meet the high utilization rate. The increased catabolic rate caused by febrile illnesses may result in deficient nutrient intake. The overproduction of insulin results in hypoglycemia. It may not be possible to ingest enough glucose to maintain blood glucose at normal levels for sufficient periods of time. These are examples of input deficiency because of failure of the adaptive mechanisms.

C. The Biochemical Abnormality

We may now inquire as to the biochemical basis of the adaptive response and the biochemical defects responsible for the failure of adaptation to environmental factors, i.e., the defects that lead to abnormal structure or amounts of enzymes and nonenzymatic proteins. Fig. 5 depicts in simplified fashion the mechanism of protein biosynthesis.

It might seem possible that defects in protein synthesis could occur as a result of a defective gene that encodes a specific messenger ribonucleic acid (mRNA), ribosomal ribonucleic acid (rRNA), or transfer ribonucleic acid (tRNA). However, upon reflection, it can be seen that the genes for rRNA and tRNA cannot be defective, since rRNA and tRNA are common to all protein synthesis. A significant defect in the

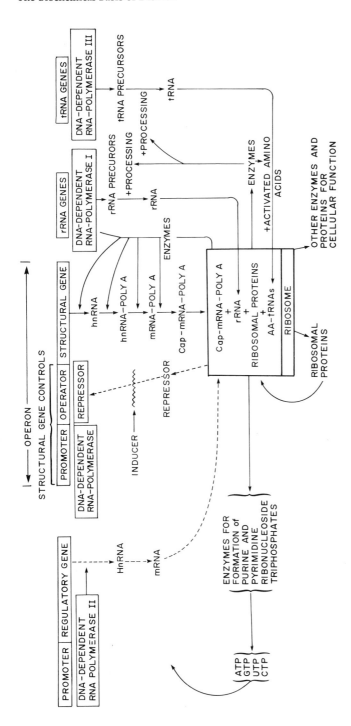

Figure 5. Simplified scheme of the mechanism of protein biosynthesis showing the intricate feedback mechanisms, the structural genes and their controls, and the regulatory gene.

genes for rRNA and tRNA would lead to a general failure of protein synthesis. Perhaps general failure of protein synthesis occurs in aging or premature aging syndromes (e.g., progeria, Werner syndrome); otherwise, it is not possible to implicate the genes for rRNA and tRNA synthesis in the production of abnormal proteins. Therefore, it seems likely that a specific defect in a specific structural gene would be responsible for the production of a specific abnormal protein.

We cannot implicate the genes for DNA-dependent RNA polymerase (types I, II and III), for ribosomal proteins, for enzymes that process rRNA and tRNA precursors, or for enzymes that activate amino acids. All of these genes encode products that are utilized for all protein synthesis. A significant defect in any of these genes would lead to a general failure of protein synthesis (see Fig. 5). Again, we are compelled to conclude that defects in protein synthesis result from a defect in the structural gene or its controls for the specific protein that contains the abnormality.

1. The Gene and Its Controls

The current concept of the gene is that it is composed of a structural gene and adjacent control regions (Fig. 5). Many of the details of the gene and its controls have been derived from the study of microorganisms (Jacob and Monod, 1961; Stent, 1964; Beckwith, 1967; Dickson et al., 1975). It is unlikely that this model, derived from investigations with prokaryotes, applies in an unmodified fashion to mammalian organisms. Indeed, as discussed in Chapters 5 and 7, the genome of eukaryotic cells is a complex structure composed of DNA and proteins, whose full informational potential is never expressed at one time. What is clear is that, concomitant with the augmentation of information, there has been a proliferation of controls, many occurring at the translational as well as transcriptional levels (Chambon, 1978).

The structural gene in prokaryotes is composed of DNA and serves as the template for the synthesis of messenger RNA (mRNA), which contains the specific information for the ordering of amino acids in the sequence in which polypeptides are synthesized at the ribosomal level (see Chapter 5). A region adjacent to the structural gene is termed the promoter and appears to be a site for binding DNA-dependent RNA polymerase, which is the enzyme that synthesizes mRNA using the structural gene as a template. The operator is a region between the promoter and the structural gene and is a control mechanism that allows the synthesis of the mRNA by the DNA-dependent RNA polymerase. A separate gene, the regulatory gene, contains the information for the synthesis of a regulatory protein called the repressor. The repressor protein binds to the operator and inhibits synthesis of the specific mRNA from the corresponding structural gene template. The repressor binds specific substances termed inducers. The repressor–inducer complex is not able to inhibit the operator, and mRNA synthesis is thus able to proceed. These sequences of events are shown in Fig. 5. There are also aporepressors which bind substances such as end-products of a metabolic pathway in such a way that the aporepressor–end-product complex acts as a complete repressor and inhibits protein synthesis. This is not shown in Fig. 5. There are other factors that have been shown to affect protein synthesis at the gene level, such as cAMP and a specific receptor protein (Zubay et al., 1970). The relevance of such mechanisms in mammalian systems remains to be demonstrated. Whether mammalian genes have operator and promoter regions is uncertain. Nevertheless, eukaryotes do appear to have regulatory genes.

A number of abnormalities of protein synthesis are known in terms of defects of the structural gene. Specific abnormalities of the regulatory genes in mammalian cells have not been demonstrated. One can only speculate at this time about the consequences of defects in the promoter and operator.

Despite the uncertainties, we can use the model depicted in Fig. 5 to examine the known defects in protein synthesis in a variety of disease entities. Table 3 summarizes some of the known defects in the structural gene which give rise to defects in protein synthesis, and lists some of the hypothetical consequences of defects in the control portions of the structural gene which might be predicted if such regions exist.

2. Defects in the Structural Gene: Synthesis of Structurally Abnormal Protein

Defects of structural genes lead to the production of correspondingly abnormal mRNAs, which may in turn lead to the synthesis of correspondingly abnormal proteins. A great number of abnormalities result from amino acid substitutions in proteins. If the substitution alters the structure significantly, there will be a change in the function of the protein. In the case of enzymes, loss of catalytic function to a varying degree may result, although other possibilities such as failure to bind a cofactor or an alteration in a regulatory site may also occur.

A mutation in a single purine or pyrimidine base in the structural gene will alter the genetic code and may lead to an amino acid substitution in the protein corresponding to the structural gene. Many abnormal proteins arise by this mechanism. For example, sickle cell hemoglobin—S hemoglobin—has a valine substituted for a glutamic acid in amino acid residue 6 in the β chain. The genetic code in mRNA for

Table 3. Possible Types of Biochemical Defects Which Give Rise to Inborn Errors of Metabolism

Gene type	Gene defect	Protein defect
Structural	1. Point mutation: change in a single purine or pyrimidine base in DNA codon	Amino acid substitution
	2. Unequal crossing over of DNA strands	
	a. Loss of one or more DNA codons	One or more amino acid deletions with preservation of identity of the protein
	b. Loss of a large number of DNA codons	Markedly abnormal protein; protein may not be detectable
	c. Fusion of different genes	Fusion of different portions of separate proteins
	d. Loss of entire gene	Failure of protein synthesis
	e. Loss of one or two purine or pyrimidine bases in a DNA codon: frameshift mutation	Abnormal protein past the point of the frameshift
	3. Mutation in termination codon	Amino acid additions
Structural gene	1. Promoter[a] is unable to bind DNA-dependent RNA polymerase	No protein synthesis
	2. Operator binds repressor permanently	No protein synthesis
	3. Operator is unable to bind repressor	Protein overproduction
Regulatory	1. No repressor[b]	Protein overproduction
	2. Excess amount of repressor	No protein synthesis or failure of adaptive response; lack of sensitivity to inducer
	3. Deficient amount of repressor	Increased sensitivity to inducer
	4. Abnormal repressor	
	a. Fails to bind inducer	No protein synthesis
	b. Binds inducer permanently	Protein overproduction or no protein synthesis
	c. Binds operator permanently	No protein synthesis
	d. Cannot bind to operator	Protein overproduction

[a] Defects in promoter and operator may be point mutations or occur as a consequence of unequal crossing over with loss or gain of a portion of the DNA.
[b] Defects in the regulatory gene may be any of those that occur for the structural gene so that the amount of repressor synthesized may vary or the structure of the repressor may be altered so that its function is abnormal.

glutamic acid is GAA and for valine is GUA, where G is guanine, A is adenine, and U is uracil. In the gene, the corresponding codes are CTT and CAT, where T is thymine and C is cytosine. A mutation transforming thymine into adenine in the structural gene for the β chain of hemoglobin changes the corresponding genetic code in the mRNA so that it translates for valine instead of glutamic acid at the ribosomal level.

A deletion of an entire codon can cause deletion of the corresponding amino acid. It is believed that unequal crossing-over between DNA strands during meiosis can result in the deletion of one or more triplet codons and the consequent loss of one or more amino acids in the corresponding protein.

If a mutation arises in a portion of the gene that regulates chain termination (UAG in mRNA, ATC in the gene), amino acid additions may result. Hemoglobin Constant Spring, which has 31 additional amino acids at the C-terminus of the α chain, is believed to have arisen because of a defect in the chain termination codon of DNA (Milner et al., 1971; Clegg et al., 1971).

Crossing-over between DNA strands may give rise to a protein containing portions of two different proteins. Hemoglobin Lepore has a structure which consists of part of the structure of the hemoglobin δ chain and part of the structure of the β chain. This suggests that there was a fusion of the genes for the δ and β chains of hemoglobin (Barnabas and Muller, 1962). It is believed that chromosomal misalignment and unequal crossing-over of the structural genes is the mechanism of formation of the Lepore group of hemoglobins. A number of different types of Lepore hemoglobins have been described, with the nature of the variant depending on the crossover site within the molecules (Badr et al., 1972). A so-called anti-Lepore variant has been described which consists of β fusion with a crossover point occurring between $\beta 22$ and $\delta 50$ and with a deletion of valine that normally occupies $\delta 137$ (Honig et al., 1978). This may have resulted from two separate crossover events. Restriction endonuclease analysis of human β- and δ-globin genes has recently shown that the β- and δ-globin genes are separated by about 7000 base pairs and that both genes are transcribed from the same DNA strand. The β-globin gene contains an intervening sequence of 800 to 1000 base pairs, and the same sequence appears to exist also in the δ gene (Flavell et al., 1978). Analysis of DNA from a homozygous individual with hemoglobin Lepore demonstrated that the Lepore gene was a fusion of part of the δ- and part of the β-globin genes (Flavell et al., 1978; Mears et al., 1978a).

Crossing-over of genes can thus give rise to fusion of proteins, amino acid deletions, and in some cases, both conditions. The loss of one or two purine or pyrimidine bases can cause a frameshift mutation. In this case the resulting protein may be totally different from the original protein and have markedly altered function. The portion of the protein from the point of the frameshift may be a nonsense sequence. If the point of the frameshift occurs early enough in the peptide sequence, a totally inactive protein may be synthesized. The loss of a sizeable portion of a gene gives rise to a markedly abnormal protein which may be indetectable in terms of its function or as a cross-reacting protein to antibodies (Boyer et al., 1973). This situation may be indistinguishable from the absence of synthesis of that specific protein.

In certain disorders, there is complete loss of either enzymatic activity or nonenzymatic protein function. In these disorders, the homozygote lacks enzymatic and antigenic activity. These conditions have been considered to be due to "silent genes." Disorders that may be considered in this category include myeloperoxidase deficiency

(Salmon et al., 1970), pseudocholinesterase deficiency (Hodgkin et al., 1965), myophosphorylase deficiency (Robbins, 1960), acatalasia (Nishimura et al., 1961; Takahara et al., 1962), a variant of factor VIII deficiency (hemophilia) (Feinstein et al., 1969), and several others (Boyer et al., 1973). Because of the absence of any identifiable protein in these conditions, it has been cautioned that enzyme or nonenzyme replacement be carefully weighed, since the patient may treat such replacement as foreign protein (Boyer et al., 1973).

In thalassemia there is a failure of protein synthesis of the various chains of hemoglobin, especially of the α and β chains of hemoglobin. The spectrum of disease ranges from a lethal condition through varying degrees of severity to little or no disease. This is related to the total amount of normally functioning hemoglobin that is synthesized.

In α-thalassemia-1, there is deletion of two α-chain genes in whole or in large part, as demonstrated by genetic analysis (Clegg and Weatherall, 1978; Weatherall, 1978). Deletion of only one α-chain gene results in α-thalassemia-2, in which there is a reduced rate of α-chain synthesis. In α-thalassemia-1, if there is continued synthesis of the γ chain, (which is unable to form F hemoglobin), γ_4 hemoglobin forms (hemoglobin Bart). In α-thalassemia-2, excess production of the β chain leads to the formation of β_4 hemoglobin (H hemoglobin). Different degrees of severity of thalassemia develop depending on whether the individual is heterozygous or homozygous for the α-thalassemia-1 or -2 genes or has both α-thalassemia-1 and -2 genes. A defect in the termination codon for the α-chain gene gives rise to hemoglobin Constant Spring, which has a very low rate of synthesis (see Table 4). The hemoglobin Constant Spring gene behaves similarly to the α-thalassemia-2 gene in that a combination of both genes produces hemoglobin H disease.

The situation is more complex in the case of β and $\delta\beta$ thalassemic diseases. In β^+ thalassemia, there appears to be reduced production of the β-chain mRNA, although there is no major deletion of the β-chain gene (Kan et al., 1975; Tolstoshev et al., 1976).

In β^0 thalassemia there is absence of β-chain synthesis. There are different forms of β^0 thalassemia. In two patients of Italian and Pakistani ethnic origin, no β-chain mRNA could be detected (Tolstoshev et al., 1976). In two patients of Chinese ethnic origin, β-chain mRNA could be detected, but was inactive in cell-free systems (Kan et al., 1975). In β^0 thalassemia in the Ferrara region of Italy, β-chain mRNA is present but is inactive in in vitro protein-synthesizing systems unless a soluble extract from normal reticulocytes is present (Conconi et al., 1972). In $\delta\beta$ thalassemia, there is deletion of both δ and β genes (Mears et al., 1978b).

A number of inborn errors have been described in which enzymatic or other function is absent but immunologically cross-reacting protein is present: glucose-6-phosphate dehydrogenase deficiency (Marks and Tsutsui, 1963), factor VIII deficiency (hemophilia variant) (Feinstein et al., 1969), and mouse erythrocyte catalase deficiency (Feinstein et al, 1968). Hence, absence of a functional protein does not necessarily indicate absence of protein synthesis. The immunologic determinants of a protein may be quite distinct from its binding or catalytic function.

The nature of the functional change in a protein will depend on the nature of the amino acid change and position in the protein structure where the change occurs. For any given protein, an enormous number of variants are possible. Glucose-6-phosphate dehydrogenase has more than 80 known variants, which results in a spectrum of clinical diseases (Yoshida, 1973). In some cases, a structural variant causes no functional impairment and no disease results. Hemoglobin has a large number of variants in its structural components. There are many amino acid substitutions, deletions, additions, and fusions. Functionally, these hemo-

Table 4. Some Examples of Hemoglobin Variants and Their Functional Change

Hemoglobin	Structural change	Functional alteration [a]
S	β6 Glu → Val	Sickling with decreased oxygen tension (1)
C-Harlem	β6 Glu → Val β73 Asp → Asn	Sickling with decreased oxygen tension, two amino acid substitutions (2)
Korle-Bu	β73 Asp → Asn	No clinical manifestations (3)
Bethesda	β145 Tyr → His	Increased oxygen affinity and erythrocytosis (4)
Fort de France	α45 His → Arg	Increased oxygen affinity (5)
St. Louis	β28 Leu → Gln	β heme in permanent ferric state, methemoglobinemia (hemoglobin M group) (6)
St. Etienne	β92 His → Gln	Unstable, β chain lacks heme, tetramer is a semihemoglobin (7)
Hope	β136 Gly → Asp	No clinical manifestations (8)
Beth Israel	β102 Asn → Ser	Low oxygen affinity and cyanosis (9)
J. Singapore	α78 Asn → Asp α79 Ala → Gly	Two amino acid substitutions (10)
Leiden	β7 Glu → 0	Amino acid deletion, mild hemolysis (11)
Tochigi	β56–59 → 0	Deletion of Gly-Asn-Pro-Lys (12)
Tours	β87 Thr → 0	Unstable, spontaneous loss of heme (13)
Constant Spring	α 142–172 (addition)	Has 31 additional amino acids attached to C-terminal end (14,15)
Lepore	δ1–87, β116–146	Fusion of δ and β genes (sequence 88–115 is common to both hemoglobins) (16)

[a] References: (1) Dean and Schechter (1978); (2) Bookchin et al. (1966); (3) Konotey-Ahulu et al. (1968); (4) Schmidt et al. (1976); (5) Braconnier et al. (1977); (6) Thillet et al. (1976); (7) Godeau et al. (1976); (8) Steinberg et al. (1974); (9) Nagel et al. (1976); (10) Blackwell et al. (1972); (11) De Jong et al. (1968); (12) Shibata et al. (1970); (13) Wajcman et al. (1973); (14) Milner et al. (1971); (15) Clegg et al. (1971); (16) Labie et al. (1966).

globins may not give rise to clinical disease, or they may produce mild to severe disease. Table 4 gives some examples of hemoglobins with their structural changes and the nature of their functional impairment.

3. Speculations about Defects in Gene Controls

Table 3 lists a number of defects known to occur in the structural gene, and the type of protein defect that results as a consequence. Also shown in Table 3 are a number of possible defects that conceivably could occur using the Jacob and Monod (1961) type of model. There is a paucity of evidence to support the possibility that defects in promoters, operators, and regulatory genes exist in human beings. Although it is believed that a promoter or its equivalent does exist for human structural genes, and there is evidence that regulatory genes exist (Tomkins et al., 1969; Gelehrter, 1976; Paigen et al., 1975), there is little evidence for the existence of an operator. It may be useful, however, to briefly consider the consequences of possible defects in promoter, operator, and regulatory genes. On the basis of these speculations we can make certain predictions, which can then be tested when the appropriate experimental techniques become available.

a. Defects in Structural Gene Controls. We can conceive of the possibility that the promoter and operator develop mutations, lose portions of their structure, or undergo other types of alterations so that their structure—and consequently their functions—is altered. Regardless of the type of structural change we envision, the result can be a change in function. If the promotor is altered such that it cannot be recognized or bound by the DNA-dependent RNA polymerase, no protein synthesis will occur. Similarly, if the operator has greatly increased affinity for its repressor and binds the repressor permanently, there will be no protein synthesis. On the other hand, if the operator is unable to bind its repressor, protein overproduction would result.

As already noted, in β^+ thalassemia the β-globin gene is present, although there is decreased synthesis of β-globin mRNA, and hence decreased production of the β chain of hemoglobin (Kan et al., 1975; Tolstoshev et al., 1976). This type of situation is compatible with a defect in a promoter or with the permanent binding of a repressor to the operator portion of the β-globin gene.

Currently, there are no known situations in which overproduction of an enzyme or protein can be reasonably attributed to the failure of repression of a gene owing to a defect in its operator. Although there are conditions characterized by excessive enzyme activity, these are related to alterations in enzyme structure such that intrinsic enzyme activity is increased or such that a defect exists in the regulatory controls at the enzyme level. For example, the variant enzyme, glucose-6-phosphate dehydrogenase Hektoen, which has a substitution of histidine for tyrosine, has an increased rate of synthesis (Yoshida, 1970) but has normal enzyme properties (Dern, 1966). Phosphoribosyl-pyrophosphate synthetase may have increased activity. In the three types of this enzyme that have been described, one has increased intrinsic enzyme activity (Becker et al., 1973), one has decreased affinity for its nucleotide allosteric modifiers (Sperling et al., 1973), and one has increased affinity for its substrate, ribose-5-phosphate (Becker et al., 1974). In acute intermittent hepatic porphyria, there is increased activity of δ-aminolevulinic acid synthetase owing to the failure of heme synthesis because of deficient activity of uroporphyrinogen I synthetase (Strand et al., 1970). Heme is able to repress the synthesis of δ-aminolevulinic acid synthetase, although the mechanism is not clear (Meyer and Schmid, 1978).

b. Defects in the Regulatory Gene: Possible Failure of Adaptation of Protein Synthesis. Of a more subtle nature are those possible defects of protein biosynthesis involving a failure of adaptation. In these conditions protein is structurally normal. However, there is a failure of protein synthesis to change in response to a suitable stimulus. Conceivably, the adaptive failures represent defects in the regulatory gene so that an abnormal repressor protein is produced. As listed in Table 3, there may be defects in the amount of repressor synthesized. If no repressor is synthesized there should be protein overproduction. If there is an excess amount of repressor, induction would be difficult unless very large amounts of an inducer are present. Excess amounts of a repressor could be expected to decrease the sensitivity of protein synthesis to an inducer. Inversely, a deficient amount of repressor should increase the sensitivity of protein synthesis to an inducer.

We can also envision the possibility of abnormal repressor protein being produced. If the repressor fails to bind its inducer, or binds the operator permanently, no protein synthesis should take place. If the repressor cannot bind the operator, protein overproduction would be expected. If the repressor binds the inducer with high affinity, protein overproduction should occur except in the absence of the inducer, when no protein synthesis would take place.

Many of the speculative possibilities could give rise to less drastic clinical abnormalities than are seen with structural gene defects. These conditions are difficult to study because there is no defect in protein structure, and in some cases a normal amount of protein may be produced. The adaptive defect is expressed mainly as a failure of protein synthesis to undergo a change in response to a stimulus.

4. Summary of the Biochemical Abnormality

The different types of gene defects are summarized in Table 3. We can summarize the implications as follows: All disease is expressed by some defect in the synthesis of a specific protein or group of proteins.

There are three theoretical categories of defects in protein synthesis. There may be defects in: (1) the structural genes; (2) the control portions of the structural genes (operator, promoter); or (3) in regulatory genes. These defects may manifest themselves in proteins with abnormal structure, in lack of protein synthesis, or in a failure of the adaptive response to environmental factors. In the last case, there is a protein which has a normal structure but which fails to change in response to a given stimulus, although the enzyme or protein may be present in normal amounts. Definitive proof for the defects in man exists only for defects in structural genes.

5. Deranged Biochemistry at the Organ Level

By referring to Table 5, it is possible to recapitulate, in a logical manner, the sequential levels of function which a particular organ carries out, and to thereby understand the myriad clinical abnormalities encountered in diseases involving a particular organ or multiple organs. Each organ must maintain its own structural and functional integrity before it can interact with the rest of the body to carry out organismic functions. Thus, it is clear that damage to a particular organ will engage the intermediary metabolism of that organ in repair, renewal, and resynthesis of damaged elements as well as removal of toxic or offending agents where necessary. These activities often require a significant expenditure of energy, and, indeed, Zotin and Zotina (1967) have demonstrated a large increase in heat production during cell renewal after injury. However, as emphasized repeatedly in this book, it is the exquisitely fine-tuned integration of structure and function that permits the cell to carry out its biochemical functions, and likewise it is such intimate organization, emerging as an expressed function of the integrated biochemistry of millions of cells, that allows an organ to carry out its functions. Disruption of the structural or functional integrity of an organ results, then, in disordered physiology. However, it is necessary to underscore the enormous functional reserve possessed by the body's organs, such that disease often does not manifest itself until much of an organ has been damaged. One should never lose sight of the fact that deranged physiology—the level at which the physician treats the patient—is the expression of deranged biochemistry.

D. Enzyme Deficiency States

A metamorphosis has occurred in medical thinking since the landmark discovery by Pauling et al. in 1949 that a defect in the hemoglobin molecule was the basis for sickle cell anemia. With the accumulated knowledge regarding metabolic pathways and the description of the double-helical structure of DNA (Watson and Crick, 1953), medicine entered into a period of discovery of inborn errors of metabolism that justified Garrod's prescient view (Garrod, 1908) that certain diseases were caused by inborn errors of metabolism. As knowledge of metabolic regulation augmented our knowledge of the metabolic pathways, there emerged an awareness that all disease, both acquired and inherited, ultimately results from a defect in the organization of the biochemistry of the cell or, in a phrase, from incoherent behavior (Chapter 2).

As classically viewed, the inborn errors of metabolism involve a mutation in a structural gene, causing an amino acid substitution or deletion affecting the active site of the protein. Defects in enzyme regulation can lead to a loss of modulation of metabolic pathways, with an overproduction of the terminal product without gross accumulation of

Table 5. Levels of Organ Function

I. Maintenance of function to provide structural and functional integrity
   A. Integration of energy, matter and information input
      1. Metabolic processes
         a. Membrane stabilization
         b. Membrane transport
         c. Resynthesis and renewal of structural elements and enzymes
         d. Maintenance of critical mass of cells for organ function to be sustained
            i. Cellular renewal
II. Maintenance of function of other organs
   A. Secretion of an informational molecule (hormone or signal)
   B. Metabolic functions: secretion of essential metabolite or its precursors
      1. Liver: protein synthesis (albumin, coagulation proteins), synthesis of cholesterol, triglycerides, and bile salts
      2. Liver and kidney: gluconeogenesis
      3. Gastrointestinal tract: absorption of nutrients
      4. Kidney: maintenance of *milieu interieur*, fluid and electrolytes, acid–base balance
      5. Lungs: O_2–CO_2 exchange, acid–base balance
      6. Blood: O_2–CO_2 transport, coagulation, conduit for transmittal of substrates (matter and energy) and hormones (information)
   C. Storage functions
      1. Adipose tissue: fat (energy)
      2. Liver: glycogen, fat soluble vitamins
      3. Muscle: glycogen, protein reserve during starvation
      4. Bone: calcium, phosphorus
   D. Detoxification
      1. Liver: urea cycle
III. Integrated functions: result of supercellular level of integration (spatiotemporal relationship of complex circuit)
   A. Central nervous system
      1. Motor activity
      2. Sensation
      3. Memory
      4. Thought

intermediary metabolites, as is often encountered with the inborn errors of metabolism. For example, a mutation might affect the ligand binding site of an enzyme without affecting the catalytic site and might thus allow the catalytic function of the enzyme to proceed normally while impairing the response of the enzyme to modifiers. The presence of such an abnormal enzyme at the rate-determining step in a biosynthetic pathway would then lead to unregulated synthesis of the end product. Of course, the amino acids comprising the allosteric site are also under the aegis of the genome and would thus also constitute an inherited defect. Thus, it is possible that defects termed "inborn errors of metabolism" may be caused by amino acid changes (deletions or substitutions) affecting either the catalytic site or a modifier site. Inherited defects may also affect enzymes that carry out covalent modification of the enzyme responsible for the particular catalytic step which appears to be affected. In this instance, the phosphorylase kinase deficiency variant of glycogen storage disease (Howell, 1978) serves as an example.

Since proteins are the functional devices of living tissue, compromise of their normal function produces a disease state. In inherited disease, the primary defect may involve either the structure of the protein—thus affecting its ability to function—or the rate of synthesis or degradation of the protein, determining the amount of a particular functional entity present. In acquired disease, however, the disorder may involve multiple pathways, dysfunction occurring as a consequence of damage to an organ. Multiple levels of damage to the structural and functional proteins comprising the organ then lead to a constellation of events, initiated because of a defect in the biological expression of the function of the particular organ.

For purposes of discussion, inborn errors of metabolism may be divided into several groups based on the consequences of the deficiency of an enzyme, a receptor, or a transport protein in a membrane. One category would include transport disorders due to suspected abnormalities of receptor or transport proteins in the membrane. A virtual plethora of such disorders have been described, and they have recently been reviewed by Roth and Segal (1979).

A number of inborn errors of metabolism are due to an enzyme deficiency which leads to the accumulation of the precursor of that particular pathway. This type of clinical expression was among the first to be described, since simple clinical screening tests often uncovered the increased excretion in the urine, or accumulation in the blood, of early metabolites of the pathway which had been blocked. An important example of such a defect is homocystinuria, where both methionine and homocystine accumulate as the result of the deficiency of activity of cystathionine synthase (Mudd and Levy, 1978). A variant of the foregoing occurs when both the precursor of the main pathway and metabolites from alternative pathways accumulate. Such an example is provided by classical phenylketonuria, caused by a deficiency of phenylalanine hydroxylase, in which one observes mental retardation and mousy-smelling urine, among other manifestations (Tourian and Sidbury, 1978). Følling (1934), using the ferric chloride spot test followed by isolation procedures, showed that phenylpyruvic acid accumulated in this disorder. Accumulation resulted from the utilization of an alternate pathway as a consequence of the blockade of the usual pathway for catabolism of phenylalanine. Another example is galactose-1-phosphate uridyl transferase deficiency (galactosemia), in which certain alternate routes are used in an attempt to relieve the abnormal accumulation of galactose. The galactitol and galactose-1-phosphate which accumulate in these disorders have been implicated in the pathogenesis of the mental retardation and liver dysfunction characteristic of this disorder (Cohn and Segal, 1977).

Another variation that results from the deficiency of a functional protein is the deficiency of an end product of a pathway. Examples of such deficiency include albinism, which was one of the original inborn errors of metabolism described by Garrod. Another example is that encountered in the adrenogenital syndrome, owing to a defect of hydroxylation at the C-21 position of the steroid nucleus, which leads to impaired production of cortisol with hypernatremia (Bongiovanni, 1978). Defective binding of the coenzyme to the apoenzyme is yet another cause of an inborn error and is encountered in forms of methylmalonic aciduria (Rosenberg, 1978). Still another variant relates to the possibility of a defect in an allosteric site on an enzyme. Such a defect has been described in the purine synthetic pathway. Here PRPP synthetase in certain patients with gout has been shown (Sperling et al., 1973) to be apparently insensitive to feedback control by the end product of the pathway. In such a circumstance, excessive production of purines is encountered, with consequent formation of stones.

The final groups of mutations affect circulating proteins rather than enzymes or membrane-structural components. These include analbuminemia, thyroxine-binding globulin deficiency, and abetalipoproteinemia.

III. Therapeutic Approaches to Disease

A. Treatment of Inborn Errors of Metabolism

Several approaches to the therapy of inborn errors of metabolism have been attempted, including those which are well defined in terms of both their rationale and their feasibility and those which are only dimly perceived as possible. In the first category we find manipulation of the diet, administration of vitamin precursors to the coenzymes, drug administration, and organ transplantation to provide a missing enzyme. In the second category are enzyme replacements both with purified enzymes and with liposomes, and genetic engineering.

The various approaches are summarized in Table 6 from a different point of view. The arrangement of Table 6 utilizes different levels of therapy which coincide with the different levels of function within the biological system. The categories of treatment range from manipulation of the environment, through symptomatic treatment, treatment at the level of the enzymatic reaction, treatment at the level of the enzyme itself, and with a final consideration of treatment at the gene level (genetic engineering).

Restriction of protein intake or development of special formulas, restricted in their concentrations of amino acids involved in certain inborn errors of metabolism, have been attempted in a number of aminoacidopathies, including phenylketonuria, homocystinuria, hyperammonemic states (urea cycle defects), and branched-chain amino acid disorders. As regards inborn errors of carbohydrate metabolism, interdiction of galactose and fructose have been successful in treating uridyl transferase deficiency (galactosemia) and hereditary fructose intolerance, respectively, while a dietary decrease in lactose-containing products has been successful in treating lactase deficiency.

Scriver (1973) has compiled a list of vitamin-responsive inborn errors of metabolism. Particular success has been achieved in those disorders responsive to pyridoxine and vitamin B_{12}.

Administration of drugs has been successful in a number of inborn errors of metabolism, including Wilson disease (Sass-Kortsak and Bearn, 1978) and cystinuria (Thier and Segal, 1978), where the chelating agent penicillamine has been effective in removing copper and cystine, respectively, which accumulate in these disorders. Allopurinol, a drug which inhibits the activity of xanthine oxidase, has been somewhat successful in treating at least the accumulation of urate in the Lesch–Nyhan syndrome, which is manifested by stone formation, arthritis, nephropathy, and tophi (Kelley and Wyngaarden, 1978). Recently, glycine has been used successfully in pharmacological amounts to treat isovaleric acidemia, since there is a glycine-N-acylase in the liver which can detoxify isovaleric acid by converting it to the glycine conjugate (Krieger and Tanaka, 1976; Yudkoff et al., 1978). Provision of large amounts of glycine in the diet has provided sufficient substrate for the enzyme to accomplish its detoxifying effect. Barbiturates have been used in one form of Crigler–Najjar disease, which is associated with hyperbilirubinemia (Schmid and McDonagh, 1978), and uridine has been used to treat orotic aciduria, since individuals so affected are pyrimidine auxotrophs (Kelley and Smith, 1978).

The Biochemical Basis of Disease

Table 6. Therapeutic Approaches to Inborn Errors of Metabolism[a]

A. Treatment at the environmental level: Simple avoidance of or protection from environmental factors and physiological states that are deleterious

Environmental factor or physiological state	Inborn error of metabolism
Sunlight and ultraviolet irradiation	Congenital erythropoietic porphyria (1), erythropoietic protoporphyria (2), xeroderma pigmentosum (3), albinism (4)
Drugs	
Barbiturates	Acute intermittent hepatic porphyria (5)
Primaquine	Glucose-6-phosphate dehydrogenase deficiency (6)
Nutrients	
Galactose	Galactokinase deficiency (7), galactosemia (8)
Fructose	Hereditary fructose intolerance (9), fructose-1, 6-diphosphatase deficiency (10)
Sucrose and isomaltose	Sucrase-isomaltase deficiency (11)
Glucose and galactose	Glucose-galactose malabsorption (12)
Toxic dietary factor	
Fava bean	Glucose-6-phosphate dehydrogenase deficiency (6)
Exercise	Myoglobinuria (13), familial periodic paralysis (14)
High altitude	Sickle cell anemia (15)

B. Treatment at the physiologic level: Compensatory or symptomatic therapy. Measures used to correct abnormal physiologic state since biochemical abnormality is inaccessible or unknown.

Symptomatic treatment	Inborn error of metabolism
Protease inhibitors	Muscular dystrophy in chickens (16) and mice (17)
Buffers (Shohl's solution)	Renal tubular acidosis (18)
Vitamin D	Fanconi syndrome (19)
Phosphate	X-linked genetic hypophosphatemia with rickets (20)
Ascorbic acid and/or methylene blue (nonenzymatic reduction of methemoglobin)	Methemoglobin reductase deficiency (21)

C. Treatment at the level of the enzymatic reaction—Substrate restriction and/or product replacement and/or removal of toxic metabolite. Substrate restriction involves special preparation of low substrate diets to prevent accumulation of toxic metabolite(s).

Substrate restricted	Inborn error of metabolism
Phenylalanine	Phenylketonuria (22)
Protein	Urea cycle defects (23), other hyperammonemic states (24,25)

Product replaced	Inborn error of metabolism
Uridine	Orotic aciduria (26)
Cortisol	Adrenogenital syndromes (27)
Blood transfusion	Thalassemia (28)
Anti-hemophiliac globulin	Hemophilia (29)

(continued)

Table 6 (*continued*)

Product replaced	Inborn error of metabolism
Fibrinogen	Afibrinogenemia (29)
Fructose	Glucose–galactose malabsorption (30)
Zinc	Acrodermatitis enteropathica (31)

Toxic metabolite removed	Method of removal	Inborn error of metabolism
Isovaleric acid	Glycine administration	Isovaleric acidemia (32,33)
Bilirubin	Phototherapy	Neonatal hyperbilirubinemia (34)
Iron	Phlebotomy	Hemochromatosis (35)
Copper	Penicillamine	Wilson disease (36)
Acetylcholine receptor antibodies	Plasmapheresis	Myasthenia gravis (37)

D. Treatment at the level of the defective enzyme or protein

Agent used to enhance enzyme or protein activity	Inborn error of metabolism
Barbiturates	Crigler–Najjar syndrome (38)
1-Deamino-8-D-arginine vasopressin (DDAVP)	Mild and moderate forms of hemophilia A and von Willebrand disease (39)
Danazol	Anti-thrombin III deficiency (40), α_1-antitrypsin deficiency (40), hereditary angioedema (41)
Folic acid	Fructose-1,6-diphosphatase deficiency (10)
Fructose	Sucrase–isomaltase deficiency (42)
Pyridoxine	Pyridoxine-responsive anemia (43)
Antibody to fructose-1-phosphate aldolase	Hereditary fructose intolerance (in vitro only) (44)

Enzyme or protein replaced	Method	Inborn error of metabolism
Glycosidase	Purified fungal enzyme	Glycogenosis, type II (45)
α-Acetylglucosaminidase	Purified human placental enzyme	Sanfilippo disease, type B (46)
Glucocerebrosidase	Purified human placental enzyme injected intravenously (47,48) or entrapped in erythrocytes (47)	Gaucher disease (47,48)
Hexosaminidase	Purified enzyme	Tay–Sachs disease (49)
α-Galactosidase	Purified enzyme	Fabry disease (50)
Glucocerebrosidase	Spleen (51) and kidney (32) transplant	Gaucher disease (51,52)

(*continued*)

Table 6 (*continued*)

Enzyme or protein replaced	Method	Inborn error of metabolism
Ceramidetrihexosidase	Kidney transplant	Fabry disease (53)
Unknown	Kidney transplant	Cystinosis (54)
Immunoglobulins (IgG, IgM)	Bone marrow transplant	Severe combined immunodeficiency (55,56)
Amyloglucosidase	Liposomes	Glycogenosis, type II (57)
Glucocerebroside: β-Glucosidase	Liposomes	Gaucher disease (58)
β-Galactosidase	Liposomes	Feline G_{M1} gangliosidosis fibroblasts (59)

Defective protein	Method or agent of protection	Inborn error of metabolism
S hemoglobin	Carbamylation of red cells extracorporeally and other agents	Sickle cell anemia (15,60)
Possible abnormal linkage between red cell membrane glycophorin and spectrin	Dimethyl adipimidate	Hereditary stomatocytosis (61)

E. Gene replacement (genetic engineering)—theoretical
   Chromosome-mediated gene transfer (62,63)
   Virus-mediated gene transfer (64)

References: (1) Haining et al. (1968); (2) Donaldson et al. (1967); (3) Wilkins (1974); (4) Witkop et al. (1978); (5) Stein and Tschudy (1970); (6) Yoshida (1973); (7) Montelone et al. (1971); (8) Segal (1972); (9) Hue (1974); (10) Taunton et al. (1978); (11) Rosenthal et al. (1962); (12) Burke and Danks (1966); (13) Wheby and Miller (1960); (14) McDowell et al. (1963); (15) Dean and Schechter (1978); (16) Stracher et al. (1978); (17) Libby and Goldberg (1978); (18) Seldin and Wilson (1978); (19) Leaf (1966); (20) Fraser and Scriver (1976); (21) Schwartz and Jaffé (1978); (22) Knox (1972); (23) Scriver and Rosenberg (1973); (24) Awrich et al. (1975); (25) Columbo et al. (1967); (26) Kelley and Smith (1978); (27) Bongiovanni (1978); (28) Weatherall (1978); (29) Ratnoff (1978); (30) Lindquist and Meeuwisse (1962); (31) Moynahan (1974); (32) Krieger and Tanaka (1976); (33) Yudkoff et al. (1978); (34) Wu (1974); (35) Knauer et al. (1965); (36) Richmond et al. (1964); (37) Dau et al. (1977); (38) Schmid and McDonagh (1978); (39) Theiss and Sauer (1977); (40) Walker et al. (1975); (41) Gelfand et al. (1976); (42) Greene et al. (1972); (43) Harris and Horrigan (1964); (44) Gitzelmann et al. (1974); (45) Lauer et al. (1968); (46) O'Brien et al. (1973); (47) Beutler et al. (1977); (48) Brady et al. (1975); (49) Johnson et al. (1973); (50) Brady et al. (1973); (51) Groth et al. (1973); (52) Desnick et al. (1973); (53) Clarke et al. (1972); (54) Malekzadeh et al. (1977); (55) Anderson (1975); (56) O'Reilly et al. (1977); (57) Tyrrell et al. (1976); (58) Belchetz et al. (1977); (59) Reynolds et al. (1978); (60) Diederich et al. (1976); (61) Mentzer et al. (1976); (62) Athwal and McBride (1977); (63) Mukherjee et al. (1978); (64) Rogers (1971).

Enzyme replacement therapy has been attempted by several methods, including organ transplantation, administration of peripheral cells and plasma, administration of purified enzymes, and providing enzymes in liposome packages. Organ transplantation has been used in Fabry disease (angiokeratoma corporis diffusum), which is a defect in ceramidetrihexosidase and which predominantly affects the kidney. Here, replacement of or transplantation of the kidney provides both a source of enzyme and a normal functioning kidney (Clarke et al., 1972). In cystinosis, whose fundamental biochemical defect is not understood but whose pathologic hallmark appears to be accumulation of cystine in the tubules of the kidney—leading eventually to renal failure—transplantation of the kidney seems to be particularly ameliorative (Malekzadeh et al., 1977).

The administration of peripheral cells and plasma has been attempted in a number of mucopolysaccharidoses, but the final verdict appears to be that the procedure is relatively unsuccessful (Cantz and Gehler, 1976). The administration of purified enzymes has been attempted in Fabry disease (Brady et al., 1973) and Gaucher disease (Brady et al., 1974) with biochemical improvement, but long-term studies are needed to determine whether or not clinical improvement occurs. Finally, attempts have been made to package purified enzymes into liposomes (Gregoriadis, 1976). These bodies are tiny spheres made from various combinations of lipids, resembling those found in natural cell membranes, and have an onion-like structure in cross section. To permit the placement of macromolecules into these bodies, it was necessary to increase the proportion of charged molecules in the original lipid layers. This represented a technological problem which was not solved until the early 1970s. This clearly is an area likely to be the focus of intensive investigation over the next few years, since, in principle, it offers the promise of providing the enzymes missing in the cerebral lipidoses.

Finally, one comes to genetic engineering and recombinant genetics. Genetic engineering involves the insertion of normal genes in place of defective or inoperative genes. This would involve not only the synthesis of the complete gene together with its control mechanisms but also the insertion of the gene into the appropriate cells without destruction or further alteration that would impair its proper function. The problem is further complicated by the fact that many genes contain interrupted sequences which do not appear in the mature (finished) mRNA. Whether it would be sufficient to use a synthetic gene without interrupted sequences, with the expectation that the correct mRNA would be synthesized, is presently unknown. Genetic engineering remains a future possibility, but far too little is presently known for this to be a practical therapeutic approach.

Recombination is a genetic term referring to the incorporation of DNA from two separate bacteria into one strand. It comes about through breakage of the parental DNA molecules and subsequent rejoining to form recombinants. The ability to carry out such a process makes it possible to join DNA molecules of unrelated organisms in vitro. The role of the restriction endonucleases is a key to this process, since these enzymes are able to cut DNA at sequence specific sites.

Baltimore (1977) has stressed the need to distinguish between recombinant DNA research and genetic engineering, defining the latter as replacement of defective DNA with material which encodes the synthesis of the normal enzyme. He argues that, because such therapy provides the promise for treating otherwise untreatable diseases, we must continue our research while recognizing the obligation to use such powerful technology wisely. This presentation was part of a forum on recombinant DNA research sponsored by the National Academy of Science in 1977. It represents a variegated presentation of both scientists and laymen which is, regretably, of unequal quality. However, on balance it presents the spectrum of viewpoints on this subject, which is one of the major ethical and scientific considerations of our generation, if not of this century.

B. The Classical Modes of Therapy in Medicine

In medicine, just a few modes of therapy are possible. Classically, patients can be treated by psychotherapy, physical methods, nutrition, drugs, and surgery. In many cases it is not possible to treat the cause of the disease, but only possible to treat the patient symptomatically. Much of the difficulty in treating patients revolves around the problem of etiology. Unless the specific etiology is known, it is not possible to eradicate the disease. The second problem has to do with access to the disease process. There are many disease entities which are inaccessible to our present therapeutic methods.

All of the approaches for the treatment of inborn errors of metabolism must be applied within the context of the recognized modes of therapy utilized in medicine. Unfortunately, not all of the modes of treatment are useful in treating inborn errors of metabolism. Psychotherapy and physical methods are not sufficiently specific and thus cannot be employed for the treatment of inborn errors except in a supportive manner.

Psychotherapy utilizes communication between the patient and the physician to engender insight in the patient as to his subconscious drives and emotions. Seemingly, the verbal expression of one's feelings and thoughts feeds back into the central nervous system, altering those feelings and thoughts, and thereby relieving the distress that has resulted from one's subconscious drives. More recently, biofeedback techniques have been used in an attempt to control physiological parameters, such as blood pressure. Psychotherapy is a useful method of therapy, often employed in conjunction with other methods. For example, relief of anxiety about one's illness, the projected course of therapy, or both, is a simple form of psychotherapy involving communication between the patient and his physician.

Physical methods employ such techniques as heat, cold, immobilization, exercise, ultrasound, radiation, electroshock, and acupuncture. These are important modalities of treatment but are gross methods which do not deal directly with the specific biochemistry of a disease.

Nutrition is utilized to provide general support for the debilitated patient with a chronic illness. Often, patients are deficient in protein, calories, potassium, vitamins, etc. The healing process is impaired by poor nutrition and is enhanced by proper nutrition. In some cases enteral nutrition is not practical or is impossible. With the techniques of total parenteral nutrition, it is now possible to provide nutritional support for almost all types of patients. Specific nutritional therapy is necessary in certain types of metabolic and related diseases. Nutritional therapy is essential in conditions where specific food intolerances are the basis for a disease.

The use of drugs is one of the most widely employed modes of therapy. It is essential that the minimum number of appropriate medications, in the proper dose, to be given at the proper times via the proper route, be prescribed. Because of the ease of administering drugs, the possibilities for error are enhanced. There is a wide range of sensitivities to various drugs among patients and a variable degree of effectiveness of different drugs. The large number of drugs available attests to the individual variation that occurs among patients and diseases.

Surgery is carried out at the macro level, often involving removal of a damaged or damaging body component. The goal of creating tissue anew is thus far not possible. Therefore, although organ transplants and plastic-surgical procedures are common, the tissues and organs used are either derived from the patient or from another person. Often the best substitutes are inert substances used to replace a particular tissue (e.g., hip joint, aorta).

The goal of medicine is to prevent all disease. As we learn more about the basic biochemical causes of disease, we can employ increasingly specific therapeutic modes via the classical methods of medicine to prevent and treat those diseases we presently understand poorly and for which we have imperfect treatments.

References

Anderson, I. M., 1975, Treatment of primary immunodeficiency, *Proc. R. Soc. Med.* **68**:577.

Athwal, R. S., and McBride, O. W., 1977, Serial transfer of a human gene to rodent cells by sequential chromosome-mediated gene transfer, *Proc. Natl. Acad. Sci. USA* **74**:2943.

Awrich, A. E., Stackhouse, W. J., Cantrell, J. E., Patterson, J. H., and Rudman, D., 1975, Hyperdibasic aminoaciduria, hyperammonemia, and growth retardation: Treatment with arginine, lysine, and citrulline, *J. Pediatr.* **87**:731.

Badr, F. M., Lorkin, P. A., and Lehmann, H., 1972, Haemoglobin P-Nilotic containing a β-δ chain, *Nature (New Biol.)* **242**:107.

Baltimore, D., 1977, quoted in: *Research with Recombinant DNA*, p. 237, National Academy of Sciences, Washington, D. C.

Barnabas, J., and Muller, C. J., 1962, Haemoglobin-Lepore$_{Hollandia}$, *Nature* **194**:931.

Becker, M. A., Kostel, P. J., Meyer, L. J., and Seegmiller, J. E., 1973, Human phosphoribosylpyrophosphate synthetase: Increased enzyme specific activity in a family with gout and excessive purine synthesis, *Proc. Natl. Acad. Sci. USA* **70**:2749.

Becker, M. A., Meyer, L. J., Kostel, P. J., and Seegmiller, J. E., 1974, Increased 5-phosphoribosyl-1-pyrophosphate (PRPP) synthetase activity and gout: Diversity and structural alterations of the enzyme, *J. Clin. Invest.* **53**:4a.

Beckwith, J. R., 1967, Regulation of the Lac operon, *Science* **156**:597.

Belchetz, P. E., Braidman, I. P., Crawley, J. C. W., and Gregoriadis, G., 1977, Treatment of Gaucher's disease with liposome-entrapped glucocerebroside: β-Glucosidase, *Lancet* **2**:116.

Beutler, E., Dale, G. L., Guinto, E., and Kuhl, W., 1977, Enzyme replacement therapy in Gaucher's disease: Preliminary clinical trial of a new enzyme preparation, *Proc. Natl. Acad. Sci. USA* **74**:4620.

Blackwell, R. Q., Boon, W. H., Liu, C.-S., and Weng, M.-I., 1972, Hemoglobin J Singapore: $\alpha78$ Asn $\rightarrow$ Asp; $\alpha79$ Ala $\rightarrow$ Gly, *Biochim. Biophys. Acta* **278**:482.

Bongiovanni, A. M., 1978, Congenital adrenal hyperplasia and related conditions, in: *The Metabolic Basis of Inherited Disease* (J. B. Stanbury, J. B. Wyngaarden, and D. S. Fredrickson, eds.), pp. 868–893, 4th ed., McGraw-Hill, New York.

Bookchin, R. M., Nagel, R. L., Ranney, H. M., and Jacobs, A. S., 1966, Hemoglobin C$_{Harlem}$: A sickling variant containing amino acid substitutions in two residues of the β-polypeptide chain, *Biochem. Biophys. Res. Commun.* **23**:122.

Boyer, S. H., Siggers, D. C., and Krueger, L. J., 1973, Caveat to protein replacement therapy for genetic disease. Immunological implications of accurate molecular diagnosis, *Lancet* **2**:654.

Braconnier, F., Gacon, G., Thillet, J., Wajcman, H., Soria, J., Maigret, P., Labie, D., and Rosa, J., 1977, Hemoglobin Fort De France (α_2^{45}(CD3) His $\rightarrow$ Arg β_2). A new variant with increased oxygen affinity, *Biochim. Biophys. Acta* **493**:228.

Brady, R. O., Tallman, J. F., Johnson, W. G., Gal, A. E., Leahy, W. R., Quirk, J. M., and Dekaban, A. S., 1973, Replacement therapy for inherited enzyme deficiency. Use of purified ceramidetrihexosidase in Fabry's disease, *N. Engl. J. Med.* **289**:9.

Brady, R. O., Pentchev, P. G., Gal, A. E., Hibbert, S. R., and Dekaban, A. S., 1974, Replacement therapy for inherited enzyme deficiency: Use of purified glucocerebrosidase in Gaucher's disease, *N. Engl. J. Med.* **291**:989.

Brady, R. O., Pentchev, P. G., and Gal, A. E., 1975, Investigation in enzyme replacement therapy in lipid storage diseases, *Fed. Proc.* **34**:1310.

Burke, V., and Danks, D. M., 1966, Monosaccharide malabsorption in young infants, *Lancet* **1**:1177.

Cantz, M., and Gehler, J., 1976, The mucopolysaccharidoses: Inborn errors of glycosaminoglycan catabolism, *Hum. Genet.* **32**:233.

Chambon, P., 1978, Summary: The molecular biology of the eukaryotic genome is coming of age, *Cold Spring Harbor Symp. Quant. Biol.* **42**:1209.

Clarke, J. T. R., Guttman, R. D., Wolfe, L. S., Beudorn, J. G., and Morehouse, D. D., 1972, Enzyme replacement therapy by renal allotransplantation in Fabry's disease, *N. Engl. J. Med.* **287**:1215.

Clegg, J. B., and Weatherall, D. J., 1978, Molecular basis of thalassaemia, *Br. Med. Bull.*, **32**:262.

Clegg, J. B., Weatherall, D. J., and Milner, P. F., 1971, Haemoglobin Constant Spring—a chain termination mutant?, *Nature* **234**:237.

Cohn, R. M., and Segal, S., 1977, Disorders of galactose metabolism, in: *Scientific Approaches to Clinical Neurology* (E. Goldensohn, and S. Appel, eds.), p. 99, Lea and Febiger, Philadelphia.

Columbo, J. P., Vassella, F., Humbel, R., and Buergi, W., 1967, Lysine intolerance with periodic ammonia intoxication, *Am. J. Dis. Child.* **113**:138.

Conconi, F., Rowley, P. T., Del Senno, L., Pontremoli, S., and Volpato, S., 1972, Induction of β-globin synthesis in the β-thalassaemia of Ferrara, *Nature (New Biol.)* **238**:83.

Dau, P. C., Lindstrom, J. M., Cassel, C. K., Denys, E. H., Shev, E. E., and Spitler, L. E., 1977, Plasmapheresis and immunosuppressive drug therapy in myasthenia gravis, *N. Engl. J. Med.* **297**:1134.

Dean, J., and Schechter, A. N., 1978, Sickle-cell anemia: Molecular and cellular bases of therapeutic approaches, *N. Engl. J. Med.* **299**:752.

De Jong, W. W. W., Went, L. N., and Bernini, L. F., 1968, Hemoglobin Leiden: Deletion of β6 or 7 glutamic acid, *Nature* **220**:788.

Dern, R. J., 1966, A new hereditary quantitative variant of glucose-6-phosphate dehydrogenase characterized by a marked increase in enzyme activity, *J. Lab. Clin. Med.* **68**:560.

Desnick, S. J., Desnick, R. J., Brady, R. O., Pentchev, P. G., Simmon, R. L., Najarian, J. S., Swaisnan, K., Sharp, H. L., and Krivit, W., 1973, Renal transplantation in type II, Gaucher's disease, in: *Enzyme Therapy in Genetic Diseases* (D. Bergsma, ed.), pp. 109–119, Williams and Wilkins, Baltimore.

Dickson, R. C., Abelson, J., Barnes, W. M., and Reznikoff, W. S., 1975, Genetic regulation: The Lac control region, *Science* **187**:27.

Diederich, D. A., Trueworthy, R. C., Gill, P., Cader, A. M., and Larsen, W. E., 1976, Hematologic and clinical responses in patients with sickle cell anemia after chronic extracorporeal red cell carbamylation, *J. Clin. Invest.* **58**:642.

Donaldson, E. M., Donaldson, A. D., and Rimington, C., 1967, Erythropoietic protoporphyria: A family study, *Br. Med. J.* **1**:659.

Feinstein, D., Chong, M. N. Y., Kasper, C. K., and Rapaport, S. I., 1969, Hemophilia A: Polymorphism detectable by a factor VIII antibody, *Science* **163**:1071.

Feinstein, R. N., Suter, H., and Jaroslow, B. N., 1968, Blood catalase polymorphism: Some immunological aspects, *Science* **159**:638.

Flavell, R., Kooter, J. M., De Boer, E., Little, P. F. R., and Williamson, R., 1978, Analysis of the β-δ-globin gene loci in normal and Hb Lepore DNA: Direct determination of gene linkage and intragene distance, *Cell* **15**:25.

Følling, A., 1934, Über Ausscheidung von Phenylbrenztraubensaure in den Harn als Stoffwechselanomalie in Verbindung mit Imbezillitat, *Z. Physiol. Chem.* **227**:169.

Fraser, D., and Scriver, C. R., 1976, Familial forms of vitamin D-resistant rickets revisited. X-linked hypophosphatemia and autosomal recessive vitamin D dependency, *Am. J. Clin. Nutr.* **29**:1315.

Garrod, A. E., 1908, Inborn errors of metabolism (Croonian Lectures), *Lancet* **2**:1, 73, 142, and 214.

Gelehrter, T. D., 1976, Enzyme induction, *N. Engl. J. Med.* **294**:646.

Gelfand, J. A., Sherins, R. J., Alling, D. W., and Frank, M. M., 1976, Treatment of hereditary angioedema with Danazol. Reversal of clinical and biochemical abnormalities, *N. Engl. J. Med.* **295**:1444.

Gitzelmann, R., Steinmann, B., Bally, C., and Lebherz, H. G., 1974, Antibody activation of mutant human fructosediphosphate aldolase B in liver extracts of patients with hereditary fructose intolerance, *Biochem. Biophys. Res. Commun.* **59**:1270.

Godeau, J. F., Beuzard, Y. G., Cacheleux, J., Brizard, C. P., Gibaud, A., and Rosa, J., 1976, Association of hemoglobin Saint Etienne ($\alpha_2\beta_2{}^{92}$F8 His → Gln) with hemoglobins A and F. Synthesis and subunit exchange in vitro, *J. Biol. Chem.* **251**:4346.

Greene, H. L., Stifel, F. B., and Herman, R. H., 1972, Dietary stimulation of sucrase in a patient with sucrase-isomaltase deficiency, *Biochem. Med.* **6**:409.

Gregoriadis, G., 1976, The carrier potential of liposomes in biology and medicine, *N. Engl. J. Med.* **295**:704 and 765.

Groth, C. G., Blomstrand, R., Dreborg, S., Hagenfeldt, L., Lofstrom, B., Ockerman, P., Samuelson, K., and Svernerholm, L., 1973. Splenic transplantation in Gaucher's disease, in: *Enzyme Therapy in Genetic Disease* (D. Bergsma, ed.), pp. 102–105, Williams and Wilkins, Baltimore.

Gudmand-Høyer, E., 1971, *Specific Lactose Malabsorption in Adults*, Fadl's Forlag, Copenhagen.

Haining, R. G., Cowger, M. L., Shurtleff, D. B., and Labbe, R. F., 1968, Congenital erythropoietic porphyria. I. Case report, special studies and therapy, *Am. J. Med.* **45**:624.

Harris, J. W., and Horrigan, D. D., 1964, Pyridoxine responsive anemia: Prototype and variations on theme, *Vitam. Horm.* **22**:721.

Hodgkin, W. E., Giblett, E. R., Levine, H., Bauer, W., and Motulsky, A. G., 1965, Complete pseudocholinesterase deficiency: Genetic and immunologic characterization, *J. Clin. Invest.* **44**:486.

Honig, G. R., Shamsuddin, M., Mason, R. G., and Vida, L. N., 1978, Hemoglobin Lincoln Park: A $\beta\delta$ fusion (anti-Lepore) variant with an amino acid deletion in the δ chain-derived segment, *Proc. Natl. Acad. Sci. USA* **75**:1475.

Howell, R. R., 1978, The glycogen storage diseases, in: *The Metabolic Basis of Inherited Disease* (J. B. Stanbury, J. B. Wyngaarden, and D. S. Fredrickson, eds.), pp. 137–159, 4th ed., McGraw-Hill, New York.

Hue, L., 1974, The metabolism and toxic effects of fructose, in: *Sugars in Nutrition* (H. L. Sipple and K. W. McNutt, eds.), pp. 357–371, Academic Press, New York.

Jacob, F., and Monod, J., 1961, Genetic regulatory mechanisms in the synthesis of proteins, *J. Mol. Biol.* **3**:318.

Johnson, W. G., Desnick, R. J., Long, D. M., Sharp, H. L., Krivit, W., Brady, B., and Brady, R. O., 1973, Intravenous injection of purified hexosaminidase A into a patient with Tay-Sachs disease, in: *Enzyme Therapy in Genetic Disease* (D. Bergsma, ed.), p. 120, Williams and Wilkins, Baltimore.

Kan, Y. W., Holland, J. P., Dozy, A. M., and Varmus, H. E., 1975, Demonstration of nonfunctional β-globin mRNA in homozygous β^0-thalassemia, *Proc. Natl. Acad. Sci. USA* **72**:5140.

Kelley, W. N., and Smith, L. H., Jr., 1978, Hereditary orotic aciduria, in: *The Metabolic Basis of Inherited Disease* (J. B. Stanbury, J. B. Wyngaarden, and D. S. Fredrickson, eds.), pp. 1045–1071, 4th ed., McGraw-Hill, New York.

Kelley, W. N., and Wyngaarden, J. B., 1978, The Lesch–Nyhan syndrome, in: *The Metabolic Basis of Inherited Disease* (J. B. Stanbury, J. B. Wyngaarden, and D. S. Fredrickson, eds.), pp. 1011–1036, 4th ed., McGraw-Hill, New York.

Knauer, C. M., Gamble, C. N., and Monroe, L. S., 1965, The reversal of hemochromatotic cirrhosis by multiple phlebotomies, *Gastroenterology* **49**:667.

Knox, W. W., 1972, Phenylketonuria, in: *The Metabolic Basis of Inherited Disease* (J. B.

Stanbury, J. B. Wyngaarden, and D. S. Fredrickson, eds.), pp. 266–295, 3rd ed., McGraw-Hill, New York.

Konotey-Ahulu, F. I. D., Gallo, E., Lehmann, H., and Ringelhann, B., 1968, Haemoglobin Korle-Bu (β73 Asp → Asn) showing one of the two amino acid substitutions of haemoglobin C Harlem, *J. Med. Genet.* **5**:107.

Krieger, I., and Tanaka, K., 1976, Therapeutic effects of glycine in isovaleric acidemia, *Pediatr. Res.* **10**:25.

Labie, D., Schroeder, W. A., and Huisman, T. H. J., 1966, The amino acid sequence of the δ-β chains of hemoglobin Lepore$_{Augusta}$ = Lepore$_{Washington}$, *Biochim. Biophys. Acta* **127**:428.

Lauer, R. M., Mascarinas, T., Racela, A. S., and Diehl, A. M., 1968, Administration of a mixture of fungal glycosidases to a patient with type II glycogenosis (Pompe's disease), *Pediatrics* **42**:672.

Leaf, A., 1966, The syndrome of osteomalacia, renal glycosuria, aminoaciduria, and increased phosphorus clearance (Fanconi syndrome), in: *The Metabolic Basis of Inherited Disease* (J. B. Stanbury, J. B. Wyngaarden, and D. S. Fredrickson, eds.), pp. 1205–1220, 2nd ed., McGraw-Hill, New York.

Libby, P., and Goldberg, A. L., 1978, Leupeptin, a protease inhibitor, decreases protein degradation in normal and diseased muscles, *Science* **199**:534.

Lindquist, B., and Meeuwisse, G. W., 1962, Chronic diarrhea caused by monosaccharide malabsorption, *Acta Paediatr. Scand.* **51**:674.

Malekzadeh, M. H., Neustein, H. B., Schneider, J. A., Pennisi, A. J., Ettenger, R. B., Uittenbogaart, C. H., Kogut, M. D., and Fine, R. N., 1977, Cadaver renal transplantation in children with cystinosis, *Am. J. Med.* **63**:525.

Marks, P. A., and Tsutsui, E. A., 1963, Human glucose-6-P dehydrogenase: Studies on the relation between antigenicity and catalytic activity—The role of TPN, *Ann. N. Y. Acad. Sci.* **103**:902.

McDowell, M. K., Herman, R. H., and Davis, T. E., 1963, The effect of a high and low sodium diet in a patient with familial periodic paralysis, *Metabolism* **12**:388.

Mears, J. G., Ramirez, F., Leibowitz, D., and Bank, A., 1978a, Organization of human δ- and β-globin genes in cellular DNA and the presence of intragenic inserts, *Cell* **15**:15.

Mears, J. G., Ramirez, F., Leibowitz, D., Nakamura, F., Bloom, A., Konotey-Ahulu, F., and Bank, A., 1978b, Changes in restricted human cellular DNA fragments containing globin gene sequences in thalassemias and related disorders, *Proc. Natl. Acad. Sci. USA* **75**:1222.

Mentzer, W. C., Lubin, B. H., and Emmons, S., 1976, Correction of the permeability defect in hereditary stomatocytosis by dimethyl adipimidate, *N. Engl. J. Med.* **294**:1200.

Meyer, U. A., and Schmid, R., 1978, The porphyrias, in: *The Metabolic Basis of Inherited Disease* (J. B. Stanbury, J. B. Wyngaarden, and D. S. Fredrickson, eds.), pp. 1166–1220, 4th ed., McGraw-Hill, New York.

Milner, P. F., Clegg, J. B., and Weatherall, D. J., 1971, Haemoglobin-H disease due to a unique haemoglobin variant with an elongated α-chain, *Lancet* **1**:729.

Monteleone, J. A., Bautler, E., Monteleone, P. L., Utz, C. L., and Casey, E. C., 1971, Cataracts, galactosuria, and hypergalactosemia due to galactokinase deficiency in a child, *Am. J. Med.* **50**:403.

Moynahan, E. J., 1974, Acrodermatitis enteropathica: A lethal inherited human zinc deficiency disorder, *Lancet* **2**:399.

Mudd, S. H., and Levy, H. L., 1978, Disorders of transsulfuration, in: *The Metabolic Basis of Inherited Disease* (J. B. Stanbury, J. B. Wyngaarden, and D. S. Fredrickson, eds.), pp. 458–503, 4th ed., McGraw-Hill, New York.

Mukherjee, A. B., Orloff, S., Butler, J. DeB., Triche, T., Lalley, P., and Schulman, J. D.,

1978, Entrapment of metaphase chromosomes into phospholipid vesicles (lipo-chromosomes): Carrier potential in gene transfer, *Proc. Natl. Acad. Sci. USA* **75**:1361.

Nagel, R. L., Lynfield, J., Johnson, J., Landau, L., Bookchin, R. M., and Harris, M. B., 1976, Hemoglobin Beth Israel. A mutant causing clinically apparent cyanosis, *N. Engl. J. Med.* **295**:125.

Nishimura, E. T., Kobara, T. Y., Takahara, S., Hamilton, H. B., and Madden, S. C., 1961, Immunologic evidence of catalase deficiency in human hereditary acatalasemia, *Lab. Invest.* **10**:333.

O'Brien, J. S., Miller, A. L., Loverde, W. A., and Veath, M. L., 1973, Sanfilippo disease type B: Enzyme replacement and metabolic correction in cultured fibroblasts, *Science* **181**:753.

O'Reilly, R. J., Dupont, B., Pahwa, S., Grimes, E., Smithwick, E. M., Pahwa, R., Schwartz, S., Hansen, J. A., Siegal, F. P., Sorell, M., Svejgaard, A., Jersild, C., Thomsen, M., Platz, P., L'Esperance, P., and Good, R. A., 1977, Reconstitution in severe combined immunodeficiency by transplantation of marrow from an unrelated donor, *N. Engl. J. Med.* **297**:1311.

Paigen, K., Swank, R. T., Tomino, S., and Ganschow, R. E., 1975, The molecular genetics of mammalian glucuronidase, *J. Cell. Physiol.* **85**:379.

Pauling, L., Itano, H. A., Singer, S. J., and Wells, I. C., 1949, Sickle cell anemia: A molecular disease, *Science* **110**:543.

Ratnoff, O. D., 1978, Hereditary disorders of hemostasis, in: *The Metabolic Basis of Inherited Disease* (J. B. Stanbury, J. B. Wyngaarden, and D. S. Fredrickson, eds.), pp. 1755–1791, 4th ed., McGraw-Hill, New York.

Reynolds, G. D., Baker, H. J., and Reynolds, R. H., 1978, Enzyme replacement using liposome carriers in feline G_{MI} gangliosidosis fibroblasts, *Nature* **275**:754.

Richmond, J., Rosenoer, V. M., Tompsett, S. L., Draper, I., and Simpson, J. A., 1964, Hepato-lenticular degeneration (Wilson's disease) treated by penicillamine, *Brain* **87**:619.

Robbins, P. W., 1960, Immunological study of human muscle lacking phosphorylase, *Fed. Proc.* **19**:193.

Rogers, S., 1971, Gene therapy: A potentially invaluable aid to medicine and mankind, *Res. Commun. Chem. Path. Pharmacol.* **2**:587.

Rosenberg, L. E., 1978, Disorders of propionate, methylmalonate, and cobalamin metabolism, in: *The Metabolic Basis of Inherited Disease* (J. B. Stanbury, J. B. Wyngaarden, and D. S. Fredrickson, eds.), pp. 411–429, 4th ed., McGraw-Hill, New York.

Rosenthal, I. M., Cornblath, M., and Crane, R. K., 1962, Congenital intolerance to sucrose and starch presumably caused by hereditary deficiency of specific enzymes in the brush border membrane of the small intestine, *J. Lab. Clin. Med.* **60**:1012.

Roth, K. S., and Segal, S., 1979, Tubular aspects of hereditary and developmental disorders of the kidney, in: *Nephrology* (J. P. Grunfeld and J. Hamberger, eds.), pp. 949–975, John Wiley and Sons, New York.

Salmon, S. E., Cline, M. J., Schultz, J., and Lehrer, R. I., 1970, Myeloperoxidase deficiency. Immunologic study of a genetic leukocyte defect, *N. Engl. J. Med.* **282**:250.

Sass-Kortsak, A., and Bearn, A. G., 1978, Hereditary disorders of copper metabolism, in: *The Metabolic Basis of Inherited Disease* (J. B. Stanbury, J. B. Wyngaarden, and D. S. Fredrickson, eds.), pp. 1098–1126, 4th ed., McGraw-Hill, New York.

Schmid, R., and McDonagh, A. F., 1978, Hyperbilirubinemia, in: *The Metabolic Basis of Inherited Disease* (J. B. Stanbury, J. B. Wyngaarden, and D. S. Fredrickson, eds.), pp. 1221–1257, 4th ed., McGraw-Hill, New York.

Schmidt, R. M., Jue, D. L., Lyonnais, J., and Moo-Pen, W. F., 1976, Hemoglobin$_{Bethesda}$, β_{145} (HC2) Tyr → His, in a Canadian family, *Am. J. Clin. Pathol.* **66**:449.

Schwartz, J. M., and Jaffé, E. R., 1978, Hereditary methemoglobinemia with deficiency of

NADH dehydrogenase, in: *The Metabolic Basis of Inherited Disease* (J. B. Stanbury, J. B. Wyngaarden, and D. S. Fredrickson, eds.), pp. 1452–1464, 4th ed., McGraw-Hill, New York.

Scriver, C. R., 1973, Vitamin-responsive inborn errors of metabolism, *Metabolism* **22**:1319.

Scriver, C. R., and Rosenberg, L. E., 1973, Urea cycle and ammonia, in: *Amino Acid Metabolism and Its Disorders* pp. 234–249, W. B. Saunders, Philadelphia.

Segal, S., 1972, Disorders of galactose metabolism, in: *The Metabolic Basis of Inherited Disease* (J. B. Stanbury, J. B. Wyngaarden, and D. S. Fredrickson, eds.), pp. 174–195, 3rd ed., McGraw-Hill, New York.

Seldin, D. W., and Wilson, J. D., 1978, Renal tubular acidosis, in: *The Metabolic Basis of Inherited Disease* (J. B. Stanbury, J. B. Wyngaarden, and D. S. Fredrickson, eds.), pp. 1618–1633, 4th ed., McGraw-Hill, New York.

Shibata, S., Miyaji, T., Ueda, S., Matsuoka, M., Iuchi, I., Yamada, K., and Shinkai, N., 1970, Hemoglobin Tochigi (β56-59 deleted). New unstable hemoglobin discovered in a Japanese family, *Proc. Jap. Acad.* **46**:440.

Sperling, O., Persky-Brosh, S., Boer, P., and De Vries, A., 1973, Human erythrocyte phosphoribosylpyrophosphate synthetase mutationally altered in regulatory properties, *Biochem. Med.* **7**:389.

Stein, J. A., and Tschudy, D. P., 1970, Acute intermittent porphyria. A clinical and biochemical study of 46 patients, *Medicine* **49**:1.

Steinberg, D., Vroom, F. Q., Engel, W. K., Cammermeyer, J., Mize, C. E., and Avigan, J., 1967, Refsum's disease—A recently characterized lipidosis involving the nervous system, *Ann. Intern. Med.* **66**:365.

Steinberg, M. H., Adams, J. G., Thigpen, J. T., Morrison, F. S., and Dreiling, B. J., 1974, Hemoglobin Hope ($\alpha_2\beta_2^{136-\text{gly}\rightarrow\text{asp}}$)-S disease: Clinical and biochemical studies, *J. Lab. Clin. Med.* **84**:632.

Stent, G. S., 1964, The operon: on its third anniversary. Modulation of transfer RNA species can provide a workable model of an operator-less operon, *Science* **144**:816.

Stracher, A., McGowan, E. B., and Shafiq, S. A., 1978, Muscular dystrophy: Inhibition of generation in vivo with protease inhibitors, *Science* **200**:50.

Strand, L. J., Felsher, B. F., Redeker, A. G., and Marver, H. S., 1970, Heme biosynthesis in intermittent acute porphyria: Decreased hepatic conversion of porphobilinogen to porphyrins and increased delta-amino-levulinic acid synthetase activity, *Proc. Natl. Acad. Sci. USA* **67**:1315.

Takahara, S., Ogata, M., Kobara, T. Y., Nishimura, E. T., and Brown, W. J., 1962, The "catalase protein" of acatalasemic red blood cells, *Lab. Invest.* **11**:782.

Taunton, O. D., Greene, H. L., Stifel, F. B., Hofeldt, F. D., Lufkin, E. G., Hagler, L., Herman, Y., and Herman, R. H., 1978, Fructose-1,6-diphosphatase deficiency, hypoglycemia, and response to folate therapy in a mother and her daughter, *Biochem. Med.* **19**:260.

Theiss, W., and Sauer, E., 1977, DDAVP: Alternative to replacement treatment in mild hemophilia A and von Willebrand-Jürgens syndrome, *Dtsch. Med. Wochschr.* **102**:1769.

Thier, S. O., and Segal, S., 1978, Cystinuria, in: *The Metabolic Basis of Inherited Disease* (J. B. Stanbury, J. B. Wyngaarden, and D. S. Fredrickson, eds.), pp. 1578–1592, 4th ed., McGraw-Hill, New York.

Thillet, J., Cohen-Solal, M., Seligmann, M., and Rosa, J., 1976, Functional and physicochemical studies of hemoglobin St. Louis β28 (B10) Leu $\rightarrow$ Gln. A variant with ferric β heme iron, *J. Clin. Invest.* **58**:1098.

Tolstoshev, P., Mitchell, J., Lanyon, G., Williamson, R., Ottolenghi, S., Comi, P., Giglioni, B., Masera, G., Modell, B., Weatherall, D. J., and Clegg, J. B., 1976, Presence of gene for β globin in homozygous β_0 thalassaemia, *Nature* **259**:95.

Tomkins, G. M., Gelehrter, T. D., Granner, D., Martin, D., Jr., Samuels, H. H., and Thompson, E. B., 1969, Control of specific gene expression in higher organisms, *Science* **166**:1474.

Tourian, A. Y., and Sidbury, J. B., 1978, Phenylketonuria, in: *The Metabolic Basis of Inherited Disease* (J. B. Stanbury, J. B. Wyngaarden, and D. S. Fredrickson, eds.), pp. 240–255, 4th ed., McGraw-Hill, New York.

Tyrrell, D. A., Ryman, B. E., Keeton, B. R., and Dubowitz, M., 1976, Use of liposomes in treating type II glycogenosis, *Br. Med. J.* **2**:88.

Wajcman, H., Labie, D., and Schapira, G., 1973, Two new hemoglobin variants with deletion. Hemoglobin Tours: Thr β87 (F3) deleted and hemoglobin St. Antoine: Gly-Leu β74-75 (E18-19) deleted. Consequences for oxygen affinity and protein stability, *Biochim. Biophys. Acta* **295**:495.

Walker, I. D., Davidson, J. F., Yound, P., and Conkie, J. A., 1975, Effect of anabolic steroids on plasma antithrombin III, α_2-macroglobulin and α_1-antitrypsin levels, *Thromb. Diath. Haemorrh.* **34**:106.

Watson, J. D., and Crick, F. H. C., 1953, Molecular structure of nucleic acids: A structure for deoxyribose nucleic acid, *Nature* **171**:737.

Weatherall, D. J., 1978, The thalassemias, in: *The Metabolic Basis of Inherited Disease* (J. B. Stanbury, J. B. Wyngaarden, and D. S. Fredrickson, eds.), pp. 1508–1523, 4th ed., McGraw-Hill, New York.

Wheby, M. S., and Miller, H. S., Jr., 1960, Idiopathic paroxysmal myoglobinuria. Report of two cases occurring in sisters. Review of the literature, *Am. J. Med.* **29**:599.

Wilkins, R. J., 1974, DNA repair, a molecular process of medical relevance, *N. Z. Med. J.* **80**:210.

Witkop, C. J., Jr., Quevado, W. C., Jr., and Fitzpatrick, T. B., 1978, Albinism, in: *The Metabolic Basis of Inherited Disease* (J. B. Stanbury, J. B. Wyngaarden, and D. S. Fredrickson, eds.), pp. 283–316, 4th ed., McGraw-Hill, New York.

Wu, P. Y. K., 1974, Immediate and long-term effects of phototherapy on preterm infants, in: *Phototherapy in the Newborn: An Overview* (G. B. Odell, R. Schaffer, A. P. Simopoulos, eds.), pp. 150–160, National Academy of Sciences, Washington, D. C.

Yoshida, A., 1970, Amino acid substitution (histidine to tyrosine) in a glucose-6-phosphate dehydrogenase variant (G6PD Hektoen) associated with overproduction, *J. Mol. Biol.* **52**:483.

Yoshida, A., 1973, Hemolytic anemia and G6PD deficiency, *Science* **179**:532.

Yudkoff, M., Cohn, R. M., Puschak, R., and Segal, S., 1978, Glycine therapy for isovaleric acidemia, *J. Pediat.* **92**:830.

Zotin, A. I., and Zotina, R. S., 1967, Thermodynamic aspects of developmental biology, *J. Theoret. Biol.* **17**:57.

Zubay, G., Schwartz, D., and Beckwith, J., 1970, Mechanism of activation of catabolite-sensitive genes: A positive control system, *Proc. Natl. Acad. Sci. USA* **66**:104.

Index

Abetalipoproteinemia, 477
Acanthosis nigricans, 563
Acetyl-CoA, 25
Acetyl CoA carboxylase
 and citrate, 525
 degradation, 244, 246
 regulation, 147, 159, 300
Acid–base catalysis, 106, 107
Active site, 101. *See also* Enzyme
  mechanisms
Active transport, 407–418
 ion pumps, 410–418
 primary, 408
 secondary, 408
Adaptive response, 221, 629–631
Adenine nucleotide compartmentation,
  505, 506
Adenosine triphosphate. *See* ATP
Adenylate energy charge, 145, 146
Adenyl cyclase
 gangliosides in receptor, 567
 hormonal activation of, 378–380, 555–558, 564–579
 hormone receptor model, 565, 572, 573
 protein inhibitor, 574
Adrenal medulla hormones, 552, 553
Adrenogenital syndrome, 643
Aging
 biochemical characteristics, 22–24, 625
 cellular replacement and, 39, 42
 premature, 632
Albinism, 643
Allosterism, 140, 141
 attributes for regulation, 141
 in enzyme regulation, 359, 360
 kinetics of, 142, 143

Allosterism (*cont.*)
 models of, 143, 297
 sequential, 144
 symmetry, 143
 site, allosteric, 295, 297
Amino acids
 in proteins, 100, 101
 transport of, 412–418
 coupled transport, 414, 415
 electrogenic nature, 415
 γ-glutamyl cycle, 413
 relation to charge, 413
 specificity of systems, 413, 414
Aminoacyl tRNA synthetase
 mechanism of action, 193
 specificity, 180, 191
Anchor principle, 105
Arrhenius plots for membrane-bound
  enzymes, 330, 341, 346–349
Aspartate transcarbamylase, 141
Atherosclerosis, 302, 624
ATP (adenosine triphosphate)
 coenzyme role, 113–115
 translocation across inner
  mitochondrial membrane, 503, 504
ATP/ADP ratio, 505, 506
ATPase pumps
 $(Na^+ + K^+)$-, 410–412
 Ca^{2+}-, 412
Autosteric regulation, 139, 152

Binding reactions, 3, 4
 examples of, 8–11
 extracellular, 4–11
 intracellular, 4–11

659

Biotin, 123, 124
Blood coagulation, 149
Bohr protons, 327
Boltzmann order principle, 68, 72, 89
Bond energy, 95, 96
Boundary lipids, 349
Bound water, 86
Brain function, 626
Bulk water, 85

Calcitonin, 547
Calcium
 ATPase pump and, 412
 as "second messenger," 584, 585
 secretory role, 483–485
 transport, 412
Calcium-binding protein, 597
Calcium-dependent regulator protein, 280, 585, 586
cAMP. *See* Cyclic AMP
Carbanion, 98
Carbene, 98
Carbohydrate
 composition of membranes, 391–393
 Golgi apparatus and glycoprotein synthesis, 391, 392
Carbonium ion, 97, 98
Carbon radical, 98
Carbon tetravalence, 95
Carboxypeptidase, 112
Carnitine acyltransferases, 528
Carrier-mediated transport, 403–418
 active, 407–418
 carrier
  chemiosmotic theory, 501
  nature of, 403, 404
 cotransport systems, 405
 dicarboxylate carrier, 507
 exchange diffusion, 407
 facilitated diffusion, 404–406
  identification criteria, 405
 of organic acids, 506, 507
 relation to electronic charge, 506, 507
Catalysis
 covalent, 107, 108
 evolution of, 16–18, 38, 40, 129
 metals in, 126
Catecholamine secretion, 549
Cell cycle, 269–272

Cell membranes. *See also* Membrane
 definitions, 381
 functions of, 373, 374
 and secretion, 439
Cell surface receptors
 antigenic, 377
  blood group specificities, 377, 378
 drug receptors, 380
 hormonal, 378–380
Cellular components, 15, 16
Centrioles, 278
Centrosome, 278
cGMP. *See* Cyclic GMP
cGMP phosphodiesterase deficiency, 583
Chalones, 270, 537
Charge transfer in hydrogen bonding, 84
Chemical bonding, 94
Chemical compartmentation, 496
Chemical reactions, 97, 98
 electrophilic, 98
 nucleophilic, 98
Chemiosmotic theory, 500, 501, 326–328, 404
Cholecystokinin, 547
Cholera toxin (choleragen), 567
 as ADP-ribosyl transferase, 552, 578
 binding to gangliosides, 578
 receptors, 568, 570, 576–579, 584
 stimulation of adenyl cyclase, 576, 577
Cholesterol
 and membrane structure, 344
 synthesis, 526, 527
  control of, 301, 302
Chorionic gonadotropin, 570
Chromaffin granules and secretion, 481
Chromatids, 279
Chromatin, 265, 266
Chromosome components, 15
Chymotrypsin, 105
Citrate, 299
Citrate cleavage enzyme, 525
Citric acid cycle intermediates, 506, 507
Coenzymes, 113–128
 A, 117, 118
 B_{12}, 126, 127
Coherent behavior, 89
Cold-sensitive enzymes, 25
Compartmentation, 495–530
 and metabolic regulation, 529, 530
 of metabolites, 505–507, 512–517

Index

Compartmentation (*cont.*)
  species differences, 509, 510
  as zymogens, 496, 498
Conformational hypothesis of oxidative phosphorylation, 326–328
Contact inhibition
  and glycoproteins, 375, 376
  lectins and aggregation factors, 375, 376
Controllability of enzymes, 163. *See also* Enzyme activity, regulation of
Cooperativity, 139, 295
  of hormone binding, 560
  kinetics of, 142–144, 298
Covalent bond, 94–96
  orbitals, 94–96
  strength of, 81
Covalent catalysis, 107, 108
Covalent modification of enzymes, 268
  as compartmentation, 497
  as regulatory mechanism, 136, 153–155
Crossing-over, 635
Crossover theorem in metabolic pathways, 162, 163
Cyclic AMP (cAMP)
  intracellular concentration, 574
  mechanism of action, 575
  in morphogenesis of *Dictyostelium*, 307
  as "second messenger," 379, 538, 551–553
  in secretion, 483–485
Cyclic GMP (cGMP), 579–584
  phosphodiesterase deficiency, 583
Cyclic nucleotides, hormone action, 579–583
Cytochrome *c* oxidase (complex IV), 324
Cytochromes, 323
Cytoplasm, 15

Degenerative diseases, 625
Dehydrogenases, 116
Deoxyribonucleic acid. *See* DNA
Desolvation, 112, 121
Dictyostelium, 307
Differential calculus, 88
2,3-Diphosphoglyceric acid, 296, 297
Disease
  biochemical basis, 621, 627–629
  categories, 623–625

Disease (*cont.*)
  degenerative, 625
  enzyme overload and, 630
  treatment of, 649, 650
Dissipative structures, 64, 70, 74, 89, 296
  Belusov–Zhabotinski reaction, 71, 76
  Benard instability, 70
  Brusselator, 77
  information transfer, 79, 80
  primitive "memory" effect, 78
DNA (deoxyribonucleic acid), 15
  functions of, 173
  and histones, 15
  informational content, 44, 45, 173, 182, 183
  instability and evolution, 43
  ligases, 276
  and mitosis, 42
  mutations of, 29, 30
  and nuclear acidic protein, 15
  polymerases, 274–276
  replicase, 276
  structure, 266, 267
  synthesis, 272–277
  transcriptional activity, 268
  unwinding protein, 276

Editing in protein synthesis, 195
Einstein's theory of fluctuations, 74
Elasticity of enzymes in regulation, 163
Electrogenic transport
  of amino acids, 415
  of organic acids, 506
Electron transport, 322, 323
Electrostatic interactions, 82
  in hydrogen bonding, 83
Electrostriction, 105
Elongation and translocation in protein synthesis, 175, 200–203
Emiocytosis, 548, 549. *See also* Exocytosis
Encounter complex, 103, 106
Endocytosis, 562
Endoplasmic reticulum, 15, 16
  enzymes of, 316, 317
  microsomes, 15
Enthalpy, 67, 68, 90
Entropy
  components of, 72, 74
  definition, 65, 68, 90

Entropy (*cont.*)
  negative, 21, 64, 73
  and water displacement, 87
Entropy flow, 71–73
Entropy production, 71–73
Enzymatic chemical reactions, 3, 4
  complete enzyme dependence, 3
  partial enzyme dependence
    (photochemical), 3, 5
Enzyme activity
  effect of lipids, 339–342, 351
  regulation of
    by compartmentation, 522
    conformation changes, 295
  role of metals, 126
Enzyme deficiency diseases
  cGMP phosphodiesterase, 583
  galactosemia, 629, 643
  glucose-6-phosphate dehydrogenase,
    637
  glycogen storage disease, 642
  hereditary fructose intolerance, 629
  homocystinuria, 642
  lactase deficiency, 283
  Lesch–Nyhan syndrome
    (hypoxanthine-guanine
    phosphoribosyltransferase), 26
  methemoglobin reductase, 13
  phenylketonuria, 643
  Pompe's (acid maltase), 250
  PRPP synthetase, 643
Enzyme degradation
  aging, 238
  half-lives, 234–241
  and lysosomes, 249–251
  and mammalian systems, 224
  rate of, 225, 229, 232, 234, 242
    abnormal enzymes, 239
    and coenzymes, 242, 243, 257
    growth and, 245
    and hormones, 244
    and ligands, 242, 243, 257
    and nutrients, 244
    reactions, 254–256
  subcellular localization, 249–253
    experimental approach, 253, 254
  sulfhydryl groups and, 256, 257
Enzyme mechanisms
  approximation–orientation, 102, 103
  catalysis by protein functional groups,
    108

Enzyme mechanisms (*cont.*)
  covalent intermediates, 102, 106
  cryoenzymology, 102, 112
  double displacement (ping-pong), 106
  enthalpy, 102
  entropy changes, 102, 103, 105
  intrinsic binding energy, 104
  minimization of charge separation,
    105
  modern rack theory, 103
  noncovalent interactions and, 87
  strain theory, 103
Enzymes
  adaptive responses, 222
    activation–inactivation, 223
    kinetic parameters, 223
    replacement, 222–224
  function, evolution of, 16–18, 38–40,
    129
  glucose-6-phosphatase, 354–357
  lock-and-key theory, 93
  membrane-bound
    hydrophobic interactions, 337, 338
    isolation and lipid removal, 353–360
    phospholipases, 338
    in vivo lipid environment, 362–364
  mode of action, 93
  multifunctional, 32–35
  nicking of, 225
  rate-controlling, 161, 162
  semistability, 171, 173
  sensitivity in regulation, 163
  specificity, 32, 33, 101
  stabilizing factors, 24–26
  UDP-glucuronyltransferase, 357–360
Equilibrium structures, 90
  crystallization, 69
Essential hypertension, 624
Eukaryotic gene controls, 632, 633
  possible defects, 638–640
Eukaryotic protein synthesis, 206
  operon, 212–214
Evolution
  implications, 46
  of metabolic control functions, 40, 42–
    44
Evolutionary feedback, 43, 80, 81, 309
  quantized phenomena, 81
  succession of instabilities, 64
Excess entropy, 74, 90
Excess entropy production, 74, 77, 90

Index

Exchange repulsion in hydrogen bonding, 84
Exocytosis, 457, 480, 486, 487, 548–553, 560, 584, 586
Extracellular enzymes, 223, 224

Far-from-equilibrium state, 2, 14
Fatty acid oxidation, 527, 528
Fatty acid synthesis, 300, 301, 525
Fatty acid synthetase, 301
  degradation of, 244, 246, 248
Fatty infiltration of liver, 477, 478
  choline deficiency, 477
Feedback inhibition, 295–297, 303
  nonlinearity, 71, 74
Feedforward activation, 296
Flavins, 122, 123
Folate coenzymes, 124, 125
Free energy, 90
Free-radical theory for ion movement, 412
Freeze clamp methodology, 54

Galactosemia, 629, 643
Gangliosides
  cholera toxin binding, 578
  in hormone receptors, 567–571
Gastrin, 546
Gastrointestinal hormones, 539
Gene
  activation and inactivation, 281–284
  amplification, 183, 208
  cloning, 208
  defects, 633–635
  repression in eukaryotes, 209
  structure, 175
Genetic code, 180–184
Genome
  discontinuous sequences, 183, 190, 207
  eukaryotic, 632, 633, 638–640
  information content, 44, 45, 173, 176, 177, 182, 183
Global stability criteria, 90
Glucagon, 251
Gluconeogenesis
  compartmentation, 518–520
  energetic requirements, 524
  fatty acid oxidation and, 520–522
  inhibitors of, 508, 515, 518–520

Gluconeogenesis (*cont.*)
  malate–aspartate shuttle, 508, 519
  substrate cycles in, 159, 160
  transport, 524
Glucose-6-phosphate dehydrogenase
  in aging red cells, 22
  deficiency of, 637
Glutamyl cysteine synthetase, 138, 139
Glyceraldehyde-3-phosphate dehydrogenase, 140
  in glycolysis, 298
  half-of-the-sites reactivity, 140
Glycogen phosphorylase/synthase, 157–159
Glycogen storage disease, 642
Glycolysis, 298, 299
Glycosyltransferase and secretion, 450
Golgi apparatus, 15, 16
  enzymes of, 320, 321
  membrane turnover, 450, 451
Growth hormone, 590, 591
Guanylate cyclase, 580
Guanyl nucleotide receptor, 571–573
  interaction with adenylate cyclase, 571

H^+-ATPase
  composition, 324, 325
  function in oxidative phosphorylation, 325, 326, 502, 503
Heisenberg uncertainty principle, 66, 94
Hemin, 214–216
Hemoglobin
  binding reactions, 5–7
  effects of amino acid substitutions, 27, 28
  functions of, 37
  genetic control of synthesis, 282
  Lepore, 635
  unstable, 27, 29
Hemoglobinopathies, 6–8, 27–29, 282, 634–637
Hereditary fructose intolerance, 629
Heterogenous nuclear RNA, 188
Histones, 266, 267
  cell cycle, control of synthesis during, 283
  covalent modification, 268
  eukaryotic gene structure, 15, 175, 213
  synthesis of, 273
Homeostasis, 2, 3, 47

Homocystinuria, 642
Hormones, 535–597
 antagonistic actions, 542, 543
 biosynthesis
  leader sequence, 544
  nonpolypeptide, 544
  polypeptide, 543
  posttranslational modification, 545–548, 554, 555
 classification, 537, 538
 criteria for hormonal action, 538, 539
 and cyclic nucleotides, 579–583
 definition, 536
 degradation, 251, 560, 561
 enzyme action of, 564
 imbalance and disease, 542
 in information transfer, 535, 539, 540
 and protein synthesis, 210–212
 receptors
  antibodies to, 563
  autoregulation of, 561–563
  interaction with, 558–560
  intracellular, 591–593
  on plasma membranes, 378–380, 555, 558
 secretion
  nonpolypeptide, 549
  polypeptide, 548
 sites of action, 540, 541
  extranuclear actions, 596
  steroid mechanism of action on DNA, 594, 595
 tissues of origin, 540, 541
 transport in plasma, 553, 554
Hydrogen bonding, 82–87
 components of, 83, 84
 and protein structure, 100
Hydrogen-ion concentration, control in metabolism, 150, 151
Hydrogen-ion gradient and oxidative phosphorylation, 501, 502
Hydrolytic reactions, 109
Hydrophilic interactions
 in enzymes, 343
 in formation of cell, 20, 21
 in membranes, 343
Hydrophobic interactions, 85
 entropy and, 85
 in formation of cell, 19–21, 385–387
 membrane-bound enzymes, 340
Hydroxymethylglutaryl-CoA reductase, 301, 527

Hypercycles, 296
 in biochemical evolution, 307, 308
Hypertension, essential, 624

Inborn errors of metabolism, 640–643
 therapeutic approaches, 643–649
Induced-fit theory, 104, 142
Inflammation, 624
Informational component in life
 and entropy, 66
 genome, 44, 45, 173, 182, 183
 protein, 177
Inherited rod-cone dysplasia, 583
Initiator codon, 197
 and Met-tRNA$_f$, 199
Insulin
 degradation, 560, 561
 mechanism of action, 587–590
 secretion, 548–550
  and glucose receptor, 551
  and hormones, 550
Intercellular communication, 376, 393
Intermediary metabolism, 135, 627
Intervening DNA sequences, 183, 189, 190, 207
Intestinal absorption, 451–454
 of fat, 454–457, 459, 468
 morphology, 451–454
Intestinal secretion, 446–451
 Golgi complex, 447, 449
 morphology, 447
Ions in metabolic control, 151, 152
 pumps, 410–419
Irrotationally bound water, 86
Isolation of membranes
 criteria of purity, 387
 effect of methods on properties, 390
 methods, 388, 389
Isozymes, 33–37

Keto–enol tautomerization, 497
Ketone body formation, 528

Lactase deficiency, 283
Lactose synthetase, 148, 149
LATS, 563
LDL receptors, 302, 624
Le Chatelier–Braun principle, 70, 90
Lectins, 375, 376

Index

Limit cycles, 14, 76, 79, 90, 306
Lipid biosynthesis, 525
Lipid liquid crystals, 329
Lipids, effect on enzyme activity, 330, 331, 339–345, 351
Lipoic acid, 120
Lipoproteins, 302
Liposomes
   in studying oxidative phosphorylation, 326
   in treatment of inborn errors, 648
Living systems
   characteristics, 65
   instability, 22–32
      causal factors, 22, 24
      stabilizing factors, 24, 25
Local equilibrium assumption, 73, 90
Low-density-lipoprotein (LDL) receptors, 302, 624
Lyotropic mesomorphism, 329
Lysosomal storage diseases, 250
Lysosomes, 15, 16
   cycle, 250
   enzyme degradation, 239, 249–252, 257
Lysozyme, 110

Malate–aspartate shuttle, 508, 519
Median metabolites, 146
Membrane. *See also* Isolation of membranes
   components, 100, 101, 314, 387–400
   fluidity, 340, 341, 344–346, 361, 362
   fusion, 445, 469
   lipids, 395
      classification, 396, 399
      composition, 398, 399
      functions, 400
   proteins
      amino acid composition, 395
      classification, 393–395
      receptors. *See* Cell surface receptors
      structural models, 381–387
         Danielli–Davson model, 382, 383
         fluid mosaic model, 385–387
         lipid–protein interaction model, 361, 362
         unit membrane model, 383, 384
Membrane-bound enzymes, 313–331. *See also* Enzymes
   Arrhenius plots for, 330, 341, 346–349
   isolation of, 315, 316

Membrane flow and differentiation
   experimental evidence, 471, 472
   versus reutilization, 474, 475
   in secretion, 468–472
Membrane reutilization
   versus membrane flow, 474, 475
   in secretion, 472–474
Merocrine secretion, 487
Messenger RNA (mRNA)
   capping and methylation, 188, 189
   posttranslational modification, 188
   processing (maturation), 179
Metabolic control, 30–32
   coupling and coherent behavior, 63–67
   and enzyme regulation, 135–164
Metabolic control principles, 18–20
Metabolism, inborn errors of, 640–643
   therapeutic approaches, 643–649
Metabolite ratios, 144
Metals in catalysis, 126
Microsomal acyl-CoA desaturation system, 317–319
Microsomal hydroxylation system, 319
Microtubules, 15, 16
   in mitosis, 280
   in secretion, 478–481
Minimum entropy production theorem, 74, 81, 90
   evolution criterion, 74
   maintenance of living systems, 81
Mitochondria, 15, 16
   enzymes of, 321, 322
   degradation, 240
Mitosis, 277–281
Mixed-function oxidases, 527
Models in science, 94
Molecular events
   cooperative phenomena, 71
   coupling and amplification, 63, 66, 81
Molecular orbital theory, 94, 95
Morphogenesis and oscillatory phenomena, 309
mRNA. *See* Messenger RNA
Multienzyme complexes, 32, 34, 149, 150
Multifunctional enzymes, 32–35
Mutations, 634, 635

NAD/NADH ratio, 146, 147
NADH-Q reductase (complex I), 324
Negative cooperativity, 139–144
   half-of-the-sites, 140

Nematic mesophase, 329
Neoplasia, 624
Neurophysins, 552
Neurotransmitters, 537
Nicking of enzymes, 225
Nicotinamide nucleotides
  coenzymatic role, 115
  mechanism, 116
  stereospecificity, 115, 116
Noncovalent interactions, 81–87, 96, 265
  flexibility, 12, 64, 81, 82
  transferability, 81
  types, 82, 97
Noncovalent regulatory mechanisms, 136, 137, 152
  kinetic inhibition, 137
  thermodynamic inhibition, 137
Nonenzymatic chemical reactions, 3, 4
  examples, 3, 4
  photobiochemical, 3, 7
  spontaneous, 3, 6
Nonequilibrium thermodynamics, 69
  local equilibrium assumption, 73, 90
  thermodynamic branch, 77, 91
  thermodynamic threshold (bifurcation point), 70, 72, 92
Nonlinear kinetics, 296
Nucleic acids
  evolution
    of DNA, 177
    of RNA, 177
  information content, 44, 45, 173, 176, 177, 182, 183
  processing, 174
Nucleosomes, 176, 266, 267
Nucleus, 15
Nutritional diseases, 624

Onsager reciprocity principles, 69, 74
Open systems
  characteristics, 69, 72
  definition, 63, 69, 91
  occurrence (living systems), 63, 69
Order through fluctuations, 64, 70–76, 80, 91
Order in living systems, 20–32, 38, 39
  emergence, 64, 66, 69, 71, 73
  functional, 64
  structural, 64
  water, 85–87

Organ function, 640, 641
Oscillatory phenomena, 14, 66, 295, 296, 304–310
  examples in biological systems, 306, 307
  physiologic significance, 309, 310
Oxaloacetate, 511, 512
Oxidation–reduction reactions, 4, 12, 13
Oxidative phosphorylation, 324–328
  compartmentation, 500

Palindromes, 178, 274
Pancreatic secretion
  exocrine cell
    morphology, 440, 442
    secretion by, 442–445
    metabolic requirements, 445, 446
Paracrines, 537
Parathormone, 546
Pasteur effect, 298, 299
Peroxisomes, 15, 16
Phenylketonuria, 643
Phi bodies, 15, 16
Phosphodiesterase, 574, 575
Phosphoenolpyruvate carboxykinase
  compartmentation, 510, 511
  degradation, 246, 248
  synthesis, 246, 248
Phosphofructokinase, 299
Phosphoribosyl pyrophosphate
  synthetase, deficiency of, 643
Physical reactions, 3, 4. *See also* Binding reactions
Physiology, 621, 622
Pinocytosis, 402, 403
Plasma membrane, 16, 328. *See also* Membrane
Polarization interactions in hydrogen bonding, 83, 84
Poly(ADP–ribose) synthetase, 273
Polysomes, 197
Pompe's disease, 250
Positive cooperativity, 139, 143, 153, 306
Posttranslational modification of protein, 183, 203, 204, 225, 546, 547
Prehormones, 545, 592
  preprohormones, 544, 545
  prohormones, 545
Product inhibition, 138
Progesterone, 593
Proinsulin, 545, 546

Prokaryotic protein synthesis, 205
Propanolol, 572
Protein
　denaturation, 22. *See also* Enzyme
　　degradation
　　factors causing instability, 26–29
　function
　　general categories, 32, 44
　　special categories, 32–40, 44
　information content, 177
　modifiers, 148
　posttranslational modification, 183,
　　203, 204, 225, 546, 547
　specifiers, 148, 149, 153
　structure, 6, 7, 100
　synthesis, 171–216
　　cap structures and initiation, 197–199
　　distinction from degradation, 224
　　editing in, 195
　　elongation and translocation, 200–203
　　in eukaryotes, 206, 212–214
　　factors, 197, 198
　　fidelity, 175, 180
　　hemin and, 214–216
　　initiation, 197–199
　　necessity for, 171, 172
　　in neoplastic cells, 282, 283
　　noncovalent forces, 175
　　in prokaryotes, 205
　　protein factors, 197, 198
　　termination and release, 203
　　transcriptional control, 174, 206–214
　　translational control, 214–216
　　translocation, 200–203
Protein kinase, 575
　cyclic-nucleotide-independent, 586
　regulatory subunit, 575, 576
Protein–lipid interactions, 338
Protein–nucleic acid interactions, 176–179
Protein–protein interactions, 147
Proticity, 326
Psychiatric disorders, 625
Pyridoxal phosphate, 118, 119
Pyrophosphate, 194
Pyruvate carboxylase
　acetyl CoA as activator, 521
　modifiers of, 523
Pyruvate dehydrogenase, 155

QH2-cytochrome *c* reductase (complex III), 324
Quantized phenomena, 81, 82

Rate-controlling enzymes, 161, 162
Reaction types, 98, 99
Reconstituted enzyme systems, 352, 353
Redox pump, 412
Reducing equivalents, 526
Reductionism and biology, 65
Regulable enzymes, 135
　kinetic properties, 135, 153
Regulation. *See* Metabolic control
Renin
　enzyme action, 564
　posttranslational modification, 546
Repression of enzyme synthesis, 297
Repressor, 206
Respiratory chain and electron transport, 322, 323
Restriction endonucleases, 207
Ribonuclease, 111, 112
Ribosomes, 15, 16
　function, 195, 196
　processing, 196
　structure, 178, 195
RNA polymerases, 177, 184–187
　complex with DNA, 185
　in eukaryote, 187

$S_{0.5}$, 141
Sarcoplasmic reticulum, 319, 320
Secretion, 439–489
　activation of, 483, 486
　chylomicra, 457, 467, 475
　granules, 15, 16, 481
　by hepatocytes, 482, 483
　membrane role, 457–460, 486, 487
　metabolic requirements, 445, 446
Sensitivity of enzymes in regulation, 163
Serine proteases, 109, 110
Serotonin, 569
Servomechanisms, defined, 295, 296
Sialic acid in membranes, 391, 392
Signal peptide theory in secretion, 498
Smectic mesophase, 329
Solvent availability in cell, 149
Stability
　asymptotic, 75, 76
　asymptotic orbital, 76

Stability (cont.)
  domains, 78
  entropy, 72
  far-from-equilibrium, 71
  global criteria, 66, 75, 90
  Lyapounov, 76
  marginal, 75
  orbital, 76
  relation to second law of thermodynamics, 68
Steady state, 13–15
  definition, 91
  thermodynamic view, 73, 74, 91
Steroid hydroxylation, 527
Stimulus–secretion coupling for hormones, 549–551
  cAMP and Ca^{2+}, 551, 553
Structural gene defects, 633–635
Substrate cycles (futile cycles), 159
  compartmentation, 523
Substrate destabilization, 105
Succinate-Q reductase (complex II), 324
Sugar transport, 414–418
  cotransport, 416, 417
  electrogenic nature, 415
Superoxide anion, 23, 25
Swivel enzyme (untwisting), 276
Synexin, 549

Temperature in regulation, 151
Termination and release in protein synthesis, 203
Testosterone binding protein, 592
Tetanus toxin, 570, 571
Thalassemia, 636
Theoretical biology, 1
  fundamental theorem
    first, 1
    second, 42, 43
    third, 46, 173
    fourth, 46, 622
    fifth, 46, 626
Thermodynamic branch, 77, 91
Thermodynamics. *See also* Nonequilibrium thermodynamics
  classical, 64, 68
  first law, 67
  second law, 68, 72
  and fluid mosaic model, 386, 387

Thermodynamics (cont.)
  linearity, 72
  linear nonequilibrium, 69
  nonlinear nonequilibrium, 69
Thermodynamic threshold (bifurcation point), 70, 72, 91
Thiamine pyrophosphate, 120, 121
Thioesters, 117, 118
Thyroglobulin secretion, 482
Thyroid hormones
  binding protein, 592
  extranuclear actions, 596
Transcription
  posttranslational modification, 187
  Rho factor, 186
  RNA polymerase, 187
  termination, 186
Transferability, 81
Transfer RNA (tRNA)
  function, 180, 190, 191
  processing, 192–194
  structure, 191, 193
Transition state, 98
Transition-state analog, 104
Transmembrane H^+-ion gradient, 325–328
Transmembrane transport
  carrier mediated, 403–418. *See also* Carrier-mediated transport
  diffusion, 401, 402
  Fick's law, 401, 402
  membrane pores, 402
  pinocytosis, 402, 403
Transport defects, 642
Treatment of diseases, 649, 650
Triglyceride lipase, 156
tRNA. *See* Transfer RNA
Tryptophan oxygenase, 248
Tubulin, 280, 281
Turnover rates, cellular, 41

Ubiquinones (coenzyme Q), 323
Ubiquitin, 268
Unwinding protein, 268

Valence bond theory, 94, 95
Van der Waals interactions, 82, 83, 105
Vasopressin secretion, 552

Index

Viscotropic control of cholesterol synthesis, 301
Vitamin D, 597

Water
   in biological systems, 1, 85–87, 93
   condensation reactions, 85
   dielectric constant, 94
   relation to protein native state, 86
   short-time structures, 85, 86
      bound water, 86
      bulk water, 85

Water (*cont.*)
   short-time structures (*cont.*)
      irrotationally bound water, 86
      transport of, 418

X-ray diffraction, 101

Yin-yang theory, 580

Zymogen, 28
   activation of, 109, 224, 496, 497